OMICS

Biomedical Perspectives and Applications

OMICS

Biomedical Perspectives and Applications

Edited by
Debmalya Barh
Kenneth Blum
Margaret Anne Madigan

CRC Press
Taylor & Francis Group
Boca Raton London New York

CRC Press is an imprint of the
Taylor & Francis Group, an **informa** business

CRC Press
Taylor & Francis Group
6000 Broken Sound Parkway NW, Suite 300
Boca Raton, FL 33487-2742

First issued in paperback 2017

Version Date: 20111004

ISNB 13: 978-1-4398-5008-4 (hbk)
ISNB 13: 978-1-138-07474-3 (pbk)

Library of Congress Cataloging-in-Publication Data

Barh, Debmalya.
OMICS : biomedical perspectives and applications / Debmalya Barh, Kenneth Blum, and Margaret A. Madigan.
p. ; cm.
Includes bibliographical references and index.
ISBN 978-1-4398-5008-4 (hardcover : alk. paper)
1. Bioinformatics. 2. Computational biology. 3. Genomics. 4. Proteomics. 5. Glycomics. I. Blum, Kenneth. II. Madigan, Margaret A. III. Title.
[DNLM: 1. Computational Biology. 2. Biomedical Research. 3. Biotechnology. 4. Genomics. 5. Molecular Biology. QU 26.5]

QH324.2.B386 2012
572.80285--dc23 2011027104

Visit the Taylor & Francis Web site at
http://www.taylorandfrancis.com

and the CRC Press Web site at
http://www.crcpress.com

Purnendu Bhusan Barh (February 22, 1940–February 27, 2008)

This book is dedicated to Purnendu Bhusan Barh (S/O Ambika Bhusan Barh), an eminent academician, philosopher, career master, and transformator who is the soul and inspiration behind the establishment of the Institute of Integrative Omics and Applied Biotechnology (IIOAB) and all its activities.

Contents

Foreword

The term *omics* refers to a field of study in biological sciences ending in -omics, such as genomics, proteomics, or metabolomics. Basically, genomics is the discipline in genetics concerning the study of the genomes of organisms and fine-scale genetic mapping efforts. Proteomics is a well-accepted term for studying proteins on a large scale, and metabolomics is a term denoting investigation of the chemical fingerprints of small molecules' metabolite profiles.

Overall, one of the major challenges of systems biology and functional genomics is to integrate proteomic, transcriptomic study of gene expression at the RNA level and metabolomic information to give a more complete picture of living organisms. Bioinformatics, another key term coined by Paulien Hogeweg and Ben Hesper in 1978, is an integrated term demonstrating the application of statistics and computer science to the field of molecular biology. The primary objective of bioinformatics is to increase and enhance the understanding of both biological and biochemical processes.

The list of authors and topics covered in this book is impressive. The editors are to be congratulated for bringing together such a unique group of experts from various fields of cutting-edge omics research. The book has twenty-seven chapters that deal with several cutting-edge features of novel technology.

The book starts with a chapter entitled "Overview of Omics" by Dr. Raghavachari that provides an overview of omics and omic technologies such as cellomics, glycomics, and lipidomics. The second chapter by Drs. Singh and Somvanshi focuses on bioinformatics and demonstrates how this can be an essential tool in omics. The third chapter provides a new twist and demonstrates the association of omics technology with nutrigenomics and nutraceuticals. Drs. López-Corrales, Stutzman, Miyoshi, Barh, and Azevedo discuss the various approaches of omics technology in toxicology research and applications in biomedical sciences in the fourth chapter. The fifth chapter covers the basic and versatile therapeutic applications of stem cells; it was written by Drs. Arya and Tripathi. Dr. Sandhiya provides an excellent chapter on the emerging trends of nanotechnology in omics-based drug discovery and development. This chapter provides a vivid description on how the integration of nanotechnology in the drug delivery system has the potential to improve specific drug targeting, drug release and interaction, and enhanced efficacy. The seventh chapter, by Drs. Zhang and Olin, discusses the biomedical applications of magnetic nanoparticles. Dr. Zaki demonstrates the usefulness of these state-of-the-art technologies on high-throughput screening in medical diagnosis and prognosis in Chapter 8. Drs. Kolukisaoglu and Thurow emphasize the applications of high-throughput omics technology in systems biology. Drs. Visaria, Prakash, and Shrivastava extensively discuss the safety aspects in diagnostic imaging techniques used in omics in Chapter 10. Dr. Gope and collaborators demonstrate the molecular genetics of human cancers in Chapter 11. Dr. Chatterjee highlights the intricate aspects on the functional identification of unknown genes in Chapter 12. Dr. Carranza-Cereceda and collaborators discuss their interesting research findings on the proteomics of phagosomal pathogens. In Chapter 14, Drs. Selvarajoo and Tsuchiya explore the governing principles of cellular networks from the perspective of systems biology. Dr. Fukunishi demonstrates the salient features of intermolecular interaction in biological systems. In Chapter 16, Dr. Zheng and collaborators demonstrate the application of neuromics and highlight how implanted brain machines interface in rats. Drs. Sharma and Munshi exhibit their concept on pharmacogenomics in the development of disease specific therapeutic strategy. Drs. Dhawan and Padh discuss the aspects of omics approaches in cancer drug discovery in Chapter 18. Drs. Ohdaira and Yoshida highlight the use of microRNA expression in the therapeutic strategy for tumors. Dr. Pereira and his collaborators extensively discuss marine metabolomics in cancer chemotherapy. Drs. Hong, Xu, Mendrick, and Tong highlight the important findings and the present status of type 2 diabetes. Dr. Viero highlights the applications of genomics and proteomics in cardiac therapies. Drs. Davies and Flower demonstrate the applications of omics in the treatment of infectious diseases in Chapter 23. Dr. Verma and collaborators highlight their findings on AIDS and HIV with omics technologies. Dr. Archer highlights aspects of epigenetics in neuropsychiatry in Chapter 25, and in Chapter 26, Dr. Blum reviews the neurogenetics and nutrigenomics of reward deficiency syndrome. Finally, Dr. Barh et al. summarize these intricate aspects and issues together and project the future pathology.

Overall, this book will be intensively useful to scientists from both academia and industry, teachers and professors, health professionals, and mostly students, who should be encouraged to study and learn from its wisdom.

Debasis Bagchi, PhD, MACN, CNS, MAIChE
Department of Pharmacological and Pharmaceutical Sciences
University of Houston College of Pharmacy
Houston, Texas

REFERENCES

Bagchi, D., Bagchi, M., and Lau, F. C. (Eds.) (2010). *Genomics, proteomics and metabolomics in nutraceuticals and functional foods*. New York: Wiley Blackwell.

Baxevanis, A. D., Petsko, G. A., Stein, L. D., and Stormo, G. D. (Eds.) (2007). *Current protocols in bioinformatics*. New York: Wiley Blackwell.

Bagchi, D., Lau, F. C., and Ghosh, D. (Eds.) (2010). *Biotechnology in functional foods and nutraceuticals*. Boca Raton, Florida: Taylor & Francis/CRC Press.

Hartung, T. (2009). Toxicology for the 21st century. *Nature* 460, 208–212.

Preface

This book, *Omics: Biomedical Perspectives and Applications*, illustrates the direction that this rapidly emerging discipline is taking. Applications of omics technologies in the postgenomics era have swiftly expanded from rare monogenic disorders to multifactorial common complex diseases, pharmacogenomics, and personalized medicine.

Omics informally refers to a field of study in biology ending in -omics, such as genomics and proteomics. The related suffix -ome is used to address the objects of such explosive fields of study as the genome and protome, respectively. The field combines different omics techniques such as transcriptomics and proteonomics. The suffix -ome as used in molecular biology refers to a totality or systems biology. The -ome suffix originated as a variant of -oma and became productive in the last quarter of the 19th century. The *Oxford English Dictionary* suggests that the third definition originated as a backformation from *mitome*, which was later also reinforced by *chromosome*. Early attestations include *biome*, first used in 1916, and *genome*, first coined as the German *Genom* in 1920. Because *genome* refers to the *complete* genetic makeup of an organism, the new suffix *-ome* suggested itself as referring to *wholeness* or *completion*.

Interestingly, bioinformaticians and molecular biologists are considered the first scientists to start to apply the -ome suffix widely. Some early advocates were bioinformaticians in Cambridge, United Kingdom, where there were many early bioinformatics labs such as the Sanger Center and European Bioinformatics Institute. One such center run by the Medical Research Council is where the first genome and proteome projects were carried out. Many -omes beyond the original *genome* have become useful and have widely adopted by research scientists. *Proteomics* has become well established as a term for studying proteins at a large scale. *Omics* can provide an easy handle to encapsulate a field; for example, an interactomics study is clearly recognizable as relating to large-scale analysis of gene-gene, protein-protein, or protein-ligand interactions. Researchers have been rapidly taking up omes and omics, as shown by the explosion of the use of these terms in PubMed since the mid-1990s, making this exciting field relatively new.

Omics research now encompasses an assortment of technologies and academic disciplines aspiring to analyze the mysteries involved in cellular function at a molecular level within organisms. Genomics, transcriptomics, pharmacogenomics, toxicogenomics, epigenomics, lipidomics, glycomics, immunomics, and proteomics are all addressed in this book, whereas the technologies covered include bioinformatics, high-throughput sequencing involving DNA and protein microarrays and mass spectrometry, stem cell research, nanoparticle drug design, the uses of magnetic nanoparticles, and diagnostic imaging.

The study of omics has become increasingly important as a specialty area within medical genetics and systems biology. This domain, originally restricted to a few researchers, has now become a vast uncharted arena where scientists from very diverse fields, including biology, biochemistry, pharmacology, pathology, toxicology, botany, neurology, psychiatry, medical and population genetics, anthropology, molecular biology, and even to some degree medical ethics converge to explore biological systems.

The increased interest stems principally from advances in molecular genetic techniques, bioinformatics, the genome project, neurosciences, nutrition science, mathematics, particle physics, and other related disciplines. Many of the dedicated scientists in this emerging field have been encouraged by enhanced public awareness of the role of genes in somatic diseases like cancer, diabetes, and HIV and complex mental diseases like bipolar depression, schizophrenia,

Alzheimer's disease, reward deficiency syndrome, and addictive, impulsive, and compulsive behaviors. The announcement of genes associated with such devastating genetically based single-gene disorders such as Huntington's disease, cystic fibrosis, and muscular dystrophy, as well as complex polygenic diseases, such as lung cancer, breast cancer, diabetes, and most recently aging, have profoundly aroused the interest of professors, students, and people all over the globe.

This book serves as an important resource and review especially to students and researchers interested in the field of *integrative omics*. The volume is also addressed to basic scientists, clinicians, and other professionals who have a specialized or even a peripheral interest in not only molecular genetics and proteomics but the field of systems biology.

In a review volume of this size, it is not possible to convey every aspect of the subject; however, we as editors have attempted to compile an outline that is comprehensive and that could serve as a state-of-the-art framework for a rather new discipline. Every effort has been made to provide an informative, basic text that presents as wide a view as possible of the current status of integrative omics.

The omics overview provides an organizational framework upon which the "Methodology and Application" section is founded. This section includes works that introduce many of the omics fields and provide background technical information and expertise. These areas include: bioinformatics, nutrigenomics, toxicology, stem cell research, magnetic nanoparticles, high-throughput screening, and safety in diagnostic imaging.

The second section, "Empirical Research," includes omics research into such diverse areas as a "Forward Genetics Approach in Genomics: Functional Identification of Unknown Genes" and "Proteomics of Phagosomal Pathogens: Lessons from *Listeria monocytogens* and New Tools in Immunology."

The third section, "Computational and Systems Biology," provides very timely topics, such as "In a Quest to Uncover Governing Principles of Cellular Networks: A Systems Biology Perspective" and "Intermolecular Interaction in Biological Systems." This section also includes an interesting topic: "Implanted Brain Machine Interfaces in Rats: A Modern Application of Neuromics."

The fourth section focuses on the application of specific omics technologies to the discovery of omics-based diagnostic and therapeutic modalities for disease treatment. These processes' challenges and successes are described in chapters that look at, for example, metabolomic research into the development of chemotherapy and the application of larger-scale high-density genome-wide association studies in type 2 diabetes to shed light on the genetic etiology and explain the difficulties involved in replicating for biomarkers. This section also covers omics-based diagnosis and treatment approaches in cardiovascular disease and cancer.

Integrative applications in various omics fields have been responsible for moving the work forward. The recent exponential growth in omics is based on the explosion of bioinformatics and other biotechnologies and the integrative multi-omics approaches being applied to research. The fifth section, "Future Perspective," deals with these issues.

The original idea for this compendium came from Dr. Debmalya Barh, who convinced CRC Press to engage all of us to edit and publish the first text in this subject area. It is our wish that the contents of this compendium will be of use to researchers and students of biology, including technologists and scientists from all disciplines, by providing both a basic platform of methods and applications and a resource for enhanced cross-pollination in a multiomics approach to future endeavors in the fertile fields of omics research. We hope that from within these chapters, these estimable researchers will impart their great appreciation of the general principles of rigorous and arduous research that can lead to appropriate and productive approaches in the study of systems biology, leading to clinical strategies and potential disease cures.

The book is an initiative of the Institute of Integrative Omics and Applied Biotechnology towards fulfillment of the mission of promoting higher education. The book is dedicated to Purnendu Bhusan Barh, Dr. Barh's beloved father.

Debmalya Barh, Kenneth Blum, and Margaret A. Madigan, Editors

Editors

Debmalya Barh, MSc, MTech, MPhil, PhD, PGDM, is a scientist, consultant biotechnologist, and intellectual property rights specialist. He is the founder and president of Institute of Integrative Omics and Applied Biotechnology, India—a global platform for multideciplinary research and advocacy. He is a pioneer researcher in male breast cancer and cardiac myxoma disease signaling pathways and drug targets. His expertise includes bioinformatics, phamacogenomics, and integrative omics-based biomarker and targeted drug discovery. Since 2008, he has authored more than 30 international peer-reviewed publications with first authorship, contributed more than 10 book chapters, and written 2 books. During this period, he has also edited three books in the area of omics and biotechnology. Because of his significant contribution towards science and higher education, he was selected by *Who's Who in the World in 2010* (p. 168). He is one of the founding members and executive editors of *The IIOAB Journal* and *IIOAB Letters* and also serves as an editorial and review board member for several professional international research journals of global repute.

Kenneth Blum, PhD, is currently Chairman of the Board and Chief Scientific Officer of LifeGen, Inc., San Diego, California, and a managing partner of Reward Deficiency Solutions, LLC, San Diego, California. He serves as a consultant and senior scientific advisor for many companies and a foundation.

Dr. Blum was for 23 years a full professor of pharmacology at the University of Texas, San Antonio, Texas. Following his service as research professor at the Wake Forest College of Medicine, Winston-Salem, North Carolina, he is currently a full professor of the Department of Psychiatry and McKnight Brain Institute University of Florida College of Medicine, Gainesville, Florida. He has received numerous awards, including the NIDA Career Teacher Award, the American Chemical Society Speakers Award, the Gordon Conference Research Award, the Presidential Excellence Award (National Council of Alcoholism and Drug Abuse), and the 2011 Lifetime Achievement Award (National Association of Holistic Addiction Studies). Dr. Blum has authored and edited 12 books, published over 400 peer-reviewed papers, and coined the terms *brain reward cascade* and *reward deficiency syndrome*. He is credited with codiscovering the first gene to associate not only with alcoholism but reward dependence in general and was the lead author in the first association study of the dopamine D2 receptor gene with severe alcoholism (*Journal of the American Medical Association*, 1990). He is considered by many to be the father of psychiatric genetics. He serves on nine editorial boards and is the associate editor on two boards, including coeditor-in-chief of the *BMC IIOAB Journal*, the official journal of the Institute of Integrative Omics and Applied Biotechnology. He is also executive editor of *Journal of Genetic Syndromes and Gene Therapy* and is an ad hoc reviewer for 40 journals worldwide. He is also the inventor of neuroadaptagen amino-acid therapy for the recovery field.

Dr. Blum's research has been covered by major newspapers all over the world, and he has made numerous television and radio appearances. In 1984, his textbook *Handbook on Abusable Drugs* was a book of the month selection. His books *Alcohol and The Addictive Brain* (with James Payne) and *Overload* (with David Miller) have received high ratings from Amazon. His work has been cited by Allen King and the Australian Broadcasting Company and on *The NY Science Show* and *Law & Order*. He has appeared on the *TODAY*, *Good Morning America*, and *Sonja Live*, to name a few. His work has received both silver and bronze medals from the Natural Products Association in 2006 and 2007. Dr. Blum has chaired three Gordon Research Conferences on alcohol and psychiatric genetics. Dr. Blum has published in almost every major scientific journal worldwide: *Science*, *Lancet*, *Nature*, *Proceedings of the National Academy of Sciences*, and *Journal of the American Medical*

Association, among others. He is actively investigating the role of natural dopamine agonists as an anticraving DNA-directed therapeutic target for prevention of relapse.

Margaret A. Madigan, BSN, is a nursing practitioner by training and is currently a senior editor and assistant to the Chairman of the Board of the LifeGen, Inc. research center located in San Diego, California. She is a native of Sydney, Australia, and has been a long-time resident of Honolulu, Hawaii. She is a graduate of the University of Sydney New South Wales having fulfilled requirements earning her a Bachelor of Health Sciences (Nursing). She served as a registered nurse at Sutherland Hospital in Sydney working on an ICU step-down ward. Ms. Madigan is a member of The New South Wales College of Nursing. She has served as a registered nurse at the Kapi'olani Medical Center, Honolulu, Hawaii, and Palomar Pomerado Health Care System Hospitals in San Diego, California. She has been certified in numerous nursing specialties, including infection control, palliative care, advanced nursing interventions, basic critical care, oncology, cancer chemotherapy, pain management, advanced fetal monitoring, neonatal resuscitation, and basic EKG monitoring, among other disciplines. Ms. Madigan is licensed in Hawaii, California, Texas, and New South Wales, Australia. She has also completed courses in brain repair for addictive disorders and has experience in psychiatric nursing. Ms. Madigan has published in the fields of neuropharmacology, neurogenetics, nutrigenomics, clinical neurology, neuroimaging, and psychiatric genetics in peer-reviewed journals. She is a graphic artist and photographer and is credited with the cover art for an issue of *The IIOAB Journal.*

Contributors

Carmen Alvarez-Dominguez, MS, PhD
Servicio de Inmunologia, Fundación Marqués de Valdecilla-IFIMAV
Hospital Santa Cruz de Liencres
Liencres, Spain

Paula B. Andrade, PhD
EQUIMTE/Laboratório de Farmacognosia
Departamento de Química
Faculdade de Farmácia
University of Porto
Porto, Portugal

Trevor Archer, PhD
Department of Psychology
University of Gothenburg
Gothenburg, Sweden

Awadhesh Kumar Arya, PhD
Department of Medicine
Banaras Hindu University
Varanasi, India

Vasco Azevedo, DVM, MSc, PhD, FESC
Departamento de Biologia Geral
ICB/UFMG
Belo Horizonte, Brazil

Debmalya Barh, MSc, MTech, MPhil, PhD, PGDM
Institute of Integrative Omics and Applied Biotechnology
Nonakuri, India

Kenneth Blum, PhD
Department of Psychiatry
University of Florida College of Medicine and McKnight Brain Institute
Gainesville, Florida

Judith Boateng, PhD
Department of Food and Animal Sciences
Alabama Agricultural and Mechanical University
Normal, Alabama

Carlos Carranza-Cereceda, BS
Servicio de Inmunologia, Fundación Marqués de Valdecilla-IFIMAV
Hospital Santa Cruz de Liencres
Liencres, Spain

Bishwanath Chatterjee, PhD
Laboratory of Developmental Biology
National Heart, Lung, and Blood Institute
National Institutes of Health
Bethesda, Maryland

Weidong Chen, PhD
Qiushi Academy for Advanced Studies
College of Computer Science
Zhejiang University
Hangzhou, China

Georgina Correia-da-Silva, PhD
Laboratório de Bioquímica
Departamento de Ciências Biológicas
Faculdade de Farmácia
University of Porto
Porto, Portugal
and
IBMC–Instituto de Biologica Molecular e Celular
University of Porto
Porto, Portugal

Jianhua Dai, PhD
College of Computer Science
Zhejiang University
Hangzhou, China

Matthew N. Davies, PhD
Social, Genetic, and Developmental Psychiatry Centre
Institute of Psychiatry
London, United Kingdom

Dipali Dhawan, PhD
Department of Cellular and Molecular Biology
B. V. Patel Pharmaceutical Education and Research Development Centre
Ahmedabad, India

Premendra Dhar Dwivedi, MSc, PhD
Food Toxicology Division
Indian Institute of Toxicology Research
Lucknow, India

Lorena Fernandez-Prieto, PhD
Servicio de Inmunologia
Fundación Marqués de Valdecilla-IFIMAV
Hospital Santa Cruz de Liencres
Liencres, Spain

Darren R. Flower, PhD
Aston Pharmacy School
School of Life and Health Sciences
Aston University
Birmingham, United Kingdom

Yoshifumi Fukunishi, PhD
Biomedicinal Information Research Center
National Institute of Advanced Industrial Science and Technology
Tokyo, Japan

Mohan L. Gope, PhD
Department of Biotechnology
Presidency College
Bangalore, India

Rajalakshmi Gope, PhD
Department of Human Genetics
National Institute of Mental Health and Neurosciences
Bangalore, India

Huixiao Hong, PhD
Division of Systems Biology
National Center for Toxicological Research
U.S. Food and Drug Administration
Jefferson, Arkansas

Xiaoling Hu, PhD
Department of Health Technology and Informatics
The Hong Kong Polytechnic University
Hung Hom, China

Üner Kolukisaoglu, PhD
Center for Life Science Automation
University of Rostock
Rostock, Germany

Nestor Luis López-Corrales, DVM, MSc, PhD
Departamento de Biologia Geral
ICB/UFMG
Belo Horizonte, Brazil

Margaret A. Madigan, BSN
LifeGen, Inc.
San Diego, California

Priyadarshini Mallick, MSc
Amity Institute of Biotechnology
Amity University
Noida, India

Donna L. Mendrick, PhD
Division of Systems Biology
National Center for Toxicological Research
U.S. Food and Drug Administration
Jefferson, Arkansas

Rohan Mitra, MSc
Department of Human Genetics
National Institute of Mental Health and Neurosciences
Bangalore, India

Anderson Miyoshi, MSc, PhD
Laboratório de Genética Celular e Molecular
Instituto de Ciĕncias Biológicas
Universidade Federal de Minas Gerais (ICB/UFMG)
Belo Horizonte, Brazil

Anjana Munshi, MPhil, PhD
King Saud University
Riyadh, Saudi Arabia

Hiroaki Ohdaira, MS
Department of Life Sciences
Meiji University
Kawasaki, Japan

Håkan Olin, PhD
Department of Natural Sciences, Engineering, and Mathematics
Mid Sweden University
Sundsvall, Sweden

Harish Padh, PhD
B. V. Patel Pharmaceutical Education and Research Development Centre
Ahmedabad, India

David M. Pereira, PhD
REQUIMTE/Laboratório de Farmacognosia
Departamento de Química
Faculdade de Farmácia
University of Porto
Porto, Portugal

Surya Prakash, MD
Tej Bahadur Saproo Hospital
Allahabad, India

Nalini Raghavachari, PhD
Genetics and Development Biology Center
The National Institutes of Health
Bethesda, Maryland

José Ramós-Vivas, PhD
Servicio de Inmunologia, Fundación Marqués de Valdecilla-IFIMAV
Hospital Santa Cruz de Liencres
Liencres, Spain

Estela Rodriguez-Del Rio, BS
Servicio de Inmunologia, Fundación Marqués de Valdecilla-IFIMAV
Hospital Santa Cruz de Liencres
Liencres, Spain

Sandhiya Selvarajan, MD
Division of Clinical Pharmacology
Jawaharlal Institute of Post-Graduate Medical Education and Research
Pondicherry, India

Kumar Selvarajoo, PhD
Institute for Advanced Biosciences
Keio University
Tsuruoka, Japan

Sanjeev Sharma, MD
Apollo Hospital
Hyderabad, India

Devashish Shrivastava, PhD
Department of Radiology
University of Minnesota
Minneapolis, Minnesota

Abhishek Angelo Singh, MTech
Department of Bioinformatics-BiGCaT
Maastricht University
Maastricht, the Netherlands

Anchal Singh, MSc, PhD
Department of Microbiology and Immunology
Kirksville College of Osteopathic Medicine
A.T. Still University of Health Sciences
Kirksville, Missouri

Udai Pratap Singh, BTech, MTech
Amity Institute of Biotechnology
Amity University
Noida, India

Pallavi Somvanshi, PhD
Bioinformatics Centre
Lucknow, India

Tina Stutzman, BS
Department of Biological Engineering
Massachusetts Institute of Technology
Cambridge, Massachusetts

Natércia Teixeira, PhD
Laboratório de Bioquímica
Departamento de Ciências Biológicas
Faculdade de Farmácia
University of Porto
Porto, Portugal
and
IBMC–Instituto de Biologica Molecular e Celular
University of Porto
Porto, Portugal

Kerstin Thurow, PhD
Center for Life Science Automation
University of Rostock
Rostock, Germany

Weida Tong, PhD
Division of Systems Biology
National Center for Toxicological Research
U.S. Food and Drug Administration
Jefferson, Arkansas

Kamlakar Tripathi, MD, DM
Department of Medicine
Institute of Medical Sciences
Banaras Hindu University
Varanasi, India

Masa Tsuchiya, PhD
Institute for Advanced Biosciences
Keio University
Tsuruoka, Japan

Patrícia Valentão, PhD
REQUIMTE/Laboratório de Farmacognosia
Departamento de Química
Faculdade de Farmácia
University of Porto
Porto, Portugal

J. Thomas Vaughan, PhD
Department of Radiology
University of Minnesota
Minneapolis, Minnesota

Ashish Swarup Verma, MSc, PhD
Amity Institute of Biotechnology
Amity University
Noida, India

Martha Verghese, PhD
Department of Food and Animal Sciences
Alabama Agricultural and Mechanical University
Normal, Alabama

Cedric Viero, PhD
Department of Cardiology
Wales Heart Research Institute
School of Medicine
Cardiff University
Cardiff, United Kingdom

Rachana Visaria, PhD
Division of Testing Services
MR Safe Devices LLC
Burnsville, Minnesota

Lei Xu, PhD
Division of Systems Biology
National Center for Toxicological Research
U.S. Food and Drug Administration
Jefferson, Arkansas

Kenichi Yoshida, DVM, PhD
Department of Life Sciences
Meiji University
Kawasaki, Japan

Maysaa El Sayed Zaki, PhD, MD
Faculty of Medicine
Mansoura University
Mansoura City, Egypt

Renyun Zhang, PhD
Department of Natural Sciences, Engineering and Mathematics
Mid Sweden University
Holmsgatan, Sweden

Shaomin Zhang, PhD
Qiushi Academy for Advanced Studies
College of Biomedical Engineering and Instrument Science
Zhejiang University
Hangzhou, China

Ting Zhao, PhD
Qiushi Academy for Advanced Studies
Zhejiang University
Hangzhou, China

Xiaoxiang Zheng, MD
Qiushi Academy for Advanced Studies
College of Biomedical Engineering and Instrument Science
Zhejiang University
Hangzhou, China

1 Overview of Omics

Nalini Raghavachari
National Heart, Lung, and Blood Institute/National Institutes of Health
Bethesda, Maryland

CONTENTS

1.1 INTRODUCTION TO THE FIELD OF OMICS

The central dogma of molecular biology (Figure 1.1a), enunciated by Crick (Crick, 1954), specified that the instruction manual is DNA (encoding genes) and that genes were transcribed into RNA to ultimately produce the basic operational elements of cellular biology, proteins whose interactions, through many levels of complexity, result in functioning living cells. This was the first description of the action of genes. After an enormous experimental effort spanning the last half-century, made possible by the development of many assays and technological advances in computing, sensing, and imaging, it has become apparent that the basic instruction manual and its processing are vastly more sophisticated than what was imagined in the 1950s. With the advent of these novel technologies, the primary focus of modern biology has shifted to link genotype to phenotype, interpreted broadly, from the level of the cellular environment to links with development and disease, and the central dogma has now been viewed as an integration of the -ome studies as depicted in Figure 1.1b.

In this context, biomedical research has been transformed recently by an exponential increase in the ability to measure biological variables of interest in grand scale (Abraham, Taylor et al., 2004). Diverse methods of large-scale measurements of biological processes have emerged in the past 15 years, and the list is growing rapidly. Remarkable technologies such as microarrays and their descendants, high-throughput sequencing, *in vivo* imaging techniques, and many others have enabled biologists to begin to analyze function at molecular and higher scales. The various aspects of these analyses have coalesced as *omics* (Wild, 2010). Omics is an emerging and exciting area in the field of science and medicine. Technologies that measure some characteristics of a large family of cellular molecules, such as genes, proteins, or small metabolites, have been named by appending

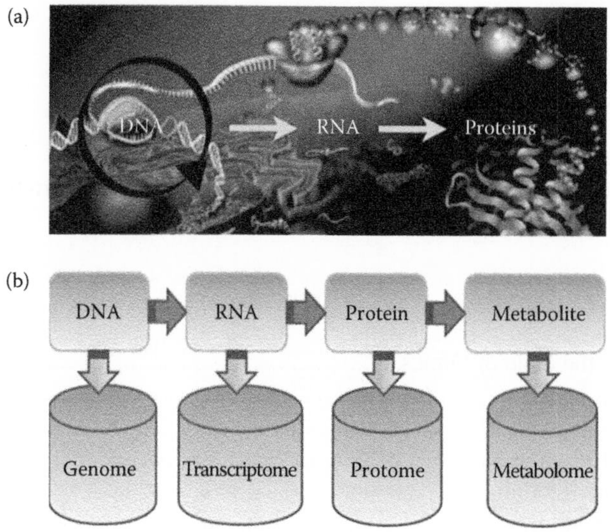

FIGURE 1.1 **(See color insert.)** (a) The central dogma (past and present) as explained by Crick. (b) Integration of the -ome studies.

the suffix *-omics*, as in *genomics*. Omics refers to the collective technologies used to explore the roles, relationships, and actions of the various types of molecules that make up the cells of an organism. These technologies encompass the following four major fields of study:

1. Genomics: the study of genome that stores the information in a cell to predict what can happen.
2. Transcriptomics: the study of mRNA or transcript that would depict what is really happening in a cell.
3. Proteomics: the study of protein molecules that would illustrate the functional roles of molecules in cellular function.
4. Metabolomics: the study of molecules involved in cellular metabolism that would eventually depict the phenotype of an organism.

Numerous promising developments have been elucidated using genomics, transcriptomics, epigenomics, proteomics, metabolomics, interactomics, cellomics, and bioinformatics (Wild, 2010). The omics technology that has driven these new areas of research consists of DNA and protein microarrays, mass spectrometry, and a number of other instruments that enable high-throughput analyses (Bier, von Nickisch-Rosenegk et al., 2008). Likewise, the field of bioinformatics has grown in parallel and with the help of the internet, rapid data analysis and information exchange are now possible. With these advancements, not only will omics have an impact on our understanding of biological processes, but the prospect of more accurately diagnosing and treating disease will soon become a reality. In an effort to understand the complex interplay of genes/proteins in disease processes, comparative genetic, transcriptomic, proteomic, and metabolomic analyses for individuals and populations are required. In particular, systems biology, more than the simple merge of omics technologies, aims to understand the biological behavior of cellular systems and enhance the capacity to test the probability of disease as early as possible through a noninvasive method of diagnosis as illustrated in Figure 1.2. The omics technologies are believed to open a new road to the field of personalized medicine in this postgenomic era. Understanding the existing and emerging technologies of genomics, transcriptomics, proteomics, and metabolomics is critical for widespread application of these technologies in the field of medicine.

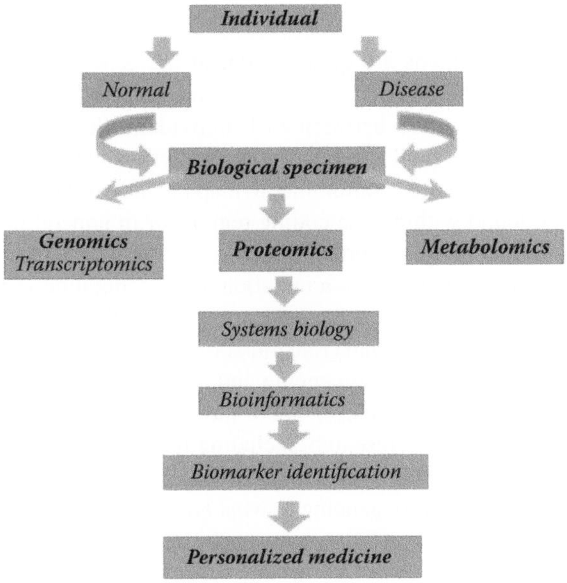

FIGURE 1.2 Flow chart for omics-based biomedical research.

Although the field of omics is ever expanding, currently genotyping, gene expression profiling, epigenomics, proteomics, and metabolomics are well established and widely used by scientists in clinical research. Technologies such as cellomics, glycomics, and lipidomics are now emerging as powerful tools for medical research. Almost invariably, these advances in omics have been associated with major expectations of transforming not only biological knowledge but also medicine and health. This chapter will provide valuable information about these powerful omics technologies.

1.2 GENOMICS

Genomics may be described as the comprehensive analysis of DNA structure and function and broadly refers to the analysis of all the genes and transcripts included within the genome (Bier, von Nickisch-Rosenegk et al., 2008). Understanding biological diversity at the whole-genome level will yield insight into the origins of individual traits and disease susceptibility. The aim of genomics is to analyze or compare the entire genetic complement of a species. Important areas of genomics are:

1. Structural genomics for the analysis of macromolecular structure using computational tools and theoretical frameworks
2. Comparative genomics (genomics for study of a species) by comparisons with model organisms
3. Functional genomics, a field of genomics attempting to make use of the vast wealth of data produced by genome sequencing projects to describe genome function
4. Pharmacogenomics, which aims to study how genes influence the response of humans to drugs, from the population to the molecular level, and uses genomic approaches and technologies for the identification of drug targets

These major fields of genomics are subclassified into genotyping, transcriptomics, pharmacogenomics, toxicogenomics, and epigenomics.

1.2.1 Genotyping

Although organisms such as humans are quite similar at the genetic level, differences exist at a frequency of about one in every 1000 nucleotide bases (Barron, 2008). This translates into approximately three million base differences between each individual. Such changes are referred to as single-nucleotide polymorphisms (SNPs), and a significant effort collectively referred to as genotyping is now underway in the research community to map the individual SNPs in humans and other organisms. SNPs may be found within gene coding regions or in noncoding regions. Their effects may be subtle, yielding slight changes in protein function, or profound, leading to the development of disease. A polymorphism is distinct from a mutation, in that mutation is considered rare, affecting less than 1% of the species, whereas polymorphism is relatively common, and its prevalence is no different from what is considered normal (Barron, 2008). Over the past decade, there has been an unprecedented surge of data directed at sequencing and categorizing all of the genes in the human genome, as well as those of other organisms. There has also been a concomitant acceleration in the technology dedicated to genomics research, including instrumentation, reagents, software, and databases. Since the introduction of array-based genotyping techniques, it has become possible to cover with varying resolution the entire genome in what are now commonly referred to as genome-wide association studies (GWAS). The GWAS have uncovered and will uncover in the future interesting and previously unknown polymorphic variants that are associated with a variety of chronic diseases (Seshadri, Fitzpatrick et al., 2010).

1.2.2 Transcriptomics

The abundance of specific mRNA transcripts in a biological specimen is a reflection of the magnitude of the expression levels of the corresponding genes. Gene expression profiling is the identification and characterization of the mixture of mRNA that is present in a biological sample. An important application of gene expression profiling is to associate differences in mRNA mixtures originating from different groups of individuals with phenotypic differences between the groups. In contrast to genotyping, gene expression profiling allows characterization of the level of gene expression. A gene expression profile provides a quantitative overview of the mRNA transcripts that were present in a sample at the time of collection (Ness, 2007). Therefore, gene expression profiling can be used to determine which genes are differentially expressed in disease conditions; these genes would then serve as disease biomarkers.

Recent advances in bioinformatics and high-throughput technologies such as microarray analysis are bringing about a revolution in our understanding of cell biology and the molecular mechanisms underlying normal and dysfunctional biological processes. This field of omics is also stimulating the discovery of new targets for the treatment of disease, which is aiding drug development, immunotherapeutics, and gene therapy. Gene expression profiling has enabled the measurement of thousands of genes in a single RNA sample. There are a variety of microarray platforms from companies such as Affymetrix, Agilent, NimbleGen, and Illumina that have been developed to accomplish this. The basic idea for each platform is simple: a glass slide or membrane is spotted or arrayed with DNA fragments or oligonucleotides that represent specific gene coding regions (Ness, 2007; Bier, von Nickisch-Rosenegk et al., 2008). Purified RNA is then fluorescently or radioactively labeled and hybridized to the slide/membrane. In some cases, hybridization is done simultaneously with reference RNA to facilitate comparison of data across multiple experiments. After thorough washing to remove nonspecific hybridization, the data can be analyzed by a variety of statistical algorithms by comparing the gene expression pattern of samples tested to identify differentially expressed genes (Holland, Smith et al., 2003) that could potentially serve as disease biomarkers.

The most popular platform is the short oligonucleotide chips produced by Affymetrix. The second major platform consists of printed cDNA fragments or a long oligonucleotide (45–80-mers) on glass slides or other types of solid support. Dissection of global changes in gene expression during predisease states, during disease progression, and following clinical treatment can provide great insight into disease mechanism and treatment management. For example, early investigations using microarrays distinguished acute myeloid and acute lymphoblastic cell gene expression patterns (Golub, Slonim et al., 1999). Subsequent studies have used microarray technology to predict outcomes in breast and ovarian cancers (Berchuck, Iversen et al., 2005; Huang, Song et al., 2003). Additionally, it has been shown that classification of diffuse large B-cell lymphomas on the basis of gene expression profiles can identify clinically significant subtypes of cancer, and the new classification has significant prognostic implications (Alizadeh, Eisen et al., 2000; Alizadeh and Staudt, 2000). Examination of systemic lupus erythematosus using microarray technology identified a subgroup of patients who may benefit from new therapeutic options (Baechler, Batliwalla et al., 2003). Novel treatments for diseases, such as multiple sclerosis, have also been suggested by gene expression profiling (Chabas, Montfort et al., 2001; Chabas, Baranzini et al., 2001). Genomic biomarkers are currently being identified in cardiovascular diseases in a large-scale study using microarray technology (unpublished data).

1.2.3 Pharmacogenomics

Pharmacogenomics is the study of how an individual's genetic inheritance affects the body's response to drugs (Evans and Relling, 1999). The field of pharmacogenomics is an intersection of pharmaceuticals and genetics and specifically studies the variability in drug response caused by heredity. The way a person responds to a drug (in both a positive and negative manner) is a complex trait that is influenced by many different genes. Without knowing all of the genes involved in drug response, scientists have found it difficult to develop genetic tests that could predict a person's response to a particular drug. A person's response to a particular drug is the result of inherited variations in genes that dictate drug response and omics researchers are exploring the ways in which these variations can be used to predict whether a patient will have a good response or a bad response or no response at all to a particular drug. For example, in their study, Johnson and Evans et al. (2001) examined the influence of genetic variation on drug response in patients by correlating gene expression or single-nucleotide polymorphisms with a drug's efficacy or toxicity. Pharmacogenomics is believed to be immensely helpful in reducing drug-caused morbidity and mortality (Algeciras-Schimnich, O'Kane et al., 2008). Pharmacogenomics has gained considerable momentum with the advent of new methods and technologies for genome analysis and is widely believed to play a major role in predictive and personalized medicine (Roden, Altman et al., 2006). It will have the most impact in areas such as oncology, where many therapies are available, but each one works only for a small percentage of cancer patients. Pharmacogenomics is also expected to help physicians and patients by enabling pharmaceutical companies to bring more drugs into the market that are targeted at those patients who are most likely to benefit from them. Pharmacogenomics holds great promise in personalized medicine by providing physicians an opportunity to individualize drug therapy for patients based on their genetic make-up.

1.2.4 Toxicogenomics

The field of toxicogenomics is used in the study of structure and output of the genome as it responds to adverse xenobiotic exposure and is very closely related to pharmacogenomics. Toxicology has traditionally been evaluated by the dosing of animals to define well-established cytologic, physiologic, metabolic, and morphologic endpoints (Ferrer-Dufol and Menao-Guillen, 2009; Ge and He, 2009; Luch, 2009) . The evaluation of the risk to humans cannot be performed in human individuals initially and thus must be derived from studies performed in other species. Typically, rodents are used to identify toxic substances such as carcinogens, reproductive toxins, and neurotoxins. Follow-up studies in nonrodent species (species extrapolation) can then be used to further define the effects of low doses and mechanism of action.

Although it is well recognized that intact animals are needed to reflect physiologic changes and mirror the effects of chronic dosing, such studies have disadvantages (Guguen-Guillouzo and Guillouzo, 2010; Mei, Fuscoe et al., 2010; Moreira, Yu et al., 2010; Pettit, des Etages et al., 2010; Thompson, 2010; Van Aggelen, Ankley et al., 2010). Experiments with animals may not be fully predictive of the response in humans because of species variation in physiology, anatomy, and metabolism. Also, toxicology studies require large numbers of animals to allow statistically significant conclusions to be drawn. Nevertheless, these numbers are still very small compared to the human population potentially at risk. In order to compensate for this relatively small sample size in these animal studies, the future risk to humans at therapeutic dosages is inferred by giving large doses of compound to these groups of animals. Finally, depending on the anticipated duration of exposure in the population, studies of up to 2 years are currently mandated to determine the carcinogenic potential. Thus, the traditional approach to toxicologic testing is costly, in terms of time, labor, and compound synthesis and, not least, the large numbers of animals.

Technological advances have now enabled scientists to simultaneously analyze thousands of genes of several species, including humans and rodents, quickly and in a reproducible manner. Current toxicogenomics applies genomics concepts and technologies to study adverse effects of chemicals. These studies use global gene expression analyses to detect expression changes that influence, predict, or help define drug toxicity. In essence, toxicogenomics combines the tools of traditional toxicology with those of genomics and bioinformatics (Zarbl, 2007). By evaluating and characterizing differential gene expression after exposure to drugs, it is possible to use complex expression patterns to predict toxicologic outcomes and to identify mechanisms involved with or related to the toxic event. Toxicogenomics thus combines conventional toxicology with the emerging technologies of genomics and bioinformatics. Gene and protein expression respond specifically to external stimuli such as pathological conditions or exposure to drugs. The corresponding genomic and proteomic technologies thus provide a new way of understanding biological systems and their response to toxic insult. This leads to a better understanding of the mechanisms of toxicity by the identification of toxicity-related gene expression signatures and the prediction of the toxic potential of unknown compounds by comparing their gene expression profiles to the fingerprints of known, similar compounds (Gant, 2003; Shostak, 2005). In addition, the identification of toxicity-related genes, together with the rapidly growing understanding of the human genome, is providing a basis for identifying and characterizing sequence variations in genes that might affect responses to chemicals. This is already having a great impact in pharmacology and toxicology, because it allows the prediction/differentiation of species-specific responses and also the identification of populations of responders and nonresponders (Mei, Fuscoe et al., 2010). The most optimistic estimates predict that the replacement of traditional methods of toxicology by toxicogenomics could eventually shorten the safety assessment of a new chemical entity from years to days and reduce costs by an estimated factor of four to six times. A more realistic picture with the data currently available suggests that toxicogenomics will reduce failure rates by helping select the right compounds for development early on and by accelerating toxicology testing and identifying suitable biomarkers amenable to screening using the generated data (Pettit, des Etages et al., 2010; Choudhuri, 2009). Toxicogenomics represents an exciting new approach to toxicology and has a great potential to influence the predictability and speed of preclinical safety assessments (Choudhuri, 2009). Published results so far show that genome-wide gene expression analysis is a powerful tool for compound classification and for the detection of new, specific, and sensitive markers for given mechanisms of toxicity (Gallagher, Tweats et al., 2009; Ge and He, 2009; Hirode, Omura et al., 2009; Smirnov, Morley et al., 2009). In addition, preliminary results support the theory that gene expression might be more sensitive than conventional toxicology endpoints. Therefore, compound classification could be performed during early, short-term (i.e., single-dose) animal studies. Hence, time, cost, and number of animals needed to identify the toxic potential of a compound would be greatly minimized. The potential identification and validation of possible marker genes are also gaining momentum. Such markers could be employed in automated, high-throughput assay systems that will provide indications regarding toxicity potential that are fast and accurate, without incurring the high costs commonly associated with microarray analysis. Appropriately chosen markers are amenable to being tested in cell-based assays that will allow scientists to evaluate compounds much earlier in the developmental process, improving clinical candidate selection. The understanding of the molecular mechanisms underlying toxicity obtained through gene expression analysis after exposure of model systems (animals or cell cultures) to test compounds will also provide more insight into species-specific response to drugs regarding efficacy and toxicity. Hence, it is expected that extrapolation across species will become more accurate by enhancing the interpretation of preclinical observations and their meaning for the human situation. This should immensely increase the predictability of toxic liabilities and of potential risk accumulation for drug combinations or drug-disease interactions.

1.2.5 Epigenomics

Epigenomics, the merged science of epigenetics and genomics, has arisen as a new discipline with the aim of understanding genetic regulation and its contribution to cellular growth and differentiation, disease, and aging. Epigenetics is the study of heritable changes other than those in the DNA sequence and encompasses two major modifications of DNA or chromatin: DNA methylation; the covalent modification of cytosine; and post-translational modification of histones, including methylation, acetylation, and phosphorylation (Banerjee and Verma, 2009). Functionally, epigenetics acts to regulate gene expression, silence the activity of transposable elements, and stabilize adjustments of gene dosage, as seen in X inactivation and genomic imprinting (Herceg, 2007). The focus of epigenomics is to study epigenetic processes on a genome-wide scale. Epigenetic processes are mechanisms other than changes in DNA sequences that are involved in gene transcription and gene silencing (Schubeler, 2009). Epigenetic studies are currently based mainly on DNA methylation, histone modification interference by noncoding RNAs such as microRNA, and small interference RNA mechanisms (Schubeler, 2009). Generally, gene silencing is observed during genomic imprinting, x-chromosome inactivation, and tissue-specific gene expression. Alteration to these patterns of gene silencing by epigenetic modification is believed to play an important role in human disease (Herceg, 2007).

Historically, technology has limited large-scale approaches to epigenomics, but the emergence of highly reproducible quantitative high-throughput microarray technology has allowed virtually all epigenomics research to be read on microarray platforms, although the substrates, preprocessing, and data analysis differs substantially depending on the modification that is being addressed (Adorjan, Distler et al., 2002). Multiple complementary technologies are emerging now to analyze DNA methylation, protein binding patterns, and chromatin regulation on a genome-wide level. Early efforts are providing glimpses into the epigenetics of gene regulation and the mechanism of cancer and aging. It is hoped that the development of high-throughput technologies will continue to unravel the enigma of the epigenome. Early approaches to epigenomics used custom-made slide-based arrays of CpG-rich regions corresponding to methylated or unmethylated DNA (Adorjan, Distler et al., 2002). There has been a shift toward commercial high-density oligonucleotide arrays because of their greater precision and potential quantitative character. These include the photolithographic masked arrays of Affymetrix, photolithographic adaptive optics arrays of NimbleGen, inkjet arrays of Agilent, and, recently, the adaptation of bead arrays for epigenetic applications of Illumina. Each of these approaches offers potential advantages and disadvantages, but as yet, no direct comparison of epigenomic technology has been performed across platforms. An advantage of a flexible design for epigenomics is that one can tailor arrays to genomic targets of interest, such as imprinted genes, differentially methylated regions, and imprinting control regions.

An example of an early step in approaching the epigenome comes from recent studies by Fraga et al. (Fraga and Esteller, 2007; Fraga, Agrelo et al., 2007) that address the relationship between epigenetics and age. Another exciting work by Beth Israel Deaconess Medical Center and the Broad Institute created a map of histone modifications in fat cells, which led to the discovery of two new factors that regulate fat formation, a key step on the road to better understanding obesity, diabetes, and other metabolic disorders (Mikkelsen, Thomsen et al., 2010). Epigenetics thus appears to be an exciting area of investigation with the potential for effective new therapies in areas of unmet medical need and the development of new diagnostic, screening, or pharmacogenomic tests.

1.3 PROTEOMICS

Proteomics is the study of proteins, including their location, structure, and function. Proteomics involves the systematic study of proteins in order to provide a comprehensive view of the structure, function, and regulation of biological systems (Patterson and Aebersold, 2003). Although all proteins are based on mRNA precursors, post-translational modifications and environmental interactions make it impossible to predict the abundance of specific proteins based on gene expression analysis alone (Patterson, 2003). In contrast to the genome, the proteome is highly variable over time between cell types and will change in response to its environment. A major challenge is the high variability in proteins and protein abundance in biological specimens (Patterson and Aebersold, 2003). Advances in instrumentation and methodologies have fueled an expansion of the scope of biological studies from simple biochemical analysis of single proteins to measurement of complex protein mixtures. Coupled with advances in bioinformatics, this approach to comprehensively describe biological systems will undoubtedly have a major impact on our understanding of the phenotypes of both normal and diseased cells. Initially, proteomics focused on the generation of protein maps using two-dimensional polyacrylamide gel electrophoresis (PAGE) (Patterson, 2003). The field has since expanded to include not only protein expression profiling, but also the analysis of post-translational modifications and protein-protein interactions. Protein expression, or the quantitative measurement of the global levels of proteins, may still be done with two-dimensional gels; however, mass spectrometry has been incorporated to increase sensitivity and specificity and to provide results in a high-throughput format (Domon and Aebersold, 2006). A variety of platforms such as mass spectrometry, tandem mass spectrometry (MS/MS), and protein microarrays are now available to conduct proteome analysis on a cellular, subcelluar, and organ level (Yates, Gilchrist et al., 2005; Cox and Mann, 2007). The study of protein-protein interactions has been revolutionized by the development of protein microarrays. Analogous to DNA microarrays, these biochips are printed with antibodies or proteins and probed with a complex protein mixture (Ressine, Marko-Varga et al., 2007). The intensity or identity of the resulting protein-protein interactions may be detected by fluorescence imaging or mass spectrometry. Other protein capture methods may be used in place of arrays, including the yeast two-hybrid system or the isolation of protein-protein complexes by affinity chromatography or other separation techniques (Ralser, Goehler et al., 2005).

Although DNA microarray technology provides a wealth of information about the expression and roles of RNA transcripts in states of disease, it is critically important to associate the events at the level of transcription with the actual proteins that are being encoded, translated, and modified. Using multidimensional gel electropheresis, high-throughput mass spectroscopy, various low density arrays for protein-protein interactions, or protein-specific antibody arrays, it is possible to study the proteomes of cells, tissues, and body fluids in search of disease-linked proteins. At the molecular and cellular level, biological functions are carried out by proteins rather than DNA or RNA (with the possible exception of ribozymes) (Kurian, Kirk et al., 1998). Thus, information obtained by proteomic analysis greatly complements data obtained from DNA microarrays.

A major technical challenge for proteomics is the significant increase in the complexity of the proteome, representing several hundred thousand or more proteins, as compared to the RNA transcriptome, which represents about 20,000–30,000 genes total. A major cause for this increased proteomic complexity is splice variants of genes that are manifested as different protein products. Another mechanism is that protein function and activity is regulated or restricted by post-translational and covalent modifications of protein structure (i.e., phosphorylation, sulfation, methylation, and glycosylation), as well as other protein-protein interactions or protein-small molecule interactions. Thus, it is equally important to develop technologies to study the post-translational events of proteins that dictate the biological microenvironment of the cells and tissues and, thus, the entire organism (Sellers and Yates, 2003; Pan, Chen et al., 2008; Pan, Kumar et al., 2009; Pan, Aebersold et al., 2009).

Proteomic analysis is expected to have wide application in the field of medicine by providing unique information about cells and tissues and eventually creating noninvasive tests to monitor biomarkers in body fluids, such as urine or blood, that would correlate clinical analysis. A proteomics application to monitor transplantation acceptance was reported (Pan, Chen et al., 2008; Pan, Kumar et al., 2009; Pan, Aebersold et al., 2009) using 2D PAGE and matrix-assisted laser desorption ionization time-of-flight (MALDI-TOF) in a rat model of liver transplantation. The authors found that haptoglobin, which has been associated with inhibition of T-cell proliferation in studies of cancer patients and some *in vitro* culture assays, was up-regulated following liver transplantation. As additional proof, the level of RNA transcript expression and intracellular localization of haptoglobin correlated with the immune events in the liver, a good example of how proteomics can complement genomics. In the field of kidney transplantation, one of the earliest searches to identify potential biomarker candidates from the urine was performed with surface enhanced laser desorption ionization (SELDI)-TOF mass spectroscopy (Clarke, Silverman et al., 2003). A study in human kidney transplantation using the same technology, SELDI-TOF mass spectroscopy, profiled urinary protein spectra from five groups of subjects: acute rejection, acute tubular necrosis, recurrent or de novo glomerulopathy, stable transplant patients with excellent function, and normal urine donor controls (Schaub, Rush et al., 2004; Schaub, Wilkins et al., 2004a; Schaub, Wilkins et al., 2004b). Two distinct urine protein patterns were observed when comparing the normal controls and stable transplant groups to the acute rejection group. A more recent study looked at the differentiation of BK virus-associated nephropathy from acute allograft rejection in kidney-transplant recipients (Jahnukainen, Malehorn et al., 2006). A plethora of biomarkers exist for diagnosis of nutritional status, metabolic diseases (carbohydrate, amino acid, and fatty acid metabolism), inflammation (C-reactive protein, haptoglobin, orosomucoid, and anti-trypsin) (Agarwal, Binz et al., 2005; Sadrzadeh and Bozorgmehr, 2004; Kanikowska, Grzymislawski et al., 2005; Kanikowska, Hyun et al., 2005), hormonal imbalance (insulin, thyroxine, adrenaline, and pituitary hormones), tissue damage (aspartate transaminase and alanine transaminase for liver and heart, collagen for joints) (Collier, Lecomte et al., 2002; Conigrave, Davies et al., 2003; Collier and Bassendine, 2002; Poole, 2003), cancer (CA15.3, CA27.29, CEA, PSA, S100-β, and hCG) (Rosai, 2003; Shitrit, Zingerman et al., 2005), neurodegeneration (amyloid plaques, and β-amyloid peptide) (Aslan and Ozben, 2004; Bossy-Wetzel, Schwarzenbacher et al., 2004; Teunissen, de Vente et al., 2002), and autoimmune diseases (autoantibodies) (Pender, Csurhes et al., 2000; Masaki and Sugai, 2004; Weetman, 2004a; Weetman, 2004b).

Candidate biomarkers have been identified for a number of diseases, including cancers of different origins (e.g., ovary, breast, and prostate) (Rapkiewicz, Espina et al., 2004), neurological disorders (Austen, Frears et al., 2000), and pathogenic organisms (Lancashire, Schmid et al., 2005), and Alzheimer's disease and diabetes.

1.4 METABOLOMICS

The metabolome consists of small molecules that are involved in the energy transmission in the cells by interacting with other biological molecules following metabolic pathways. In cells, the rate of enzymatic reactions is also regulated by metabolites. The metabolome is highly variable and time dependent and consists of a wide range of chemical structures (Fridman and Pichersky, 2005). It is also important to point out here that metabolomics and metabonomics are generally interchangeable terms. Metabolic phenotypes are the by-products that result from the interaction between genetic, environment, lifestyle, and other factors (Fridman and Pichersky, 2005). Metabolomics, as a method to define the small molecule diversity in the cell and to display differences in small molecule abundance, shows many advantages in terms of metabolic analyses because metabolites are the functional entities within the cells, and their concentration levels vary as a consequence of genetic or physiological changes. An important challenge of metabolomics is to acquire qualitative and quantitative information concerning the metabolites that are perturbed because of changes in environmental factors. Metabolomics analysis is typically performed by employing gas chromatography time-of-flight mass spectrometry, high performance liquid chromatography-mass spectrometry, or capillary electrophoresis mass spectrometry instruments, nuclear magnetic resonance spectroscopy, and more recently vibrational spectroscopy (Robertson, Reily et al., 2005). Metabolome analysis can also be performed through combined application of several technologies together in order to achieve wide coverage and better identification. Compared with transcriptomics and proteomics, improvements in instrumentation and data analysis software are still needed for metabolomic studies.

In animals and humans, metabolic profiling of body fluids to characterize metabolic disorders has been ongoing since the introduction of gas chromatography and mass spectrometry. Nuclear magnetic resonance (NMR) techniques have also been applied for a wide range of components of blood and urine. Current metabolomic studies are making use of technologies such as mass spectrometry (MS), gas chromatography/MS, and NMR to produce metabolic profiles or signatures of toxicity, disease, and drug efficacy. A major aspect of organismal biology is the metabolism and elimination of proteins, hormones, and exogenous molecules, including drugs. In fact, if a given drug therapy resulted in a set of molecular events that created a unique metabolome detected in blood plasma, for example, these metabolic biomarkers could be highly specific as metrics for therapeutic efficacy but actually not be comprised of any of the metabolites of the drug. In other settings, it is hoped that metabolomic profiles of drugs will also correlate with unwanted and dangerous side effects and could therefore be used to enhance the safety of drug therapy.

Metabolic signatures provide prognostic, diagnostic, and surrogate markers for a disease state. For example, NMR spectroscopy of urine and plasma samples was used to examine early graft dysfunction in a pig ischemia/reperfusion model (Holland, Smith et al., 2003) in order to assess and predict early graft dysfunction (Kurian, Flechner et al., 2005). In another study, NMR spectroscopy in combination with pattern recognition tools was used to investigate the composition of organic compounds in urine from patients with multiple sclerosis, patients with other neurological diseases, and healthy controls (Holland, Pfleger et al., 2005). Using the marmoset monkey model of experimental autoimmune encephalomyelitis, the relation of disease progression and alteration of the urine composition was investigated and compared with the measurements obtained with the human patient samples. A recent study has led to the development of a new statistical paradigm to coanalyze NMR and ultra performance liquid chromatography combined with orthogonal acceleration TOF-MS data (Heverhagen, Hartlieb et al., 2002; Hutcheson, Canning et al., 2002) across different samples of urine. Application of these tools has been shown to improve the efficiency of biomarker identification. Finally, another source for metabonomic biomarkers is the low-molecular-weight range serum proteome, the peptidome, which may also contain disease-specific information (Hu, Ye et al., 2009). This seems to be an untapped resource of candidates for new and specific biomarkers, because it is comprised of a multitude of small protein fragments that present a recording or snapshot of events taking place at the level of disease-associated microenvironments. Because intact tissue proteins are too large to passively diffuse through the cell and across the endothelial basement membranes into the circulation, the peptidome could provide an accessible portal to identify and quantify a wide range of protein changes that are taking place in all of the cells and tissues (Hu, Ye et al., 2009). Therefore, metabolomics appears to be a valuable platform for studies of complex diseases and for the development of new therapies both in nonclinical disease model characterization and in clinical settings.

1.5 CELLOMICS

The field of cellomics was driven by the need to define the functions of genes and the proteins that they encoded. It was apparent by the mid-1990s that knowing the human genome was the start not the end of the biological challenge for basic research and drug discovery. Light microscopy, especially digital imaging fluorescence microscopy on living cells, was chosen as the best approach to defining the functions of genes and proteins (Yasuda, 2010). Human interactive imaging methods were pretty well developed by the 1980s, and fundamental information about the temporal and spatial dynamics of cells and their constituents was being published by a growing academic community. However, the human interactive imaging tools in the absence of automated imaging methods and informatics tools to archive, mine, and display complex imaging data made the process of studying cells time consuming and complicated (Yasuda, 2010). Similar to the field of genomics, there was a need for the development of an automated system to acquire, process, analyze, display, and mine massive amounts of cellular data derived from arrays of cells treated in various ways. This need for high-content screening of cells has paved the way for developing novel technologies such as automated digital microscopy and flow cytometry and Arrayscan, to offer a complete solution for single cell analysis. These technologies are currently being put to use in biomedicine.

1.6 LIPIDOMICS

Lipidomics, the systems-level analysis of lipids and their interacting partners, can be viewed as a subdiscipline of metabolomics. An enormous number of chemically distinct molecular species arise from the various combinations of fatty acids with backbone structures (Blanksby and Mitchell, 2010; Shevchenko and Simons, 2010). Lipidomics is the emerging field of systems-level analysis of lipids and factors that interact with lipids (Wenk, 2005). Although important, the study of lipids has been hampered by analytical limitations. Lipids are molecules that are highly soluble in organic solvents. It is clear, however, that without special precautions many classes of lipid molecules (such as the very polar phosphoinositides) will escape into the aqueous milieu during phase partitioning (Brown and Murphy, 2009).

Lipids, the fundamental components of biological membranes, play multiple important roles in biological systems. The most important functions are creating in the cell a subsystem in the context of the whole and relatively independent of the exterior environment through lipid bilayer structures, providing an appropriate hydrophobic medium for the functional implementations of membrane proteins and their interactions and producing second messengers by enzyme reactions (Brown and Murphy, 2009). Abnormal lipid metabolism has been observed in numerous human diseases such as diabetes, obesity, atherosclerosis, and Alzheimer's disease, leading to tremendous interest in lipid research in biomedical research (Aukrust, Muller et al., 1999; Hjelmesaeth, Hartmann et al., 2001). Current research on lipids tends to shift from determining the individual molecular structures of single lipids in biological samples to characterizing global changes of lipid metabolites in a systems-integrated context in order to understand the crucial role of lipids in physiopathology (Wenk, 2005). Traditional strategies for lipid analysis usually prefractionate lipids into classes using thin-layer chromatography normal-phase liquid chromatography, or solid-phase extraction and then separate particular classes of lipids into individual molecular species by high-performance liquid chromatography coupled with either ultraviolet or evaporative light-scattering detector. However, such classical techniques often either lack sensitivity or require large sample volumes and multi-step procedures for sample preparation, and the resolution is limited, i.e. only a limited set of individual molecular species are analyzed. Recent advancements in mass spectrometry and innovations in chromatographic technologies have largely driven the development of high-throughput analysis of lipids. With the advent of soft ionization, technologies such as matrix-assisted laser desorption/ionization, electrospray ionization, and atmospheric pressure chemical ionization for MS, possibly coupled to liquid chromatography (LC) rapid and sensitive analysis of the majority or a substantial fraction of lipids possible in one analysis, is currently possible. Most common strategies currently used in lipidomics include direct-infusion electrospray ionization (ESI)-MS and ESI-MS/MS, LC coupled with ESI-MS or MS/MS, and MALDI combined with Fourier transform ion cyclotron resonance MS or MALDI-TOF-MS) (van Meer, 2005; Wenk, 2005).

Despite all advances recently made, the diversity of structures and properties and the wide range of concentrations of lipids provide a huge and almost impossible challenge for analytical methodology when aiming at a single technological platform capable of measuring and identifying all lipids in a single sample simultaneously. As a consequence, multiple, often complementary, analytical approaches are currently used in the field of lipidomics (van Meer, 2005).

1.7 GLYCOMICS

The term glycomics is derived from the chemical prefix for sweetness or a sugar, *glyco*, and was formed to follow the naming convention established by genomics and proteomics (Liang, Wu et al., 2008). Glycomics is an integrated approach to study structure-function relationships of complex carbohydrates or glycans such as glycolipids, glycoproteins, lipopolysaccharides, peptidoglycans, and proteoglycans. Comparative studies of specific carbohydrate chains of glycoproteins can provide useful information for the diagnosis, prognosis, and immunotherapy of tumors. Glycan-based drugs have generated much excitement and provided important insight into the power of glycan-based therapeutics. However, the ultimate promise of glycans as drugs is only beginning to be exploited. The emerging omics domain of glycomics has lagged behind that of genomics and proteomics, mainly because of the inherent difficulties in analysis of glycan structure and functions. A wide variety of technologies are now being brought to bear on the technically difficult problems of glycan structural analysis and investigation of functional roles. Enabling technologies such as high-throughput mass spectroscopy, glycan microarrays, aminoglycoside antibiotic microarrays and glycan sequencing, quantum dots, and gold nano particles are currently helping to unravel the complexity resulting from diverse glycans in biological systems (Liang, Wu et al., 2008). In an effort to harness the promise of glycans as therapeutics, advances have been made in analyzing glycan structures in a rapid manner using a minimum of material, in synthesizing glycan structures *in vitro*, and in harnessing endogenous glycosylation pathways *in vivo* to create new reproducible glycan structures. Recently, there has been a marked increase in the reporting of techniques that have been successfully applied to the analysis of complex glycans and glycoconjugates, including MS (Kaji, Saito et al., 2003; Zhang, Cocklin et al., 2003) and capillary electrophoretic (Que, Mechref et al., 2003) techniques. Many of these technologies have distinct advantages compared with traditional analytical methodologies, including the ability to analyze minute amounts of biologically based material.

Comparative studies of glycans can provide useful information for the diagnosis, prognosis, and therapy for several diseases. For example, glycoproteomics analysis was used to discover serum markers in liver cancer; GP73, a glycoprotein, was found to be elevated in hepatocellular carcinoma, and this marker have been successfully used as a positive predictor of diagnosis and treatment (Marrero and Lok, 2004).

1.8 FUTURE PROSPECTS OF OMICS: INTEGRATION OF OMICS TECHNOLOGIES AND APPLICATIONS IN CLINICAL MEDICINE

The omics field is now transforming biomedical research where one gene or protein was studied at a time to a world in which whole organelles and pathways are studied simultaneously using less biological material. An integrative approach using the data collected by various omics platforms developed by companies shown in Table 1.1 is expected to fulfill the dream of specific disease biomarkers, individualized care, and treatment of human diseases. Whereas high-throughput omics approaches to analyze molecules at different cellular levels are rapidly becoming available, it is

TABLE 1.1
List of Biotechnology Companies and Their Resources

Biotechnology Companies	Technology	Tools
Affymetrix	Genomics, microarrays, sequencing	Gene chips
Agilent Technologies	Genomics, protomics, microarrays, sequencing	Glass arrays
Applied Biosystems/Life Technologies	Genomics, protomics, microarrays, sequencing	Arrays
Axcell Biosciences	Informatics	Proteomic database
Bio-Rad	Genomics	Arrays
Celera	Genomics	Informatics
Cellomics Inc.	Cellomics	High content screening
Ciphergen Biosystems	Proteomics, toxicogenomics	Proteinchip systems
Decode Genetics	Genomics, proteomics	Bioinformatics
Epigenomics	Epigenomics, pharmacogenomics	Arrays
Global Lipidomics	Lipidomics	Service
Human Metabolome Technologies	Metabolomics	CE-MS tool
Illumina	Genomics	Bead arrays
Incyte Genomics	Proteomics, toxicogenomics	Informatics
Large Scale Biology Corp	Proteomics, toxicogenomics	2-DE
Nimblegen Systems	Genomics	Arrays
PerkinElmer	Genomics, proteomics	Reagents
Proteome Inc.	Proteomics, toxicogenomics	Service
Proteome Sciences plc.	Proteomics, toxicogenomics	2D gel electrophoresis
Sequenom Inc.	Genomics	DNA analysis tools
Sigma-Aldrich	Genomics	Reagents, arrays

also becoming clear that any single omics approach may not be sufficient to characterize the complexity of biological systems (Gygi, Han et al., 1999; Gygi, Rochon et al., 1999). For example, the expression level of a given gene does not indicate the amount of protein produced nor its location, biological activity, or functional relationship with metabolomes. Moreover, in cells, many levels of regulation occur after genes have been transcribed, such as post-transcriptional, translational, and post-translational regulation and all forms of biochemical control such as allosteric or feedback regulation. For example, in a study by ter Kuile and Westerhoff (2001), control of glycolysis was shown to be shared between metabolic, proteomic, and genomic levels, thereby suggesting that the functional genomics cannot stop at the mRNA level or any single level of information. Integrated multiomics approaches have been applied recently, and the studies have enabled researchers to unravel global regulatory mechanisms and complex metabolic networks in various eukaryotic organisms (Hegde, White et al., 2003; Mootha, Bunkenborg et al., 2003; Ray, Mootha et al., 2003; Alter and Golub, 2004). These early studies have clearly demonstrated that integrated omics analysis may be a key to decipher complex biological systems. It is widely believed that the application of the currently available omics technologies as depicted in Figure 1.3 will not only have an impact on our understanding of biological processes, but will also improve the prospect of more accurately diagnosing and treating diseases. The reality of applying omics technologies to unravel disease processes, identify disease biomarkers, and finally make a recommendation for personalized medicine is expected to revolutionize medical practice. Several examples of different omics technologies that have been integrated and methods used in studying diseases in human and animal models have been summarized in Table 1.2.

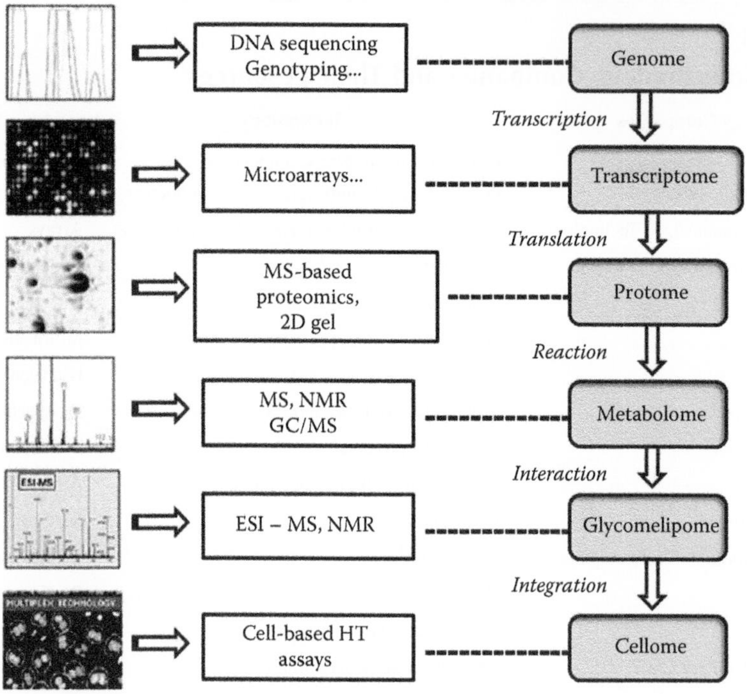

FIGURE 1.3 Technologies for omics analysis.

TABLE 1.2
Integration of Omics Technologies and Applications in Clinical Medicine

Disease Pathology	Omics Technology	References
Alzheimer's disease, Parkinson's disease, and multiple sclerosis	Metabolomics, mass spectrometry	Alimonti, Ristori et al. (2007)
Coronary disease	Lipidomics, liquid chromatography-mass spectrometry	Bergheanu, Reijmers et al. (2008)
Parkinson disease	Metabolomics, high-performance liquid chromatography, electrochemical colorimetric array detection	Bogdanov, Matson et al. (2008)
Biomarkers/kideney transplanation	SELDI-TOF, mass spectrometry	Clarke, Silverman et al. (2003); Schaub, Wilkins et al. (2004)
Diabetes, obesity, coronary heart disease	Functional genomics, metabonomics, NMR spectroscopy, mass spectrometry	Brindle (2002); Griffin and Vidal-Puig (2008)
Muscular dystrophy in mice	Metabolomics, NMR	Griffin (2006)
Crohn's disease and ulcerative colitis	Proteomics, genomics, protein microarrays	Kader, Tchernev et al. (2005)
Ovarian cancer	Glycomics, mass spectrometry (MALDI-FTMS)	Leiserowitz, Lebrilla et al. (2008)
Leigh syndrome, mitochondrial complex I deficiency	Proteomics, PAGE, LC-MS/MS, genomics, homozygosity mapping, Affymetrix GeneChip mapping	Pagliarini, Calvo et al. (2008)
Various cancers	Genomics, transcriptomics, RNA interference	Pai, Lin et al. (2006)
Organ transplantation/rejection	Proteomics/2D PAGE/MALDI-TOF	Pan, Jain et al. (2004)

TABLE 1.2 (CONTINUED)
Integration of Omics Technologies and Applications in Clinical Medicine

Disease Pathology	Omics Technology	References
Kidney cancer	Proteomics, metabolic profiling, PAGE, MS, immunoblotting	Perroud, Lee et al. (2006)
Type II diabetes and dyslipidemia	Metabolomics, biofluid NMR spectroscopy	Ringeissen, Connor et al. (2003)
Amyotrophic lateral sclerosis in a mouse model	Genomics, proteomics, immunochemistry, genetic engineering, gene silencing	Saito, Yokota et al. (2005)
Gene expression in human liver	Genomics, expression profiling, genotyping	Schadt, Molony et al. (2008)
Phenylketonuria	Genomics, population genetics, metabolomics	Scriver (2007)
Alzheimer's disease	Genotyping, genomics	Seshadri, Fitzpatrick et al. (2010)
Normal glucose metabolism, homeostasis, insulin sensitivity	Metabolic profiling, metabolomics, LC-MS/MS, radioimmunoassay, hexokinase assay	Shaham, Wei et al. (2008)
Rheumatoid arthritis, hypertension, Crohn's disease, coronary artery disease, bipolar disorder, diabetes	Genomics, genome-wide association, genotyping, GeneChip arrays	The Wellcome Trust Case Control Consortium (2007)
Crohn's disease and ulcerative colitis	Genomics, expression microarrays, quantitative reverse transcription-polymerase chain reaction	Wu, Dassopoulos et al. (2007)
Plant storage proteins, allergens	Proteomics, affinity columns, PAGE	Yano and Kuroda (2008)

REFERENCES

Abraham, V. C., Taylor, D. L., et al. (2004). High content screening applied to large-scale cell biology. *Trends Biotechnol.* 22: 15–22.

Adorjan, P., Distler, J., et al. (2002). Tumour class prediction and discovery by microarray-based DNA methylation analysis. *Nucleic Acids Res.* 30: e21.

Agarwal, R., Binz, T., et al. (2005). Analysis of active site residues of botulinum neurotoxin E by mutational, functional, and structural studies: Glu335Gln is an apoenzyme. *Biochemistry* 44: 8291–8302.

Algeciras-Schimnich, A., O'Kane, D. J., et al. (2008). Pharmacogenomics of tamoxifen and irinotecan therapies. *Clin. Lab. Med.* 28: 553–567.

Alimonti, A., Ristori, G., et al. (2007). Serum chemical elements and oxidative status in Alzheimer's disease, Parkinson disease and multiple sclerosis. *Neurotoxicol.* 28: 450–456.

Alizadeh, A. A., Eisen, M. B., et al. (2000). Distinct types of diffuse large B-cell lymphoma identified by gene expression profiling. *Nature* 403: 503–511.

Alizadeh, A. A., and Staudt, L. M. (2000). Genomic-scale gene expression profiling of normal and malignant immune cells. *Curr. Opin. Immunol.* 12: 219–225.

Alter, O., and Golub, G. H. (2004). Integrative analysis of genome-scale data by using pseudoinverse projection predicts novel correlation between DNA replication and RNA transcription. *Proc. Natl. Acad. Sci. U.S.A.* 101: 16577–16582.

Aslan, M., and Ozben, T. (2004). Reactive oxygen and nitrogen species in Alzheimer's disease. *Curr. Alzheimer Res.* 1: 111–119.

Aukrust, P., Muller, F., et al. (1999). Enhanced levels of soluble and membrane-bound CD40 ligand in patients with unstable angina: Possible reflection of T lymphocyte and platelet involvement in the pathogenesis of acute coronary syndromes. *Circulation* 100: 614–620.

Austen, B. M., Frears, E. R., et al. (2000). The use of seldi proteinchip arrays to monitor production of Alzheimer's betaamyloid in transfected cells. *J. Pept. Sci.* 6: 459–469.

Baechler, E. C., Batliwalla, F. M., et al. (2003). Interferon-inducible gene expression signature in peripheral blood cells of patients with severe lupus. *Proc. Natl. Acad. Sci. U.S.A.* 100: 2610–2615.

Banerjee, H. N., and Verma, M. (2009). Epigenetic mechanisms in cancer. *Biomark Med.* 3: 397–410.

Barron, A. (2008). DNA sequencing and genotyping. *Electrophoresis* 29: 4617.

Berchuck, A., Iversen, E. S., et al. (2005). Patterns of gene expression that characterize long-term survival in advanced stage serous ovarian cancers. *Clin. Cancer Res.* 11: 3686–3696.

Bergheanu, S. C., Reijmers, T., et al. (2008). Lipidomic approach to evaluate rosuvastatin and atorvastatin at various dosages: Investigating differential effects among statins. *Curr. Med. Res. Opin.* 24: 2477–2487.
Bier, F. F., von Nickisch-Rosenegk, M., et al. (2008). DNA microarrays. *Adv. Biochem. Eng. Biotechnol.* 109: 433–453.
Blanksby, S. J., and Mitchell, T. W. (2010). Advances in mass spectrometry for lipidomics. *Annu. Rev. Anal. Chem.* 3: 433–465.
Bogdanov, M., Matson W. R., et al. (2008). Metabolomic profiling to develop blood biomarkers for Parkinson's disease. *Brain* 131: 389–396.
Bossy-Wetzel, E., Schwarzenbacher, R., et al. (2004). Molecular pathways to neurodegeneration. *Nat. Med.* 10(Suppl.): S2–S9.
Brindle, K. M. (2002). Detection of apoptosis in tumors using magnetic resonance imaging and spectroscopy. *Adv. Enzyme Regul.* 42: 101–112.
Brown, H. A., and Murphy, R. C. (2009). Working towards an exegesis for lipids in biology. *Nat. Chem. Biol.* 5: 602–606.
Chabas, A., Montfort, M., et al. (2001). Mutation and haplotype analyses in 26 Spanish Sanfilippo syndrome type A patients: Possible single origin for 1091delC mutation. *Am. J. Med. Genet.* 100: 223–228.
Chabas, D., Baranzini, S. E., et al. (2001). The influence of the proinflammatory cytokine, osteopontin, on autoimmune demyelinating disease. *Science* 294: 1731–1735.
Choudhuri, S. (2009). Looking back to the future: From the development of the gene concept to toxicogenomics. *Toxicol. Mech. Methods* 19: 263–277.
Clarke, W., Silverman, B. C., et al. (2003). Characterization of renal allograft rejection by urinary proteomic analysis. *Ann. Surg.* 237: 660–665.
Collier, J., and Bassendine, M. (2002). How to respond to abnormal liver function tests. *Clin. Med.* 2: 406–409.
Collier, T. L., Lecomte, R., et al. (2002). Assessment of cancer-associated biomarkers by positron emission tomography: Advances and challenges. *Dis. Markers* 18: 211–247.
Conigrave, K. M., Davies, P., et al. (2003). Traditional markers of excessive alcohol use. *Addiction* 98(Suppl. 2): 31–43.
Cox, J., and Mann, M. (2007). Is proteomics the new genomics? *Cell* 130: 395–398.
Crick, F. H. (1954). The Complementary Structure of DNA. *Proc. Natl. Acad. Sci. U.S.A.* 40: 756–758.
Domon, B., and Aebersold, R. (2006). Mass spectrometry and protein analysis. *Science* 312: 212–217.
Evans, W. E., and Relling, M. V. (1999). Pharmacogenomics: Translating functional genomics into rational therapeutics. *Science* 286: 487–491.
Ferrer-Dufol, A., and Menao-Guillen, S. (2009). Toxicogenomics and clinical toxicology: An example of the connection between basic and applied sciences. *Toxicol. Lett.* 186: 2–8.
Fraga, M. F., Agrelo, R., et al. (2007). Cross-talk between aging and cancer: The epigenetic language. *Ann. N.Y. Acad. Sci.* 1100: 60–74.
Fraga, M. F., and Esteller, M. (2007). Epigenetics and aging: The targets and the marks. *Trends Genet.* 23: 413–418.
Fridman, E., and Pichersky, E. (2005). Metabolomics, genomics, proteomics, and the identification of enzymes and their substrates and products. *Curr. Opin. Plant Biol.* 8: 242–248.
Gallagher, W. M., Tweats, D., et al. (2009). Omic profiling for drug safety assessment: Current trends and public-private partnerships. *Drug Discov. Today* 14: 337–342.
Gant, T. W. (2003). Application of toxicogenomics in drug development. *Drug News Perspect.* 16: 217–221.
Ge, F., and He, Q. Y. (2009). Genomic and proteomic approaches for predicting toxicity and adverse drug reactions. *Expert Opin. Drug Metab. Toxicol.* 5: 29–37.
Golub, T. R., Slonim, D. K., et al. (1999). Molecular classification of cancer: Class discovery and class prediction by gene expression monitoring. *Science* 286: 531–537.
Griffin, J. L. (2006). Understanding mouse models of disease through metabolomics. *Curr. Opin. Chem. Biol.* 10: 309–315.
Griffin, J. L., and Vidal-Puig, A. (2008). Current challenges in metabolomics for diabetes research: A vital functional genomic tool or just a ploy for gaining funding? *Physiol. Genom.* 34: 1–5.
Guguen-Guillouzo, C., and Guillouzo, A. (2010). General review on in vitro hepatocyte models and their applications. *Methods Mol. Biol.* 640: 1–40.
Gygi, S. P., Han, D. K., et al. (1999). Protein analysis by mass spectrometry and sequence database searching: Tools for cancer research in the post-genomic era. *Electrophoresis* 20: 310–319.
Gygi, S. P., Rochon, Y., et al. (1999). Correlation between protein and mRNA abundance in yeast. *Mol. Cell. Biol.* 19: 1720–1730.

Hegde, P. S., White, I. R., et al. (2003). Interplay of transcriptomics and proteomics. *Curr. Opin. Biotechnol.* 14: 647–651.

Herceg, Z. (2007). Epigenetics and cancer: Towards an evaluation of the impact of environmental and dietary factors. *Mutagenesis* 22: 91–103.

Heverhagen, J. T., Hartlieb, T., et al. (2002). Magnetic resonance cystometry: Accurate assessment of bladder volume with magnetic resonance imaging. *Urology* 60: 309–314.

Hirode, M., Omura, K., et al. (2009). Gene expression profiling in rat liver treated with various hepatotoxic-compounds inducing coagulopathy. *J. Toxicol. Sci.* 34: 281–293.

Hjelmesaeth, J., Hartmann, A., et al. (2001). Metabolic cardiovascular syndrome after renal transplantation. *Nephrol. Dial. Transplant.* 16: 1047–1052.

Holland, N. T., Pfleger, L., et al. (2005). Molecular epidemiology biomarkers: Sample collection and processing considerations. *Toxicol. Appl. Pharmacol.* 206: 261–268.

Holland, N. T., Smith, M. T., et al. (2003). Biological sample collection and processing for molecular epidemiological studies. *Mutat. Res.* 543: 217–234.

Hu, L., Ye, M., et al. (2009). Recent advances in mass spectrometry-based peptidome analysis. *Expert Rev. Proteomics* 6: 433–447.

Huang, G., Song, Y., et al. (2003). [The bystander effect of HSV-tk/GCV system on human cervical carcinoma cell line ME180]. *Sichuan Da Xue Xue Bao Yi Xue Ban* 34: 51–54.

Hutcheson, J. C., Canning, D. A., et al. (2002). Magnetic resonance imaging of fetal urinoma. *Urology* 60: 697.

Jahnukainen, T., Malehorn, D., et al. (2006). Proteomic analysis of urine in kidney transplant patients with BK virus nephropathy. *J. Am. Soc. Nephrol.* 17: 3248–3256.

Johnson, W. E., Evans, H., et al. (2001). Immunohistochemical detection of Schwann cells in innervated and vascularized human intervertebral discs. *Spine* 26: 2550–2557.

Kader, H. A., Tchernev, V. T., et al. (2005). Protein microarray analysis of disease activity in pediatric inflammatory bowel disease demonstrates elevated serum PLGF, IL-7, TGF-beta1, and IL-12p40 levels in Crohn's disease and ulcerative colitis patients in remission versus active disease. *Am. J. Gastroenterol.* 100: 414–423.

Kaji, H., Saito, H., et al. (2003). Lectin affinity capture, isotope-coded tagging and mass spectrometry to identify N-linked glycoproteins. *Nat. Biotechnol.* 21: 667–672.

Kanikowska, D., Grzymislawski, M., et al. (2005). Seasonal rhythms of acute phase proteins in humans. *Chronobiol. Int.* 22: 591–596.

Kanikowska, D., Hyun, K. J., et al. (2005). Circadian rhythm of acute phase proteins under the influence of bright/dim light during the daytime. *Chronobiol. Int.* 22: 137–143.

Kurian, E., Kirk, W. R., et al. (1998). Affinity of fatty acid for rRat intestinal fatty acid binding protein: Further examination. *Biochemistry* 37: 6614.

Kurian, S. M., Flechner, S. M., et al. (2005). Laparoscopic donor nephrectomy gene expression profiling reveals upregulation of stress and ischemia associated genes compared to control kidneys. *Transplantation* 80: 1067–1071.

Lancashire, L., Schmid, O., et al. (2005). Classification of bacterial species from proteomic data using combinatorial approaches incorporating artificial neural networks, cluster analysis and principal components analysis. *Bioinformatics* 21: 2191–2199.

Leiserowitz, G. S., Lebrilla, C., et al. (2008). Glycomics analysis of serum: A potential new biomarker for ovarian cancer? *Int. J. Gynecol. Cancer* 18: 470–475.

Liang, P. H., Wu, C. Y., et al. (2008). Glycan arrays: Biological and medical applications. *Curr. Opin. Chem. Biol.* 12: 86–92.

Luch, A. (2009). Preface in *Molecular, clinical and environmental toxicology*. Berlin: Department for Product Safety and ZEBET, Federal Institute for Risk Assessment.

Marrero, J. A., and Lok, A. S. (2004). Newer markers for hepatocellular carcinoma. *Gastroenterology* 127(Suppl. 1): S113–S119.

Masaki, Y., and Sugai, S. (2004). Lymphoproliferative disorders in Sjogren's syndrome. *Autoimmun. Rev.* 3: 175–182.

Mei, N., Fuscoe, J. C., et al. (2010). Application of microarray-based analysis of gene expression in the field of toxicogenomics. *Methods Mol. Biol.* 597: 227–241.

Mikkelsen, J. D., Thomsen, M. S., et al. (2010). Use of biomarkers in the discovery of novel anti-schizophrenia drugs. *Drug Discov. Today* 15: 137–141.

Mootha, V. K., Bunkenborg, J., et al. (2003). Integrated analysis of protein composition, tissue diversity, and gene regulation in mouse mitochondria. *Cell* 115: 629–640.

Moreira, E. G., Yu, X., et al. (2010). Toxicogenomic profiling in maternal and fetal rodent brains following gestational exposure to chlorpyrifos. *Toxicol. Appl. Pharmacol.* 245: 310–325.

Ness, S. A. (2007). Microarray analysis: basic strategies for successful experiments. *Mol. Biotechnol.* 36: 205–219.

Pagliarini, D. J., Calvo, S. E., et al. (2008). A mitochondrial protein compendium elucidates complex I disease biology. *Cell* 134: 112–123.

Pai, S. I., Lin, Y. Y., et al. (2006). Prospects of RNA interference therapy for cancer. *Gene Ther.* 13: 464–477.

Pan, C., Jain, A., et al. (2004). Analysis of causes of late mortality in liver transplant recipients. *Chinese Critical Care Med.* 16: 547–551.

Pan, C., Kumar, C., et al. (2009). Comparative proteomic phenotyping of cell lines and primary cells to assess preservation of cell type-specific functions. *Mol. Cell Proteomics* 8: 443–450.

Pan, J., Chen, H. Q., et al. (2008). Comparative proteomic analysis of non-small-cell lung cancer and normal controls using serum label-free quantitative shotgun technology. *Lung* 186: 255–261.

Pan, S., Aebersold, R., et al. (2009). Mass spectrometry based targeted protein quantification: methods and applications. *J. Proteome Res.* 8: 787–797.

Patterson, S. D. (2003). Proteomics: Evolution of the technology. *Biotechniques* 35: 440–444.

Patterson, S. D., and Aebersold, R. H. (2003). Proteomics: The first decade and beyond. *Nat. Genet.* 33(Suppl.): 311–323.

Pender, M. P., Csurhes, P. A., et al. (2000). Surges of increased T cell reactivity to an encephalitogenic region of myelin proteolipid protein occur more often in patients with multiple sclerosis than in healthy subjects. *J. Immunol.* 165: 5322–5331.

Perroud, B., Lee, J., et al. (2006). Pathway analysis of kidney cancer using proteomics and metabolic profiling. *Mol. Cancer* 5: 64.

Pettit, S., des Etages, S. A., et al. (2010). Current and future applications of toxicogenomics: Results summary of a survey from the HESI Genomics State of Science Subcommittee. *Environ. Health Perspect.* 118: 992–997.

Poole, A. R. (2003). Biochemical/immunochemical biomarkers of osteoarthritis: Utility for prediction of incident or progressive osteoarthritis. *Rheum. Dis. Clin. North Am.* 29: 803–818.

Que, A. H., Mechref, Y., et al. (2003). Coupling capillary electrochromatography with electrospray Fourier transform mass spectrometry for characterizing complex oligosaccharide pools. *Anal. Chem.* 75: 1684–1690.

Ralser, M., Goehler, H., et al. (2005). Generation of a yeast two-hybrid strain suitable for competitive protein binding analysis. *Biotechniques* 39: 165–166, 168.

Rapkiewicz, A. V., Espina, V., et al. (2004). Biomarkers of ovarian tumours. *Eur. J. Cancer* 40: 2604–2612.

Ray, H. N., Mootha, V. K., et al. (2003). Building an application framework for integrative genomics. *AMIA Annu. Symp. Proc.* 981.

Ressine, A., Marko-Varga, G., et al. (2007). Porous silicon protein microarray technology and ultra-/superhydrophobic states for improved bioanalytical readout. *Biotechnol. Annu. Rev.* 13: 149–200.

Ringeissen, S., Connor, S. C., et al. (2003). Potential urinary and plasma biomarkers of peroxisome proliferation in the rat: Identification of N-methylnicotinamide and N-methyl-4-pyridone-3-carboxamide by 1H nuclear magnetic resonance and high performance liquid chromatography. *Biomarkers* 8: 240–271.

Robertson, D. G., Reily, M. D., et al. (2005). Metabonomics in preclinical drug development. *Expert Opin. Drug Metab. Toxicol.* 1: 363–376.

Roden, D. M., Altman, R. B., et al. (2006). Pharmacogenomics: Challenges and opportunities. *Ann. Intern. Med.* 145: 749–757.

Rosai, J. (2003). Immunohistochemical markers of thyroid tumors: Significance and diagnostic applications. *Tumorigenesis* 89: 517–519.

Sadrzadeh, S. M., and Bozorgmehr, J. (2004). Haptoglobin phenotypes in health and disorders. *Am. J. Clin. Pathol.* 121(Suppl.): S97–S104.

Saito, Y., Yokota, T., et al. (2005). Transgenic small interfering RNA halts amyotrophic lateral sclerosis in a mouse model. *J. Biol. Chem.* 280: 42826–42830.

Schadt, E. E., Molony, C., et al. (2008). Mapping the genetic architecture of gene expression in human liver. *PLoS Biol.* 6: e107.

Schaub, S., Rush, D., et al. (2004a). Proteomic-based detection of urine proteins associated with acute renal allograft rejection. *J. Am. Soc. Nephrol.* 15: 219–227.

Schaub, S., Wilkins, J., et al. (2004b). Urine protein profiling with surface-enhanced laser-desorption/ionization time-of-flight mass spectrometry. *Kidney Int.* 65: 323–332.

Schaub, S., Wilkins, J. A., et al. (2004). Proteomics in renal transplantation: Opportunities and challenges. *Clin. Transpl.* 253–260.

Schubeler, D. (2009). Epigenomics: Methylation matters. *Nature* 462: 296–297.

Scriver, C. R. (2007). The PAH gene, phenylketonuria, and a paradigm shift. *Hum. Mutat.* 28: 831–845.

Sellers, T. A., and Yates, J. R. (2003). Review of proteomics with applications to genetic epidemiology. *Genet. Epidemiol.* 24: 83–98.

Seshadri, S., Fitzpatrick, A. L., et al. (2010). Genome-wide analysis of genetic loci associated with Alzheimer disease. *JAMA* 303: 1832–1840.

Shaham, O., Wei, R., et al. (2008). Metabolic profiling of the human response to a glucose challenge reveals distinct axes of insulin sensitivity. *Mol. Syst. Biol.* 4: 214.

Shevchenko, A., and Simons, K. (2010). Lipidomics: Coming to grips with lipid diversity. *Nat. Rev. Mol. Cell Biol.* 11: 593–598.

Shitrit, D., Zingerman, B., et al. (2005). Diagnostic value of CYFRA 21-1, CEA, CA 19-9, CA 15-3, and CA 125 assays in pleural effusions: Analysis of 116 cases and review of the literature. *Oncologist* 10: 501–507.

Shostak, S. (2005). The emergence of toxicogenomics: A case study of molecularization. *Soc. Stud. Sci.* 35: 367–403.

Smirnov, D. A., Morley, M., et al. (2009). Genetic analysis of radiation-induced changes in human gene expression. *Nature* 459: 587–591.

ter Kuile, B. H., and Westerhoff, H. V. (2001). Transcriptome meets metabolome: Hierarchical and metabolic regulation of the glycolytic pathway. *FEBS Lett.* 500: 169–171.

Teunissen, C. E., de Vente, J., et al. (2002). Biochemical markers related to Alzheimer's dementia in serum and cerebrospinal fluid. *Neurobiol. Aging* 23: 485–508.

Thompson, K. (2010). Toxicogenomics and studies of genomic effects of dietary components. *World Rev. Nutr. Diet* 101: 115–122.

Van Aggelen, G., Ankley, G. T., et al. (2010). Integrating omic technologies into aquatic ecological risk assessment and environmental monitoring: hurdles, achievements, and future outlook. *Environ. Health Perspect.* 118: 1–5.

van Meer, G. (2005). Cellular lipidomics. *EMBO J.* 24: 3159–3165.

Weetman, A. P. (2004a). Autoimmune thyroid disease. *Autoimmunity* 37: 337–340.

Weetman, A. P. (2004b). Cellular immune responses in autoimmune thyroid disease. *Clin. Endocrinol.* 61: 405–413.

The Wellcome Trust Case Control Consortium. (2007). Genome-wide association study of 14,000 cases of seven common diseases and 3,000 shared controls. *Nature* 447: 661–678.

Wenk, M. R. (2005). The emerging field of lipidomics. *Nat. Rev. Drug Discov.* 4: 594–610.

Wild, C. P. (2010). OMICS technologies: An opportunity for two-way translation from basic science to both clinical and population-based research. *Occup. Environ. Med.* 67: 75–76.

Wu, F., Dassopoulos, T., et al. (2007). Genome-wide gene expression differences in Crohn's disease and ulcerative colitis from endoscopic pinch biopsies: Insights into distinctive pathogenesis. *Inflamm. Bowel Dis.* 13: 807–821.

Yano, H., and Kuroda, S. (2008). Introduction of the disulfide proteome: Application of a technique for the analysis of plant storage proteins as well as allergens. *J. Proteome Res.* 7: 3071–3079.

Yasuda, K. (2010). [On-chip Cellomics technology for drug screening system using cardiomyocyte cells from human stem cell]. *Yakugaku Zasshi* 130: 545–557.

Yates, J. R., 3rd, Gilchrist, A., et al. (2005). Proteomics of organelles and large cellular structures. *Nat. Rev. Mol. Cell Biol.* 6: 702–714.

Zarbl, H. (2007). Toxicogenomic analyses of genetic susceptibility to mammary gland carcinogenesis in rodents: Implications for human breast cancer. *Breast Dis.* 28: 87–105.

Zhang, Y., Cocklin, R. R., et al. (2003). Rapid determination of advanced glycation end products of proteins using MALDI-TOF-MS and PERL script peptide searching algorithm. *J. Biomol. Tech.* 14: 224–230.

Section I

Methodology and Application

2 Bioinformatics
A Brief Introduction to Changing Trends in Modern Biology

Abhishek Angelo Singh
Maastricht University
Maastricht, the Netherlands

Pallavi Somvanshi
Bioinformatics Centre
Lucknow, India

CONTENTS

2.1 INTRODUCTION

Computers have become ubiquitous in the field of biological science. They serve a range of purposes, including collecting and processing signals detected by DNA sequencers, storing data in public repositories, and annotating data and performing simulations on the stored data. Biological molecules like DNA, RNA, and proteins are information carriers, instructing the system when and where a particular biological process needs to take place. In each human cell, approximately 5,000 different proteins are expressed (Celis et al., 1991); efficient handling of this massive amount of data requires strong computational resources. In the world of computer science, data consist of a numerical value or a value that can be processed by a computer, whereas in the biological world, data consist of raw knowledge and observations and can be seen as pieces of a puzzle that needs to be put in correct perspective for a clear vision. Contributions from the fields of computer science and mathematics have been very significant in expanding our understanding of cell mechanisms at the molecular level. The continuous advancements in computing technologies, supercomputers, and computer clusters are bringing significant changes to modern biology. The past few decades have witnessed steady progress in the field of bioinformatics. The late 1960s witnessed the development of algorithms responsible for the construction of phylogenetic trees and protein sequence alignment (Fitch and Margoliash, 1967; Cantor, 1968); the 1970s witnessed the development of algorithms for secondary structure prediction of RNA and proteins (Tinoco et al., 1971; Chou and Fasman, 1974); the 1980s witnessed development in sequence analysis, protein structure prediction (tertiary), molecular evolution, and database development; while the past decade witnessed development in methods of *in silico* drug designing.

The role of bioinformatics is pivotal in the postgenomic era. Today it is hard to imagine designing a biological experiment without taking into account biological databases. Bioinformatics aids in terms of the application of knowledge to a biological problem and cuts down on cost and time for experimental design. Put simply, bioinformatics provides necessary computational tools and databases for efficient and successful running of a biological project. Good examples are developments in the field of *omics* involving handling, processing, and analyzing large-scale genomic data arising from various genome-sequencing projects. Bioinformatics precision extends beyond the study of genes and pathways to include the study of drug targets and therapeutic drugs.

The strength of bioinformatics resides in its ability to link diverse research and academic fields such as molecular biology, genetics, biochemistry, clinical genetics, molecular diagnostics, pharmacogenomics, biomedical informatics, mathematics, statistics, informatics, artificial intelligence, physics, chemistry, medicine, and biology, making it an interdisciplinary field (Figure 2.1). Hence, the field of bioinformatics can be seen as a fine amalgamation of various disciplines. Bioinformatics is a dynamic and rapidly developing branch of modern science that has the capability to change the rule of thumb of biology: predictions are not based on general principles (Lake and Moore, 1998; Howard, 2000; Rashidi and Buehler, 2000).

Bioinformatics tools are widely used by the scientific community for a variety of tasks including comparison of biological sequences, establishing ancestral relationship, structure prediction of a biomolecule, primer designing, genome-map construction, restriction-map construction, high-throughput data analysis, pathway analysis, and *in silico* drug designing.

2.2 UNIQUENESS OF BIOINFORMATICS

Successful running of an *in silico* analysis (bioinformatics-based analysis) to decode the language of biomolecules requires modest hardware, and most bioinformatics tools and biological repositories are freely accessible. The results obtained through the *in silico* analysis unarguable cut down in laboratory time and cost. The efficiency of the bioinformatics analysis is increased several-fold when linked to different repositories. Bioinformatics acts as a knowledge bridge in comprehending the consequences of mutations in DNA, revealing gene structure, establishing ancestral relationships, and determining the consequences of a structural disorder. Bioinformatics provides its selfless services (Foster, 2005) to the scientific community, for example, the myGrid e-science project (http://www.mygrid.org.uk) (Hey and Trefethen, 2005).

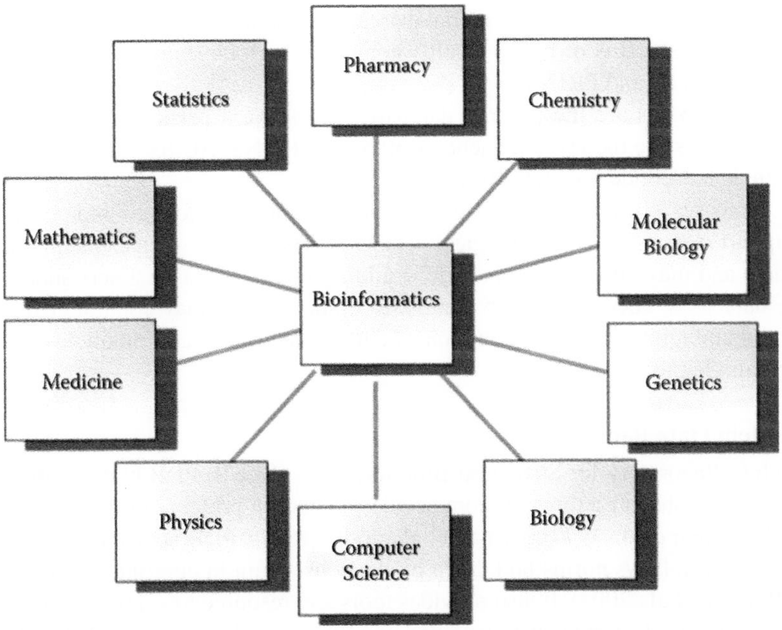

FIGURE 2.1 The strength of bioinformatics resides in its ability to link diverse research and academic fields.

2.3 AIMS OF BIOINFORMATICS

The fundamental aims of bioinformatics are as follows:

1. Construction of biological databases: for example, the Protein Data Bank (PDB) for 3D macromolecular structures (Bernstein et al., 1977; Berman et al., 2000).
2. Development of algorithms for identification of relationships among the members of data set: for example, in order to identify homologous sequence, programs like Basic Local Alignment Search Tool are required (Altschul et al., 1997).
3. Comprehensive analysis and interpretation of biological data.

It is evident that computers are the core part of bioinformatics, but it is necessary that computers are fed with accurate data input.

2.3.1 Biological Databases

Biological databases store and organize large biological data sets. The biological data sets are composed of information from scientific experiments, scientific literature, and analysis on these datasets. These may hold genomics, proteomics, phylogenetic, and gene expression information (Altman, 2004). Biological databases can be mined for information associated with gene function, structure, localization, clinical effects of mutations, and similarity between biological sequences and structures; this information helps us in building and understanding a large number of biological phenomena ranging from the structure and interactions of biomolecules to the evolution of species.

Based on the type of data, biological databases can be grouped into primary, secondary, and specialized databases. Primary databases contain unprocessed biological sequence or structural data submitted by research groups across the world. Examples of primary databases are the European Molecular Biology Laboratory Nucleotide Sequence Database (EMBL-Bank), DNA Data Bank of Japan (DDBJ), GenBank of the National Center for Biotechnology Information (NCBI), Swiss-Prot, and Protein Data Bank.

Secondary databases are well-curated databases. Although the information content is derived from primary databases, it is of higher quality compared with its source. Examples of secondary databases are PROSITE and PRINTS.

Specialized databases are the tailored databases and serve a particular interest. Examples of specialized databases are the HIV Sequence Database, TRANSFAC for transcription factors, and dbSNP for single nucleotide polymorphism.

Primary databases act as central storing and distribution hubs for raw biological information and are a support system for many biological databases. They can be seen as the first level of information holders and may provide templates for building other databases. Users should be cautious while using databases, because primary databases (data sources) may contain errors, duplication of records, and ambiguity in the presentation of data. Some of the commonly used databases are discussed in this chapter.

2.3.1.1 Protein Data Bank

The Research Collaboratory for Structural Bioinformatics (RCSB) PDB houses information about experimentally validated structures of biomolecules, including protein, nucleic acids, and complex assemblies (Bernstein et al., 1977; Berman et al., 2000). RCSB PDB is also member of worldwide PDB (wwPDB) and follows norms laid down by the consortium in curating and annotating RCSB PDB data (Berman et al., 2003). It also provides tools and resource to support users in performing advanced searches based on sequence, structure, and function. The homepage of RCSB PDB (http://www.pdb.org/pdb/home/home.do) can be mined for more information.

2.3.1.2 Molecular Modeling Database

The Molecular Modeling Database (MMDB) stores experimentally determined structures of proteins and nucleic acids derived from the PDB with some additional features to give more information about the structures (Wang et al., 2007). The added features include chemical graphs, computationally identified 3D domains, links to literature, and links to similar sequences, etc. More details can be seen on the home page of MMDB (http://www.ncbi.nlm.nih.gov/Structure/MMDB/mmdb.shtml).

2.3.1.3 PRINTS

The PRINTS database accumulates conserved protein motifs that act as fingerprint marker to describe a particular protein family (Attwood et al., 2004). Fingerprints can determine protein folds and functionalities comprehensively as compared to single motif. More details can be seen on the home page (http://www.bioinf.manchester.ac.uk/dbbrowser/PRINTS/index.php).

2.3.1.4 CATH

CATH (Orengo et al., 1997; Pearl et al., 2001) uses protein domain information to classify hierarchically protein structures in PDB. The four levels of hierarchy are: class, architecture, topology, and homologous superfamily. Crystal structures having resolution better than 4.0 angstron are considered together with nuclear magnetic resonance (NMR) structures (Orengo et al., 1997, 1998, 1999). These resources can be found on the home page (http://www.cathdb.info).

2.3.1.5 SCOP

The Structural Classification of Protein (SCOP) database comprises protein structures classified according to their evolutionary and structural relationships, i.e., protein domains from all species are classified into families, superfamilies, folds, and classes (Murzin et al., 1995). More information can be accessed from the home page (http://scop.mrc-lmb.cam.ac.uk/scop/).

2.3.1.6 GenBank

The GenBank database has DNA sequences (Benson et al., 2008). It also contains bibliographic and biological annotations. GenBank, along with DDBJ and EMBL, is a member of the International Nucleotide Sequence Database Collaboration. These organizations try to maintain uniformity in their content by exchanging data on a regular basis. More information can be acquired at http://www.ncbi.nlm.nih.gov/genbank.

2.3.1.7 dbSNP

dbSNP houses data on single-nucleotide polymorphisms (SNPs) and helps in identifying genotype-phenotype associations (Sherry et al., 2001). It contains SNP data (location in gene, nucleotides, and affected amino acids) and also information about the organisms from which SNPs are derived, including population details. The data from dbSNP can aid researchers in identifying SNPs in a gene, in determining SNPs that influence phenotype, and in obtaining functional information of gene product. More information can be found at http://www.ncbi.nlm.nih.gov/projects/SNP.

2.3.1.8 KEGG

Kyoto Encyclopedia of Genes and Genomes, commonly known as KEGG, is an integrated database resource of 16 databases roughly grouped into systems information, genomic information, and chemical information (Kanehisa et al., 2006). It contains data on genome sequences, metabolic pathways, orthologs, and compound structures. KEGG aids in providing biological understanding of large-scale datasets generated by high-throughput experimental technologies. More information can be obtained at http://www.genome.jp/kegg.

2.3.1.9 TRANSFAC

TRANSFAC is a database of transcription factors along with their experimentally verified binding sites and regulated genes (Matys et al., 2006). Academic and nonprofit organizations have free access to these data. More information can be obtained at the TRANSFAC home page (http://www.biobase-international.com/pages/index.php?id=transfac).

2.3.1.10 OMIM

The Online Mendelian Inheritance in Man (OMIM) is a database of genes and genetic disorders in humans (Hamosh et al., 2002). It is focused on inherited or heritable genetic diseases. It acts as a phenotypic resource to the human genome project. A lot of information is available at http://www.ncbi.nlm.nih.gov/omim.

2.3.2 Algorithms and Biology

Algorithms are a set of instructions, implemented using computers to perform complex tasks in bioinformatics analysis. The range of tasks that algorithms perform in bioinformatics includes comparative analysis of sequences, building phylogenetic trees, predicting gene and protein structure, visualization of structures, probing databases, simulation of biological processes, and image processing. The efficiency of any bioinformatics analysis depends upon the grade of algorithm being used.

The field of bioinformatics has not yet reached its adolescence, as most of the algorithms being used to solve confounding problems in modern biology are not comprehensive enough to reflect a clear picture of the biological system in question. Furthermore, the algorithms that intend on producing a good-enough picture of *in vivo* processes have very high computational demands.

There is a huge demand for developing algorithms that may give a comprehensive picture of *in vivo* biological processes. However, the biggest challenge for algorithm developers is to find a middle road, i.e., an algorithm that is computationally less demanding and at the same time comprehensive enough. Researchers are striving to overcome these hurdles posed by the complexities of biological systems.

2.3.3 Analysis and Interpretation of Experimental Data

The combination of biological data with a statistical or mathematical model illustrates a more comprehensive picture of cellular activities. A strong statistical model is important for characterization of various analyses measured in an experiment. Integration of various biological data resources significantly contributes to the accuracy of analysis and can provide insights into complex diseases and traits, for example:

1. Integrating nonsynonymous single-nucleotide polymorphism data with the native structure of protein can help in identifying deleterious nonsynonymous single-nucleotide polymorphisms (Singh et al., 2011a).
2. In molecular biomarker studies, the proficient integration of molecular, clinical, and imaging data may aid in reaching a conclusive result.

The projects in biological sciences vary in their demand for analysis and integration of resources; hence statistical models and analysis are customized according to the need of the project. The correct analysis and interpretation of results is also important for successfully running a wet lab protocol. In this area, researchers from the bioinformatics discipline can provide aid by integrating multiple sources of biological knowledge (databases and tools), which in turn helps in correct interpretation and elucidation of biological processes (Gerstein, 2000).

2.4 BIOINFORMATICS AND RESEARCH AREAS IN MODERN BIOLOGY

Bioinformatics methods are indispensable in biological science. Below we describe a few of the major research areas in bioinformatics; a review of all research areas is out of the scope of this chapter.

2.4.1 Genomics

The successful completion of large numbers of genome-sequencing projects has resulted in an ocean of genomic data. The precise analysis of these data for the correct identification of the gene, its location, genetic control elements, tagging structural and functional information, and identification of genetic variation(s) is the greatest challenge to the scientific community. Bioinformatics helps in the organization and storage of high-throughput data and provides tools for analyzing these data. It was through the availability of computational tools and databases that the field of comparative genomics has made progress in leaps and bounds. The analysis of a single sequence reveals lots of information, but there are still certain regions in genomes that remain unexplored, and the functionality of these regions can only be explored by comparing them with other genomes that are less divergent or homologous in nature (Kellis et al., 2003). The availability of freely accessible biological databases and computational tools for the mining and analysis of genomes and their genes has made it possible to answer a few of the many confounding questions of modern biology. Biological databases such as dbSNP (Sherry et al., 2001) can be mined for genetic variations of a given gene, whereas TRANSFAC (Matys et al., 2006) is for transcription factors, and a computational tool like Sorting Intolerant from Tolerant (SIFT) (Ng and Henikoff, 2001, 2002, 2003; Kumar et al., 2009) can be used for predicting deleterious SNPs. Some web servers such as SNPnexus (Chelala et al., 2009) are completely dedicated to the analysis of mutations in the regulatory regions of genes, and Regulatory Sequence Analysis Tools (RSAT), available at http://rsat.ulb.ac.be, provides several computational tools for detecting regulatory signals in genomes (Turatsinze et al., 2008). These are some of the prominent bioinformatics tools used in the field of genomics. Ongoing projects for genome annotation include ENCODE, the ENCyclopedia of DNA Elements (http://www.genome.gov/10005107); Entrez Gene (http://www.ncbi.nlm.nih.gov/gene); Ensembl (http://www.ensembl.org/index.html); the Gene Ontology Consortium (http://www.geneontology.org); RefSeq (http://www.ncbi.nlm.nih.gov/projects/RefSeq); and the Vertebrate and Genome Annotation Project (http://vega.sanger.ac.uk/index.html).

2.4.2 Proteomics

Proteomics refers to a comprehensive study of proteins, particularly their structures, functions, and interactions (Anderson and Anderson, 1998). Structural and functional information about the proteins is vital in elucidating their role in the regulation of cellular processes and in biological pathways (Morel et al., 2004). Bioinformatics provides a pool of databases and computational tools for studies related to proteomics, such as for protein expression analysis. Bioinformatics techniques can be employed to match a large amount of data against predicted data from protein sequence databases followed by statistical analysis. Examples of bioinformatics tools and databases used in protein expression analysis are as follows:

1. ExPASy 2D PAGE databases and services (http://expasy.org/ch2d/2d-index.html) (Hoogland et al., 1999), which contain references to known 2D PAGE database servers and to 2D PAGE-related services
2. GelScape (http://www.gelscape.ualberta.ca) (Young et al., 2004), a platform-independent program for analyzing standard 1D and 2D protein gels
3. NCI Flicker (http://www-lmmb.ncifcrf.gov/flicker) (Lemkin et al., 1979), which compares 2D sample gels against 2D gel database maps, possibly suggesting putative protein spot identification

Bioinformatics methods help in providing a snapshot of proteins present in biological samples. Also in protein structure prediction, correct determination of secondary structure is important because it governs the shape acquired by the polypeptide chain. Some of the commonly used methods for secondary structure prediction are the Chou and Fasman method (Chou and Fasman 1974), the GOR method (Garnier et al., 1978), and the PHD program (Rost and Sander, 1993a, 1993b). Tertiary or native structure of protein is responsible for its biological activity. Hence, a lot of focus has been on developing tools or protocols for correct or approximate prediction of tertiary structure. Computationally predicted structures serve as templates to build hypotheses for further research in structural and functional biology. The gold standards in structure prediction are set by NMR and X-ray crystallography methods.

2.4.3 Transcriptomics

Only 5% of human DNA is transcribed; the rest of the genome controls and regulates expression of this 5% of the genome. The expression information of individual genes at the mRNA level can help in establishing the relationship between gene expression patterns and its biological significance (Carulli et al., 1998; Scheel et al., 2002; Morel et al., 2004). Bioinformatics can be used for mining of the transcriptomics data followed by its analysis, this could help in creating an understanding of how certain genes are activated and deactivated and how the levels of gene products produced. Understanding transcriptomics data and its analysis is particularly important to researchers studying the process of cellular differentiation and carcinogenesis.

2.4.4 Metabolomics

Metabolomics refers in particular to the study of metabolite profiles, the study of chemical fingerprints left after a completion of cellular processes (Daviss, 2005). Metabolic profiling is important because it can give us a snapshot of the cellular physiology that can augment the information about the cellular physiology that mRNA gene expression data and proteomic analysis does not reveal. The field of bioinformatics with its data management, data processing, statistical analysis, data mining, data integration, and the mathematical modeling of metabolic networks' abilities can assist in the development of metabolomics (Shulaev, 2006).

2.4.5 Cytomics

Cytomics refers to single cell study of cell-system heterogeneity using image or flow cytometry (Gomase and Tagore, 2008a). By studying cell phenotypes, we can establish a correlation between the disease process as the sum of the respective genotype and exposure influences. The cell phenotype contains the information about the cell health (disease status) and helps in predicting therapy-dependent future developments. The integration of information from cytomics with proteomics may assist in identifying cells with a specific set of phenotype characteristics; this may aid in the identification of tumor markers (Bernas et al., 2006). Bioinformatics tools can be exhaustively used in extracting molecular cell phenotype(s).

2.4.6 Physiomics

Physiomics refers to an integrative study of genome, proteome, and metabolome (Gomase and Tagore, 2008b). It comprehensively uses experimental databases and computer algorithms for correct identification of physiological phenotypes of genes and proteins. Overall, bioinformatics tools and databases may aid the field of physiomics in constructing physiological features associated with genes, proteins, and interactions among and between them. The field of physiomics is extremely useful in the development of drugs and biochips (Gomase and Tagore, 2008b).

2.4.7 Glycomics

Glycomics refers to a comprehensive study of the entire sugar compliment of an organism, encompassing its genetic, pathologic, physiologic, and other aspects (Aoki-Kinoshita, 2008). The integration of glycomics data with proteomic and genomics data can help in elucidating the relationship between the glycome and genome (Pilobello and Mahal, 2007). This in turn will help in determining the role of carbohydrates or sugar moloecules in various pathways.

2.4.8 Lipidomics

In biological systems, lipidomics refers to comprehensive study of the pathways and networks of cellular lipids (Wenk, 2005; Watson, 2006). Lipids are important because they perform structural, energy storage, and signaling roles (Wenk, 2005). Lipidomics, along with the data integration and data analysis capabilities of bioinformatics, can help elucidate important biological phenomena and their influence(s) on biological pathways.

2.4.9 Interactomics

Interactomics refers to the study of interactions between and among protein(s) and other biomolecules inside cells (Kiemer and Cesareni, 2007). A network of such interactions is called an *interactome*. Comparison of interactomes among and between species may help reveal the traits of such networks.

2.5 FREQUENTLY USED BIOINFORMATICS TOOLS

A list of commonly used bioinformatics tools, which can be accessed freely over the World Wide Web, is presented. A comprehensive list of bioinformatics tools used in modern biology is beyond the scope of this chapter.

2.5.1 Electronic Polymerase Chain Reaction

Electronic polymerase chain reaction (e-PCR) aids in the identification of sequence-tagged sites within DNA sequences (Schuler, 1997). The e-PCR, aids in searching subsequences that closely match PCR primers and have the right order, orientation, and spacing. The program employs a fuzzy matching strategy that improves search sensitivity of the program and also allows incorporating gaps in primer alignment. The latest release of e-PCR provides a search mode using a query sequence against a sequence database (Rotmistrovsky et al., 2004). More information about this program can be accessed on http://www.ncbi.nlm.nih.gov/Tools, and the program can be accessed through http://www.ncbi.nlm.nih.gov/projects/e-pcr. Researchers have used the *in silico* designing of primers and validated using PCR. For example, *in silico* primers have been designed for the amplification of all the structural regions of the quasi species of dengue viruses that could be useful in molecular diagnostic (Somvanshi and Seth, 2008). Using this technique, numbers of *in silico* primers have been designed, synthesized, and validated in wet laboratories for the detection of other bacterial pathogens, especially *Aeromonas hydrophila* (Singh et al., 2009; Singh et al., 2010b; Singh et al., 2011a; Singh and Somvanshi, 2009c).

2.5.2 Map Viewer

Map Viewer displays integrated chromosomal maps of many organisms including vertebrates, invertebrates, fungi, protozoa, and plants (Wheeler et al., 2008). The tool aids in locating genes along with other biological features. More information about the program can be found at http://

www.ncbi.nlm.nih.gov/Tools, and the program can be accessed through http://www.ncbi.nlm.nih.gov/mapview.

2.5.3 Model Maker

Through Model Maker, the user can view the evidence that was used to construct a gene model on an assembled genomic sequence, and it can be accessed from sequence maps (Wheeler et al., 2008). Moreover, Model Maker enables the user to construct his version of the model by selecting desired exons.

2.5.4 Open Reading Frame Finder

Open Reading Frame (ORF) Finder detects all open reading frames of selected minimum size in a query sequence (Wheeler et al., 2008). The program detects open reading frames using standard or alternative genetic codes. ORF Finder may assist in preparing complete and precise sequence submissions and is packaged in Sequin (http://www.ncbi.nlm.nih.gov/projects/Sequin), a sequence submission software. The program can be accessed at http://www.ncbi.nlm.nih.gov/projects/gorf. It has already been used to find the ORF that could be used for the expression of genes (Singh et al., 2009, 2011a; Singh and Somvanshi, 2009c).

2.5.5 TaxPlot

TaxPlot is a three-way comparison of genomes based on the protein encoded by them (Wheeler et al., 2008). The tool selects a reference genome, compares two other genomes to it, and uses pre-computed BLAST results to plot a point for each predicted protein in reference genome. The program can be accessed at http://www.ncbi.nlm.nih.gov/sutils/taxik2.cgi.

2.5.6 VAST Search

VAST Search is a structure similarity search program; compares 3D coordinates of a recently ascertained protein structure with the existing ones in the MMDB/PDB databases (Wheeler et al., 2008). The program computes and generates a list of structure neighbors that could be browsed interactively. The program can be accessed at http://structure.ncbi.nlm.nih.gov/Structure/VAST/vastsearch.html.

2.5.7 Basic Local Alignment Search Tool

Basic Local Alignment Search Tool (BLAST) is a set of similarity searching programs intended to search all of the accessible sequence databases regardless of whether the query is protein or DNA; it seeks local alignments and identifies relationships between sequences sharing isolated regions of similarity (Altschul et al., 1990, 1997). Variants of BLAST are as follows:

1. BLASTn: Input nucleotide sequence is compared with nucleotide sequence database to find sequences containing regions homologous to input sequence.
2. BLASTp: Input protein sequence is compared with protein sequence database to find sequences containing regions homologous to input sequence.
3. BLASTx: Input nucleotide sequence is translated and compared with protein sequence database to find sequences containing regions homologous to input sequence.
4. tBLASTn: Input protein sequence is compared with translated nucleotide sequence database to find sequences containing regions homologous to input sequence.
5. tBLASTx: Input nucleotide sequence is translated and compared with translated nucleotide sequence database to find sequences containing regions homologous to input sequence.

Several applications of this tool are used to find the homology of genes and proteins present in available organisms such as influenza virus (Somvanshi et al., 2008a), in genes (Somvanshi et al., 2008b), hemolysin (Singh et al., 2009), and aerolysin (Singh and Somvanshi, 2009c; Singh et al., 2010b). BLAST can be accessed through NCBI (Johnson et al., 2000) at http://blast.ncbi.nlm.nih .gov/Blast.cgi.

2.5.8 SIFT

SIFT is used for identifying potential deleterious nonsynonymous single-nucleotide polymorphism(s) (nsSNPs) (Ng and Henikoff 2001, 2002, 2003; Kumar et al., 2009). SIFT uses sequence homology to predict deleterious nsSNPs; amino acids at specific positions that are important for protein function must remain conserved in the alignment of homologous sequences. The tool (and more information about it) can be accessed at http://sift.jcvi.org.

2.5.9 Clustal

Clustal is a multiple sequence alignment program and is available in command line (ClustalW) as well as in graphical interface (ClustalX) (Chenna et al., 2003; Larkin et al., 2007). The program is written in C++ programming language. The program can be executed either on default parameters or on customized parameters; the main parameters of the program that could be adjusted are gap-opening penalty and gap-extension penalty. Precompiled executables for most of the operating systems can be downloaded from http://www.clustal.org.

2.5.10 PHYLIP

PHYLogeny Inference Package, commonly known as PHYLIP (Felsenstein, 1989), is a freely available computational package of programs for inferring phylogenies. The package consists of 35 portable programs (http://bioweb2.pasteur.fr/docs/phylip/doc/main.html). The source code is written in C programming language, and precompiled executables are available for Windows, Mac OS, and Linux systems and can be downloaded from http://evolution.genetics.washington .edu/phylip.html.

2.5.11 FoldX

FoldX program provides a quick and quantitative estimate of the effect of mutation on the stability of proteins. The program exploits atomic description in the structure of proteins (Schymkowitz et al., 2005). The FoldX program (and more information about it) can be obtained from http://foldx .crg.es.

2.5.12 MODELLER

MODELLER is commonly used for homology modeling of the tertiary structure of proteins (Eswar et al., 2006; Marti-Renom, 2000). Target and template sequence alignment is fed as an input into the program, and then it automatically computes a model containing all of the nonhydrogen bond atoms. Other tasks performed by MODELLER include: *de novo* modeling of loop structure, optimization of protein model, comparison of protein structures, etc. The strength of MODELLER lies in its ability to model structures by satisfying spatial constraints (Sali and Blundell, 1993; Fiser et al., 2000). MODELLER has been frequently used to generate the 3D structure of uncrystallized protein from different origins such as bacterial, viral, human, and so on (Somvanshi and Singh, 2008, 2010; Singh et al., 2009; Singh and Somvanshi, 2009a, 2009b,

2009c, 2009d, 2009e, 2010). The tool (and more information about it) can be obtained from http://www.salilab.org/modeller.

2.5.13 GROMACS

Groningen Machine for Chemical Simulations (GROMACS) is a package written in ANSI C programming language and performs molecular dynamics simulation. The program does not have its own force field, but it is compatible with the following force fields: GROMOS, OPLS, AMBER, and ENCAD (Van Der Spoel et al., 2005). The program offers flexibility to customize force routines and tabulated functions. The program can be downloaded from http://www.gromacs.org.

2.5.14 AutoDock

AutoDock (Goodsell et al., 1996; Morris et al., 1998, 2009) is a cluster of automated docking tools. It assists in understanding the binding of a ligand molecule to a receptor of known three-dimensional structure. The docking tool consists of two main programs:

1. AutoDock, which performs docking of ligand molecule to grid(s) characterizing target protein
2. AutoGrid, which computes the grid(s) significant for docking

AutoDock can assist organic synthetic chemists in designing better binders. Some of the application areas of AutoDock are structure-based drug design, lead optimization, virtual screening, protein-protein docking, and chemical mechanism study. This tool became very popular, and researchers are continuously using it for screening drug molecules on the active region of 3D proteins (Somvanshi and Singh, 2008; Singh and Somvanshi, 2009a, 2009b, 2009c, 2009d, 2009e). The tool (and more information about it) can be obtained from http://autodock.scripps.edu/downloads.

2.5.15 Gene Designer for Gene Designing and Codon Optimization

Gene Designer (https://www.dna20.com/index.php?pageID=220) is commonly used for designing genes in a given expression host. Codon optimization technique is used to improve the protein expression in living organisms by increasing the translational efficiency of a gene of particular interest (Mani et al., 2010). The nucleotides of a DNA sequence are separated into triplets (codons), and then the codon is replaced with a new (degenerate codon), generated with a given frequency distribution. This amino acid will be the same, but codon with the low frequency of an amino acid will be replaced with a codon of high frequency, according to the desired species frequency distribution. Optimizer software is used for optimization and calculation of CAI, G+C, and A+T (Puigbo et al., 2007). CAIcal and MrGene are used for optimization of DNA sequences at maximum suitable threshold level (Sharp and Li, 1987). It has become popular to improve the level of expression in the host (Singh et al., 2010a; Mani et al., 2010, 2011). It is also useful for overexpression in the host and provides new insights into emerging research in synthetic biology.

2.6 PRACTICAL APPLICATIONS OF BIOINFORMATICS

The field of bioinformatics is very vast and consequentially has a vast area of application. Below we have listed some of the practical applications of bioinformatics.

2.6.1 Identification of Drug Targets

A comprehensive understanding of disease mechanisms and an efficient use of computational tools can lead to the identification and validation of potential novel drug targets. This will lead to development of more specific medicine(s) that act on cause and have fewer side effects.

2.6.2 Personalized Medicine

Developments in the field of pharmacogenomics will lead to a smooth transition of clinical medicines into personalized medicines. Knowledge about an individual genetic profile can help in prescribing the best possible drug therapy and dosage.

2.6.3 Preventive Medicine

Advances in the field of genomics have unraveled many specific details about the genetic basis of a disease. Preventive measures such as change in lifestyle or medication at an early stage could shield an individual from many diseases.

2.6.4 Gene Therapy

Comprehensive knowledge and understanding of gene expression can aid us in targeting and manipulating expression of genes that lead to disease. This will lead to treatment, cure, or even prevention of disease. Simulation programs can be run in order to find the most probable gene to be targeted for manipulation. Knowledge from simulated gene expression could be obtained before implementation of any manipulation in a real biological system.

2.6.5 Evolutionary Study

Over time biologists have been aided by bioinformatics in studying various aspects of evolution, including:

1. Measuring changes in DNA during the evolution of a certain organism
2. Studying complex evolutionary events such as speciation and gene duplication
3. Building computational models to study a population and making predictions for the population

2.6.6 Agriculture

The agricultural community has benefited vastly from genome sequencing programs for plants. With the aid of bioinformatics tools, specific genes can be targeted in genomes to produce insect- and disease-resistant plants with healthier more productive offspring.

2.6.7 Veterinary Science

The information revealed through sequencing projects of farm animals such as, cattle, pigs, and sheep has helped veterinary scientists immensely in improving the production and health of livestock. More details about the practical applications of bioinformatics can be found at http://www.ebi.ac.uk/2can/bioinformatics/bioinf_realworld_1.html.

2.7 LIMITATIONS

Bioinformatics analysis greatly depends upon the quality and quantity of data generated by the biologist(s) and the comprehensiveness of the algorithm being used to analyze data. The field of bioinformatics is still in its infancy, and many algorithms are in the developmental stage and thus are not comprehensive enough to reflect the complete *in vivo* picture of a biological system.

2.8 WEB RESOURCE FOR BIOINFORMATICS

Through the internet, it is possible to access any database and any computational tool (most of the databases and tools are freely accessible) needed to perform *in silico* analysis. Table 2.1 provides

TABLE 2.1
Web Servers Used in Bioinformatics and Computational Biology Research

Name	Description	Web Link
	Important bioinformatics web resource locators	
ArrayExpress	Public curated repository of array data such as gene expression, comparative genome hybridization data, and chip-on-chip data	http://www.ebi.ac.uk/arrayexpress
ExPASy	Dedicated to *in silico* protein study	http://expasy.org
miRBase	Resource of microRNA	http://www.mirbase.org
RNA World	Lists web links on related RNA topics	http://www.imb-jena.de/RNA.html
Pasteur	Miscellaneous links	http://bioweb.pasteur.fr/intro-en.html
	Important biological databases	
GenBank/DDBJ/EMBL	Nucleotide sequences	http://www.ncbi.nlm.nih.gov/nuccore
PDB	Protein structures	http://www.pdb.org/pdb/home/home.do
Swiss-Prot	Protein sequences	http://expasy.org/people/swissprot.html
OMIM	Genetic diseases and genetic disorders	http://www.ncbi.nlm.nih.gov/omim
KEGG	Biological pathways	http://www.genome.jp/kegg
dbSNP	Single nucleotide polymorphism	http://www.ncbi.nlm.nih.gov/projects/SNP
PubMed	Literature references	http://www.ncbi.nlm.nih.gov/pubmed
	Important bioinformatics tools	
BLAST	Homology search	http://blast.ncbi.nlm.nih.gov/Blast.cgi
FASTA	Homology search	http://www.ebi.ac.uk/Tools/sss
Entrez	Database search	http://www.ncbi.nlm.nih.gov/gene
Clustalw2	Multiple sequence alignment	http://www.ebi.ac.uk/Tools/msa/clustalw2
GenScan	Gene structure prediction	http://genes.mit.edu/GENSCAN.html
PredictProtein	Protein structure prediction	http://www.predictprotein.org
Mfold	RNA structure prediction	http://mfold.rna.albany.edu/?q=mfold
PHYLIP	Phylogeny inference package	http://www.phylip.com
RasMol	Structure visualization	http://www.umass.edu/microbio/rasmol

a list of major resource locators for bioinformatics, important biological databases, and important software used in the field. Although this is not a comprehensive list, and the use of databases and tools varies from analysis to analysis, the databases and tools listed may provide a starting point to those entering the field of bioinformatics.

REFERENCES

Altman, R. B. (2004). Building successful biological databases. *Briefings in Bioinformatics*, 4–5.

Altschul, S. F., Gish, W., Miller, W., et al. (1990). Basic local alignment search tool. *Journal of Molecular Biology*, 403–410.

Altschul, S. F., Madden, T. L., Schäffer, A. A., et al. (1997). Gapped BLAST and PSI-BLAST: A new generation of protein database search programs. *Nucleic Acids Research*, 3389–3402.

Anderson, N. L., Anderson, N. G. (1998). Proteome and proteomics: New technologies, new concepts, and new words. *Eloectrophoresis*, 1853–1861.

Aoki-Kinoshita, K. F. (2008). An introduction to bioinformatics for glycomics research. *PLOS Computational Biology*, e1000075.

Attwood, T. K., Bradley, P., Gaulton, A., et al. (2004). *The PRINTS protein fingerprint database: Functional and evolutionary applications.* New York: John Wiley & Sons.

Benson, D. A., Karsch-Mizrachi, I., Lipman, D. J., et al. (2008). GenBank. *Nucleic Acids Research,* D26–D31.

Berman, H. M., Henrick, K., Nakamura, H. (2003). Announcing the worldwide Protein Data Bank. *Nature Structural Biology*, 98.

Berman, H. M., Westbrook, J., Febg, Z., et al. (2000). The Protein Data Bank. *Nucleic Acids Research*, 235–242.

Bernas, T., Gregori, G., Asem, E. K., et al. (2006). Integrating cytomics and proteomics. *Molecular Cell Proteomics*, 2–13.

Bernstein, F. C., Koetzle, T. F., Williams, G. J., et al. (1977). The Protein Data Bank: A computer-based archival file for macromolecular structures. *Journal of Molecular Biology*, 535–542.

Cantor, C. R. (1968). The occurrence of gaps in protein sequences. *Biochemical and Biophysical Research Communications*, 410–416.

Carulli, J. P., Artinger, M., Swain, P. M., et al. (1998) High throughput analysis of differential gene expression. *Journal of Cell and Biochemistry Supplement*, 30–31.

Celis, J. E., Leffers, H., Rasmussen, H. H., et al. (1991). The master two-dimensional gel database of human AMA cell proteins: Towards linking protein and genome sequence and mapping information. *Electrophoresis*, 765–770.

Chelala, C., Khan, A., Lemoine, N. R. (2009). SNPnexus: A web database for functional annotation of newly discovered and public domain single nucleotide polymorphisms. *Bioinformatics*, 655–661.

Chenna, R., Sugawara, H., Koike, T., et al. (2003). Multiple sequence alignment with the Clustal series of programs. *Nucleic Acids Research*, 3497–3500.

Chou, P. Y., Fasman, G. D. (1974). Prediction of Protein Conformation. *Biochemistry*, 222–245.

Darwin, C. (1859). *On the Origin of Species by Means of Natural Selection, or the Preservation of Favoured Races in the Struggle for Life*, New York: D. Appleton and Company.

Daviss, B. (2005). Growing pains for metabolomics. *The Scientist*, 25.

Eswar, N., Marti-Renom, M. A., Webb, B., et al. (2006). Comparative protein structure modeling with MODELLER. *Current Protocols in Bioinformatics*, 5.6.1–5.6.30.

Felsenstein, J. (1989). PHYLIP: PHYLogeny Inference Package. *Cladistics*, 164–166.

Fiser, A., Do, R. K., Sali, A. (2000). Modeling of loops in protein structures. *Protein Science*, 1753–1773.

Fitch, W. M., Margoliash, E. (1967). Construction of phylogenetic trees. *Science*, 270–284.

Foster, I. (2005). Service-oriented science. *Science*, 814–817.

Garnier, J., Osguthorpe, D. J., Robson, B. (1978). Analysis of the accuracy and implications of simple methods for predicting the secondary structure of globular proteins. *Journal of Molecular Biology*, 97–120.

Gerstein, M. (2000). Integrative database analysis in structural genomics. *Nature Structural Biology*, 960–963.

Gomase, V. S., Tagore, S. (2008a). Cytomics. *Current Drug Metabolism*, 263–266.

Gomase, V. S., Tagore, S. (2008b). Physiomics. *Current Drug Metabolism*, 259–262.

Goodsell, D. S., Morris, G. M., Olson, A. J. (1996). Automated docking of flexible ligands: Applications of AutoDock. *Journal of Molecular Recognition*, 1–5.

Hamosh, A., Scott, A. F., Amberger, J., et al. (2002). Online Mendelian Inheritance in Man (OMIM): A knowledgebase of human genes and genetic disorders. *Nucleic Acids Research*, 52–55.

Hey, T., Trefethen, A. E. (2005). Cyberinfrastructure for e-Science. *Science*, 817–821.

Hoogland, C., Sanchez, J. C., Tonella, L., et al. (1999). The SWISS-2DPAGE database: What has changed during the last year. *Nucleic Acids Research*, 289–291.

Howard, K. (2000). The bioinformatics gold rush. *Scientific American*, 58–63.

Johnson, M. Zaretskaya, I. Raytselis, Y., et al. (2000). NCBI BLAST: A better web interface. *Nucleic Acids Research*, W5–W9.

Kanehisa, M., Goto, S., Hattori, M., et al. (2006). From genomics to chemical genomics: New developments in KEGG. *Nucleic Acids Research*, D354–D357.

Kellis, M., Patterson, N., Endrizzi, M., et al. (2003). Sequencing and comparison of yeast species to identify genes and regulatory elements. *Nature*, 241–254.

Kiemer, L., Cesareni, G. (2007). Comparative interactomics: Comparing apples and pears? *Trends in Biotechnology*, 448–454.

Kumar, P., Henikoff, S., Ng, P. C. (2009). Predicting the effects of coding non-synonymous variants on protein function using the SIFT algorithm. *Nature Protocol*, 1073–1082.

Lake, J. A., Moore, J. E. (1998). Phylogenetic analysis and comparative genomics. *Trends in Biotechnology*, 22–23.

Larkin, M. A., Blackshields, G., Brown, N. P., et al. (2007). ClustalW and ClustalX version 2.0. *Bioinformatics*, 2947–2948.

Lemkin, P., Merril, C., Lipkin, L., et al. (1979). Software aids for the analysis of 2D gel electrophoresis images. *Computer and Biomedical Research*, 517–544.

Mani, I., Singh, V., Chaudhary, D. K., et al. (2011). Codon optimization of the major antigenic genes of diverse strains of influenza A virus. *Interdisciplinary Sciences: Computational Life Sciences*, 36–42.

Mani, I., Chaudhary, D. K., Somvanshi, P., et al. (2010). Codon optimization of the potential antigens encoding genes from Mycobacterium tuberculosis. *International Journal of Applied Biology and Pharmaceutical Technology*, 292–301.

Marti-Renom, M. A., Stuart, A., Fiser, A., et al. (2000). Comparative protein structure modeling of genes and genomes. *Annual Reviews in Biophysics and Biomolecular Structures*, 291–325.

Matys, V., Kel-Margoulis, O. V., Fricke, E., et al. (2006). TRANSFAC and its module TRANSCompel: Transcriptional gene regulation in eukaryotes. *Nucleic Acids Research*, D108–D110.

Morel, N. M., Holland, J. M., van der Greef, P., et al. (2004). Primer on medial genomics: Part XIV. Introduction to system biology: A new approach to understanding disease and treatment. *Mayo Clinic Proceedings*, 651–658.

Morris, G. M., Goodsell, D. S., Halliday, R. S., et al. (1998). Automated docking using a Lamarckian genetic algorithm and and empirical binding free energy function. *Journal of Computational Chemistry*, 1639–1662.

Morris, G. M., Huey, R., Lindstrom, W., et al. (2009). AutoDock4 and AutoDockTools4: Automated docking with selective receptor flexibility. *Journal of Computational Chemistry*, 2785–2791.

Murzin, A. G., Brenner, S. E., Hubbard, T., et al. (1995). SCOP: A structural classification of proteins database for the investigation of sequences and structures. *Journal of Molecular Biology*, 536–540.

Ng, P. C., Henikoff, S. (2002). Accounting for human polymorphisms predicted to affect protein function. *Genome Research*, 436–446.

Ng, P. C., Henikoff, S. (2001). Predicting deleterious amino acid substitutions. *Genome Research*, 863–874.

Ng, P. C., Henikoff, S. (2003). SIFT: Predicting amino acid changes that affect protein function. *Nucleic Acids Research*, 3812–3814.

Orengo, C. A., Martin, A. M., Hutchinson, G., et al. (1998). Classifying a protein in the CATH database of domain structures. *Acta Crystallographica D Biological Crystallography*, 1155–1167.

Orengo, C. A., Michie, A. D., Jones, S., et al. (1997). CATH: A hierarchic classification of protein domain structures. *Structure*, 1093–1108.

Orengo, C. A., Pearl, F. M., Bray, J. E., et al. (1999). The CATH Database provides insights into the protein structure/function relationships. *Nucleic Acids Research*, 275–279.

Pearl, F. M. G., Martin, N., Bray, J. E., et al. (2001). A rapid classification protocol for the CATH domain database to support structure genomics. *Nucleic Acids Research*, 223–227.

Pilobello, K. T., Mahal, L. K. (2007). Deciphering the glycode: The complexity and analytical challenge of glycomics. *Current Opinion in Chemical Biology*, 300–305.

Puigbo, P., Guzman, E., Romeu, A., et al. (2007). OPTIMIZER: A web server for optimizing the codon usage of DNA sequences. *Nucleic Acids Research*, 1–6.

Rashidi, H. H., Buehler, L. K. (2000). *Bioinformatics Basics: Applications in Biological Science.* Boca Raton: CRC Press.

Rost, B., Sander, C. (1993a). Improved prediction of protein secondary structure by use of sequence profiles and neural networks. *Proceedings of the National Academy of Sciences*, 7558–7562.

Rost, B., Sander, C. (1993b). Prediction of protein secondary structure at better than 70% accuracy. *Journal of Molecular Biology*, 584–599.

Rotmistrovsky, K., Jang, W., Schuler, G. D. (2004). A web server for performing electronic PCR. *Nucleic Acids Research*, W108–W112.

Sali, A., Blundell, T. L. (1993). Comparative protein modelling by satisfaction of spatial restraints. *Journal of Molecular Biology*, 779–815.

Scheel, J., Von Brevern, M. C., Horlein, A., et al. (2002). Yellow pages to the transcriptome. *Pharmacogenomics*, 791–807.

Schuler, G. D. (1997). Sequence mapping by electronic PCR. *Genome Research*, 541–550.

Schymkowitz, J., Borg, J., Stricher, F., et al. (2005). The FoldX web server: An online force field. *Nucleic Acids Research*, W382–W388.

Sharp, P. M., Li, W. H. (1987). The codon adaptation index: A measure of directional synonymous codon usage bias, and its potential applications. *Nucleic Acids Research*, 1281–1295.

Sherry, S. T., Ward, M. H., Kholodov, M., et al. (2001). dbSNP: The NCBI database of genetic variation. *Nucleic Acids Research*, 308–311.

Shulaev, V. (2006). Metabolomics technology and bioinformatics. *Briefings in Bioinformatics*, 128–139.

Singh, V., Chaudhary, D. K., Mani, I., et al. (2010a). Molecular identification and codon optimization analysis of major virulence encoding genes of Aeromonas hydrophila. *African Journal of Microbiology Research*, 952–957.

Singh, V., Mani, I., Chaudhary, D. K., et al. (2011a). Molecular detection and cloning of thermostable hemolysin gene from Aeromonas hydrophila. *Molecular Biology*, 551–560.
Singh, V., Somvanshi, P., Rathore, G., et al. (2010b). Gene cloning, expression and characterization of recombinant aerolysin from Aeromonas hydrophila. *Applied Biochemistry & Biotechnology*, 1985–1991.
Singh, A., Sivakumar, D., Somvanshi, P. (2011b). Cataloguing functionally relevant polymorphisms in gene DNA ligase I: A computational approach. *3 Biotech*, 47–56.
Singh, V., Somvanshi, P., Rathore, G., et al. (2009). Gene cloning, expression and homology modeling of hemolysin gene from Aeromonas hydrophila. *Protein Expression & Purification*, 1–7.
Singh, V., Somvanshi, P. (2009a). Homology modeling of adenosine A2A receptor and molecular docking for exploration of appropriate potent antagonists for treatment of Parkinson's disease. *Current Aging Science*, 127–134.
Singh, V., Somvanshi, P. (2009b). Homology modeling of 3-oxoacyl-acyl carrier protein synthase II (KAS II) from Mycobacterium tuberculosis H37Rv and molecular docking for exploration of drugs. *Journal of Molecular Modeling*, 453–460.
Singh, V., Somvanshi, P. (2009c). Inhibition of oligomerization of aerolysin from Aeromonas hydrophila: Homology modeling and docking approach for exploration of hemorrhagic septicemia. *Letters in Drug Design & Discovery*, 215–223.
Singh, V., Somvanshi, P. (2009d). Structural modeling of the NS 3 helicase of tick-borne encephalitis virus and their virtual screening of potent drugs using molecular docking. *Interdisciplinary Sciences: Computational Life Sciences*, 168–172.
Singh, V., Somvanshi, P. (2009e). Targeting the peptide deformylase of Mycobacterium tuberculosis leads to drug discovery. *Letters in Drug Design & Discovery*, 487–493.
Singh, V., Somvanshi, P. (2010). Toward the virtual screening of potential drugs in the homology modeled NAD+ dependent DNA ligase from Mycobacterium tuberculosis. *Protein & Peptide Letters*, 269–276.
Smith, T. F., Waterman, M. S. (1981). Identification of common molecular subsequences. *Journal of Molecular Biology*, 195–197.
Somvanshi, P., Seth, P. K. (2008). *In silico* primer designing of structural region of dengue virus for molecular diagnostic. *Internet Journal of Genomics and Proteomics*.
Somvanshi, P., Singh, V., Seth, P. K. (2008a). Phylogenetic and computational proteomes analysis of influenza A virus subtype H5N1. *Internet Journal of Genomics & Proteomics*.
Somvanshi, P., Singh, V., Seth, P. K. (2008b). Phylogenetic investigation of *Lin* genes involved in degradation of hexachlorocyclohexane (HCH). *Internet Journal of Toxicology*.
Somvanshi, P., Singh, V. (2010). Homology modeling and identification of catalytic residue in the nucleoprotein of influenza A virus. *International Journal of Medical Engineering & Informatics*, 26–36.
Somvanshi, P., Singh, V. (2008). Homology modeling and prediction of catalytic amino acid in the neurotoxin from Indian cobra (Naja naja). *Open Bioinformatics Journal*, 97–102.
Tinoco, I., Uhlenbeck, O. C., Levine, M. D. (1971). Estimation of secondary structure in ribonucleic acids. *Nature*, 362–367.
Turatsinze, J. V., Thomas-Chollier, M., Defrance, M., et al. (2008). Using RSAT to scan genome sequences for. *Nature Protocols*, 1578–1588.
Van Der Spoel, D., Lindahl, E., Hess, B., et al. (2005). GROMACS: Fast, flexible, and free. *Journal of Computational Chemistry*, 1701–1718.
Wang, Y., Addess, K. J., Chen, J., et al. (2007). MMDB: Annotating protein sequences with Entrez's 3D-structure database. *Nucleic Acids Research*, 298–300.
Watson, A. D. (2006). Thematic review series: Systems biology approaches to metabolic and cardiovascular disorders. Lipidomics: A global approach to lipid analysis in biological systems. *Journal of Lipid Res*, 2101–2111.
Wenk, M. R. (2005). The emerging field of lipidomics. *Nature Reviews Drug Discovery*, 594–610.
Wheeler, D. L., Barrett, T., Benson, D. A., et al. (2008). Database resources of the National Center for Biotechnology. *Nucleic Acids Research*, D13–D21.
Young, N., Chang, Z., Wishart, D. S. (2004). GelScape: A web-based server for interactively annotating, manipulating, comparing and archiving 1D and 2D gel images. *Bioinformatics*, 976–978.

3 Nutrigenomics

Martha Verghese and Judith Boateng
Alabama A&M University
Normal, Alabama

CONTENTS

3.1 INTRODUCTION

Conceptually, modern nutrition involves not just providing optimal nutrients for nourishment and improving health but, in addition, understanding the mechanisms, protection, and identification of the biologically active molecules and their demonstrated efficacy (Kussman and Affolte, 2006). A growing number of studies have demonstrated that nutrients and other bioactive compounds in food can regulate gene expression in many ways (Mead, 2007; García-Cañas, 2010). This emerging science, termed nutrigenomics, aims to describe nutrients in one of their biological roles, that is, as signaling molecules that are recognized by cellular sensing mechanisms and result in translation of these dietary signals into changes in gene, protein, and metabolite expression (Afman and Müller, 2006). One of the principal roles of nutrigenomics research is to identify the genes that influence the risk of diet-related diseases on a genome-wide scale and to understand the mechanisms that underlie these genetic predispositions (Müller & Kersten, 2003; Kolehmainen et al., 2005). These disorders are complex and multifactorial in their origin, involving not only genetic factors but also a number of behavioral and environmental factors such as exposure to certain food components (Ordóvas, 2007; García-Cañas, 2010). Nutrigenomics constitute a well-integrated analytical approach including the latest developments in high-throughput omics techniques such as genomics, genotyping, transcriptomics, proteomics, and metabolomics, as bioinformatics for the comprehensive study of different aspects of this biological complexity, which is also referred to as systems biology (van Ommen and Stierum, 2002; García-Cañas, 2010).

3.2 SINGLE-NUCLEOTIDE POLYMORPHISM IN NUTRIGENOMICS

The crux of nutrigenomics is that individuals for metabolic, environmental, or genetic reasons respond differently to nutrients and foods. Genetic factors known as single-nucleotide polymorphisms (SNPs) may be the key factor in human genetic variations that provide the molecular basis for these differences between individuals (Hesketh et al., 2006). However, the presence or absence of a particular gene is not an indicator to a particular disease process. The potential that this could increase or decrease a disease process depends on a complex interaction between the human genome and environmental and behavioral factors. Single-nucleotide polymorphisms are single-base pair differences or variations in a genetic sequence (Lander et al., 1998). Approximately 10 million SNPs that exist occur in about 1% of the population. According to Hinds et al. (2004), some SNPs occur in 5–50% of the population, and most humans are heterozygous for more than 50,000 SNPs across their genes.

3.2.1 Evidence from Epidemiological Studies

With numerous ongoing efforts to identify these SNPs, there is also focus on studying associations between disease risk and these genetic variations using a molecular epidemiological approach (Zhu et al., 2004). An example of a classic SNP known to influence nutrient requirements is 677 (C > T). This polymorphism is common in the gene for methylene-tetrahydrofolate (THF) reductase the enzyme responsible for reducing methylene-THF (the cofactor for methylating dUMP to dTMP in deoxynucleotide synthesis) to methyl-THF (the cofactor in methylating homocysteine to methionine) (Ames, 1999; Kaput et al., 2006; Hesketh, et al. 2006; Zeisel, 2007). It has been reported that individuals with this polymorphism tend to exhibit lower plasma-folate levels, which may eventually affect plasma-homocysteine levels, DNA synthesis, and methylation. Several population studies have looked at SNP frequency and folate status in relation to colorectal cancer risk.

Another pertinent example is the ApoE gene, from among several different classes of apolipoprotein, that transports lipid in plasma and other body fluids. At least three common apolipoprotein E (APOE) genetic alleles, ε2, ε3, and ε4, accounting for more than 99% of the variations of APOE, have been identified. Although more than half of the general populations carry the ε3 (60–90%) allelic variations, the occurrences of these variants vary among populations (Gerdes et al., 1992, 1996; Singh et al., 2006; Svobodova et al., 2007; Barzegar et al., 2008; Hubacek et al., 2010). It has been reported that individuals carrying the ε4 allele (Cys112 → Arg, rs429358) have been observed to predispose to high total cholesterol levels, whereas those expressing the ε2 allele (Arg158 → Cys, rs7412) have lower cholesterol than those carrying the most common ε3ε3 genotype (Hubacek et al., 2010). In particular, the ε4 allelic variant is associated with an increased risk for cognitive impairment and Alzheimer's disease (Schmechel et al., 1993; Poirier, 2003; Valenza et al., 2010). Individuals who inherit a single copy of the e4 allele have an increased chance of developing the disease, and those who inherit two copies of the allele are at greater risk.

Recently, a link was made between the fat mass and obesity (FTO) and INSIG2 (insulin-induced gene-2) genes and the incidence of obesity (Frayling et al., 2007; Lovegrove and Gitau, 2008; Andreasen et al., 2009). FTO is expressed in several tissues such as the adipose, pancreatic islets, skeletal muscle, and brain tissues. The identification of FTO, the major susceptibility gene for polygenic obesity, was initially found to predispose to diabetes through an effect on body-mass index (BMI). It was observed that individuals who were carriers of the A allele (rs9939609) of this SNP had higher BMI compared to carriers of the T-allele (Yajnik et al., 2009; Tan et al., 2008; Andreasen et al., 2008, 2009). As indicated by Orkunoglu-Suer et al. (2008), it also appears that the FTO variant may have a role in the control of food intake and hence food choice. This is a clear indication that the risk of obesity is dependent on environmental factors.

INSIG2 rs756605, which lies 10 kilobases 5' to the first exon of the insulin-induced gene-2, encodes a protein located in the membrane of the endoplamic reticulum. It is reported to regulate the processing of several proteins associated with fatty-acid synthesis and adipogenesis (Kim et al., 1996; Fajas et al., 1999; Yabe et al., 2002; Gong et al., 2006; Andreasen et al., 2009). Although some have reported a link between INSIG2 and the risk of obesity among certain populations (Hotta et al., 2008), a series of publications have found no association (Smith et al., 2007; Bressler et al., 2009; Cha et al., 2009). However, individual carriers of this gene variant are less able to inhibit the synthesis of fatty acids and cholesterol and are thus prone to accumulation of body fat (Orkunoglu-Suer et al., 2008). Even though no significant association has yet been observed between INSIG2 and obesity, Andreasen et al. (2008) stressed that physically inactive individuals who are carriers of this polymorphism may be prone to developing obesity.

3.3 ROLE OF NUTRIENTS IN GENE EXPRESSION

Over the years, nutrition research has dealt with providing nutrients to nourish the population. Nowadays, it focuses on improving the health of individuals through diet. It has become known through a series of population and experimental studies that diet can modulate health by either improving health or contributing to the risk of chronic diseases. Identifying diet as a primary causative factor for disease risk should be paramount when directing attention toward nutrition research. In nutrition research, the key scientific objective is to determine the underlying factors associated with diseases such as obesity, diabetes, cardiovascular disease, and cancer. Research has shown that these disorders are partly mediated by chronic exposure to certain food components, and thus, a critical part of the prevention strategy concerns changing food habits (Afman and Müller, 2006). The consensus is that nutrition for the most part can significantly contribute to undernutrition, indicating that many disease and disorders are correlated to suboptimal nutrition in terms of insufficiency of essential nutrients, excess of macronutrients, or even toxic concentrations of certain food components (van Ommen, 2004). With that in mind, it is essential to note that the concept of modern nutrition lies not just in providing optimal nutrients for nourishment and improving health but, in addition, in understanding the mechanisms, protection, and identification of the biologically active molecules and their demonstrated efficacy (Kussman and Affolte, 2006). The emerging science of nutrigenomics has demonstrated that diet may not only be a determinant factor in reducing the risk of chronic diseases but may and, justifiably so, mitigate the expression of several critical genes, thus modifying their activities and effect on diseases. Hence, the new era of nutritional research translates this rather empirical knowledge of evidence based molecular science, because food components interact with our body at the systemic, organ, cellular, and molecular levels (Kussmann et al., 2008; Kussmann and Affolter, 2009).

3.4 BIOACTIVE COMPONENTS

Bioactive compounds are non-nutrient or extranutritional constituents that typically occur in small quantities in plant and lipid-rich foods (Kris-Etherton et al., 2002; Liu, 2004). These compounds show great promise as disease fighters in the body by acting as antioxidants, activating liver detoxification enzymes, blocking the activity of bacterial or viral toxins, inhibiting cholesterol absorption, decreasing platelet aggregation, destroying harmful ROS, suppressing malignant cells, or interfering with the processes that can cause these diseases (Kris-Etherton et al., 2002; Pennington, 2002; Liu, 2004). Some of the better-known bioactive components are polyphenols, which include flavonoids, carotenoids, glucosinolates, oligosaccharides, polyunsaturated fatty acids, plant sterols (phytosterols), phytates, and lecithins. Nutrigenomics is now accepted as a multidisciplinary field of research that aims to determine how bioactive compounds in the diet can influence human health by analyzing their molecular interactions with several genes and their effect on transcription factors, protein expression, and metabolic profile (Chankvetadze and Cifuentes, 2010).

3.5 IMPACT OF BIOACTIVE COMPONENTS ON GENE EXPRESSION

3.5.1 Flavonoids

Flavonoids are one of the major groups of polyphenols. They constitute a large family of food constituents, many of which alter metabolic processes and have a positive impact on health (Beecher, 2003). It is the general term given to nearly 4,000 compounds that impart the colorful pigment to fruits, vegetables, and herbs, and they are also found in legumes, grains, and nuts (Nijveldt et al., 2001).

Generally, flavonoids are a subclass of polyphenols consisting of two aromatic rings. Each ring is composed of at least one hydroxyl, connected through a three-carbon bridge becoming part of a six-member heterocyclic ring (Beecher, 2003). The flavonoids are further divided into subclasses based on the connection of an aromatic ring to the heterocyclic ring, the molecular structure, as well as the oxidation state and functional groups of the heterocyclic ring (Beecher, 2003; Nijveldt et al., 2001). The major groups include the most common flavonols, flavones, anthocyanins, and flavan-3-ols such as catechin, whereas the minor groups include the chalcones and aurones, isoflavonoids, flavanones, and other phenolic pigments. As a group, flavonoids are universally distributed among vascular plants, and most are in the form of glycosides. Structurally, flavonoid is a 2-phenyl- benzo[~]pyrane or flavane nucleus comprising two benzene rings (A and B) (Figure 3.1) that is linked through a heterocyclic pyrane C ring.

3.5.2 Carotenoids

Carotenoids are isoprenoid molecules belonging to the tetraterpenes family. Carotenoids are represented by over 600 structures and consist primarily of at least eight-isoprene units (Fraser and Bramley, 2004; Tapiero et al., 2004). Carotenoids are responsible for most of the yellow to red colors in photosynthetic organisms. Carotenoids cannot be synthesized by humans and thus must be obtained through the diet. Some of the prominent sources are from plants, algae, and bacteria. In plants, they can be found in the roots, leaves, shoots, seeds, fruits, and flowers (Fraser and Bramley, 2004).

Carotenoids have varied functions in relation to human health because of their reported antioxidant activities and their ability to suppress chronic diseases (Fraser and Bramley, 2004). They have also been found to aid in immune and genome functions. Of late, research has primarily focused on individual carotenoids, such as b-carotene, lycopene, lutein, and zeaxanthine and their role as biological antioxidants, anticarcinogenic properties, and regulators of the immune system (Figure 3.2). Studies using human and animal cells have identified a gene, connexin43, whose expression is upregulated by chemopreventive carotenoids and that allows direct intercellular gap junctional communication (GJC). Since GJC is lost in cancer cells, its restoration through stabilization of connexin43 mRNA may be considered a cancer-preventive property of carotenoids (Zhang et al., 1992; Bertram, 1999; Aust et al., 2003). Another antiproliferative property of carotenoids is shown by inhibited expression of the N-*myc* gene (Okuzumi et al., 1990; De Flora et al., 1999)

3.5.3 Polyunsaturated Fatty Acids

Polyunsaturated fatty acids (PUFA) are an abundant source of dietary fat that occur throughout the animal and plant kingdom. They are long-chain fatty acids containing two or more double bonds. Omega-3 fatty acids (*n*-3) and omega-6 fatty acids (*n*-6) PUFA are the two main classes of PUFA encountered in the diet, and both are essential for normal processes in the body. (Spector, 1999; Jump, 2002). The metabolism of a-linolenic acid (ALA) and linoleic acid (LA) is well documented. The steps convert LA and ALA to their higher unsaturated derivatives; the long chain eicosapentaenoic acid (EPA, 20:5*n*-3) is a precursor of prostaglandins, thromboxanes, and leukotrienes (Tapiero et al., 2004) and docosahexaenoic acid (DHA, 22:6*n*-3). These two major types of omega-3 fatty acids are more readily used by the body (Connor, 1999), and arachidonic acid, the primary metabolite of LA, entails the activities of successive desaturation and elongation reactions (Bezard et al., 1994; Makni et al., 2008, Innis, 1991; Linscheer and Vergroesen, 1994; Russo, 2009). Because the two families of fatty acids compete for the same metabolic enzymes for desaturation and elongation, ALA is usually present at much lower levels in the diet and in the tissues of the body than LA, the primary essential fatty acid of the *n*-6 family usually consumed in the Western diet (Verghese et al., 2011).

Although omega-3 PUFAs; ALA, EPA, and DHA; and omega-6 PUFA 18:2w6, LA collectively protect against coronary heart disease, LA is the major dietary fatty acid regulating low-density

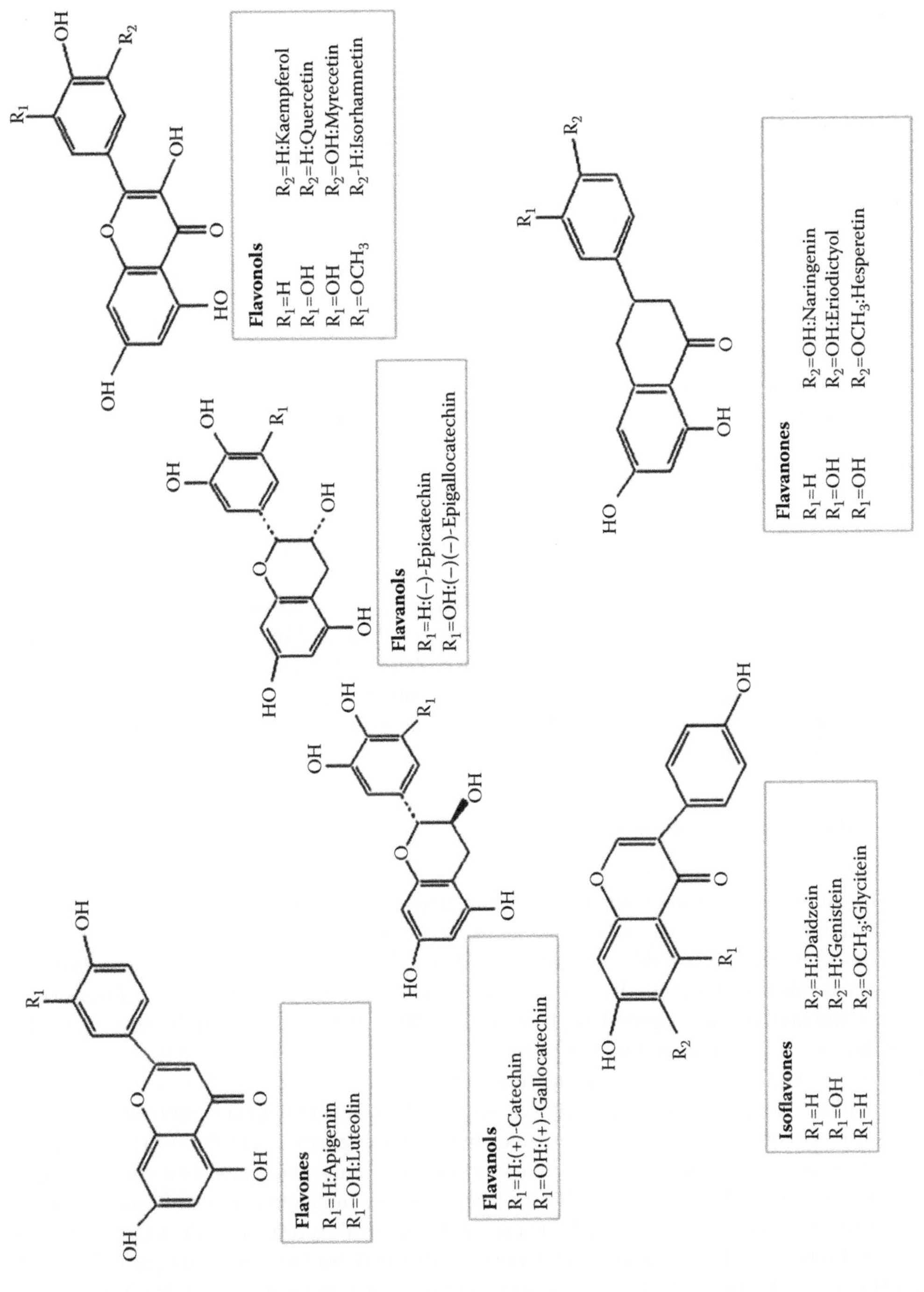

FIGURE 3.1 Chemical structures of flavonoids.

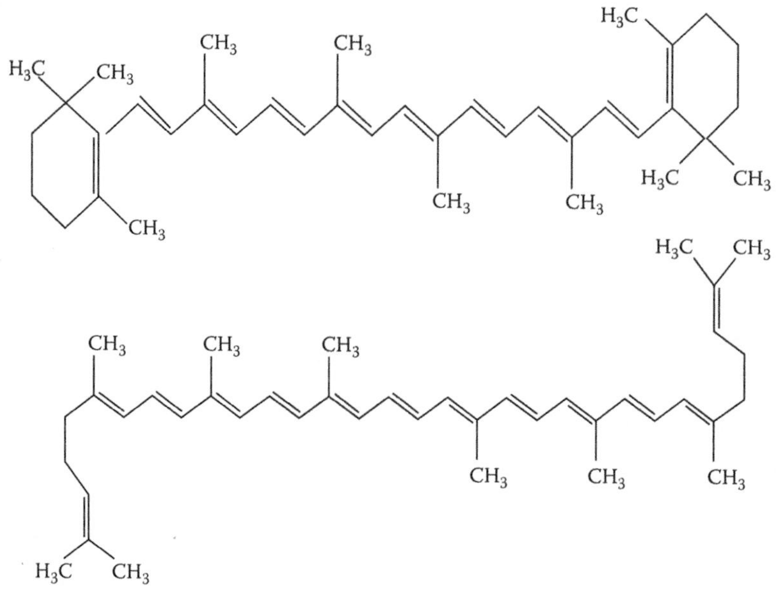

FIGURE 3.2 Structure of β-carotene and lycopene.

lipoprotein C (LDL-C) metabolism by downregulating LDL-C production and enhancing its clearance. There is evidence from research that has documented these PUFAs; especially *n*-3 possesses anticancer, anti-inflammatory, antiobesity, and antidiabetic properties among others. Others have reported the possibility that *n*-3 long-chain PUFA may be implicated in the regulation of mood and behavior and, hence, its role in neurological disorders. PUFAs play a significant role in the prevention of these diseases primarily by affecting cellular membrane lipid composition, metabolism, and signal-transduction pathways and by direct control of gene expression (Table 3.1) (Clarke and Jump, 1996; Sessler and Ntambi, 1998; Ntambi and Bené, 2001).

3.5.4 Glucosinolates/Isothiocyanates

Isothiocyanates (ITCs) are sulfur-containing phytochemicals and are the biologically active products of glucosinolates. They are plant-derived compounds that are commonly known as mustard oil and are found in cruciferous vegetables such as broccoli, brussels sprouts, and cauliflower (Tseng et al., 2004). Cruciferous vegetables contain a variety of glucosinolates, and each forms a different isothiocyanate when hydrolyzed. The conversion or breakdown of glucosinolates to ITCs occurs by the enzyme myrosinase (Figure 3.3) (Tseng et al., 2004). However, even in the absence of this enzyme, human intestinal microflora can cause the formation and absorption of ITCs.

The four most prominent ITCs include phenethyl isothiocyanate, which is a tissue specific inducer of phase II enzymes found in the liver; allyl isothiocyanate; benzyl isothiocyanate, which increases the activity of detoxification enzymes in the small intestine and the liver; and sulforaphane, which increases detoxification activity in the liver, lungs, stomach, and small intestine (Conaway et al., 2002). Isothiocyanates have a protective role in the prevention of cancers such as breast, prostate, and lung (Zhang, 2004). These phytochemicals aid in detoxification of carcinogens in the body as well as reducing oxidative stress that may lead to DNA damage and possibly cancer (Figure 3.4). ITCs can also help induce apoptosis or programmed cell death, which can help prevent the formation of tumors and further reduce the potential of cancer (Bonnesen et al., 2001). In normal breast cells, 17β-estradiol is irreversibly metabolized to 16 α-hydroxyestrone or 2-hydroxyestrone.

TABLE 3.1
Functions and Effects of Various Dietary Fatty Acids

Fatty Acid	Functions and Effects
Medium-chain	Rapid sources of calories (energy)
Saturated	
Lauric (12:0)	Hyperlipidemic, hypercholesterolemic, prothrombotic
Myristic (14:0)	
Palmitic (16:0)	
Stearic (18:1*n*-9)	Neutral or hypolipidemic, precursor of oleic acid
Monounsaturated	
Oleic (18:1*n*-9)	Hypolipidemic/hypocholesterolemic, precursor of eicosatrienoic acid (20:3*n*-9) in essential fatty acid insufficiency
Elaidic (18:1 trans)	Analogous to 18:0
Erucic (22:1*n*-9)	Impaired fatty acid oxidation in heart of rat
n-6 polyunsaturated	
Linoleic (18:2*n*-6)	Essential fatty acid (45 mg/kg/day), component of acylglucoceramides, precursor of arachidonic acid, hypolipidemic compared with saturated fatty acid, increases membrane fluidity with hypertension
γ-Linolenic (18:3*n*-6)	Precursor or eicosatrienoic acid and arachidonic acid, modifies eicosanoid levels
γ-Homolinolenic (20:3*n*-6)	Precursor of prostaglandin E_1 series of eicosanoids
Arachidonic (20:4*n*-6)	Membrane fluidity, precursor of eicosanoids
n-3 polyunsaturated	
α-Linolenic (18:3*n*-3)	Hypolipidemic, membrane fluidity, precursor of EPA and DHA (essential?), reduces eicosanoid synthesis
Eicosapentanoic (20:5*n*-3)	Hypolipidemic, reduces arachidonic acid synthesis and eicosanoids, precursor of prostaglandin E_1 and TXA_3, precursor of TXB_3
Docosahexanoic (22:6*n*-3)	Hypolipidemic, reduces arachidonic acid synthesis; reduces eicosanoid in some cells, essential for vision and neural membranes?

TXA_3, thromboxane A3; TXB_3, thromboxane B3.

Perhaps the important modulation of the metabolism of chemical carcinogens is to limit the availability of their genotoxic metabolites. This can be attained either through impairment of the cytochrome P450-mediated generation and/or increased detoxification of genotoxic metabolites resulting from upregulation of enzyme systems such as glutathione *S*-transferase and quinone reductase (Lampe and Peterson, 2002; Zhang, 2004; Fimognari and Hrelia, 2007; Higdon et al., 2007). ITCs, specifically sulforafane, play roles in the prevention of diseases because of their ability to inhibit cytochrome P450 enzymes in phase I biotransformation in addition to being inducers of phase II enzymes such as glutathione *S*-transferase (GST) (Bonnesen et al., 2001). The mode of these actions by the compound *Cruciferae* is for the most part dependent on their structures, with effects of indole derivatives and ITC being distinct (Lampe and Peterson, 2002).

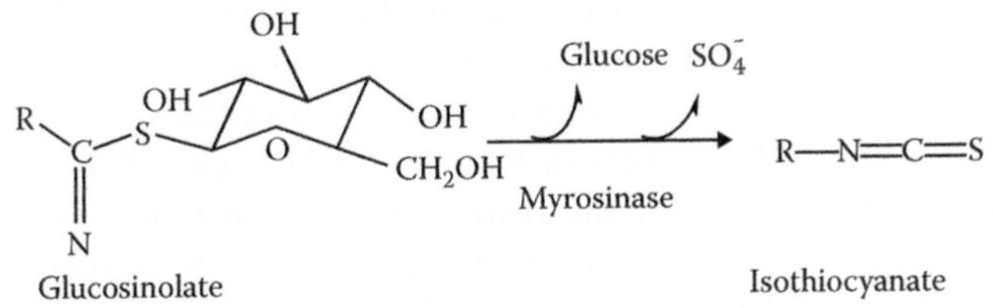

FIGURE 3.3 Conversion of glucosinolate to isothiocyanate via myrosinase.

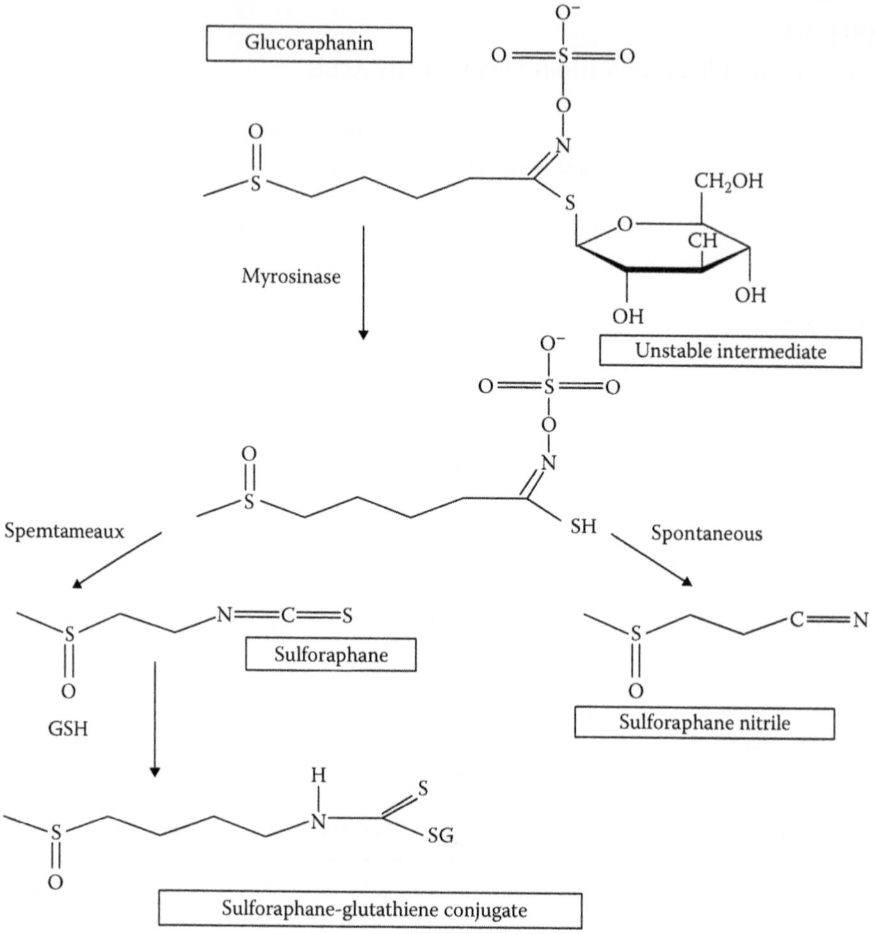

FIGURE 3.4 Conversion of glucosinolate to isothiocyanate via myrosinase.

Induction of these enzymes occurs at the transcriptional level and may be largely mediated by the antioxidant several transcription of genes. There are some ITCs that induce either phase I or II enzymes or both. Those that induce phase I contain a specific DNA sequence called xenobiotic response element (XRE), whereas those that induce phase II contain a sequence called antioxidant response element (ARE) (Slattery et al., 2000; Murray et al., 2001; Talalay and Fahey, 2001; Lampe and Peterson, 2002; Zhang, 2004). It is known that compounds that induce both phase I (e.g., XRE-driven) and phase II (e.g., ARE-driven) speed carcinogenic compounds through the metabolic pathway toward elimination, whereas those that induce XRE-driven gene expression without stimulating ARE-driven expression speed up, rather than retard, chemical carcinogenesis (Bonnesen et al., 2001; Lampe and Peterson, 2002).

The protective effects of ITCs on cancer risk may be influenced by inherited differences in individuals. These genetic polymorphisms affect expression of transcription factors or ligand binding affinity of receptors associated with the induction of the GST enzymes (Lampe and Peterson, 2002; Higdon et al., 2007). Null variants for GSTM1 and GSTT1 genes contain large deletions that result in the absence of the respective GST enzymes. Because both of these enzymes are involved in the metabolism of xenobiotics and reactive oxygen species, individuals who are homozygous for the *GSTM1*-null or *GSTT1*-null gene are at greater risk of developing cancer.

REFERENCES

Afman, L., Müller, M. (2006). Nutrigenomics: From molecular nutrition to prevention of disease. *J. Am. Diet Assoc.* 106: 569–576.

Ames, B. N. (1999). Cancer prevention and diet: Help from single nucleotide polymorphisms. *Proc. Natl. Acad. Sci. U.S.A.* 96: 12216–12218.

Andreasen, C. H., Mogensen, M. S., et al. (2008). Non-replication of genome-wide based associations between common variants in INSIG2 and PFKP and obesity in studies of 18,014 Danes. 3: e2872.

Andreasen, C. H., Mogensen, M. S., et al. (2009). Studies of CTNNBL1 and FDFT1 variants and measures of obesity: Analyses of quantitative traits and case-control studies in 18,014 Danes. *BMC Med. Genet.* 10: 17.

Aust, O., Ale-Agha, N., et al. (2003). Lycopene oxidation product enhances gap junctional communication. *Food Chem. Toxicol.* 41: 1399–1407.

Barzegar, A., Moosavi-Movahedi, A. A., et al. (2008). The mechanisms underlying the effect of alpha-cyclodextrin on the aggregation and stability of alcohol dehydrogenase. *Biotechnol. Appl. Biochem.* 49: 203–211.

Beecher, G. R. (2003). Review: Overview of dietary flavonoids: Nomenclature, occurrence and intake. *J. Nutr.* 133: 3248S–3254S.

Bertram, J. S. (1999). Carotenoids and gene regulation. *Nutr. Rev.* 57: 182–191.

Bezard, J., Blond, J. P., et al. (1994). The metabolism and bioavailability of essential fatty acids in animal and human tissues. *Reprod. Nutr. Dev.* 34: 539–568.

Bonnesen, C., Eggleston, I. M., et al. (2001). Dietary indoles and isothiocyanates that are generated from cruciferous vegetables can both stimulate apoptosis and confer protection against DNA damage in human colon cell lines. *Cancer Res.* 61: 6120–6130.

Bressler, J., Fornage, M., et al. (2009). The INSIG2 rs7566605 genetic variant does not play a major role in obesity in a sample of 24,722 individuals from four cohorts. *BMC Med. Genet.* 10: 56.

Cha, S., Koo, I., et al. (2009). Association analyses of the INSIG2 polymorphism in the obesity and cholesterol levels of Korean populations. *BMC Med. Genet.* 10: 96.

Chankvetadze, B., Cifuentes, A. (2010). Natural bioactive compounds and nutrigenomics. *J. Pharm. Biomed. Anal.* 51: 289.

Clarke, S. D., Jump, D. B. (1996). Polyunsaturated fatty acid regulation of hepatic gene transcription. *J. Nutr.* 126 (Suppl. 4): 1105S–1109S.

Conaway, C. C., Yang, Y. M., et al. (2002). Isothiocyanates as cancer chemopreventive agents: Their biological activities and metabolism in rodents and humans. *Curr. Drug Metab.* 3: 233–255.

Connor, W. E. (1999). Alpha-linolenic acid in health and disease. *Am. J. Clin. Nutr.* 69: 827–828.

De Flora, .S., Bagnasco, M., et al. (1999). Modulation of genotoxic and related effects by carotenoids and vitamin A in experimental models: Mechanistic issues. *Mutagenesis*. 14: 153–172.

Fajas, L., Schoonjans, K., et al. (1999). Regulation of peroxisome proliferator-activated receptor gamma expression by adipocyte differentiation and determination factor 1/sterol regulatory element binding protein 1: Implications for adipocyte differentiation and metabolism. *Mol. Cell. Biol.* 19: 5495–5503.

Fimognari, C., Hrelia, P. (2007). Alpha-linolenic acid in health and disease. *Mutat. Res.* 635: 90–104.

Fraser, P. D., Bramley, P. M. (2004). Review: The biosynthesis and nutritional uses of carotenoids. *Prog. Lipid Res.* 43: 228–265

Frayling, T. M., McCarthy, M. I. (2007). Genetic studies of diabetes following the advent of the genome-wide association study: Where do we go from here? *Diabetologia* 50: 2229–2233.

García-Canas, V., Simo, C., et al. (2010). Advances in nutrigenomics research: Novel and future analytical approaches to investigate the biological activity of natural compounds and food functions. *J. Pharm. Biomed. Anal.* 51: 290–304.

Gerdes, L. U., Gerdes, C., et al. (1996). The apolipoprotein E polymorphism in Greenland Inuit in its global perspective. *Hum. Genet.* 98: 546–550.

Gerdes, L. U., Klausen, I. C., et al. (1992). Apolipoprotein E polymorphism in a Danish population compared to findings in 45 other study populations around the world. *Genet. Epidemiol.* 9: 155–167.

Gong, X. W., Wei, D. Z., et al. (2006). Lowry method for the determination of pegylated proteins: The error, its reason, and a method for eliminating it. *Anal. Biochem.* 354: 157–158.

Hesketh, J., I. Wybranska, et al. (2006). Nutrient-gene interactions in benefit-risk analysis. *Br. J. Nutr.* 95: 1232–1236.

Higdon, J. V., Delage, B., et al. (2007). Cruciferous vegetables and human cancer risk: Epidemiologic evidence and mechanistic basis. *Pharmacol. Res.* 55: 224–236.

Hinds, D. A., Seymour, A. B., et al. (2004). Application of pooled genotyping to scan candidate regions for association with HDL cholesterol levels. *Hum. Genomics* 1: 421–434.

Hotta, K., Nakamura, M., et al. (2008). INSIG2 gene rs7566605 polymorphism is associated with severe obesity in Japanese. *J. Hum. Genet.* 53: 857–862.

Hubacek, J. A., Pelikanova, T., et al. (2010). A polymorphism in the cyclooxygenase 2 gene in type 1 diabetic patients with nephropathy. *Physiol. Res.* 60: 377–380.

Innis, S. M. (1991). Essential fatty acids in growth and development. *Prog. Lipid Res.* 30: 39–103

Jump, D. B. (2002). Dietary polyunsaturated fatty acids and regulation of gene transcription. *Curr. Opin. Lipididol.* 13: 155–164.

Kaput, J., Astley, S., et al. (2006). Harnessing Nutrigenomics: Development of web-based communication, databases, resources, and tools. *Genes Nutr.* 1: 5–11.

Kim, J. B., Spiegelman, B. M. (1996). ADD1/SREBP1 promotes adipocyte differentiation and gene expression linked to fatty acid metabolism. *Genes Dev.* 10: 1096–1107.

Kolehmainen, M., Poutanen, K., et al. (2005). Nutrigenomics: Key for identification of molecular effects of nutritional factors. *Duodecim* 121: 2139–2141

Kris-Etherton, P. M., Hecker, K. D., et al. (2002). Bioactive compounds in foods: Their role in the prevention of cardiovascular disease and cancer. *Am. J. Med.* 113 (Suppl. 9B): 71S–88S.

Kussmann, M., Affolter, M. (2006). Proteomic methods in nutrition. *Curr. Opin. Clin. Nutr. Metab. Care* 9: 575–583.

Kussmann, M., Affolter, M. (2009). Proteomics at the center of nutrigenomics: Comprehensive molecular understanding of dietary health effects. *Nutrition* 25: 1085–1093.

Kussmann, M., Rezzi, S., et al. (2008). Profiling techniques in nutrition and health research. *Curr. Opin. Biotechnol.* 19: 83–99.

Lampe, J. W., Peterson, S. (2002). Brassica, biotransformation and cancer risk: Genetic polymorphisms alter the preventive effects of cruciferous vegetables. *J. Nutr.* 132: 2991–2994.

Lander, E. S. (1998). Scientific commentary: The scientific foundations and medical and social prospects of the Human Genome Project. *J. Law Med. Ethics* 26: 184–188, 178.

Linscheer, W. G., and Vergroesen, A. J. (1994). Lipids. In M. E. Shils, J. Olson, and M. Shike (Eds.), *Modern nutrition in health and disease* (8th ed.) (pp. 47–88). Philadelphia: Lea and Febiger.

Liu, R. H. (2004). Review: Potential synergy of phytochemicals in cancer prevention: Mechanism of action. *J. Nutr*. 134: 3479S–3485S.

Lovegrove, J. A., Gitau, R. (2008). Personalized nutrition for the prevention of cardiovascular disease: A future perspective. *J. Hum. Nutr. Diet* 21: 306–316.

Makni, M., Fetoui, H., et al. (2008). Hypolipidemic and hepatoprotective effects of flax and pumpkin seed mixture rich in omega-3 and omega-6 fatty acids in hypercholesterolemic rats. *Food Chem. Toxicol.* 46: 3714–3720.

Mead, M. N. (2007). Nutrigenomics: the genome-food interface. *Environ. Health Perspect.* 115: A582–A589.

Müller, M., Kersten, S. (2003). Nutrigenomics: Goals and strategies. *Nat. Rev. Genet.* 4: 315–322.

Murray, A. E., Lies, D., et al. (2001). DNA/DNA hybridization to microarrays reveals gene-specific differences between closely related microbial genomes. *Proc. Natl. Acad. Sci. U.S.A.* 98: 9853–9858.

Nijveldt, R. J., van Nood, E., et al. (2001). Flavonoids: A review of probable mechanisms of action and potential applications. *Am. J. Clin. Nutr.* 74: 418–425.

Ntambi, J. M., Bené, H. (2001). Polyunsaturated fatty acid regulation of gene expression. *J. Mol. Neurosci.* 16: 273–288.

Okuzumi, J., Nishino, H., et al. (1990). Inhibitory effects of fucoxanthin, a natural carotenoid, on N-myc expression and cell cycle progression in human malignant tumor cells. *Cancer Lett.* 55: 75–81.

Ordóvas, J. (2007). Diet/genetic interactions and their effects on inflammatory markers. *Nutr. Rev.* 65: S203–S207.

Orkunoglu-Suer, F. E., Gordish-Dressman, H., et al. (2008). INSIG2 gene polymorphism is associated with increased subcutaneous fat in women and poor response to resistance training in men. *BMC Med. Genet.* 9: 117.

Pennington, C. R. (2002). Nutritional management: When and how should we become involved? *Clin. Nutr.* 21:191–194.

Poirier, M. F. (2003). Schizophrenia and drug abuse: Genetic aspects. *Encephale* 29: S23–S27.

Russo, G. L. (2009). Dietary n-6 and n-3 polyunsaturated fatty acids: From biochemistry to clinical implications in cardiovascular prevention. *Biochem. Pharmacol.* 77: 937–946.

Schmechel, D. E., Saunders, A. M., et al. (1993). Increased amyloid beta-peptide deposition in cerebral cortex as a consequence of apolipoprotein E genotype in late-onset Alzheimer disease. *Proc. Natl. Acad. Sci. U.S.A.* 90: 9649–9653.

Sessler, A. M., Ntambi, J. M. (1998). Polyunsaturated fatty acid regulation of gene expression. *J. Nutr.* 128: 923–926.

Singh, P. P., Singh, M., et al. (2006). APOE distribution in world populations with new data from India and the UK. *Ann. Hum. Biol.* 33: 279–308.

Slattery, M. L., Kampman, E., et al. (2000). Interplay between dietary inducers of GST and the GSTM-1 genotype in colon cancer. *Int. J. Cancer* 87: 728–733.

Smith, A. J., Cooper, J. A., et al. (2007). INSIG2 gene polymorphism is not associated with obesity in Caucasian, Afro-Caribbean and Indian subjects. *Int. J. Obes.* 31: 1753–1755.

Spector, A. A. (1999). Essentality of Fatty Acids. *Lipids* 34: 1–4.

Svobodova, H., Kucera, F., et al. (2007). Apolipoprotein E gene polymorphism in the Mongolian population. *Folia Biol.* 53: 138–142.

Talalay, P., Fahey, J. W. (2001). Phytochemicals from cruciferous plants protect against cancer by modulating carcinogen metabolism. *J. Nutr.* 131 (Suppl. 11): 3027S–3033S.

Tan, J. T., Dorajoo, R., et al. (2008). FTO variants are associated with obesity in the Chinese and Malay populations in Singapore. *Diabetes* 57: 2851–2857.

Tapiero, H., Townsend, D. M., et al. (2004). The role of carotenoids in the prevention of human pathologies. *Biomed. Pharmacother.* 58: 100–110.

Tseng, E., Scott-Ramsay, E. A., et al. (2004). Dietary organic isothiocyanates are cytotoxic in human breast cancer MCF-7 and mammary epithelial MCF-12A cell lines. *Exp. Biol. Med.* 229: 835–842.

Valenza, A., Bizzarro, A., et al. (2010). The APOE-491 A/T promoter polymorphism effect on cognitive profile of Alzheimer's patients. *Neurosci. Lett.* 472: 199–203.

van Ommen, B. (2004). Nutrigenomics: Exploiting systems biology in the nutrition and health arenas. *Nutrition* 20: 4–8.

van Ommen, B., Stierum, R. (2002). Nutrigenomics: Exploiting systems biology in the nutrition and health arena. *Curr. Opin. Biotechnol.* 13: 517–521.

Verghese, M., Boateng, J., Walker, L. T. (2011). Flax seed (*Linum usitatissimum*) fatty acids. In V. R. Preedy, R. R. Watson, and V. B. Patel (Eds.), *Nuts & seeds in health and disease prevention* (1st ed.) (pp. 487–498). London: Academic Press.

Yabe, D., Brown, M. S., et al. (2002). Insig-2, a second endoplasmic reticulum protein that binds SCAP and blocks export of sterol regulatory element-binding proteins. *Proc. Natl. Acad. Sci. U.S.A.* 99: 12753–12758.

Yajnik, C. S., Janipalli, C. S., et al. (2009). FTO gene variants are strongly associated with type 2 diabetes in South Asian Indians. *Diabetologia* 52: 247–252.

Zeisel, S. H. (2007). Nutrigenomics and metabolomics will change clinical nutrition and public health practice: Insights from studies on dietary requirements for choline. *Am. J. Clin. Nutr.* 86: 542–548.

Zhang, L. X., Cooney, R. V., et al. (1992). Carotenoids up-regulate connexin43 gene expression independent of their provitamin A or antioxidant properties. *Cancer Res.* 52: 5707–5712.

Zhang, Y. (2004). Cancer-preventive isothiocyanates: Measurement of human exposure and mechanism of action. *Mutat. Res.* 555: 173–190.

Zhu, Y., Spitz, M. R., et al. (2004). An evolutionary perspective on single-nucleotide polymorphism screening in molecular cancer epidemiology. *Cancer Res.* 64: 2251–2257.

4 Omics Approaches in Toxicology Research and Biomedical Applications

Nestor Luis López-Corrales, Anderson Miyoshi, and Vasco Azevedo
Universidade Federal de Minas Gerais
Belo Horizonte, Brazil

Tina Stutzman
Massachusetts Institute of Technology
Cambridge, Massachusetts

Debmalya Barh
Institute of Integrative Omics and Applied Biotechnology
Nonakuri, India

CONTENTS

4.1 INTRODUCTION AND BASIC CONCEPTS

Toxicology studies measure the effects of an agent on an organism's food consumption and digestion, on its body and organ weight, on microscopic histopathology, and on cell viability, immortalization, necrosis, and apoptosis (Waters and Fostel, 2004). A specific objective is to identify possible hazards, know their mechanisms and process, conduct risk assessment, and understand the implications of gene, protein, and metabolite function, mediating any possible toxic effect. The improvement and access to new methodologies in genomics, proteomics, and metabolomics have increased the speed with which gathered data can be used to understand and prevent the effects of environmental and chemical toxins (stressors) on biological systems. The joint application of all these technologies has lead to the development of new toxicological subdisciplines through the incorporation of the so-called *omics* technologies (methodologies related to DNA, RNA, proteins, and metabolites), focused on the elucidation of the adverse effects of compounds in organisms. Toxicogenomics was the original name of the new field. In order to get a proper idea about this new area, it is important to consider first that toxicogenomics is a multidisciplinary approach arising from the combination of methodological advances achieved in different scientific areas and second that the integration of omics technologies (genomics, proteomics, metabolomics, epigenomics with bioinformatics, and toxicology) will originate subfields that stand on their own right.

4.1.1 Genomics in Toxicology: Toxicogenomics

As mentioned above, toxicogenomics makes use of all of the omics technologies in the study of toxicology. At same time, it is possible to make distinctions between subfields: toxicogenomics combines genomics and toxicology; toxicoproteomics combines proteomics and toxicology; toxicometabolomics combines metabolomics and toxicology; and toxicoepigenomics combines epigenomics and toxicology. The integration and interrelationships among fields are key issues in the conjunction of omics-toxicology, because it can be very difficult to achieve scientific goals evaluating techniques separately. Figure 4.1 provides a view of the hierarchy and relationships among DNA, RNA, protein, metabolites, and includes epigenetics modifications. In any case, and to avoid possible misunderstandings, this chapter focuses mainly on the genomics and proteomics technological approaches and their integration with toxicology.

4.1.2 Toxicogenomics and Its Implications: Main Objectives in Toxicogenomics Studies

Nuwaysir et al. (1999) defined the use of omics technologies in toxicology as a new field of research, toxicogenomics (TGx). This area emerged with the expectation that it would provide ways to

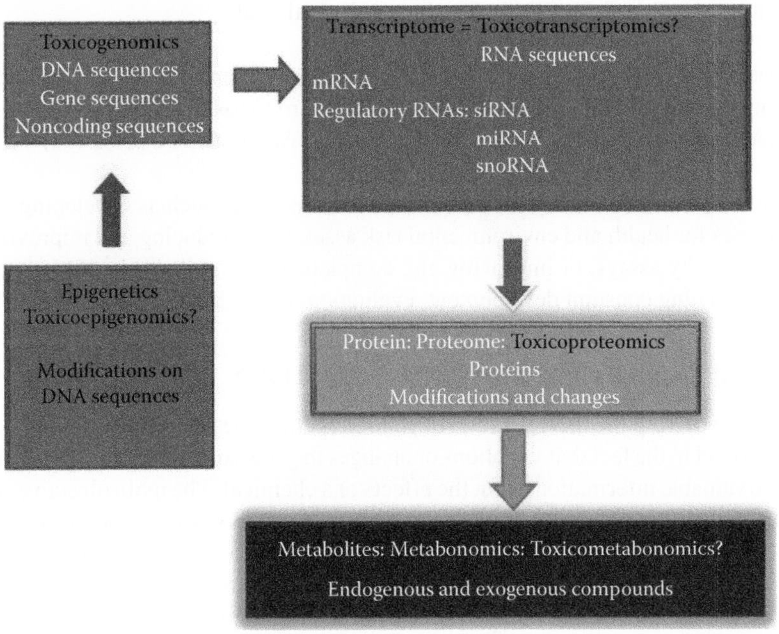

FIGURE 4.1 Hierarchy and relationships among DNA, RNA, protein, and metabolites, including epigenetic modifications. (Adapted from the Committee to Evaluate and Determine the Applications of Toxicogenomics Technologies to Predictive Toxicology and Risk Assessment at the National Research Council, National Institutes of Environmental Health Sciences, Research Triangle Park, NC.)

identify and characterize the molecular mechanisms underlying toxicity and play a role in all stages of compound safety evaluation, and more importantly, it is applicable to both drugs and chemicals at same time.

For a better understanding of toxicogenomics and the impact of omics in toxicology, we should consider two different aspects:

1. Involvement of environmental factors in common pathologies such as cancer, diabetes, hypertension, or asthma (to name a few)
2. A lack of knowledge about molecular mechanisms linking pathologies and environmental factors

One basis for the environment-disease relationship can be attributed to the paradigm that explains that chemicals may be distributed or metabolized in cells and interact with, disrupt, or damage target genes. Subsequently, pathologies arise as the result of disruption of genes that regulate essential biological process like cycle control, DNA repair, or differentiation (Mattingly et al., 2004). At the same time, factors such as the size of the genome and the diversity of hazardous chemicals, as well as the circumstances of exposure and genetic variability, make the understanding of links between chemical and disease a challenging process. If issues like the lack of information on genetic interactions, regulation of gene expression, or epigenetics are added to the original challenge, we will get a better idea about the task ahead and the scope of the toxicogenomics field. According to the report recently published by the Health and Environmental Sciences Institute (HESI) (Pettit et al., 2010), there are three main fields in which TGx has a direct impact on the understanding of the biological mechanisms of toxic compounds, drugs, and chemicals that affect public and environmental health; the determination of biomarker candidates (for example, associated with risk status and onset of disease); and increased knowledge of differences among species, which would allow for an improvement in the use of animal models.

In this context, the TGx field would have three principal goals:

1. To understand the relationship between environmental stress and human disease susceptibility
2. To identify useful biomarkers of disease and exposure to toxic substances
3. To elucidate the molecular mechanisms of toxicity (Waters and Fostel, 2004)

Under current circumstances, it is possible to add more objectives such as developing complementing methodologies for health and environmental risk assessment, reducing and improving the use of animals in the toxicity assays, or improving and complementing traditional tests in toxicology. All of these are undergoing constant development, evaluation, and validation.

4.2 GENE-EXPRESSION ANALYSIS IN TOXICOLOGY

What is the reason driving the incorporation of genomics and genetics methods into toxicology? The answer can be found in the fact that variations or changes in gene expression associated with pathways would produce valuable information about the effects of a chemical. The main objective of using gene expression techniques to understand the effects of environmental hazards, chemicals, or drugs is to analyze not only the simultaneous expression of thousands of genes but also their interactions.

At present, gene expression microarrays are the technology of choice in TGx studies. The use of microarrays allows the comparison of genome-wide gene expression patterns in dose and time contexts and thus is indicated to detect any modification in the expression profiles that may occur as a consequence of the effects of drugs and environmental chemicals. Different types of microarrays have been developed since they were first described. For specific information about the use of DNA microarrays in toxicology, we recommend the review by Letieri (2006), which describes principles and applications of arrays in toxicology and ecotoxicology. Early gene expression profiling experiments carried out for toxicogenomics studies used cDNA microarrays. Although this cDNA technology is rapidly being replaced by synthetic-oligonucleotide—short and long—microarrays (Figure 4.2), the technological concepts underlying the two approaches are largely analogous.

4.2.1 Oligonucleotide Arrays

In general, oligonucleotide arrays (trademarked as a GeneChip by Affymetrix) consist of short base-pair gene fragments as the DNA located (spotted) on the array. Specific details on the array production can be found in the work of Arteaga-Salasa et al. (2007). Once constructed, an oligonucleotide array is used for one of two things:

1. Quantification of the amount of mRNA in a single sample (e.g., to determine the amount of mutated versus nonmutated mRNA)
2. The comparison of two different samples hybridized to two separate arrays

4.2.2 cDNA Arrays

cDNA arrays are based on the same principle, but in this case the probes are larger pieces of DNA that are complementary to the genes of interest. Microarrays have certain limitations. These measurements are only semiquantitative due to a number of factors, including crosshybridization and sequence-specific binding anomalies. Other limitations include the number of samples that can be processed efficiently at a time, validation of the analysis, interlaboratory differences, and differences between platforms used. The issues concerning variability in the data obtained by microarrays are the subject of intensive research. A description of sources of variation in genome profiles (focused on toxicogenomic studies) can be found in the work of Boedigheimer et al. (2008). Specific guidelines such as Minimum Information about a Microarray Experiment (MIAME) have been

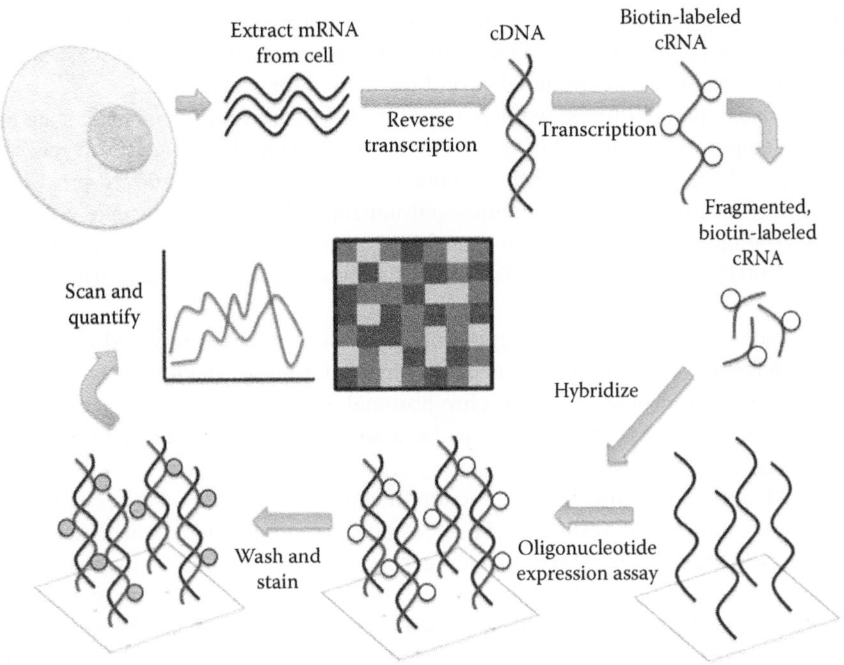

FIGURE 4.2 The use of oligonucleotide arrays. mRNA is extracted from cells and amplified through a process that labels the RNA for analysis. The sample is then applied to an array and any bound RNA stained. (Extracted from http://www.scq.ubc.ca/spot-your-genes-an-overview-of-the-microarray.)

developed to work with arrays. This package provides key information to enable the unambiguous interpretation of the results of a microarray-based experiment (Brazma, 2009).

4.2.3 Quantitative Polymerase Chain Reaction Complementing Arrays Methodologies

Quantitative polymerase chain reaction (QPCR) is a technique commonly used in gene-expression analysis. The QPCR assay works by amplifying large (typically 10–15 kb) stretches of genomic DNA. Under quantitative conditions, any damage to the DNA that is able to stop or significantly inhibit the progression of the DNA polymerase used in the PCR results in reduced amplification.

This method (as well as variations such as TaqMan®) can also test gene expression in a set of sequences previously determined. QPCR allows more quantitative measurements to be to obtained than microarrays. The literature is rich in examples of microarray technology coupled with quantitative reverse transcription (RT)-PCR offering a high-throughput method to explore the number of differentially expressed genes. The combination of arrays-RT-PCR is used routinely on toxicogenomics with very different objectives. Some examples are described by Cheng et al. (2006), who studied the effect of kojix (a food additive and preservative, skin-whitening agent used in cosmetics and plant growth); Nawarak et al. (2009), who analyzed the anticancer effect of arbutin treatment on melanoma cells; Sexton et al. (2008), who identified early molecular markers for events in pulmonary injury resulting from tobacco-smoke component exposure; and Zhao et al. (2007), who studied gene-expression profiles between strains of mice treated with HIV-1-based vectors for gene therapy. The limitations and advantages of this technique, as well as its potential for biomonitoring and mechanistic ecotoxicology, are reviewed by Meyer (2010).

4.3 EPIGENETICS AND EPIGENOMICS IN TOXICOLOGY

The term epigenetics describes the study of changes in the function of DNA (reversible and heritable) that occur without modifications in the sequence. These changes are included into the group of variations affecting the regulation of gene expression. Because the variations in gene expression caused by epigenetic factors are part of the individual response to drugs and disease susceptibility, epigenetics variations are also considered part of the field of toxicogenomics.

The *epigenome*, a parallel to the word *genome*, refers to the overall epigenetic state of the cell. Normally, the epigenome is built during development to shape the diversity of gene expression programs in the different cell types of the organism through a highly organized process. It is considered as an interface between the dynamic environment and the inherited static genome. Indeed, epigenetic aberrations can have effects on gene function similar to those based on genetic polymorphisms (Szyf, 2007). Interindividual differences in the epigenetic state could affect susceptibility to xenobiotics.

Efforts to understand the role of the epigenetic modifications have revealed their relationships in defined pathologies. Some types of cancer, mental retardation, neurodegenerative symptoms, imprinting disorders, syndromes involving chromosomal instabilities, and a great number of human life-threatening diseases have been found to be associated with aberrant epigenetic regulation (Santos-Reboucas and Pimentel, 2007).

DNA methylation, imprinting, and histone modifications (the balance between histone acetylation/deacetylation, which is governed by the opposite actions of histone desacetylases and histone acetyl transferases) are considered the main epigenetics mechanisms. DNA methylation (the addition of a methyl group to the 5-carbon of cytosine in a CpG island by the enzyme DNA methyltransferase) is the most easily measured by means of toxicogenomic technologies. Methylation of DNA plays a key role in processes like parental imprinting, development, X-chromosome inactivation, silencing of foreign DNA, proper centromeric segregation, or T-cell function. This process has also been presented as a prominent factor in aging and long-term memory storage (Hayashi et al., 2007).

In addition to the modification of core histones and DNA methylation, three other types of epigenetic control have been recently described:

1. Posttranscriptional gene silencing transcripts based on noncoding RNAs
2. Methylation-dependent sensitive insulators (such as the insulin-like growth factor 2 and H19 genes)
3. Action of the polycomb and trithorax groups, chromatin modifiers that control chromatin accessibility and maintain transcription in the first stages of embryonic life, throughout development, and in adulthood

One of the objectives of this chapter is to give a general view of the features and applications of omics in toxicology using different perspectives. In this context, it is important to mention the recently published review by Trosko and Upham (2010), which focuses on the epigenetic effects of chemicals and their correlation with end-point diseases. Essentially, the authors challenge the paradigm that chemicals are directly responsible for DNA damage in the genomic-nuclear DNA of relevant cells of the human body. Without denying the direct effect of compounds on DNA, they contend that chemicals also contribute to human diseases via either epigenetic or cytotoxic mechanisms. There are three basic end points of toxicity: mutagenesis (genotoxicity), cell death (cytotoxicity by necrosis, apoptosis, autophagy, or anoikis), and alteration of the expression of genes at the transcriptional, translational, or posttranslational levels (toxicoepigenomics). These authors hypothesize that chemicals act preferentially by inducing epigenetic changes.

4.4 PROTEOMICS AND TOXICOLOGY: TOXICOPROTEOMICS

Gene-expression profiles can be considered the primary level of integration between environmental factors and the genome. The characterization of the disease process is the main difference between classical genomic approaches and proteomics. The latter is performed directly by capturing proteins that participate in the disease. Proteomics in toxicogenomics necessarily involves the comprehensive functional annotation and validation of proteins in response to toxicant exposure. Understanding the functional characteristics of proteins and their activity requires a determination of cellular localization and quantification, tissue distribution, posttranslational modification state, domain modules, and their effect on protein interactions, protein complexes, ligand-binding sites, and structural representation. In addition, an important consideration is that each protein may be present in multiple modified forms, and sometimes the variations in modification status can be critical both for the function and expression of the protein.

It is also possible to observe effects that result from the formation of protein adducts through reactive chemical intermediates (originated by toxic substances and/or endogenous oxidative stress). The mentioned alterations on proteins caused by intermediates can also perturb regulatory networks.

All of this complexity adds to the challenge of proteome analysis itself. In contrast to the microarray technologies applied to gene expression, most analytical proteomic methods require elaborate serial analyses.

At present, the most commonly used technologies for proteomics research involve 2D gel electrophoresis for protein separation followed by mass spectrometry analysis of proteins of interest. Gel-based analyses generate an observable map of the proteome analyzed, although liquid separation and can be coupled with the use of appropriate software to achieve analogous results.

Other methodology is *shotgun proteomics* (which for some authors is, to a certain extent, analogous to the genome sequencing strategy of the same name). A key advantage of shotgun proteomics is its unsurpassed performance in the analysis of complex peptide mixtures (Wolters et al., 2001; Washburn et al., 2002).

The application of quantitative analyses has become a critical element of proteome analyses, and methods have been developed for application to both gel-based and shotgun-proteomic analyses. The most effective quantitative approach for gel-based analyses is difference gel electrophoresis. A description about its technical features and applications can be found in the work of Larbi and Jefferies (2009).

In addition to the above, there are two important methodologies that account for the use of proteomic analysis applied to toxicogenomics or toxicoproteomics. The first is the use of bioinformatics tools to search nucleotide and protein sequences in the database. In one example, the measured peptide masses from matrix-assisted laser desorption ionization (MALDI)-mass spectrometry (MS) spectra of tryptic peptide digests are searched against databases to identify the corresponding proteins. In proteomics, an important issue is the standardization of data analysis methods, data representation, and reporting formats. Problems regarding the variety of MS instruments, data analysis algorithms, and software used in proteomics, as well as differences in protein and peptide identifications (attributable to the use of different database search algorithms) have been reported in several sources. The second important methodology is proteome profiling, in which MALDI time-of-flight (MALDI-TOF-MS is used to acquire a spectral profile from a tissue or biofluid sample. Acquiring profiles/patterns from biological fluids like plasma is an enormous advance. Both the MALDI-TOF-MS and surface-enhanced laser desorption ionization time-of-flight mass spectrometry (SELDI, also known as SELDI-TOF-MS) are becoming more popular in clinical medicine and very recently in environmental toxicology. Two examples of using SELDI-based toxicoproteomics to identify biomarkers for specific environmental exposures or stressors are given by Benninghoff (2007). These studies demonstrate the application of the proteomics approach to wide-scale biomonitoring of numerous wildlife species in which contaminant concentration and the presence (or absence) of low-molecular-weight biomarkers in blood plasma are assessed concurrently.

4.5 BIOINFORMATICS AND DATA ANALYSIS IN TOXIGENOMICS

Besides the already described impediments shared by all omics methodologies used in toxicology, one of the greatest challenges is the effective collection, management, analysis, and interpretation of data. Although the various toxicogenomic technologies (genomics, transcriptomics, proteomics, and metabolomics) explore different aspects of cellular responses, the approaches to experimental design and high-level data analysis are universal.

The first step in any essay is the experimental design. Currently, and despite the fact that costs have decreased, economic factors still remain an obstacle for large population-based studies. On the subject of experimental design, a model of the workflow for a toxicogenomic experiment has been proposed by the Committee on Applications of Toxicogenomic Technologies to Predictive Toxicology and Risk Assessment of the National Research Council. The model is focused on the main steps of an experiment, emphasizing elements common to any assay regardless of the objective of the analysis and considering that at the end the underlying experimental hypothesis will indicate specific details for each step. In toxicogenomics, the data analysis is mainly focused on gene-expression profiles obtained from microarrays. The analysis can be classified in different types according to the recommendations of the committee regarding class analysis/class discovery, class comparisons, and class prediction in either unsupervised or supervised methods. It is important to consider the amount of bioinformatics tools specifically developed to deal with the increasing amount of data. These resources are providing intelligent ways to store and analyze information. Some of these tools are described in Table 4.1.

4.5.1 Functional and Network Inference for Mechanistic Analysis

In addition to providing the means for predicting the response to a compound, it would be of great interest to obtain information about the underlying mechanisms of action. Understanding the mechanisms of actions of drugs and environmental agents is a process that implicates the translation of the toxicogenomic-based hypotheses to validated findings. Bioinformatic tools play a key role in developing those hypotheses by integrating information that can facilitate interpretation, including gene ontology terms, which describe gene products (proteins), functions, processes, and cellular locations; pathway database information; genetic mapping data; structure-activity relationships; dose-response curves; phenotypic or clinical information; genome sequence and annotation; and other published literature. It is possible to access public software, developed using different methodological approaches, that will help to detect any possible mechanisms, but there is an important amount of work ahead to convert the prescriptive models to predictive ones.

See the original report of the Committee on Applications of Toxicogenomic Technologies to Predictive Toxicology and Risk Assessment of the National Research Council for a detailed description. Also see references and the works by Eisen et al. (1998). Raychaudhuri et al. (2000), Wang et al. (2002), and the Significance Analysis of Microarrays (SAM) package by Tusher et al. (2001) are key references.

Bild et al. (2009) provide an interesting example of how data sets from different sources but on the same pathology (in this case a type of breast tumor) can be integrated to produce valuable information and to identify relations between breast cancer subtypes, pathway deregulation, and drug sensitivity. Integration is useful to estimate pathway mechanisms or to predict responses to therapies. The study integrates independent gene-expression data sets to measure an individual phenotype or type of tumor, combining independent methods of genomic analyses to highlight the complexity of signaling pathways underlying different phenotypes and to identify optimal therapeutic opportunities. Interestingly, the method also allows for the prediction of sensitivity to commonly used cytotoxic chemotherapies to provide additional specificity to subtypes that might be better aligned with the characteristics of the individual patient.

TABLE 4.1
Databases and Bioinformatics Tools of Interest in Toxicogenomics

Database	Web Site
ActoR	http://actor.epa.gov/actor/faces/ACToRHome.jsp
Comparative Toxicogenomics Database (CTD)	http://ctd.mdibl.org
ChemBank	http://chembank.broad.harvard.edu
MMSINC	http://mms.dsfarm.unipd.it/MMsINC/search
Chemical Effects in Biological Systems (CEBS)	http://www.niehs.nih.gov/cebs-df
DrugBank	http://www.drugbank.ca
Kyoto Encyclopedia of Genes and Genomes (KEGG)	http://www.genome.jp/kegg
PharmGED	http://bidd.cz3.nus.edu.sg/phg
Pharmacogenetics and Pharmacogenomics Knowledge Base (PharmGKB)	http://www.pharmgkb.org
Potential Drug Target Database (PDTD)	http://www.dddc.ac.cn/pdtd
PubChem	http://pubchem.ncbi.nlm.nih.gov
Search Tool for Interactions of Chemicals (STITCH)	http://stitch.embl.de
ToxRTool (Germany)	http://ecvam.jrc.ec.europa.eu or http://ecvam.jrc.it (ToxRTool made publicly available from the ECVAM website)
SuperToxic	http://bioinformatics.charite.de
dbZach: A MIAME-Compliant Toxicogenomic Supportive	http://dbzach.fst.msu.edu
ArrayTrack	http://www.fda.gov/ScienceResearch/BioinformaticsTools/Arraytrack

Source: Adapted from Mattingly, C. J., *Toxicol Lett*, 186, 62–65, 2009; Burgoon, L. D., Boutros, P. C., Dere, E., et al., *Toxicol Sci*, 90, 558–568, 2006; and Schneider, K., Schwarz, P., Burkholder, I., et al., *Toxicol Lett*, 189, 138–144, 2009.

4.5.2 The Importance of Interacting Information in Toxicogenomics Approaches

One of the most important questions in biology and medicine is how to identify genetic mutations affecting human traits such as blood pressure, longevity, and onset of disease. The same challenges apply to toxicogenomics. Numerous scientific teams and consortiums are currently examining the genomes of thousands of people in an attempt to find mutations present only in individuals with certain traits. Until now, mutations have been largely examined in isolation, without regard to how they work together inside the cell. Pathway maps are now available describing in detail the network of genes and proteins that underlies cell function.

How to take advantage of these pathway maps to better identify relevant mutations and to show how these mutations work mechanistically? This basic approach of combining genetic information with known maps of the cell will have wide-ranging applications in understanding and treating disease.

Interactions amongst loci in mammalian species have been mapped through two main forward genetic approaches: genome-wide linkage and genome-wide association studies (GWAS). Recently Hannum et al. (2009) developed a methodology through the inclusion of a bi-cluster structure in the data and integrating genetic interactions derived from GWAS with protein complexes and functional annotations. The result is a map of protein complexes and pathways interconnected by dense

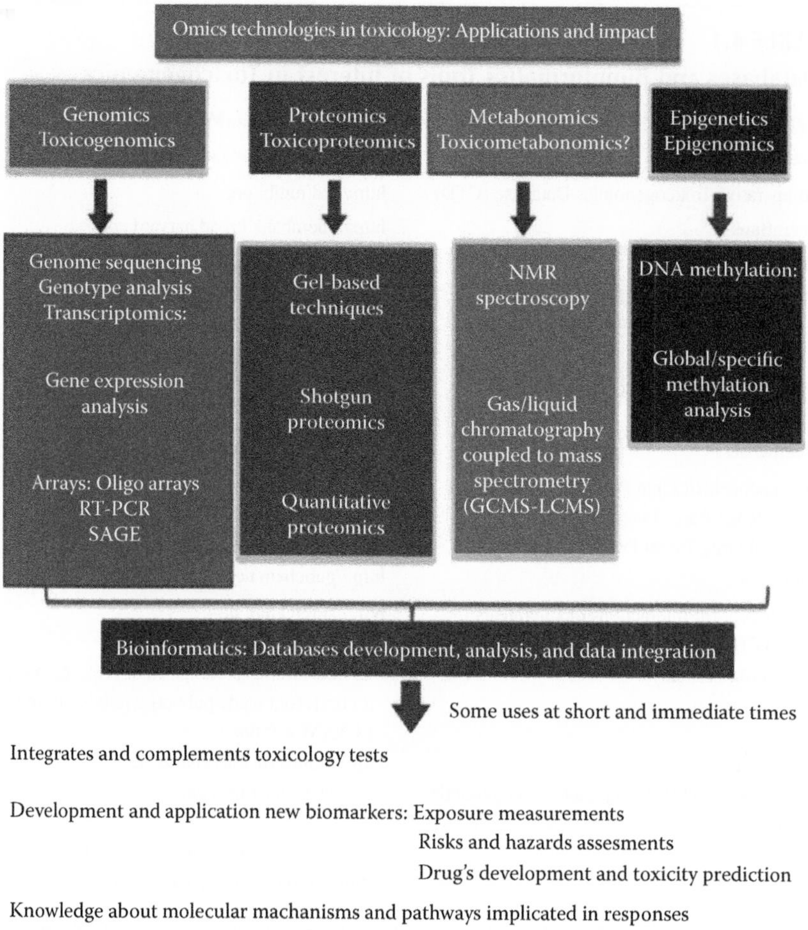

FIGURE 4.3 Omics methodologies in toxicology. NMR, nuclear magnetic resonance spectroscopy; GCMS, gas chromatography and mass spectrophotometry; LCMS, liquid chromatography and mass spectrophotometry.

bundles of genetic interactions, which raises statistical power and provides biological context to the genetic interactions uncovered in natural populations.

4.5.2.1 Concluding Remarks

The above methodologies provide different alternatives to predict and quantify effects of toxins and chemical and environmental compounds. One important issue is that the described approaches would bring data from different areas (genomics, proteomics, and metabonomics). The final objective will be to join them with the goal of generating information to understand biological responses to a specific compound. Consequently, a key factor is the integration of data to estimate responses in a complete biological system level. Figure 4.3 summarizes the main methods involved in the methodological big picture of omics technologies in toxicology. Readers should be aware that techniques are constantly developing and improving. Consequently, advances are expected, and the data integrations from different techniques are one of the most important issues to consider.

4.6 DATABASES IN TOXICOGENOMICS

There is a general agreement about the vital importance of databases in biological sciences. In toxicogenomics, the challenges and opportunities for a rigorous database are the capture, storage, and integration of a large volume of diverse data (Mattes et al., 2004). TGx always has a need for accurate tools to handle, store, and categorize data from different sources. Besides the diversity and complexity of data, as mentioned already, the interaction of the information becomes a fundamental issue when TGx databases are developed. How databases are built, what kind of data they gather, what is the degree of interactions, and what tools are included are all important factors to take into account to make the right decisions, predictions, and estimations from the stored data.

Statistics indicate the existence of more than 1000 databases in the public domain, but only a few are of interest to TGx studies. This small group of tools holds information about chemicals, drugs, and mechanisms of action and comprises gene profiling and toxic data. Table 4.1 summarizes some databases of interest in toxicogenomics. For a detailed description of each, it is recommended to see the original publications.

The survey of Williams-Devane et al. (2009), developed on public genomic sources across to more than 40 public Internet resources housing microarray data (Microarray World list of databases, http://www.microarrayworld.com/DatabasePage.html), identifies a group of databases showing relevance to toxicogenomics. In addition, this work lists other public cheminformatic resources potentially useful for supporting a public toxicogenomics capability, including a brief description of survey results for each data resource below, followed by chemical-indexing results for the two major resources. It is also important to consider private databases and tools produced by private companies: for example, Affymetrix (http://www.affymetrix.com), PharmaADME (http://www.pharmaadme.org), and Entelos, Inc. (http://www.entelos.com/index.php).

It is impossible to fully describe the characteristics of the main toxicogenomics databases in this chapter. For detailed descriptions of contents, utilities, and possibilities, a search of the bibliography and websites indicated is strongly recommended. There are two important resources that deserve special consideration: the Comparative Toxicogenomics Database and the Profiles of Chemical Effects on Cells Database.

4.6.1 The Comparative Toxicogenomics Database

The Comparative Toxicogenomics Database (CTD) is one of the most well known and used sources in the field. A few years ago Mattingly et al. (2004, 2006) describe the progress achieved by the development of a tool to support cross-species comparative studies of genes and proteins important from a toxicological point of view. It was complementary to the original report about the database published also by Mattingly et al. in 2003 and subsequently in 2004. The authors developed the CTD at the Mount Desert Island Biological Laboratory (MDIBL) in Salisbury Cove, Maine, USA, providing the following:

1. Annotated associations among genes, proteins, references, and toxic agents, with a focus on annotating data from aquatic and mammalian organisms
2. Nucleotide and protein sequences from diverse species
3. A range of analysis tools for customized comparative studies
4. Information to investigators on available molecular reagents

The process of data curation in the CTD is described by Davis et al. (2008) and Davis et al. (2009), who give detailed descriptions of each step in the process.

4.6.2 Profiles of Chemical Effects on Cells Database

The Profiles of Chemical Effects on Cells (pCEC) database is another bioinformatics resource of special interest. A detailed description has been published recently by Sone et al. (2010). This TGx database has a system to classify chemicals that have an effect on human health. The pCEC compiles data organized into a variety of chemical groups classified according to the type of pathway or type of toxicity and provides a visualization of the gene expression information by means of clustering techniques such as self-organizing maps and minimum-spanning trees. These characteristics, according to the authors, make pCEC a unique tool to predict toxicity and find biomarkers. Besides, pCEC could potentially include information about the health effects of chemicals combined with gene alteration profiles in specific tissues and cell types.

4.7 MAIN APPLICATIONS OF TOXICOGENOMICS METHODOLOGIES

One of the main objectives in toxicogenomics is to incorporate the above technologies into the toxicological routine assays and programs, focusing on the following goals:

1. Reduction of times and cost to evaluate drugs and chemicals
2. Validate findings
3. Reduce the use of animal models

4.7.1 Risk Assessment and Toxicogenomics

There are different ways in which global analysis of gene expression can be useful in assessing the risk of chemical agents, particularly considering the concentrations of compounds at which populations are exposed. Methodologies for gene expression evaluation are sensitive, and the fact that almost any toxic response is accompanied by modifications in the expression of genes indicates that it is possible to use gene-expression profiles to determine whether responses at molecular levels continue to occur at dose levels below those that produce detected adverse effects (Daston, 2008).

4.7.1.1 Evaluating Dose Response

There are many works using gene expression to evaluate risk and dose-response. For example, experiments to correlate formaldehyde doses with histological changes in rats grouped genes using gene ontology terms. This allowed not only for the identification of groups of genes acting/responding together but also for the interactions between them, which in turn can be used as a means of analysis (Andersen et al., 2008).

As mentioned above, any toxic response almost invariably results in changes in the expression of genes. The observation that specific mechanisms of toxicity result in specific changes in gene expression is currently supported by different lines of work. Boverhof et al. (2009a, 2009b), cited mentioned by Daston (2008), show that the toxic response is driven by the responses to ligands of nuclear receptors, because a large component of the signal transduction cascade for this pathway depends on gene expression.

At same time there is evidence supporting the analysis of gene-expression profiles to identify potential adverse effects in the absence of toxicity and overtoxicity signs, as indicated by Heinloth et al. (2004). This work establishes a clear relationship between subtle injuries and gene-expression profiles, produced by nontoxic doses of a product classified originally as a toxicity producer. Another example is the work by Kiyosawa et al. (2007), who attempted to identify candidate biomarker gene sets to evaluate the potential risk of chemical-induced glutathione depletion in liver using a dose-response and time-course assay.

Microarray platforms now span the entire genome in many species. This, coupled to the possibility of interlaboratory validation as well as the use of public databases and repository, is a clear indication that current methodologies are good enough to detect small effects at minimal levels of exposure to toxins. Besides, if pleiotropic effects act at low exposure levels to produce differential gene-expression profiles, the responsible sequences can be also identified.

Taken together, all the above advances enable us today to estimate the effects of low dosage levels and conduct research on the linearity of dose-response curves. The lowest-observed adverse-effect level (LOAEL) is defined as the lowest concentration or amount of a substance found by experiment or observation that causes an adverse alteration of morphology, functional capacity, growth, development, or life span of a target organism distinguishable from normal (control) organisms of the same species and strain under defined conditions of exposure. Thus the comparison between doses above and below the LOAEL will address hypothesized dose-dependent transitions in toxic response.

4.7.1.2 Integrating Toxicological Assays and Omics Methods in Risk Assessment: Some Examples

It is generally accepted that genomic technologies have the potential to enhance and complement existing toxicology end points. However, this can be achieved through a systematic evaluation and robust experimental design with genomic end points anchored to traditional toxicology end points.

The software GO-Quant was developed few years ago in an attempt to address the interest in the integration of toxicogenomics technologies into risk assessment. The objective was to extract and use quantitative gene expression values obtained with microarrays and determine the degree to which functional gene systems change based on dose or time (Yu et al., 2006).

Another example of the many possibilities afforded by the integration of traditional toxicological assays and toxicogenomics technology is shown again in the work of Boverhof et al. (2009a, 2009b). These experiments systematically illustrate the potential utility of genomic end points to enhance the local lymph node assay (LLNA). The conclusions support further exploration of this approach through the examination of a more diverse array of chemicals, trying to validate and complement a traditional toxicological approach.

4.7.2 Biomarker Discovery in Toxicogenomics

A biomarker is defined as "a characteristic that is objectively measured and evaluated as an indicator of normal biologic processes, pathogenic processes, or pharmacologic responses to a therapeutic intervention." End point, on the other hand, is defined as "a characteristic or variable that reflects how a patient feels, functions, or survives," and it is distinguished from a surrogate end point, which is defined as "a biomarker that is intended to substitute for a clinical end point" (National Institutes of Health). The main attributes of a high-quality biomarker for a specific chemical class or mechanism of action have been described by Benninghoff (2007).

The utility of gene expression biomarkers has been already acknowledged for many years (Van Leeuwen et al., 1986). These authors studied the use of transcriptional biomarkers and effects of smoking. More recently, Kim et al. (2005) used putative transcriptional marker to identify genotoxic effects (but not carcinogenesis), and Amin et al. (2004) studied renal toxicity.

One of the great expectations about toxicogenomics is that it will decrease the time it takes to determine new biomarkers useful in drug development, clinical trials, clinical monitoring, and toxicity assessment. The transcriptional fingerprint can reflect an accumulative response that represents complex interactions within the organism, including pharmacologic and toxicologic effects. If these interactions are reproducible and significantly correlate to an end point, the molecular fingerprint would potentially qualify as a predictive biomarker, as it happens *de facto* in the drug industry.

There is also a place for exposure biomarkers, as it has been pointed out in studies to detect the presence of environmental contaminants. An important issue is that the responses are changes able to be detected at different levels: either on the expression of genes and proteins or on the metabolic profiles. Consequently, the spectrum to detect and use new biomarkers to evaluate dose responses, exposure hazards, risk assessment, or drug toxicity is wide. Until now, most of the efforts have been focused on using single technologies to determine biomarkers, but again, the integration of different methodological approaches to provide multidomain biomarkers looks much more promising and predictive than those derived from a single method. In fact, there are strong recommendations to include the existing predictive measures on the new approaches as well as to pay attention to those biomarkers not specifically attached to an end point.

Finally, as stated by the Committee on Applications of Toxicogenomic Technologies to Predictive Toxicology and Risk Assessment of the National Research Council (National Institutes of Health), it is important to take into account toxicogenomics projects under development in different countries like the Netherlands [where a breast cancer study gives an example of a biomarker reaching Phase 4 (Bogaerts et al., 2006)] or Japan [the Japanese toxicogenomics project reviewed and recently published by Uehara et al. (2010)].

4.7.2.1 Concluding Remarks

1. There is a relationship between modification of gene-expression profiles and nontoxic doses.
2. Data must be gathered about the relationship between subtle injuries, nontoxic doses, and gene expression.
3. Methods have to be sensitive enough to detect small changes in gene expression through microarray methods.
4. It is possible to group genes involved in nontoxic dose responses.
5. It is possible to integrate toxicogenomics in risk assessment projects.

4.7.3 Toxicogenomics and Nanomaterials

Nanomaterials are defined as materials with at least one dimension of 100 nanometers or less and having nanostructure-dependent properties (e.g., chemical, mechanical, electrical, optical, magnetic, and biological), with a commercial or medical application. There are many applications for nanomaterials at present. One of the problems is that their properties may potentially lead to nanostructure-dependent biological activity that differs from and is not directly predicted by the bulk properties of the constitutive chemicals and compounds (Poma and Di Giorgio, 2008). There is not enough information about how man-made nanostructures can interact with or influence biological system and how the exposure to nanoparticles would affect the living systems. In addition, the literature indicates that manufactured nanoscale materials may spread in the body in unpredictable ways. Nanoscale materials have been observed to preferentially accumulate in particular organelles (Mishra et al., 2005). Studies about the effect of inhalation of nanoparticles show their accumulation in organs like the kidney and liver (Oberdorster et al., 2002). Moreover, nanoparticle toxicity can be attributed to the release of toxic ions (for example CdSe/ZnS nanoparticles), nonspecific interaction with biological structures facilitated by their shape [as in the case of nanotubes (Lam et al., 2004)], and specific interaction with biomacromolecules through surface modifications.

Biocompatibility, toxicity, and the ability to penetrate cells are three critical factors that will determine the utility of nanoparticles in clinical applications. Here is where current toxicogenomics methodologies would fit. An example is presented in a recent study on the ecotoxicity of silver nanoparticles (AgNPs) investigated in *Caenorhabditis elegans* (Roh et al., 2009). The authors integrate gene-expression analysis with organism and population-level end points using *C. elegans* functional genomics tools to test the ecotoxicological relevance of AgNPs-induced gene expression.

An interesting view about the toxic effects of nanomaterials and synthetic polymers used in drug delivery is given by Kabanov (2006). The author coined the concept of *polymer genetics* to describe a new field or subarea focused on the understanding of the pharmacological and toxicological effects of polymer formulations of biological agents.

Nevertheless, the problem of toxicity of nanoscale materials is not new. It was encountered by researchers several decades ago in the context of the toxicity of water-soluble synthetic polymers, liposomes, nanoparticles, and other nanoscale objects.

As indicated by Kabanov (2006), the toxicity problem is addressed again in cases such as the development of biodegradable polymers, the optimization of polymer chemical composition, and the surface coating of liposomes and nanoparticles, among other approaches. Considering current nanodrugs or nanomedicines in use (liposomal doxorubicin, Doxil, or albumin-bound paclitaxel, Abraxane), as well as in those under clinical trials, it is possible to agree with the author on the fact that even safe polymers and nanomaterials developed for human use may affect the responses to biological agents in the body, and these interactions are currently not well understood. Poma and Di Giorgio (2008) described the utility of the toxicogenomics methodologies to analyze effects of nanoparticles in different *in vivo* and *in vitro* systems, including data on the documented negative effects of different nanomaterials on health.

4.7.3.1 Concluding Remarks

The following issues should be considered when addressing the relationship between toxicogenomics and nanomaterials:

1. Expectations of the use of nanomaterials
2. Concerns about the safety of nanomaterials
3. Physicochemical properties of nanomaterials
4. Possibility of toxic effects
5. Need to develop safe nanomaterials

4.7.4 Genotoxicity, Carcinogenesis, and Toxicogenomics

The starting point is the very well known association between DNA damage and cancer development. Based on this association, the preclinical safety evaluation paradigm for drugs and chemicals consists of assessing their genotoxicity, e.g., their potential to cause DNA damage and carcinogenicity (generation of tumors in animals upon long-term exposure). There is a standard genotoxicity testing battery, coupled to the analysis of potential of chemicals to induce tumors. Some of these tests involve the use of transgenic mice such as $p53^{+/-}$ Hras2 and Tg.AC.

A thorough study of the literature allows us to gather a series of facts that support the use of TGx assays/information in genotoxic and carcinogenetic assays:

1. The genotoxicity testing battery is highly sensitive for the detection of carcinogens. However, the testing paradigm has low specificity (Kirkland et al., 2005, 2006). This is supported by the fact that 50% of the noncarcinogens among marketed pharmaceuticals had some positive genotoxicity (Snyder and Green, 2001).
2. Carcinogenicity assessment for compounds negative for genotoxicity is typically required for marketing approval of pharmaceuticals.
3. Only one-third of marketed drugs found to be positive or equivocal for carcinogenicity were also positive for genotoxicity, indicating the existence of a considerable number of nongenotoxic carcinogens.
4. It is important to have a better knowledge of events and mode of action (MOA) in assays as indicated by International Life Sciences Institute and Risk Science Institute.
5. There is a need to develop alternative short-term experimental approaches with better cancer predictive potential (even if it is partial), including the investigation of MOA(s).
6. There is a need to discriminate between distinct subgroups: genotoxic and nongenotoxic carcinogens according to mechanism of action. This is an important step in the cancer-risk assessment for chemicals and drugs.

Besides, there is evidence that some genotoxic chemicals may exhibit a threshold dose-response curve in *in vitro* systems. It is of paramount importance to differentiate a true threshold dose-response curve from an assay detection limit. To do this, it is critical to understand the underlying mechanisms (Kirkland and Muller, 2000). Currently, mechanistic research requires lengthy experimental follow-up strategies with uncertain outcome, which may lead to significant delays in the introduction of new drugs to the market.

An increasing body of evidence indicates that toxicogenomic analysis of cellular stress responses provides an insight into mechanisms of action of genotoxicants. The features of this response in mammalian cells are detailed by Aubrecht and Caba (2005).

4.7.4.1 Concluding Remarks

There is substantial evidence to support the use of TGx methods in carcinogenesis and genotoxicity: there is a possibility of determining DNA-reactive versus DNA-nonreactive genotoxins *in vivo*, as well as genotoxic versus nongenotoxic carcinogens *in vitro*, through the use of toxicogenomics methodologies. A need to redefine and differentiate classes of carcinogenetic products based on mechanisms of action has driven an unparalleled development of TGx tools coupled to biostatistical methods. Finally, the omics technologies to analyze transcriptional regulation (epigenetics and micro-RNA), as well as proteomics and metabolomics to evaluate the mechanisms of genotoxicity and carcinogenicity, have also flourished over the past few years.

As indicated by Oberemm et al. (2009), practical applications of TGx techniques provide a mechanistic context for results obtained with both genotoxicity and carcinogenicity assays. Considering that gene-expression profiles can help determine non-DNA reactive mechanisms in the trials of genotoxic and carcinogenetic compounds, it is clear that if not a fact already, it is just a matter of time before TGx's methodologies are incorporated into the genotoxicity and carcinogenesis fields

Table 4.2 presents a summary of some of the main toxicogenomics projects under development around the world. Specific websites for each provide details about the systems and methodologies in use.

4.7.5 Toxicogenomics in Drug Discovery and Development

The cost of drug discovery has increased to alarming levels. The amount of money to bring a new candidate drug to the market is calculated to be about 1 billon US$. (Barros, 2005). A substantial portion of the cost is related to the ratio of candidate drug losses (30–60%) because of failure to detect clinical side effects and nonclinical toxicological findings. There are many reasons for these problems, and there is also a general agreement on the need to improve the toxicological tests to reduce them. Of course, the objective is to reach 100% accuracy when testing toxicological liabilities, but in the real world, tests cannot perfectly predict and instead are fallible and able to produce errors. In this context, achieving precise and accurate TGx tests to distinguish real toxic effects versus real nontoxic ones is one of the main objectives in the drug development process.

Other important economical factors are the *in vivo* toxicological testing programs. Besides the ethical considerations (which are leading efforts to reduce the use of animals), the cost of animal testing is one of the largest in the drug development industry. There is a specific interest to implement alternative methodologies for assessing the safety of new and old compounds ensuring human health but simultaneously reducing the reliance on animal testing. The general idea is that application of modern technologies such as transgenic, *in silico*, and *in vitro* systems along with novel and predictive biomarkers should decrease the use of large, expensive, and sometimes unethical animal tests.

Genomics technologies can provide tools to decrease the length and cost of drug development and compound attrition. Supporting this concept, we can mention the real possibility of identifying specific gene-expression profiles linked to the dose and response to chemicals and compounds. In addition, obtaining and combining gene-expression profiles with proteomics and metabolomics profiles would improve and complement the detection of negative/toxic effects, which in turn would reduce costs and increase the speed of drug development. The growing speed of toxicogenomics in

TABLE 4.2
Features of Some Toxicogenomics Projects under Development and Institutions Involved

Project Institutions	Features/Objectives	Web Site/Reference
National Center for Toxicogenomics	Three different areas: 1. Environmental genetics. 2. Environmental stress and cancer. 3. Chemical Effects in Biological Systems (CEBS).	http://www.niehs.nih.gov/research/atniehs/labs/lrb/enviro-gen/index.cfm http://www.niehs.nih.gov/research/atniehs/labs/ltp/esc/index.cfm http://www.niehs.nih.gov/research/resources/databases/cebs/index.cfm
Toxicogenomics Research Consortium NIH	- Research on environmental stress responses. - Standards and practices for the analysis of gene expression data. - Development of relational databases. - Risk detection and earlier intervention in disease processes.	http://www.niehs.nih.gov/research/supported/centers/trc
European Molecular Biology Laboratory (EMBL) and European Bioinformatics Institute (EBI)	- Maximize interactions between nutrigenomics, environmental genomics and toxicogenomics (NET) communities. - Provide data and bioinformatics services to scientific community. - Basic research in bioinformatics. - Provide advanced bioinformatics training at all levels. - Disseminate cutting-edge technologies to industry.	http://www.ebi.ac.uk/net-project
Mount Desert Island Biological Laboratory	Home of the Comparative Toxicogenomics Database (CTD), a public database that helps researchers discover relationships between chemicals in the environment, genes, and disease.	http://www.mdibl.org
Netherlands Toxicogenomics Centre (NTC)	- Develop new methods that better chart the risks of chemical compounds and simultaneously offer an alternative to the current practice of animal testing.	http://www.toxicogenomics.nl
OECD Organization for Economic Co-operation and Development	The OECD works in cooperation with the International Programme on Chemical Safety (IPCS). The IPCS focuses its work on new biomarkers by exploring the science and evidence basis for toxicogenomics, and the OECD is responsible in other two items: 1. Molecular screening for characterizing individual chemicals and chemical categories. 2. Survey on the available omics tools, focusing more on defining the needs and possibilities for the application of toxicogenomics in a regulatory context. Works by both programs are being coordinated closely and will feed into each other.	http://www.oecd.org

continued

TABLE 4.2 (CONTINUED)
Features of Some Toxicogenomics Projects under Development and Institutions Involved

Project Institutions	Features/Objectives	Web Site/Reference
The Japanese Toxicogenomics Project (TJP)	The TJP was a 5-year project (2002–2007) performed by National Institute of Health Sciences, 15 pharmaceutical companies, and the National Institute Biomedical Innovation (NIBIO), which was the core institute where the actual work was performed.	http://www.nibio.go.jp/english
The Korean Toxicogenomics Projects	The National Institute of Toxicological Research (NITR) in Korea is in charge to develop toxicogenomics-based toxicity and safety assessment techniques. NITR is conducting various research and development projects, such as construction of toxicoinformatics infrastructure called the Korea Toxicoinformatics Integrated System (KOTIS), which is carried out jointly with ISTECH Inc. KOTIS is modeled against ArrayTrack, and CEBS is composed of a database system and its analysis programs. Details are depicted in Chung et al. (2009).	Chung et al. (2009)
carcinoGENOMICS	The aim of the project called carcinoGENOMICS is to develop *in vitro* methods for assessing the carcinogenic potential of compounds, as an alternative to current rodent bioassays for genotoxicity and carcinogenicity. The major goal is to develop a battery of mechanism-based *in vitro* tests accounting for various modes of carcinogenic action. The novel assays will be based on the application of omics technologies. Thereby also exploring stem cell technology, to generate omic responses from a well defined set of model compounds causing genotoxicity and carcinogenicity.	http://www.carcinogenomics.eu
INTARESE	The INTARESE Consortium comprises a total of 33 partners, from many of the leading research and user organizations in Europe, including 10 universities, 17 national research institutions/centers, 4 national governmental agencies, 1 intergovernmental organization, and 1 representative from industry. Expertise covers all the relevant areas of science, including environmental science, epidemiology, toxicology, information technology, geography and GIS, statistics, and modeling. The project is coordinated at Imperial College London with administrative coordination done through ICON.	http://www.intarese.org
COMET Consortium for Metabonomics Toxicology	Formed between five major pharmaceutical companies and Imperial College London, UK.	http://bc-comet.sk.med.ic.ac.uk

the last few years has created expectations about its utility in the understanding and prediction of unwanted chemical and physical effects in organisms and ecosystems. Although the bibliography indicates the potential in this area, toxicogenomics has not been established either in the drug development process nor in drug safety evaluation on a routine basis. Despite the advantages in terms of cost and speed, there are also limitations as a result of the overlapping of the new techniques with traditional preclinical assessment (Wills and Mitchell, 2009).

The use of microarray analysis in toxicology has generated specific gene-expression profiles for various chemical substances that can be related to dose and response. This relation makes possible the identification of sensitive biomarkers for early deleterious effects, and examples are well depicted in the work of Van Hummelen and Sasaki (2010). The authors highlight specific issues that need improvement (data analysis and interpretation and accessible data repositories) to achieve a more efficient incorporation of genomics in drug development and environmental toxicology research.

Another interesting view of the use of biomarkers in safety testing is given by Hewitt and Herget (2009). They use, for example, kidney toxicity (nephrotoxicity) produced by cisplatin. This drug causes deregulation in several important genes. These changes in gene expression, according to the authors, are well correlated with the mechanisms producing tubular damage, tissue regeneration, and remodelation by the cisplatin.

In the rat animal model, biomarkers of early cardiotoxicity have been evaluated recently by Mori et al. (2010). Using genome profile analysis from microarrays, the work determines the relationships between a set of genes and histopathological features like degeneration of myocardium and inflammation. The authors identified five genes as candidate genomics biomarkers for the detection of early cardiotoxicity.

One specific area in which scientists are focusing their efforts is to anchor the histopathological findings/parameters to toxicogenomic profiles. Oberemm et al. (2009) examine the utility of proteomic and transcriptomic analysis with this objective. The authors describe, through the functional analysis, a set of genes and corresponding proteins useful as biomarkers for early hepatocarcinogenesis.

Other efforts can be observed in the analysis of drug-induced hepatotoxicity, a frequent cause of liver disease and acute liver failure, particularly in patients treated with multiple drugs. There are many common substances in medicine with possible hepatotoxicity effects including tetracyclines, sulfonamides, tuberculostatic agents, macrolides, quinolones, and beta-lactams. As a consequence, drug induced liver injury, can be considered a hot spot for an early detection using toxicogenomics approaches.

An interesting strategic application of genomics to increase the efficiency of developing safer medicines has been recently developed. Based on previously published data, the researchers identified three different steps in the drug development process (target choice, molecule selection, and clinical benefit/risk) in which toxicogenomics could interact through different methods. This work provides good examples illustrating the value of toxicogenomics in early drug discovery. The authors advocate the use of gene expression from *in vitro* models to assist in compound or chemical-scaffold selection at a stage where the structure-activity relationship is broad and malleable. The authors also suggest that the value of expression analysis in primary rat hepatocytes indicates that target organ toxicity can be predicted if predictions are based on a large, robust database. Likewise, gene expression changes from short-term studies can serve as more sensitive markers of a toxicity that manifests pathologically *in vivo* only with longer duration treatment when evaluated against the context of a large reference database (Ryan et al., 2008).

An example of enhancing and complementing toxicological end points with genomics approaches is demonstrated in the previously mentioned work by Boverhof et al. (2009a, 2009b). This work evaluates the sensitivity of genomic responses compared with the traditional LLNA end points on analysis of lymph-node cell proliferation and evaluation of the responses for their ability to provide insights into mode of action. The authors consider anchoring genomic end points to the traditional toxicology end points as a first step to implement genomic technologies.

There is a general agreement about the importance of achieving consistency in gene expression responses across different microarray platforms, in mouse strains (animal models in general), and in laboratories in order to successfully apply these end points as potential biomarkers in toxicological evaluations.

4.7.5.1 Toxicogenomics in Drug Discovery and Development: What Can We Expect?

According to Registration Evaluation Authorization and Restriction of Chemicals (REACH), indications are that in the next 10 years around 30,000 chemical substances will be registered by a process for which manufacturers and importers need to produce data for all chemical substances created or imported into the European Union. It will seriously impact the industry including increasing costs, scientific efforts, and logistical problems. It will also impact animal welfare, because large numbers of laboratory animals are likely to be required. To reduce this impact, REACH Guidance suggested that the European Commission, member states, industry, and other stakeholders should continue to contribute to the promotion of alternative test methods on an international and national level, which would include toxicogenomics methodologies (Van Hummelen and Sasaki, 2010).

Commercial tools to obtain gene-expression profiles and analyze them help to answer specific questions in toxicology and pharmacogenomics safety. However, there are still some challenges, including: the coupling of academic findings and toxicology assays for routine industrial use, the implementation of toxicogenomics in screening during early drug discovery phases, and effective use of this information. In order to obtain the efficient incorporation of genomics into drug development and environmental research issues with data analysis, data interpretation tools and accessibility of data repositories need to be solved.

4.7.5.2 Concluding Remarks

Both ethical and economical aspects support the full implication of toxicogenomics in the earliest stages of the drug discovery and development process. Toxicogenomics would offer the possibility of identifying the potential safety problems earlier, thus reducing the time in the process, and eventually, reducing animal testing programs. These features will improve the efficiency of costly and lengthy *in vivo* studies, with respect to the pharmaceutical productivity dilemma of long cycle times and high attrition in drug discovery and development.

4.8 FUTURE PERSPECTIVES IN TOXICOGENOMICS

To have a correct idea about of the future impact of toxicogenomics technologies, it is important to consider the information provided in different surveys.

The National Academy of Sciences National Research Council published a report in 2007 describing the potential of toxicogenomics, specially the use of genomic profiles obtained with microarrays. This report indicated that the main hurdles to the implementation of arrays would be the costs, analysis, and storage of data. Problems such as the integration and validation of data can obviously be added at this point.

More recently the survey promoted by the HESI Committee on the Application of Genomics to Mechanism-based Risk Assessment (USA) focused on the analysis of the status and future use of TGx to evaluate drug and chemical compounds in nonclinical and experimental models. Their report indicates that toxicogenomic data will have a moderate to high impact in a variety of areas of safety and risk assessment over the next 2–5 years. Two areas in which a major potential impact of TGx is expected are in the understanding the biological mechanism of action of the toxic effect and in the identification of candidate biomarkers for toxicity.

The search was mainly focused on gene expression techniques and quantification of gene responses to xenobiotic exposure in experimental and preclinical models. An important conclusion was that in comparison with other areas such as biomarker detection (in which the genomics technologies play a crucial role), in toxicology gene expression techniques are still under development and validation.

Additional areas in which this report considers that TGx will contribute are compound selection with respect to safety potential, the identification of species differences in toxicity, and the identification of drug target and off-target effects. This report is less optimistic about TGx data influencing the decision-making process in safety assessment, influencing regulatory decisions, and contributing to intellectual property for products or for a class of products.

There is a general agreement on the fact that the future of genomics technologies is absolutely dependent on both data quality and quality of interpretation. Thus accuracy and perception of data are both key areas for focus in the coming years. In this context, there is a special emphasis on the need of trained human resources in bioinformatics and on the fact that a lack of standardization will remain a barrier for the effective sharing of microarray data and the development of public databases.

Another important consideration is the intensification of the use of *in vitro* models in TGx assays in order to simplify data extraction and interpretation, reduce costs, and increase screening speed using smaller amounts of material. An interesting source is the Combined Challenge Maps for the period 2010–2020 (Smith et al., 2008). This is an exercise promoted by HESI with two main objectives: firstly to identify issues with significant and relevant challenges during the next decades that can be addressed as part of the HESI scientific portfolio and secondly to provide information from private companies, research institutions, government agencies, and regulatory authorities in the developed world about strategies on these issues.

Several issues regarding the use of toxicogenomics are considered of high importance in these maps. Genomics, for instance, is predicted to have a high impact in the short term (2010–2020). Risk assessment through biomonitoring populations follows immediately afterwards, with assessment of risk in sensitive/vulnerable populations predicted to have a middle/long-term impact. Animal welfare considerations are also mentioned, especially in the context of the "three Rs" (replace, reduce, and refine) in the use of animals in research. Indeed, risk assessment (one of the major subfields in toxicogenomics methodologies) is directly affected in all the periods in the map: regulatory frameworks, computational tools for toxicology, omics for risk assessment, risk assessment of coexposures, individual susceptibility, strategies to improve testing, epigenetics in risk assessment, exposure risk-based assessment, and improving biomonitoring through the biomarkers will be implicated in the middle/long-term with different impacts. In summary, there is a great need to solve the current problems, to extend the use of toxicogenomics in combination with improvements in data analysis and protocol standarization. In this context, the indications by the National Institute of Environmental Health Sciences (NIEHS) about the future technological developments in the field of toxicogenomics, as well as what can be expected in the next 10 years, are of great interest.

4.8.1 Technology Developments Affecting the Growth of Toxicogenomics in the Next Decade

According to NIEHS and its committee about technologies in toxicogenomics, the most critical aspects of technology development that will drive toxicogenomics in the next decade will be:

1. New sequencing technologies that offer the prospect of cost-effective individual whole-genome sequencing and comprehensive genotype analysis.
2. Array-based whole-genome scanning for variations in individual genes, known as *single-nucleotide polymorphisms*, will dramatically increase throughput for genotyping in population studies.
3. Advances in NMR and MS instrumentation will enable high-sensitivity analyses of complex collections of metabolites and proteins and quantitative metabolomics and proteomics.
4. New bioinformatic tools, database resources, and statistical methods will integrate data across technology platforms and link phenotypes and toxicogenomic data.

REFERENCES

Amin, R. P., Vickers, A. E. F., Sistare, K. L., et al. (2004). Identification of putative gene based markers of renal toxicity. *Environ. Health Perspect.* 112: 465–479.

Andersen, M. E., Clewell, H. J., III, Bermudez, E., et al. (2008). Genomic signatures and dose-dependent transitions in nasal epithelial responses to inhaled formaldehyde in the rat. *Toxicol. Sci.*105: 368–383.

Arteaga-Salasa, J. M., Zuzana, H., Langdon, W. B., et al. (2007). An overview of image-processing methods for Affymetrix GeneChips. *Brief. Bioinform.* 9, 25–33.

Aubrecht, J., Caba, E. (2005). Gene expression profile analysis: An emerging approach to investigate mechanisms of genotoxicity. *Pharmacogenomics* 6, 419–428.

Barros, S. (2005). The importance of applying toxicogenomics to increase the efficiency of drug discovery. *Pharmacogenomics* 6: 547–550.

Benninghoff, A. D. (2007). Toxicoproteomics: The next step in the evolution of environmental biomarkers. *Toxicol. Sci.* 95: 1–4.

Bild, A. H., Parker, J. S., Gustafson, A. M., et al. (2009). An integration of complementary strategies for gene-expression analysis to reveal novel therapeutic opportunities for breast cancer. *Breast Cancer Res.* 11; R55.

Boedigheimer, M. J., Wolfinger, R. D., Bass, M. B., et al. (2008). Sources of variation in baseline gene expression levels from toxicogenomics study control animals across multiple laboratories. *BMC Genomics* 9: 285.

Bogaerts, J., Cardoso, F., Buyse, M., et al. (2006). Gene signature evaluation as a prognostic tool: Challenges in the design of the MINDACT trial. *Nat. Clin. Pract. Oncol.* 3: 540–551.

Boverhof, D. R., Bhaskar Gollapudi, B., Hotchkiss, J. A., et al. (2009a). Evaluation of a toxicogenomic approach to the local lymph node assay (LLNA). *Toxicol. Sci.* 107: 427–439.

Boverhof, D. R., Gollapudi, B. B., Hotchkiss, J. A., et al. (2009b). Evaluation of a toxicogenomic approach to the local lymph node assay (LLNA). *Toxicol. Sci.* 107: 427–439.

Burgoon, L. D., Boutros, P. C., Dere, E., et al. (2006). dbZach: A MIAME-compliant toxicogenomic supportive relational database. *Toxicol. Sci.* 90: 558–568.

Brazma, A. (2009). Minimum Information About a Microarray Experiment (MIAME)—successes, failures, challenges. *The Scientific World Journal* 9: 420–423.

Cheng, S. L., Huang-Liu, R., Sheu, J. N., et al. (2006). Toxicogenomics of kojic acid on gene expression profiling of a375 human malignant melanoma cells. *Biol. Pharm. Bull.* 29: 655–669.

Chung, T.-H., Yoo, J.-H., Ryu, J.-C., et al. (2009). Recent progress in toxicogenomics research in South Korea. *BMC Proceedings* 3 (Suppl. 2): S6.

Daston, G. P. (2008). Gene expression, dose-response, and phenotypic anchoring: Applications for toxicogenomics in risk assessment. *Toxicol. Sci.* 105: 233–234.

Davis, A. P., Murphy, C. G., Rosenstein, M. C., et al. (2008). The Comparative Toxicogenomics Database facilitates identification and understanding of chemical-gene-disease associations: Arsenic as a case study. *BMC Med. Genomics* 1: 48.

Davis, A. P., Murphy, C. G., Saraceni-Richards, C. A., et al. (2009). Comparative Toxicogenomics Database: A knowledgebase and discovery tool for chemical-gene-disease networks. *Nucleic Acids Res. 37:* D786–D792.

Eisen, M. B., Spellman, P. T., Brown, P. O., et al. (1998). Cluster analysis and display of genome-wide expression patterns. *Proc. Natl. Acad. Sci. U.S.A.* 95: 14863–14868.

Hannum, G., Srivas, R., Guénole, A., et al. (2009). Genome-wide association data reveal a global map of genetic interactions among protein complexes. *PLoS Genet.* 5: e1000782.

Hayashi, H., Nagae, G., Tsutsumi, S., et al. (2007). High-resolution mapping of DNA methylation in human genome using oligonucleotide tiling array. *Hum. Genet.* 120: 701–711.

Heinloth, A. N., Irwin, R. D., Boorman, G. A., et al. (2004). Gene expression profiling of rat livers reveals indicators of potential adverse effects. *Toxicol. Sci.* 80: 193–202.

Hewitt, P., Herget, T. (2009). Value of new biomarkers for safety testing in drug development. *Expert Rev. Mol. Diagn.* 9: 531–536.

Kabanov, A. V. (2006). Polymer genomics: An insight into pharmacology and toxicology of nanomedicines. *Adv. Drug Deliv. Rev.* 58: 1597–1621.

Kim, J. Y., Kwon, J., Kim, J.E., et al. (2005). Identification of potential biomarkers of genotoxicity and carcinogenicity in L5178Y mouse lymphoma cells by cDNA microarray analysis. *Environ. Mol. Mutagen.* 45: 80–89.

Kirkland, D., Aardema, M., Henderson, L., et al. (2005). Evaluation of the ability of a battery of three in vitro genotoxicity tests to discriminate rodent carcinogens and noncarcinogens: I. Sensitivity, specificity and relative predictivity. *Mutat. Res.* 584: 1–256.

Kirkland, D., Aardema, M., Muller, L., et al. (2006). Evaluation of the ability of a battery of three in vitro genotoxicity tests to discriminate rodent carcinogens and noncarcinogens: II. Further analysis of mammalian cell results, relative predictivity and tumour profiles. *Mutat. Res.* 608: 29–42.

Kirkland, D. J., Muller, L. (2000). Interpretation of the biological relevance of genotoxicity test results: The importance of thresholds. *Mutat. Res.* 464: 137–147.

Kiyosawa, N., Uehara, T., Gao, W., et al. (2007). Identification of glutathione depletion-responsive genes using phorone-treated rat liver. *J. Toxicol. Sci.* 32: 469–486.

Lam, C. W., James, J. T., McCluskey, R., et al. (2004). Pulmonary toxicity of single-wall carbon nanotubes in mice 7 and 90 days after intratracheal instillation. *Toxicol. Sci.* 77: 126–134.

Larbi, N. B., Jefferies, C. (2009). 2D-DIGE: Comparative proteomics of cellular signalling pathways. *Methods Mol. Biol.* 517: 105–132.

Leitner, J. M., Graninger, W., Thalhammer, F. (2010). Hepatotoxicity of antibacterials: Pathomechanisms and clinical. *Infection* 38: 3–11.

Letieri, T. (2006). Recent applications of DNA microarray technology to toxicology and ecotoxicology. *Environ. Health Perspect.* 114: 4–9.

Mattes, W. B., Pettit, S. D., Sansone, S. A., et al. (2004). Database development in toxicogenomics: Issues and efforts. *Environ. Health Perspect.* 112: 495–505.

Mattingly, C. J. (2009). Chemical databases for environmental health and clinical research. *Toxicol. Lett.* 186: 62–65.

Mattingly, C. J., Colby, G. T., Forrest, J. N., et al. (2003). The Comparative Toxicogenomics Database (CTD). *Environ. Health Perspect.* 111: 793–795.

Mattingly, C. J., Colby, G. T., Forrest, J. N., et al. (2004). Promoting comparative molecular studies in environmental health research: An overview of the comparative toxicogenomics database (CTD). *Pharmacogenomics* 4: 5–8.

Mattingly, C. J., Rosenstein, M. C., Colby, G. T., et al. (2006). The Comparative Toxicogenomics Database (CTD): A resource for comparative toxicological studies. *J. Exp. Zool. A. Comp. Exp. Biol.* 305: 689–692.

Meyer, J. N. (2010). QPCR: A tool for analysis of mitochondrial and nuclear DNA damage in ecotoxicology. *Ecotoxicology* 19: 804–811.

Mishra, S. R., Dubenko, I., Losby, J., et al. (2005). Magnetic properties of magnetically soft nanocomposite Co-SiO2 prepared *via* mechanical milling. *J. Nanosci. Nanotechnol.* 5: 2082–2087.

Mori, Y., Kondo, C., Tonomura, Y., et al. (2010). Identification of potential genomic biomarkers for early detection of chemically induced cardiotoxicity in rats. *Toxicology* 271: 36–44.

Nawarak, J., Huang-Liu, R., Kao, S. H., et al. (2009). Proteomics analysis of A375 human malignant melanoma cells in response to arbutin treatment. *Biochim. Biophys. Acta* 1794: 159–67.

Nuwaysir, E. F., Bittner, M., Trent, J., et al. (1999). Microarrays and toxicology: The advent of toxicogenomics. *Mol. Carcinog.* 24, 153–159.

Oberdorster, G., Sharp, Z., Atudorei, V., et al. (2002). Extrapulmonary translocation of ultrafine carbon particles following whole-body inhalation exposure of rats. *J. Toxicol. Environ. Health* 65: 1531–1543.

Oberemm, A., Ahr, H. J., Bannasch, P., et al. (2009). Toxicogenomic analysis of N-nitrosomorpholine induced changes in rat liver: Comparison of genomic and proteomic responses and anchoring to histopathological parameters. *Toxicol. Appl. Pharmacol.* 241: 230–245.

Pettit, S., des Etages, S. A., Mylecraine, L., et al. (2010). Current and future applications of toxicogenomics: Results summary of a survey from the HESI Genomics State of Science Subcommittee. *Environ. Health Perspect.* 118: 992–997.

Poma, M. L., Di Giorgio, L. (2008). Toxicogenomics to improve comprehension of the mechanisms underlying responses of in vitro and in vivo systems to nanomaterials: A review. *Curr. Genomics* 9: 571–585.

Wills, Q., Mitchell, C. (2009). Toxicogenomics in drug discovery and development: Making an impact. *ATLA* 37 (Suppl. 1): 33–37.

Raychaudhuri, S., Stuart, J. M., Altman, R. B. (2000). Principal components analysis to summarize microarray experiments: Application to sporulation time series. *Pac. Symp. Biocomput.* 455–466.

Roh, J. Y., Sim, S. J., Yi, J., et al. (2009). Ecotoxicity of silver nanoparticles on the soil nematode *Caenorhabditis elegans* using functional ecotoxicogenomics. *Environ. Sci. Technol.* 43: 3933–3940.

Ryan, T. P., James, S. L., Craig, T. E. (2008). Strategic applications of toxicogenomics in early drug discovery. *Curr. Opin. Pharmacol.* 8: 654–660.

Santos-Reboucas, C. B., Pimentel, M. M. G. (2007). Implication of abnormal epigenetic patterns for human diseases. *Eur. J. Human Genet.* 15: 10–17.

Schneider, K., Schwarz, P., Burkholder, I., et al. (2009). ToxRTool: A new tool to assess the reliability of toxicological data. *Toxicol. Lett.* 189: 138–144.

Sexton, K., Balharry, D., BéruBé, K. A. (2008). Genomic biomarkers of pulmonary exposure to tobacco smoke components. *Pharmacogenet. Genomics* 18: 853–860.

Smith, L. L., Brent, R. L., Cohen, S. M., et al. (2008). Predicting future human and environmental health challenges: The Health and Environmental Sciences Institute's scientific mapping exercise. *Crit. Rev. Toxicol.* 38: 817–845.

Snyder, R. D., Green, J. W. (2001). A review of the genotoxicity of marketed pharmaceuticals. *Mutat. Res.* 488, 151–169.

Sone, H., Okura, M., Zaha, H., et al. (2010). Profiles of chemical effects on cells, pCEC. *J. Toxicol. Sci.* 35: 115–125.

Szyf, M. (2007). The dynamic epigenome and its implications in toxicology. *Toxicol. Sci.* 100: 7–23.

Trosko, J., Upham, B. L. (2010). A paradigm shift is required for the risk assessment of potential human health after exposure to low level chemical exposures: A response to the toxicity testing in the 21st century report. *Int. J. Toxicol.* 29: 344–357.

Tusher, V. G., Tibshirani, R., Chu, G. (2001). Significance analysis of microarrays applied to the ionizing radiation response. *Proc. Natl. Acad. Sci. U.S.A.* 98: 5116–5121.

Uehara, T., Ono, A., Maruyama, T., et al. (2010). The Japanese toxicogenomics project: Application of toxicogenomics. *Mol. Nutr. Food Res.* 54: 218–227.

Van Hummelen, P., Sasaki, J. (2010) State-of-the-art genomics approaches in toxicology. *Mutat. Res.* 3: 165–171.

van Leeuwen, A., Schrier, P. I., Giphart, M. J., et al. (1986). TCA: A polymorphic genetic marker in leukemia and melanoma cell lines. *Blood* 67: 1139–1142.

Wang, J., Delabie, J., Aasheim, H., et al. (2002). Clustering of the SOM easily reveals distinct gene expression patterns: Results of a reanalysis of lymphoma study. *BMC Bioinformatics* 3: 36.

Washburn, M. P., Ulaszek, R., Deciu, C., et al. (2002). Analysis of quantitative proteomic data generated via multidimensional protein identification technology. *Anal. Chem.* 74: 1650–1657.

Waters, M. D., Fostel, J. M. (2004). Toxicogenomics and systems toxicology: Aims and prospects. *Nat. Rev.* 5: 936–948

Williams-Devane, C. R., Wolf, M. A., Richard, A. M. (2009). Toward a public toxicogenomics capability for supporting predictive toxicology: Survey of current resources and chemical indexing of experiments in GEO and ArrayExpress. *Toxicol. Sci.* 109: 358–371.

Wolters, D. A., Washburn, M. P., Yates, J. R., III (2001). An automated multidimensional protein identification technology for shotgun proteomics. *Anal. Chem.* 73: 5683–5690.

Yu, X., Griffith, W. C., Hanspers, K., et al. (2006). A system-based approach to interpret dose- and time-dependent microarray data: Quantitative integration of gene ontology analysis for risk assessment toxicological sciences 92: 560–577.

Zhao, Y., Keating, K., Thorpe, R. (2007). Comparison of toxicogenomic profiles of two murine strains treated with HIV-1-based vectors for gene therapy. *Toxicol. Appl. Pharmacol.* 225: 189–197.

5 Stem Cells
Basics and Therapeutic Applications

Awadhesh Kumar Arya and Kamlakar Tripathi
Banaras Hindu University
Varanasi, India

CONTENTS

5.1 INTRODUCTION

The stem cell is a unique and essential cell type found in animals. There are many different types of stem cells found in the body, with some more differentiated, or committed, to a particular function than others. In other words, when stem cells divide, some of the progeny mature into cells of a specific type (e.g., heart, muscle, blood, or brain cells), whereas others remain stem cells, ready to repair some of the everyday wear and tear undergone by our bodies. These stem cells are capable of continually reproducing themselves and serve to renew tissue throughout an individual's life. For example, they constantly regenerate the lining of the gut, revitalize skin, and produce a whole range of blood cells. Over the past two decades, scientists have been gradually deciphering the processes by which unspecialized stem cells become the many specialized cell types in the body. If these unique properties can be understood and harnessed, stem cells hold great potential as tools for medical research and as therapeutic agents (The President's Council on Bioethics, 2004). In medical research, stem cells are commonly used to produce large numbers of genetically uniform cultures of organ tissues: for example, liver, muscle, or neural tissue. This would allow controlled comparison of the effects of drugs or chemicals on these tissues (Rohwedel et al., 2001). The use of human stem cell cultures reduces the use of animals for research and testing purpose. In addition, stem cell cultures are also beneficial for testing pharmaceuticals that can be tailored to provide greater benefits and with fewer side effects in patients (Evans and Kaufman, 1981) (Figure 5.1).

5.2 SOURCES AND CHARACTERISTICS OF STEM CELLS

5.2.1 The Embryonic Stem Cell

The embryonic stem cell (ES) is defined by its origin, which is from the earliest stages of the development of the embryo, called the blastocyst. Specifically, embryonic stem cells are derived from the inner cell mass of the blastocyst at a stage before it would implant in the uterine wall. The embryonic stem cell can self-replicate and is pluripotent; it can give rise to cells derived from all three germ layers. These cell types have characteristics of neuronal ganglia, lung epithelia, gut tissue, muscle cells, bone, and cartilage (Figure 5.2).

Studies of ES cells derived from mouse blastocysts became possible 20 years ago with the discovery of techniques that allowed the cells to be grown in the laboratory. The isolation, culture, and partial characterization of stem cells isolated from human embryos were reported in November of 1998 (Thomson et al., 1998). Human embryonic stem cells could be produced either by *in vitro*

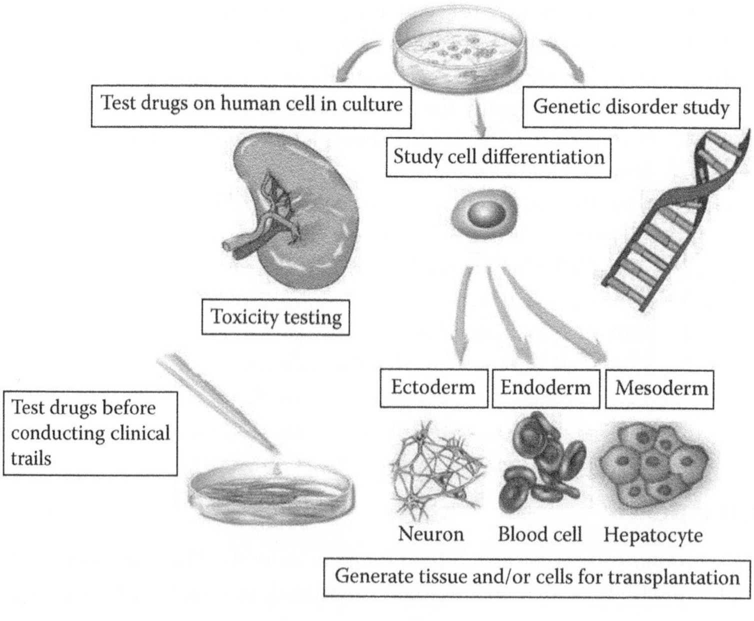

FIGURE 5.1 Role of stem cells in basic and clinical research.

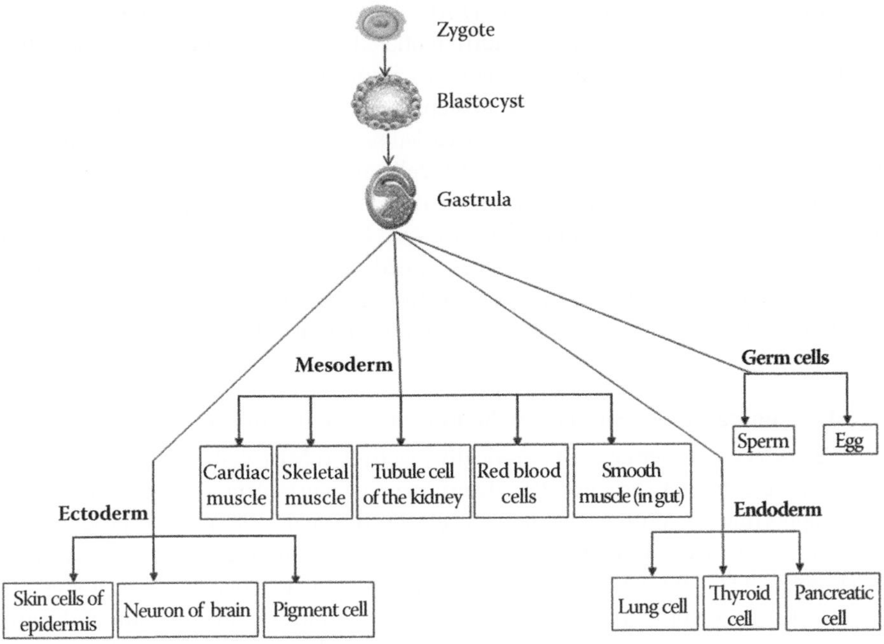

FIGURE 5.2 Development and differentiation of embryonic stem cell.

fertilization (IVF) or by transfer of an adult nucleus to an enucleated egg cell or oocyte (somatic cell nuclear transfer) (Richard et al., 2006). The stem cells that are created by somatic cell nuclear transfer are therefore copies or clones of the original adult cell because their nuclear DNA matches that of the adult cell.

The process of IVF requires the retrieval of a woman's eggs via a surgical procedure after undergoing an intensive regimen of fertility drugs that stimulate her ovaries to produce multiple mature eggs. When IVF is used for reproductive purposes, all of the donated eggs are fertilized in order to maximize their chance of producing a viable blastocyst that can be implanted in the womb. Because not all the fertilized eggs are implanted, this has resulted in a large bank of excess blastocysts that are stored in freezers. The blastocysts stored in IVF clinics could prove to be a major source of embryonic stem cells for use in medical research. This would also facilitate the isolation of stem cells with specific genetic traits necessary for the study of particular diseases. The creation of stem cells specifically for research using IVF is, however, ethically problematic for some people because it involves intentionally creating a blastocyst that will never develop into a human being.

5.2.1.1 Mouse Embryonic Stem Cell Culture

Mouse embryonic stem (mES) cells are pluripotent stem cells derived from the epiblast cells within the inner cell mass of a blastocyst (Brook and Gardner, 1997; Evans and Kaufman, 1981; Martin, 1981). The techniques for culturing mES cells from the inner cell mass of the preimplantation blastocyst were first reported 20 years ago (Evans and Kaufman, 1981; Martin, 1981), and versions of these standard procedures are still used in laboratories throughout the world.

For the culture of mouse ES cells, the inner cell mass of a preimplantation blastocyst is removed from the trophectoderm and taken into small plastic culture dishes that contain growth medium supplemented with fetal calf serum and that are sometimes coated with a *feeder* layer of nondividing cells [mouse embryonic fibroblast (MEF) cells that are chemically inactivated so they will not divide are used as feeder cells]. These cells can also be grown *in vitro* without feeder layers if the cytokine leukemia inhibitory factor (LIF) is added to the culture medium. After several days to a week, the cells, having undergone proliferation, are removed and dispersed into new culture dishes containing MEF feeder layer. Under these *in vitro* conditions, the ES cells aggregate to form colonies. Some colonies consist of dividing, nondifferentiated cells; in other colonies, the cells may be differentiating. Depending on the culture conditions, it may be difficult to prevent the spontaneous differentiation of mouse ES cells. In the final step, the individual, nondifferentiating colonies are dissociated and replated into new dishes, a step called passage. After passage, only those cells that have properties that enable them to grow into ES cells are transferred into new plates to establish an ES cell line. Following some version of this fundamental procedure, human and mouse ES cells can be grown and passaged for 2 or more years, through hundreds of population doublings, and still maintain a normal complement of chromosomes, called a karyotype (Smith, 2001; Thomson et al., 1998). Recent development of ES cell culture has focused on the potential of ES cells to undergo directed differentiation *in vitro* in order to provide a source of cells for therapeutic alleviation of disease (Figure 5.3).

5.2.1.2 Maintaining Mouse Embryonic Stem Cell in Their Undifferentiated State

ES cells are grown on a feeder layer of mitotically inactivated MEF cells in order to keep them pluripotent. Culture condition is an important determinant of pluripotent stem cell phenotype. LIF is essential to supporting the self-renewal of mES cells (Smith et al., 1988). LIF is produced by feeder cells and, in their absence, allows mouse ES cells *in vitro* to continue proliferating without differentiating (Rathjen et al., 1990). LIF exerts its effects by binding to a two-part receptor complex that consists of the LIF receptor and the gp130 receptor. The amino-terminal domain of the gp130 and other cell-surface receptor stimulated extracellular regulated kinase (ERK). ERK is a component of a signal-transduction pathway that counteracts the proliferative effects of STAT3 activation with SHP-2. ERK activation in response to various stimuli appears attenuated in ES cells relative to other cell types, despite the presence of comparable levels of ERK proteins. This may result in part from the specific expression in ES cells of an altered form of the Gabl scaffold protein that suppresses

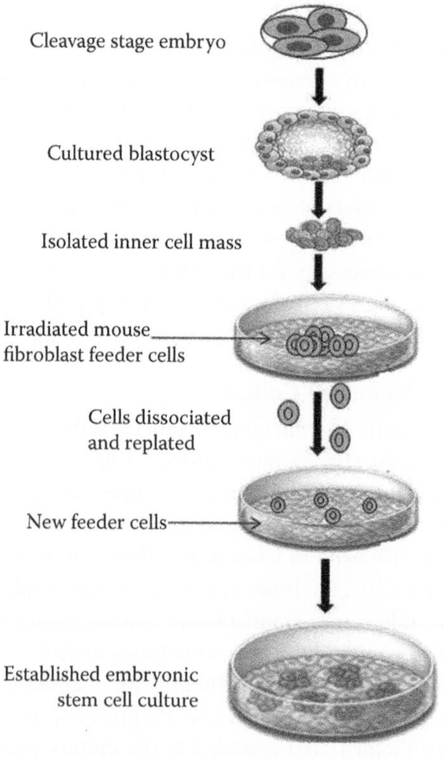

FIGURE 5.3 Establishment of embryonic stem cell culture.

linkage of certain receptors to the Ras-ERK cascade (Burdon et al., 1999). Further, the expression of this altered form of Gabl may be promoted by the transcription factor Octamer-4 (Oct-4). Oct-4 is a homeodomain transcription factor of the POU family. This protein is critically involved in the self-renewal of undifferentiated embryonic stem cells. In mouse ES cells, Oct-4 expression and increased synthesis of Gabl may help suppress induction of differentiation. Overall, the self-renewal of ES cells appears to depend on the balance between conflicting intracellular signals.

5.2.1.3 Human Embryonic Stem Cell Culture and Maintenance

The successful culture and maintenance of human embryonic stem cells (HESCs) offer great promise for furthering our basic understanding of many regulatory and developmental processes and for developing effective clinical applications for treating, even correcting, genetic disorders. HESCs are pluripotent in nature and are generated from the inner cell mass of a human blastocyst, the culture of which involves a multistep process. The pluripotent cells of the inner cell mass are separated from the surrounding trophectoderm by immunosurgery followed by antibody-mediated dissolution of the trophectoderm. The inner cell masses are grown in culture dishes containing growth medium supplemented with fetal bovine serum on feeder layers of mouse embryonic fibroblasts, which were gamma-irradiated to prevent replication. After 9–15 days, when inner cell masses divided and formed clumps of cells, cells from the periphery of the clumps are chemically or mechanically dissociated and replated in the same culture conditions. Colonies of apparently homogeneous cells are selectively removed, mechanically dissociated, and replated. These homogeneous cells are expanded and passaged to create a cell line. Thomson and Odorico (2000) performed human ES cell culture from fresh or frozen embryos generated in IVF laboratories from couples undergoing treatment for infertility. From the embryos that developed to the blastocyst stage, they established five human ES cell lines: H1, H7, H9, H13, and H14 (Figure 5.4).

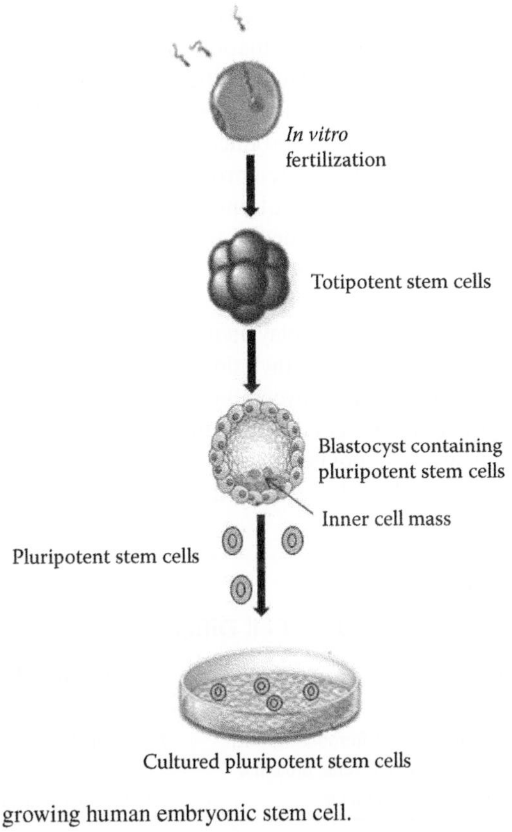

FIGURE 5.4 Method for growing human embryonic stem cell.

The five original human ES cell lines continued to divide without differentiating for 5–6 months. Since then, the H9 line has divided for nearly 2 years *in vitro*, for more than 300 population doublings, and has yielded two subclones, H9.1 and H9.2. The human ES cell lines expressed high levels of telomerase activity. Telomerase is a ribonucleoprotein that adds telomere repeats to chromosome ends and is involved in maintaining telomere length, which plays an important role in replicative life span (Harley, 1991; Harley et al., 1992). Telomerase expression is highly correlated with immortality in human cell lines, and reintroduction of telomerase activity into some diploid human somatic cell lines extends the replicative life span (Bodnar et al., 1998). The human ES cell lines expressed cell-surface markers that characterize undifferentiated nonhuman primate ES and human EC cells, including stage-specific embryonic antigen (SSEA)-3, SSEA-4, TRA-1–60, TRA-1–81, and alkaline phosphatase (Andrews et al., 1987). Mouse ES cells do not express SSEA-3 or SSEA-4; they express SSEA-1, which human and monkey ES cells do not, suggesting basic species differences between early mouse and human development. Human ES cells also express the transcription factor Oct-4 (Roach et al., 1993), as mouse ES cells do. Oct-4 is expressed in stem cells and downregulated during differentiation, strongly indicating that stem cell selection using drug resistance genes driven by the Oct-4 promoter will be a useful avenue for manipulating human ES cells. Furthermore, it is hoped that by elucidating the mechanisms behind which cells differentiate into tissues, scientists can develop a better understanding of human developmental biology.

5.2.1.4 Directed Differentiation of Human Embryonic Stem Cells

As with cultures of mouse ES cells, human ES cells begin to differentiate if they are removed from feeder layers and grown in suspension culture on a nonadherent surface. The human ES cells form embryoid bodies, which, in the early stages, may be simple or cystic and filled with fluid. Although human embryoid bodies vary in their cellular content, many include cells that look like neurons and heart muscle cells (Itskovitz et al., 2000; Reubinoff et al., 2000; Roach et al., 1993). After human embryoid body formation, the cells are dissociated and replated in monolayer cultures, which are then exposed to specific growth factor by transfection with a gene that will regulate differentiation via a specific pathway or by coculture with companion cells. For example, dopaminergic neurons can be produced when HESCs are cocultured with mouse stromal PA6 cells, and neuronal cells have been produced after treatment of HESCs with nerve growth factor (NGF) and retinoic acid (Schuldiner et al., 2001). Other growth factors, such as activin-A and transforming growth factor-β1, trigger the differentiation of mesodermally derived cells, whereas hepatocyte growth factor and NGF promote differentiation into all three germ layers, including endoderm (Table 5.1).

Environmental and epigenetic factors also play an important role in regulating the differentiation of pluripotent HESCs. For example, DNA methylation is required for differentiation, and together with the chromatin regulators, such as the polycomb group proteins, they are important for epigenetic modifications. Among the environmental factors that influence the state of potency, is oxygen concentration. At low oxygen levels, hypoxia has been shown to promote more pluripotent and multipotent cell type at the expense of their differentiated progeny (Ezashi et al., 2005).

TABLE 5.1
Growth Factors Used for Cell Differentiation

Growth Factors	Cell Differentiation
TGF-β1, Activin-A	Mesodermal differentiation
Epidermal growth factor, fibroblast growth factor, RA, bone morphogenic protein 4	Ectodermal differentiation
NGF, hepatocyte growth factor	All three germ layers

5.2.2 Embryonic Carcinoma Cells

Embryonic carcinoma (EC) cells are the stem cells that occur in unusual germ cell tumors, also called teratocarcinomas, and provide a striking paradigm of the stem cell concept of cancer. EC cells have several distinctive features in particular; they are often capable of multilineage differentiation. Crucially, this capacity is retained by clonal isolates (Kleinsmith and Pierce, 1964; Martin and Evans, 1975), formally establishing the presence of pluripotent stem cells.

Teratocarcinomas are of particular interest because they contain EC cells, which in many ways resemble normal ES cells. Like human ES cells, human EC cells proliferate extensively *in vitro* and in teratomas formed *in vivo* after injection into immunocompromised mice. Both cell types express a panel of surface markers, including the embryonic stage-specific antigens SSEA-3 and SSEA-4. Unlike human ES cells and human EC cells, mouse ES and EC cells express SSEA-1. Conversely, mouse EC and ES cells do not express SSEA-3 or SSEA-4. Another set of antigens that have been identified in human EC cells are epitopes associated with keratan sulfate, notably TRA-1–60 and TRA-1–81 (Andrews et al., 1984; Badcock et al., 1999), as well as GCTM2, K21, and K4 (Rettig et al., 1985; Pera and Herzfeld, 1998). It appears that these epitopes are commonly expressed by human EC cells, and indeed some, notably TRA-1–60, have been shown to be useful serum markers in germ cell tumor patients because they are shed by EC cells (Marrink et al., 1991; Mason et al., 1991; Gels et al., 1997).

5.2.3 Embryonic Germ Cells

Embryonic-like stem cells, called embryonic germ (EG) cells, can be derived from primordial germ (PG) cells (the cells of the developing fetus from which eggs and sperm are formed) of the mouse and human fetus. Germ cells are responsible for the fidelity of DNA inheritance from one generation to the next. In November of 1998, the isolation, culture, and partial characterization of germ cells derived from the gonadal ridge of human tissue obtained from abort uses were reported. (Shamblott et al., 1998). PG cells cluster outside the embryo before migrating through the hindgut endoderm and into the genital ridges and were cultured in the presence of LIF, stem cell factor, and basic fibroblast growth factor to obtain pluripotent stem cells (Matsui et al., 1992; Resnick et al., 1992). These pluripotent stem cells are referred to as EG cells. EG cells share several important characteristics with embryonic stem cells (ESCs), such as morphology, pluripotency, self-renewal, high levels of intracellular AP activity, presentation of specific cell surface antigens, capacity to differentiate into three germ layer in chimeras and teratomas, and germ line transmission (Knowles et al., 1978; Labosky et al., 1994; Solter and Knowles, 1978; Stewart et al., 1994). Furthermore, the behavior of these cells *in vivo* is not well understood; significant research will be required to avoid unwanted outcomes, including ectopic tissue formation (additional, unwanted tissue), tumor induction, or other abnormal development (Kato et al., 1999).

5.2.4 Adult Stem Cells

Stem cells that are found in developed tissue, regardless of the age of the organism at the time, are referred to as adult stem cells (Audrey et al., 1999). Adult stem cells, like all stem cells, share at least two characteristics. First, they can make identical copies of themselves for long periods of time; this ability to proliferate is referred to as long-term self-renewal. Second, they can give rise to mature cell types that have characteristic morphologies (shapes) and specialized functions. Unlike embryonic stem cells, which are defined by their origin (the inner cell mass of the blastocyst), adult stem cells share no such definitive means of characterization. In fact, no one knows the origin of adult stem cells in any mature tissue. Some have proposed that stem cells are somehow set aside during fetal development and restrained from differentiating. The list of adult tissues reported to contain stem cells is growing and includes bone marrow, peripheral blood, brain, spinal cord, dental pulp, blood vessels, skeletal muscle, epithelia of the skin and digestive system, cornea, retina, liver, and pancreas (Figure 5.5).

An adult stem cell should also be able to give rise to fully differentiated cells that have mature phenotypes, are fully integrated into the tissue, and are capable of specialized functions that are appropriate for the tissue. The most well known example of adult stem cell is the hematopoietic stem cells (HSCs) of blood. More recently, mesenchymal stem cells required for the maintenance of bone, muscle, and other tissues have been discovered (Pittenger et al., 1999). At present, there is, however, a paucity of research, with few notable exceptions, in which researchers were able to conduct studies of genetically identical (clonal) stem cells. In order to fully characterize the generating and self-renewal capabilities of the adult stem cell and therefore to truly harness its potential, it will be important to demonstrate that a single adult stem cell can, indeed, generate a line of genetically identical cells, which then gives rise to all of the appropriate, differentiated cell types of the tissue in which it resides.

5.2.4.1 Hematopoietic Stem Cells

A HSC is a cell isolated from the blood or bone marrow that can renew itself, can differentiate to a variety of specialized cells, can mobilize out of the bone marrow into circulating blood, and can undergo programmed cell death, called apoptosis, a process by which cells that are detrimental or unneeded self-destruct (Weissman, 2000). HSCs are primarily found in the bone marrow; they are also present in a variety of other tissues including peripheral blood and umbilical cord blood and are found in lesser numbers in the liver, spleen, and perhaps many organs (Holyoake et al., 1999). HSCs generate the multiple hematopoietic lineages through a successive series of intermediate progenitors. These include common lymphoid progenitors, which can generate only B, T, and NK cells, and common myeloid progenitors, which can generate only red cells, platelets, granulocytes, and monocytes (Akashi et al., 2000; Kondo et al., 1997) (Figure 5.6).

A major thrust of basic HSC research since the 1960s has been identifying and characterizing these stem cells. Because HSCs look and behave in culture like ordinary white blood cells, it is

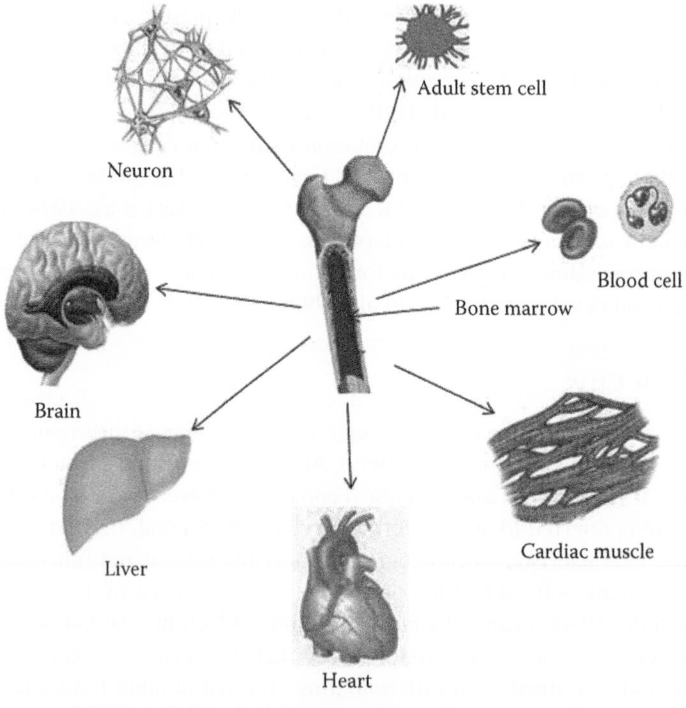

FIGURE 5.5 Source and differentiation of adult stem cell in various other cells.

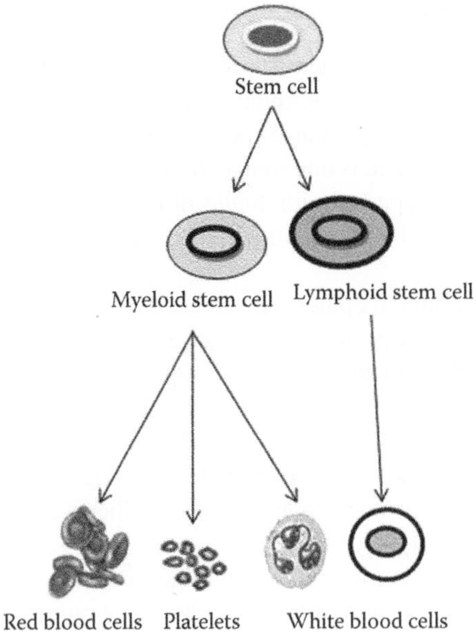

FIGURE 5.6 Differentiation of hematopoietic stem cell.

extremely difficult to identify them by morphology (size and shape). Even today, scientists must rely on cell surface proteins, which serve, only roughly, as markers of white blood cells. Currently, the phenotype of human HSCs and progenitors is not completely defined. Studies suggest that human HSCs and progenitors are small quiescent cells that express the surface glycoprotein CD34 (Berenson et al., 1988; Civin et al., 1996).

5.2.4.2 Sources of Hematopoietic Stem Cell

The HSCs are mesodermal in origin, derived early in embryogenesis, and become deposited in the bone marrow, liver, and yolk sac of the embryo. HSCs can be purified using monoclonal antibodies, and recently, common lymphoid progenitor and myeloid-erythroid progenitor cells have been isolated and characterized. Bone marrow is the classic source for HSCs. It may be more plastic and versatile than expected because they are multipotent and can be differentiated into many cell types both *in vitro* and *in vivo*. The recent evidence shows that patients receiving peripherally harvested cells have higher survival rates than bone marrow recipients do. The peripherally harvested cells contain twice as many HSCs as stem cells taken from bone marrow and engraft more quickly. This means patients may recover white blood cells, platelets, and their immune and clotting protection several days faster than they would with a bone marrow graft. For clinical transplantation of human HSCs, it is now preferred to harvest donor cells from peripheral, circulating blood. It has been known for decades that a small number of stem and progenitor cells circulate in the bloodstream, but in the past 10 years, researchers have found that they can coax the cells to migrate from marrow to blood in greater numbers by injecting the donor with a cytokine, such as granulocyte-colony stimulating factor. To collect HSCs, an intravenous tube is inserted into the donor's vein, and blood is passed through a filtering system that pulls out $CD34^+$ white blood cells and returns the red blood cells to the donor.

Along with the use of bone marrow tissue as a source of hematopoietic stem cells, scientists are also using blood from human umbilical cord and placenta. This tissue supports the developing fetus during pregnancy, is delivered along with the baby, and is usually discarded. The first successful umbilical cord blood transplant was used to treat children suffering from Fanconi anemia (Laughlin, 2001).

5.3 STEM CELL MARKERS

With the development of technology, scientists have found a wide array of stem cells that have unique capabilities to self-renew, grow indefinitely, and differentiate or develop into multiple types of cells and tissues. In the human body, many type of stem cells exist, but they all are found in very small populations, in some cases there is one stem cell in 100,000 cells in circulating blood. So, how do scientists identify these rare types of cells found in many different cells and tissues, a process that is much akin to finding a needle in a haystack? The answer is rather simple thanks to stem cell markers.

The first surface marker of human epidermal stem cells to be described was β1 integrins, receptors that bind extracellular matrix proteins (Jones et al., 1995; Moll, 1995; Lyle et al., 1998). β1 integrins can be used to enrich stem cells, either in culture or directly from the epidermis, by fluorescence-activated cell sorting (FACS) or differential adhesiveness to extracellular matrix-coated dishes, and to visualize the stem cells using confocal microscopy (Jones and Watt, 1993; Jensen et al., 1999). Other proposed surface markers include the antigen recognized by monoclonal antibody 10G7 (Li et al., 1998), low surface expression of E-cadherin (Moles and Watt, 1997), and high expression of Delta1 (Lowell et al., 2000).

FACS is commonly used with markers to sort out the rare stem cells from the millions of other cells. With this technique, a suspension of tagged cells (i.e., bound to the cell-surface markers are fluorescent tags) is sent under pressure through a very narrow nozzle—so narrow that cells must pass through one at a time. Upon exiting the nozzle, cells then pass, one-by-one, through a light source, usually a laser, and then through an electric field. The fluorescent cells become negatively charged, while nonfluorescent cells become positively charged. The charge difference allows stem cells to be separated from other cells. After this step, a population of cells that have all of the same marker characteristics can be collected (Figure 5.7).

Genetic and molecular biology techniques are extensively used to study how cells become specialized in the organism's development. In this regard, many genes and transcription factors

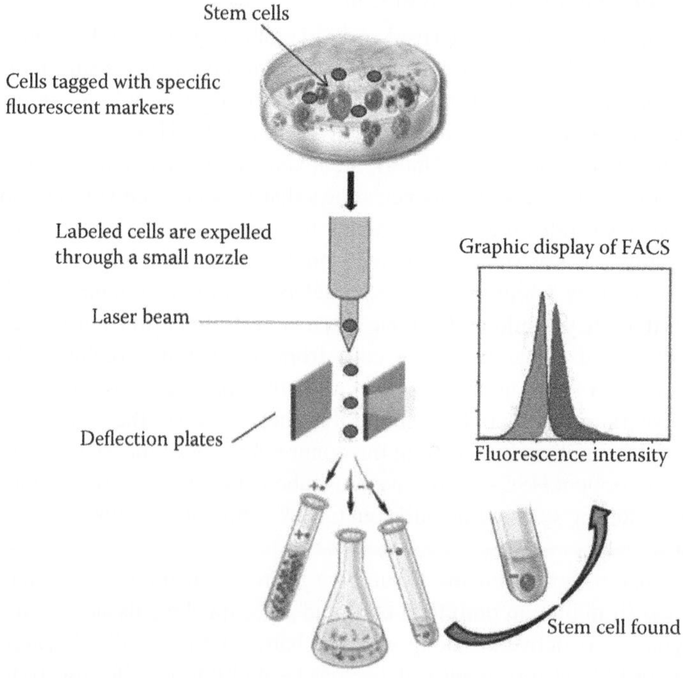

FIGURE 5.7 Separation of stem cells by FACS.

(proteins found within cells that regulate a gene's activity) that are unique in stem cells are identified. The presence of genes that are *active* and play a role guiding the specialization of a cell are detected by polymerase chain reaction. This technique is helpful in identifying *genetic markers* that are characteristic of stem cells. For example, a gene marker called PDX-1 is specific for a transcription factor protein that initiates activation of the insulin gene. This marker is used to identify cells that are able to develop islet cells in the pancreas. The importance of this new technique is that it allows the tracking of stem cells as they differentiate or become specialized (Table 5.2).

5.4 POTENTIALS OF STEM CELL THERAPIES

Stem cell treatments are a type of genetic medicine that introduces new cells into damaged tissue in order to treat a disease or injury. Stem cells are now opening the vistas of a new era and an alternative for curing various diseases. Presently, donated organs and tissues are used to replace ailing or destroyed tissue, but the need for transplantable tissues and organs far exceeds the available supply. The science of stem cell therapies has the potential to lead to treatments for major degenerative diseases by providing healthy cells to replace diseased tissues and organs (Chapman et al., 1999). Stem cells, directed to differentiate into specific cell types, offer the possibility of a renewable source of replacement cells and tissues to treat diseases including Alzheimer's disease, Parkinson's disease, spinal cord injury, stroke, burns, heart disease, diabetes, osteoarthritis, and rheumatoid arthritis (Figure 5.8).

5.4.1 Stem Cell Approaches for the Treatment of Autoimmune Diseases

When the immune system fails to recognize its own constituent parts as self and mistakenly attacks them, the result is known as an autoimmune disease. Common autoimmune diseases include

TABLE 5.2
Markers Commonly Used to Identify Stem Cells

Marker Name	Cell Type
Oct-4	Embryonic stem cell
SSEAs	Teratocarcinoma stem cells
Stem cell antigen, Thy-1, c-Kit	HSC, mesenchymal stem cell (MSC)
Fetal liver kinase-1	Endothelial
CD4 and CD8	White blood cell (WBC)
CD34	HSC, satellite (muscle stem cell), endothelial progenitor
$CD34^+$ $Sca\text{-}1^+$	MSC
CD38	Absent on HSC, present on WBC lineage
CD44	Mesenchymal
CD133	Neural stem cell, HSC
Muc-18 (CD146)	Bone marrow fibroblast, endothelial
Hoechst dye	Absent on HSC
β1 integrin, albumin	Hepatocyte
Microtubule-associated protein-2	Neuron
Cytokeratin 19	Pancreatic epithelium
Insulin-promoting factor-1 (PDX-1), somatostatin, Insulin	Pancreatic islet
Alkaline phosphatise	ES, EC

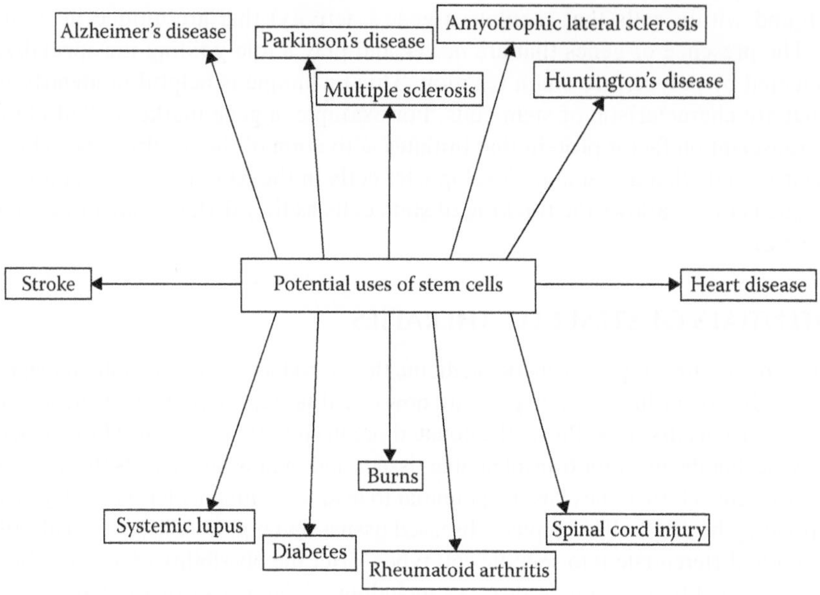

FIGURE 5.8 Potential uses of stem cells in therapies.

rheumatoid arthritis, systemic lupus erythematosis (lupus), type 1 diabetes, multiple sclerosis, Sjogren's syndrome, and inflammatory bowel disease. The immune-mediated injury in autoimmune diseases can be organ-specific, such as type 1 diabetes, which is the consequence of the destruction of the pancreatic β-islet cells or multiple sclerosis, which results from the breakdown of the myelin covering of nerves. These autoimmune diseases are amenable to treatments involving the repair or replacement of damaged or destroyed cells or tissue.

5.4.1.1 Systemic Lupus Treatment with Stem Cells

Lupus is a non-organ-specific autoimmune disease, characterized by widespread injury caused by immune reactions against many different organs and tissues. Although the specific cause of systemic lupus erythematosis (SLE) is unknown, multiple factors are associated with the development of the disease, including genetic, racial, hormonal, and environmental factors (Cooper et al., 1998; Rahman and Isenberg, 2008). At present, there is no permanent cure for SLE. The treatment is aimed at preventing complications, as well as treating the symptoms and signs of the disease. The objective of hematopoietic stem cell therapy for lupus is to destroy the mature, long-lived, and auto reactive immune cells and to generate a new, properly functioning immune system. In most cases, the patient's own stem cells (autologous stem cells) have been used (Marmount et al., 1995). Hematopoietic stem cell transplantation was first reported for patients with systemic lupus erythematosus in 1997 (Richard and Traynor, 2003). The treatment started with injecting the patient with a growth factor, which coaxes large numbers of hematopoietic stem cells to be released from the bone marrow into the blood stream. These cells are harvested from the blood, purified away from mature immune cells, and stored. After sufficient quantities of these cells are obtained, the patient undergoes a regimen of cytotoxic (cell-killing) drug and/or radiation therapy, which eliminates the mature immune cells. Then the hematopoietic stem cells are returned to the patient via a blood transfusion into the circulation where they migrate to the bone marrow and begin to differentiate to become mature immune cells. The body's immune system is then restored. However, the recovery phase, during which the immune system is reconstituted, represents a period of dramatically increased susceptibility to bacterial, fungal, and viral infection, making this a high-risk therapy.

5.4.1.2 Diabetes Treatment with Stem Cells

Diabetes mellitus (DM) is actually a group of diseases characterized by abnormally high levels of the sugar glucose in the bloodstream. This excess glucose is responsible for most of the complications, which include blindness, kidney failure, heart disease, stroke, neuropathy, and amputations. DM in all its forms currently affects at least 200 million people in the world, and this number is expected to rise to more than 350 million by the year 2030 (Wild et al., 2004). About 10% of total diabetic patients suffer from type 1 diabetes, which develops when the body's immune system sees its own cells as foreign and attacks and destroys them. As a result, the islet cells of the pancreas, which normally produce insulin, are destroyed (Atkinson and Eisenbarth, 2001). In the absence of insulin, glucose cannot enter cells, and glucose accumulates in the blood. Type 2 diabetes tends to affect older, sedentary, and overweight individuals with a family history of diabetes. Type 2 diabetes occurs when the body cannot use insulin effectively. This is called insulin resistance, and the result is the same as with type 1 diabetes: a build up of glucose in the blood (DeFronzo, 1997).

There is currently no cure for diabetes. People with type 1 diabetes must take insulin several times a day to keep their blood glucose concentrations as close to normal as possible, whereas people with type 2 diabetes can often control their blood glucose concentrations through a combination of diet, exercise, and oral medication. Replacing these missing islet cells of the pancreas could be a means to treat the insulin-delivery malfunction characteristic of this form of DM as well. Stem cell therapy is one of the most promising treatments for the near future. The discovery of methods to isolate and grow human embryonic stem cells in 1998 renewed the hopes of doctors, researchers, and diabetes patients and their families that a cure for type 1 diabetes, and perhaps type 2 diabetes as well, may be within striking distance. ESCs proliferate indefinitely in defined culture conditions and have the potential to differentiate into any of the more than 200 cellular types present in the human body, including β-cells (Odorico et al., 2001). With a ready supply of cultured stem cells at hand, a line of embryonic stem cells could be grown up as needed for anyone requiring a transplant. The cells could be engineered to avoid immune rejection. Before transplantation, they could be placed into nonimmunogenic material so that they would not be rejected, and the patient would avoid the devastating effects of immunosuppressant drugs.

A method for inducing differentiation of human ES cells into insulin-producing β-cell-like cells by signals from the embryonic mouse pancreas was suggested by Brolen et al. in 2005. These human ES cell-derived insulin-producing cells share important features with normal β-cells such as synthesis (proinsulin) and processing (C-peptide) of insulin and nuclear localization of key β-cell transcription factors (Brolen et al., 2005). Further improvement of this method may lead to the formation of an unlimited source of cells suitable for transplantation. In 2000, Shapiro and colleagues described the successful cure of type 1 diabetes in a small number of patients using a procedure known as the *Edmonton Protocol*. In this procedure, a large number of islet cells were transplanted in type 1 diabetic patients combined with a glucocorticoid-free immunosuppressive drug. The main advantage of islet transplantation includes the achievement of physiologic insulin secretion in those patients who are able to achieve insulin independence. This procedure cannot provide long-term treatment, and there was complete graft loss 1 year after the final transplantation (Shapiro et al., 2006).

Insulin-producing cells derived from stem cells and islet cells extracted from donor pancreas have some similarities in transplantation for type 1 diabetes therapy. They all provide the hope of sufficiently tight control of blood glucose to avoid the diabetic late complications that with current diabetes drug therapies have been almost impossible to avoid (Ping et al., 2007).

5.4.1.3 Stem Cell Therapy for Rheumatoid Arthritis

Rheumatoid arthritis (RA) is an autoimmune disorder of unknown etiology characterized by symmetric, erosive synovitis and, in some cases, extra-articular involvement. It can affect anyone at any age, but it usually develops between the ages of 30 and 50. More than 75% of those affected are women. According to the National Institutes of Health, more than 2 million people in the United States have RA. Most patients experience a chronic fluctuating course of disease that, despite therapy, may result in progressive joint destruction, deformity, disability, and even premature death. A permanent cure for rheumatoid arthritis is not available. Drugs are used to reduce joint inflammation and pain, maximize joint function, and prevent joint destruction and deformity. Stem cell transplantation now offers a potentially useful treatment for rheumatoid arthritis. Recently it has been suggested that high-dose immunosuppressive treatment with autologous stem cell transplantation can be used for RA (Joske et al., 1997; Snowden et al., 1999; Pavletic et al., 1998; Durez et al., 1998).

5.4.1.4 Hematopoietic Stem Cell Transplantation Procedures for Autoimmune Disease

In the majority of patients undergoing hematopoietic stem cell transplantation (HSCT), peripheral blood stem cells are used as the source of stem cells. These cells are mobilized with cyclophosphamide in combination with granulocyte-colony stimulating factor (G-CSF) or with G-CSF alone according to local protocols. *In vitro* purging before autologous HSCT (44%) was performed using either $CD34^+$-positive selection (92%) or by negative purging of lymphocyte subsets by monoclonal antibodies, particularly anti-CD52 (CAMPATH 1), anti-CD3, anti-CD19, or anti-CD20 (8%). The conditioning regimen consists of either total-body irradiation (7%) or various combinations of chemotherapy alone (93%), including combinations based on cyclophosphamide (52%), busulfan (4%), and BEAM (carmustine, cytarabine, melphalan, and etoposide) (34%). Antithymocyte globulins are used in 55% of the patients (Dominique et al., 2010).

5.4.2 Stem Cells for the Treatment of Neurological Disorders

In humans, many neurological disorders such as Parkinson's disease, Huntington's disease, amyotrophic lateral sclerosis (ALS), Alzheimer's disease, multiple sclerosis (MS), stroke, and spinal cord injury are caused by a loss of neurons and glial cells in the brain or spinal cord. Successful generation of neurons and glia from stem cells in culture are fueling efforts for the development of potentially powerful new therapeutic strategies for a broad spectrum of human neurological diseases. Adult stem cell therapy could offer a safe and effective treatment for diseases that were previously considered to be irreversible and provide a healthy and efficacious alternative to drug-related therapies not only for Alzheimer's disease but for other neurological disorders. In the mid-1990s, neuroscientists learned that some parts of the adult human brain do, in fact, generate new neurons, at least under certain circumstances. Moreover, they found that the new neurons arise from *neural stem cells* in the fetal as well as the adult brain. These undifferentiated cells resemble cells in a developing fetus that give rise to the brain and spinal cord. The researchers also found that these neural stem cells could generate many, if not all, types of cells found in the brain. This includes neurons, the main message carriers in the nervous system, which use long and thin projections. The discovery of a regenerative capacity in the adult central nervous system holds out the promise that it may eventually be possible to repair damage from terrible degenerative diseases such as Parkinson's disease and amyotrophic lateral sclerosis as well as from brain and spinal cord injuries resulting from stroke or trauma. To utilize this discovery, scientists are pursuing two fundamental strategies. One is to grow differentiated cells in a laboratory dish that are suitable for implantation into a patient by starting with undifferentiated neural cells. The idea is either to treat the cells in culture to nudge them toward the desired differentiated neuronal cell type before implantation or to implant them directly and rely on signals inside the body to direct their maturation into the right kind of brain cell. A variety of stem cells might be used for this task, including so-called neural precursor cells that are inwardly committed to differentiating into a particular cell type but are outwardly not yet changed or pluripotent embryonic stem cells. The other repair strategy relies on finding growth hormones and other trophic factors that can fire up a patient's own stem cells and endogenous repair mechanisms to allow the body to cope with damage from disease or injury. Researchers are vigorously pursuing both strategies to find therapies for central nervous system disorders that involve axons to transmit signals over long distances as well as crucial neural-support cells called oligodendrocytes and astrocytes.

5.4.2.1 Alzheimer's Disease Treatment by Stem Cells

Alzheimer's disease (AD) was first described in 1906, when German psychiatrist and neuropathologist Alois Alzheimer observed the pathological hallmarks of the disease in the brain of a female patient who had experienced memory loss, language problems, and unpredictable behavior and who at postmortem had abnormal clumps of protein (β-amyloid plaques) and tangled bundles of protein fibers (neurofibrillary tangles) in the brain. AD is characterized by degeneration and loss of neurons and synapses throughout the brain, particularly in the basal forebrain, amygdala, hippocampus, and cortical area. Memory and cognitive function of patients progressively decline, and patients become demented and die prematurely (Whitehouse et al., 1981; Bartus et al., 1982; Coyle et al., 1983). Currently up to 5.3 million Americans suffer from AD, and the number will jump to 11–16 million by the year 2050 unless a prevention method or an effective cure is found (Alzheimer's Association, 2010). Alzheimer's disease is complex, and there is no single *magic bullet* to prevent or cure it. That is why current treatments focus on several different aspects, including helping people maintain mental function; managing behavioral symptoms; and slowing, delaying, or preventing the disease. Four medications are approved by the U.S. Food and Drug Administration to treat Alzheimer's. Donepezil, rivastigmine, memantine, and galantamine are used to treat mild to moderate Alzheimer's. These drugs work by regulating neurotransmitters. They may help maintain thinking, memory, and speaking skills and help with certain behavioral problems. However, these drugs do not change the underlying disease process and may help only for a few months to a few years.

Regeneration of new neurons from neural stem cell in certain regions of the mature brain holds promise for structural brain repair. Stem cell therapy has been suggested as a possible strategy for replacing damaged circuitry and restoring learning and memory abilities in patients with Alzheimer's disease (Zhongling et al., 2009). Mesenchymal stem cells (MeSCs) are a subset of stromal cells present in bone marrow capable of producing multiple mesenchymal cell lineages, including bone, cartilage, fat, tendons, and other connective tissues, (Majumdar et al., 1998; Pereira et al., 1995; Prockop, 1997; Pittenger et al., 1999). Recent reports have shown that human MeSCs (HMeSCs) also have the ability to differentiate into a diverse family of cell types that may be unrelated to their phenotypical embryonic origin, including muscle and heptocytes (Ferrari et al., 1998; Makino et al., 1999; Petersen et al., 1999; Mackenzie and Flake, 2001; Imasawa et al., 2001; Liechty et al., 2006). *In vivo* transplantation studies showed that neural and glial cells differentiated from HMeSCs (Schwarz, 1999; Chopp et al., 2000; Chen et al., 2000; Li et al., 2000; Kopen et al., 1999) have potential therapeutic use in the central nervous system. Spontaneous fusion of stem cells has been reported by two groups (Terada et al., 2002; Ying et al., 2002). These reports show that stem cells acquired phenotypes from other cells by fusion, which may occur when these stem cells directly touch other cells after transplantation. This result indicates that HMeSCs are capable of becoming neurons and astrocytes when cocultured with differentiated neural stem cells (NSCs). Because no exogenous differentiation factors, such as retinoic acid or brain-derived neurotrophic factor, were added to these cultures, and no cell-to-cell contact existed in this coculture system, it is reasonable to hypothesize that the membrane-permeable endogenous factor(s) released from differentiating human neural stem cells (HNSCs) altered the cell fate decisions of HMeSCs. These results indicate that the brain environment may produce factor(s) that allow the differentiation of not only NSCs but also MeSCs into neurons. These studies suggest that HMeSCs may serve as an alternative to HNSCs for potential therapeutic use in neuroreplacement.

5.4.2.2 Stem Cell Therapy for Parkinson's Disease

Parkinson's disease was first described by the British physician James Parkinson, who wrote about a "shaking palsy," which later came to bear his name, in a paper published in 1817. It is estimated that

6.3 million people have Parkinson's worldwide, affecting all races and cultures. The age of onset is usually over 60, but it is estimated that one in ten are diagnosed before the age of 50, with slightly more men than women affected. Anyone can get Parkinson's, but it is more common in older people. It is not contagious and cannot be spread from one person to another. PD is a disorder of the central nervous system that occurs as a result of gradual loss of a specific type of dopamine-producing nerve cell, located in an area of the brain called the substantia nigra. Because dopamine is responsible for transmitting the electrical signals, its deficiency makes patients have difficulty in moving freely, holding a posture, talking, and writing. Other symptoms may include depression and other emotional changes; difficulty in swallowing, chewing, and speaking; urinary problems or constipation; skin problems; and sleep disruptions. PD is both chronic and progressive; it lasts throughout one's lifetime and worsens over time. There are currently no blood or laboratory tests that have been proven to help in diagnosing sporadic PD. Therefore, the diagnosis is based on medical history and a neurological examination.

At present, there is no cure for PD, but a variety of medications provide dramatic relief from the symptoms. Usually, patients are given levodopa combined with carbidopa. Carbidopa delays the conversion of levodopa into dopamine until it reaches the brain. Nerve cells can use levodopa to make dopamine and replenish the brain's dwindling supply. In some cases, surgery may be appropriate if the disease does not respond to drugs. A therapy called deep brain stimulation (DBS) has now been approved by the U.S. Food and Drug Administration. In DBS, electrodes are implanted into the brain and connected to a small electrical device called a pulse generator that can be externally programmed. DBS can reduce the need for levodopa and related drugs.

Because of degeneration of dopamine-producing cells of the brain that takes place in PD, it can be treated successfully by regenerating these dopamine-producing cells. In this regard, stem cell therapy would be the only truly effective treatment of PD. For neuroreplacement therapy, the most logical approach is to produce neurons from HNSCs *in vitro* and then transplant them into the brain tissue. Dopaminergic cells were produced *in vitro* in the case of PD and then transplanted into the basal ganglia as dopamine-release vehicles in the target area (Clarkson, 2001). The first effective treatment for PD was L-dihydroxyphenyl alanine (L-DOPA), a precursor of dopamine, but long-term administration of L-DOPA consequently produced grave side effects (Lang and Lozano, 1998a, 1998b). Based on the same hypothesis, the tissue isolated from human embryos or other animals that produce dopamine have been directly transplanted into PD patient striatum to increase dopamine content (Borlongan and Sanberg, 2002; Clarkson, 2001; Freeman et al., 2001; Fricker et al., 2001; Larsson and Widner, 2000; Lindvall, 2000; Lindvall and Hagell, 2000, 2001; Subramanian, 2001). The major hurdle in the transplantation of HESC-derived cells is ethical concern over the source of tissues and safety measures. Some transplantation studies have reported an incidence of tumors or teratomas following transplantation of predifferentiated HESCs in the brain (Brederlau et al., 2006; Erdo et al., 2003). This presents an unacceptable risk for patients and currently hampers the development of HESC-derived cell therapies. These tumors are thought to arise from residual undifferentiated ESCs or precursor cells that maintain their proliferative capacity *in vivo*. To overcome these risks, a number of different strategies are currently being tested. Some of these include the use of genetic modification to knock out tumor-inducing activity in HESC (Parish et al., 2005), purifying cells prior to transplantation using sorting technology (Pruszak et al., 2007) and inducing selective apoptosis of tumor-inducing cells (Bieberich et al., 2004). The establishment of a pure, differentiated HESC population that is safe and risk-free is an important translational research goal for all areas of regenerative medicine. Overcoming this hurdle will push HESCs one step closer to the clinic.

5.4.2.3 Huntington's Disease

The first complete description of Huntington's disease (HD) was reported by George Huntington among the population of Long Island, New York in 1872 (Huntington, 1872). HD affects males and females in relatively equal numbers. The disorder occurs in various geographic and ethnic populations worldwide. The prevalence of Huntington's disease in European populations is estimated at 4–8 cases per 100,000 people, and the disorder may also be frequent in India and parts of central Asia (Harper, 1992). HD is an autosomal dominant neurodegenerative disorder characterized by involuntary choreiformic movements, cognitive impairment, and emotional disturbances (Greenmayre and Shoulson, 1994; Harper, 1996). Its cause has been linked to abnormal expansion in a length of a CAG triplet repeat sequence in a gene on chromosome 4p, now called the Huntington gene. Huntington's disease is usually apparent in middle adult life, with development of abnormal movements that are at first most evident in the hands and face. The person has difficulty in performing a sequence of hand movements. Movement disorder is usually slowly progressive and eventually may become a disability. Attention, judgement, awareness, and executive functions may be seriously deficient at an early stage, but memory is frequently not impaired until late in the disease. These abnormal movements accompanied by disturbance of mood, particularly depression, are common, leading to a high suicide rate (Farrer, 1986). The U.S. Food and Drug Administration has approved Xenazine (tetrabenazine) for the treatment of in people with HD. Xenazine is a new drug and is the first treatment of any kind approved in the United States for any symptom of HD.

Stem cell therapy offers a hope for treatment of HD in the future because of its ability to restore or preserve brain function by replacing and protecting striatal neurons. An ideal source of cell transplantation in HD would be NSCs, which could participate in normal central nervous system (CNS) development and differentiate into regionally appropriate cell types in response to environmental factors. Transplantation of NSCs to replace degenerated neurons with genetically modified NSCs producing neurotrophic factors has been used to protect striatal neurons against excitotoxic insults (Bjorklund and Lindvall, 2000). In animal models of HD, cell replacement using grafts of fetal striatal neurons promotes functional recovery, and some evidence from clinical trials indicates that this can also occur in patients (Lindvall et al., 2004). Recent studies shows that the human NSCs implanted into the brains of rats reduce motor impairments in HD through trophic mechanisms (Ryu et al., 2004; McBride et al., 2004). The first clinical study of cellular transplantation in HD patients was conducted as a pilot study in Mexico in 1990 (Madrazo et al., 1995). These studies involved tissue implantation within 1–2 h of spontaneous abortion of the donated human fetal tissue, whereas most subsequent studies have been based on tissue donation from elective abortion. In this case, there are sensitive ethical and social issues associated with using human fetal tissues for transplantation, in particular for embryonic and fetal tissues collected from voluntary or spontaneous terminations of pregnancy, and it is imperative that such studies are undertaken only within a framework of strict ethical appraisal and informed consent (Boer, 1994).

There are many reports about successful application of stem cell-based therapy in animal models of HD with functional recovery. However, there are still many obstacles to be overcome before clinical application of cell therapy in HD. With the recent explosive progress in the field of stem cell biology and its applications, one might hope that stem cell-based cell therapy would be one of the life-saving cures for HD patients in the near future.

5.4.2.4 Amyotrophic Lateral Sclerosis

ALS is a relentlessly progressive, adult-onset neurodegenerative disorder characterized by degeneration and loss of motor neurons in the cerebral cortex, brainstem, and spinal cord, leading to muscle wasting and weakness. When the motor neurons die, the ability of the brain to initiate and control muscle movement is lost. With voluntary muscle action progressively affected, patients in the later stages of the disease may become totally paralyzed; the disease often leads to death within 5 years after the onset of clinical symptoms (Hudson, 1990; Rowland and Shneider, 2001). Currently there is no effective treatment available to save patients with ALS. The U.S. Food and Drug Administration has approved the drug riluzole for the treatment of ALS. Riluzole prolongs survival in some patients for up to several months. However, there is no standard medical treatment that can reverse the debilitating effects of ALS or offer patients any medium-term or long-term improvements.

A stem cell-based therapy can be used for treatment if it can restore or preserve the function of both upper and lower motor neurons, and new neurons must be integrated into existing neural circuitries. In this regard, some studies have indicated that it is possible to generate motor neurons in culture from stem cells that include ESCs and NSCs (Wichterle et al., 2002; Harper et al., 2004; Miles et al., 2004; Li et al., 2005). Mouse ESC-derived motor neurons establish functional synapses, with muscle fibers transplanted into motor neuron-injured rat spinal cord surviving and extending axons into ventral root. Whereas neuronal replacement in ALS patients seems a distant goal, using stem cells to prevent motor neurons from dying is a more realistic and shorter-term clinical approach. Transplantation of NSCs isolated from fetal spinal cord (Xu et al., 2006) or neurons generated from the NT2 human teratocarcinoma cell line (Garbuzova et al., 2002) in spinal cord is also effective in delaying disease progression in a mouse ALS model. These cell transplantation studies have shown functional improvement in animal models of ALS. Previously, the XCell-Center in Germany focused on reducing symptoms and slowing down the chronic consequences of ALS in order to improve each patient's quality of life. They used the procedure of intrathecal injection or lumbar puncture, in which patients are treated by injecting the stem cells into the cerebrospinal fluid that flows within the spinal canal. The entire treatment consists of three steps: bone marrow collection, laboratory processing, and stem cell implantation. On the first day, bone marrow is collected from the patient's iliac crest (hip bone) using thin-needle mini-puncture under local anesthesia. Although some pain is felt when the needle is inserted, most patients do not find the bone marrow collection procedure particularly painful. The entire procedure normally takes about 30 minutes. The day after the collection of bone marrow, the stem cells are processed from the bone marrow in a state-of-the-art, government-approved (cGMP) laboratory. In the lab, both the quantity and quality of the stem cells are measured. These cells have the potential to transform into multiple types of cells and are capable of regenerating or repairing damaged tissue. On the third day, the stem cells are implanted back into the patient by lumbar puncture. A spinal needle is inserted between the L4 and L5 vertebrae under local anesthesia, and a small amount of spinal fluid is removed. A portion of that spinal fluid is mixed with the stem cell solution, which is then injected into back into the patient's spinal fluid, not the spinal cord. Normally, the lumbar puncture procedure is not painful. In a pilot study, almost 40% of the ALS patients treated with stem cells at the XCell-Center showed improvement.

5.4.2.5 Stem Cell Therapy for Multiple Sclerosis

MS is a chronic neurological disorder that affects the central nervous system (brain and spinal cord). It is caused by the destruction of oligodendrocytes (OLs) and myelin sheath by an inflammation-mediated mechanism. OLs and myelin sheath surrounds axons. Their destruction leads to conduction deficits and a variety of neurological symptoms and, in some patients, major disability. Currently there is no cure for MS. There are treatments available that may slow its progression and alleviate associated symptoms. Naturally, treatment approaches for MS focus on targeting the immune system, either in a nonspecific way (systemic immunosuppression with cytotoxic agents) or through immunomodulation (to specifically downregulate the myelin-reactive autoimmune lymphocytes or to enhance the regulatory immune networks) in order to control the inflammatory process, which causes demyelination (Steinman, 2001). Unfortunately, currently existing treatments for MS (both the immunosuppressive ones and the immunomodulating, i.e., glatiramer acetate and interferon-β) are only partially effective. Therefore, for MS, the use of stem cells may provide a logical solution, because these cells can migrate locally into the areas of white-matter lesions (plaques) and have the potential to support local neurogenesis and rebuilding of the affected myelin. This is achieved both by support of the resident CNS stem cell repertoire and by differentiation of the transplanted cells into neurons and myelin-producing cells (oligodendrocytes).

5.5 CONCLUSION

In the past millennium, scientists have come to understand the nature of disease and developed antibiotics, nutritional supplements, and biologics to cure many diseases. However, disease that results from degenerative pathology, malignancy, and immune-mediated disorders have continued to remain a challenge, and stem cell therapy has emerged as a new ray of hope for patients suffering from these illnesses. Understanding the biological behavior of stem cells from embryos and their potential to differentiate into various tissues that can regulate the function of organs has given an entirely new dimension to stem cell research. We have had recent successes in engineering stem cells to morphologically differentiate into hematopoietic cells, β-cell, neuronal cell, fibroblast, hepatocyte, and angioblastic cells. There are bigger challenges in the identification of various transcription factors to modulate the signals and make these cell lines physiologically viable and able to undertake the job of the endocrine, autocrine, and paracrine systems independently and perform the task to which they have been assigned by nature. Furthermore, the role of stem cell therapy in degenerative neurological disorders like motor-neuron disease, chorea, Alzheimer's disease, and Parkinsonism has yet to be determined. However, the success of stem cell therapy has already been established in leukemia and myelodisplastic syndrome, whereas its role in other diseases is still at the experimental stage. The purpose of biotechnology and genetic engineering is not, after all, to create another life in a test tube but to provide a life with value, virtue, and dignity to living human beings. Stem cell research is a humble beginning in this endeavor.

REFERENCES

Akashi, K., Traver, D., Miyamoto, T., et al. (2000). A clonogenic common myeloid progenitor that gives rise to all myeloid lineages. *Nature* 404: 193–197.

Alzheimer's Association (2010). Alzheimer's disease facts and figures. *Alzheimer's & Dementia* 6: 158–194.

Andrews, P. W., Banting, G. S., Damjanov, I., et al. (1984). Three monoclonal antibodies defining distinct differentiation antigens associated with different high molecular weight polypeptides on the surface of human embryonal carcinoma cells. *Hybridoma* 3: 347–361.

Andrews, P. W., Oosterhuis, J., Damjanov, I. (1987). Cell lines from human germ cell tumours. In E. Robertson (Ed.), *Teratocarcinomas and embryonic stem cells: A practical approach* (pp. 207–248). Oxford: IRL.

Atkinson, M. A., Eisenbarth, G. S. (2001). Type I diabetes: New perspectives on disease pathogenesis and treatment. *Lancet* 358: 221–229.

Audrey, R. C., Mark, S. F., Michele, S. G. (1999). *Stem cell research and applications monitoring the frontiers of biomedical research.* New York: American Association for the Advancement of Science and Institute for Civil Society.

Badcock, G., Pigott, C., Goepel, J., et al. (1999). The human embryonal carcinoma marker antigen TRA-1–60 is a sialylated keratan sulphate proteoglycan. *Cancer Res.* 59: 4715–4719.

Bartus, R., Dean, R. L., Beer, B., et al. (1982). The cholinergic hypothesis of geriatric memory dysfunction. *Science* 217: 408–411.

Berenson, R. J., Andrews, R. G., Bensinger, W. I., et al. (1988). Antigen CD34+ marrow cells engraft lethally irradiated baboons. *J. Clin. Invest.* 81: 951–955.

Bieberich, E., Silva, J., Wang, G., et al. (2004). Selective apoptosis of pluripotent mouse and human stem cells by novel ceramide analogues prevents teratoma formation and enriches for neural precursors in ES cell-derived neural transplants. *J. Cell Biol.* 167: 723–734.

Bjorklund, A., Lindvall, O. (2000). Cell replacement therapies for central nervous system disorders. *Nat. Neurosci.* 3: 537–544.

Bodnar, A. G., Ouellette, M., Frolkis, M., et al. (1998). Extension of life-span by introduction of telomerase into normal human cells. *Science* 279: 349–352.

Boer, G. J. (1994). Ethical guidelines for the use of human embryonic or fetal tissue for experimental and clinical neurotransplantation and research. *J. Neurol.* 242: 1–13.

Borlongan, C. V., Sanberg, P. R. (2002). Neural transplantation for treatment of Parkinson's disease. *Drug Discov. Today* 7: 674–682.

Brederlau, A., Correia, A. S., Anisimov, S. V., et al. (2006). Transplantation of human embryonic stem cell-derived cells to a rat model of Parkinson's disease: Effect of in vitro differentiation on graft survival and teratoma formation. *Stem Cells* 24: 1433–1440.

Brolen, G. K., Heins, N., Edsbagge, J., Semb, H. (2005). Signals from the embryonic mouse pancreas induce differentiation of human embryonic stem cells into insulin-producing beta-cell-like cells. *Diabetes* 54: 2867–2874.

Brook, F. A., Gardner, R. L. (1997). The origin and efficient derivation of embryonic stem cells in the mouse. *Proc. Natl. Acad. Sci. U.S.A.* 94: 5709–5712.

Burdon, T., Stracey, C., Chambers, I., et al. (1999). Suppression of SHP-2 and ERK signalling promotes self-renewal of mouse embryonic stem cells. *Dev. Biol.* 210: 30–43.

Chapman, A. R., Frankel, M. S., Michele, S. G. (1999). *Stem cell research and applications monitoring the frontiers of biomedical research.* New York: American Association for the Advancement of Science and Institute for Civil Society.

Chen, J., Li, Y., Chopp, M. (2000). Intracerebral transplantation of bone marrow with BDNF after MCAo in rat. *Neuropharmacology* 39: 711–716.

Chopp, M., Zhang, X. H., Li, Y., et al. (2000). Spinal cord injury in rat: Treatment with bone marrow stromal cell transplantation. *Neuroreport* 11: 3001–3005.

Civin, C. I., Trischmann, T., Kadan, N. S., et al. (1996). Highly purified CD34- positive cells reconstitute hematopoiesis. *J. Clin. Oncol.* 14: 2224–2233.

Clarkson, E. D. (2001). Fetal tissue transplantation for patients with Parkinson's disease: A database of published clinical results. *Drugs Aging* 18: 773–785.

Cooper, G. S., Dooley, M. A., Treadwell, E. L., et al. (1998). Hormonal, environmental, and infectious risk factors for developing systemic lupus erythematosus. *Arthritis Rheum.* 41: 1714–1724.

Coyle, J. T., Price, D. L., DeLong, M. R. (1983). Alzheimer's disease: A disorder of cortical cholinergic innervation. *Science* 219: 1184–1190.

DeFronzo, R. A. (1997). Pathogenesis of type 2 diabetes: Metabolic and molecular implications for identifying diabetes genes. *Diabetes Rev.* 5: 178–269.

Dominique, F., Myriam L., Alan, T., et al. (2010). Autologous hematopoietic stem cell transplantation for autoimmune diseases: An observational study on 12 years' experience from the European Group for Blood and Marrow Transplantation Working Party on Autoimmune Diseases. *Hematologia* 95: 284–292.

Durez, P., Toungouz, M., Schandene, L., et al. (1998). Remission and immune reconstitution after T-cell-depleted stem-cell transplantation for rheumatoid arthritis. *Lancet* 352: 881.

Erdo, F., Buhrle, C., Blunk, J., et al. (2003). Host-dependent tumorigenesis of embryonic stem cell transplantation in experimental stroke. *J. Cereb. Blood Flow Metab.* 23: 780–785.

Evans, M. J., Kaufman, M. H. (1981). Establishment in culture of pluripotential cells from mouse embryos. *Nature* 292: 154–156.

Ezashi, T., Das, P., Roberts, R. M. (2005). Low O2 tensions and the prevention of differentiation of hES cells. *Proc. Natl. Acad. Sci. U.S.A.* 102: 4783–4788.

Farrer, L. A. (1986). Suicide and attempted suicide in Huntington disease: Implication for preclinical testing of persons at risk. *Am. J. Med. Genet.* 24: 305–311.

Ferrari, G., Cusella, A. G., Coletta, M., et al. (1998). Muscle regeneration by bone marrow-derived myogenic progenitors. *Science* 279: 1528–1530.

Freeman, T. B., Willing, A., Zigova, T., et al. (2001). Neural transplantation in Parkinson's disease. *Adv. Neurol.* 86: 435–445.

Fricker, R. A., Lundberg, C., Dunnett, S. B. (2001). Neural transplantation: Restoring complex circuitry in the striatum. *Restor. Neurol. Neurosci.* 19: 119–138.

Garbuzova, D. S., Willing, A. E., Milliken, M. (2002). Positive effect of transplantation of hNT neurons (NTera 2/D1 cell-line) in a model of familial amyotrophic lateral sclerosis. *Exp. Neurol.* 174: 169–180.

Gels, M. E., Marrink, J., Visser, P. (1997). Importance of a new tumor marker TRA-1–60 in the follow-up of patients with clinical state I nonseminomatous testicular germ cell tumors. *Ann. Surg. Oncol.* 4: 321–327.

Greenmayre, J. T., Shoulson, I. (1994). Huntington's disease. In D. B. Calne (Ed.), *Neurodegenerative Diseases* (pp. 685–704). Philadelphia, W.B. Saunders.

Harley, C. B. (1991). Telomere loss: Mitotic clock or genetic time bomb. *Mutat. Res.* 256: 271–282.

Harley, C. B., Vaziri, H., Counter, C. M., et al. (1992). The telomere hypothesis of cellular aging. *Exp. Gerontol.* 27: 375–382.

Harper, P. S. (1992). The epidemiology of Huntington's disease. *Hum. Genet.* 89: 365–376.

Harper, P. S. (1996). *Huntington's disease.* 2nd Ed. Philadelphia: W.B. Saunders.

Harper, J. M., Krishnan, C., Darman, J. S. (2004). Axonal growth of embryonic stem cell-derived motoneurons in vitro and in motoneuron-injured adult rats. *Proc. Natl. Acad. Sci. U.S.A.* 101: 7123–7128.

Holyoake, T. L., Nicolini, F. E., Eaves, C. J. (1999). Functional differences between transplantable human hematopoietic stem cells from fetal liver, cord blood, and adult marrow. *Exp. Hematol.* 27: 1418–1427.

Hudson, A. J. (1990). *Amyotrophic lateral sclerosis: concepts in pathogenesis and etiology*. Toronto: University of Toronto Press.

Huntington, G. (1872). On chorea. *Med. Surg. Reporter Philadelphia* 26: 317–321.

Imasawa, T., Utsunamiya, Y., Kawamura, T., et al. (2001). The potential of bone marrow-derived cells to differentiate to glomerular mesangial cells. *J. Am. Soc. Nephrol.* 12: 1401–1409.

Itskovitz, E. J., Schuldiner, M., Karsenti, D., et al. (2000). Differentiation of human embryonic stem cells into embryoid bodies comprising the three embryonic germ layers. *Mol. Med.* 6: 88–95.

Jensen, U. B., Lowell, S., Watt, F. M. (1999). The spatial relationship between stem cells and their progeny in the basal layer of human epidermis: A new view based on whole mount labelling and lineage analysis. *Development* 126: 2409–2418.

Jones, P. H., Watt, F. M. (1993). Separation of human epidermal stem cells from transit amplifying cells on the basis of differences in integrin function and expression. *Cell* 73: 713–724.

Jones, P. H., Harper, S., Watt, F. M. (1995). Stem cell patterning and fate in human epidermis. *Cell* 80: 83–93.

Joske, D. L., Ma, D. T., Langlands, D. R., et al. (1997). Autologous bone marrow transplantation for rheumatoid arthritis. *Lancet* 350: 337–338.

Kato, Y., Rideout, W. M., Hilton, K., et al. (1999). Developmental potential of mouse primordial germ cells. *Development* 126: 1823–1832.

Kleinsmith, L. J., Pierce, G. B. (1964). Multipotentiality of single embryonal carcinoma cells. *Cancer Res.* 24: 1544–1552.

Knowles, B. B., Aden, D. P., Solter, D. (1978). Monoclonal antibody detecting a stage-specific embryonic antigen (SSEA-1) on preimplantation mouse embryos and teratocarcinoma cells. *Curr. Top. Microbiol. Immunol.* 81: 51–53.

Kondo, M., Weissman, I. L., Akashi, K. (1997). Identification of clonogenic common lymphoid progenitors in mouse bone marrow. *Cell* 91: 661–672.

Kopen, G. C., Prockop, D. J., Phinney, D. G. (1999). Marrow stromal cells migrate throughout forebrain and cerebellum, and they differentiate into astrocytes after injection into neonatal mouse brains. *Proc. Natl. Acad. Sci. U.S.A.* 96: 10711–10716.

Labosky, P. A., Barlow, D. P., Hogan, B. L. (1994). Embryonic germ cell lines and their derivation from mouse primordial germ cells. *CIBA Found. Symp.* 182: 157–168.

Lang, A. E., Lozano, A. M. (1998a). Parkinson's disease: First of two parts. *N. Engl. J. Med.* 339: 1044–1053.

Lang, A. E., Lozano, A. M. (1998b). Parkinson's disease: Second of two parts. *N. Engl. J. Med.* 339: 1130–1143.

Larsson, L. C., Widner, H. (2000). Neural tissue xenografting. *Scand. J. Immunol.* 52: 249–256.

Laughlin, M. J. (2001). Umbilical cord blood for allogeneic transplantation in children and adults. *Bone Marrow Transplant.* 27: 1–6.

Li, A., Simmons, P. J., Kaur, P. (1998). Identification and isolation of candidate human keratinocyte stem cells based on cell surface phenotype. *Proc. Natl. Acad. Sci. U.S.A.* 95: 3902–3907.

Li, P., Tessler, A., Han, S. S. (2005). Fate of immortalized human neuronal progenitor cells transplanted in rat spinal cord. *Arch. Neurol.* 62: 223–229.

Li, Y., Chopp, M., Chen, J., et al. (2000). Intrastriatal transplantation of bone marrow nonhematopoietic cells improves functional recovery after stroke in adult mice. *J. Cereb. Blood Flow Metab.* 20: 1311–1319.

Liechty, K. W., MacKenzie, T. C., Shaaban, A. F., et al. (2000). Human mesenchymal stem cells engraft and demonstrate site-specific differentiation after in utero transplantation in sheep. *Nat. Med.* 6: 1282–1286.

Lindvall, O. (2000). Neural transplantation in Parkinson's disease. *Novartis Found. Symp.* 231: 110–123.

Lindvall, O., Hagell, P. (2000). Clinical observations after neural transplantation in Parkinson's disease. *Prog. Brain Res.* 127: 299–320.

Lindvall, O., Hagell, P. (2001). Cell therapy and transplantation in Parkinson's disease. *Clin. Chem. Lab. Med.* 39: 356–361.

Lindvall, O., Kokaia, Z., Martinez, S. A. (2004). Stem cell therapy for human neurodegenerative disorders: How to make it work. *Nat. Med.* 10 (supplement): S42–S50.

Lowell, S., Jones, P., Le, R. I., et al. (2000). Stimulation of human epidermal differentiation by Notch/Delta signalling at the boundaries of stem cell clusters. *Curr. Biol.* 10: 491–500.

Lyle, S., Christofidou, S. M., Liu, Y., et al. (1998). The C8/144B monoclonal antibody recognizes cytokeratin 15 and defines the location of human hair follicle stem cells. *J. Cell Sci.* 111: 3179–3188.

Mackenzie, T. C., Flake, A. W. (2001). Human mesenchymal stem cells persist, demonstrate site-specific multipotential differentiation, and are present in sites of wound healing and tissue regeneration after transplantation into fetal sheep. *Blood Cells Mol. Dis.* 27: 601–604.

Madrazo, I., Franco-Bourland, R. E., Castrejon, H., et al. (1995). Fetal striatal homotransplantation for Huntington's disease: First two case reports. *Neurol. Res.* 17: 312–315.

Majumdar, M. K., Thiede, M. A., Mosca, J. D., et al. (1998). Phenotypic and functional comparison of cultures of marrow-derived mesenchymal stem cells (MSCs) and stromal cells. *J. Cell Physiol.* 176: 57–66.

Makino, S., Fukuda, K., Miyoshi, S., et al. (1999). Cardiomyocytes can be generated from marrow stromal cells in vitro. *J. Clin. Invest.* 103: 697–705.

Marmount, A., Tyndall, A., Gratwohl, A., et al. (1995). Hematopoietic precursor-cell transplants for autoimmune disease. *Lancet* 345: 978.

Marrink, J., Andrews, P. W., van Brummenm P. J. (1991). TRA-1-60: A new serum marker in patients with germ cell tumors. *Int. J. Cancer* 49: 368–372.

Martin, G. R. (1981). Isolation of a pluripotent cell line from early mouse embryos cultured in medium conditioned by teratocarcinoma stem cells. *Proc. Natl. Acad. Sci. U.S.A.* 78: 7634–7638.

Martin, G. R., Evans, M. J. (1975). The formation of embryoid bodies in vitro by homogeneous embryonal carcinoma cell cultures derived from isolated single cells. In M. I. Sherman and D. Solter (Eds.), *Teratomas and differentiation* (pp. 169–187). New York: Academic Press.

Mason, M. D., Pera, M. F., Cooper, S. (1991). Possible presence of an embryonal carcinoma-associated proteoglycan in the serum of patients with testicular germ cell tumors. *Eur. J. Cancer* 27: 300.

Matsui, Y., Zsebo, K., Hogan, B. L. (1992). Derivation of pluripotential embryonic stem cells from murine primordial germ cells in culture. *Cell* 70: 841–847.

McBride, J. L., Behrstock, P. S., Er-yun, C., et al. (2004). Human neural stem cell transplants improve motor function in a rat model of Huntington's disease. *J. Comp. Neurol.* 475: 211–219.

Miles, G. B., Yohn, D. C., Wichterle, H. (2004). Functional properties of motoneurons derived from mouse embryonic stem cells. *J. Neurosci.* 24: 7848–7858.

Moles, J. P., Watt, F. M. (1997). The epidermal stem cell compartment: Variation in expression levels of E-cadherin and catenins within the basal layer of human epidermis. *J. Histochem. Cytochem.* 45: 867–874.

Moll, I. (1995). Proliferative potential of different keratinocytes of plucked human hair follicles. *J. Invest. Dermatol.* 105: 14–21.

Odorico, J. S., Kaufman, D. S., Thomson, J. A. (2001). Multilineage differentiation from human embryonic stem cell lines. *Stem Cells* 19: 193–204.

Parish, C. L., Parisi, S., Persico, M. G., et al. (2005). Cripto as a target for improving embryonic stem cell-based therapy in Parkinson's disease. *Stem Cells* 23: 471–476.

Pavletic, Z. S., O'Dell, J. R., Bishop, M. R., et al. (1998). Treatment of refractory rheumatoid arthritis utilizing an outpatient autologous blood stem cell transplantation protocol. *Blood* 92 (Suppl. 1): 370b.

Pera, M. F., Herzfeld, D. (1998). Differentiation of human pluripotent teratocarcinomas stem cells induced by bone morphogenetic protein-2. *Reprod. Fertil. Dev.* 10: 551–555.

Pereira, R. F., Halford, K. W., O'Hara, M. D., et al. (1995). Cultured adherent cells from marrow can serve as long-lasting precursor cells for bone, cartilage, and lung in irradiated mice. *Proc. Natl. Acad. Sci. U.S.A.* 92: 4857–4861.

Petersen, B. E., Bowen, W. C., Patrene, K. D., et al. (1999). Bone marrow as a potential source of hepatic oval cells. *Science* 284: 1168–1170.

Ping, L., Fang, L., Lei, Y. (2007). Stem cells therapy for type 1 diabetes. *Diabetes Res. Clin. Pract.* 78: 1–7.

Pittenger, M. F., Mackay, A. M., Beck, S. C., et al. (1999). Multilineage potential of adult human mesenchymal stem cells. *Science* 284: 143–147.

The President's Council on Bioethics (2004). http://www.bioethics.gov.

Prockop, D. J. (1997). Marrow stromal cells as stem cells for nonhematopoietic tissues. *Science* 276: 71–74.

Pruszak, J., Sonntag, K. C., Aung, M. H., et al. (2007). Markers and methods for cell sorting of human embryonic stem cell-derived neural cell populations. *Stem Cells* 25: 2257–2268.

Rahman, A., Isenberg, D. A. (2008). Systemic lupus erythematosus. *N. Engl. J. Med.* 358: 929–939.

Rathjen, P. D., Toth, S., Willis, A., et al. (1990). Differentiation inhibiting activity is produced in matrix-associated and diffusible forms that are generated by alternate promoter usage. *Cell* 62: 1105–1114.

Resnick, J. L., Bixler, L. S., Cheng, L., et al. (1992). Long-term proliferation of mouse primordial germ cells in culture. *Nature* 359: 550–551.

Rettig, W. J., Cordon-Cardo, C., Ng, J. S., et al. (1985). High-molecular-weight glycoproteins of human teratocarcinomas defined by monoclonal antibodies to carbohydrate determinants. *Cancer Res.* 45: 815–821.

Reubinoff, B. E., Pera, M. F., Fong, C. Y. (2000). Embryonic stem cell lines from human blastocysts: somatic differentiation *in vitro*. *Nat. Biotechnol* 18: 399–404.
Richard, H., Shanks, P., Darnovsky, M., et al. (2006). *Stem cells and public policy*. New York: The Century Foundation.
Richard, K. B., Traynor, E. A. (2003). Hematopoietic stem cell transplantation for systemic lupus erythematosus. *Arthritis Res. Ther.* 5: 207–209.
Roach, S., Cooper, S., Bennett, W., et al. (1993). Cultured cell lines from human teratomas: Windows into tumour growth and differentiation and early human development. *Eur. Urol.* 23: 82–87.
Rohwedel, J., Guan, K., Hegert, C., et al. (2001). Embryonic stem cells as an in vitro model for mutagenicity, cytotoxicity, and embryotoxicity studies: Present state and future prospects. *Toxicol. In Vitro* 15: 741–753.
Rowland, L. P., Shneider, N. A. (2001). Amyotrophic lateral sclerosis. *N. Engl. J. Med.* 344: 1688–1700.
Ryu, J. K., Jean, K., Sung, J. C., et al. (2004). Proactive transplantation of human neural stem cells prevents degeneration of striatal neurons in a rat model of Huntington disease. *Neurobiol. Dis.* 16: 68–77.
Schuldiner, M., Eiges, R., Eden, A., et al. (2001). Induced neuronal differentiation of human embryonic stem cells. *Brain Res.* 913: 201–205.
Schwarz, E. J., Alexander, G. M., Prockop, D. J., et al. (1999). Multipotential marrow stromal cells transduced to produce L-dopa: Engraftment in a rat model of Parkinson disease. *Hum. Gene Ther.* 10: 2539–2549.
Shamblott, M. J., Axelman, J., Wang, S. (1998). Derivation of pluripotent stem cells from cultured human primordial germ cells. *Proc. Natl. Acad. Sci. U.S.A.* 95: 13726–13731.
Shapiro, A. M., Lakey, J., Ryan, E., et al. (2000). Islet transplantation in seven patients with type 1 diabetes mellitus using a glucocorticoid-free immunosuppressive regimen. *N. Engl. J. Med.* 343: 230–238.
Shapiro, A. M., Ricordi, C., Hering, B. J., et al. (2006). International trial of the Edmonton protocol for islet transplantation. *N. Engl. J. Med.* 355: 1318–1330.
Smith, A. G. (2001). Embryo-derived stem cells: of mice and men. *Annu. Rev. Cell Dev. Biol.* 17: 435–462.
Smith, A. G., Heath, J. K., Donaldson, D. D. (1988). Inhibition of pluripotential embryonic stem cell differentiation by purified polypeptides. *Nature* 336: 688–690.
Snowden, J. A., Biggs, J. C., Milliken, S. T., et al. (1999). A phase I/II dose escalation study of intensified cyclophosphamide and autologous blood stem cell rescue in severe, active rheumatoid arthritis. *Arthritis Rheum.* 42: 2286–2292.
Solter, D., Knowles, B. B. (1978). Monoclonal antibody defining a stage-specific mouse embryonic antigen (SSEA-1). *Proc. Natl. Acad. Sci. U.S.A.* 75: 5565–5569.
Steinman, L. (2001). Multiple sclerosis: A two-stage disease. *Nat. Immunol.* 2: 762–764.
Stewart, C. L., Gadi, I., Bhatt, H. (1994). Stem cells from primordial germ cells can reenter the germ line. *Dev. Biol.* 161: 626–628.
Subramanian, T. (2001). Cell transplantation for the treatment of Parkinson's disease. *Semin. Neurol.* 21: 103–115.
Terada, N., Hamazaki, T., Oka, M., et al. (2002). Bone marrow cells adopt the phenotype of other cells by spontaneous cell fusion. *Nature* 416: 542–545.
Thomson, J. A., Odorico, J. S. (2000). Human embryonic stem cell and embryonic germ cell lines. *Trends Biotechnol.* 18: 53–57.
Thomson, J. A., Waknitz, M. A., Swiergiel, J. J., et al. (1998). Embryonic stem cell lines derived from human blastocysts. *Science* 282: 1061–1062.
Weissman, I. L. (2000). Translating stem and progenitor cell biology to the clinic: Barriers and opportunities. *Science* 287: 1442–1446.
Whitehouse, P. J., Price, D. L., Clark, A. W., et al. (1981). Alzheimer disease: Evidence for selective loss of cholinergic neurons in the nucleus basalis. *Ann. Neurol.* 10: 122–126.
Wichterle, H., Lieberam, I., Porter, J. A., et al. (2002). Directed differentiation of embryonic stem cells into motor neurons. *Cell* 110: 385–397.
Wild, S., Roglic, G., Green, A., et al. (2004). Global prevalence of diabetes: Estimates for the year 2000 and projections for 2030. *Diabetes Care* 27: 1047–1053.
Xu, L., Yan, J., Chen, D. (2006). Human neural stem cell grafts ameliorate motor neuron disease in SOD1 transgenic rats. *Transplantation* 82: 865–875.
Ying, Q. L., Nichols, J., Evans, E. P., et al. (2002). Changing potency by spontaneous fusion. *Nature* 416: 545–548.
Zhongling, F., Gang, Z., Lei, Y. (2009). Neural stem cells and Alzheimer's disease: Challenges and hope. *Am. J. Alzheimer's Dis. Other Dementias* 24: 52–57.

6 Emerging Trends of Nanotechnology in Omics-Based Drug Discovery and Development

Sandhiya Selvarajan
Jawaharlal Institute of Post-Graduate Medical Education and Research
Pondicherry, India

CONTENTS

6.1 INTRODUCTION

Nanotechnology deals with designing particles and devices, 1–100 nm in size, with unique characteristic features at the cellular, atomic, and molecular levels. The application of nanotechnology concepts for enhancing diagnostics, prophylaxis, and therapy has given rise to an emerging branch of science known as nanomedicine (Linkov, Satterstrom, & Corey, 2008). One of the major aims of nanomedicine is to devise therapeutically functional nanoparticles and nanodevices to probe inside the cellular organelles for diagnostic purposes as well as to deliver drugs at the desired site for targeted drug delivery. Thus nanostructures may facilitate the understanding of molecular level mechanisms involved

TABLE 6.1
Currently Available Nanodrugs and Their Uses

Drugs	Nanoformulations	Uses
Amphotericin B	Liposomal amphotericin B (AmBisome)	Fungal infections, visceral leishmaniasis
Aprepitant	Micronized, nanosized NK1 receptor antagonist (aprepitant)	Cancer chemotherapy-induced emesis
Doxorubicin	Pegylated liposomal doxorubicin (Doxil/Caelyx/Myocet)	Recurrent breast cancer
Paclitaxel	Solvent-free, albumin-bound paclitaxel (Abraxane, Genexol-PM)	Breast carcinoma
Sirolimus	Nano suspension of sirolimus (Rapamune)	Immunosuppression

in both physiological and pathological processes (Teli, Mutalik, & Rajanikant, 2010). Moreover with the advent of nanodevices that can interact with DNA, RNA, proteins, and enzymes, the pathological changes can be detected at an early stage, leading to early intervention and cure of the disease.

In addition to diagnosis, nanoparticles are also being investigated for the targeted delivery of drugs, DNA, RNA, protein, and contrast agents in the desired site of action (Cortivo et al., 2010; Mukherjee, Ray, & Thakur, 2009). Among the various nanocarriers, liposomes, polymers, micelles, and dendrimers have been shown to enhance the solubility and biocompatibility of poorly soluble lipid drugs. Nanodevices like nanopores, nanotubes, nanoshells, nanowires, and nanocantilevers also offer an opportunity to achieve specific distribution of drug in the required site. Moreover, with the growing advances in the field of surface modification using different moieties, the desirable functionalities required for targeted drug delivery can be imparted on the nanocarriers (Bhaskar et al., 2010). Thus, nanotechnology aids in the designing and development of nanodrug-delivery system with the advantage of maximum therapeutic efficacy with fewer adverse effects (Rawat et al., 2007). The currently available nanoformulations include Abraxane, AmBisome, aprepitant, and Rapamune, as shown in Table 6.1 with their indications (Cortivo et al., 2010; Wang & Thanou, 2010).

Nanoparticles possess a unique feature attributed to the leaky vasculature and impaired lymphatic drainage of the malignant tumors known as the enhanced permeability and retention (EPR) effect (Putz et al., 2010). The EPR phenomenon is used in the treatment of cancer because nanoformulations can penetrate tumor vascular endothelium owing to their smaller size and remain within the tumor cells, resulting in prolonged duration of action because they are not drained by the lymphatic system. Thus the application of nanotechnology in cancer chemotherapy seems to be infinite, with the ability to detect cancer-related pathogenesis at an early stage, release of therapeutic agents exactly at the desired site of action, and monitoring the effectiveness of the delivered drug (Wang & Thanou, 2010).

In the present scenario, liposomes, polymers, and dendrimers have been successfully investigated to boost the development of targeted drug delivery. In addition, the biodegradable polymers have widened the availability of drugs for brain tumors and neurodegenerative diseases like Alzheimer's disease. Nanotechnology is spreading its wings in various branches of medicine with more emphasis towards the design and development of newer drugs. Hence this chapter will focus on the emerging trends of nanotechnology in drug discovery, development, and drug-delivery methods.

6.2 THE ROLE OF NANOTECHNOLOGY IN DRUG DISCOVERY AND DEVELOPMENT

The major problems associated with conventional drug-delivery systems are enzymatic degradation of drug, first-pass metabolism, and the presence of physiological barriers like the blood-brain

TABLE 6.2
Major Techniques of Nanotechnology Used in Diagnosis, Drug Discovery, and Development

Techniques of Nanomedicine	Applications
1. Biosensors	Monitor and administer correct doses of drug *in vivo*
2. Biomimetics	Diagnosis and drug delivery
3. Drug carriers (e.g., dendrimer, liposomes, micelles, polymer)	Delivery of contrast agents, lipid-insoluble drugs, and genes

barrier (BBB), restricting the entry of drugs into the desired site of action. This has led to the intense hunt for the discovery of alternative drug-delivery methods through intranasal, intrathecal, and intraventricular routes (Patel et al., 2009). The recent advances in the field of nanotechnology, life sciences, and biotechnology have paved the way for targeted drug delivery to the diseased area. They help to achieve maximum therapeutic efficacy at low dose, resulting in fewer side effects and improved patient adherence (Majidi et al., 2009). (The newer techniques of nanotechnology used in drug discovery and development are shown in Table 6.2.)

Nanoparticles (NPs) with diameters less than 100 nm, created by chemical and physical processes, show novel properties owing to their larger surface to mass ratio. The potential application of the nanoparticles originates from the catalytic reactions promoted with the ability to adsorb other compounds, also known as quantum phenomena. This phenomenon, along with other added features like surface modification of nanoparticles, helps to target specific organs, prolong the duration of action, and limit the uptake by macrophages (Nobs et al., 2004). The large functional surface area of the nanoparticles designed to bind, adsorb, and deliver drugs can be used to enhance the absorption of lipid insoluble drugs, resulting in prolonged duration of action (Alam et al., 2010; Su et al., 2010). The inherent problems associated with denaturation of peptide drugs can be overcome with the help of nanocarriers because they escape first-pass metabolism owing to their small size (Cho et al., 2008). The distinct properties of nanoparticles like nanosphere, nanocapsule, nanoshell, polymeric micelles, liposomes, gold colloids, and dendrimers have been receiving a lot of attention in the fields of bioengineering, drug delivery, and drug discovery. Studies have shown that polymers, liposomes, solid lipid nanoparticles (SLN), and micelles increase the concentration of drug in the brain by inhibiting ATP-binding cassette transporters expressed at the blood-brain barrier. A study, conducted in HIV patients with antiretroviral drugs loaded in nanocarriers, showed a higher concentration in the brain (Wong et al., 2010). Another study conducted on the murine C6 glioma cells *in vitro* and the brain glioma-bearing rats *in vivo*, with a functionalized liposomal nanoconstruct, composed of epirubicin along with tamoxifen and transferrin has shown an increased transport of drugs across the BBB. This design can be used as a novel strategy for delivering chemotherapeutic agents to the brain tumor (Tian et al., 2010). (The ideal characteristics expected from the design and development of nanoformulations is shown in Table 6.3.) Among the various nanocarriers,

TABLE 6.3
Characteristic Features of Ideal Nanodrugs

1. More efficacious than the usual drug
2. Less adverse reactions than the conventional drug
3. No drug interactions or interference with the pharmacological aspects of other drugs
4. Cost-effective
5. Stable following administration through different routes
6. No toxic effects on the environment

liposomes, nanocrystals, solid lipid nanoparticles, polymers, silver nanoparticles, dendrimers, virosomes, and carbon nanotubes have aroused considerable interest in drug delivery, and these nanocarriers will be discussed briefly under this section.

6.2.1 Liposomes for Drug Delivery

Liposomes are small vesicles composed of phospholipids and have the capacity to incorporate any drug, independent of its lipid solubility. Based on the number of concentric bilayers present in the structure, the liposomes can be classified into unilamellar and multilamellar (Wu & Chang, 2010). Because liposomes have the potential for targeting the mononuclear phagocytic system (MPS), they serve as a favourable drug carrier for amphotericin B, an antifungal drug used in the treatment of fungal infections as well as visceral leishmaniasis (Hamill et al., 2010; Vigna et al., 2010). The advantage of liposomal formulation of amphotericin B in these conditions includes enhanced efficacy with less nephrotoxicity. Currently the duration of action of liposomes has been shown to be prolonged by coating with polyethylene glycol (PEG). A coating of PEG material helps to prevent the interaction of nanoformulations with plasma components by forming a protective hydrophilic layer on their surface. This feature of PEG coating is used to shield the nanoparticles from the MPS, resulting in reduced clearance and a relatively prolonged half-life of the therapeutic agent (Lipka et al., 2010). Such PEGylated liposomes are termed as stealth liposomes, and they show dose-independent pharmacokinetics (Putz et al., 2010). The polymer coating of nanocarriers may occasionally delay the release of drug at the desired site, resulting in reduced efficacy. However, this may be overcome with sterically stabilized nanoparticles designed to unmask the coating on their arrival at the target site (Garnett et al., 2009).

Currently PEGylated liposomes are also being explored for delivery of drug via pulmonary, oral, and nasal routes for the administration of drugs like insulin (Smith, 2007). Studies using insulin-loaded liposomes as dry powder for pulmonary delivery have been found to show relatively high bioavailability with good hypoglycemic effect and long-lasting action. Moreover, PEGylation of liposomes have been found to increase the stability of drug, especially in lung cancer, because it avoids interaction with the surfactant lining of lungs compared with liposomes. In addition, after being taken up by the tumor cells of the lung, 80% of the drug was found to be retained for 48 hours (Anabousi et al., 2006). Similarly, trials conducted in patients with locally advanced carcinoma of breast, cervix, bronchus, glioma, head, and neck to study the biodistribution and pharmacokinetics of (111)In-diethylentriaminepentaacetate-labeled PEGylated liposomes showed that the uptake was greater with PEGylated liposomes in solid tumors, especially in head and neck cancer compared with conventional liposomes (Harrington et al., 2001). A phase I trial, carried out in patients with lung carcinoma to investigate the safety and pharmacokinetics of aerosolized sustained release lipid inhalation targeting cisplatin, demonstrated tolerability without any dose-limiting toxicity even at the maximum dose administered. Moreover, the safety data from this trial did not show any hematologic toxicity, nephrotoxicity, ototoxicity, or neurotoxicity. Moreover, this study has shown a novel drug-delivery method for the safe and effective treatment of lung cancer with aerosolized liposomal cisplatin (Wittgen et al., 2007). In a study on indinavir, a protease inhibitor targeted against MPS via mannosylated liposomes has shown a significant increase in the concentration of indinavir in macrophage-loaded tissues like liver, spleen, and lungs as compared with conventional liposomes and free drug formulation. Such studies suggest the potential use of indinavir loaded mannosylated liposomes for anti-HIV therapy (Dubey et al., 2010). At present, the potential application of surface-linked liposomal antigens for developing vaccines with minimal allergic effects is also being investigated (Smith, 2007). In addition, a surface-linked liposomal antigen could be used to build viral vaccines that induce cytotoxic T-cell response by presenting tumor antigens to antigen-presenting cells and inducing effective antitumor responses.

6.2.2 Nanocrystals for Drug Delivery

Nanocrystals are nanosized aggregates of thousands of drug molecules in crystalline form, with a thin surfactant coating. They are used in the delivery of poorly lipid soluble drugs without the need of a vehicle. It allows for safe and effective passage of lipid-insoluble therapeutic agents through capillaries, eliminating the potential toxicity associated with carrier molecules. Their potential use to deliver drugs has given rise to the U.S. Food and Drug Administration approval of Rapamune, also known as sirolimus, an immunosuppressant drug used in the prevention of graft rejection in liver-transplantation patients. This has been proven in a study done with a sirolimus-eluting stent that showed an increased sirolimus release achieved with the help of nanovehicles present in the eluting materials (Luderer et al., 2010). Another nanodrug called Emend, also known as aprepitant, a NK_1 receptor antagonist, approved for the prevention of cancer chemotherapy-induced nausea and vomiting, has been found to be more effective in the crystalline form and seems to be a promising formulation for the near future (Shono et al., 2010; Wu et al., 2004).

6.2.3 Polymeric Nanoparticles

Polymeric nanoparticles are nanosized colloidal materials with the ability to encapsulate, adsorb, and covalently bind drugs. Water-soluble polymers like poly(ethylene glycol) or poly(*N*-vinyl-2-pyrrolidinone) are preferred to increase the solubility of anticancer drugs. However, the use of naturally occurring polymers, especially polysaccharides like chitosan and alginate, in the place of synthetic polymers has been shown to be safer.

Polymeric micelles have been used in the targeted delivery of anticancer agents in both preclinical and clinical studies (Kedar et al., 2010). A study done on diabetic rats, with oral administration of alginate/chitosan nanoparticles loaded with insulin, showed a fall in basal serum-glucose levels by more than 40% (Sarmento et al., 2007). In another study, carried out in Sarcoma 180-bearing mice and rats, following intravenous administration of 5-fluorouracil-loaded *N*-succinyl-chitosan nanoparticles, the duration of action was found to be increased (Yan et al., 2010). Similarly *in vitro* studies carried out using albumin particles loaded with anti-inflammatory drugs like celecoxib have revealed sustained release of drug from the nanocarrier (Seedher & Bhatia, 2009). These findings suggest that polymeric micelles composed of amphiphilic copolymers are suitable agents for the encapsulation of poorly water-soluble drugs (Yokoyama, 2010). They can be used to target the tumor site by both passive and active mechanisms owing to their inherent properties like size, stability, and EPR effect (Jones & Leroux, 1999). Moreover, the other characteristics of polymeric micelles like functionality at the outer shell are used in the specific targeting of anticancer drugs.

Polymeric micelles conjugated with ligands like antibodies, epidermal growth factors, a2-glycoprotein, transferrin, and folate have been found to be more efficient in targeting specific tumor cells. In addition, external stimuli including heat, ultrasound, and pH can be used to enhance drug accumulation within the cancer cells (Torchilin, 2004; Wu et al., 2010). The novel drug-delivery systems using soluble polymeric micelles for the targeted delivery of anticancer drugs like paclitaxel and docetaxel have shown minimal toxic effects in normal cells with enhanced efficacy (Wang, Petrenko, & Torchilin, 2010). Moreover, study with docetaxel-loaded nanoparticles (Doc-NP) has been found to be more effective in inhibiting the progression of human ovarian cancer growth without much toxicity in normal cells. Thus intratumoral delivery of Doc-NP could become a clinically potential therapeutic regimen for carcinoma of ovary in the future (Zheng et al., 2010).

However, while choosing the polymer, care should be taken to opt for particles with nontoxic and nonimmunogenic properties subsequent to their degradation. Moreover, they should be rapidly eliminated from the body following their degradation without getting accumulated. In the present scenario, the high cost and scarcity of safe polymers with regulatory approval are the major drawbacks associated with their widespread application in nanomedicine. This has been found to be overcome with the use of lipid nanocarriers like solid lipid nanoparticles.

6.2.4 Solid Lipid Nanoparticles

Solid lipid nanoparticles are composed of a solid lipid matrix, 50 nm to 1 μm in size, in which the drug is incorporated. These are a new generation of submicron-sized lipid emulsions where the liquid lipid has been substituted by a solid lipid (Mukherjee, Ray, & Thakur, 2009). They are made of lipids like cholesterol, glycerol, triglycerides, stearic acid, and palmitic acid that are well tolerated by the body. Their biocompatibility, biodegradability, flexibility, and smaller size makes the SLN particles more suitable for drug delivery because they show combined advantages of emulsions and liposomes (Bimbo et al., 2010). They have been developed as an alternative carrier system for emulsions, liposomes, and polymeric nanoparticles in controlled drug delivery.

SLNs have unique features like small size, large surface area, and high drug-loading capacity. Moreover, because they remain solid at room temperature, the mobility of incorporated drugs is reduced, a feature more desirable for controlled drug release. In addition to avoiding aggregation of drug particles within solid lipid complex, different surfactants and polymers are used in SLN with an accepted generally recognized as safe status (Mukherjee, Ray, & Thakur, 2009). Apart from being administered through the oral route, they are also investigated for administration through intravenous, pulmonary, and dermal routes. Thus solid lipid nanoparticles are almost guaranteed to achieve the goal of controlled and site-specific drug delivery, which may open a new approach in the treatment of complex diseases (Mukherjee, Ray, & Thakur, 2009). Moreover, the surface of SLNs can be modified using polymers or copolymers to make the particles steal towards the reticuloendothelial system. PEG stearate-modified SLNs have shown reduced uptake by macrophages in proportion to the length of PEG chain following intraperitoneal injection in mice. Apart from this, SLNs appears to be a suitable nanocarrier for delivering drug to the brain because they have affinity and adhere to the endothelial cells of the BBB (Bondi et al., 2010).

In vitro experiments with insulin-loaded SLNs and wheat germ agglutinin (WGA)-*N*-glutarylphosphatidylethanolamine-modified SLNs have been found to inhibit degradation of insulin from digestive enzymes. The insulin-stabilizing effect of WGA-modified SLNs can be used in the future to design nanocarriers for oral administration of insulin (Zhang et al., 2006). Pharmacokinetic studies following intravenous administration of doxorubicin incorporated in SLNs in rats have revealed higher plasma concentration compared with doxorubincin alone. Similarly a study done on mice following intravenous administration of camptothecin-incorporated SLNs has proven to be a potential targeted drug-delivery system for sustained drug release in the brain because it allows a reduction in the dose along with reduced systemic toxicity (Yang et al., 1999).

6.2.5 Dendrimers

Dendrimers, a unique class of polymers, are highly branched macromolecules with convenient size and shape. They are made of monomers with huge surface areas to which therapeutic agents and other biologically active molecules can be attached. Their well defined structure, size, surface functionalization, and stability make them an attractive vehicle for drug delivery. Drug molecules incorporated in the dendrimers by means of complexation or encapsulation are being investigated for the delivery of antibiotics like penicillin and anticancer drugs (Iwamura, 2006).

Dendrimers with application in drug delivery use polymers like polyamidoamine, melamine, poly(l-glutamic acid), polyethyleneimine, poly(propyleneimine), PEG, and chitin (Labieniec & Watala, 2010; Nomani et al., 2010). In the present scenario, measures are being taken to design a dendrimer with the ability to carry a ligand that recognizes cancer cells, a suitable drug to kill tumor cells, and a contrast agent to detect the progress of the disease. Likewise, investigations are being carried out to design dendrimers to release their contents in the presence of certain trigger factors like a change in pH specific to the tumor environment. Currently, in xenograft models, an amphiphilic telodendrimer system with the ability to self-assemble as multifunctional micelles in aqueous solution has been used in the delivery of the water-insoluble anticancer drug, paclitaxel. This study has demonstrated enhanced uptake of smaller paclitaxel-loaded micelles by the tumor cells, whereas larger micelles are taken up by the liver and lungs. Moreover, paclitaxel-loaded micelles have shown enhanced antitumor efficacy compared with that of paclitaxel (Luo et al., 2010). Such developments in the drug discovery process help the pharmaceutical industry design existing older drugs in newer formulations to reduce the time and cost spent in the discovery of newer molecules.

6.2.6 Virosomes

The inherent capacity of the virus to enter and infect a specific cell in addition to the ability of the viral shell to target cell membrane receptors is being exploited to deliver drugs into the host cytoplasm. The use of virus as a nanocarrier in drug-delivery methods requires the isolation of viral genetic information from the viral shell to avoid infection in the host. One of the methods explored to circumvent infection from the use of viral vectors is to use empty viral shells known as virosomes. The virosomes are usually obtained from influenza virus by solubilizing the viral membrane along with ultracentrifugation and reconstitution of the viral envelope. Because viral derivatives show high immunogenic properties, their synthetic analogues can be used in combination with liposomes to minimize the immune response (Cavanagh et al., 2008; Moser et al., 2007). However, the lipid-based emulsion systems have been shown to be more effective as gene carriers compared with viral vectors because they lack the potential to cause infection in the host. The emulsion-DNA complexes remain stable in the plasma because their surfaces are poorly recognized by the immune-related cells and serum proteins. Moreover, the surfaces of the emulsion complexes can be readily modified by varying the lipid composition (Nam et al., 2009).

6.2.7 Carbon Nanostructures

Carbon nanotubes (CNTs), also known as fullerenes, are hollow, carbon-based, cage-like architectures of nanosize. They have unique electrical and mechanical properties in combination with biological molecules like proteins. Fullerenes exhibit a variety of pharmacological activities and have diverse applications in drug-delivery methods. The most common and stable form of fullerene is Buckminster fullerene, which is composed of 60 carbon atoms. Nanotubes, being about half the diameter of a DNA molecule, are smaller than nanopores and help to identify altered genes and other DNA changes associated with cancer. The common configurations of nanotubes are designed in the form of single-walled nanotubes and multiwall nanotubes (Cveticanin et al., 2010). Their size as well as surface characteristics make them attractive agents for drug delivery (Cranford & Buehler, 2010). Surface-functionalized CNTs linked to peptides can be internalized within mammalian cells for vaccine delivery at the required site. CNTs can be used to transport DNA with the application of molecular dynamics simulations and can potentially become a gene delivery tool in the future. Moreover, tissue-selective targeting and intracellular targeting of mitochondria have been demonstrated with the use of fullerene structures. In addition, experiments with fullerenes have also shown that they exhibit antioxidant and antimicrobial behavior, which offers an opportunity to discover newer drugs to treat oxidative stress and infectious diseases (Markovic & Trajkovic, 2008).

Single-walled carbon nanotubes can be used for the analysis of cellular changes during surface-protein binding and protein-protein binding. Carbon nanotubes provide highly specific electronic sensors for detecting clinically important biomolecules like antibodies associated with human autoimmune diseases. However, in a study conducted in the lung of C57BL/6 mice, it has been shown that inhalation of single-walled CNT can cause inflammatory responses, oxidative stress, collagen deposition, fibrosis, and mutations of K-ras gene (Shvedova et al., 2008). Hence, further studies need to be carried out to explore the possible adverse effects of carbon nanotubes before using them for diagnosis and therapy.

6.2.8 Nanocantilevers

The nanocantilevers are nanostructures with the potential to transform a chemical reaction into mechanical motion in nanoscale. The mechanical motion thus produced can be measured directly by deflecting a beam of light from the surface of the cantilever. The static mode of the cantilever is used to acquire information regarding the presence of certain target molecules in the analyzing sample. The surface stress caused by the adsorption of these molecules results in minute deflections of the cantilever, which correlate directly to the concentration of the target substance. By attaching specific antibodies to cantilevers, the simultaneous imaging of target antigens and identification of antigen-antibody interactions can be achieved (Kooser et al., 2003).

6.2.9 Nanopores

Nanopores are nanodevices with small holes to allow only one strand of DNA to pass through at a time. Apart from confirming DNA sequencing, they monitor the shape and electrical properties of each base of the DNA strand. Because the properties are unique for each of the four bases that make up the genetic code, the passage of DNA through a nanopore can be used to interpret the encoded information, including errors in the genetic code associated with cancer. Better understanding of such errors may help in the discovery and development of new drugs to rectify mutations in the genetic material (Krems, Pershin, & Di, 2010).

6.2.10 Quantum Dots

Quantum dots are tiny crystals that glow and emit light in accordance with their size on being stimulated by ultraviolet light. By combining different-sized quantum dots in a single bead, probes can be created to release distinct colors and intensities of light. On being stimulated by UV light, the crystals of each bead will emit light as a sort of spectral bar code that can recognize a particular region of the DNA. The diversity of quantum dots allows for the creation of unique labels, which may be used to identify the various regions of DNA simultaneously. This may pave the way for a new mode of cancer diagnosis in the future, eliminating the need for biopsy. Currently quantum dots have a major role to play in many biomedical imaging applications. Combining the high sensitivity and specificity of antibodies with the brightness of quantum dots, the ability to detect pathological changes in the tissues can be improved (Hsu & Huang, 2004; Jokerst et al., 2009).

6.3 TARGETED DRUG DELIVERY USING NANOTECHNOLOGY

The practical problems associated with most of the currently available conventional drug-delivery systems include poor bioavailability, degradation of drug, short duration of action, and unwanted side effects. These problems can be overcome by targeting the diseased cells without disturbing the normal cells, which has been one of the major goals of drug discovery and delivery (Jatariu, Popa, & Peptu, 2010). Recent advances in nanotechnology offer the opportunity to get closer to this target by delivering the drug in the right place at the right time (De Jong & Borm, 2008).

Nanostructures possess the competence to protect encapsulated drugs from enzymatic degradation in the gastrointestinal tract and deliver the drug to the required site in the body (Tang & Singh, 2010). They enhance the oral bioavailability of drugs because of their specialized uptake mechanisms called absorptive endocytosis and following the absorption remain in circulation for a longer time. The uptake of nanoparticles has been found to be 15–250 times more than that of microparticles. Thus, by manipulating the characteristics of polymers, the release of drug from nanostructures can be controlled to achieve the desired therapeutic concentration for the required duration.

The delivery of drug to the desired site could be achieved through passive and active targeting mechanisms. A passive targeting approach is achieved owing to the leaky vasculature of tumor cells, which facilitates the accumulation of chemotherapeutic agents in solid tumors compared with the healthy tissue. However, passive targeting with nanoparticles encounters obstacles like mucosal barriers, nonspecific uptake, and degradation before systemic absorption of the drug. Active targeting is different in that it involves surface functionalization of drug carriers with ligands that are selectively recognized by receptors on the surface of the target cells (Torchilin, 2010). Currently, the most important aspects of drug delivery include specific targeting of the diseased tissue with nanoparticles of appropriate size, functionalization with antibodies, and timed release of the drug (Krishna et al., 2009). However, during targeted drug therapy until it binds to the target and releases the drug, care should be taken to avoid diffusion of drug out of the nanoparticle.

Recently, site-specific targeted drug delivery has been needed for finding the effective dose and for disease control (Alexis et al., 2008). Nanomaterials used in targeted drug delivery improve the bioavailability of poorly soluble drugs, especially those with severe adverse effects like paclitaxel, doxorubicin, 5-fluorouracil, and dexamethasone (Gan et al., 2010). Polylactic/glycolic acid- and polylactic acid-based nanoparticles have been formulated to encapsulate dexamethasone, a corticosteroid with antiproliferative and anti-inflammatory effects. This drug binds to the cytoplasmic receptors, and the subsequent drug-receptor complex is transported to the nucleus, resulting in the expression of certain genes that control cell proliferation.

6.4 THE ROLE OF GENOMICS AND PROTEOMICS WITH NANOTECHNOLOGY IN DRUG DISCOVERY

Currently, the pharmaceutical industry is focusing on newer methods of drug discovery and development to develop highly efficacious drugs with reduced time to reach the market. With the advent of nanotechnology, there is great scope for reaching this goal by merging nanotechnology with life sciences and biotechnology. The postgenomic era has revealed many potentially important drug targets for various diseases. However, to exploit their value to the full extent, the efficiency of screening and validation processes needs improvement. Analyses of signaling pathways by nanotechniques might provide new insights into the pathogenesis of disease, identify more efficient biomarkers, and recognize the mechanisms of action of drugs.

Recent advances in nanotechnology include the use of nanobiosensors, nanobiochips, and nanoscale assays in various areas of drug discovery including target identification, validation, assay development, and lead optimization. Similarly, these advances are also used in absorption, distribution, metabolism, excretion, and toxicity (ADME-Tox) studies. Biosensors are better-suited for cell-based assays and the study of receptors because they do not require the receptor to be removed from the cell membrane. The primary application of the biosensors in the present scenario is for the optimization of limited-scope drug libraries against specific targets (Hansen & Thundat, 2005).

A novel nanobiosensor, designed using magnetic nanoparticles has demonstrated rapid screening of telomerase activity in biological samples (annealing telomerase-synthesized TTAGGG repeats) and the ability to switch to their magnetic state. This phenomenon of can be readily measured using magnetic detectors. A high-throughput version of this technique uses magnetic resonance imaging and allows processing of hundreds of samples within a few minutes with ultra high sensitivity. These studies have established and validated a novel, powerful tool for rapidly sensing telomerase activity and for developing similar magnetic nanoparticles. Because elevated telomerase levels are detected in many malignancies, this technique may provide an easy method for diagnosis and therapeutic interventions used in cancer treatment (Xie et al., 2009).

One of the applications of nanobiotechnology in drug discovery involves the use of nanoparticles and nanoscale miniaturizations of microfluidic technologies in proteomics. The development of miniaturized devices enables rapid and direct analysis of the specific binding of small molecules to proteins. These miniaturized devices play a significant role in the screening and discovery of new drugs. In a study, Abl, a protein-tyrosine kinase related to chronic myelogenous leukemia, covalently linked with the silicon nanowires within microfluidic channels was used to assess the concentration-dependent binding of ATP and its inhibition by Gleevec along with four additional small molecules (Wang et al., 2005). These studies have shown that silicon nanowire devices can be used to differentiate the affinities of distinct, small-molecule inhibitors to a receptor. This may pave a way for a newer technology in drug discovery and development.

Presently, nanomedicine is concerned with increasing the understanding of cell biology and providing insight into the processes involved in endocytosis, clathrin-mediated endocytosis (CME), receptor downregulation, nutrient uptake, and the maintenance of signal transmission across nerve-cell junctions (Goh et al., 2010). Studies have shown that disturbances in CME may have a major impact in cancer and neurodegenerative diseases. Live cell imaging, along with a novel fluorescence assay, used to visualize the formation of clathrin-coated vesicles at single clathrin-coated pits within seconds, is a novel assay that is being explored for drug screening. Moreover, such techniques, along with the use of electron microscopy, have been used to detect the newer cellular structures used by viruses to gain entry into the host. An understanding of this process could provide a better strategy for the development of new antiviral drugs (Ting, Chang, & Wang, 2009).

Similarly optical biosensors capable of exploiting surface plasmon resonance (SPR) wave have been used over the past decade to analyze biomolecular interactions. These sensors determine the affinity and kinetics of a wide variety of molecular interactions in real time, without the need for a molecular tag or label. Conventional SPR is applied in specialized biosensing instruments that use expensive sensor chips of limited reusable capacity and require complex chemistry for ligand or protein immobilization. SPR has been successfully applied with colloidal gold particles in buffered solution. This application offers many advantages over conventional SPR because it is cost effective, is easily synthesized, and can be coated with various proteins or protein-ligand complexes or by charge adsorption. Using colloidal gold, the SPR phenomenon can be monitored in any UV-visible spectrophotometer. For high-throughput applications, this technology has been modified in an automated clinical-chemistry analyzer. Currently, the use of metal nanocolloids offers enhanced throughput and flexibility for real-time biomolecular-recognition monitoring, at a sensible cost.

The gold nanoparticle-choline complex, a negatively charged, direct ligand-gated nicotinic ion-channel blocker, has been found to block the current evoked by acetylcholine in nicotinic acetylcholine receptors (nAChRs). Studies carried out in adrenal-gland perfusion with this complex have been found to block nAChRs and diminish the release of catecholamines by about 75%. In an *in vivo* study, following injection of this complex, muscle relaxation was produced in rats, thus establishing their role on nicotinic receptors. Such results may help in the application of gold nanoparticle-choline complex as a newer direct ion-channel blocker for muscle relaxation (Chin et al., 2010).

The surface modification of gold nanoparticles using a biocompatible polymer like chitosan may improve its binding properties with biomolecules. This has been proven in a study with chitosan-reduced gold nanoparticles designed to deliver oral and nasal administration of insulin to improve the pharmacodynamic activity. Such chitosan-reduced gold nanoparticles loaded with insulin may play a promising role in controlling the postprandial hyperglycemia (Bhumkar et al., 2007). Similarly, a study done to evaluate the toxic effects of different doses of gold nanoparticles following intraperitoneal injection in mice for 8 days showed that they were able to cross the blood-brain barrier and accumulate in the neural tissue with no evidence of toxicity (Lasagna-Reeves et al., 2010). These findings promise a potential role for gold-nanoparticle complex in drug discovery over the coming years.

Nanofluidics, the latest nanotechnique, is being used to detect the kinetic parameters of protease mixtures, protease-substrate interactions, and high-throughput screening reactions. Similarly, nanoarrays, the modified miniaturization of microarrays, are being used to assess the interactions between individual molecules at a cellular level. The major aim of such arrays is to produce so-called laboratories-on-a-chip, with a million test reactors, each capable of screening an individual drug. The chips could significantly increase the number of experiments possible with a minimal amount of protein available (Yue, Stachowiak, & Majumdar, 2004). This device may be used to study the membrane proteins and in future may pave the way for designing drugs that act by controlling proteins in cell membranes (Tian et al., 2005; Yue, Stachowiak, & Majumdar, 2004).

The chips can also be used to establish bioaffinity tests for proteins, nucleic acids, and receptor-ligand pairs (Tian et al., 2005). In the present scenario, such miniature devices are being constructed to study synthetic cell membranes in an effort to hasten the discovery of new drugs for various infectious diseases and cancer. Investigations are also being carried out to discover drugs with the ability to deactivate the efflux pumps present in the cell membrane, resulting in increased intracellular concentration as well as duration of action of chemotherapeutic agents.

Currently, progress with biochips, a nanotechnology-enabled electrical detection system, is the most interesting field in drug discovery and development. Because electrical detection systems permit the detection of a binding process directly via an electrode, this avoids the need for a complicated optical instrument. In the present scenario, with nanoarrays, nanowire sensors, surface plasmon resonance sensors, and cantilever sensors all in the early stages of development, there is a long way to go in the field of biomedical research and drug discovery (Barnett & Goldys, 2010; Live, Bolduc, & Masson, 2010; Shukla et al., 2010).

6.5 NANOTECHNOLOGY IN CANCER DIAGNOSIS AND CHEMOTHERAPY

The available therapeutic modalities for cancer, including surgical removal, radiotherapy, and chemotherapy, are not always associated with a complete cure. In addition, they may cause serious adverse effects and reduce the quality of life among cancer patients. Various nanoparticles like nanoshells and gold nanoparticles are being investigated for the early detection and treatment of cancer.

Nanoshells are tiny beads coated with gold designed to absorb a specific wavelength of light depending on their thickness. The most useful nanoshells are those that absorb near infrared light that can easily penetrate several centimeters in human tissues and create an intense heat that is lethal to cells. This has been proven in cell culture studies, in which the heat generated by the light-absorbing nanoshells have been shown to kill tumor cells while leaving adjacent normal cells intact (Zhang et al., 2010). Administered systemically, nanoshells have been shown to absorb light in the near-infrared (NIR) region and accumulate in the tumor because of the EPR effect. They also induce photothermal ablation of the tumor following irradiation with an NIR laser. Their tumor specificity can be increased by functionalizing the nanoshell surface with tumor-targeting moieties. Nanoshells can also be made to scatter light and hence may be used in various imaging modalities such as dark-field microscopy and optical coherence tomography (Morton et al., 2010).

Compared with the current methods of detecting cancer, gold nanoparticles, used as ultrasensitive fluorescent probes to detect cancer biomarkers in blood, have been found to be many times more sensitive. They could also be employed for direct detection of viral or bacterial DNA and promise a suitable probe for biomedical applications because they can be easily primed without getting burnt following chronic exposure to light. They also have the potential to detect carcino embryonic antigen and alpha feto protein, the most important biomarkers in the diagnosis of various malignancies, including liver, lung, and breast cancers.

6.6 CURRENT TRENDS OF NANOTECHNOLOGY IN THERAPEUTICS

Cancer, one of the leading causes of death worldwide, is likely get newer modes of therapies in the near future with the progress of nanotechnology, which raises new possibilities in the diagnosis and treatment of human cancers. The treatment of cancer using nanoparticles by means of targeted drug delivery is one of the latest achievements in the field of medicine. With more than 10 million new cases every year, cancer has become one of the most overwhelming diseases worldwide. The potential for nanoparticles in cancer-drug delivery is infinite, with novel applications like multifunctional nanoparticles being explored to play a significant role.

The most common cancer seen in women, breast carcinoma, is currently treated with chemotherapeutic agents like anthracyclines and taxanes during the early and advanced stages of the disease. However, the major problem with the use of taxanes like paclitaxel is their poor lipid solubility, which is enhanced by the addition of cremophor solvent. The cremophor solvent causes infusion-related hypersensitivity reactions, one of the major adverse effects of paclitaxel infusion in ovarian cancer. This adverse reaction has been overcome with Abraxane (ABI-007), a solvent-free albumin bound paclitaxel of nanosize, which has shown a better response rate along with safety profile compared with the use of paclitaxel alone. Currently abraxane has been approved by U.S. Food and Drug Administration for treatment of metastatic breast cancer following the failure of first line chemotherapeutic drugs (Petrelli, Borgonovo, & Barni, 2010; Vishnu & Roy, 2010). Similarly, in a study done to evaluate the safety and pharmacokinetics of albumin-bound paclitaxel in patients with advanced solid tumors and hepatic dysfunction, an acceptable tolerability was demonstrated compared with cremophor-based paclitaxel (Biakhov et al., 2010). Recently Genexol-PM, a novel Cremophor EL-free polymeric micelle formulation of paclitaxel, has been found to be a promising nanodrug for metastatic breast cancer. In a phase II trial, Genexol-PM was found to be superior to paclitaxel because the delivery of a high dose of drug at the desired site resulted in less toxicity (Lee et al., 2008). Similarly, another trial, designed to evaluate the efficacy and safety of the combination of Genexol-PM and cisplatin for the treatment of advanced non-small-cell lung cancer (NSCLC), showed significant antitumor activity. Moreover, with the use of Genexol-PM, higher doses of paclitaxel can be administered without any significant toxicity (Kim et al., 2007). However, this drug has been found to be associated with other dose-dependent adverse effects like myalgia, neuropathy, and neutropenia (Lee et al., 2008).

Ovarian cancer, which can spread all over the peritoneal cavity, is currently being treated by direct injection of anticancer drugs into the peritoneal cavity. The administration of drug through an intraperitoneal route carries the risk of infection leading to peritonitis if aseptic precautions are not strictly followed. The nanoparticles equipped with positively charged, biodegradable polymers injected into or near the tumor cells have the potential to treat ovarian cancer and carcinoma of the prostate (Yeo & Xu, 2009).

Apart from cancer, the most common illness seen among the elderly is diabetes mellitus, and the major obstacle in its treatment is the subcutaneous administration of insulin, resulting in pain at the injection site. This results in decreased compliance and quality of life in diabetic patients because the injection has to be administered once or twice daily for the remainder of the patient's life. To overcome this drawback, the possibility of administering insulin via oral and nasal routes has been investigated over the past few years, but without success because of enzymatic degradation and reduced mucosal absorption. With the development of biocompatible and biodegradable polymers, oral bioavailability of insulin can be enhanced by promoting its uptake via a transcellular or paracellular pathway (Ramesan & Sharma, 2009).

Biodegradable nanoparticles loaded with insulin-phospholipid complex have shown controlled drug release characterized by an initial burst and subsequent delayed release in both acidic and basic mediums (Cui et al., 2006). Recently chitosan and its salts have been tried as functional recipients for delivering insulin via oral, nasal, and transdermal routes (Wong, 2009). Nanoparticles loaded with insulin like Actrapid and NovoRapid using biodegradable polyester (polyepsilon-caprolactone) and a polycationic nonbiodegradable acrylic polymer (Eudragit RS) showed a burst release of insulin from nanoparticles. This study has shown decreased glycemia with orally administered NovoRapid-loaded nanoparticles in diabetic rats. This finding may be of interest for the design and discovery of oral insulin formulation in the near future (Socha et al., 2009).

Similarly, in another study, nanoparticles made of biodegradable polyester (poly(epsilon-caprolactone)) and a polycationic nonbiodegradable acrylic polymer (Eudragit RS) were able to release about 70% of the short-acting insulin analogue aspart-insulin following their oral administration in a neutral medium for more than 24 h. Given orally in diabetic rats, insulin-loaded nanoparticles were found to reduce glycemia for a prolonged period with improved glycemic response for 12–24 hours. Moreover, they were able to preserve the biological activity of the insulin analogue aspart. However, the postprandial peak suppression was prolonged for more than 24 hours in comparison with 6–8 hours with regular insulin because of their monomeric configuration (Damge et al., 2010).

6.7 ADVERSE EFFECTS OF NANOPARTICLES

The recent developments in the field of nanotechnology have the potential to introduce huge amounts of nanoparticles in the atmosphere, leading to the pollution of air and water. There is a growing concern regarding the health hazards posed to human beings and animals following exposure to inhalation of nanoparticles. Moreover, inhaling nanoparticles may show an occupational hazard, especially among the workers of industries using techniques for designing and manufacturing nanoparticles. Inhalation being the major route of exposure to nanoparticles, pulmonary functions may be affected with chronic exposure. Moreover, the absorption of deposited nanoparticles from the lung into the blood may result in systemic side effects. Apart from inhalation, nanoparticles may also be absorbed through skin, eye, and gastrointestinal tract from the surroundings (Mostofi et al., 2010).

Previous studies on ultrafine particles, which are the same size as nanoparticles, have revealed acute and chronic toxic effects. Studies conducted on animals to detect the acute toxicity effects of nanoparticles have shown damage to various organs. The toxicity studies conducted in rats and mice have shown increases in pulmonary and cardiovascular diseases along with reduced immunity following exposure to nanoparticles (Chen et al., 2009; Kreyling et al., 2009; Li et al., 2009).

Similarly, studies done in nanoparticles have shown that a particle of nanosize is more toxic compared with that of its counterpart in micro size. This is attributed to the larger surface area of the nanoparticles because the increase in their surface reactivity leads to accentuated toxic effects. Moreover, it has been found that the toxicity of the nanoparticles is influenced by their physicochemical properties, structure, aggregating property, and surface coatings (Mostofi et al., 2010). Among the various toxicities associated with exposure to nanoparticles, cardiovascular complications like tachycardia, vasoconstriction, raised systolic blood pressure, and increased plasma viscosity have been found to be more fatal. Patients with a history of ischemic heart disease, arrhythmias, and congestive heart failure are prone to such risks.

6.8 SUMMARY

In the present scenario, nanomedicine has emerged as one of the major applications of nanotechnology with more emphasis towards drug-delivery methods. Among the various branches of nanomedicine, targeted drug delivery has become an essential part of drug discovery, to target specifically the diseased cells while shielding the healthy cells from their harmful effects. Currently, nanoparticles are being used extensively because of their particle size and surface characteristics, which help them to achieve specific drug targeting. They aid in the controlled as well as sustained release of drugs during absorption and transport by means of passive as well as active transport. Nanoparticles can be administered through oral, nasal, parenteral, and intraocular routes to deliver drugs to the unreachable areas in the body. In addition, the nanosize of the drug carrier allows the drug to gain access into the cell and various cellular organelles including the nucleus.

The use of nanotechnology in the medical field, especially in drug discovery and newer drug delivery, is set to extend rapidly. Currently, many substances are under investigation for drug delivery with an emphasis towards cancer chemotherapy. The substances under investigation for the design of nanoparticles for drug delivery include albumin, gelatin liposomes, polymers, and solid metals like gold, iron, and silica. Nanostructures like polymers can be used in the management of neurodegenerative, cerebrovascular, and inflammatory diseases because of their ability to penetrate the blood-brain barrier.

One of the advantages of drug-delivery methods using nanocarriers is the potential to reformulate the existing old drugs with little efficacy and more side effects into more efficacious, potent, safer drugs with decreased adverse effects. In due course, nanotechnology may be used to enhance the performance of drugs that are unable to get approval in the clinical trials because of their poor solubility and size. The surface properties of nanoparticles can be modified in such a way that they are not recognized by the immune system and remain in circulation for a long time. In the present scenario, nanocarriers hold the promise to deliver drugs, genes, and nucleic acids to various areas of the body, overcoming the physiologic barriers to delivery. However, the storage, handling, and administration of nanoformulations may be difficult because of their smaller size. Another disadvantage of nanoformulations is that they may not be of much use in designing less potent drugs. Moreover, they may cause genetic damage and mutations following chronic exposure.

6.9 CONCLUSION

Recent progress and advances in the field of nanotechnology have already added nanoparticles to the environment; the amount is expected get higher in the future. Moreover, nanoparticles used to reduce drug toxicity and side effects may themselves impose newer unknown risks in the patients. These risks may manifest in the form of cardiovascular, respiratory, malignant, and genetic disorders.

Epidemiological as well as toxicological studies have provided sufficient proof for the association of increased cardiovascular morbidity and mortality with elevated levels of particulate matter in the atmosphere. Cardiovascular complications like tachycardia, vasoconstriction, increased systolic blood pressure, and increased plasma viscosity were found to be related to the exposure to nanoparticles. This may cause worsening of cardiac function, especially in patients with ischemic heart disease, arrhythmias, and congestive heart failure. Because nanoparticles mimic particulate matter in size, studies are being carried out to find out their adverse effects, especially with regard to cardiovascular side effects. Current investigations provide evidence for the transfer of nanosized particles from the lungs to liver, heart, spleen, and brain through circulation. Hence, to protect living organisms from the potential dangers of nanomaterials, more studies need to be carried out in the future to detect the toxicity of nanoparticles following acute and chronic exposure. To conclude, in the present scenario, the medical fraternity has a major responsibility to protect society from the harmful effects of nanoparticles while reaping the benefits of nanotechnology.

REFERENCES

Alam, M. I., Beg, S., Samad, A., Baboota, S., Kohli, K., Ali, J., Ahuja, A., Akbar, M. (2010). Strategy for effective brain drug delivery. *Eur. J. Pharm. Sci.* 40: 385–403.

Alexis, F., Pridgen, E., Molnar, L. K., Farokhzad, O. C. (2008). Factors affecting the clearance and biodistribution of polymeric nanoparticles. *Mol. Pharmacol.* 5: 505–515.

Anabousi, S., Kleemann, E., Bakowsky, U., Kissel, T., Schmehl, T., Gessler, T., Seeger, W., Lehr, C. M., Ehrhardt, C. (2006). Effect of PEGylation on the stability of liposomes during nebulisation and in lung surfactant. *J. Nanosci. Nanotechnol.* 6: 3010–3016.

Barnett, A., Goldys, E. M. (2010). Modeling of the SPR resolution enhancement for conventional and nanoparticle inclusive sensors by using statistical hypothesis testing. *Opt. Express.* 18: 9384–9397.

Bhaskar, S., Tian, F., Stoeger, T., Kreyling, W., de la Fuente, J. M., Grazu, V., Borm, P., Estrada, G., Ntziachristos, V., Razansky, D. (2010). Multifunctional nanocarriers for diagnostics, drug delivery and targeted treatment across blood-brain barrier: Perspectives on tracking and neuroimaging. *Part. Fibre Toxicol.* 7: 3.

Bhumkar, D. R., Joshi, H. M., Sastry, M., Pokharkar, V. B. (2007). Chitosan reduced gold nanoparticles as novel carriers for transmucosal delivery of insulin. *Pharm. Res.* 24: 1415–1426.

Biakhov, M. Y., Kononova, G. V., Iglesias, J., Desai, N., Bhar, P., Schmid, A. N., Loibl, S. (2010). nab-Paclitaxel in patients with advanced solid tumors and hepatic dysfunction: A pilot study. *Expert Opin. Drug Saf.* 9: 515–523.

Bimbo, L. M., Sarparanta, M., Santos, H. A., Airaksinen, A. J., Makila, E., Laaksonen, T., Peltonen, L., Lehto, V. P., Hirvonen, J., Salonen, J. (2010). Biocompatibility of thermally hydrocarbonized porous silicon nanoparticles and their biodistribution in rats. *ACS Nano* 4: 3023–3032.

Bondi, M. L., Craparo, E. F., Giammona, G., Drago, F. (2010). Brain-targeted solid lipid nanoparticles containing riluzole: Preparation, characterization and biodistribution. *Nanomedicine* 5: 25–32.

Cavanagh, D. R., Remarque, E. J., Sauerwein, R. W., Hermsen, C. C., Luty, A. J. (2008). Influenza virosomes: A flu jab for malaria? *Trends Parasitol.* 24: 382–385.

Chen, J., Dong, X., Zhao, J., Tang, G. (2009). In vivo acute toxicity of titanium dioxide nanoparticles to mice after intraperitioneal injection. *J. Appl. Toxicol.* 29: 330–337.

Chin, C., Kim, I. K., Lim, D. Y., Kim, K. S., Lee, H. A., Kim, E. J. (2010). Gold nanoparticle-choline complexes can block nicotinic acetylcholine receptors. *Int. J. Nanomedicine* 5: 315–321.

Cho, K., Wang, X., Nie, S., Chen, Z. G., Shin, D. M. (2008). Therapeutic nanoparticles for drug delivery in cancer. *Clin. Cancer Res* 14: 1310–1316.

Cortivo, R., Vindigni, V., Iacobellis, L., Abatangelo, G., Pinton, P., Zavan, B. (2010). Nanoscale particle therapies for wounds and ulcers. *Nanomedicine* 5: 641–656.

Cranford, S. W., Buehler, M. J. (2010). In silico assembly and nanomechanical characterization of carbon nanotube buckypaper. *Nanotechnology* 21: 265706.

Cui, F., Shi, K., Zhang, L., Tao, A., Kawashima, Y. (2006). Biodegradable nanoparticles loaded with insulin-phospholipid complex for oral delivery: Preparation, in vitro characterization and in vivo evaluation. *J. Control Release* 114: 242–250.

Cveticanin, J., Joksic, G., Leskovac, A., Petrovic, S., Sobot, A. V., Neskovic, O. (2010). Using carbon nanotubes to induce micronuclei and double strand breaks of the DNA in human cells. *Nanotechnology* 21: 015102.

Damge, C., Socha, M., Ubrich, N., Maincent, P. (2010). Poly(epsilon-caprolactone)/eudragit nanoparticles for oral delivery of aspart-insulin in the treatment of diabetes. *J. Pharm. Sci.* 99: 879–889.

De Jong, W. H., Borm, P. J. (2008). Drug delivery and nanoparticles: Applications and hazards. *Int. J. Nanomedicine* 3: 133–149.

Dubey, V., Nahar, M., Mishra, D., Mishra, P., Jain, N. K. (2010). Surface structured liposomes for site specific delivery of an antiviral agent-indinavir. *J. Drug Target.* 19: 258–269.

Gan, L., Han, S., Shen, J., Zhu, J., Zhu, C., Zhang, X., Gan, Y. (2010). Self-assembled liquid crystalline nanoparticles as a novel ophthalmic delivery system for dexamethasone: Improving preocular retention and ocular bioavailability. *Int. J. Pharm.* 396: 179–187.

Garnett, M. C., Ferruti, P., Ranucci, E., Suardi, M. A., Heyde, M., Sleat, R. (2009). Sterically stabilized self-assembling reversibly cross-linked polyelectrolyte complexes with nucleic acids for environmental and medical applications. *Biochem. Soc. Trans.* 37: 713–716.

Goh, L. K., Huang, F., Kim, W., Gygi, S., Sorkin, A. (2010). Multiple mechanisms collectively regulate clathrin-mediated endocytosis of the epidermal growth factor receptor. *J. Cell Biol.* 189: 871–883.

Hamill, R. J., Sobel, J. D., El-Sadr, W., Johnson, P. C., Graybill, J. R., Javaly, K., Barker, D. E. (2010). Comparison of 2 doses of liposomal amphotericin B and conventional amphotericin B deoxycholate for treatment of AIDS-associated acute cryptococcal meningitis: A randomized, double-blind clinical trial of efficacy and safety. *Clin. Infect. Dis.* 51: 225–232.

Hansen, K. M., Thundat, T. (2005). Microcantilever biosensors. *Methods* 37: 57–64.

Harrington, K. J., Mohammadtaghi, S., Uster, P. S., Glass, D., Peters, A. M., Vile, R. G., Stewart, J. S. (2001). Effective targeting of solid tumors in patients with locally advanced cancers by radiolabeled pegylated liposomes. *Clin. Cancer Res.* 7: 243–254.

Hsu, H. Y., Huang, Y. Y. (2004). RCA combined nanoparticle-based optical detection technique for protein microarray: A novel approach. *Biosens. Bioelectron.* 20: 123–126.

Iwamura, M. (2006). [Dendritic systems for drug delivery applications]. *Nippon Rinsho* 64: 231–237.

Jatariu, A. N., Popa, M., Peptu, C. A. (2010). Different particulate systems: Bypass the biological barriers? *J. Drug Target.* 18: 243–253.

Jokerst, J. V., Raamanathan, A., Christodoulides, N., Floriano, P. N., Pollard, A. A., Simmons, G. W., Wong, J., Gage, C., Furmaga, W. B., Redding, S. W., McDevitt, J. T. (2009). Nano-bio-chips for high performance multiplexed protein detection: Determinations of cancer biomarkers in serum and saliva using quantum dot bioconjugate labels. *Biosens. Bioelectron.* 24: 3622–3629.

Jones, M., Leroux, J. (1999). Polymeric micelles: A new generation of colloidal drug carriers. *Eur. J. Pharm. Biopharm.* 48: 101–111.

Kedar, U., Phutane, P., Shidhaye, S., Kadam, V. (2010). Advances in polymeric micelles for drug delivery and tumor targeting. *Nanomedicine.* 6: 714–729.

Kim, D. W., Kim, S. Y., Kim, H. K., Kim, S. W., Shin, S. W., Kim, J. S., Park, K., Lee, M. Y., Heo, D. S. (2007). Multicenter phase II trial of Genexol-PM, a novel Cremophor-free, polymeric micelle formulation of paclitaxel, with cisplatin in patients with advanced non-small-cell lung cancer. *Ann. Oncol.* 18: 2009–2014.

Kooser, A., Manygoats, K., Eastman, M. P., Porter, T. L. (2003). Investigation of the antigen antibody reaction between anti-bovine serum albumin (a-BSA) and bovine serum albumin (BSA) using piezoresistive microcantilever based sensors. *Biosens. Bioelectron.* 19: 503–508.

Krems, M., Pershin, Y. V., Di, V. M. (2010). Ionic memcapacitive effects in nanopores. *Nano. Lett.* 10: 2674–2678.

Kreyling, W. G., Semmler-Behnke, M., Seitz, J., Scymczak, W., Wenk, A., Mayer, P., Takenaka, S., Oberdorster, G. (2009). Size dependence of the translocation of inhaled iridium and carbon nanoparticle aggregates from the lung of rats to the blood and secondary target organs. *Inhal. Toxicol.* 21 (Suppl. 1): 55–60.

Krishna, A. D., Mandraju, R. K., Kishore, G., Kondapi, A. K. (2009). An efficient targeted drug delivery through apotransferrin loaded nanoparticles. *PLoS One* 4: e7240.

Labieniec, M., Watala, C. (2010). Use of poly(amido)amine dendrimers in prevention of early nonenzymatic modifications of biomacromolecules. *Biochimie* 92: 1296–1305.

Lasagna-Reeves, C., Gonzalez-Romero, D., Barria, M. A., Olmedo, I., Clos, A., Sadagopa, R. V., Urayama, A., Vergara, L., Kogan, M. J., Soto, C. (2010). Bioaccumulation and toxicity of gold nanoparticles after repeated administration in mice. *Biochem. Biophys. Res. Commun.* 393: 649–655.

Lee, K. S., Chung, H. C., Im, S. A., Park, Y. H., Kim, C. S., Kim, S. B., Rha, S. Y., Lee, M. Y., Ro, J. (2008). Multicenter phase II trial of Genexol-PM, a Cremophor-free, polymeric micelle formulation of paclitaxel, in patients with metastatic breast cancer. *Breast Cancer Res. Treat.* 108: 241–250.

Li, C., Liu, H., Sun, Y., Wang, H., Guo, F., Rao, S., Deng, J., Zhang, Y., Miao, Y., Guo, C., Meng, J., Chen, X., Li, L., Li, D., Xu, H., Wang, H., Li, B., Jiang, C. (2009). PAMAM nanoparticles promote acute lung injury by inducing autophagic cell death through the Akt-TSC2-mTOR signaling pathway. *J. Mol. Cell. Biol.* 1: 37–45.

Linkov, I., Satterstrom, F. K., Corey, L. M. (2008). Nanotoxicology and nanomedicine: Making hard decisions. *Nanomedicine* 4: 167–171.

Lipka, J., Semmler-Behnke, M., Sperling, R. A., Wenk, A., Takenaka, S., Schleh, C., Kissel, T., Parak, W. J., Kreyling, W. G. (2010). Biodistribution of PEG-modified gold nanoparticles following intratracheal instillation and intravenous injection. *Biomaterials* 31: 6574–6581.

Live, L. S., Bolduc, O. R., Masson, J. F. (2010). Propagating surface plasmon resonance on microhole arrays. *Anal. Chem.* 82: 3780–3787.

Luderer, F., Lobler, M., Rohm, H. W., Gocke, C., Kunna, K., Kock, K., Kroemer, H. K., Weitschies, W., Schmitz, K. P., Sternberg, K. (2010). Biodegradable sirolimus-loaded poly(lactide) nanoparticles as drug delivery system for the prevention of in-stent restenosis in coronary stent application. *J. Biomater. Appl.* 25: 851–875.

Luo, J., Xiao, K., Li, Y., Lee, J. S., Shi, L., Tan, Y. H., Xing, L., Holland, C. R., Liu, G. Y., Lam, K. S. (2010). Well-defined, Size-tunable, multifunctional micelles for efficient paclitaxel delivery for cancer treatment. *Bioconjug. Chem.*

Majidi, J., Barar, J., Baradaran, B., Abdolalizadeh, J., Omidi, Y. (2009). Target therapy of cancer: Implementation of monoclonal antibodies and nanobodies. *Hum. Antibodies* 18: 81–100.

Markovic, Z., Trajkovic, V. (2008). Biomedical potential of the reactive oxygen species generation and quenching by fullerenes (C60). *Biomaterials* 29: 3561–3573.

Morton, J. G., Day, E. S., Halas, N. J., West, J. L. (2010). Nanoshells for photothermal cancer therapy. *Methods Mol. Biol.* 624: 101–117.

Moser, C., Amacker, M., Kammer, A. R., Rasi, S., Westerfeld, N., Zurbriggen, R. (2007). Influenza virosomes as a combined vaccine carrier and adjuvant system for prophylactic and therapeutic immunizations. *Expert Rev. Vaccines* 6: 711–721.

Mostofi, R., Wang, B., Haghighat, F., Bahloul, A., Jaime, L. (2010). Performance of mechanical filters and respirators for capturing nanoparticles: Limitations and future direction. *Ind. Health* 48: 296–304.

Mukherjee, S., Ray, S., Thakur, R. S. (2009). Solid lipid nanoparticles: A modern formulation approach in drug delivery system. *Indian J. Pharm. Sci.* 71: 349–358.

Nam, H. Y., Park, J. H., Kim, K., Kwon, I. C., Jeong, S. Y. (2009). Lipid-based emulsion system as non-viral gene carriers. *Arch. Pharm. Res.* 32: 639–646.

Nobs, L., Buchegger, F., Gurny, R., Allemann, E. (2004). Current methods for attaching targeting ligands to liposomes and nanoparticles. *J. Pharm. Sci.* 93: 1980–1992.

Nomani, A., Haririan, I., Rahimnia, R., Fouladdel, S., Gazori, T., Dinarvand, R., Omidi, Y., Azizi, E. (2010). Physicochemical and biological properties of self-assembled antisense/poly(amidoamine) dendrimer nanoparticles: The effect of dendrimer generation and charge ratio. *Int. J. Nanomedicine* 5: 359–369.

Patel, M. M., Goyal, B. R., Bhadada, S. V., Bhatt, J. S., Amin, A. F. (2009). Getting into the brain: Approaches to enhance brain drug delivery. *CNS Drugs* 23: 35–58.

Petrelli, F., Borgonovo, K., Barni, S. (2010). Targeted delivery for breast cancer therapy: The history of nanoparticle-albumin-bound paclitaxel. *Expert Opin. Pharmacother.* 11: 1413–1432.

Putz, G., Schmah, O., Eckes, J., Hug, M. J., Winkler, K. (2010). Controlled application and removal of liposomal therapeutics: Effective elimination of pegylated liposomal doxorubicin by double-filtration plasmapheresis in vitro. *J. Clin. Apher* 25: 54–62.

Ramesan, R. M., Sharma, C. P. (2009). Challenges and advances in nanoparticle-based oral insulin delivery. *Expert Rev. Med. Devices* 6: 665–676.

Rawat, A., Vaidya, B., Khatri, K., Goyal, A. K., Gupta, P. N., Mahor, S., Paliwal, R., Rai, S., Vyas, S. P. (2007). Targeted intracellular delivery of therapeutics: An overview. *Pharmazie* 62: 643–658.

Sarmento, B., Ribeiro, A., Veiga, F., Sampaio, P., Neufeld, R., Ferreira, D. (2007). Alginate/chitosan nanoparticles are effective for oral insulin delivery. *Pharm. Res.* 24: 2198–2206.

Seedher, N., Bhatia, S. (2009). Competition between COX-2 inhibitors and some other drugs for binding sites on human serum albumin. *Drug Metabol. Drug Interact.* 24: 37–56.

Shono, Y., Jantratid, E., Kesisoglou, F., Reppas, C., Dressman, J. B. (2010). Forecasting in vivo oral absorption and food effect of micronized and nanosized aprepitant formulations in humans. *Eur. J. Pharm. Biopharm.* 76: 95–104.

Shukla, S., Kim, K. T., Baev, A., Yoon, Y. K., Litchinitser, N. M., Prasad, P. N. (2010). Fabrication and characterization of gold-polymer nanocomposite plasmonic nanoarrays in a porous alumina template. *ACS Nano.* 4: 2249–2255.

Shvedova, A. A., Kisin, E., Murray, A. R., Johnson, V. J., Gorelik, O., Arepalli, S., Hubbs, A. F., Mercer, R. R., Keohavong, P., Sussman, N., Jin, J., Yin, J., Stone, S., Chen, B. T., Deye, G., Maynard, A., Castranova, V., Baron, P. A., Kagan, V. E. (2008). Inhalation vs. aspiration of single-walled carbon nanotubes in C57BL/6 mice: Inflammation, fibrosis, oxidative stress, and mutagenesis. *Am. J. Physiol. Lung Cell Mol. Physiol.* 295: L552–L565.

Smith, N. B. (2007). Perspectives on transdermal ultrasound mediated drug delivery. *Int. J. Nanomedicine* 2: 585–594.

Socha, M., Sapin, A., Damge, C., Maincent, P. (2009). Influence of polymers ratio on insulin-loaded nanoparticles based on poly-epsilon-caprolactone and Eudragit RS for oral administration. *Drug Deliv.* 16: 430–436.

Su, Y. C., Chen, B. M., Chuang, K. H., Cheng, T. L., Roffler, S. R. (2010). Sensitive quantification of PEGylated compounds by second-generation anti-poly(ethylene glycol) monoclonal antibodies. *Bioconjug. Chem.* 21: 1264–1270.

Tang, Y., Singh, J. (2010). Thermosensitive drug delivery system of salmon calcitonin: In vitro release, in vivo absorption, bioactivity and therapeutic efficacies. *Pharm. Res.* 27: 272–284.

Teli, M. K., Mutalik, S., Rajanikant, G. K. (2010). Nanotechnology and nanomedicine: Going small means aiming big. *Curr. Pharm. Des.* 16: 1882–1892.

Tian, F., Hansen, K. M., Ferrell, T. L., Thundat, T., Hansen, D. C. (2005). Dynamic microcantilever sensors for discerning biomolecular interactions. *Anal. Chem.* 77: 1601–1606.

Tian, W., Ying, X., Du, J., Guo, J., Men, Y., Zhang, Y., Li, R. J., Yao, H. J., Lou, J. N., Zhang, L. R., Lu, W. L. (2010). Enhanced efficacy of functionalized epirubicin liposomes in treating brain glioma-bearing rats. *Eur. J. Pharm. Sci.* 197: 3–53.

Ting, G., Chang, C. H., Wang, H. E. (2009). Cancer nanotargeted radiopharmaceuticals for tumor imaging and therapy. *Anticancer. Res.* 29: 4107–4118.

Torchilin, V. P. (2004). Targeted polymeric micelles for delivery of poorly soluble drugs. *Cell. Mol. Life Sci.* 61: 2549–2559.

Torchilin, V. P. (2010). Passive and active drug targeting: Drug delivery to tumors as an example. *Handb. Exp. Pharmacol.* 3: 53.

Vigna, E., De Vivo, A., Gentile, M., Morelli, R., Lucia, E., Mazzone, C., Recchia, A. G., Vianelli, N., Morabito, F. (2010). Liposomal amphotericin B in the treatment of visceral leishmaniasis in immunocompromised patients. *Transpl. Infect. Dis.* 12: 428–431.

Vishnu, P., Roy, V. (2010). nab-paclitaxel: A novel formulation of taxane for treatment of breast cancer. *Womens Health* 6: 495–506.

Wang, M., Thanou, M. (2010). Targeting nanoparticles to cancer. *Pharmacol. Res.* 62: 90–99.

Wang, T., Petrenko, V. A., Torchilin, V. P. (2010). Paclitaxel-loaded polymeric micelles modified with MCF-7 cell-specific phage protein: Enhanced binding to target cancer cells and increased cytotoxicity. *Mol. Pharm.* 7: 1007–1014.

Wang, W. U., Chen, C., Lin, K. H., Fang, Y., Lieber, C. M. (2005). Label-free detection of small-molecule-protein interactions by using nanowire nanosensors. *Proc. Natl. Acad. Sci. U.S.A.* 102: 3208–3212.

Wittgen, B. P., Kunst, P. W., van der Born, K., van Wijk, A. W., Perkins, W., Pilkiewicz, F. G., Perez-Soler, R., Nicholson, S., Peters, G. J., Postmus, P. E. (2007). Phase I study of aerosolized SLIT cisplatin in the treatment of patients with carcinoma of the lung. *Clin. Cancer Res.* 13: 2414–2421.

Wong, H. L., Chattopadhyay, N., Wu, X. Y., Bendayan, R. (2010). Nanotechnology applications for improved delivery of antiretroviral drugs to the brain. *Adv. Drug Deliv. Rev.* 62: 503–517.

Wong, T. W. (2009). Chitosan and its use in design of insulin delivery system. *Recent Pat. Drug Deliv. Formul.* 3: 8–25.

Wu, H. C., Chang, D. K. (2010). Peptide-mediated liposomal drug delivery system targeting tumor blood vessels in anticancer therapy. *J. Oncol.* 2010: 723–798.

Wu, X. L., Kim, J. H., Koo, H., Bae, S. M., Shin, H., Kim, M. S., Lee, B. H., Park, R. W., Kim, I. S., Choi, K., Kwon, I. C., Kim, K., Lee, D. S. (2010). Tumor-targeting peptide conjugated pH-responsive micelles as a potential drug carrier for cancer therapy. *Bioconjug. Chem.* 21: 208–213.

Wu, Y., Loper, A., Landis, E., Hettrick, L., Novak, L., Lynn, K., Chen, C., Thompson, K., Higgins, R., Batra, U., Shelukar, S., Kwei, G., Storey, D. (2004). The role of biopharmaceutics in the development of a clinical nanoparticle formulation of MK-0869: A beagle dog model predicts improved bioavailability and diminished food effect on absorption in human. *Int. J. Pharm.* 285: 135–146.

Xie, M., Hu, J., Long, Y. M., Zhang, Z. L., Xie, H. Y., Pang, D. W. (2009). Lectin-modified trifunctional nanobiosensors for mapping cell surface glycoconjugates. *Biosens. Bioelectron.* 24: 1311–1317.

Yan, C., Gu, J., Guo, Y., Chen, D. (2010). In vivo biodistribution for tumor targeting of 5-fluorouracil (5-FU) loaded N-succinyl-chitosan (Suc-Chi) nanoparticles. *Yakugaku Zasshi* 130: 801–804.

Yang, S. C., Lu, L. F., Cai, Y., Zhu, J. B., Liang, B. W., Yang, C. Z. (1999). Body distribution in mice of intravenously injected camptothecin solid lipid nanoparticles and targeting effect on brain. *J. Control Release* 59: 299–307.

Yeo, Y., Xu, P. (2009). Nanoparticles for tumor-specific intracellular drug delivery. *Conf. Proc. IEEE Eng. Med. Biol. Soc.* 2009: 2403–2405.

Yokoyama, M. (2010). Polymeric micelles as a new drug carrier system and their required considerations for clinical trials. *Expert Opin. Drug Deliv.* 7: 145–158.

Yue, M., Stachowiak, J. C., Majumdar, A. (2004). Cantilever arrays for multiplexed mechanical analysis of biomolecular reactions. *Mech. Chem. Biosyst.* 1: 211–220.

Zhang, J., Fu, Y., Jiang, F., Lakowicz, J. R. (2010). Metal nanoshell: Capsule for light-driven release of small molecule. *J. Phys. Chem. C Nanomater. Interfaces* 114: 7635–7659.

Zhang, N., Ping, Q., Huang, G., Xu, W., Cheng, Y., Han, X. (2006). Lectin-modified solid lipid nanoparticles as carriers for oral administration of insulin. *Int. J. Pharm.* 327: 153–159.

Zheng, D., Li, D., Lu, X., Feng, Z. (2010). Enhanced antitumor efficiency of docetaxel-loaded nanoparticles in a human ovarian xenograft model with lower systemic toxicities by intratumoral delivery. *Oncol. Rep.* 23: 717–724.

7 Magnetic Nanoparticles in Biomedical Applications

Renyun Zhang and Håkan Olin
Mid Sweden University
Sundsvall, Sweden

CONTENTS

7.1 MAGNETISM AND MAGNETIC NANOPARTICLES

7.1.1 Types of Magnetism

Magnetism is a term to describe how materials respond to an applied magnetic field. Common forms of magnetism are paramagnetism, diamagnetism, ferromagnetism, anti-ferromagnetism, and ferrimagnetism (Okuhata, 1999). In a magnetic field, all materials are influenced, more or less, and express different magnetism. There are two magnetic fields, the B-field and the H-field. The B-field is usually called the magnetic flux density but goes also under different names, such as the magnetic field and magnetic induction. The H-field or magnetic field is also known under other names: magnetizing field, magnetic field strength, and magnetic field intensity. In this paper, we use either B-field or magnetic flux density for the B-field and H-field and magnetic field for the H-field. The B-field is influenced by matter, and the relationship between B and H is

$$B = \mu \cdot H = \mu_r \cdot \mu_0 \cdot H \tag{7.1}$$

(Benenson et al., 2001), where μ is the permeability, μ_r *is* relative permeability, and μ_0 is vacuum permeability (also called magnetic constant or permeability of free space). The magnetic susceptibility (x_m) is

$$x_m = \mu_r - 1 \tag{7.2}$$

The different forms of magnetism and their magnetic susceptibility and relative permeability are shown in Table 7.1.

TABLE 7.1
Characteristics of Different Magnetizations and Their Magnetic Susceptibility and Relative Permeability

	Magnetic Susceptibility (x_m)	Relative Permeability (μ_r)	Direction of Magnetization (*M*) and Magnetic Field Strength (*H*)	Structures of Magnetic Substances
Diamagnetism	<0 (-10^{-4} to -10^{-9})*	<1	Opposite	
Paramagnetism	>0 (10^{-6} to 10^{-4})*	>1	Parallel	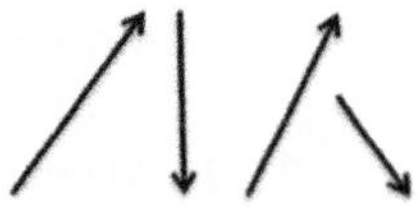
Ferromagnetism	>0 (10^{2} to 10^{7})**	>>1	Parallel	
Antiferromagnetism	Small**	>1	—	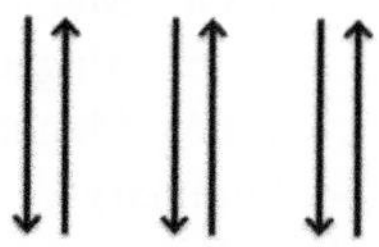
Ferrimagnetism	>0 (10^{1} to 10^{4})**	>1	—	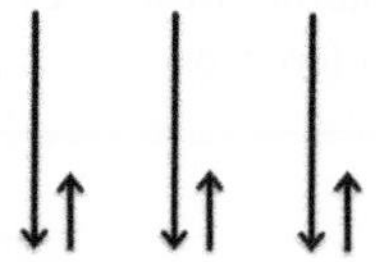

* Data from Benenson et al. (2001).

** Data from Judy and Myung (2002) showing the value of x/μ_0, where $x = M/H$, and M is magnetization.

7.1.2 Magnetic Nanoparticles

Magnetic nanoparticles normally refer to those materials that are made of iron, cobalt, or nickel elements or those materials that contain these elements. Iron and magnetism were written about in a Chinese book named *Book of the Devil Valley Master*, where it is stated, "the lodestone makes iron come or it attracts it" (Li, 1954). The magnetic moment of the iron atom is due to its four unpaired electrons in its 3*d* orbital. The compounds exhibit different forms of magnetism depending on the crystal structures.

Iron nanoparticles include iron and iron-oxide nanoparticles and some nanomaterials that contain iron or iron oxide. Among these nanoparticles, iron-oxide nanoparticles are of great interest because of their superparamagnetic properties and their wide application. For example, in the biomedical field, iron-oxide nanoparticles are commonly used because of their low toxicity and lower susceptibility to oxidation than other magnetic nanoparticles (Gupta and Gupta, 2005a, 2005b; Teja and Koh, 2009). Magnetite and maghemite are the two main forms of magnetic iron-oxide nanoparticles. A third form is hematite, which is weakly ferromagnetic or antiferromagnetic, while magnetite is ferromagnetic and maghemite is ferrimagnetic (Teja and Koh, 2009). Structurally, the difference between magnetite and maghemite is the charge state; in magnetite there are Fe^{3+} and Fe^{2+}, whereas there is only Fe^{3+} in the maghemite structure.

Cobalt is another magnetic material that is ferromagnetic and is commonly used for making magnetic and high-strength alloys. The magnetic moment of cobalt is twice that of magnetite or maghemite (Dobson, 2006b), and hexagonal close-packed cobalt shows the largest magnetic anisotropy energy among the 3*d* ferromagnetic elements (Gambardella et al., 2003).

Nickel is also a ferromagnetic material and has the lowest magnetization among these three ferromagnetic elements. The Curie temperature of nickel is about 360°C, which is lower than that for iron (768°C), iron oxide (622°C), and cobalt (1150°C) (Dewhurst et al., 1988).

There exist different compositions and phases of magnetic nanoparticles. In addition to pure metals (iron, cobalt, and nickel) and their oxides (magnetite and maghemite), alloys like Fe-Co alloy (Turgut et al., 1997) and iron-platinum (FePt) (Sun et al., 2000) and spinel-type ferromagnets like $CuFe_2O_4$ (Altincekic et al., 2010) and $CoFe_2O_4$ (Artus et al., 2008) exhibit magnetic properties (Lu et al., 2007). All of these magnetic materials can be synthesized as nanosized particles.

7.1.3 Magnetisms of Magnetic Nanoparticles

In bulk magnetic materials, the magnetization is the unit vector sum of magnetic moments of atoms. Because bulk magnetic materials have multiple magnetic domains, and all the domains are not necessarily aligned at the same time, the real magnetization is less than the sum (Teja and Koh, 2009). However, when the size of a magnetic material is nanoscaled, single domains can be achieved. For spherical magnetic nanoparticles, the critical diameter to form a single domain state is when the magnetostatic energy is equal to the domain-wall energy. This critical diameter can be evaluated by

$$D_C \approx 18\frac{\sqrt{AK_{\text{eff}}}}{\mu_0 M^2} \tag{7.3}$$

(Lu et al., 2007), where A is the exchange constant, M is the saturation magnetization, and K_{eff} is an anisotropy.

The single domain limit is one of the two finite size effects of magnetic nanoparticles. The other one is the superparamagnetic limit. Superparamagnetism is common in small, single-domain magnetic nanoparticles, like magnetite and maghemite nanoparticles, that are smaller than 10 nm. In small nanoparticles, the magnetization of the particle can randomly flip direction because of the thermal fluctuations. When the size of a magnetic nanoparticle is small enough, the thermal energy kT can exceed the energy barrier $K_{eff}V$ (where k is Boltzmann's constant, T is the absolute temperature, and V is the volume of the particle), causing fluctuation of the magnetization direction (Lu et al., 2007; Neuberger et al., 2005). Here, the small nanoparticle, the superparamagnet, is considered as a single giant magnetic moment rather than contribution by individual magnetic moments in the nanoparticles. Using a Néel-Arrhenius equation, the Néel relaxation time, τ_N, between two flips is given by:

$$\tau_N = \tau_0 \exp\left(K_{eff}V / kT\right) \tag{7.4}$$

where τ_0 is the attempt time characteristic of the material.

7.2 SYNTHESIS AND FUNCTIONALIZATION OF MAGNETIC NANOPARTICLES

To use magnetic nanoparticles in biology or medicine, small particles with uniform size and shape are desired. Different procedures have been developed to obtain proper magnetic nanoparticles, including physical vapor deposition (PVD), mechanical attrition, chemical methods, etc. PVD and chemical methods are bottom-up routines, where the nanoparticles are the assembly of single atoms. The mechanical attrition is a top-down method, where larger-grained materials are fractured down to nano-sized particles (Willard et al., 2004). However, based on the phases, the methods can also be classified as gas-phase methods, liquid-phase methods, or two-phase methods, etc. (Teja and Koh, 2009).

7.2.1 Synthesis

7.2.1.1 Gas-Phase Methods

Gas-phase methods commonly depend on thermal decomposition, reduction, hydrolysis, and other reactions to produce solid materials from the gas phase (Pierson, 1999). Among the gas-phase methods, PVD and chemical vapor deposition (CVD) are the most commonly used. Common PVD methods include radio frequency (RF) sputtering and reactive evaporation (Fujii et al., 1995). Nowadays, CVD is preferred for making magnetic nanoparticles rather than PVD. Unlike PVD methods that evaporate atoms into gas vapor from bulk materials, CVD methods use chemical reactions to form nanostructures with the participation of precursors. For synthesis of magnetic nanoparticles, metallo-organics containing magnetic elements are usually used as precursors. The reason for using metallo-organics is their low reaction temperature and low pressure limit (Teja and Koh, 2009). Alternatively, by using laser pyrolysis of organometallic precursors, one can also synthesize magnetic nanoparticles. For example, using iron pentacarbonyl as a precursor and infrared laser pyrolysis, one can form γ-phase iron-oxide nanoparticles (Alexandescu et al., 2005).

Generally, gas-phase methods always contain a thermal process, which can also be defined as thermal deposition methods. By using these methods, one can synthesize both metallic oxide magnetic nanoparticles and pure metallic nanoparticles. Moreover, by controlling the experimental parameters, the size and shape of the magnetic nanoparticles can be easily controlled. The disadvantages of these methods are the requirements for vacuum environment and high temperatures.

7.2.1.2 Liquid-Phase Methods

Liquid-phase methods utilize both aqueous and nonaqueous solutions. Compared with gas-phase methods, liquid-phase methods are commonly less expensive and are give higher yields. The disadvantages of the liquid-phase methods are the need for separation and purification of the products.

Liquid-phase methods include microemulsion (Langevin, 1992), coprecipitation (Liu et al., 2004), and sol-gel methods (Ennas et al., 1998; Teja and Koh, 2009). In microemulsion methods, both a water phase and an oil or organic phase are used; it is thus called a two-phase method.

Nanosized water droplets are used for the growth of magnetic nanoparticles, where the droplets are in the oil phase, and they are covered with surfactants to limit the size of magnetic nanoparticles (Solla-Gullón et al., 2009; Chin and Yaacob, 2007). To control the size of the nanoparticles, one should control the size of the droplet by adjusting the ratio of water and surfactant. These methods are widely used to synthesize iron oxide magnetic nanoparticles.

Coprecipitation synthesis of magnetic nanoparticle normally happens in a mixture of magnetic elements containing salt and a base or a mild oxidant. For example, magnetite nanoparticle can be synthesized in the mixture of Fe^{2+} salt, base, and nitrate ions in aqueous solutions (Sugimoto and Matijevic, 1980). The size and shape of the magnetic nanoparticles can be controlled by adjusting the type of the salts, temperature, pH value, and ionic strength (Lu et al., 2007). Moreover, because the reaction condition is not restricted, the reproducibility of these methods is quite high.

Sol-gel methods generally use water as solvent, whereas the precursors can be hydrolyzed by acid or base. The methods involve the hydrolysis and condensation of metal alkoxides, and the reaction process can be expressed as (Gash et al., 2001):

$$[Fe(OH_2)_6]^{3+} + H_2O \xrightarrow{K=10^{-3.5}} [Fe(OH)(OH_2)_5]^{2+} + H_3O^+$$

$$[Fe(OH)(OH_2)_5]^{2+} \Leftrightarrow [(H_2O)FeOFe(H_2O)_5]^{4+} + H_2O \tag{7.5}$$

7.2.1.3 Hydrothermal Methods

In hydrothermal methods, also called hydrothermal synthesis, nanoparticles are grown based on their solubility in hot water under high pressure. Fine magnetic nanoparticles can be obtained using these methods, because high supersaturations can be achieved during the synthesis processes that are important for the formation of magnetic nanoparticles. Parameters such as concentration and temperature are used to adjust the size and shape of the products. These methods are environmentally benign, because there are no organic reagents, and there is no need for posttreatment.

7.2.1.4 Mechanical Attrition

Mechanical attrition, also called ball-milling, is another method to produce magnetic nanoparticles, by the structural decomposition of coarser-grained structures. The advantages of this method are the simplicity and low cost, whereas the disadvantage is the mix-milling materials in the products that come from the attrition balls (Koch, 1997).

7.2.2 Modification and Functionalization of Magnetic Nanoparticles

Because of the development of synthetic techniques, it is now possible to produce fine magnetic nanoparticles with controlled size, shape, crystal structure, and dispersion. However, to apply these nanoparticles, further steps are needed to alter the nanoparticles for specific functions. For example, if we use the magnetic nanoparticles as drug carriers, we need to load the drug onto the nanoparticles. However, how can we load the drug? This section covers this problem.

In addition, small nanoparticles easily aggregate if not stabilized. This aggregate is one of the major obstacles for the application of magnetic nanoparticles in biological and biomedical applications. As shown in Figure 7.1, if we use the magnetic nanoparticles as a drug carrier but they are aggregated, the efficiency will decrease. Also, the aggregate problem will influence the loading process of the drug; for these reasons, further surface modifications of the synthesized magnetic nanoparticles are necessary.

7.2.2.1 Stabilization of Magnetic Nanoparticles

To stabilize and disperse magnetic nanoparticles, a protection layer is needed. The most common method is to coat the nanoparticles with molecules such as surfactants, polymers, carbon, or silica.

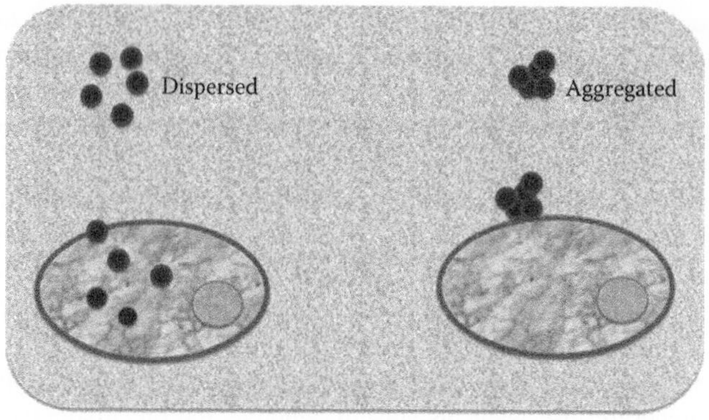

FIGURE 7.1 Influence of the dispersion of magnetic nanoparticles as drug carrier to the drug delivery. In the left part, dispersed nanoparticles can enter into the cell and deliver the drug inside the cell; this can achieve high drug concentration inside the cell. In the right part, aggregated nanoparticles cannot enter the cell, which reduces the drug delivery efficiency very much.

Surfactant and polymer protection commonly generates steric repulsion or electrostatic repulsion of the particles. For example, tetraheptylammonium 2-propanol can be used to protect magnetic nanoparticles (Zhang et al., 2006, 2009; Wang et al., 2006) through steric repulsion. The molecular interaction with magnetic nanoparticles through electrostatic interaction and the carbon chains of tetraheptylammonium 2-propanol generate steric repulsion between adjacent magnetic nanoparticles. Molecules like sodium oleate, dextran, and poly(vinyl alcohol) (PVA) are frequently used to protect magnetic nanoparticles. Generally, surfactant and polymer coating consist of molecules that can be chemically or physically attached to magnetic nanoparticles.

Silica coatings also generate steric repulsion and are used to prevent the aggregation of magnetic nanoparticles. The advantage is that the silica layer prevents the direct contact of magnetic core with additional agents, which avoids unwanted interactions (Lu et al., 2007).

The advantage with carbon coating is that the carbon layer can be effective against environmental degradation because the carbon layer is chemically stable. This carbon layer structure is also called a carbon nanocapsule and can be produced by using arc-discharge and ion-beam sputtering methods and others (Tomita et al., 2000).

7.2.2.2 Functionalization of Magnetic Nanoparticles

Although the modification as described above can stabilize dispersed magnetic nanoparticles, it is not enough for biomedical use, because the coating layer does not contain any specific function. To move to the next application step, the functional group or molecules or other nanoparticles need to be attached to the magnetic nanoparticles.

Four kinds of routine can be used to functionalize the magnetic nanoparticles.

1. Covalent bonding, which forms covalent bonds between the functional molecules and the magnetic nanoparticles that have –OH groups on the surface. The –OH groups from the base that is used in the synthesis can form covalent bonds with, for example, the silane group in 3-aminopropyltriethoxysilane (APTES) (Giri et al., 2005; Zhang et al., 2010).
2. Electrostatic interaction, which is based on the surface properties of the synthesized magnetic nanoparticles. Different synthetic methods can form a positively or negatively charged surface. By using the surface charge, one can coat molecules with opposite charge (Wang et al., 2007).
3. Adsorption, by which the functional molecules are physically adsorbed on magnetic nanoparticles (Sousa et al., 2001).

4. Hydrophobic or hydrophilic interaction, which is based on the surface properties of synthesized magnetic nanoparticles. For example, liposomes can be used to coat magnetic nanoparticles based on the hydrophobicity of the particles.

Figure 7.2 shows a schematic drawing of three kinds of interaction modes. All these methods can directly immobilize functional molecules or drug on magnetic nanoparticles for biomedical use or offer opportunities for drug or biomolecules loading. For example, the modification of APTES offers a $-NH_2$ group that can be used to link with other molecules through the interaction of the amino group and carboxylate group.

The functional coatings on magnetic nanoparticles for biomedical applications include lipids, liposomes, proteins, dendrimers, polyacrylamide, oligonucleotides sequences, RNA, and anticancer drugs (McCarthy et al., 2007). The nanoparticles can be capsulated in polymers or liposomes, form core-shell structures, be end grafted by polymers or other biomolecules (Sun et al., 2008).

7.2.2.2.1 Liposome Coating

Liposomes are a kind of amphiphilic molecule that carries both hydrophobic and hydrophilic terminals, and because it can form capsules for drug carrying has been applied in drug-delivery investigation for a long time. The coating of liposomes on magnetic nanoparticles is commonly based on the surface properties of the nanoparticles that are generated during synthetic processes. Liposome-coated magnetic nanoparticles are called magnetoliposome (Dobson, 2006a; Mornet et al., 2004; Ito et al., 2005). Different hydrophobicity of the surface shows different affinity to the hydrophobic tail or the hydrophilic head. Biomedically, the advantages of liposome coatings, their behavior, and their capsulate efficiency are well studied in vivo.

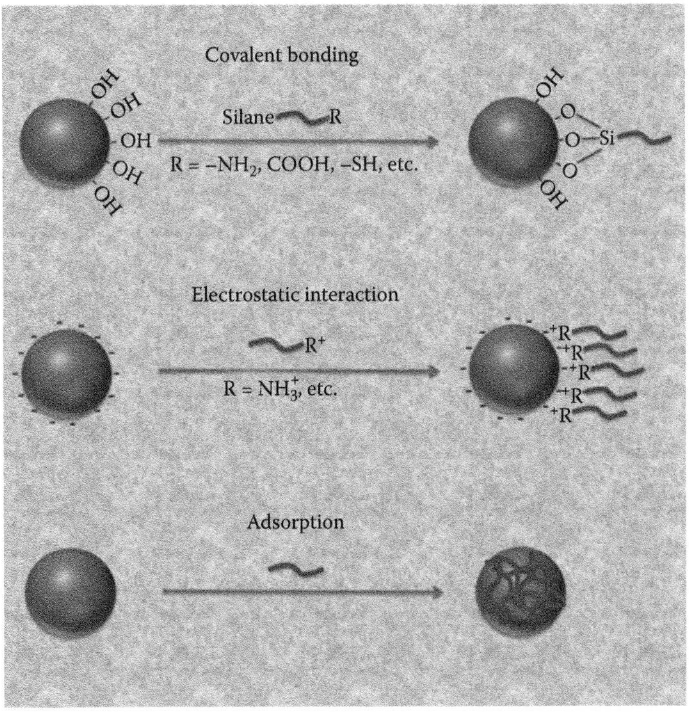

FIGURE 7.2 Binding modes of the functionalization of magnetic nanoparticles showing the covalent bonding through silane group, the electrostatic interaction, and the physical adsorptions.

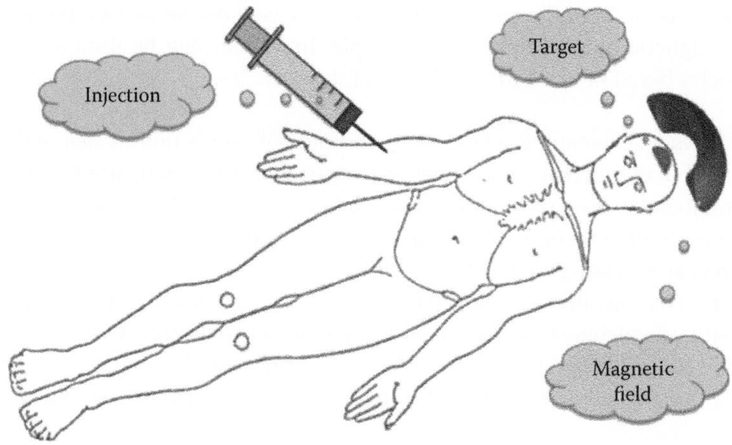

FIGURE 7.3 *In vivo* target accumulation of magnetic nanoparticles for magnetic resonance imaging by external magnetic field.

7.2.2.2.2 Polymer Coating

Polymer coatings can be attached to magnetic nanoparticles through covalent bonding, electrostatic interaction, or adsorption. Polymers that contain carboxylic acid, phosphates, and sulfates can easily be immobilized on magnetic nanoparticles through covalent bonding (Cornell and Schertmann, 1996). Charged polymers or long-chain polymers can be attached on magnetic nanoparticles through electrostatic interaction or adsorption. Owing to their chemical variety, polymers have been used to coat magnetic nanoparticles for *in vitro* and *in vivo* biomedical applications successfully. However, polymer coatings may also influence the performance of the magnetic nanoparticles (Sun et al., 2008). Commonly used polymers include polysaccharide dextran, poly(ethylene glycol), and PVA.

7.2.2.2.3 Ligand Loading

Liposomes and polymer coatings need targeting agents, optical dyes, or therapeutic agents to acquire specific functions. These functional molecules could be directly coated either on the nanoparticles or on the protective polymer or liposome layers. Ligands such as targeting agents can be used to deliver the magnetic nanoparticles to the local area, dyes can be used to image the location of nanoparticles, and therapeutic agents can be used for curing diseases (Figure 7.3).

7.3 BIOMEDICAL APPLICATIONS OF MAGNETIC NANOPARTICLES

The synthesis, stabilization, modification, and functionalization of magnetic nanoparticles are designed for application in biomedicine. The applications are mainly concerning magnetic resonance (MR) imaging, gene delivery, drug delivery, biological separation, and hyperthermia.

7.3.1 Characteristics of Magnetic Nanoparticles That Affect Their Applications in Biomedical Areas

There are several characteristics of magnetic nanoparticles that affect their application in biology or biomedicine including the size of the particles, magnetic properties, toxicity, surface charge, surface hydrophobicity, and protein binding affinity.

7.3.1.1 Size

The size of the magnetic nanoparticles is of importance because it determines how far that the particles can enter the body. As we know, the human body has many barriers to prevent the entry of foreign matter. For example, particles in the air that are larger than 4 μm are filtered by the lungs when we breathe. Smaller nanoparticles, less than 100 nm, can be phagocytosed through liver cells, and those particles larger than 200 nm will be filtered by the venous sinuses of the spleen (Neuberger et al., 2005). Depending on the size, magnetic nanoparticles can be taken up through phagocytosis (all particle sizes) or pinocytosis (particles smaller than 150 nm) (Muller et al., 1997).

7.3.1.2 Magnetism

The magnetism of magnetic nanoparticles is also important in applications. In both MRI and alternating magnetic-field treatment of tumor tissues, the magnetism of the nanoparticles is a main factor. Suitable magnetism can improve the imaging effect and increase the efficiency of killing tumor cells.

7.3.1.3 Toxicity

The toxicity matters because it is of crucial importance to evaluate the safety of a ligand or a drug for *in vivo*/animal tests or real use, and all pharmaceutical substances that are intended for human use have to be tested for toxic and side effects. All original synthesized magnetic nanoparticles and the modified or functionalized magnetic nanoparticles for *in vivo* use or study should be biocompatible or have a reasonable level of toxicity. To evaluate the cytotoxicity of magnetic nanoparticles, a common method is the intraperitoneal application of magnetic nanoparticles in mice, which can give the LD50 dosage (Neuberger et al., 2005).

7.3.1.4 Surface Charge

Surface charge should also be considered since it influences the endocytosis behavior. This is due to the negative-charge property of cell membrane, which has stronger affinity to positively-charged magnetic nanoparticles than for negatively-charged ones (Neuberger et al., 2005; Muller et al., 1997).

7.3.1.5 Protein Binding Affinity

Protein binding affinity is important for the distribution of the magnetic nanoparticles before drug delivery and for the degradation and elimination of the particles after drug delivery. The binding of protein to the particles can be by physical adsorption or enzyme-substrate interaction. The physical adsorption depends on the size and the surface properties of the magnetic nanoparticles, whereas the enzyme-substrate interaction depends on the reagents that are immobilized on the nanoparticles.

7.3.2 Magnetic Resonance Imaging

7.3.2.1 Principles

Usually, pharmacokinetic modeling is used for prediction of the behavior of drugs *in vivo*, but it is less useful for predicting the dynamics of nanoparticles, because they have different parameters controlling their distribution (Jain et al., 2008). In contrast to pharmacokinetic modeling, MR imaging can provide real-time, target monitoring of the delivery of nanoparticle *in vivo* (Kohler et al., 2006; Son et al., 2005; Mornet et al., 2004; Nitin et al., 2004).

Magnetic resonance imaging (MRI) is a noninvasive medical imaging technology to visualize the internal structure of bodies. It is recognized as a better technique than computed tomography (CT) for neurological, musculoskeletal, and oncological imaging. The method is based on the counterbalance between the small magnetic moment on a proton and protons present in biological tissue (Pankhurst et al., 2003). Simply, MRI is based on the nuclear magnetic resonance signal of protons from the water in organisms, through the combined effect of a strong static magnetic field B_0 and a transverse RF field (Mornet et al., 2004). Two relaxation times, named longitudinal (T_1) and transverse (T_2) relaxation times (which refer to spin-lattice and spin-spin relaxation times), are the most important parameters in MR imaging.

Based on the T_1 and T_2, one can do T_1-weighted MR imaging and T_2-weighted MR imaging. T_1-weighted MR imaging is operated by utilizing short repetition time TR (elapsed time between successive RF excitation pulses) and short delay time TE (time interval between the RF pulses and the measurement of the first signal). The short utilization allows the full recovery of tissues within short T_1 and meanwhile allows partial recovery of tissues with long T_1 (Mornet et al., 2004). By contrast, T_2-weighted MR imaging uses long TR and TE.

7.3.2.2 Magnetic Nanoparticles in Magnetic Resonance Imaging

MRI shows higher contrast in some part of the body than CT, but the contrast can be further improved by the addition of contrast agents. For T_1-weighted MR imaging, high-spin paramagnetic ions are usually used as contrast agents (Mornet et al., 2006). These contrast agents assisting in MR imaging commonly target extracellular fluid, the blood pool, and capillary permeability (Brasch, 1991).

The aim of magnetic nanoparticles in MR imaging is to increase the contrast of T_2-weighted MR imaging. The presence of magnetic nanoparticles can reduce the transverse time and shorten the longitudinal relaxation time (Alexiou et al., 2001). Paramagnetic chelates and ferromagnetic and superparamagnetic nanoparticles are commonly used for enhancing the contrast, and to date the superparamagnetic nanoparticles are the most promising. The advantage of superparamagnetic nanoparticles, compared with paramagnetic ions, is the high molar relaxivities, which is of importance at low concentrations (Ito et al., 2005).

Iron-oxide nanoparticles are the most sensitive superparamagnetic nanoparticle for contrast MR imaging (Choi et al., 2004). Iron oxide has been approved by the U.S. Food and Drug Administration for using as contrast agent in MR imaging (Kohler et al., 2006).

Superparamagnetic nanoparticles are commonly used for T_2-weighted MR imaging. As for animal tests, the relaxivity of T_2, R_2, is measured to obtain the quantitative information on nanoparticle accumulation (Chertok et al., 2008). The R_2 maps are usually calculated from the signal intensities using the following equation (Chertok et al., 2008):

$$R_2(t) = \frac{1}{T_2(t)} = \frac{\ln[S_1(TE_1,t)/S_2(TE_2,t)]}{TE_2 - TE_1} \tag{7.6}$$

where $S_1(TE_1, t)$ and $S_2(TE_2, t)$ are the signal intensities at time point t. For quantitative analysis of the MR imaging data, one can compare the change of R_2 at time t with the value of the initial relaxivity value by:

$$\Delta R_2(\%) = \frac{R_2(t) - R_2(0)}{R_2(0)} \times 100\% \tag{7.7}$$

The magnetic nanoparticles, with or without surface modifications, can be used as either positive or negative contrast agents. The contrast agents can be defined as matter that is introduced to the imaging region, to increase the differences between the cell and its surroundings for *in vitro* imaging of the differences between different tissues. Theoretically, negative contrast agents produce predominantly spin-spin relaxation effects, shortening both T_1 and T_2 relaxation times. However, positive contrast agents just shorten the T_1 time (http://www.mr-tip.com).

7.3.2.2.1 *In Vitro* Magnetic Resonance Imaging

Magnetic nanoparticles, such as superparamagnetic iron-oxide nanoparticles (SPION), have been demonstrated to be of low toxicity and have shown promise as a means to visualize labeled cells. The SPION can be synthesized smaller than 50 nm, which permits the nanoparticles to be easily transported across the cell membrane. Moreover, the low toxicity enables large loading of nanoparticles without significant cell death (Bowen et al., 2002). To perform *in vitro* MR imaging, an incubation procedure is first employed before the imaging step, where the cells and magnetic nanoparticles are mixed and cultured for a certain time. The incubation allows the uptake of magnetic nanoparticles into cell lines.

In addition to their role as contrast agents, magnetic nanoparticles like fluorescent reagent-labeled magnetic nanoparticles can be used to image the cell lines by both MRI and fluorescent microscopic imaging (Choi et al., 2004). Moreover, if the fluorescent reagent is also a drug, it will offer a treatment effect, or if the fluorescent reagent is a substrate of proteins inside or on the membrane of a cell, it will offer opportunities to monitor the distribution and the binding of the nanoparticles to proteins.

7.3.2.2.2 In Vivo *Magnetic Resonance Imaging*

In vitro MRI is ordinarily performed in scientific research, whereas *in vivo* MRI is applied toward medicine, e.g., clinical diagnoses. *In vivo* MR imaging can give information about tissue structures of the body, such as the distribution of water and fat, which could tell, for example, where a cancer is located.

In the human body, hemoglobin is an iron-containing molecule that can be used to map brain functions. The paramagnetic and diamagnetic properties of deoxygenated and oxidized hemoglobin afford high and low relaxivity, allowing imaging of the brain (Okuhata, 1999). However, the shortcoming of using this biomolecule is the low contrast in many clinical situations. Thus, exogenous contrast agents, like magnetic nanoparticles, that can improve the diagnostic value of MRI are desired (Mornet et al., 2004).

Magnetic nanoparticles, for *in vivo* MRI, are normally introduced by intravenous injection (Chertok et al., 2008). The magnetic nanoparticles can provide high MR contrast by shortening the longitudinal and transverse relaxation of surrounding protons. For T_1 imaging, the shortening processes are based on the close interaction of T_1 agents with protons, whereas the interaction could be hindered by the thickness of modified surface molecules on the magnetic nanoparticles (Sun et al., 2008). For T_2 imaging, the shortening is based on the susceptibility difference between the surrounding medium and the magnetic nanoparticles.

By increasing the injection amount of magnetic nanoparticles, the T_1 and T_2 relaxation times can be significantly reduced, and the relaxation rates R_1 and R_2 are found to be proportional to the concentration of magnetic nanoparticles (Jain et al., 2008). The effective concentration of magnetic nanoparticle as contrast agents for MRI can be estimated by the equation below. Post and pre means the postcontrast and precontrast.

$$1/T_{1\text{post}} = 1/T_{1\text{pre}} + R_1[C] \quad (7.8)$$

$$1/T_{2\text{post}} = 1/T_{2\text{pre}} + R_1[C] \quad (7.9)$$

The injected magnetic nanoparticles can be delivered to the target tissue by an external magnetic field. The delivered magnetic nanoparticles then act as contrast agents and improve the imaging qualities as described above. Moreover, magnetic nanoparticles afford more opportunities to monitor the dynamics of the nanoparticle distribution within the brain, in a noninvasive way (Chertok et al., 2008). To do target observation, three steps can be used (Figure 7.3):

1. Inject suspension with magnetic nanoparticles by intravenous injection.
2. Apply an external magnetic field around the target tissue or organism to accumulate the magnetic nanoparticles.
3. Monitor the tissue or organism with MRI.

7.3.3 Magnetic Nanoparticles in Biological Separation

Magnetic nanoparticles can be collected by a magnet. This property offers a way to separate biomolecules and cells. Before the application of magnetic nanoparticles in biological separations, magnetic beads with diameters of 1–5 μm were used. However, the magnetic beads lack some of the advantages of magnetic nanoparticles, for example, large surface/volume ratios and easy entry into cells (Gu et al., 2003). Three steps are commonly used to do bioseparation (see also Figure 7.4):

1. *Modification of the magnetic nanoparticles*: To separate the target biomolecules or cells, the magnetic nanoparticles should have the ability to specifically interact with the targets. Otherwise, if the magnetic nanoparticles cannot bind to the targets or the magnetic nanoparticles can bind to other molecules beside the targets separation would not succeed.
2. *Incubation of the modified magnetic nanoparticles with samples*: This step contains the mixture of the magnetic nanoparticles with the sample and the control of the incubation conditions like temperature and pH value.
3. *Separation of the targets*: This step can be easily done by adding a magnet on the place you want to attract the magnetic nanoparticles that are covered on the targets. (In the case of cell separation, magnetic nanoparticles could be on the surface of the cell or inside the cell.) Then the liquid is removed, and the collection is purified by other liquids such as water or buffers.

Magnetic-nanoparticle–based separation has been demonstrated in several systems including DNA/RNA separation (Ito et al., 2005), immunomagnetic separation (Gupta and Gupta, 2005a), protein separation (Bao et al., 2007), and cell separation (McCloskey et al., 2003). The DNA/RNA separation can be done based on the hybridization of the nucleotide sequence and the disassociation temperature of the hybridized sequence. For example, one can modify a sequence on magnetic nanoparticles and then add the modified nanoparticles in the DNA sequences mixture to let the sequence hybridize. After that, it is easy to separate the target sequence that is fully matched with the sequence on the magnetic nanoparticles. Moreover, one can separate multitargets at the same

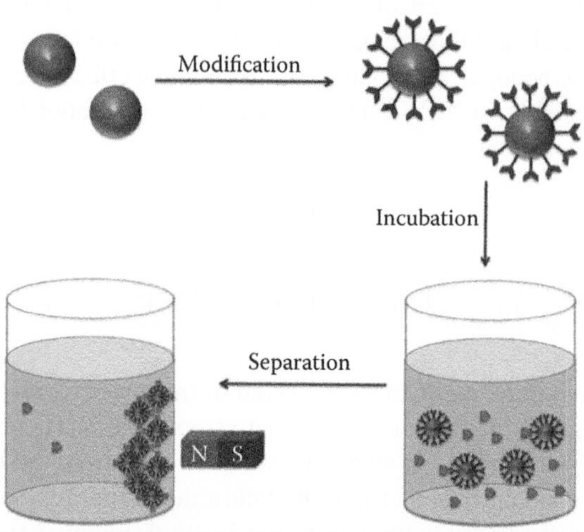

FIGURE 7.4 Procedures for bioseparation. The magnetic nanoparticles are first modified with some molecules that can specifically bind to the target cell or biomolecules; then the modified magnetic nanoparticle are incubated with the mixture of target molecule and other molecules; after the incubation, the target can be separated by magnet. To purify to separation, one can redisperse the collected sample and do magnetic separation again.

time (Zhao et al., 2003) and then further separate the sequences by the disassociation temperature differences. Protein separation uses a process similar to DNA separation, but the binding is based on the interaction of protein with its substrate on the modified magnetic nanoparticles (Bao et al., 2007). Cell separation can be performed with two different protocols: one is based on the interaction of the molecules on the magnetic nanoparticles with the biomolecules on cell membrane (McCloskey et al., 2003), and the other one is to let the magnetic nanoparticles enter the cells (Zhang et al., 2006).

7.3.4 Magnetic Nanoparticles in Gene Delivery

Gene delivery based on magnetic nanoparticles is usually performed by transfection methods, which are based on the principles developed by Widder et al. (1978). The delivery of genes by using magnetic carriers was first demonstrated by Mah et al. (2000) at the University of Florida. In their experiments, magnetic microspheres covered with adeno-associated virus were used both *in vitro* and *in vivo*. This method is called magnetofection and is of increasing importance (Dobson, 2006a; Gersting et al., 2004; Gupta and Gupta, 2005a; Sahoo and Labhasetwar, 2003).

DNA cannot cross the cellular membranes by diffusive transport like other low–molecular-weight molecules (Plank et al., 2003). Therefore, a substrate is needed to bring the gene sequence into the cell and reach the nuclei similar to virus mechanisms. In magnetofection, there are two rules to ensure the function: one is that the association of vectors with the magnetic nanoparticles must not interfere with the transport processes used; the other is that the association should be stable enough to survive from the defense mechanisms of the host and blood fluids (Plank et al., 2003).

The process of magnetic-nanoparticle–based gene delivery includes the steps of gene-sequence loading, release of gene sequences, and transfection. First, gene sequences should be loaded on the magnetic nanoparticles, and covalent bonding or noncovalent interactions can be used for this as described in the previous section. However, an important point is that the bonding should be strong enough to not break until inside the cell and at the same time weak enough to release the gene sequence when at target. A balance between loading efficiency and release effect is always a topic in gene delivery.

To do magnetofection, labeled gene sequences can be delivered to target cells through nonviral magnetofection, adenoviral magnetofection, and retroviral magnetofection (Scherer et al., 2002). Nonviral magnetofection is not based on viral particles during the transfection processes. Nonviral gene transfection agents are favorable because of their safety profile and the ability to be readministered. However, the efficiency of nonviral protocols is usually lower than virus-participated transfection (Xenariou et al., 2006), leading to the requirement for new approaches to improve efficiency. In one practical solution, magnetofection efficiency is improved by coating the magnetic nanoparticles with plasmids DNA (Scherer et al., 2002). Other vectors, including lipids, polymers, gold nanoparticles, quantum dots, silica nanoparticles, fullerenes, and carbon nanotubes, can be used in transfection (Ragusa et al., 2007). The advantage of magnetic nanoparticles (e.g., superparamagnetic nanoparticles) as vectors is that the whole process can be monitored using MRI at real time. Plasmid DNA can be electrostatically adsorbed on superparamagnetic nanoparticles, for example, iron-oxide nanoparticles (Gersting et al., 2004). In nonviral magnetofection, the magnetic nanoparticles can be coated with polyethylenimine (commonly named transMAG-PEI) to positively charge the nanoparticles, which allow the binding of negative charged DNA molecules to the nanoparticles (Jahnke et al., 2007). In addition to polyethylenimine (PEI), lipids can also be used to attach gene sequences to the magnetic nanoparticles for nonviral magnetofection. Magnetofection has hundreds to thousands times higher transgene expression than standard nonviral transfection, polyfection, and lipofection (Gersting et al., 2004). Figure 7.5 shows a schematic representation of the nonviral magnetofection process, where the DNA is loaded on the magnetic nanoparticles through electrostatic interaction between DNA and the PEI layer on magnetic nanoparticle surface.

Adenoviral transfection is a standard way to study gene expression *in vitro* and *in vivo*. The transfection is dependent on the coxsackie and adenovirus receptor (CAR) status. However, in some tissues, like lung epithelium and tumor tissue, the CAR status is changing, small, or even absent

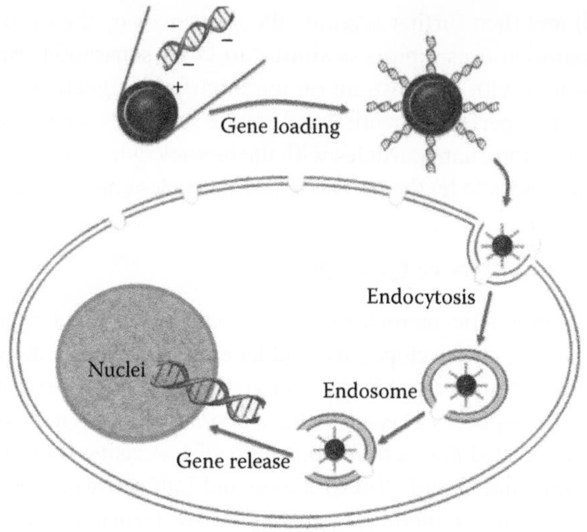

FIGURE 7.5 Schematic drawing of the pathway of magnetofection. The gene sequence is first loaded on magnetic nanoparticles through electrostatic interaction between negatively-charged DNA and positively-charged PEI or other polymers on magnetic nanoparticles. After that, the gene-loaded magnetic nanoparticles will approach the cell membrane, probably with the assistance of the magnetic field, and then enter the cell through endocytosis. Subsequently, after the endosome process, the gene sequence will be released to the nuclei and accomplish the transfection process.

(Scherer et al., 2002). To overcome this obstacle, mediators have been introduced, e.g., magnetic nanoparticles, polymers. Similar to nonviral magnetofection, the viruses for adnoviral magnetofection can also be attached to magnetic nanoparticles through electrostatic interaction using PEI- or chitosan-modified magnetic nanoparticles (Scherer et al., 2002; Bhattarai et al., 2008). The attachment procedure can be performed by simply mixing the magnetic nanoparticles with adenoviruses (Gliddon et al., 2008). The enhancement effect of magnetic nanoparticles in adenoviral transfection depends on the ratio of adenovirus and magnetic nanoparticles (Plank et al., 2003).

Retroviral transfection is another commonly used method, where the retroviral vectors are derived from RNA viruses and the vectors can be integrated into a gene in the host cell lines. The advantage of retroviral transfection is the broad range of host specificity. However, there are also some disadvantages to this method; for example, the preparation of the viruses is difficult (Ito et al., 2009). Magnetofection is recently found to be useful in retroviral transfection, because it can increase the gene expression after retroviral transfection. The most common magnetic vector is the magnetite cationic liposomes (Akiyama et al., 2010), which can be simply produced by mixing magnetite nanoparticles with lipid mixtures (Ito et al., 2009).

7.3.5 Magnetic Nanoparticles in Drug Delivery

In addition to gene delivery, drug delivery is another important application of magnetic nanoparticles. Clinical trials for systems based on iron-oxide nanoparticles have been investigated since 1996, when Lubbe et al. did the first phase I clinical trials. Later, Koda et al. (2002) did another phase I trail for targeting drug delivery, and 30 of 32 patients were found targeted in their trail (McBain et al., 2008). These two examples indicate that the application iron oxide for *in vivo* drug targeting and delivery can be achieved and can be applied in real treatments.

The advantages of magnetic nanoparticles as drug carrier, e.g., iron-oxide nanoparticles, comparing with other nanoparticles, are their biocompatible properties and their magnetism. The biocompatible properties ensure the *in vivo* applications, and the magnetism makes the targeting drug delivery externally controllable. However, to make a drug carrier, there are other parameters that should be considered: the size of the particles, the cytotoxicity, the surface properties, the loading capacity of drugs, the drug release kinetics, the stability, etc.

There are three kinds of targeting drug-release modes: passive, active, and physical. Drug delivery in passive mode is not specific to any particular place in the body, whereas active mode is based on the specific binding of drugs to cells or tissues. These two modes are depending on the properties of the drug deliver systems themselves. The physical targeting adds external control to the system, which is considered as the advantages of magnetic nanoparticles that can be controlled by an external magnetic field (Neuberger et al., 2005). Using magnetic nanoparticles, one can also combine the active mode with the physical mode by modifying the magnetic nanoparticles with some molecules or antibody that can specifically bind to cells. Based on this protocol, one can perform first and second order targeting (Neuberger et al., 2005) at the same time, because the external magnetic field applied on a magnetic nanoparticle can bring the delivering system to the target site as close as possible, and the surface modification of the magnetic nanoparticle can perform receptor oriented targeting.

Figure 7.6 illustrates a schematic drawing of the application of magnetic nanoparticles in targeted drug delivery. The delivery system consists of a magnetic field and drug-loaded magnetic nanoparticles, which are also labeled with a molecule that specifically binds to some units on cell membranes. During step I, drug-loaded magnetic nanoparticles are moving to the target site under magnetic field, which is magnetic-field–driven first-order targeting, to bring the nanoparticle close to the target cell. Then the labeled molecule will immobilize the nanoparticle on the cell through the specific interaction between the molecule and the unit on cell membrane, which is step II, a second-order targeting step. At step III, when the nanoparticles are immobilized, the drug molecules that loaded on the magnetic nanoparticles will release and diffuse into the cell.

To perform an accurate targeting drug delivery, surface modification of magnetic nanoparticles is one of the essential steps. Polymers, liposomes, emulsion, and other molecules have been applied to modify magnetic nanoparticles. The delivery system benefits from these modifications, but there are also some obstacles (Jain et al., 2008):

1. Modification of a polymer, like starch, or block copolymer needs complex conjugations, which limit the drug association efficiency.
2. The incorporation of liposomes and emulsion reduce the drug-loading capacities.
3. The dispersion of magnetic nanoparticles in polymers increased the size of the delivery systems, which will influence the targeting efficiency.

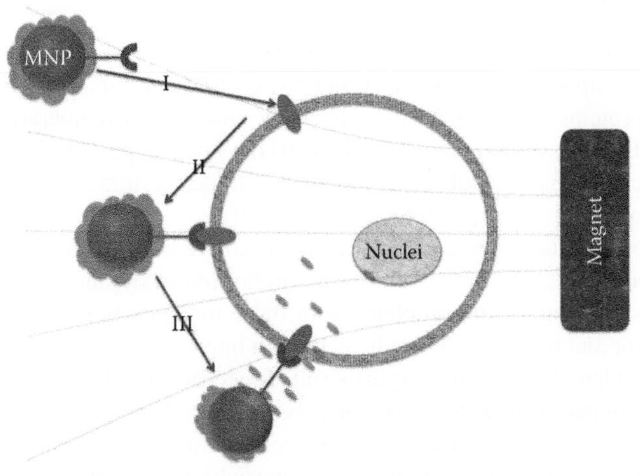

FIGURE 7.6 Application of magnetic nanoparticles in target drug delivery, with the aid of labeled molecules and external magnetic field.

There are also other obstacles related to the body and the external magnetic field:

1. The application of external magnetic field might cause the accumulation of the magnetic nanoparticles in the body.
2. The external magnetic field can just be applied to the superficial organs.
3. Phagocytosis of the body will remove the magnetic nanoparticles by the coating of plasma proteins after the magnetic nanoparticles enter bloodstream (Kumar et al., 2010; Namdeo et al., 2008).

In addition to targeting drug delivery, other modified magnetic nanoparticles have been developed. For example, tetraheptylammonium capped iron-oxide nanoparticles are found to overcome the drug resistance of leukemia K562 cell lines (Zhang et al., 2006, 2009; Wang et al., 2006). After the iron-oxide nanoparticles were modified with tetraheptylammonium, the surface of the nanoparticle became hydrophobic, making the iron nanoparticle a suitable substrate for P-glycoprotein. P-glycoprotein is the most important protein on the cancer cell membrane, which contributes for some time to the multidrug resistance of the cells after drug treatment. The multidrug resistance pumps out the drugs from inside to outside, causing low drug concentration inside the cancer cells, and subsequently low treatment effect. Zhang and co-workers reported how to overcome the multidrug resistance of Leukemia K562 cells (Zhang et al., 2006, 2009; Wang et al., 2006), where anticancer drugs doxorubicin and daunorubicin were loaded on the tetraheptylammonium capped iron-oxide nanoparticles. After the iron-oxide nanoparticles enter the cell lines, the anticancer drugs will release and diffuse inside the cell. Meanwhile, the tetraheptylammonium capped iron-oxide nanoparticles are also diffusing. Because both the anticancer drug- and tetraheptylammonium-capped iron-oxide nanoparticles can be recognized by the P-glycoprotein, there is a competition between the capped nanoparticles and the anticancer drug. The binding of capped nanoparticles to the protein will block the recognition of P-glycoprotein to the drugs, leading to less pumping out effect and resulting in high drug concentration inside the cells compared with the absence of magnetic nanoparticles. This is of importance for the treatment of cancers, because the drug resistance is one of the main obstacles in cancer treatment. Traditionally, to overcome the drug resistance, researchers have to develop a new generation of drugs, but the speed of the development of new drugs is far behind the generation of drug resistance. However, with the aid of modified magnetic nanoparticles, this problem might be reduced.

7.3.6 Magnetic Nanoparticles in Hyperthermia

Hyperthermia is a powerful physical method to damage tumor or cancer cells by heating certain organs or tissues to temperatures between 41 and 46 °C (Jordan et al., 1999), because most of the biomolecules like proteins are influenced by hyperthermia at this temperature range. There are three domains in clinically applied hyperthermia: whole body hyperthermia, regional hyperthermia, and local hyperthermia. One of the advantages of hyperthermia is the low toxicity, which is commonly high in other treatments (Falk and Issels, 2001).

To perform hyperthermia treatments, hot water, capacitive heating, and induction heating have previously been used (Kim et al., 2009). Microwave, radio frequency, and ultrasound are commonly used to heat local sites. One of the problems with these methods is nonselective heating, which means both the tumor and normal cells are heated; both inter- and intracellular places are affected. To overcome this problem, intracellular hyperthermia is needed, which only heats the mutated cells.

For intracellular hyperthermia, the heat must be generated inside the tumor cells. Magnetic nanoparticles can be used for this, because they are small enough to enter cells and can be heated by an external alternative magnetic field. Magnetic nanoparticles for hyperthermia, which is also called magnetic fluid hyperthermia (Johannsen et al., 2005), is based on the power of absorption of magnetic nanoparticles by an alternative magnetic field. The power absorption is commonly expressed by a parameter named specific absorption rate (SAR), which can be calculated for a given material as SAR = Af, where A is the area of the hysteresis loop (with unit J/g, also called specific loss), and f is the frequency of the magnetic field. Because the temperature of hyperthermia is in a range of 41 to 46°C, the nanoparticles should be have a corresponding suitable power loss (Pollert et al., 2007).

On the other hand, in local hyperthermia, the heterogeneous temperature distribution and the inability of heating smaller visceral masses hindered the employment of localized treatments, pointing toward the design of specific magnetic nanoparticles with an appropriate Curie temperature range that can be found, for example, in copper-nickel magnetic nanoparticles (Chatterjee et al., 2005).

Another factor that influences the accurate hyperthermia is the selective binding of magnetic nanoparticles to tumor cells. With high selectivity, the magnetic nanoparticles can specifically bind to and enter tumor cells and subsequently kill the tumor cell by intracellular hyperthermia. To increase the selectivity of the magnetic nanoparticles, surface modification is the most appropriate way to go. For example, chitosan-coated magnetic nanoparticles can more specifically bind to KB carcinoma cell than starch-coated nanoparticles (Kim et al., 2009). So far, the most common coating materials are polymers, because there are several advantages for polymer coating, including the overcoming of the reticuloendothelial system clearance problem, reduction of toxicity, reduction of the heat damage to normal tissues (Kim et al., 2009), and increase of the selectivity. However, the size of the magnetic nanoparticles should be considered, because some modification may greatly increase the size of the nanoparticles. The size is of importance because it influences the penetration of the nanoparticles into cells and the transport dynamics caused by the external magnetic field.

The combination of the external magnetic field with radiation can also increase the treatment effect of hyperthermia. One example is the combination of hyperthermia and ionizing radiation. This combination can increase the response rate of cells to 70% (Falk and Issels, 2001). The combination of these two methods can also overcome thermotolerance, which quickly appears after hyperthermia (Jordan et al., 1999).

Another example is to change the waveforms of the applied magnetic field. Sinusoidal waveforms can maximize the effect of treatment, and square waveforms can increase the heat by at least 50% (Morgan and Victora, 2010).

7.4 SUMMARY

Magnetic nanoparticles, including metallic nanoparticles (iron, cobalt, nickel), metallic oxide nanoparticles (magnetite, maghemite nanoparticles), alloy nanoparticles (copper-nickel and iron-nickel), and complex nanoparticles ($CoFe_2O_4$), are increasingly used for bioapplications. Many kinds of methods have been developed to synthesize these nanoparticles with specific structures and for different applications. Moreover, to embed more properties or to change the surface properties, methods have been developed to modify the surface of magnetic nanoparticles. Biomedically, these magnetic nanoparticles can be used in several applications, including:

1. Bioseparation: to separate different biomolecules
2. Magnetic resonance imaging: to increase the imaging contrast
3. Gene delivery: for targeting delivery of gene to specific cells
4. Drug delivery: for targeting delivery of drugs and to simultaneously overcome the drug resistance
5. Hyperthermia: to heat the tumor cell to death by the alternating magnetic field

The most obvious advantage of magnetic nanoparticles compared with other nanoparticles is the ability to combine several functions:

1. The combination of magnetic resonance imaging and drug/gene delivery enables estimations of the distribution of the magnetic nanoparticles while performing drug/gene delivery.

2. The combination of drug delivery and hyperthermia offers synergistic application of chemotherapy and hyperthermia, leading to potentially higher treatment effect.
3. The combination of drug delivery, hyperthermia, and magnetic resonance imaging, if not too complicated, might prove to be useful in future biomedical and clinical applications.

There are also disadvantages using magnetic nanoparticles. For example, the magnetic nanoparticles can be modified to be biocompatible, but the nanoparticles themselves are not biodegradable, which might cause an accumulation of magnetic nanoparticles inside the body after treatment.

Magnetic nanoparticles are generally interesting nanomaterials in biomedicine. They offer a platform to combine different techniques to decrease human morbidity and mortality. However, in spite of all that research in this exciting field has achieved up until now, safety and efficiency can still be improved in the future.

INTERNET RESOURCES

http://www.mr-tip.com/serv1.php?type=db1&dbs=Contrast%20Agents

REFERENCES

Akiyama, H., Ito, A., Kawabe, Y., et al. (2010). Genetically engineered angiogenic cell sheets using magnetic force-based gene delivery and tissue fabrication techniques. *Biomaterials* 31: 1251–1259.

Alexandescu, R., Morjan, I., Voicu, I., et al. (2005). Combining resonant/non-resonant processes: Nanometer-scale iron-based material preparation via CO laser pyrolysis. *Applied Surface Science* 248: 138–146.

Alexiou, C., Arnold, W., Hulin, P., et al. (2001). Magnetic mitoxantrone nanoparticle detection by histology, X-ray and MRI after magnetic tumor targeting. *Journal of Magnetism and Magnetic Materials* 225: 187–193.

Altincekic, T. G., Boz, I., Baykal, A., et al. (2010). Synthesis and characterization of $CuFe_2O_4$ nanorods synthesized by polyol route. *Journal of Alloys and Compounds* 493: 493–498.

Artus, M., Ammar, S., Sicard, L., et al. (2008). Synthesis and magnetic properties of ferrimagnetic $CoFe_2O_4$ nanoparticles embedded in an antiferromagnetic NiO matrix. *Chemistry of Materials* 20: 4861–4872.

Bao, J., Chen, W., Liu, T., et al. (2007). Bifunctional Au-Fe_3O_4 nanoparticles for protein separation. *ACS Nano* 1: 293–298.

Benenson, W., Harris, J. W., Stocker, H., et al. (2001). *Handbook of physics*, New York: Springer-Verlag.

Bhattarai, S. R., Kim, S. Y., Jang, K. Y., et al. (2008). N-hexanoyl chitosan-stabilized magnetic nanoparticles: Enhancement of adenoviral-mediated gene expression both in vitro and in vivo. *Nanomedicine: Nanotechnology, Biology, and Medicine* 4: 146–154.

Bowen, C. V., Zhang, X., Saab, G., et al. (2002). Application of the static dephasing regime theory to superparamagnetic iron-oxide loaded cells. *Magnetic Resonance in Medicine* 48: 52–61.

Brasch, R. C. (1991). Rationale and applications for macromolecular Gd-based contrast agents. *Magnetic Resonance in Medicine* 22: 282–287.

Chatterjee, J., Bettge, M., Haik, Y., et al. (2005). Synthesis and characterization of polymer encapsulated Cu-Ni magnetic nanoparticles for hyperthermia applications. *Journal of Magnetism and Magnetic Materials* 293: 303–309.

Chertok, B., Moffat, B. A., David, A. E., et al. (2008). Iron oxide nanoparticles as a drug delivery vehicle for MRI monitored magnetic targeting of brain tumors. *Biomaterials* 29: 487–496.

Chin, A., Yaacob, I. (2007). Synthesis and characterization of magnetic iron oxide nanoparticles via w/o micro-emulsion and Massart's procedure. *Journal of Materials Processing Technology* 191(1–3): 235–237.

Choi, H., Choi, S. R., Zhou, R., et al. (2004). Iron oxide nanoparticles as magnetic resonance contrast agent for tumor imaging via folate receptor-targeted delivery. *Academic Radiology* 11: 996–1004.

Cornell, R. M., Schertmann, U. (1996). *The iron oxide: Structure, properties, reactions, occurrence and uses.* Weinheim: VCH.

Dewhurst, R. J., Edwards, C., McKie, A. D. W., et al. (1988). A remote laser system for ultrasonic velocity measurement at high temperature. *Journal of Applied Physiology* 63: 1225–1227.

Dobson, J (2006a). Gene therapy progress and prospects: magnetic nanoparticle-based gene delivery. *Gene Therapy* 13, 283–287.

Dobson, J. (2006b). Magnetic nanoparticles for drug delivery. *Drug Development Research* 67: 55–60.
Ennas, G., Musinu, A., Piccaluga, G., et al. (1998). Characterization of iron oxide nanoparticles in an Fe_2O_3-SiO_2 composite prepared by a sol-gel method. *Chemistry of Materials* 10: 495–502.
Falk, M. H., Issels, R. D. (2001). Hyperthermia in oncology. *International Journal of Hyperthermia* 17: 1–18.
Fujii, E., Torii, H., Tomozawa, A., et al. (1995). Iron oxide films with spinel, corundum and bixbite structure prepared by plasma-enhanced metalorganic chemical vapor deposition. *Journal of Crystal Growth* 151: 134–139.
Gambardella, P., Rusponi, S., Veronese, M., et al. (2003). Giant magnetic anisotropy of single cobalt atoms and nanoparticles. *Science* 300: 1130–1133.
Gash, A. E., Tillotson, T. M., Satcher, J. H., et al. (2001). Use of epoxides in the sol-gel synthesis of porous iron(III) oxide monoliths from Fe(III) salts. *Chemistry of Materials* 13: 999–1007.
Gersting, S. W., Schillinger, U., Lausier, J., et al. (2004). Gene delivery to respiratory epithelial cells by magnetofection. *The Journal of Gene Medicine* 6: 913–922.
Giri, S., Trewyn, B. G., Stellmaker, M. P., et al. (2005). Stimuli-responsive controlled-release delivery system based on mesoporous silica nanorods capped with magnetic nanoparticles. *Angewandte Chemie* 117: 5166–5172.
Gliddon, B. L., Nguyen, N. V., Gunn, P. A., et al. (2008). Isolation, culture and adenoviral transduction of parietal cells from mouse gastric mucosa. *Biomedical Materials* 3: 034117.
Gu, H., Ho, P. L., Tsang, K. W. T., et al. (2003). Using biofunctional magnetic nanoparticles to capture vancomycin-resistant enterococci and other Gram-positive bacteria at ultralow concentration. *Journal of the American Chemical Society* 125: 15702–15703.
Gupta, A. K., Gupta, M. (2005a). Cytotoxicity suppression and cellular uptake enhancement of surface modified magnetic nanoparticles. *Biomaterials* 26: 1565–1573.
Gupta, A. K, Gupta, M. (2005b). Synthesis and surface engineering of iron oxide nanoparticles for biomedical applications. *Biomaterials* 26: 3995–4021.
Ito, A., Shinkai, M., Honda, H., et al. (2005). Medical application of functionalized magnetic nanoparticles. *Journal of Bioscience and Bioengineering* 100: 1–11.
Ito, A., Takahashi, T., Kameyama, Y., et al. (2009). Magnetic concentration of a retroviral vector using magnetite cationic liposomes. *Tissue Engineering: Part C* 15: 57–64.
Jahnke, A., Hirschberger, J., Fischer, C., et al. (2007). Intra-tumoral gene delivery of feIL-2, feIFN-gamma and feGM-CSF using magnetofection as a neoadjuvant treatment option for feline fibrosarcomas: A phase-I study. *Journal of Veterinary Medicine A* 54: 599–606.
Jain, T. K., Richey, J., Strand, M., et al. (2008). Magnetic nanoparticles with dual functional properties: Drug delivery and magnetic resonance imaging. *Biomaterials* 29: 4012–4021.
Johannsen, M., Gneveckow, U., Eckelt, L., et al. (2005). Clinical hyperthermia of prostate cancer using magnetic nanoparticles: Presentation of a new interstitial technique. *International Journal of Hyperthermia* 21: 637–647.
Jordan, A., Scholz, R., Wust, P., et al. (1999). Magnetic fluid hyperthermia (MFH): Cancer treatment with AC magnetic field induced excitation of biocompatible superparamagnetic nanoparticles. *Journal of Magnetism and Magnetic Materials* 201, 413–419.
Judy, J. W., Myung, N. (2002). Magnetic materials for MEMS. In *MRS workshop on MEMS materials* (pp. 23–26).
Kim, D. H., Kim, K. N., Kim, K. M., et al. (2009). Targeting to carcinoma cells with chitosan- and starch-coated magnetic nanoparticles for magnetic hyperthermia. *Journal of Biomedical Materials Research: Part A* 88: 1–11.
Koch, C. C. (1997). Synthesis of nanostructured materials by mechanical milling: Problems and opportunities. *NanoStructured Materials* 9: 13–22.
Koda, J., Venook, A., Walser, E., et al. (2002). A multicenter, Phase I/II trial of hepatic intraarterial delivery of doxorubicin hydrochloride adsorbed to magnetic targeted carriers in patients with hepatocarcinoma. *European Journal of Cancer* 38(Suppl 7): S18.
Kohler, N., Sun, C., Fichtenholtz, A., et al. (2006). Methotrexate-immobilized poly(ethylene glycol) magnetic nanoparticles for MR imaging and drug delivery. *Small* 2: 785–792.
Kumar, A., Jena, P. K., Behera, S., et al. (2010). Multifunctional magnetic nanoparticles for targeted delivery. *Nanomedicine: Nanotechnology, Biology, and Medicine* 6: 64–69.
Langevin, D. (1992). Micelles and microemulsions. *Annual Review of Physical Chemistry* 43: 341–369.
Li, S.-H. (1954). Origine de la boussole II: Aimant et boussole. *Isis* 45: 175–196.
Liu, Z. L., Wang, H. B., Lu, Q. H., et al. (2004). Synthesis and characterization of ultrafine well-dispersed magnetic nanoparticles. *Journal of Magnetism and Magnetic Materials* 283: 258–262.

Lu, A.-H., Salabas, E. L., Schüth, F. (2007). Magnetic nanoparticles: Synthesis, protection, functionalization, and application. *Angewandte Chemie International Edition* 46: 1222–1244.

Lubbe, A. S., Bergemann, C., Riess, H., et al. (1996). Experiences with magnetic drug targeting: A phase epidoxorubicin in 14 patients with advanced solid tumors. *Cancer Research* 56, 4686–4693.

Mah, C., Fraites, T. J., Zolotukhin, I., et al. (2000). Improved method of recombinant AAV delivery for systemic targeted gene therapy. *Molecular Therapy* 6, 106–112.

McBain, S. C., Yiu, H. H. P., Dobson, J. (2008). Magnetic nanoparticles for gene and drug delivery. *International Journal of Nanomedicine* 3: 169–180.

McCarthy, J. R., Kelly, K. A., Sun, E. Y., et al. (2007). Targeted delivery of multifunctional magnetic nanoparticles. *Nanomedicine*, 2: 153–167.

McCloskey, K. E., Chalmers, J. J., Zborowski, M. (2003). Magnetic cell separation: Characterization of magnetophoretic mobility. *Anal. Chem.* 75: 6868–6874.

Morgan, S. M., Victora, R. H. (2010). Use of square waves incident on magnetic nanoparticles to induce magnetic hyperthermia for therapeutic cancer treatment. *Applied Physics Letters* 97: 093705.

Mornet, S., Vasseur, S., Grasset, F., et al. (2004). Magnetic nanoparticle design for medical diagnosis and therapy. *Journal of Materials Chemistry* 14: 2161–2175.

Mornet, S., Vasseur, S., Grasset, F., et al. (2006). Magnetic nanoparticle design for medical applications. *Progress in Solid State Chemistry* 34: 237–247.

Muller, R. H., Luck, M., Garnisch, S., et al. (1997). Intravenously injected particles: Surface properties and interaction with blood proteins—the key determining the organ distribution. In U. Häfeli et al. (Eds.), *Scientific and clinical application of magnetic carriers* (pp. 135–136). New York: Plenum Press.

Namdeo, M., Sasena, S., Tankhiwale, R., et al. (2008). Magnetic nanoparticles for drug delivery applications. *Journal of Nanoscience and Nanotechnology* 8: 3247–3271.

Neuberger, T., Schöpf, B., Hofmann, H., et al. (2005). Superparamagnetic nanoparticles for biomedical applications: Possibilities and limitations of a new drug delivery system. *Journal of Magnetism and Magnetic Materials* 293: 483–496.

Nitin, N., LaConte, L. E. W., Zurkiya, O., et al. (2004). Functionalization and peptide-based delivery of magnetic nanoparticles as an intracellular MRI contrast agent. *Journal of Biological Inorganic Chemistry* 9: 706–712.

Okuhata, Y. (1999). Delivery of diagnostic agents for magnetic resonance imaging. *Advanced Drug Delivery Reviews* 37: 121–137.

Pankhurst, Q. A., Connolly, J., Jones, S. K., et al. (2003). Applications of magnetic nanoparticles in biomedicine. *Journal of Physics D: Applied Physics* 36: R167–R181.

Pierson, H. O. (1999). *Handbook of chemical vapor deposition: Principles, technology, and application.* New York: William Andrew Inc.

Plank, C., Schillinger, U., Scherer, F., et al. (2003). The magnetofection method: Using magnetic force to enhance gene delivery. *Biological Chemistry* 384, 737–747.

Pollert, E., Knizek, K., Marysko, M., et al. (2007). New Tc-tuned magnetic nanoparticles for self-controlled hyperthermia. *Journal of Magnetism and Magnetic Materials* 316: 122–125.

Ragusa, A., García, I., Penades, S. (2007). Nanoparticles as nonviral gene delivery vectors. *IEEE Transactions on NanoBioscience* 6: 319–330.

Sahoo, S. K., Labhasetwar, V. (2003). Nanotech approaches to drug delivery and imaging. *Drug Discovery Today* 8: 1112–1120.

Scherer, F., Anton, M., Schillinger, U., et al. (2002). Magnetofection: Enhancing and targeting gene delivery by magnetic force in vitro and in vivo. *Gene Therapy* 9: 102–109.

Solla-Gullón, J., Gómez, E., Valles, E., et al. (2009). Synthesis and structural, magnetic and electrochemical characterization of PtCo nanoparticles prepared by water-in-oil microemulsion. *Journal of Nanoparticle Research* 12: 1149–1159.

Son, S. J., Reichel, J., He, B., et al. (2005). Magnetic nanotubes for magnetic-field-assisted bioseparation, biointeraction, and drug delivery. *Journal of the American Chemical Society* 127: 7316–7317.

Sousa, M. H., Rubim, J. C., Sobrinho, P. G., et al. (2001). Biocompatible magnetic fluid precursors based on aspartic and glutamic acid modified maghemite nanostructures. *Journal of Magnetism and Magnetic Materials* 225: 67–72.

Sugimoto, T., Matijevic, E. (1980). Formation of uniform spherical magnetite particles by crystallization from ferrous hydroxide gels. *Journal of Colloid and Interface Science* 74: 227–243.

Sun, C., Lee, J. S. H., Zhang, M. (2008). Magnetic nanoparticles in MR imaging and drug delivery. *Advanced Drug Delivery Reviews* 60: 1252–1265.

Sun, S. H., Murray, C. B., Weller, D., et al. (2000). Monodisperse FePt nanoparticles and ferromagnetic FePt nanocrystal superlattices. *Science* 287: 1989–1992.

Teja, A. S., Koh, P.-Y. (2009). Synthesis, properties, and applications of magnetic iron oxide nanoparticles. *Progress in Crystal Growth and Characterization of Materials* 55: 22–45.

Tomita, S., Hikita, M., Fujii, M., et al. (2000). A new and simple method for thin graphitic coating of magnetic-metal nanoparticles. *Chemical Physics Letters* 316: 361–364.

Turgut, Z., Huang, M. Q., Gallagher, K., et al. (1997). Magnetic evidence for structural-phase transformations in Fe-Co alloy nanocrystals produced by a carbon arc. *Journal of Applied Physics* 81: 4039–4041.

Wang, S. H., Shi, X., van Antwerp. M., et al. (2007). Dendrimer-functionalized iron oxide nanoparticles for specific targeting and imaging of cancer cells. *Advanced Functional Materials* 17: 3043–3050.

Wang, X. M., Zhang, R. Y., Wu, C. H., et al. (2006). The application of Fe_3O_4 nanoparticles in cancer research: A new strategy to inhibit drug resistance. *Journal of Biomedical Materials Research Part A* 80A: 852–860.

Widder, K. J., Senyei, A. E., Scarpelli, D. G. (1978). Magnetic microspheres: A model system for site specific drug delivery in vivo. *Proceedings of the Society for Experimental Biology and Medicine* 58: 141–146.

Willard, M. A., Kurihara, L. K., Carpenter, E. E., et al. (2004). Chemically prepared magnetic nanoparticles. *International Materials Reviews* 49: 125–170.

Xenariou, S., Griesenbach, U., Ferrari, S., et al. (2006). Using magnetic forces to enhance non-viral gene transfer to airway epithelium in vivo. *Gene Therapy* 13: 1545–1552.

Zhang, R. Y., Wang, X. M., Wu, C. H., et al. (2006). Synergistic enhancement effect of magnetic nanoparticles on anticancer drug accumulation in cancer cells. *Nanotechnology* 17: 3622–3626.

Zhang, R. Y., Wu, C. H., Wang, X. M., et al. (2009). Enhancement effect of nano Fe_3O_4 to the drug accumulation of doxorubicin in cancer cells. *Materials Science and Engineering: C* 29: 1697–1701.

Zhang, R. Y., Hummelgård, M., Olin, H. (2010). Simple synthesis of clay-gold nanocomposites with tunable color. *Langmuir* 26: 5823–5828.

Zhao, X., Tapec-Dytioco, R., Wang, K., et al. (2003). Collection of trace amounts of DNA/mRNA molecules using genomagnetic nanocapturers. *Analytical Chemistry* 75: 3476–3483.

8 High-Throughput Screening in Medical Diagnosis and Prognosis

Maysaa El Sayed Zaki
Mansoura University
Mansoura, Egypt

CONTENTS

8.1 INTRODUCTION

Recent omics technologies have opened the door to discovering new biomarkers for the diagnosis, prognosis, therapeutic-response prediction, and population screening of various human diseases. In this article, we will focus on the recent advances and future directions in omics-based medical diagnostics methods.

In the postgenome era, efforts are focused on biomarker discovery and the early diagnosis of diseases through the application of various omics technologies like transcriptomics, proteomics, metabonomics, peptidomics, glycomics, phosphoproteomics, and lipidomics on various tissue samples and body fluids. Currently, the biological samples analyzed include blood, urine, sputum, saliva, nipple-aspirate fluid, breath, tear fluid, and cerebrospinal fluid. Diversity of components in these samples (e.g., amino acids, peptides, proteins, or metabolites) can further increase the analytical complexity.

Generally speaking, proteomics and metabolomics have the advantage of being capable of searching for proteins or metabolites in the blood or urine (rather than the primary tissue, where a disease might appear), but typically these approaches identify far fewer proteins or metabolites than can be identified with the use of microarray analysis. Microarray analysis is considered to be a more mature technique than the other approaches and has the relative ease of working with nucleic acids; microarrays remain the omics technique that is most likely to have early applications in diagnosis or prognosis. Table 8.1 summarizes the description of omics technology.

TABLE 8.1
Description of Omics Technology

Term	Definition
Functional genomics or transcriptomics	It is known as gene expression profiling and consists of analyzing patterns of gene expression and correlating these patterns with underlying disease biology. There are a wide range of techniques used, including DNA microarray analysis and serial analysis of gene expression. It is usually used in the diagnosis of stroke, Alzheimer's disease, and cancer.
Genomics	It is the study of genomes and the complete collection of the different genes that they contain. Genomics studies have proven that many other elements have important functions in the genome, such as transcription factor-binding domains, regions encoding micro RNAs, and antisense transcripts, and large, evolutionarily conserved regions. The technique usually used in genomics is high-throughput genome sequencing. These techniques are usually used in diagnosis of cancer, atherosclerosis, and brain ischemia.
Proteomics	Proteomics is the examination of a collection of proteins to determine how, when, and where they are expressed. Techniques include two-dimensional gel electrophoresis, MS, and protein microarrays. These techniques are usually used in the diagnosis of cancer, atherosclerosis, and brain ischemia.
Metabolomics or metabonomics	Metabonomics is a large-scale approach for monitoring as many as possible of the compounds involved in cellular processes in a single assay to derive metabolic profiles. Although metabolomics first referred to the monitoring of individual cells and metabonomics referred to multicellular organisms, these terms are now often used interchangeably. Techniques applied to metabolic profiling include NMR and MS. These techniques are usually used for drug development and cancer diagnosis.

MS, mass spectrometry; NMR, nuclear magnetic resonance.

8.2 TRANSCRIPTOMICS

The transcriptome includes the total complement of messenger RNA (mRNA) molecules, also called "transcripts," produced in a specific cell or the population of cells comprising a tissue of interest. The transcripts usually originate from less than 5% of the genome in humans and other mammals. Each gene (locus of expressed DNA) may produce a variety of mRNA molecules using the process of alternative splicing. Therefore, the transcriptome has a level of complexity greater than the genome that encodes it. Transcriptome varies because of underlying the wide range of biochemical, physical, and developmental differences. It also differs from cell to cell, depending on environmental conditions.

Methods measuring transcriptomics are considered a high throughput technology enabling the quantification, to some extent, of expression levels of thousands of gene transcripts. It is now possible to simultaneously assess a sufficiently large number of genes to assay almost the entire complement of transcripts within a cell or tissue with the availability of various platforms. We will try to give insight into this promising technology with multiple applications in both basic science and clinical research, where it can be used to discover novel biomarkers potentially useful for diagnostic, prognostic, and therapeutic purposes in medicine.

The use of hybridization array technology is a powerful tool for transcriptomics analysis that has been used to identify biomarkers associated with some tumor types because the patterns of gene expression can be used to classify types of tumors and predict their outcome. Hybridization array is a systemic arrangement of probes where matching known and unknown DNA samples is accomplished by using nucleotide hybridization techniques for exact base pairing (Archacki and Wang, 2004). The technique uses thousands of spotted probes that are immobilized on a solid support that can be a silicon chip, microscope glass slide, or nylon membrane. The spotted sequences consist of DNA, cDNA, or oligonucleotides. For hybridization to the array, RNA extracted from a cell or tissue of interest is submitted to reverse transcription and yfluorescent labeling with fluorophores such as phycoerythrin, cyanine-3 (Cy3), or cyanine-5 (Cy5). Each fluorophore can be chosen for a one-color array, whereas in a two-color array, two samples are labeled with two distinct dyes, usually Cy3 and Cy5, and hybridized together on a single chip. Labeled nucleic acids with unknown identity (so-called targets) bind to the complementary sequence on the array and can be identified according to their location after hybridization. Fluorescent signals are screened with scanners specifically designed for arrays. So far most scanners use lasers for excitation, but some machines apply a white light source and selectable filters. The emitted fluorescence is detected either by photomultiplier tube or charge coupled device detector. Hybridization arrays are roughly divided into macroarrays, microarrays, and microelectronic arrays (Heidecker and Hare, 2007) (Figure 8.1 represents hybridization microchip technology).

An earlier technique, macroarray, has been used for the expression analysis of tens or hundreds of genes. Probes for macroarrays are usually printed on nylon membranes containing up to 5,000 spots, each greater than 300 microns in diameter. Diffusion of the probe molecules on macroarrays restricts the number of spots per square inch of membrane and results in trapping of the probes within the pores of the membrane, slowing down the kinetics of hybridization. Microarrays, on the other hand, use an impermeable glass support that enables high spot density with a spot diameter of 150 microns. This technique facilitates better diffusion and interaction between target and probe, easier removal of the excess unbound targets, and better image quality for data acquisition (Southern et al., 1999).

Probes used for both macroarrays and microarrays can be clones, PCR products, or presynthesized oligonucleotides. The labeling of the probe can be either single-color or two-color labeling. The use of distinct and nonoverlapping wavelengths in two-color microarrays for the excitation of the two labels Cy3 and Cy5 allows separate analysis of fluorescent intensities of each tagged cDNA sample. Thus, the gene expression pattern can be compared by determining the ratio of fluorescent intensities of the two dyes after hybridization with the probes.

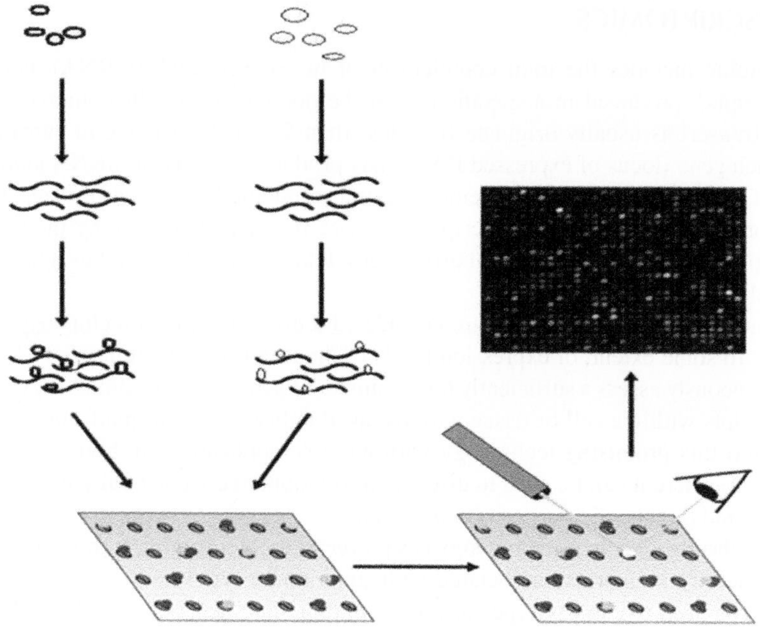

FIGURE 8.1 Hybridization microchip technology.

Another advance in the methods for transcriptomic assay is the use of microelectronic arrays. This method is still under development. They are made up of sets of microelectrodes covered by a thin layer of agarose coupled with an affinity moiety. Charged molecules (DNA, RNA, enzymes, and proteins) and nanoscale and micron scale structures can be directed to specific areas on the array under the control of electric fields both immobilized and hybridized. Microelectronic arrays are used both for analytical function and for immobilization of DNA to make the attachment of nanoparticles possible (Edman et al., 1997; Eiter et al., 2002; Zimmermann et al., 2003).

Sample processing is similar as for the two-color array with the main difference being that biotin is used for labeling. There is controversy regarding the reliability and comparability of data resulting from arrays with different types of labeling. For the use of direct labeling, possible dye bias for two-color arrays has been discussed. Because Cy3 and Cy5 dye molecules are not exactly the same size and shape, this could result in differential rates of incorporation of nucleotides into the growing nucleic acid chain during a labeling reaction. Further concerns regarding competition of the two samples during the hybridization process were excluded with comparative studies ('t Hoen et al., 2004). In this issue, a previous study compared one-color and two-color assays on three different microarray platforms (Agilent, CapitalBio, and TeleChem) for reproducibility, specificity, sensitivity, and accuracy of the two approaches (Patterson et al., 2006). The list of differentially expressed genes was highly consistent across one- and two-color microarray data and was stable within individual platforms considering *P* value and fold-change thresholds for significance. Whereas two-color arrays have a slight advantage with regard to power (sensitivity) and detection of fold changes, the benefit of single color arrays lies in experimental design simplicity and flexibility. Hybridization of only one sample per microarray chip in single color arrays facilitates comparison between various groups. The choice of commercially available microarray chips is extensive. For example, Affymetrix offers commercially available probe sets for more than 15 different species. The other type, Human Genome U133 plus 2.0 arrays, covers probes for sequences of identified genes of the whole human genome and 6,500 additional sequences of unknown function for the analysis of over 47,000 transcripts. Increased reproducibility and accuracy are achieved by using redundant probes for the same transcripts, which are reported as average signal intensity. Additionally, the chips include a further internal control, notably a perfect and mismatched probe for a given transcript. The probe, which contains the exact complementary sequence of the target, is termed perfect match (PM) probe and is paired with a mismatch (MM) probe that is comprised of a sequence that differs by a single base. Both PM and MM probes hybridizing with nucleic acids induce fluorescent signals that are measured and used to calculate the ratio of specific- and cross-hybridization (Pozhitkov et al., 2006). Exceeding a predefined and validated ratio confirms specific binding of the target with the corresponding probe.

Most array manufacturers provide gene identification based on databases of known genomic sequences. Arrays also contain transcripts not currently identified—expressed sequence tags, more commonly for species with less comprehensively characterized genomes. Initial characterization of those probes can be identified with a blast search of the unknown sequence in the genome of a species that has been analyzed to a wider extent, like human, rat, or mouse.

New arrays are designed for investigation of alternative splicing events. These so-called GeneChip Exon Arrays have the advantage of exploring a higher density and variety of exon transcripts per array. Also disease-specific chips have been created such as the LymphoChip from Stanford University (Alizadeh et al., 1999) and the CardioChip from Brigham and Women's Hospital (Barrans et al., 2001). Current technologies will likely render disease-specific chips obsolete, given the ability to screen entire genomes, eliminating potential bias in chip design.

Early gene expression studies have been plagued with variable reproducibility, but it has been shown that if strict laboratory methods are adopted, then good reproducibility can be shown, even across array platforms. The most difficult aspect of an array experiment in the clinical setting is obtaining high-quality mRNA (Irizarry et al., 2005; Larkin et al., 2005; Bammler et al., 2005a, 2005b; Baird, 2006a; Du et al., 2006; Burczynski, and Dorner, 2006). The difficulty arises from the fact that once the blood sample is drawn, RNA degrades very quickly and so must be isolated as quickly as possible. This is achievable in specialized laboratories, as used by Moore et al. (2005), where mRNA was isolated from peripheral blood mononuclear cells (PBMCs), but is not usually practical in a hospital or clinical setting. The PAXgene approach used by Tang et al. (2006) has the advantage that RNA is stabilized by certain procedure, and so sample processing can be delayed for up to 24 h. However, it has the limitation of a severe decrease in the sensitivity relative to mRNA isolated from PBMCs. The subject of which leukocyte types should be used is also an interesting one, and it is not clear to what extent this affects the results obtained in the clinical setting. The PAXgene methodology uses all of the peripheral leukocytes, whereas PBMCs as used by Moore et al. (2005) consist primarily of lymphocytes and monocytes.

The sequencing of the human genome and the advent of microarray technology are opening up new possibilities for applying genomic information in medicine by giving the chance for multiple genes to be studied simultaneously (Schena et al., 1995; Lockhart et al., 1996; Lander, 1999). Information from multiple genes seems to offer considerable hope for providing new insights into the pathogenesis of complex but common disorders, and moreover it can be used to guide diagnostics and therapeutic modalities.

8.2.1 Application of Transcriptomic in Stroke

Stroke is one example of such complex conditions. Current stroke management is based largely on the clinical examination aided by both brain and cardiovascular imaging studies, but this proved to be imperfect at best. The accuracy of early stroke diagnosis may only be as much as 60% (Hand, 2006), and moreover the etiology of ischemic stroke cannot be identified in 30–40% (Baird, 2006b). There is also at present only one treatment available for ischemic stroke, the thrombocytic agent rtPA (recombinant tissue plasminogen activator) that must be administered within 3 h of onset. Therefore, given the lack of treatment options for stroke and the need for rapid and accurate diagnosis, there is great potential to apply gene expression profiling in stroke.

The peripheral blood is a most practical source of mRNA in this clinical setting, despite being in part a surrogate tissue for cerebral ischemia. A study has found that the cell types used in diagnosis have an important impact on the accuracy of diagnosis. Du et al. (2006) have reported that the genes expressed in acute ischemic stroke are most likely to be expressed in neutrophils under physiological conditions, although this does not exclude the possibility that cerebral ischemia induces the expression of genes in lymphocytes that are not expressed under normal physiological conditions. Nevertheless, this finding also does not explain the 85% accuracy of the panel of 22 genes used by Moore et al. (2005) when tested by Tang et al. (2006), in which the blood draw was obtained within 3 h of stroke onset. It may be that at present only the most robust and prolonged gene expression changes are being detected reproducibly in stroke patients. It is currently not clear to what extent the methodology needs to be standardized in terms of cell populations and time of blood draw, whereas it is crucial that the most exact and strictest of laboratory methods should be employed.

Promising results of blood genomic profiling in stroke have been obtained in pilot studies. Moore et al. (2005) reported that a gene-expression signature of acute ischemic stroke could be identified from profiling of PBMCs using Affymetrix microarrays on which there are 22,283 gene probes. The work was validated with real-time PCR in an independent cohort of subjects. Tang et al. (2006) used the PAXgene TM approach (Hombrechtikon, Switzerland) with the whole blood RNA stabilization platform and have also come up with a gene-expression signature of acute ischemic stroke.

It is interesting that there was an overlap of approximately 20% of the genes in the two studies. The findings did not appear to be affected by different time points after stroke, the use of different arrays, different cell types, and different statistical analyses.

This amazing work leads to many new questions. Because blood is a surrogate tissue for stroke, how specific are these signatures? Furthermore, early gene expression studies have varied markedly in their quality, in particular, the quality of RNA that was used, and have been plagued by issues of reproducibility: should the methodology be standardized? Are the clinical signatures an improvement on current diagnostic and management approaches for stroke? And if clinically useful panels of genes can be identified, can they be made into cost-effective tools for use in the clinical setting?

8.2.2 Application of Transcriptomic in Diverse Diseases

The uses of transcriptomic technologies are being studied in other diverse diseases states such as heart disease, Alzheimer's disease, and diabetes.

8.2.3 Application of Transcriptomic in Cancer

Cancer literature has led the way in studies of tumor specimens to the extent that gene-expression signatures are now being used in the management of patients with breast cancer in The Netherlands (Van de Vijver et al., 2002; Quackenbush, 2006; Barry et al., 2010). Furthermore, the results from gene-expression experiments are being applied to improve diagnostic accuracy in hematological malignancies as Burkitt's lymphoma, permitting more accurate differentiation from diffuse large B-cell lymphoma (Dave et al., 2006; Hummel et al., 2009). More data indicate that transcriptomic sequencing is a powerful method to identify novel genetic alterations in acute lymphoblastic leukemia and may be used to identify novel targets for therapeutic intervention.

Numerous reports have demonstrated the potential power of expression profiling for the molecular diagnosis of human cancers (Abdullah-Sayani et al., 2006; Ginsburg and Haga, 2006; Quackenbush, 2006; Sandvik et al., 2006; Tinker et al., 2006). In particular, using large-scale meta-analysis of cancer microarrays, some common cancer biomarkers have been identified. For example, TOP2A is present in 18 types of cancers versus normal signatures, representing ten types of cancer (Rhodes et al., 2004). Similarly, seven gene pairs were identified for common cancer biomarkers (colon, melanoma, ovarian, and esophageal cancers); these biomarkers may be broadly used to increase the sensitivity and accuracy of cancer diagnosis (Basil et al., 2006).

8.3 PROTEOMICS

Proteomics is the large-scale identification and functional characterization of all the expressed proteins in a given cell or tissue, including all of the protein isoforms and modifications. There are challenges in developing analytical tools for proteome analysis. Despite the crucial role that proteomics (Weston et al., 2004) will play in medical-systems biology, the field currently faces huge technical challenges that are a consequence of the complex, dynamic, idiosyncratic, and largely uncharacterized proteome present in most samples of human fluids, cells, or tissues. Many reviews (Anderson and Anderson, 2002; Anderson et al., 2004) tried to summarize the situation for the analysis of human plasma and serum and illustrate the magnitude of the problems. After extensive recent efforts to evolve better methodologies via such approaches as fractionation schemes or the isolation of targeted protein or peptide classes, the fact remains that even after over 30 years of productive work in isolating, characterizing, and quantifying proteins, the large-scale proteomics vision of global and comprehensive protein analysis remains largely unfulfilled.

In particular, there are five critical, unsolved problems that need to be adequately addressed before proteomics can begin to realize its potential role for medical-systems biology and before significant progress can be made with second-generation proteomics areas of study such as intermolecular interactions, biological function, and structures. Figure 8.2 represents the benefits of applications of proteomic analysis in medicine.

There is evidence in the lack of accurate and reproducible quantification of all components of a requirement that is fundamental to systems biology approaches. In addition, there is a need for extensive coverage of the proteome, because most current global approaches probably detect less than 0.1% of the protein species present in a complex sample, mostly because of the dynamic range of more than 10 orders of magnitude between the most abundant protein albumin, and the least abundant proteins measured clinically (Anderson and Anderson, 2002). Furthermore, although protein catalogue information can be useful, the comparative nature of systems biology demands differential analyses, in which changes in multiple species across many samples are routinely measured. Also, it is an important requirement for identifying differentially expressed protein species at the structural level, including precise amino acid sequence as well as the nature and structure of posttranslational modifications. The widely adopted approach of attempting to map a

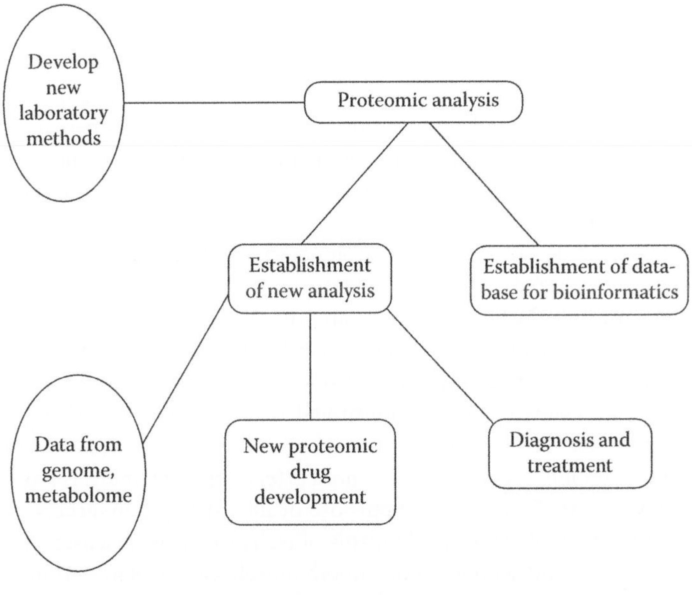

FIGURE 8.2 The benefits of applications of proteomic analysis in medicine.

limited complement of tryptic peptides to specific protein species using the currently substantially incomplete proteome databases will not meet the needs of medical-systems biology for the future.

Finally, the work flows for these primary capabilities need to be robust, cost and time effective, and capable of yielding data from hundreds of samples, even at a discovery stage, before focused and high-throughput assays can be developed for key biomarker subsets. The frequently used tools for proteomic investigations include two-dimensional gel electrophoresis (2D-gel) and mass spectrometry (MS). However, 2D-gel technology has many inherent drawbacks like poor resolution for less abundant proteins and the inability to detect proteins with extreme properties (small, large, hydrophobic, and acidic or basic properties), and identification of the proteins is difficult, time consuming, and costly.

Mass spectrometry-based proteomics technology has been considered as a promising approach for the early diagnosis of cancers. Other promising and innovative approaches include direct tissue profiling and imaging mass spectrometry, in which matrix-assisted laser desoption/ionization (MALDI) MS analysis of thin tissue sections allows a detailed assessment of the complex protein pattern within tissue samples (Chaurand et al., 2004), and the development of natural protein microarrays for diagnosing cancer based on antibody responses to tumor antigens (Qiu et al., 2004).

There are emerging several potentially breakthrough approaches in proteomics related to quantification improvements via labeling procedures. An example of these efforts is the newly presented multiplexing tagging reagent termed iTRAQ (isobaric tags for relative and absolute quantification), reported to work through an efficient global peptide labeling strategy that can derivatize all peptides, including those that are posttranslationally modified (Ross et al., 2004).

8.3.1 Application of Proteomic in Atherosclerosis

Proteomics can be used in medicine for basic or applied research, in fundamental or clinical sciences, to unravel disease processes or to discover biomarkers and therapeutic targets. So far, the value of proteomics has mainly been shown in fundamental applications such as discovery tools. It has demonstrated its power to identify and classify, with the help of bioinformatics, proteins from pathogenic microorganisms or from several cell fractions such as membrane, nuclei, or other components.

In clinical medicine, numerous applications of proteomics have been published. Nevertheless, so far, no proteomic methods have entered the field of routine clinical chemistry laboratories for routine diagnosis. Biochemical tests should fulfill at least one of the conditions listed; otherwise the tests will have little value in routine clinical practice. The proteomerics is clinically important when establishing the diagnosis of a disease, providing prognostic information about a disease guide the therapy of a disease, suggesting preventive actions when no disease has been detected yet, and/or providing predictive information on potential genetic diseases where preventive action is possible. If none of these criteria are met, a biomedical test is not justified and should therefore not be performed (Sanchez et al., 2007).

One of the promising fields of the use of proteomeric is its use in diagnosis of atherosclerosis. Diagnosis of atherosclerosis usually depends on radiological imaging, but this method remains limited by variable efficacy, high cost, and the difficulty of identifying the pathological process. Haptoglobin was considered as a suggestive biomarker for diagnosis of atherosclerosis; other researchers investigated secreted proteins from the supernatant of cultured artery segments (Kritharides, 2003). The results clearly demonstrated that the number and type of secreted proteins is directly proportional to the severity and complexity of the lesion. MALDI-time-of-flight (TOF)-MS showed that proteins secreted by cultured artery segment analysis has been widely used in the field of cardiovascular disease for the purpose of identifying biomarkers of specific disease state and understanding the molecular mechanisms of ischemic heart diseases.

A study on the proteome and secretome of smooth muscle cells isolated from segments of arteries from patients undergoing coronary artery bypass grafting was published (Dupont et al., 2005).

Reference two-dimensional gel electrophoresis (2-DE) maps of secreted and intracellular proteins were built, leading to the identification of 83 intracellular and 18 secreted proteins. Most of these proteins appeared to be of cytoskeletal origin, but others were involved in a wide range of functions such as protein biosynthesis and proteolysis, cellular defense, or metabolic pathways. This confirmed the complex pathogenesis of heart disease.

8.3.2 Application of Proteomic in Brain Ischemia

Another important disease entity with promising use of proteamers is in the arena of brain ischemia. A number of groups have used proteomics with cellular and animal models to study ischemia of the brain. For instance, one study reported a combination of affinity enrichment of phosphoproteins, isotopic labeling, 2-DE, and MS to investigate chemical ischemia on neural embryonic stem cells. The study identified differential isoforms of seven proteins from neurons exposed to chemical ischemia. Some of these are known to be responsive to oxidative stress, and others are part of the immediate molecular response to chemical ischemia induced in cultured cells (Dupont et al., 2005).

Animal models of brain damage and ischemia have also been developed. Murine transient or permanent middle cerebral artery occlusion (MCAO) is the most widely used animal model of focal ischemia. An original approach combining surface-enhanced laser desorption/ionization-TOF-MS, 2-DE, and peptide mass fingerprinting allowed Suzumaya et al. (2004) to discover cerebrospinal fluid (CSF) protein patterns after transient brain ischemia.

They highlighted the rapid increase of monomeric transthyretin in rat CSF after transient MCAO. Brain damage has also been studied using spontaneously hypertensive stroke-prone rats (SHSPR), an inbred animal model of cerebrovascular pathology that resembles human stroke. Unlike the MCAO model, the SHSPR model displays a vasogenic ischemia that refers to the breakdown of the cerebral vasculature's autoregulatory system or the blood-brain barrier, with no evidence of cytotoxic oedema (transfer of water from the extracellular space into the cells) (Schrattenholz et al., 2005). To investigate the pathogenic mechanism leading to cerebrovascular disease in SHSPR, Gianazza and Sironi (2004) monitored the appearance of brain damage and the alteration of the permeability of the blood-brain barrier by magnetic resonance imaging (MRI) and 2-DE. They highlighted impairment of the blood-brain barrier simultaneously with the detection of brain abnormalities by MRI, followed by the appearance of plasma proteins in the CSF. These proteins include thiostatin, a marker of inflammation, and high molecular weight proteins that indicate gross alteration in the permeability of the barrier.

The role that proteomics can play in the field of research on neurodegenerative disorders is well described in the goals of the Human Proteome Brain Project launched by the Human Proteome Organization (Gianazza and Sironi, 2004; Hamacher et al., 2004).

8.3.3 Application of Proteomic in Cancer Diagnosis

Another field in the use of proteomerics is cancer diagnosis. Up to now, this technology has been applied to many types of cancer for biomarker discovery and diagnosis, including ovarian, prostate, breast, bladder, renal, lung, pancreas, and astroglial tumors (Coombes et al., 2005; Kolch et al., 2005; Posadas et al., 2005; Thadikkaran et al., 2005; Veenstra et al., 2005; Ebert et al., 2006). Recently, many new strategies have been developed for cancer biomarker discovery. For example, the combination of multi-dimensional liquid chromatography and 2D-gels has been applied to plasma proteomics of lung cancer (Okano et al., 2006); 2D-gel coupled with MALDI-TOF-TOF was used to screen biomarker candidates in serum samples of breast cancer patients (Huang et al., 2006); coupling 2D liquid chromatography (LC)/MS/MS with automated genome-assisted spectra interpretation enables the direct, high-throughput and high-sensitivity identification of biomarkers from complex biological samples (Powell et al., 2006).

8.4 METABONOMICS

We usually use two terms: metabolomics and metabonomics. Metabolomics means using measurement of all metabolite concentrations in cells and tissues. Metabonomics is the quantitative measurement of the metabolic responses of multicellular systems to pathophysiological stimuli or genetic modification.

Recent technological progress in nuclear magnetic resonance (NMR) spectroscopy and MS, the two most accepted methods used in the measurement of metabolites, has improved the sensitivity and spectral resolution of analytic assays on metabolomic samples in attempt to achieve a comprehensive biochemical assessment. Extended developments in both NMR and hyphenated MS technology play crucial roles, although alternative approaches with great potential are also now emerging. Major improvements in NMR include the introduction of extremely high field magnets (currently up to 900 MHz), cryoprobe technology, and the measurement of microsamples. In high field NMR (800 MHz) signal-to-noise ratios, the cryoprobe has been improved by a factor of 15–20, as compared with a standard probe in a 600-MHz apparatus. Following this improvement, acquisition times can be reduced dramatically. This has paved the way to including ^{13}C NMR spectroscopy as a promising tool for metabolic profiling (Veenstra et al., 2005). For this, spectrum acquisition now requires 30 min per sample instead of overnight measurements.

Nanoprobe technology allows the analysis of samples in volumes as small as 25 ml and has been applied in the metabolic profiling of cerebrospinal fluid microdialysates obtained from several brain regions of freely moving rats. Additionally, high-resolution magic-angle spinning (HR-MAS) NMR spectroscopy provides information on metabolite concentration and compartmentalization in intact tissue samples. The use of HR-MAS produces narrow-line-width spectra comparable with those from solid-state NMR (Yang et al., 2004; Huang et al., 2006; Powell et al., 2006).

The latest generation of Fourier transform ion cyclotron resonance mass spectrometers represents a quantum leap forward in the robust capacity of mass spectrometers for high-resolution metabolite analysis. In standard analytical experiments, resolutions of >100,000 (50% valley definition), yielding accurate masses within 0.5 ppm, are easily obtainable without compromising the chromatographic conditions (including nano-LC) and without the use of internal calibrants (Odunsi et al., 2005; Seidel et al., 2006). The accurate masses obtained give elemental compositions that provide a major step towards metabolite identification.

Although such high mass resolution might minimize the need for chromatographic separation before introduction to the mass spectrometer, this is probably not possible. Direct infusion of the sample into the mass spectrometer would be predicted to lead to suppression of ionization in the electrospray inlet and the formation of adducts.

Interestingly, however, an optimized protocol including the use of the non-salt-based buffer Tricine for metabolome analysis in yeast by direct infusion has been reported (Castrillo et al., 2003). On the other hand, enhanced chromatographic resolution from the use of a long (90 cm) monolithic capillary column effectively reduced ionization suppression so that several hundred peaks could be detected in an *Arabidopsis thaliana* extract (Tolstikov et al., 2003). The limited sample availability has led to reduction of the dimensions of the LC columns to enable increased sensitivity.

Capillary LC connected to a quadrupole-time-of-flight mass spectrometer generated about 2000 different mass signals in extracts of *Arabidopsis* (von Roepenack-Lahaye et al., 2004). Highly polar compounds, normally missed by standard reversed phase chromatography, are detected by hydrophilic interaction chromatography. In combination with electrospray ionization ion-trap MS (in positive and negative ion mode), oligosaccharides, aminosugars, and sugar nucleotides were detected (Tolstikov and Fiehn, 2002).

The highest separation efficiency before MS detection is obtained with the use of capillary electromigration methods. From *Bacillus subtilis*, 1692 metabolites were analyzed by using a combination of three methods of capillary electrophoresis (CE)-MS. To maximize detection sensitivity, the mass spectrometer scan range was narrowed to 30 m/z; about 30 injections were necessary to cover the necessary mass range (Soga et al., 2003). Pressure-assisted CE-MS allowed the analysis of multivalent ions as nucleotides and CoA-derivatives (Soga et al., 2002). By applying air pressure to the inlet of a noncharged polymer-coated capillary, a conductive liquid junction between the capillary and the MS was maintained. Selectivity originating from the mass spectrometer detector (i.e., by precursor ion scanning for specific fragment ions) was applied in the CE-MS analysis of sugar nucleotides. Furthermore, CE is an excellent tool for the on-line preconcentration of metabolites. Dynamic pH junction, sweeping, and dynamic pH junction sweeping are three recent, complementary methods for electrokinetic focusing (Soo et al., 2004). A further increase in separation efficiency is obtained by using CE in a two-dimensional approach (Britz-McKibbin et al., 2003): hyphenated micro-LC (with a monolithic column) and CE. Sweeping and a dynamic pH junction were employed to interface the two dimensions. The method was evaluated for 54 standard metabolites and applied to a *Bacillus subtilis* extract.

As an alternative to the commonly used MS and photodiode array detection systems, the use of an electrochemical detector (ECD) sheds new light on the metabolome by the very sensitive detection of redox active compounds. Recent developments in this area are the coulometric array technology (16 channels), including the option of combining it with gradient elution (so far, ECD is mainly used in isocratic systems) (Shi et al., 2002; Jia et al., 2002).

Another approach is the use of Fourier transform infrared spectroscopy. This economical and rapid technique was applied in an animal model study for idiosyncratic toxicity (Ellis et al., 2003; Kaddurah-Daouk et al., 2004). For body fluid, tissue, and whole organism profiling, Raman spectroscopy has several applications (Harrigan et al., 2004). Stable isotope-based dynamic metabolic profiling (SiDMAP) proved to be a powerful approach, enabling a better understanding of changes in substrate flow as a basis for drug mechanisms and disease (Clarke and Goodacre, 2003; Boros et al., 2003).

In conclusion, progress has been made in the dynamic range and coverage capacity of metabolite profiling. Nevertheless, despite the availability of such high technologies, the comprehensive identification of metabolites remains a key challenge that awaits the better integration of NMR and MS/MS data, as well as the development of novel approaches.

In medical-systems biology, the metabolome coverage of body fluids is still limited, and the integration of techniques in a platform is mandatory (Van der Greef et al., 2003).

8.4.1 Applications of Metabolomic Assay in Diagnosis of Cancer

Metabolomics allows for a global assessment of a cellular state within the context of the immediate environment, taking into account genetic regulation, altered kinetic activity of enzymes, and changes in metabolic reactions (Mendes et al., 1992, 1996; Griffin and Shockcor, 2004). Thus, compared with genomics or proteomics, metabolomics reflects changes in phenotype and therefore function. The omics sciences are, however, complementary because "upstream" changes in genes and proteins are measured "downstream" as changes in cellular metabolism (Griffin and Shockcor, 2004). The converse, however, is that metabolomics is also a terminal view of the biological system, not allowing for representation of the genes and proteins that are increased or decreased. Other features of metabolomics are similar to those of proteomics and transcriptomics, including the ability to assay biofluids or tumor samples and the relatively inexpensive, rapid, and automated techniques once start-up costs are taken into account.

Interestingly, modern interest in metabolomics in oncology originally stemmed from the claim in the late 1980s that cancer could be identified by NMR spectra of blood samples (Fossel et al., 1986). Unfortunately, these data were later found to be falsified, and the field of metabolomics was tainted. Despite this, the notion has persisted that correctly applying metabolomics on patient specimens may affect oncologic practice. Metabolomics should be used for identifying multivariate biomarkers, including fingerprints, profiles, or signatures, the patterns that characterize a state of cancer. We might eventually be able to diagnose cancer earlier when it is still amenable to cure by using this technology, determine the aggressiveness of cancer to help direct prognosis and therapy, and predict drug efficacy. These signatures can be practical and accurate, although they also require sophisticated analytic techniques (Van der Greef et al., 2004; Ardekani et al., 2002).

The use of metabolomics as a diagnostic tool has been validated using citrate and choline in prostate and breast cancer, respectively, both of which are now covered by health insurance providers (Scheidler et al., 1999; Serkova et al., 2007; Bartella et al., 2007). The precedent for this type of omics technology includes colon cancer diagnosis and prognosis by gene microarrays and ovarian cancer diagnosed with serum protein profiles (Buckhaults et al., 2001; Petricoin et al., 2002). Conceivably, metabolomics may play a role in investigating tumors posing a diagnostic challenge. Evidence suggests that metabolomic profiles are already used in diagnosing ovarian cancer by analyzing either serum or tumor tissue (Denkert et al., 2006; Odunsi et al., 2005). Detection of pancreatic cancer has also been possible *in vivo* by analyzing cellular glucose use via GS/MS (Boros et al., 2005). Future studies should evaluate the use of body fluids other than blood, for example ascitic fluid in ovarian cancer, pancreatic secretions in pancreatic cancer, and/or bronchoalveolar or pleural fluid in lung cancer. If pathognomonic profiles can be identified and validated in these fluids, metabolomics may save time, cost, and effort in obtaining a definitive diagnosis in situations where no other test can provide answers. Additionally, there could be a future role for metabolomics as a screening tool, particularly in those tumors that readily produce or secrete easily accessible fluid.

8.4.2 Applications of Metabolomic Assay in Development of Drugs

Similar to other strategies currently being investigated to individualize therapy, such as the assessment of mutations or amplification of receptor tyrosine kinase genes in Global Innovation through Science and Technology (GIST), metabolomic studies should be integrated into preclinical and clinical research and assessed for predictive value (Prenen et al., 2006; Heinrich et al., 2006). A form of *in vivo* metabolomics, position emission tomography (PET) imaging, with the use of radioactive glucose, choline, or thymidine as metabolic end points, has already been evaluated as a predictor of drug efficacy in some tumor types. In recurrent GIST studies, compared with standard computed tomography scanning, [18F]fluordeoxyglucose (FDG) PET was superior in predicting an early response to imatinib therapy when evaluated in 56 patients before and after initiating imatinib therapy (Gayed et al., 2004; Holdsworth et al., 2007). Furthermore, changes on [^{18}F]FDG PET have been predictive of the response to standard cytotoxic treatments in patients with breast cancer, locally advanced or metastatic nonsmall cell lung cancer, ovarian cancer, after high-dose salvage chemotherapy in relapsed germ-cell cancer, and in treatment-naïve patients with cervical cancer (Pio et al., 2006; Kidd et al., 2007). In hematological malignancies not amenable to [^{18}F]FDG PET imaging, metabolomic analysis on circulating tumor cells after [^{13}C]glucose administration could be used in assessing treatment effects, thereby providing biological response information noninvasively. This could also be applied to circulating tumor cells from solid tumors.

The principal objectives of early clinical trials are to determine the maximum tolerated dose of new drugs or drug combinations while also collecting information on drug tolerability, pharmacokinetics, and pharmacodynamics. Increasingly, biomarkers are being used preclinically and in early clinical development to identify, validate, and optimize therapeutic targets, to confirm mechanism of drug action, and as pharmacodynamic end points (Park et al., 2004). As discussed earlier, metabolomics is already being assessed as a pharmacodynamic marker of novel agents, whereas another application is in the characterization of toxic effects. For example, metabolomics can be used as a biomarker of hepatic, renal, and lung toxicity with various metabolites, including glucose, lactate, lipoproteins, and amino acids, increasing or decreasing providing a recognizable pattern associated with organ dysfunction (Robertson et al., 2000; Beckwith-Hall et al., 1998). Much of these data have not been validated, and there is some overlap between various toxins but the pattern, temporal rate of change, and the extent of change in metabolites can still provide toxicity assessments (Serkova and Niemann, 2006). Such patterns may be used for preclinical drug screening or as a means of following a patient clinically to monitor target organ effects.

Although metabolomic technology has improved and evidence is accumulating to support its use in clinical decision making, the discipline is still in its infancy, and metabolomics has somewhat lagged behind other omics sciences because of technical limitations, database challenges, and costs. Future development and application will be dependent on several factors, such as the establishment of spectral databases of metabolites and associated biochemical identities, as well as cross-validation of NMR- or MS-obtained metabolites and correlation with other quantitative assays. Lastly, it will be important to integrate the results of metabolomic assessments with other omics technology so that the entire spectrum of the malignant phenotype can be characterized.

8.5 PEPTIDOMICS

The use of peptidomics highlights both the visualization and identification of the whole peptidome of a cell or tissue, that is, all expressed peptides with their posttranslational modifications. In general, there are two sources of the candidate peptidome biomarkers: one is the peptides and fragments derived from parental protein molecules, and the other is the cleavage products generated, ex vivo, after blood clotting. Both can make contributions to diagnosis.

Several researchers adopted methods for measurement of peptidomic. Some researches (Villanueva et al., 2004) developed an automated procedure for serum peptide profiling that uses magnetic, reverse-phase beads to capture peptides, followed by MALDI-TOF-MS analysis.

8.5.1 Application of Peptidomic in Diagnosis of Cancer

The low molecular weight of both CSF and serum, although a promising reservoir for biomarker discovery, remains largely uncharacterized. Previous studies have demonstrated that a pattern of 274 peptides can be used to correctly predict 96.4% of samples from patients with or without brain tumors. Another laboratory achieved success by adopting the use of highly optimized peptide extraction and MALDI-TOF-MS (Villanueva et al., 2006). It was evaluated by profiling 106 serum samples from patients with advanced prostate cancer, breast cancer, and bladder cancer. The study identified 61 signature peptides that provide accurate prediction between the cancerous patients and controls. Obviously, the peptides identified as cancer-type-specific markers proved to be the products that were generated from enzymatic breakdown after patient blood collection.

8.5.2 Application of Peptidomic in Diagnosis of Neurodegenerative Disorders

Another promising field for use of peptidomic analysis is neurodegenerative disorders. Analysis of endogenous peptides produced by aberrant cleavage of proteins in the disease state not only can provide alternatives for disease diagnosis but also can highlight the pathways that may be involved in the neurodegenerative diseases. One significant advantage of using a MS-based approach is that it can unambiguously record protein fragment peaks in mining the low-mass proteome and peptidome. If a biomarker for a given disease state is a fragment of a larger protein, it may be very difficult to produce effective antibodies for conventional tests such as enzyme-linked immunoassay. Moreover, coupling LC to MS overcomes the limitation observed in 2D gel electrophoresis, because this method has very low resolution for small proteins and peptides.

Because of its proximity to the brain, CSF has been the subject of several peptidomics studies recently. Many of the CSF peptides identified so far are biologically active. One study (Stark et al., 2001) developed an organic phase extraction and MS-based profiling strategy and succeeded in identifying of a number of peptide fragments of human CSF proteins. Later on another study (Yuan and Desiderio, 2005) applied ultrafiltration with a molecular mass limit of <5 kDa and solid phase extraction, followed by LC-MS/MS to characterize human CSF peptidome. In this principle study, 20 representative peptides derived from 12 proteins were identified. The ultrafiltration method divided the CSF samples into low molecular weight *peptidome* and higher molecular weight *proteome* fractions. Although proteome fractions were subjected to tryptic digestions followed by LC-MS/MS identifications in conventional proteome mapping experiments, the endogenous CSF peptidome fractions were analyzed directly, by capillary LC coupled to tandem MS experiments. This approach has become the method of choice for peptidome profiling in CSF and other body fluid samples. By adopting a similar strategy and using a hybrid LTQ-orbitrap mass spectrometer with high accuracy and high resolution, Zougman et al. (2008) were able to enhance the CSF peptidome profiling to assure identification of 563 peptides derived from 91 precursors.

Serum is another sample that can be used in peptidomes analysis. It is recognized that the low molecular weight (LMW) fraction of the serum proteome may contain shed proteins and protein fragments emanating from physiologic and pathologic events taking place in all of the perfused tissues.

Theoretical models were proposed (Geho et al., 2006) that predict that the vast majority of LMW biomarkers exist in association with circulating high molecular mass carrier proteins. Lopez et al. (2005) have examined the carrier-protein-bound fraction of serum to discover patterns of peptide ions that provide diagnoystic signatures of adenocarcinoma. They pulled down the target peptides by affinity chromatography and analyzed them by MALDI-TOF. Although numerous LMW putative markers were detected, the amino acid sequence or identity of each marker and how each one contributed to the classification in terms of specificity and sensitivity were not investigated.

An emerging exciting field in neuroscience-related peptidomics research is imaging mass spectrometry (IMS) (McDonnell and Heeren, 2007). In a typical procedure for IMS on tissues, a thin-sliced tissue section is placed on a sample plate, and MALDI matrix is deposited either as a thin layer or as a spot pattern. The sample is then introduced into the instrument for MS analysis, and a laser is rastered across the tissue section collecting an array of mass spectra at each set of XY coordinates. This array of mass spectra can then be processed to produce individual molecular ion images, in which each pixel represents ion signals extracted from the corresponding spectrum. IMS offers several advantages over immunocytochemistry, an antibody-based alternative approach in determining the distribution of neuropeptides. These advantages include higher throughput, higher chemical specificity, and the ability to discover novel peptides. As an example, the distribution of five structurally related Aβ peptides in mouse brain sections has been determined by IMS with attomole sensitivity and 50-μm lateral resolution (Stoeckli et al., 2002). Although most MALDI-based IMS studies rely on mass measurement alone to identify peptides and proteins, DeKeyser et al. (2007) have taken advantage of the TOF/TOF mass analyzer that enables both postsource decay and collision-induced disassociation fragmentation in MS/MS and provided increased confidence for assignment of a number of neuropeptide families. The first attempt (Pierson et al., 2002) to directly profile proteins and neuropeptides in the brain tissue of a rat model of Parkinson's disease (PD) demonstrated differential expression of numerous proteins as well as some changes in posttranslational modifications such as alterations of acetylation. In a later study (Skold et al., 2006), PEP-19, a 6.7-kDa polypeptide that belongs to a family of proteins involved in calcium transduction through their ability to interact with neuronal calmodulin, was found to be significantly decreased in model PD brain. Because IMS studies enabled the localization of specific peptide or protein molecules to the diseased region of the tissues, there is greater chance for this technology to discover potentially disease-relevant biomarkers. The limitation, however, is that the tissue is not as easily available as the body fluids. Also, the tissue-based disease biomarkers are more likely indicative of a late stage of disease progression. Nonetheless, once identified in diseased tissues via IMS technology, these putative (neuro) peptide or protein markers can serve as useful candidates to develop body fluid-based diagnostic measurement (Wei and Li, 2009).

8.6 GLYCOMICS

The emerging field of glycomics has been attracting interest in the area of medical research. The aim of glycomics is to identify and study all of the glycan molecules produced by an organism, encompassing all glycoconjugates (glycolipids, glycoproteins, lipopolysaccharides, peptidoglycans, and proteoglycans), whereas glycoproteomics refers only to the characterization of the glycosylation of proteins. Comparative studies of the specific carbohydrate chains of glycoproteins can provide useful information for the diagnosis, prognosis, and immunotherapy of tumors (Kobata and Amano, 2005).

Carbohydrate recognition often consists of low affinity interactions between glycans and their associated binding partners. Therefore, a successful microarray platform must be able to probe protein-carbohydrate, carbohydrate-carbohydrate, nucleic acid-carbohydrate, and intact cell-carbohydrate interactions with sufficient specificity to detect weak binding. To achieve this, the glycan moiety should be displayed in a fashion that is readily accessed by the analyte, and the system must strive to achieve low background signal caused by nonspecific adsorption of analyte to nonbinding glycans and underivatized portions of the array. Carbohydrate-modified surfaces are ideal for interrogating these weak interactions, because monolayers of sugars on the surface of a chip are analogous to the multivalent display of glycan frequently found on the cell surface. Figure 8.3 represents functions of glycomics.

This multivalency serves to amplify the binding event, increasing the level of the desired signal. Several factors must be taken into consideration when designing a microarray. First, a suitable surface must be selected for displaying the glycan library. This can include microtitre plates, functionalized glass slides, nitrocellulose, and gold surfaces. Second, a linking chemistry, whether covalent or noncovalent, must be adopted. This requires careful consideration of the source of the glycans (synthetic versus isolated) and whether a spacer will be incorporated to increase the accessibility of the sugar in solution above the surface. Numerous linking chemistries are possible, such as lipid-modified glycans for noncovalent adsorption to nitrocellulose and hydrophobic surfaces,

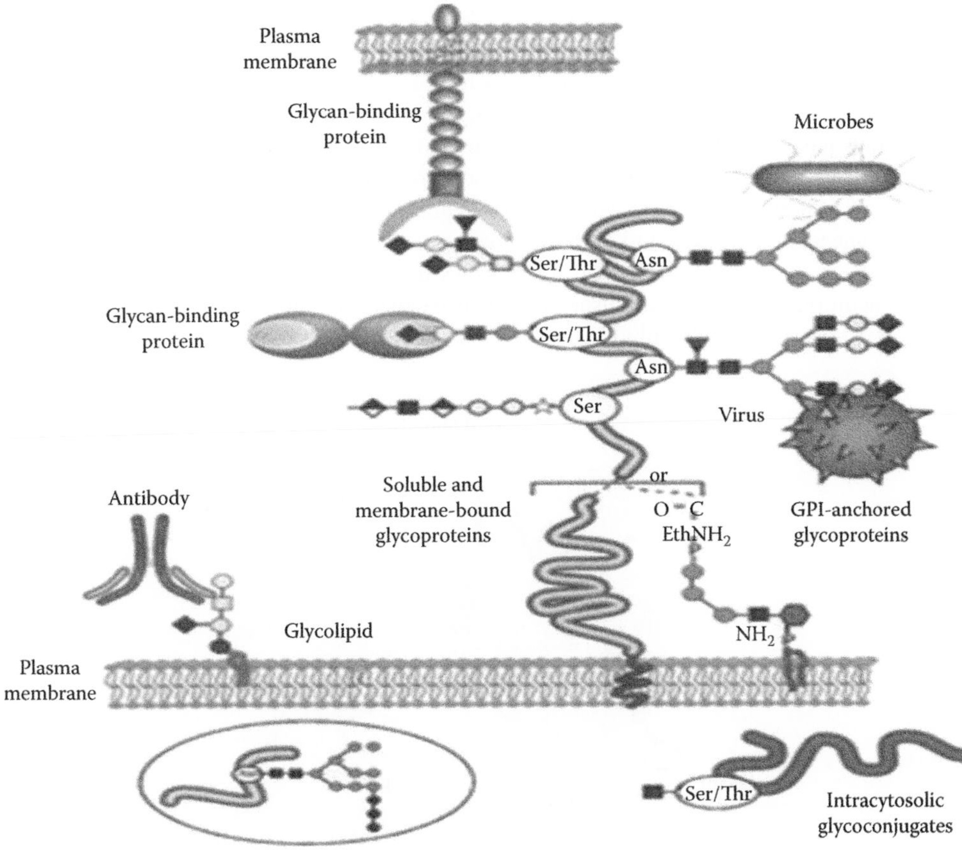

FIGURE 8.3 Functions of glycomics.

and sulfydryl, amino, and maleimido derivative sugars, for covalent attachment to functionalized surfaces. Third, a protocol must also be developed for blocking potential sites for nonspecific binding of analyze to the surface. Finally, a method must be established to detect weak binding events on the surface of the array. Most often, this will entail a fluorescent probe, although some groups have had success with surface plasmon resonance (SPR) and MALDI-TOF-MS.

8.6.1 Application of Glycomic in Infections

While ongoing research continues to refine the existing microarray platforms, biomedical studies have already found numerous applications for the current generation of carbohydrate microarrays. Arrays of carbohydrates have been used to identify the glycan specificity of proteins, nucleic acids, and intact cells (pathogen adhesion and cell-cell recognition; Figure 8.4).

Glyochips have also been employed to examine human serum for carbohydrate reactivity and to profile glycan-dependant enzymes and enzyme inhibitors. Carbohydrate microarrays have found particular application in the study of HIV glycobiology (Ratner et al., 2004). The predominant surface envelope protein of HIV, glycoprotein gp120, plays a vital role in viral entry. The gp120 molecule possesses numerous sites for N-linked glycosylation, and the glycans are known to participate in host-cell adhesion and immunoevasion. Utilizing a microarray comprised of a panel of synthetic high-mannose oligosaccharides. A study was carried out to profile the mannose-binding specificities of two anti-HIV proteins (Cyanovirin-N and scytovirin): the HIV-inactivating human monoclonal antibody 2G12 and the dendritic C-type lectin DC-SIGN (Ratner et al., 2004). The Consortium for Functional Glycomics has also used their microarray to study the carbohydrate ligand-binding properties of DC-SIGN and a related endothelial cell receptor DCSIGNR (Adams et al., 2004).

Continuing efforts to develop carbohydrate-based vaccines for pathogens (including malaria and HIV) and cancer have highlighted additional uses for the glycochip. Microarray analysis of serum anticarbohydrate activity will facilitate vaccine design and dose trials.

Microarrays have already been used to show anticarbohydrate activity in healthy individuals. These results detected the expected anti-ABO blood group response (varying by the individual's blood type), as well as antibodies towards glycans representative of Gram-negative bacterial polysaccharide (suggesting past exposure to these antigens).

Seeberger's group has used their own microarrays to detect bacterial adhesion to cell surface glycans. In addition, Seeberger and Disney (2004) have also used an aminoglycoside glycochip to study antibiotic resistance and antibiotic toxicity and to detect RNA binding.

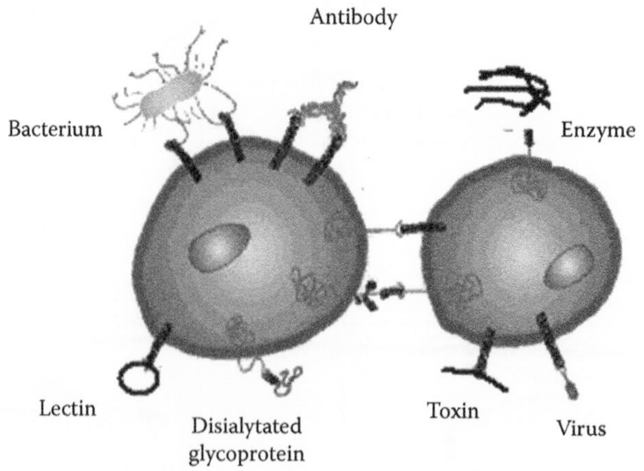

FIGURE 8.4 Glycomic receptors in host-pathogen interaction.

8.6.2 Application of Glycomics in Liver Cancer

An example application of glycomics is that glycoproteomic analysis was used to discover serum markers in liver cancer (Block et al., 2005). A glycoprotein, Golgi protein 73 (GP73), was found to be elevated in the serum of people with hepatocellular carcinoma (HCC).

Serum GP73 levels correlate with a diagnosis of HCC, with a positive predictive value equal to or greater than the currently used marker, α-fetoprotein (Marrero and Lok, 2004). In another study, 19 glycoproteins were found to be hyperfucosylated in HCC and are potential biomarkers for HCC diagnosis (Comunale et al., 2006). MS combined with lectin-based glycoprotein capture strategies for the discovery of serum glycoprotein biomarkers has been reviewed by Drake et al. (2006). Recently, a novel strategy, natural glycoprotein microarray using multi-lectin fluorescence detection, was shown to be useful for the identification of potential cancer biomarkers (Patwa et al., 2006).

8.7 PHOSPHOPROTEOMICS

Phosphoproteomics is the characterization of the phosphorylation of proteins. Is it possible to mine potential cancer biomarkers for molecular diagnosis and prognosis from the tumor phosphoproteome? Preliminary studies indicated that a distinct pattern could exist in the phosphotyrosine proteome of cancer patients.

Currently, there are several emerging strategies for the enrichment and quantification of phosphoproteins. Reinders and Sickmann (2005) reviewed the most frequently used methods for the isolation and detection of phospho-proteins and -peptides, such as specific enrichment or separation strategies as well as the localization of the phosphorylated residues by various mass spectrometric techniques. In particular, the coupling of stable-isotope labeling with amino acids in cell culture to MS is considered to be a powerful, simple,k and quantitative phosphoproteomics technology (Amanchy et al., 2005; Mann, 2006).

8.7.1 Application of Phosphoproteomics in Diagnosis of Cancer

Phosphoproteomics have been used in diagnosis of various types of cancer. Lim et al. (2004) reported that three proteins (vimentin, hsp70, and actin) were consistently hyperphosphorylated at tyrosine residues in breast tumors but not in normal tissues.

Kim et al. (2005) identified 238 phosphorylation sites in 116 proteins from the HT-29 human colon adenocarcinoma cell line, using immobilized metal-affinity chromatography combined with LC-MS/MS analysis. However, it is noted that the identification of phosphorylated proteins remains a difficult task.

8.8 LIPIDOMICS

Lipidomics is the systems-level analysis and characterization of lipids and their interacting partners. Lipids have been implicated in many human diseases, including cancer.

8.8.1 Application of Lipidomics in Diagnosis of Cancer

It is possible that a global analysis of lipid patterns could provide diagnostic information for particular cancers, although at present little research is available in this field. For example, the simplest phospholipid, lysophosphatidic acid, was found to be markedly elevated in the ascetic fluid of ovarian cancer patients (Umezu-Goto et al., 2004). Furthermore, levels of sphingolipids are altered in various types of cancers (Wenk et al., 2005). Recently, to facilitate the understanding of the role of liypid mediators in cancer, a LC-electrospray ionization-MS-MS assay was developed to conduct lipidomic analysis of 27 mediators, including prostaglandins, prostacyclines, thomboxanes, dihydroprostaglandins, and isoprostanes (Masoodi and Nicolaou, 2006).

8.9 OMICS TECHNOLOGY IN DRUG DEVELOPMENT

In simple term, drug discovery requires the identification of a disease, knowledge of the disease mechanism, and identification of a target (point of intervention). Currently it takes an estimated 7–10 years to develop and market a drug at a cost that exceeds $800 million. Figure 8.5 shows the steps in drug development.

Although some drugs may be the result of fortuitous discovery (e.g., penicillin) or from the screening of natural products (e.g., taxol prepared from an extract from the Pacific yew tree is used in the treatment of ovarian cancer), most drugs are the result of carefully designed research programs of screening, molecular modification, and mechanism-based drug design.

Most drug targets are proteins. These can be human proteins that are either defective or abnormally expressed or can be proteins within a pathogen. The site of interaction of the drug with target is often at the protein's active site.

The current trends in drug discovery are toward automation, high-throughput processes, and technologies such as miniaturization. The traditional route to drug discovery of understanding the mechanism of disease and synthesizing compounds that interact with targets has identified fewer new drugs or new molecular entities. It is important that researchers identify quickly and focus upon the most promising drug candidates at the earliest possible point in the drug discovery process.

With the completion of the Human Genome Project, many novel approaches to drug discovery have become evident. The human genome contains approximately 35,000 genes. These can be copied into mRNA, which acts as a template for the translation of the genetic code into proteins. However, the biological effect of some genes has yet to be identified. Some genes have the potential to express multiple forms of mRNA, which can lead to forms of the same protein within different tissues.

This means that the potential targets for drug intervention are approximately 100,000. Nanotechnology can enhance the drug discovery process through miniaturization, automation speed, and reliability of assays.

8.9.1 DNA Microarray Technology in Drug Development

Arrays impact many aspects of the drug discovery process, from initial identification of drug targets and identification of small molecule drug candidates, to toxicogenomic studies, which are aimed at verifying drug safety. The microarray is helping to unravel much of the genomes' complexity. The necessity of high density arrays for drug discovery has become increasingly apparent as the number of potential targets to be investigated has increased as a result of genome sequencing efforts. Since its invention, microarray information capacity has increased so that currently it is able to measure expression for nearly 50,000 transcripts or genotype more than 100,000 polymorphisms in a single experiment.

8.9.2 Impact of Nanotechnology on Arrays

Affymetrix has recently developed an automated 96-array high-throughput (HT) format of array. HT array analysis provides a method of identifying drug targets and pathways for complex disease mechanisms. Once a disease pathway is identified, it needs to be validated and then verified that pathway disruption will actually affect the disease. By using whole-genome expression profiling, a wide range of effects that result from disrupting the pathway can be understood so that potential drug targets for drug design can be evaluated. After identifying and validating a target, HT arrays can also be used to screen libraries of compounds to identify those that disrupt expression of particular disease genes.

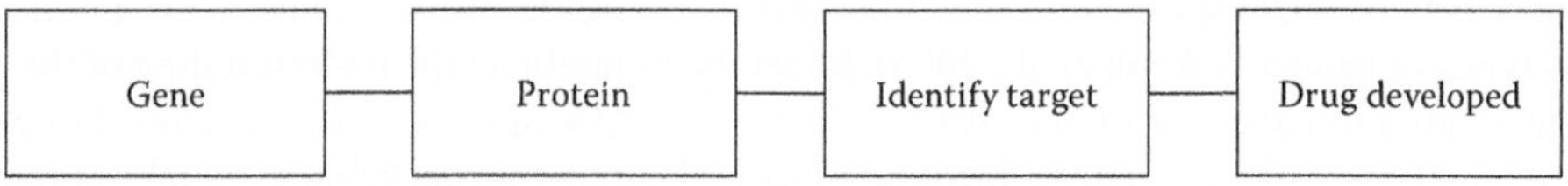

FIGURE 8.5 Steps in drug development.

8.9.3 Protein Chips in Drug Development

Although DNA microarrays can measure the levels of messenger RNA expressed in a cell, they cannot directly measure the proteins or protein functionality that those messengers produce. Protein biochips, however, can make direct measurements of the relative levels of proteins and their interactions with other molecules. This makes them ideal for use in searching for relevant targets and disease-specific marker proteins. The technique allows scientists to monitor a cell's metabolism and its response to external stimuli.

8.9.4 Impact of Nanotechnology on Protein Chips in Drug Development

Protein microarrays can be regarded as tools that will enable rapid progress in the study of proteomics. Despite the technical challenges that are associated with them, many companies are investing in them. Research is being carried out to determine the substances to which proteins bind. This has implications in the understanding of disease and the development of cures for them. Researchers in several areas of cancer research use SPR technology to study biomolecular binding events. SPR is an electron charged density wave phenomenon that occurs at the surface of a metallic film when light is reflected under certain conditions. SPR is altered by the binding of molecules to the opposite side of the film from the reflected light. This method enables data to be gathered on binding specificity, affinity, and kinetics. The technology is used by Biacore for isolation of analytes from complex mixtures and for identification of proteins and other molecules.

Celera is a company evaluating cell-surface proteins from normal and diseased cells to identify likely therapeutic targets for antibodies. The company applies techniques to capture cell-surface proteins, which it then identifies using a program based on liquid chromatography and mass spectrometry. Using this technique, the company has identified differentially expressed proteins on the surface of pancreatic cancer cells. The proteins are undergoing further validation to confirm their viability as targets.

Avalon uses both gene expression and monitoring cytogenic changes to discover drug candidates and their mechanisms of action. This means that an actual drug target does not have to be verified. Instead they are monitored through gene expression.

8.9.5 Target Validation in Drug Development

It is obvious that with the genomic and proteomic methods described above and with the complete map of the human genome, there will be no shortage of new targets to be evaluated. The problem arises in determining whether these targets are significant. This can be done by target validation.

Targeted gene disruption (TGD) can be used for target validation. TGD relates to the production of knockout or transgenic mice animals to study the effects of removing a particular gene coding for the molecular target. By removing the gene encoding a protein that has been identified as a candidate will mimic the application of a strong repressor for that protein, thus allowing investigators to determine the proteins normal cellular function. However, there are compensatory mechanisms in any organism that may invalidate the intended result of the knockout, and in addition it takes time to produce and analyze these animals.

Antisense and ribozyme technology uses molecules to hybridize to mRNA and prevent expression of the protein product of that RNA through either induced degradation or catalytic cleavage of the mRNA. The company BIOGNOSTIK offers ANTISENSE, a custom design service. ANTISENSE oligonucleotides are designed to switch off protein expression by interfering with the mRNA. They also manufacture phosphorothioate antisense oligonucleotides that are capable of entering cells in tissue culture and binding to complementary mRNA and preventing their translation.

Aptamer technology makes use of DNA molecules that bind specifically to proteins and can be employed to inhibit the function of a specific protein or as competitive binders of small molecule drugs to target proteins.

8.9.6 Lead Identification in Drug Development

Following the identification and validation of a target in a disease, a lead compound must be generated. This is a compound that demonstrates a desired biological activity on a validated molecular target. The majority of leads discovered are derived from a collection that is referred to as a library. Today, many of these libraries are commercially available. Most pharmaceutical companies have their own compilation of compounds that have been synthesized and screened against a variety of targets.

The drug company Nuevolution has generated a library comprising 100 million diverse compounds. The abundance of information that has been obtained from functional genomics and virtual screening has given rise to bioinformatics, which combines elements of biology with computer science. Bioinformatics has been used to predict the three-dimensional (3D) structures of targets and potential chemicals that could interact with them and inhibit protein function, by comparing primary amino acid sequences of novel proteins with the determined 3D structure of known proteins.

De Novo Pharmaceuticals has developed algorithms to assess how well a particular compound molecule will bind to the active site of a target. Candidates are screened *in silico* using the company's EasyDock program to determine binding affinities for a ligand.

8.9.7 Model Validation in the Real World

To judge whether a drug is safe and effective, information must be gained on how it is absorbed, distributed throughout the body, metabolized, and excreted and how it affects the action of the body's cells, tissues, and organs. This validation is summarized by the acronym ADMET, which stands for Absorption, Distribution, Metabolism, Excretion, and Toxicity. Scientists have developed studies that may be conducted outside the living body by using cell and tissue culture and computer programs that simulate human and animal systems. Computer models help to predict the properties of substances and their probable actions in living systems.

Cellomics have developed sensitive digital imaging techniques that can detect morphological changes in cells in real-time. The effects of a drug can be monitored, and any adverse reactions can be detected early.

Advances in polymer chemistry and membranes have allowed the generation of synthetic 3D tissues that can be used for the evaluation of candidate drugs. For example the MatTek Corporation has synthesised *in vitro* tissue models derived from normal human cells. The models mimic respiratory epithelium and skin; thus, the results more closely mimic the real-world interaction of drug candidate to target human organ.

The response of human tissues itself can also be probed ex vivo. This involves removing tissue from a patient to evaluate the efficacy of a drug. This can also involve transplanting human tissue into an animal model.

8.9.8 Pharmacogenomics in Drug Development

Pharmacogenomics is the study of how an individual's genetic inheritance affects the body's response to drugs. Environment, diet, age, lifestyle, and state of health can all influence a person's response to medicine. However, understanding an individual's genetic makeup is thought to be the route to creating personalized drugs with greater efficacy and safety.

Single-nucleotide polymorphisms (SNPs) can be used to determine different susceptibilities to drugs between individuals. Therefore millions of SNPs must be identified and analyzed to determine their involvement in drug response.

Perlegen, which is a spin-off company of Affymetrix, used high-density chip technology to sequence 50 genomes and found 1.5 million SNPs. It has created microarrays for comparison with clinical samples to identify disease- or drug response-related SNPs. The importance of SNPs can be realized by looking at the CYP450 family.

Polymorphisms in genes that code for these enzymes can influence their ability to metabolize certain drugs. (e.g., codeine is oxidized to morphine by CYP2D6). In different people and different populations, the activity of CYP oxidase differs. A drug that works one way in one population group may differ in another group. Many people of Ethiopian or Saudi Arabian origin have high expression of CYP2D6, which allows them to metabolize drugs recognized by the enzyme much more quickly. The interaction between CYP and new drugs is important. Predominant degradation of a drug by one of the CYPs would result in no further research on that drug. It is thought that the removal of some major medicines may not have been necessary if pharmacogenomics had been used to predict adverse events.

8.10 BIOINFORMATICS

Bioinformatics tools are required to extract the diagnostic or prognostic information from the complex data. Based on pattern recognition technologies, discriminatory patterns (a panel of gene, protein, or peptide patterns) can be identified for the diagnosis of persons with and without a disease.

8.10.1 Bioinformatics Use in Diagnosis of Cancer

In the diagnosis of cancer with the use of transcriptomics investigations, numerous approaches are used to conduct diagnostic and prognostic predictions for cancer patients. They are based on gene expression profiles such as the recently developed independent component analysis (Zhang et al., 2005). The performance of different algorithms has been compared by Dudoit et al. (2002). For proteomics experiments, the computational issues in the processing and classification of protein mass spectra have been reviewed in details by Hilario et al. (2006), and the performance of various methods has been evaluated by Shin and Markey et al. (2006).

For metabonomic data analysis, a few studies are available, such as a principle component analysis method that was used for the diagnosis of liver cancer and the statistical total correlation spectroscopy analysis method for biomarker identification from metabolic NMR data sets (Cloarec et al., 2005).

There are, however, still some bioinformatics challenges in omics research. One substantial problem is the overfitting of data when the number of parameters in a model is too great relative to the number of samples, and the outcome is that the model fits the original data but might predict poorly for independent data. Two methods are frequently used to avoid data overfitting: crossvalidation (applied to the entire data-analysis process) and validation of independent data sets.

Another major concern is that different bioinformatic analyses generate different predictive patterns. The main problem for these discrepancies could be the small number of samples. Ein-Dor et al. (2005) demonstrated that thousands of samples are needed for transcriptomic analysis to generate a robust gene list for predicting outcome in cancer. Unfortunately, no such framework is available for estimating efficient sample sizes in other omics research.

Globally, the main similarity for all of the omics technologies is that they all rely on analytical chemistry methods and generate complex data sets. In particular, MS technology is a common and powerful tool for all omics platforms except transcriptomics. Alternatively, in omics research, the development of robust data analysis approaches is important to generate discriminatory patterns (e.g., gene, protein, or metabolite). In principle, the bioinformatic tools developed for microarray analysis should be transferable to other omics, such as proteomics.

In particular, the lessons learned from analysis of DNA microarray data, including clustering, compendium, and pattern-matching approaches, should be helpful for designing new methods in proteomics (Kremer et al., 2005) but are limited because of the unique attributes of proteomics data. For example, the number of interrogations in microarrays is determined preexperimentally by the number of genes or gene-specific probes on the array, whereas the number of targets in the proteomic or metabonomic analysis of complex samples is unknown. Hence, it is difficult to estimate the degree of confidence of the findings in such omics experiments.

8.11 COMPANIES AND OMICS TECHNIQUES

- Advalytix AG has developed technology for precise electronic control of chemical reactions on the surface of a biochip. The technology uses nanopumps that utilize surface acoustic waves.
- Affymetrix has developed the GeneChip® system, which is a platform for acquiring, analyzing, and managing genetic information. It consists of a disposable DNA probe arrays containing selected gene sequences on a chip, instruments to process the arrays, and software to manage the information. The system has been developed to help understand the human genome and to improve the diagnosis, monitoring, and treatment of disease.
- Attophotonics Bioscience develops products to collect, analyze, and manage bioanalytical data and information for use in proteomics, biomedical research, clinical diagnostics, and therapeutic-drug development.
- BioCrystal Ltd has developed nanocrystalline fluorescent markers for cell and intracellular detection and analysis.
- Caliper Life Sciences uses LabChip technologies to create breakthrough tools that accelerate drug discovery and enhance the diagnosis of disease.
- Epigenomics is developing diagnostic and pharmacodiagnostic products based on DNA methylation. By detecting and interpreting DNA methylation patterns, epigenomics can create a digital readout for each cell (Digital Phenotype®).
- Immunicon Corporation has developed a diagnostic based on immunomagnetic selection and fluorescence of rare cells in blood. The system is for clinical use in the field of cancer. The company has patented magnetic nanoparticles called ferrofluids.
- Integrated BioDiagnostics is developing technology to simplify and accelerate analysis of biological systems. The Lab-on-a-slide system integrates different steps on one slide for preparation and analysis.
- Nanobiogène is a nanoengineering and nanoinstrumentation company specializing in microfluidics and nanofluidics, biochips, and miniaturization technologies. The company is developing microfluidic and nanofluidic instrumentation spotting tools involved in biochip manufacturability.
- Nanogen is a supplier of molecular diagnostic tests. The NanoChip® Molecular Biology Workstation is an automated instrument used for DNA-based analyses. The workstation is used by researchers in the fields of molecular biology and genetics to study how genes function and to understand the correlation between genetic variation and disease.
- NanoInk is a company commercializing the Dip Pen Nanolithography™ method of nanoscale manufacturing, which is a process used to build nanometre scale structures and patterns by drawing materials directly onto a surface.
- Nanoxis AB is a nanobiotechnology company developing advanced biochips within the area of proteomics and genetics
- Roche Diagnostics has developed diagnostic tests and systems that play a role in early detection, targeted screening, evaluation, and monitoring of disease. The AmpliChip is a DNA chip-based diagnostic test that provides information on a patient's metabolic status.
- Solexa is developing systems for the analysis of individual genomes. The single molecule array allows simultaneous analysis of hundreds of individual molecules. The technology is being applied to the measurement of individual genetic variation.
- Spinelix specializes in the area of human diagnostics. It provides diagnostic solutions on nanotechnology enabling the production of minute structures on the molecular level.

Company websites are shown in Table 8.2.

TABLE 8.2
Companies in Omics Technology and Their Internet Sites

Company	Link
Advalytix	http://www.advalytix.de
Affymetrix	http://www.affymetrix.com
Attophotonics Bioscience	http://www.bionanotec.org/Attophotonics
Biocrystal	http://www.biocrystal.com
Caliper Life Sciences	http://www.calipertech.com
Epigenomics	http://www.epigenomics.com
Immunicon	http://www.immunicon.com
Ibidi	http://www.ibidi.com
BioDiagnostics	http://www.biodiagnostics.net
Nanobiogène	http://www.nanobiogene.com
Nanogen	http://www.nanogen.com
NanoInk	http://www.nanoink.net
Nanoxis	http://www.nanoxis.se
Roche	http://www.roche.com
Solexa	http://www.solexa.com
Spinelix	http://www.spinelix.com

8.12 CONCLUSION

The current sciences of omics includes the study of genomics for DNA variations, transcriptomics for messenger RNA, proteomics for peptides and proteins, and metabolomics for intermediate products of metabolism. Technological high throughput and analytical tools allow simultaneous examination of thousands of genes, transcripts, proteins, and metabolites to extract information. Hypothesis-driven research and discovery-driven research (through omics methodologies) are complementary and synergistic. Modern screening technologies speed up the discovery process and give broader insight into both biochemical events that follow the exposure to harmful agents, for example, chemical substances, ionizing radiation, or electromagnetic fields, and diagnosis of various pathological conditions. Moreover, it highlights the response to therapy at a molecular level. Omics technologies have the advantage of containing methods for investigations of DNA, RNA, and protein level as well as changes in the metabolism.

INTERNET RESOURCES

http://www.researchandmarkets.com
http://www.pubs.acs.org/journals/mdd
http://www.nanoink.net/documents/GENews_reprint.pdf
http://www.news-medical.net/?id=628
http://www.nature.com/cgi-taf/DynaPage.taf?file=/nbt/journal/v22
http://www.ornl.gov/sci/techresources/Human_Genome/medicine/pharma.shtml
http://www.bcbs.com/tec/vol19/19-09.html
http://www.pubs.acs.org/cen/coverstory/8130/8130drugdiscovery2.html

REFERENCES

Abdullah-Sayani, A., Jolien, M., de-Mesquita, B., et al. (2006). Technology insight: Tuning into the genetic orchestra using microarrays: Limitations of DNA microarrays in clinical practice. *Nat. Clin. Pract. Oncol.* 501–516.

Adams, E. W., Ratner, D. M. Bokesh, H.R, et al. (2004). Oligosaccharide and glycoprotein microarrays as tools in HIV-glycobiology: Glycan dependent gp120/protein interactions. *Chem. Biol.* 11: 875.

Alizadeh, A., Eisen, M., Davis, R. E., et al. (1999). The lymphochip: A specialized cDNA microarray for the genomic-scale analysis of gene expression in normal and malignant lymphocytes. *Cold Spring Harb. Symp. Quant. Biol.* 64: 71–78.

Amanchy, R., Kalume, D. E., Iwahori, A., et al. (2005). Phosphoproteome analysis of HeLa cells and rational treatment of cancer: Realistic hope? *Ann. Oncol.* 16: 16–22.

Anderson, N. L., Anderson. N. G. (2002). The human plasma proteome: History, character, and diagnostic prospects. *Mol. Cell Proteomics* 1: 845–867.

Anderson, N. L., Polanski, M., Pieper, R., et al. (2004). The human plasma proteome: A non-redundant list developed by combination of four separate sources. *Mol. Cell Proteomics* 3: 311–326.

Archacki, S., Wang, Q. (2004). Expression profiling of cardiovascular disease. *Hum. Genomics* 1: 355–370.

Ardekani, A. M., Liotta, L. A., Petricoin, E. F. (2002). Clinical potential of proteomics in the diagnosis of ovarian cancer. *Expert Rev. Mol. Diagn.* 2: 312–320.

Baird, A. E. (2006). The blood option: Transcriptional profiling in clinical trials. *Pharmacogenomics* 7: 141–144.

Baird, A. E. (2006). Blood biologic markers of stroke: improved management, reduced cost? *Curr. Atheroscler. Rep.* 8: 267–275.

Bammler, T., Beyer, R. P., Bhattacharya, S. et al. (2005a). Electronic microarray analysis of 16S rDNA amplicons for bacterial detection. *J. Biotechnol.* 115: 11–21.

Bammler, T., Beyer, R. P., Bhattacharya, S., et al. (2005b). Standardizing global gene expression analysis between laboratories and across platforms. *Nat. Meth.* 2: 351–356.

Barrans, J. D., Stamatiou, D., Liew, C. (2001). Construction of a human cardiovascular cDNA microarray: Portrait of the failing heart. *Biochem. Biophys. Res. Commun.* 280: 964–969.

Barry, W. T., Kernagis, D. N., Dressman, H. K. (2010). Intratumor heterogeneity and precision of microarray-based predictors of breast cancer biology and clinical outcome. *J. Clin. Oncol.* 28: 2198–2206.

Bartella, L., Thakur, S. B., Morris, E. A., et al. (2007). Enhancing nonmass lesions in the breast: Evaluation with proton (1H) MR spectroscopy. *Radiology* 245: 80–87.

Basil, C. F., Zhao, Y., Zavaglia, K., et al. (2006). Common cancer biomarkers. *Cancer Res.* 66: 2953–2961.

Beckwith-Hall, B. M., Nicholson, J. K., Nicholls, A. W., et al. (1998). Nuclear magnetic resonance spectroscopic and principal components analysis investigations into biochemical effects of three model hepatotoxins. *Chem. Res. Toxicol.* 11: 260–272.

Block, T. M., Comunale, M. A., Lowman, M., et al. (2005). Use of targeted glycoproteomics to identify adenocarcinoma cells. *J. Proteome Res.* 4: 1339–1346.

Boros, L. G., Lerner, M. R., Morgan, D. L., et al. (2005). [1,2-13C2]-D-glucose profiles of the serum, liver, pancreas, and DMBA-induced pancreatic tumors of rats. *Pancreas* 31: 337–343.

Boros, L. G., Steinkamp, M. P., Fleming, J. C., et al. (2003). Defective RNA ribose synthesis in fibroblasts frompatients with thiamine-responsive megaloblastic anemia (TRMA). *Blood* 102: 3556–3561.

Britz-McKibbin, P., Terabe, S. (2003). On-line preconcentration strategies for trace analysis of metabolites by capillary electrophoresis. *J. Chromatogr. A* 1000: 917–934.

Buckhaults, P., Rago, C., St. Croix, B., et al. (2001). Secreted and cell surface genes expressed in benign and malignant colorectal tumors. *Cancer Res.* 61: 6996–7001.

Burczynski, M. E., Dorner, A. J. (2006). Pharmacogenetics and pharmacogenomics in drug development. *Pharmacogenomics* 7: 187–202.

Castrillo, J. I., Hayes, A., Mohammed, S., et al. (2003). An optimized protocol for metabolome analysis in yeast using direct infusion electrospray mass spectrometry. *Phytochemistry* 62: 929–937.

Chaurand, P., Schwartz, S. A., Caprioli, R. M. (2004). Assessing protein patterns in disease using imaging mass spectrometry. *J. Proteome Res.* 3: 245–252.

Clarke, S., Goodacre, R. (2003). Raman spectroscopy for whole organism and tissue profiling. In *Metabolic profiling: Its role in biomarker discovery and gene function analysis* (pp. 95–110). Edited by Harrigan, G. G., Goodacre, R. Boston: Kluwer Academic Publishers.

Cloarec, O., Dumas, M. E., Craig, A., et al. (2005). Statistical total correlation spectroscopy: An exploratory approach for latent biomarker identification from metabolic 1H NMR data sets. *Anal. Chem.* 77: 1282–1289.

Comunale, M. A., Lowman, M., Long, R. E., et al. (2006). Proteomic analysis of serum-associated fucosylated glycoproteins in the development of primary hepatocellular carcinoma. *J. Proteome Res.* 5: 308–315.

Coombes, K. R., Morris, J. S., Hu, J., et al. (2005). Serum proteomics profiling: A young cytoskeletal and stress proteins in primary human breast cancers: Implications for adjuvant use of kinase-inhibitory drugs: Clinical cancer data storage, analysis and integration. *Biosci. Rep.* 25: 95–106.

Dave, S. S., Fu, K., Wright, G. W., et al. (2006) Molecular diagnosis of Burkitt's lymphoma. *N. Engl. J. Med.* 354: 2431–2442.

DeKeyser, S. S., Kutz-Naber, K. K., Schmidt, J. J., et al. (2007). Imaging mass spectrometry of neuropeptides in decapod crustacean neuronal tissues. *J. Proteome Res.* 6: 1782–1791.

Denkert, C., Budczies, J., Kind, T., et al. (2006). Mass spectrometry-based metabolic profiling reveals different metabolite patterns in invasive ovarian carcinomas and ovarian borderline tumors. *Cancer Res.* 66: 10795–10804.

Drake, R. R., Schwegler, E. E., Malik, G., et al. (2006). Lectin capture strategies combined with mass spectrometry for the discovery of serum glycoprotein biomarkers. *Mol. Cell. Proteomics* 5: 1957–1967.

Du, X., Tang, Y., Xu, H., et al. (2006). Genomic profiles for human peripheral blood T cells, B cells, natural killer cells, monocytes, and polymorphonuclear cells: Comparisons to ischemic stroke, migraine, and Tourette syndrome. *Genomics* 87: 693–703.

Dudoit, S., Fridlyand, J., Speed, T. P. (2002). Comparison of discrimination methods for the classification of tumors using gene expression data. *J. Am. Stat. Assoc.* 97: 77–90.

Dupont, A., Corseaux, D., Dekeyzer, O., et al. (2005). The proteome and secretome of human arterial smooth muscle cells. *Proteomics* 5: 585–596.

Ebert, M. P., Meuer, J., Wiemer, J. C., et al. (2006). Advances, challenges, and limitations in serum: Proteome-based cancer diagnosis. *J. Proteome Res. 5:* 19–25.

Edman, C. F., Raymond, D. E., Wu, D. J., et al. (1997). Electric field directed nucleic acid hybridization on microchips. *Nucleic Acids Res.* 25: 4907–4914.

Ein-Dor, L., Kela, I., Getz, G., et al. (2005). Outcome signature genes in breast cancer: Is there a unique set? *Bioinformatics* 21: 171–178.

Eiter, T., Zimmermann, K., Scheiflinger, F. (2002). Analysis of the detection limit on a microelectronic array. *BioTechniques* 33: 494–496.

Ellis, D. I., Harrigan, G. G., Goodacre, R. (2003). Metabolic fingerprinting with Fourier transform infrared spectroscopy. In *Metabolic profiling: Its role in biomarker discovery and gene function analysis* (pp. 111–124). Edited by Harrigan, G. G., Goodacre, R. Boston: Kluwer Academic Publishers.

Fossel, E. T., Carr, J. M., McDonagh, J. (1986). Detection of malignant tumors: Water-suppressed proton nuclear magnetic resonance spectroscopy of plasma. *N. Engl. J. Med.* 315: 1369–1376.

Gayed, I., Vu, T., Iyer, R., et al. (2004). The role of 18F-FDG PET in staging and early prediction of response to therapy of recurrent gastrointestinal stromal tumors. *J. Nucl. Med.* 45: 17–21.

Geho, D. H., Liotta, L. A., Petricoin, E. F., et al. (2006). The amplified peptidome: The new treasure chest of candidate biomarkers. *Curr. Opin. Chem. Biol.* 10: 50–55.

Gianazza, E., Sironi, L. (2004). Vasculature, vascular disease and atherosclerosis. In *Biomedical applications of proteomics* (pp. 39–55). Edited by Sanchez, J. C., Corthals, G. L., Hochstrasser, D. F. Weinheim: Wiley-VCH.

Ginsburg, G. S., Haga, S. B. (2006). Translating genomic biomarkers into clinically useful diagnostics. *Expert Rev. Mol. Diagn.* 6: 179–191.

Griffin, J. L., Shockcor, J. P. (2004). Metabolic profiles of cancer cells. *Nat. Rev. Cancer* 4: 551–561.

Guo, Y., Feinberg, H., Conroy, E., et al. (2004). Structural basis for distinct ligand-binding and targeting properties of the receptors DC-SIGN and DC-SIGNR. *Nat. Struct. Mol. Biol.* 11: 591–598.

Hamacher, M., Klose, J., Rossier, J., et al. (2004). Does understanding the brain need proteomics and does understanding proteomics need brains? Second HUPO HBPP Workshop hosted in Paris. *Proteomics* 4: 1932–1934.

Hand, P. J., Kwan, J., Lindley, R. I., et al. (2006). Distinguishing between stroke and mimic at the bedside: The brain attack study. *Stroke* 37: 769–775.

Harrigan, G. G., LaPlante, R. H., Cosma, G. N., et al. (2004). Application of high-throughput Fourier-transform infrared spectroscopy in toxicology studies: Contribution to a study on the development of an animal model for idiosyncratic toxicity. *Toxicol. Lett.* 146: 197–205.

Heidecker, B., Hare, J. M. (2007). The use of transcriptomic biomarkers for personalized medicine. *Heart Failure Rev.* 12: 9004–9007.

Heinrich, M. C., Maki, R. J., Corless, C. L., et al. (2006). Sunitinib (SU) response in imatinib-resistant (IM-R) GIST correlates with KIT and PDGFRA mutation status: 2006 ASCO Annual Meeting Proceedings Part I. *J. Clin. Oncol.* 24: 9502.

Hilario, M., Kalousis, A., Pellegrini, C., et al. (2006). Processing and classification of protein mass. *Mass Spectrom. Rev.* 25: 409–449.

Holdsworth, C. H., Badawi, R. D., Manola, J. B., et al. (2007). CT and PET: Early prognostic indicators of response to imatinib mesylate in patients with gastrointestinal stromal tumor. *Am. J. Roentgenol.* 189: W324–W330.

Huang, H. L., Stasyk, T., Morandell, S. (2006). Biomarker discovery in breast cancer serum using 2D differential gel electrophoresis/MALDI-TOF/TOF and data validation by routine clinical assays. *Electrophoresis* 27: 1641–1650.

Hummel, M., Bentink, S., Berger, H., et al. (2009). A biologic definition of Burkitt's lymphoma from transcriptional and genomic profiling. *N. Engl. J. Med.* 354: 2419–2430.

Irizarry, R. A., Warren, D., Spencer, F., et al. (2005). Multiple-laboratory comparison of microarray platforms. *Nat. Methods* 2: 345–349.

Jia, L., Liu, B. F., Terabe, S., et al. (2002). Two-dimensional separation method for analysis of Bacillus subtilis metabolites via hyphenation of micro-liquid chromatography and capillary electrophoresis. *Anal. Chem.* 76: 1419–1428.

Kaddurah-Daouk, R., Beecher, C., Kristal, B. S., et al. (2004). Bioanalytical advances for metabolomics and metabolic profiling. *PharmaGenomics* January: 46–52.

Kidd, E. A., Siegel, B. A., Dehdashti, F., et al. (2007). The standardized uptake value for F-18 fluorodeoxyglucose is a sensitive predictive biomarker for cervical cancer treatment response and survival. *Cancer* 110: 1738–1744.

Kim, J. E., Tannenbaum, S. R., White, F. M. (2005). Global phosphoproteome of HT-29 human colon adenocarcinoma cells. *J. Proteome Res.* 4: 1339–1346.

Kobata, A., Amano, J. (2005). Altered glycosylation of proteins produced by malignant cells, and application for the diagnosis and immunotherapy of tumours. *Immunol. Cell Biol.* 83: 429–439.

Kolch, W., Mischak, H., Pitt, A. R. (2005). The molecular make-up of a tumour: Proteomics in cancer research. *Clin. Sci.* 108: 369–383.

Kremer, A., Schneider, R., Terstappen, G. C. (2005). A bioinformatics perspective on proteomics: Data Storage, Analysis, and Integration. *Biosci. Reports* 25: 95–106.

Kritharides, L. (2003). Haptoglobin elutes from human atherosclerotic coronary arteries: A potential marker of arterial pathology. *Atherosclerosis* 168: 389–396.

Lander, E. S. (1999). Array of hope. *Nat. Genet.* 21 (Suppl.): 3–4.

Larkin, J. E., Frank, B. C., Gavras, H. (2005). Independence and reproducibility across microarray platforms. *Nat. Methods* 2: 337–343.

Lim, Y. P., Wong, C. Y., Ooi, L. L., et al. (2004). Selective tyrosine hyperphosphorylation of cytoskeletal and stress proteins in primary human breast cancers: Implications for adjuvant use of kinase-inhibitory drugs. *Clin. Cancer Res.* 10: 3980–3987.

Lockhart, D. J., Dong, H., Byrne, M. C., et al. (1996). Expression monitoring by hybridization to high-density oligonucleotide. *Nat. Biotechnol.* 14: 1675–1680.

Lopez, M. F., Mikulskis, A., Kuzdzal, S., et al. (2005). High-resolution serum proteomic profiling of Alzheimer disease samples reveals disease-specific, carrier-protein-bound mass signatures. *Clin. Chem.* 51: 1946–1954.

Mann, M. (2006). Functional and quantitative proteomics using SILAC. *Nat. Rev. Mol. Cell Biol.* 7: 952–958.

Marrero, J. A., Lok, A. S. (2004). Newer markers for hepatocellular carcinoma. *Gastroenterology* 127 (Suppl. 1): S113–S119.

Masoodi, M., Nicolaou, A. (2006). Lipidomic analysis of twenty-seven prostanoids and isoprostanes by liquid chromatography/electrospray tandem mass spectrometry. *Rapid Commun. Mass Spectrom.* 20: 3023–3029.

McDonnell, L. A., Heeren, R. M. A. (2007). Imaging mass spectrometry. *Mass Spectrom. Rev.* 26: 606–643.

Mendes, P., Kell, D. B., Westerhoff, H. V. (1992). Channelling can decrease pool size. *Eur. J. Biochem.* 204: 257–266.

Mendes, P., Kell, D. B., Westerhoff, H. V. (1996). Why and when channeling can decrease pool size at constant net flux in a simple dynamic channel. *Biochim. Biophys. Acta* 1289: 175–186.

Moore, D. F., Li, H., Jeffries, N., et al. (2005). Using peripheral blood mononuclear cells to determine a gene expression profile of acute ischemic stroke: A pilot investigation. *Circulation* 111: 212–221.

Odunsi, K., Wollman, R. M., Ambrosone, C. B., et al. (2005). Detection of epithelial ovarian cancer using 1H-NMR-based metabonomics. *Int. J. Cancer* 113: 782–788.

Okano, T., Kondo, T., Kakisaka, T., et al. (2006). Plasma proteomics of lung cancer by a linkage of multi-dimensional liquid chromatography and two-dimensional difference gel electrophoresis. *Proteomics* 6: 3938–3948.

Park, J. W., Kerbel, R. S., Kelloff, G. J., et al. (2004). Rationale for biomarkers and surrogate end points in mechanism-driven oncology drug development. *Clin. Cancer Res.* 10: 3885–3896.

Patterson, T. A., Lobenhofer, E. K., Fulmer-Smentek, S. B., et al. (2006). Performance comparison of one-color and two-color platforms within the MicroArray Quality Control (MAQC) project. *Nat. Biotechnol.* 24: 1140–1150.

Patwa, T. H., Zhao, J., Anderson, M. A., et al. (2006). Screening of glycosylation patterns in serum using natural glycoprotein microarrays and multi-lectin fluorescence detection. *Anal. Chem.* 78: 6411–6421.

Petricoin, E. F., Ardekani, A. M., Hitt, B. A., et al. (2002). Use of proteomic patterns in serum to identify ovarian cancer. *Lancet* 359: 572–577.

Pierson, J., Norris, J. L., Aerni, H. R., et al. (2002). Molecular profiling of experimental Parkinson's disease: Direct analysis of peptides and proteins on brain tissue sections by MALDI mass spectrometry. *J. Proteome* Res. 3: 289–295.

Pio, B. S., Park, C. K., Pietras, R., et al. (2006). Usefulness of 3'-[F-18]fluoro-3'-deoxythymidine with positron emission tomography in predicting breast cancer response to therapy. *Mol. Imaging Biol.* 8: 36–42.

Posadas, E. M., Simpkins, F., Liotta, L. A., et al. (2005). Proteomic analysis for the early detection and rational treatment of cancer: Realistic hope? *Ann. Oncol.* 16: 16–22.

Powell, D. W., Merchant, M. L., Link, A. J. (2006). Discovery of regulatory molecular events and biomarkers using 2D capillary chromatography and mass spectrometry. *Expert Rev. Proteomics* 3: 63–74.

Pozhitkov, A., Noble, P. A., Domazet-Loso, T., et al. (2006). Tests of rRNA hybridization to microarrays suggest that hybridization characteristics of oligonucleotide probes for species discrimination cannot be predicted. *Nucleic Acids Res.* 34: e66.

Prenen, H., Cools, J., Mentens, N., et al. (2006). Efficacy of the kinase inhibitor SU11248 against gastrointestinal stromal tumor mutants refractory to imatinib mesylate. *Clin. Cancer Res.* 12: 2622–2627.

Qiu, J., Madoz-Gurpide, J., Misek, D. E., et al. (2004). Development of natural proteinmicroarrays for diagnosing cancer based on an antibody response to tumor antigens. *J. Proteome Res.* 3: 261–267.

Quackenbush, J. (2006). Microarray analysis and tumor classification. *N. Engl. J. Med.* 354: 2463–2472.

Ratner, D. M., Adams, E. W., Disney, M. D., et al. (2004). Tools for glycomics: Mapping interactions of carbohydrates in biological systems. *ChemBioChem* 5: 1375–1383.

Reinders, J., Sickmann, A. (2005). State-of-the-art in phosphoproteomics. *Proteomics* 5: 4052–4061.

Rhodes, D. R., Yu, J., Shanker, K., et al. (2004). Large-scale meta-analysis of cancer microarray data identifies common transcriptional profiles of neoplastic transformation and progression. *Proc. Natl. Acad. Sci. U.S.A.* 101: 9309–9314.

Robertson, D. G., Reily, M. D., Sigler, R. E., et al. (2000). Metabonomics: Evaluation of nuclear magnetic resonance (NMR) and pattern recognition technology for rapid in vivo screening of liver and kidney toxicants. *Toxicol. Sci.* 57: 326–337.

Ross, P., Huang, Y., Marchese, J., et al. (2004). Relative and absolute quantitation in yeast proteomics using multiplexed isobaric peptide tags. *Proceedings 52nd ASMS*, Nashville, May 23–27, 2004, Slot 392. http://www.inmerge.com/ASMS/DisplayAbstractList.aspx?Session=TPT.

Sanchez, J. C. Y., Lescuyer, L. A: P., Hochstrasser, D. F. (2007). Biomedical applications of proteomics. In *Proteome research: Concepts, technology and application.* Edited by Wilkins, M. R., Appel, R. D., Williams, K. L., et al. Heidelberg, Germany: Springer.

Sandvik, A. K., Alsberg, B. K., Norsett, K. G., et al. (2006). Gene expression analysis and clinical diagnosis. *Clin. Chim. Acta* 363 157–164.

Scheidler, J., Hricak, H., Vigneron, D. B., et al. (1999). Prostate cancer: Localization with three-dimensional proton MR spectroscopic imaging: Clinicopathologic study. *Radiology* 213: 473–480.

Schena, M., Shalon, D., Davis, R. W., Brown, P. O. (1995). Quantitative monitoring of gene expression patterns with a complementary DNA microarray. *Science* 270: 467–470.

Schrattenholz, A., Wozny, W., Klemm, M., et al. (2005). Differential and quantitative molecular analysis of ischemia complexity reduction by isotopic labeling of proteins using a neural embryonic stem cell model. *J. Neurol. Sci.* 229: 261–267.

Seeberger, P. H., Disney, M. D. (2004). The use of carbohydrate microarrays to study carbohydrate-cell interactions and to detect pathogens. *Chem. Biol.* 11: 1701–1707.

Seidel, A. Brunner, S., Seidel, P., et al. (2006). Modified nucleosides: An accurate tumour marker for clinical diagnosis of cancer, early detection and therapy control. *Br. J. Cancer* 94: 1726–1733.

Serkova, N. J., Niemann, C. U. (2006). Pattern recognition and biomarker validation using quantitative 1H-NMR-based metabolomics. *Expert Rev. Mol. Diagn.* 6: 717–731.

Serkova, N. J., Spratlin, J. L., Eckhardt, S. G. (2007). NMR-based metabolomics: Translational application and treatment of cancer. *Curr. Opin. Mol. Ther.* 9: 572–585.

Shi, H., Vigneau-Callahan, K. E., Matson, W. R., Kristal, B. S. (2002). Attention to relative response across sequential electrodes improves quantitation of coulometric array. *Anal. Biochem.* 302: 239–245.

Shin, H., Markey, M. K. (2006). A machine learning perspective on the development of clinical decision support systems utilizing mass spectra of blood samples. *J. Biomed. Inform.* 39: 227–248.

Skold, K., Svensson, M., Nilsson, A. (2006). Decreased striatal levels of PEP-19 following MPTP lesion in the mouse. *J. Proteome Res.* 5: 262–269.

Soga, T., Ohashi, Y., Ueno, Y. (2003). Quantitative metabolome analysis using capillary electrophoresis mass spectrometry. *J. Proteome Res.* 2: 488–494.

Soga, T., Ueno, Y., Naraoka, H., et al. (2002). Pressure-assisted capillary electrophoresis electrospray ionization mass spectrometry for analysis of multivalent anions. *Anal. Chem.* 74: 6224–6229.

Soo, E. C., Aubry, A. J., Logan, S. M., et al. (2004). Selective detection and identification of sugar nucleotides by CE-electrospray-MS and its application to bacterial metabolomics. *Anal. Chem.* 76: 619–626.

Southern, E,, Mir, K., Shchepinov, M. (1999). Molecular interactions on microarrays. *Nat. Genet.* 21: 5–9.

Stark, M., Danielsson, O., Griffiths, W. J., et al. (2001). Peptide repertoire of human cerebrospinal fluid: Novel proteolytic fragments of neuroendocrine proteins. *J. Chromatogr. B* 754: 357–367.

Stoeckli, M., Staab, D., Staufenbiel, M., et al. (2002). Molecular imaging of amyloid b peptides in mouse brain sections using mass spectrometry. *Anal. Biochem.* 311: 33–39.

Suzuyama, K., Shiraishi, T., Oishi, T., et al. (2004). Combined proteomic approach with SELDI-TOF-MS and peptide mass fingerprinting identified the rapid increase of monomeric transthyretin in rat cerebrospinal fluid after transient focal cerebral ischemia. *Brain Res. Mol. Brain Res.* 129: 44–53.

Tang, Y., Xu, H., Du, X., et al. (2006). Gene expression in blood changes rapidly in neutrophils and monocytes after ischemic stroke in humans: A microarray study. *J. Cereb. Blood. Flow Metab.* 28: 1089–1102.

Thadikkaran, L., Seigenthaler, M. A., Crettaz, D., et al. (2005). Recent advances in blood-related proteomics. *Proteomics* 5: 3019–3034.

't Hoen, P. A., Turk, R., Boer, J. M., et al. (2004). Intensity-based analysis of two-colour microarrays enables efficient and flexible hybridization designs. *Nucleic Acids Res.* 32: e41.

Tinker, A. V., Boussioutas, A., Bowtell, D. D. (2006). The challenges of gene expression microarrays for the study of human cancer. *Cancer Cell* 9: 333–339.

Tolstikov, V. V., Fiehn, O. (2002). Analysis of highly polar compounds of plant origin: Combination of hydrophilic interaction chromatography and electrospray ion trap mass spectrometry. *Anal. Biochem.* 301: 298–307.

Tolstikov, V. V., Lommen, A., Nakanishi, K., et al. (2003). Monolithic silica-based capillary reversed-phase liquid chromatography/electrospray mass spectrometry for plant metabolomics. *Anal. Chem.* 75: 6737–6740.

Umezu-Goto, M., Tanyi, M. J., Lahad, J., et al. (2004). Lysophosphatidic acid production and action: Validated targets in cancer. *J. Cell. Biochem.* 92: 1115–1140.

Van der Greef, J., Davidov, E., Verheij, E. R., et al. (2003). The role of metabolomics in systems biology. In *Metabolic profiling: Its role in biomarker discovery and gene function analysis* (pp. 170–198). Edited by Harrigan, G. G., Goodacre, R. Boston: Kluwer Academic Publishers.

Van der Greef, J., Stroobant, P., van der Heijden, R. (2004). The role of analytical sciences in medical systems biology. *Curr. Opin. Chem. Biol.* 8: 559–565.

Van de Vijver, M. J., He, Y. D., van't Veer, L. J., et al. (2002). A gene-expression signature as a predictor of survival in breast cancer. *N. Engl. J. Med.* 347: 1999–2009.

Veenstra, T. D., Conrads, T. P., Hood, B. L., et al. (2005). Biomarkers: Mining the biofluid proteome. *Mol. Cell. Proteomics* 4: 409–418.

Villanueva, J., Philip, J., Entenberg, D., et al. (2004). Serum peptide profiling by magnetic particle-assisted, automated sample processing and MALDI-TOF mass spectrometry. *Anal. Chem.* 76: 1560–1570.

Villanueva, J., Shaffer, D. R., Philip, J., et al. (2006). Differential exoprotease activities confer tumor-specific serum peptidome patterns. *J. Clin. Invest.* 116: 271–284.

von Roepenack-Lahaye, E., Degenkolb, T., Zerjeski, M., et al. (2004). Profiling of Arabidopsis secondary metabolites by capillary liquid chromatography coupled to electrospray ionization quadrupole time-of-flight mass spectrometry. *Plant Physiol.* 134: 548–559.

Wei, X., Li, L. (2009). Mass spectrometry-based proteomics and peptidomics for biomarker discovery in neurodegenerative diseases. *Int. J. Clin. Exp. Pathol.* 2: 132–148.

Wenk, M. R. (2005). The emerging field of lipidomics. *Nat. Rev. Drug* 4: 594–610.

Weston, A. D., Hood, L. (2004). Systems biology, proteomics, and the future of health care: Toward predictive, preventative, and personalized medicine. *J. Proteome Res.* 3: 179–196.

Yang, J., Xu, G., Hong, Q., et al. (2004). Diagnosis of liver cancer using HPLC-based metabonomics avoiding false-positive result from hepatitis and hepatocirrhosis diseases. *J. Chromatogr.* 813: 2559–2565.

Yuan, X., Desiderio, D. M. (2005). Human cerebrospinal fluid peptidomics. *J. Mass Spectrom.* 40: 176–181.

Zhang, X. W., Yap, Y. L., Wei, D., et al. (2005). Molecular diagnosis of human cancer type by gene expression profiles and independent component analysis. *Eur. J. Hum. Genet.* 13: 1–9.
Zimmermann, K., Eiter, T., Scheiflinger, F. (2003). Consecutive analysis of bacterial PCR samples on a single electronic microarray. *J. Microbiol. Methods* 55: 471–474.
Zougman, A., Pilch, B., Podtelejnikov, A., et al. (2008). Integrated analysis of the cerebrospinal fluid peptidome and proteome. *J. Proteome Res.* 2008; 7: 386–399.

9 High-Throughput Omics

The Application of Automated High-Throughput Methods and Systems for Solutions in Systems Biology

Üner Kolukisaoglu and Kerstin Thurow
University of Rostock
Rostock, Germany

CONTENTS

9.1 INTRODUCTION

Systems biology nowadays invades all areas of biomedical research and seems to provide a promising avenue for the understanding and diagnosis of different diseases and even for finding a cure for them (Wolkenhauer et al., 2009). Although the concept of systems biology was founded over 60 years ago (Weiner, 1948), its breakthrough came in the last decade along with the sequencing of the first whole genomes. This was the starting point for comprehensive studies of particular molecule types in one organism. The study of these omes, such as genomes, transcriptomes, proteomes, or metabolomes, led to the formation of research fields of their own, summarized under the term *omics*. In parallel, the availability of comprehensive genome data together with high-throughput measurement methods revived the idea of system biological approaches to study biomedical problems (Kitano, 2002).

Nowadays, vast arrays of omes are known, and the same numbers of omics have been developed. The results of these methods and their integration are also in use for systems biological investigation of many biomedical tasks and questions. The major problem of this development lies in the technicality involved in the generation of high-quality quantitative data from omics analyses in a reasonable time frame. This represents a major bottleneck of systems biology in medical applications (Wolkenhauer et al., 2009). Therefore systems biology has been characterized as *high-throughput reductionism* (Katagiri, 2003). A major path out of this dilemma may be the progressive automation of most analyses in the laboratories working on these approaches.

The automation of high-throughput processes in omics research, nicknamed *high-throughput omics*, and its contribution to systems biology in biomedicine will stand at the center of this chapter. Therefore, in the next section an introduction into different aspects of life science automation is given. This technology was mainly developed and applied to the discovery and development of drugs, which is summarized and described in the second section. In the third section the different omics fields and their degree and potential of automation are covered. The future prospects of omics on systems biology and the influence of automation within this process are the subject of the final section.

9.2 HIGH-THROUGHPUT SCREENING: A STORY ABOUT SMALL VOLUMES, STANDARD FORMATS, AND ROBOTS

9.2.1 The Necessity of High Throughputs in Life-Science Research

Nearly all branches of life sciences are characterized today by high numbers of experiments. Chemists may have the wish to optimize certain chemical reactions by varying parameters such as temperature, pressure, solvents, or catalysts. Medicinal chemists are interested in the establishment of compound libraries with tens to hundred thousands of compounds that can be used for drug discovery processes. Biologists and pharmacologists need to perform different numbers of assays to get as much as possible knowledge on the biological activity and toxicity of new drug candidates in order to avoid compound failures in later stages of the drug-development process and to make our drugs more safe.

Why is it necessary to perform such high numbers of experiments? Figure 9.1 shows a typical result for the optimization of a chemical reaction by varying the parameters of temperature and pressure. If the number of experiments and resulting data points are too small (wide-meshed grid), optimal parameter configurations leading to high yields will be missed. More experiments and more data points will result in the determination of these so-called hits.

Another important argument for high numbers of experiments comes from the drug discovery process. Because only one out of 100,000 initially tested compounds can successfully enter the market after a 12–15 year development process, a higher number of compounds tested increases the probability of finding more hits. If we assume that a starting compound has to be modified and

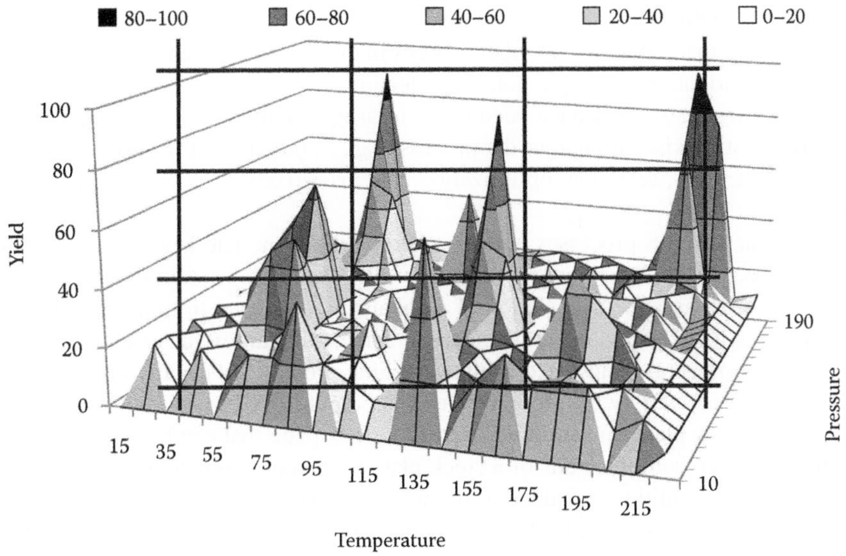

FIGURE 9.1 Yields of a typical chemical reaction varying temperature and pressure.

that one reaction will have a reaction time of 1 hour, a synthetic chemist will need up to 8 years to perform all possible reactions with five compounds, five solvents, five concentrations, five catalysts, five temperatures, five pressures, and five times. Thus the development of new methods to increase the throughput in life-science experiments and investigations is essential for modern life-science research.

9.2.2 High-Throughput Screening: A Definition

High-throughput screening (HTS) is a technology that developed over the past 20 years and is today used in biology, chemistry, and mainly in the process of drug discovery. With HTS methods, tens of thousands of compounds can be tested against a particular target daily. Active compounds, antibodies, or genes that modulate a particular pathway can be rapidly identified. Positive results from these experiments are called hits. They provide starting points for drug design and for understanding the interaction or role of a particular biological process. High-throughput screening involves advanced liquid handling, modern robotics, sensitive detection methods, and sophisticated control software.

The development of high-throughput screening methods was driven by the needs of the pharmaceutical industry and technological strategies within the *in vitro* diagnostics industry. In the meantime, HTS has been widely implemented throughout pharmaceutical research and development (HighTech Business Decisions, 2002). High-throughput screening assays are typically performed within a single well of a microplate; each well represents a single data point. High-throughput screening is the generally accepted term for processes of up to 100,000 wells per day. The term moderate-throughput screening is used for lower throughput rates (Hann and Oprea, 2004). Throughputs above 100,000 wells per day are considered to be ultrahigh-throughput screening (uHTS). High-throughput screening has traditionally applied for biochemical assays that measure how compounds bind to targeted molecules or how compounds inhibit enzyme activities. These assays can be performed as part of automated, high-throughput procedures using 96- or 384-well microwell plates and for uHTS such as 1,536-well plates or even higher density. Although these high-throughput biochemical assays rapidly produce large quantities of information about positive reactions on one system, these systems provide little information about interactions of the pathway of interest with other components of a biological system (Sannes, 2002). The assay design affects the extent to which automation can be used, the cost of the screen, the sensitivity, and the ability to find hits (Simpson, 2006). Improvements and innovations in assay development include the development of technologies that reduce the number of steps involved, have increased biological relevance, or provide more information per single point of detection.

Although significant advances have been made in drug discovery in recent years, a large number of bottlenecks still exist throughout the process (e.g., in target identification, qualification and validation, lead discovery, selection, optimization, and prioritization, as well as in preclinical and clinical testing).

The first step in all HTS processes is assay-plate preparation. The key labware of HTS is the microplate (see Section 9.2.4) with a grid of small wells. The wells contain the reagents to be processed or tested, often an aqueous solution of a chemical compound in dimethyl sulfoxide. Each well on the plate is filled with a different solution; some wells are used for experimental control purposes. The experimental plates used in the screening process are usually prepared from stock plates, which can be created by the laboratory or can be purchased from a commercial source. In the easiest setups, assay plates are simply a copy of a stock plate that can be created by pipetting a small amount of liquid from the wells of a stock plate to the corresponding wells of a completely empty plate. Biological entities are added to the wells to perform a biological reaction. These can consist of enzymes, proteins, or cells. The wells are incubated to allow for absorption, binding, or other reaction between the biological component and the compound to be tested. Measurements will follow for each well on the plate in order to visualize the experimental effect. Mainly spectroscopic methods such as absorption, luminescence, or fluorescence measurement are employed. These measurements in HTS are usually performed with specialized automated machines that can analyze numerous plates within a few seconds, generating thousands of data points. Depending on the results of this first assay, additional investigation can be required on the hit wells that gave interesting results.

9.2.3 High-Throughput Screening versus High-Content Screening

High-content screening (HCS) assays, which are generally microscopic cell-based assays, offer the potential to address many of the mentioned bottlenecks. HCS systems in contrast to HTS assays (mostly biochemical assays) provide researchers with massive amounts of biological information because they show how a compound is likely to interact in a biological system, not just about how it interacts with a potential drug target. As a result it is also possible to obtain information about other interactions that may occur within the cell and that may potentially affect the efficacy and/or safety of the compounds being evaluated (Fisler and Burke, 2004). With high-content analyses, general fields of interest, including proliferation, cell cycle, apoptosis, and cytotoxicity, can be investigated. Examples include the screening of compound libraries for particular cytostatics in cancer research and secondary screenings in a library for cytotoxicity of hit compounds. Apart from that, high-content screening also allows for more complex analysis methods such as receptor binding and internalization studies like G-protein-coupled receptor-based (GPCR) assays (Milligan, 2003), which are of specific interest in pharmaceutical research (Filmore, 2004; Felder et al., 2002). Other assay methods that were not available before the invention of high-content screening include examinations on intracellular protein translocation or phosphorylation status in individual signal components within intact cells. Besides phosphorylation of mediator and target proteins, translocations between cell compartments also play an important role in signal processes. In addition to cytoplasm-to-nucleus translocation, other translocation forms include membrane-to-cytoplasm and cell compartment-to-cytoplasm translocation (endosome, endoplasmic reticulum, and mitochondria). Translocation and phosphorylation assays are often referred to when examining signal cascades in diseases and development processes such as in stem-cell research (Borchert et al., 2005; Richards et al., 2006; Chan et al., 2005).

The main advantages of HCS compared with HTS are that the biological information yielded by the HCS approach is many times as complex and detailed. Automated microscopy allows a direct view of the metabolism in intact cells, in contrast to the more conventional cell-based assays configured for HTS, where only indirect evidence of metabolism in the cell is determined by cell-lysate analysis or biochemical assay where the target is examined in complete isolation from the cell complex. In contrast to HTS, which consists of parallel screening of many substances on a single target, the HCS process presents the advantage of enabling parallel screening of many cellular events. The main feature of HCS processes is the simultaneous quantitative recording of several parameters on cellular and subcellular level. This mostly requires the use of costly reagents such as antibodies and fluorescent dyes in several time-consuming process steps. This reduces throughput and increases costs in comparison with HTS. Although analyzing an HTS assay plate in a fluorescence, absorption, or luminescence reader—depending on the application—only takes seconds, HCS usually takes a few minutes per microplate. The increased time is a result of the variable number of individual microscopic images per well. The emphasis is on throughput in HTS, which means that one-step assays are used wherever possible with a homogeneous evaluation process and low costs per data point. HCS is therefore an orthogonal method to HTS; the future will likely see a combination of both technologies.

9.2.4 Standardization as a Success Factor in Laboratory Automation

Laboratory automation is still a quite young field strongly connected with modern drug development. The main success factor for this strong and fast development was the introduction of a standardized microplate format (Astle, 2000). The microplate (microtiter plate) was originally developed in 1951–1952 when Dr. Takatsy machined eight rows with 12 positions. Common usage started in the laboratory in the late 1950s, when John Liner (Lindor Company) introduced a molded version. The microplate has usually 96, 384, or 1,536 wells arranged in a 2:3 matrix. Starting in 1965, Cooke Labs USA produced 96-well plates in acrylic and later in polystyrene. In 1968, the company issued a patent for microplates and registered the trademark. In 1974, Allister Voller and Ken Walls started using microplates for enzyme-linked immunosorbent assays (ELISAs). The first 96-well filter plate patent was issued in 1981 to Dr. Cleveland (University of California at San Diego). With the development of HTS for small molecule drug discovery in the 1990s, the market demanded microplates with higher well densities for increased sample throughput. The 384-well microplate was developed by Genetix Corp. in 1992, which increased the number of wells by a factor of four on the same footprint as a 96-well microplate. These 384-well microplates were rapidly adopted and influenced the development of even higher well densities such as 1,536- and 3,456-well microplates by Whatman (now part of GE Healthcare) and by a collaboration between Whatman and Aurora Biosciences (now known as Vertex Pharmaceuticals), respectively, in 1996. One of the most important evolutions of the microplate was initiated by the Society for Bimolecular Screening and a key group of manufacturers who established standards for the microplate format. This standardization was necessary for the development of automation equipment to move, sort, and wash plates in future instruments. All microplate manufacturers at this time modified their production equipment to the exact specifications of the new microplate standards (SBS ANSI 2004).

The volume of the sample wells typically ranges from a few microliters to a few hundred microliters (Table 9.1). Since the 1970s, the 96-well microplate became a ubiquitous tool in scientific assays. The most common standard nowadays is the 384-well microplate; the smaller well volume of this microplate enables a lower usage of sample, solvents, and reagents and thus more cost-effective investigations. This trend continues with the 1,536-well plates; the 9,600-well format brings the microplate into nanorange, and picorange is not far behind. This requires new technologies and automation solutions for the liquid handling procedures. In the case of solid reagents, the increasing miniaturization leads to the problem of high inhomogenities combined with a low reproducibility of the results. An additional negative effect of smaller volumes is also the changing ratio of volume to surface, which is of great importance for all upscaling processes. An increasing miniaturization leads to a decreasing volume-to-surface ratio, which can result in increased surface effects with

TABLE 9.1
Volume-to-Surface Ratios for Different Well Dimensions

Wells	Volume	Surface	Volume/Surface Ratio
1,536	10 µl	35 mm^2	0.35
384	149 µl	178 mm^2	0.84
96	530 µl	371 mm^2	1.45
8	2.991 µl	1,166 mm^2	2.56
1	100 ml	121 cm^2	8.22
1	2,000 ml	444 cm^2	44.95

high impact on the reaction kinetics. Table 9.1 shows typical volumes and surfaces for different well dimensions. Thus the miniaturization for all technologically relevant methods and procedures is limited.

Today more than 15 companies produce microplates in different types including different well densities from several wells to several thousand wells per microplate; colored microplates for different optical readouts (i.e., black for fluorescence; white for luminescence); solid-bottom wells or clear-bottom wells for top- or bottom-reading fluorescence, respectively; different plastics (polystyrene, polypropylene, etc.); various coatings (i.e., reduction of nonspecific binding, tissue-culture treated, etc.); sterile plates; plates with filter bottoms; deep-well plates; and many more (Banks, 2009).

9.2.5 Automation Systems for High-Throughput Screening

Equipment suppliers have developed a wide range of laboratory-automation tools. The development of such tools has dramatically been influenced and driven by the advancement of the microplate technology. The use of highly parallel reader systems is only as feasible as sample preparation and injection into the system are suitable for high-throughput applications. The most important automation factors in high-throughput applications therefore include the pipetting and microplate transport systems. A typical setup for HTS might also include storing and stacking systems, incubators and harvesters, and detection instrumentation. Instrumentation and equipment suppliers are increasing the compatibility of their equipment with higher-density microplate formats and capabilities in handling nanoliter and picoliter volumes. The high numbers processed in HTS systems and the increasing information density from cell-based assays poses increasing demands on data processing and storage (Thurow et al., 2004). The main challenges over the next few years will especially involve the development of innovative information-technology solutions (Stoll and Thurow, 2008).

9.2.5.1 Liquid-Handling Systems

Parallel to the development of the first microplate, Dr. Takatsy also developed the first form of a microplate automation tool. The loop that mixed and transferred a predefined volume from one well to another had been used in serial dilution testing (Buie, 2010). A first parallel and more automated loop system was manufactured by Cooke Engineering in 1964: the Microtiter®. Eight to 12 loops were moved from row to row in a plate. Although this was still a manual process, it provided a significant improvement of throughput and accuracy. In 1967, the first fully automated serial-dilution instrument (Autotiter) was introduced by Astec (now TomTec). The system was used at Smith Cline & French to perform thousands of hemagglutination-inhibition tests for the rubella vaccine and became in the following years a very popular instrument in clinical laboratories. An important fact in the automation of liquid-handling systems was the development of microscale DC motors and valve technology that led to the introduction of highly accurate semiautomated motorized syringe-based pipetting devices. A semiautomated, DC motor-driven, adjustable pipetting device was introduced in 1971 (Digital Dilutor from Hamilton). The system used two of their calibrated syringes as the pipetting pistons. The programming of sequences of motor and valve functions became possible because of the development of microprocessor technology in the late 1970s and resulted in fully automated motorized syringe-based liquid-aspirating and -dispensing devices. However, only the evolution of motor and microprocessor technology in the 1980s enabled the development of the first truly automated liquid-handling workstations. With respect to the requirements of drug-discovery laboratories, TomTec expanded the functionality of the Wallac Betaplate (first automated microplate-based instrument for scintillation counting, 1986) by including possibilities for harvesting and automated pipetting; the automated pipettors Harvestor 2 and Quadra 3 were introduced to the market in 1990. The first 96-channel pipetting/liquid transfer device (Quadra96) with a positioning stage base for holding microplates, pipette tips, and reservoirs and with a 96-channel pipetting head mounted above on a linear z-axis was introduced by TomTec Inc. in 1990. The current automated liquid-handling workstations have all evolved from these early basic designs, adding more functionality and software sophistication. For a selection of current liquid-handling systems, see Table 9.2.

9.2.5.2 Detection Systems: Reader

The increasing demand for ELISAs led in 1976 with the Multiskan photometer to the development of the earliest version of the common day microplate readers by Lab Systems (now part of Thermo Fischer Scientific). The first multidetection microplate reader including fluorescence polarization was introduced in 1997 by BMG LABTECH. BioTek introduced in 2004 the Synergy™ 4 with Hybrid Technology, a multidetection system capable of performing an unlimited number of microplate-based assays. The first multidetection microplate reader on the market with UV-visible spectrometer absorbance was the POLARstar Omega (BMG Labtech) in 2007.

Detection systems for high-throughput screening include today radiometric and fluorescence-based methods, absorbance/colorimetric methods, luminescence, and others. The choice of the detection method has great impact on the speed, efficiency, and accuracy, but also on the costs of the entire system operation. Fluorescence polarization, fluorescence intensity, chemoluminescence, and Förster resonance energy transfer (FRET) are expected to show the biggest increase in use in the upcoming years. Detection systems continue to evolve. HTS users are interested in new innovations such as automated patch-clamp technology, single-molecule detection, advanced imaging systems, and label-free technologies. Drivers of detection mode changes include avoiding radioactivity and using safer methods with less disposal costs, increasing sensitivity (achieve better signal-to-noise ratios), increasing speed and throughput, and facilitating miniaturization and smaller volumes.

Optical readers are used in clinical chemistry and diagnostics and molecular and cell biology, as well as foodstuff analysis. The main application areas in pharmaceutical industry include ELISAs, protein and cell growth assays, nucleic acid quantification, molecular interactions, detection of

TABLE 9.2
Liquid-Handling Systems (Selection)

Manufacturer	System	MTP Formats	Volumes
Beckman Coulter	BioRAPTR	96–3,456	100 nl to 60 μl
Beckman Coulter	Biomek3000 workstation	384	1–200 μl
Caliper Life Sciences	Zephyr SPE workstation	96	1–200 μl
Hamilton	Star Line	96–384	0.5–5,000 μl
Hamilton	Nimbus	1,536	10–300 μl
Tecan	Freedom EVO® series	1–384	100 nl to 5,000 μl
Agilent	Bravo automated liquid handling platform	96–1,536	100 nl to 200 μl
Eppendorf	epMotion	96–384	1–1,000 μl

enzyme activity, cell toxicity, proliferation and viability, ATP quantification, immunoassays, and, of course, high-throughput screening of compounds and targets in drug discovery. Current-day plate readers are equipped with software tools for data analysis and automation.

9.2.5.3 Robotic Systems

Automated transport systems are the main driver for increasing throughputs. They connect liquid-handling systems, reader systems, and different other subsystems necessary for certain assays and thus enable fully automated unattended operation of the screening systems. Robots are defined by the Robotics Industry Association as "a manipulator designed to move materials, parts or specialized devices through variable programmed motions for the performance of a variety of tasks."

Electromechanical transport systems for industrial automation emerged in the 1950s. The use of robotics in manufacturing environments was pioneered by Unimation Inc., the world's first robotics company, in 1956. In 1961, the first robot system was commercially installed at General Motors for die-casting (Devol, 1961). Although robots had been widely used in the industry, they did not appear in the laboratory until the 1970s, when simple autosampler devices began to be developed for segmented flow analyzers and, later, high-performance liquid chromatography systems. This was mainly caused by the different requirements regarding size, flexibility, and programmability of devices for laboratory environments in contrast to manufacturing processes. The development of microprocessors and precision small DC-motor technology in the late 1970s was a major step for the implementation of robotic systems in the laboratory. Compact, microprocessor-controlled user-programmable robot arms were introduced in the early 1980s. The Zymark Corporation patented a robot arm with interchangeable hands that became the base for the development of robotic laboratory workstations capable of carrying out programmable multistep sample manipulations. These robots could be adapted to numerous assays and sample-handling approaches, and they were used in preanalytical sample preparation or potency and stability testing in the pharmaceutical industry (Godolphin, 1983). In the 1980s, Dr. Masahide Sasaki developed his vision of an automated clinical laboratory for performing medical investigations of patient samples. The laboratory was completed in the 1990s, equipped to perform all clinical laboratory testing for a 600-bed hospital with a government-mandated maximum staff of 19 employees (Sasaki et al., 1998). Similar developments started at the same time at the University of Virginia (Boyd et al., 1996). Robotic laboratory automation was pushed in the following years by automated gene sequencing in the Human Genome Project, drug discovery in pharmaceutical companies (Rutherford and Stinger, 2001), or high-throughput proteomics.

Today there are a number of electrically driven common robotic configurations available (for a summary see Table 9.3). The workspace assessable by the robot is limited, and different configurations have different workspace geometry. The main robotic configurations today include Cartesian,

TABLE 9.3
Laboratory Robots (Selection)

Robot	Vendor	Degrees of Freedom	Weight (kg)	Payload (kg)	Positioning (mm)	Interfaces	Software
XP-Robot	Zymark	4	39	1.4	+/−1	RS422 proprietär	Easylab
ORCA	Beckman	5	8	2.5	+/−0.25	RS232	SAMI
HP3JC	Motoman	6	27	3	+/−0.03	Ethernet Profibus DP	INFORM III
F3	CRS	6	53	3	+/−0.05	Polara Rapl-3	16dig. output RS232
RX 60	Stäubli	6	44	2.5	+/−0.02	Ethernet Profibus DP	VAL 3
Katana	Neuronics	6	4.8	0.4	+/−0.1	Ethernet USB	Katana 4D
RV-1AJ	Mitsubishi	6	19	1	+/−0.02	Cosimir Cosirop	Ethernet RS232

Gantry, Cylindrical, Articulated, and Selective Compliant Assembly Robot Arm. All of the configurations can be equipped with different end effectors enabling pick and place or even screwing operations to be performed by the robot. Screening systems with robots as central transport systems are usually operated multidirectionally. Multiple workstations can be accessed during the screening process. Sophisticated scheduling software is required to optimize the systems operation and throughput and to avoid failures of the systems. Robots can become a rate-limiting factor for high-throughput systems in this system concept.

Another approach for transport systems are linear devices such as belts or conveyors. They have been used in manufacturing environments for decades but only appeared in the laboratory automation environment in the 1990s. Flexible conveyor approaches have the advantage of being fast and relatively inexpensive (higher precision devices are much more costly). Their workspace can be extended to almost any length. In contrast to robots, they are only transport systems that do not offer the pick and place dexterity for samples or vessels. This capability must thus be integrated into the workstations connected by the conveyor. Conveyor systems are usually used in a linear mode (unidirectional operation) to ensure highest throughputs. In such linear systems, the rate-limiting component will be the workstation that takes the longest time to perform its processes. If different liquid handling or readout steps are integrated into the assay, the duplication of the necessary workstations is required, which will result in increasing costs for the complete system. A major advantage of such systems is that no sophisticated system-scheduling software is required.

9.3 THE NECESSITY OF AUTOMATED HIGH-THROUGHPUT SCREENS IN DRUG DISCOVERY

9.3.1 Why Automate the Drug Discovery Lab?

Pharmaceutical companies currently have a pressing demand for improvement in high-throughput screening technology. Microorganisms are adapting to classical drugs, resulting in bacterial resistance against antibiotics. Many classical drugs have negative side effects, and there are still a high number of diseases without a useful medication. Additionally numerous patents end in the near future, and new blockbusters are required to sustain the pharmaceutical industry.

The recent advances in genomics, proteomics, and systems biology have led to the possibility of simultaneous studies of high numbers of molecules. Large numbers of potential drug targets and related volumes of information and data have been created. Automation of all procedures is necessary to handle these large amounts of samples and information. In addition to the fact that laboratory automation allows more tasks to be carried out within a shorter time frame, it also makes possible higher quality and reproducibility of data combined with better documentation of all experiments. Because of the quality improvement of the data, more difficult analyses can be performed.

Many of the procedures in drug development and drug discovery include repetitive and routine tasks that are important targets for automation. Semi- or fully automated workstations can be used for delivering small reagent or solvent volumes or washing the samples in microplates. The goal of automation in pharmaceutical and biotech companies is not a reduction of labor costs, but the improvement of experimental results and work flow (Kalorama Information, 2008). Automation of the drug discovery process can free scientists from labor-intensive tasks and give them time for creative and value-added tasks such as experimental design. The drug-discovery market still requires continuous improvements in automation. After a period of strongly proprietary automation solutions, the manufacturers are now changing their strategy to open automation solutions. Especially small and medium-sized laboratories require flexible and powerful devices that can either be used as stand-alone systems or can be easily integrated with other devices including laboratory robots. Thus a rapid reconfiguration of the automation systems depending on the type of application becomes possible. Vendors offering cost-effective open automation solutions will receive a higher market share in the future compared with vendors marketing closed systems.

9.3.2 Applications of HTS in Drug Discovery

The screening of chemical compounds regarding their pharmacological activity has been ongoing in various forms for at least 40 years. When a compound interacts with a target in a productive way, that compound then passes the first milestone on its way to becoming a drug. Compounds that fail this initial screen go back into the library, perhaps to be screened later against other targets (Rubenstein and Coty, 2001). Traditional biochemical and pharmacological drug-discovery methods are working with 1-ml reaction volumes and require the use of solid compounds. The assay capacity is limited to 20–50 compounds per week, resulting in a total screening time of 1–2 years for screening 3000 selected compounds (Pereira and Williams, 2007). The screening of natural compounds was a starting point in the late 1980s for the development of HTS methods. Until 1995, HTS was mainly designed as therapeutic target HTS. With the development of high-throughput liquid chromatography (LC)-mass spectrometry (MS), HTS moved into absorption, distribution, metabolism, excretion, and toxicity HTS with cytotoxicity tetrazolium assays. First generation HTS was in the 1990s oriented on throughput and included assay adaptation, HTS screening, and lead optimization for full libraries with 50,000–200,000 compounds (Figure 9.2). An increase in the number of compounds (from 200,000 to 1.5 million) in libraries and in the number of molecular targets led to a change in the HTS strategy, shifting large parts of assay development and tool production into close proximity of the screening processes in the beginning of the 21st century. Since 2007, the drug-development process was mainly project related and usually required focused libraries as an alternative to screening full compound libraries (Mayr and Fuerst, 2008).

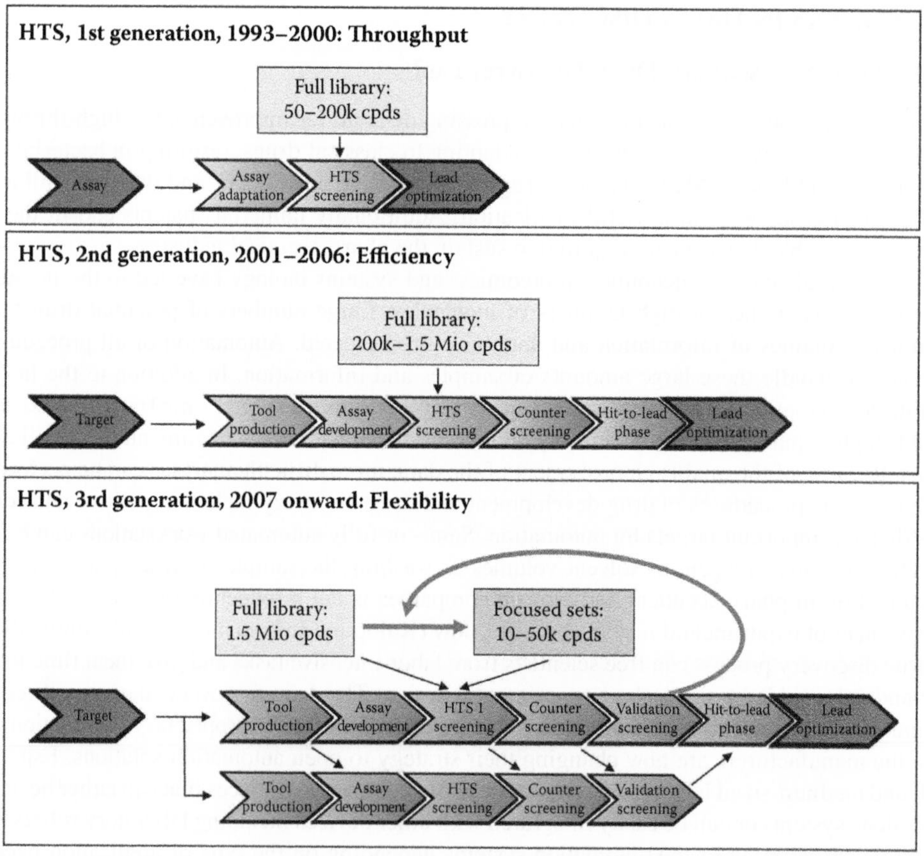

FIGURE 9.2 Generations of HTS in drug discovery. (Taken from Mayr, L. and Fuerst, P., *Journal of Biomolecular Screening*, 13, 443–448, 2008.)

Today HTS in drug discovery can be defined as the process for rapid and parallel testing of batches of compounds for biological activity or binding activity against target molecules. Test compounds act, for example, as inhibitors of target enzymes, as competitors for binding of a natural ligand to its receptor, and as agonists or antagonists for receptor-mediated intracellular processes. Positive hits of this screening process undergo further investigations and tests and can become lead compounds after this second level of triage. Leads are used as starting points for the synthesis of structurally related compounds in order to increase the biological activities and optimize toxicity or bioavailability. Thus potential failures in the drug discovery process can be eliminated as early as possible.

Since the 1980s, improvements in screening technologies have resulted in throughputs that have increased from 10,000 assays per year to current levels of more than 100,000 assays per day. HTS today is not limited to the determination of biological activity but also includes absorption, distribution, metabolism, and excretion (ADME) and toxicity testing. The majority of all assays are performed in 384 well microplates. This format has been established as the standard format for screening and compound storage in pharmaceutical industry (Fox, 2006).

High-throughput assays in drug development can be classified as functional (measuring the activity of a compound in modifying the actual function of a target protein such as ion currents of ion channels) or nonfunctional (measuring binding of a compound to the target protein or use some indirect measure of target activity). Examples of nonfunctional assays are tritiated binding assays, the measure of fluorescence activity associated with calcium signaling, and techniques such as FRET. Functional assays are preferred because they produce less false-positive hits.

Pharmaceutical companies typically screen between 100,000 and 300,000 compounds per screen to produce approximately 100–300 hits. Usually only one or two of these hits become lead compounds for further development. Improvement of lead generation depends on the diversity of compound libraries. HTS assays are also used in safety studies, e.g., screening for blockade of the hERG ion channel, which has been associated with sudden death in a number of marketed drugs (Brown, 2004). To avoid this effect, many companies are screening their libraries to eliminate these compounds before a hit is even identified.

9.3.3 Current State and Future Prospects for Drug-Discovery Technologies

The market for laboratory automation is still quite young and has emerged as recently as the 1990s. In the last 10–15 years, high-throughput screening instruments, assays, and services have emerged as a significant growth market. The growth rates for HTS were even higher than the rates of growth for pharmaceutical research and development in the last years, but they were still behind the actual needs of the industry. According to recent strategic business reports, the global market for drug-discovery technologies has been estimated to be US$33 billion for 2008 (Global Industry Analysts, 2008). With a growth rate of about 15% from 2001–2010, the market is further projected to reach a value of US$43.8 billion by 2010. The largest market with an estimated market share of nearly 55% represents the US followed by the European market with a market share of nearly 25% (in 2010). The largest segment among drug-discovery technologies is represented by bioanalytical instruments with a global-market segment of about 36% with growth rates of 13% (expected value 2010 US$15.3 billion). The HTS market as the second largest drug-discovery technology showed growth rates of 16% during the period 2001–2010 with a total market value of US$8.1 billion in 2010. The market for plate readers is expected to grow at a 4% annual rate until 2012 (Kalorama Information, 2008). Liquid-handling systems represent the largest segment of the laboratory automation with a market value of US$2.75 billion in 2005 and an annual growth rate of 8–10% (Global Industry Analysts, 2006).

An additional problem especially in the United States and in Europe is the decreasing number of lab technicians available in the next decades. Laboratory automation can help to fill this gap with innovative automation products. High growth rates are also predicted for the information-technology solution market and the bioinformatics market for the upcoming years (Global Industry Analysts, 2006).

9.3.4 Success of HTS in Drug Discovery

As mentioned before, HTS technologies have been used in chemical and pharmaceutical companies since 1995. First approaches using HTS in drug discovery were based on random screening of large corporate compound libraries. Experiences from pharmaceutical companies show that HTS success depends on the target class. Full diversity screening was successful for certain classes like ion channels and family A GPCRs but did not lead to acceptable results for family B or family C members of this receptor class (Macarron, 2006). Successful HTS screens have been performed by different companies, which have been described before (Golebiowski et al., 2001, 2003). Different clinical candidates have resulted from diversity or project-related screening, such as a complex stilbene derivative as a potential drug for the prevention of respiratory syncytial virus disease or a novel nonpeptide HIV protease inhibitor. One spectacular success of using HTS technologies was the identification of inhibitors of the severe acute respiratory syndrome (SARS) coronavirus main proteinase. Fifty thousand compounds have been tested using a quenched-FRET assay producing 69 hits and five secondary hits (Blanchard et al., 2004).

In general, it can be concluded that this approach has not yet fulfilled the initial (but unrealistic) expectations that screening a greater number of compounds would generate more leads of improved quality. The experience of many pharmaceutical companies showed that screening of numerous compounds cannot be successful without additional information on the target and the compound (Gribbon and Sewing, 2005). The HTS strategies of many companies changed to a greater use of knowledge-based approaches. The main drivers for successful HTS are thorough target analyses and careful selection of compounds. The introduction of property filters in compound selection is essential in order to get high-quality HTS results because the success of a hit or lead strongly depends on its bioavailability. Thus in addition to the selectivity and potency of a compound, multiple compound characteristics (e.g., ADME properties) are also important drivers in lead development (Alanine et al., 2003). Poor HTS results can also be caused by two additional problems: the target is not druggable (mostly because of protein-protein interactions) or the libraries do not contain the right compounds. The latter is not surprising because the chemical space available for screenings is very limited. From the 10^{40}–100^{100} possible small molecules, only a maximum of 10^7 compounds are tested in drug-discovery laboratories.

Today, many pharmaceutical companies are screening 100,000–300,000 or more compounds per screen to produce approximately 100–300 hits. On average, one or two of these become lead compound series. Larger screens of up to 1,000,000 compounds in several months may be required to generate something closer to five leads. According to previous studies performed with 44 different companies, on average 43% of the investigated targets had generated leads. Altogether, in different HTS laboratories worldwide, 326 leads were found in 2001 using HTS; 62 of these leads have been tested in humans (Sannes, 2002). These include kinases, viral proteins, GPCR, cytomegalovirus, receptors, and cancer as targets (Fox et al., 2002). One hundred four clinical candidates, leads originally found through the use of HTS technologies, were generated in 2006; four products are now on the market. The number of leads found by using HTS technologies increased from 326 in 2002 to 746 in 2004 (Fox et al., 2006).

Drug discovery is still undergoing revolutionary and very rapid changes. Additionally HTS cannot be considered in isolation from other aspects of drug discovery. These aspects will influence the long-term future of HTS.

9.3.5 The Future of HTS

The size and diversity of compound collections of pharmaceutical and biotech companies have tremendously increased within the past 10–15 years because of the development of high throughput technologies for chemical synthesis and biological testing. The sequencing of the human genome delivered numerous novel molecular targets for pharmaceutical interventions. Increased numbers of targets and compounds call for greater parallelism and/or increased throughput in screens. Furthermore, the expense of targets and compounds drives a trend toward smaller assay volumes through miniaturization. The most important factor, which is expected to have a direct influence on future developments, is cost reduction in medicine in the next decade(s) (Global Industry Analysts, 2006).

Future trends in HTS will include novel target classes such as ion channels, transporters, or protein-protein interactions. The development of adequate compound libraries for these targets will be essential for the future success of HTS in the drug discovery process (Schreiber et al., 2002). Miniaturization will be one of the development drivers for the next decade. Acceptance will be strongly connected to the value received for miniaturization versus the resource inputs required. Drop-based microfluidics has shown its potential, allowing 1,000 times faster screening (100 million reactions in 10 hours) at one millionth the cost (using 10^{-7} times the reagent volume) compared with conventional techniques (Agrestia et al., 2010). New developments will also be made for new detection methods. These methods include patch-clamp technology (Sigworth and Klemic, 2002; Fertig et al., 2002; Stett et al., 2003) and lab-on-a-chip systems (Dittrich and Manz 2006; Craighead, 2006), as well as mass-spectrometric measurements (Ozbal et al., 2004; Fleischer et al., 2009; Koehn, 2008). A silicon sheet of lenses that can be placed over microfluidic arrays to analyze 200,000 drops per second has been recently introduced (Schonbrunn et al., 2010).

One of the largest current bottlenecks is the implementation of high-content–type assays in high-throughput formats. More complex biology requires multiparametric assays to analyze a high amount of diverse data from cell imaging (Sterling, 2008). This is strongly connected with innovative possibilities of analyzing, archiving, and retrieving complex imaging data.

Decision markers for the implementation of new technologies also include the extent to which new technology provides extended value beyond the primary screening process. Implementation also depends on the information content provided by new technology and the extent to which it can be integrated into the laboratory environment without reducing or destroying flexibility.

Future developments in automation solutions for HTS will include integrated platforms with multiple reagent and detection technologies, handling of multiple-sized tubes and supporting intelligent connectivity to automation systems. The need for standardized interfaces in laboratories in order to enable the operation of instruments from different manufacturers at a common platform is steadily increasing (Global Industry Analysts, 2006). Automation and data explosion must be accompanied by the development of optimized processes for knowledge integration, information flow, and decision making (Kalorama Information, 2008).

9.4 THE GROWING PORTION OF AUTOMATED PROCESSES IN OMICS RESEARCH

According to the concept of life's complexity, pyramid molecules represent the basis of the cell's functional organization. Therefore, the bottom of this pyramid is formed by the genome, transcriptome, proteome, and metabolome of a cell (Oltvai and Barabasi, 2002). Systems biology approaches, as they have been introduced at the beginning of this chapter, are mostly used to integrate data and results from analyses for omics research. As a consequence, systems biology in biomedical disciplines would consist of genomic, transcriptomic, proteomic, and metabolomic investigations. This set of basic approaches applied on the huge variety of cells, tissues, and organs of interest would alone justify the efforts to increase the velocity of analysis by automated laboratory methods. In Table 9.4, a summary of commonly applied omics methods in systems biology is given. Although this table is far from being complete, it makes apparent that the number of omics investigations in systems biology goes far beyond these basic methods. This effect is mainly caused by the rising number of omes that have been discovered in recent years, like the epigenome, the microRNA genome, the phosphoproteome, and the ionome, just to mention a few examples. Together with the number of these fields, the number of corresponding omics has also increased, which makes the

TABLE 9.4
A Collection of Different Human Omes and Omics

Ome	Field of Study	Selected URLs and Readings
Genome	Genomics: determination of the 3-Gbp human DNA	http://www.ornl.gov/sci/techresources/Human_Genome/home.shtml; http://huref.jcvi.org; Collins et al. (2003); Mukhopadhyay (2009)
Transcriptome	Transcriptomics: qualitative and quantitative determination of all transcripts in a cell	http://h-invitational.jp; http://www.microarrayworld.com; http://www.ncbi.nlm.nih.gov/IEB/Research/Acembly/index.html; http://rfam.sanger.ac.uk; Wang et al. (2009)
Proteome	Proteomics: qualitative, quantitative, and functional analysis of all proteins in a cell	http://www.hupo.org; http://www.proteomicworld.org; http://www.proteinatlas.org; http://www.hprd.org; Tyers and Mann (2003); Patterson and Aebersold (2003); Malmström et al. (2007)
Metabolome	Metabolomics: qualitative and quantitative determination of all small molecules (metabolites) in a cell	http://www.hmdb.ca; http://humancyc.org; Ellis et al. (2007); Wishart et al. (2009)
Epigenome	Epigenomics: characterization of all inherited alterations not caused by changes of the underlying DNA sequence	http://www.epigenome.org; Lister et al. (2009); Gaulton et al. (2010)
Exome	Exomics: characterization of all protein-coding DNA and RNA (exons)	Cirulli et al. (2010); Kim et al. (2010); Ng et al. (2008, 2010)
Kinome	Kinomics: analysis of the kinase inventory	http://kinase.com/human/kinome; http://kinase.bioinformatics.tw; http://www.cellsignal.com/reference/kinase/index.html; Manning et al. (2002); Johnson and Hunter (2005)
Metabonome	Metabonomics: quantitative analysis of metabolites in response to biological alterations like disease or therapy	http://www.hmdb.ca; http://humancyc.org; Ellis et al. (2007)
Lipidome	Lipidomics: study of all lipids in a cell	http://www.lipidomics.net; http://www.lipidmaps.org; van Meer (2005); Bougnoux et al. (2006); Schittmayer and Birner-Gruenberger (2009)
Glycome	Glycomics: study of all carbohydrates in a cell	http://www.functionalglycomics.org; http://www.ncgg.indiana.edu; http://www.hupo.org/research/hgpi; http://glycosciences.de; Lee et al. (2005); Freeze (2006); Satomaa et al. (2009)
Interactome	Interactomics: investigation of all protein-protein interactions in a cell	http://www.hprd.org; http://mips.helmholtz-muenchen.de/proj/ppi; http://160.80.34.4/HomoMINT/Welcome.do; http://www.humanproteinpedia.org; http://kinase.bioinformatics.tw; Stelzl et al. (2005); Mathivanan et al. (2006, 2008); Stumpf et al. (2008)
Glycoproteome	Glycoprotemics: analysis of all proteins covalently bound to carbohydrates	http://www.ncgg.indiana.edu; http://www.hupo.org/research/hgpi; http://glycosciences.de; Lee et al. (2005); An et al. (2009)
Lipoproteome	Lipoproteomics: analysis of all proteins covalently bound to lipids	http://www.lipidmaps.org/data/proteome/index.cgi; Hoofnagle and Heinecke (2009); Vaisar (2009); Schittmayer and Birner-Gruenberger (2009)
Phosphoproteome	Phosphoproteomics: analysis of all phosphorylated proteins	http://141.61.102.18/phosida/index.aspx; http://kinase.bioinformatics.tw; http://phospho.elm.eu.org; http://www.phosphosite.org; Gnad et al. (2007); Macek et al. (2009)

TABLE 9.4 (CONTINUED)
A Collection of Different Human Omes and Omics

Ome	Field of Study	Selected URLs and Readings
Fluxome	Fluxomics: analysis of dynamic changes of small molecules (metabolic fluxes) in cells	http://www-ciwdpb.stanford.edu/frommer-lab/research/fluxomics; Cascante and Marin (2008); Frommer et al. (2009); Dauner (2010)
Regulome	Regulomics: comprehensive investigation of transcriptional regulation	http://www.internationalregulomeconsortium.ca; http://www.eutracc.eu; Stamatoyannopoulos (2004)
Metallome	Metallomics: quantification of metals and metaloids and the determination of their coordinative environment within a biological system	http://www.speciation.net; Szpunar (2005); Mounicou et al. (2009)
Toponome	Toponomics: description of the topology of all proteins within the context of cells and tissues	http://www.hprd.org; http://www.humanproteinpedia.org; Schubert (2010); Schubert et al. (2006); Pierre and Scholich (2008)
Microbiome	Microbiomics: comprehensive characterization of the human microbiota	http://www.hmpdacc.org; http://www.human-microbiome.org; Human Microbiome Jumpstart Reference Strains Consortium (2010)

employment of automation in these laboratories indispensable. In the following subsections, an overview of the state of high-throughput methods and automation in the basic fields of genomics, transcriptomics, proteomics, and metabolomics is given, concluding with a discussion of future trends and challenges in systems biology and the evolution of novel omics.

9.4.1 Genomics

The emergence and establishment of most omics approaches was largely based on the success of whole-genome sequencing from any given organism. The evolution of genomics research itself, more precisely of genome sequencing, is an ideal example to study for critical factors in the development of any omics field.

One of the major obstacles at the beginning of genome sequencing was the sequencing costs: at the beginning of the Human Genome Project (HGP) in 1988, US$3 billion over 15 years were budgeted for the sequencing of the complete human genome. The project finished successfully before the deadline at far below the calculated cost. One of the major reasons for this development was the decrease in sequencing costs. At the beginning of the HGP, the costs were around US$10 per base pair (bp). In February 2001, when the first complete sequences of the human genome were published (International Human Genome Sequencing Consortium, 2001; Venter et al., 2001), the price had dropped below 10 US cents per bp, a decrease of more than 100-fold within 13 years (Collins et al., 2003).

This logarithmic decrease of costs was caused by several different factors. A major driver of this development was the dramatic increase in high-throughput sequencing. The speed and cost efficiency of this process were enhanced by technological development on the one hand and automation on the other. This interplay of technological development and automation resulted in a quite impressive decrease in the costs for a full genome sequence since the publication of the first draft of the human genome: the price for the first genome in 2001 was about US$300,000,000–500,000,000, whereas the costs for the first personal genome published of James Watson were below US$1,000,000 (Wheeler et al., 2008). In 2010, the same service could be purchased for under US$50,000, and it is assumed that the so-called $1,000 genome can be realized by the year 2013 (Bonetta, 2010; Wade, 2009).

This development was mainly driven by technological changes to increase the throughput of sequencing. The first version of the human genome was achieved using classical dideoxy sequencing of map-based or shotgun clones (International Human Genome Sequencing Consortium, 2001; Venter et al., 2001). Afterward, novel-sequencing technologies were established, of which the 454 Life Sciences and the Solexa/SOLiD technologies are the most popular ones. These so-called next-generation sequencing (NGS) platforms are based on cycle-array–sequencing machines with specific software under full automation. In massively parallel DNA-sequencing reactions in millions of reads, several megabase pairs up to 1 gigabase pairs of sequence information can be achieved in one run (Mukhopadhyay, 2009; Rothberg and Leamon, 2008). These developments would not have been possible without technological leaps in sequencing methodology and information technology.

These NGS platforms shifted genomics approaches in biomedical fields to an advanced level. Full-genome sequencing of single individuals becomes more and more affordable, which is a step forward to personalized medicine. Comparative genomics of individual genome sequences will also contribute to the identification of genes and mutations responsible for many kinds of disorders.

The 1000 Genomes Project is one example. In this international collaboration, the full genomes of approximately 2,000 individuals from all over the world have been sequenced to determine genetic variations at a population frequency of more than 1% (Via et al., 2010).

The described progress in DNA sequencing technology was tremendous but is still not at its end. There are several efforts to reach the goal of the $1,000 genome with technologies such as nanopore sequencing or single-molecule real-time technology. Although these systems have not been commercially launched yet, it seems realistic that these platforms, sometimes referred as next-next or third-generation sequencers, will lower sequencing costs of full genomes to almost negligible levels in the next few years (Mukhopadhyay, 2009). With this step, full-genome sequencing would become a common genomics tool also for system biologists, but it is quite naive to reduce genomics to pure A, C, G, and T. It is a long known fact that organismal DNA does not occur unmodified all alone. Its methylation patterns and its chromatin structure, as well as the methylation status of accompanied histones, are inherited, influence genetic activity, and are responsible for growth syndromes, behavioral disorders, and cancer (Henckel and Amaud, 2010; Talbert and Henikoff, 2010). The mechanisms and regulation of imprinting genomics and epigenomics are still poorly understood, and their investigation is still in its infancy. Therefore it has to be stated that the progress in genome sequencing was a major step forward in genomics, but analyzing and understanding the epigenome will be one of the next large challenges in the path. The accomplishment of this task will mainly depend on the application of high-throughput methods also in this field. First approaches in this direction have been taken and successfully applied with the first sequence of the human methylome (Lister at al., 2009), but there is still a long road until epigenetics can reach the analytical capacity of classical genomics.

9.4.2 Transcriptomics

The transcriptome is defined as the comprehensive set of transcripts of a cell. It is the task of transcriptomics to identify and analyze these molecules, as well as to quantify their expression level. In the beginning, the major purpose of this methodology was the analysis of mRNAs; this means the determination of their coding sequence, the analysis of splicing variants, the determination of their 5'- and 3'-untranslated regions, and also the quantification of the different transcripts.

Generally the transcriptome can be analyzed in two ways: by sequencing or by hybridization. In former days of transcriptome research, partial sequencing of randomly chosen cDNA clones was a predominant technique. The generation of millions of such expressed sequence tags (ESTs) was realized by the use of picking robots for clone selection and liquid handlers for bacterial culture handling and plasmid preparation. Therefore the generation of EST repositories is an early example of successful lab automation in the history of omics. Later, the development of tag-based methods like the serial expression of gene expression with all its modifications extended the employment of sequence-based transcriptomics (Datson, 2008).

Parallel hybridization techniques evolved, which led to the development of microarray technology. Only 20 years have passed since the first attempts to produce high-density peptide and nucleotide slides were published (Fodor et al., 1991). Nowadays DNA microarrays, especially whole-genome tiling arrays, are almost a synonym for transcriptome analyses. Although their range of application goes far beyond simple transcript analysis, microarrays have become an indispensable part of biomedical research as the exponential increase of microarray citations shows (Hoheisel, 2006; Wheelan et al., 2008). There were several factors that led to the success of this technology. Two of them have to be pointed out concerning the subject of this review: one advantage of microarray experiments is the high degree of automation. This is mainly caused by the high density of probes on the slides. This miniaturization led to the development of fully automated workstations and to the processing of the microarrays in them. Another reason for the success of this technology is its high degree of throughput. Ten thousand transcripts can be identified, analyzed, and quantified in a single array experiment. This fast gathering of huge amounts of data made this technology an ideal tool for systems biologists.

Although the application of DNA microarrays is still in its booming phase (Wheelan et al., 2008), a novel technology evolved that will take over its role at least in transcriptomics. The NGS technologies and platforms described in the previous chapter also found their way into transcriptomics as RNA-Seq (Wang et al., 2009). Under this term (or RNA sequencing), the deep sequencing of cDNAs is understood. The possibility of getting millions of reads within one run by high-throughput sequencing offers several advantages over the use of DNA microarrays. Among these advantages the lower costs, better quantification, and higher resolution have to be pointed out (Wang et al., 2009; Cirulli et al., 2010).

The invention of RNA-Seq emphasizes the importance of NGS technology not just for the classical genome sequencing field but also for the progress of transcriptomics. Recently this method was successfully applied for individual exome sequencing (Ng et al., 2008). A similar approach was later used to find the gene responsible for Miller syndrome. Therefore, the exome of affected individuals was sequenced to sufficient depth and after filtering against several data sets (single-nucleotide polymorphisms and HapMap); mutations in a pyrimidine biosynthesis gene could be identified to cause this Mendelian disorder (Ng et al., 2010). It is very probable that the evolution of third-generation sequencers will further take over the transcriptomics field.

9.4.3 Proteomics

Nucleotide-based molecules in a biological system are called genomes or transcriptomes; the same holds true for all polypeptides, which are known as proteomes. Proteomics is a central field of systems biology because proteins are involved in almost any biological activity, and thus they are indispensable for the understanding of any biological system. Although proteins are nothing more than translations of their encoding DNAs according to the central dogma of cellular biology (DNA → RNA → protein), it is more difficult to analyze the proteome than just by simple peptide sequencing.

The idea to catalog all of the proteins of a biological system came up very early, and in 1994 it was called proteomics for the first time (Patterson and Aebersold, 2003). In early phases of proteomics, in untargeted approaches, the proteins were separated and visualized by two-dimensional electrophoresis and afterwards identified by MS. Although this was a powerful approach, which is still one of the standard procedures today, the limitations of two-dimensional electrophoresis for protein separation were recognized a decade ago (Gygi et al., 2000). To overcome this limitation, other separation methods were introduced for quantitative proteome analysis. The most prominent among them is the proteolytic digestion of unseparated protein mixtures and their separation by LC afterward (Patterson and Aebersold, 2003). Apart from the separation methods, MS is still the standard for the identification of peptides in untargeted proteomics. In targeted approaches like the investigation of the kinome (Table 9.4), other techniques including protein arrays and microfluidics are employed, but these miniaturized methodologies will not be the subject of the present review.

Currently almost five million MS spectra and about two million peptide identifications from untargeted analyses in human sources have been published (http://www.humanproteinpedia.org). Nevertheless, these figures are still just a beginning compared with the tasks of proteomics projects in the future. The gene-centric Human Proteome Project is an informative example that depicts the challenges of proteomics research and its demands for high-throughput and automation. With this project, the Human Proteome Organization (HUPO) proposed to generate a human proteome map using the approximately 20,000 protein-coding genes in the human genome as a starting point (HUPO, 2010). This should be achieved in three steps.

In the first step, at least one representative for every human protein shall be identified and characterized including its abundance and major modifications. The costs of just a part of this project chapter were calculated at US$300 million (Anderson et al., 2009). An essential element of this calculation is based on cost reduction of monoclonal antibody production against each protein by automated high-throughput processes (De Masi et al., 2005; Anderson et al., 2009). The description of the efforts to produce the initial set of antibodies for this project, their verification, and their application gives a first impression about the potential for automated high-throughput screening processes within this project (Uhlen et al., 2005). It has to be noted that this calculation of US$300 million does not include the analysis of protein modifications (e.g., phosphorylation, glycosylation, and coupling of metals) or of protein isoforms. Moreover, the majority of human genes are spliced alternatively, which results in about 2.6 protein isoforms per gene (Nakao et al., 2005). All of these considerations increase the need for higher throughput using enhanced instrumentation and also more automated processes.

In the second step, a protein distribution atlas shall be established that determines the profile of each protein in various tissues and organs to single-cell resolution. In the current version of the Human Protein Atlas with 11,200 antibodies, more than 9 million images were made for this purpose (http://www.proteinatlas.org/atlas_history.php). These figures evoke an impression about the throughput velocity needed in this project. Furthermore, this protein distribution atlas shall also include the subcellular localization of all proteins. This localization and topology of the proteins, also known as the toponome, is another dimension of proteomics that will be the subject of a further chapter.

In the third and last step, a protein pathway and network map shall be generated. This means that on one hand the protein-protein interactions have to be analyzed. Such interactomics and toponomics will also be discussed in one of the next chapters. On the other hand, the functions of the proteins have to be characterized. A major part of the protein inventory mainly acts as enzymes, and enzyme analysis is a classical field of high throughput analysis in automated lab environments. There is a high probability that also in this field of enzymatic analysis of proteomes, classical lab automation will play a substantial role.

As depicted for the Gene-centric Human Proteome Map Project, proteomics offers many opportunities for high-throughput processing. Also for the existing core analyses of human proteomics, automation and high-throughput methods are indispensable. Thus it can be stated that the progress of human proteome analysis is highly dependent on further development of instrumentation and laboratory automation.

9.4.4 Metabolomics

The term metabolomics was coined as an analogy to the analysis of the genome and proteome for the investigation of the entity of metabolites in a biological system. The measurements of particular compounds in tissue or a body fluid like in urine, blood, or saliva belong to the old traditional methods of diagnosis. Also the investigation of the causes of disease-related metabolic profiles, their elucidation, and the use of the results for therapy and drug development are fundamental methods of biomedical research. Therefore metabolomics defined as comprehensive metabolic profiling within a biological system would be a consequent continuation of these applications but would not cover the whole power of these investigations. The analysis of metabolite differences caused by biological perturbations like disease or therapy are defined as metabonomics (Nicholson et al., 2002; Ellis et al., 2007). The number of disease-linked compounds is around 1,000, whereas about 8,000 compounds could be identified in the human body (Wishart et al., 2009; http://www.hmdb.ca). This example shows that metabonomics is a rather limited approach, just targeting on a subset of metabolomic research.

Especially in the field of systems biology, metabolomics is considered as an emerging field of huge potential (Morrow, 2010). Human metabolomics have reached a highly developed stage technologically. Different methods for metabolome analysis have been established: after chromatographic separation detection methods like Fourier transform infrared spectroscopy, Raman microspectroscopy, nuclear magnetic resonance, and MS-based techniques are employed (for a summary see Ellis et al., 2007), of which the last two are currently the most popular (Morrow, 2010). Interestingly, the central role of mass spectrometry is reminiscent of different procedures in proteomics (Griffiths and Wang, 2009), but this is not the only similarity. In metabolomics, targeted and untargeted approaches can also be differentiated. In untargeted mass fingerprinting, approaches with measurement times around 1 minute, high-throughput analysis of samples gets realistic (Zhou et al., 2010). This is another step toward the characterization of the individual metabotype or metabolic phenotype, which would open several ways for personal medicine and also systems biology of health and disease (Holmes et al., 2008). In targeted approaches, the time frames for analysis are even lower. The analysis of specific metabolite classes is of special interest because of still poorly understood relationships between metabolite appearance and disease. For instance, lipidomics are of high interest because of striking relationships between diet, lipids, and different types of cancer (Bougnoux et al., 2006).

These examples demonstrate some similarities between proteomics and metabolomics and the importance of metabolome analysis for systems biology. It also holds true that research on various diseases, their diagnosis, and their therapy will be strongly influenced by metabolome analysis (van Meer, 2005; Ellis et al., 2007; Holmes et al., 2008). As a consequence of the similarity of proteomic and metabolomic methodology, both methodologies are highly dependent upon further development of instrumentation and laboratory automation.

9.5 AN OUTLOOK ON HIGH-THROUGHPUT OMICS: THE FUTURE OF AUTOMATION IN SYSTEMS BIOLOGY

As indicated in the previous chapters, together with genome sequencing and genomics, the number of other omics is still increasing. In Table 9.4, a collection of different omics is summarized. Although this list is far from being comprehensive, it gives an impression of structures and trends of omes and omics research.

In the previous chapter the basic omics fields (genomics, transcriptomics, proteomics, and metabolomics) were described, and the potential for high-throughput analysis and automation was discussed. These basic omes follow the central dogma of cellular biology (DNA → RNA → protein) and that the cellular content of small molecules is a direct consequence of protein expression. According to this view, different types of omes and omics can be differentiated. First, there are the basic omics mentioned above. Second, there are omes such as the exome, kinome, or lipidome, which are subsets of the four basic omes. In the case of the exome, it is just the protein coding part of the DNA (or RNA). Kinomics just analyzes the kinases within the entity of the proteome, and lipidomics concentrates on the lipids part of the metabolome. The transcription factors are sometimes referred as regulomes (http://www.internationalregulomeconsortium.ca). These proteins are also a subset of the proteome with the purpose of DNA-protein interaction to regulate transcript formation.

In Figure 9.3, major interactions between the four basic omes are illustrated. This scheme is suggestive of the dynamic character of DNA, RNA, proteins, and small molecules and their interactions with each other. Additionally it points out the third type of omes, which are summarized and designated as *2+omes* in this context. These are omes that evolve by the interaction of a basic ome with itself or with other molecules. The fourth type of omics, provisionally designated as *novel* omics, analyzes those molecules and phenomena that do not fit into the scheme of the first three. The 2+omes and the novel omes are subject of the following sections, and the final section focuses on the future and frontiers of laboratory automation in omics research and systems biology.

9.5.1 2+Omics as a Future Challenge for Systems Biology

As mentioned above, there are molecules and molecular clusters labeled as 2+omes that evolve due to the interaction of DNA, RNA, proteins, or metabolites with themselves or with each other. An example for 2+omes according to this definition is given by the epigenome. The major difference between the pure genome sequence and the epigenome is its methylation status. This inherited methylation of cytosine residues is catalyzed and regulated by proteins with significant consequences for cellular differentiation, development, and disease (Lister et al., 2009). In this case specific DNA-protein interactions leads to the formation of hybrid molecules of methylated DNA, sometimes also called DNA methylome. The analysis of this methylome cannot be accomplished by simple

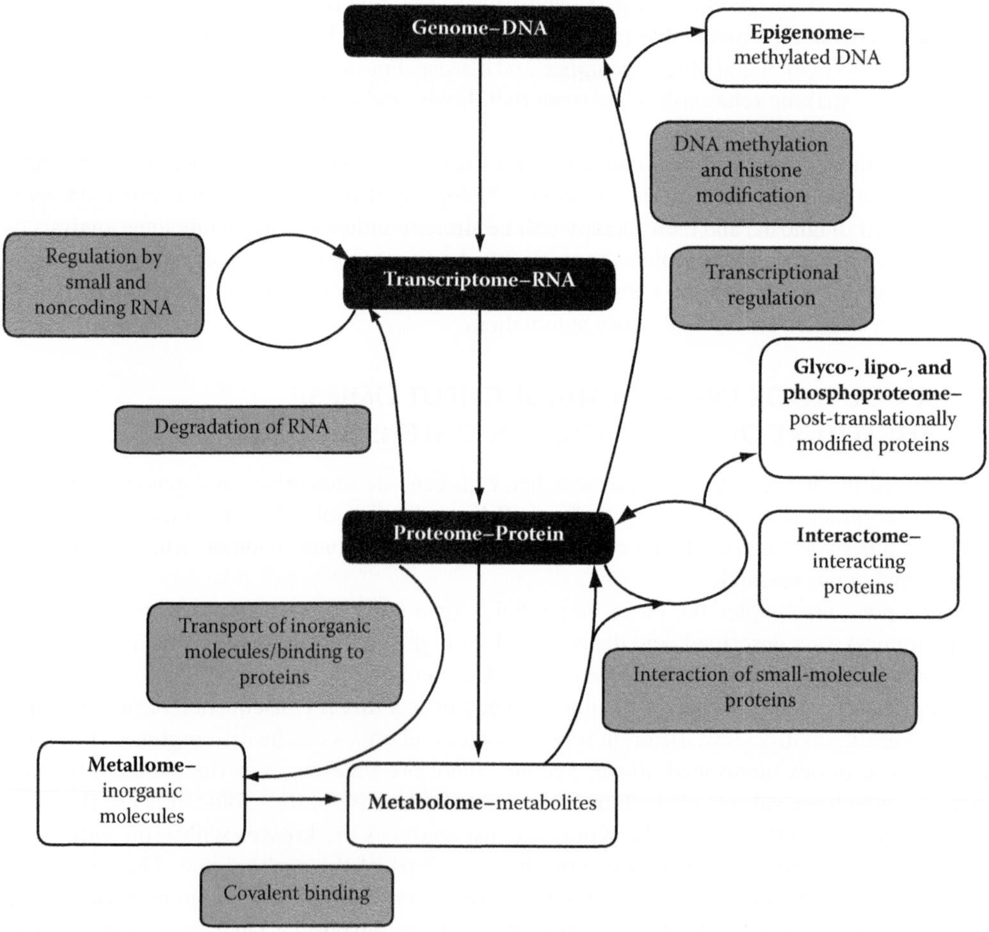

FIGURE 9.3 The four basic omes, their major interactions, and the evolution of other human omes.

sequencing but by previous chemical modification of methylated DNA. Afterwards this DNA is subjected to modern sequencing techniques for epigenome analysis. The addition of sodium bisulfate (BS) leads to the conversion of cytosine to uracil, whereas methylcytosine stays unaffected. Afterward, this DNA is used for sequencing reactions, also called BS sequencing. By application of the described method, the first human methylomes could be determined at single-nucleotide resolution (Lister et al., 2009).

The analysis of posttranslationally modified proteins, like the glyco-, lipo-, and phosphoproteomes, shows another example of 2+omics with similarities to epigenome analysis. They all come into existence by the interaction of basic molecules (proteins and small molecules), and these interactions result in hybrid molecules. Strictly speaking, the analysis of all of these modified proteins belongs in the great field of proteomics (see above). In contrast to classical proteomics, the modifications of these proteins take place posttranslationally, whereas the central dogma of cell biology predicts an orthogonal flow of information without feedback and correction on the protein level. Nevertheless, technically the major difference between classical proteomics and the analysis of modified proteins lies in different enrichment, separation, and chemical treatment procedures before MS identification (An et al., 2009; Hoofnagle and Heinecke, 2009; Macek et al., 2009). This means that high-throughput analysis of these proteins does not primarily depend on enhanced instrumentation but rather on flexible sample preparation and treatment because of the variety of applied methods.

A substantial example for a 2+ome by interaction with itself is given by the interactome: this ome evolves by the physical interaction of at least two proteins. To give a slight impression about the efforts to analyze the interactome, one experimental approach shall be considered. The simplest form of an interaction between two different proteins is the formation of heterodimers. The easiest way to test the interaction of two proteins is done by yeast two-hybrid (YTH) assays. As described previously, the human genome encodes for about 20,000 proteins with 2.6 isoforms in average (Nakao et al., 2005; HUPO, 2010). Assuming roughly 50,000 isoforms, this means that there are hypothetically 2.5×10^9 possible interactions. One can imagine that comprehensive testing of the human interactome just by YTH assays is a huge task that would be impossible to accomplish without massive application of automated processes. It has been estimated that in the human interactome, 650,000 interactions take place (Stumpf et al., 2008). Actually, according to the Human Protein Reference Database (http://www.hprd.org), almost 40,000 protein-protein interactions are described, less than a tenth of the assumed number. A major reason for this gap can be explained by experimental difficulties. One of the classical traps of YTH is the mandatory nuclear localization of the interacting proteins. Consequently, membrane proteins could not be tested with this system, which excludes up to one-third of the human proteome (Stagljar and Fields, 2002). To overcome this problem, recently a novel YTH assay on the basis of the split ubiquitin system was developed for membrane protein YTH (Snider et al., 2010).

The last example especially emphasizes the problems arising from 2+omics: Most analyses described in this text deal with tens of thousands of genes, proteins, or small molecules multiplied with spatiotemporal factors or numbers of individuals. Comparing these figures with the combinatorial possibilities of the interactome, the first numbers look rather tiny. The sample numbers to analyze protein-protein interactions rise exponentially, and the complexity is magnitudes higher than that for the analysis of single molecules. These considerations make the miniaturization as well as the automation of the analysis of the interactome indispensable. The examples described show that the basics for comprehensive analysis of 2+omes are available, now but there is still a long way to go for comprehensive coverage in this field. This goal cannot be accomplished without the extensive application of automated high-throughput processes.

9.5.2 The Evolution of Novel Omics

As described before, omes summarized under the term novel were those omes that did not fit into the first three categories. The reason for this rather uninformative definition is the subject of the first three omics. In these disciplines, DNA, RNA, proteins, or metabolites or their interactions are investigated. In the fourth category, either these molecules are investigated in different contexts or other molecules and objects are the focus of investigation.

Metallomics are an illustration of the analysis of molecules that have not been the subject of discussion yet. The metallome comprises all inorganic molecules in cells, either covalently or coordinatively bound to metabolites or to proteins. Proteins are of particular interest because up to one-third of all proteins require metal cofactors (Mounicou et al., 2009). Because of the special interest in metal-binding proteins, most analytical techniques in metallomics are derived from proteomics, which shall not be repeated here.

An interesting example for the comprehensive analysis of objects in the human body is given by microbiomics. It is a long-known fact that humans do not live in isolation but are colonized by a multitude of viruses, bacteria, and microbial eukaryotes. Therefore, *Homo sapiens* is often referred to as a *supraorganism* consisting of human cells and the microbiome (Bäckhed and Crawford, 2010). Recently, initial results of the consortium in the analysis of the human microbiome have been published. In the first approach, 178 microbial reference genomes were sequenced, and analysis of the results revealed that the human body probably inherits far more microbial strains (10,000–100,000) than assumed before (Human Microbiome Jumpstart Reference Strains Consortium, 2010). Although next-generation sequencing made a technically essential contribution to the analysis of the microbiome, there are still a lot of efforts and technological inventions needed for a better understanding of it.

The toponome is an informative example for a different context of molecular characterization and the next dimension of analysis. The set of all proteins expressed in a cell, tissue, or organism is defined as their proteome, and the interaction of these proteins is defined as the corresponding interactome. The toponome describes the spatial distribution of these proteins and protein clusters within a cell or tissue (Schubert et al., 2006; Pierre and Scholich, 2008). In some cases, the definition of toponome has been extended to the topology of all major molecules in cells or tissues (Somani and Somani, 2008), but we will concentrate on the generally accepted definition focusing on proteins.

Most interestingly, the evolution and success of toponomics is strongly coupled to an automated imaging procedure. Visual analysis of protein topology was made possible for the first time by the application of an immunochemical technique called multiepitope-ligand cartography (MELC; Schubert et al., 2006). For this process, single tissue sections are repeatedly labeled with specific antibodies, an image of fluorescence is taken, and then this fluorescence is bleached again for a new cycle of labeling. Afterward, the images are merged and compared pairwise to map colocalizing proteins also called combinatorial molecular phenotypes (CMPs). Theoretically with n specific antibodies 2^n CMPs can be tested. For 50 proteins, this would make $2^{50} = 10^{15}$ combinations (Pierre and Scholich, 2008), but even 100 proteins have been successfully tested in repetitive cycles (Schubert et al., 2006). Recently 17 tags against cell-surface proteins were tested on prostate cancer tissue sections, and about 2,100 CMPs out of 131,072 theoretical CMPs were identified (Schubert et al., 2009). These figures demonstrate the power of this method and also the level of complexity that these investigations have reached.

The analysis of the fluxome is another example for the investigation of basic molecules in different contexts. In the previous example of toponomics, the spatial distribution of the proteome was the focus. In fluxomics, previously called metabolic flux analysis, the temporal changes and distribution of metabolites, ions, and other small molecules are the center of interest. Especially in systems biology, the fluxes of metabolites are of major interest. Because of the dynamic characteristics of metabolic pathways and networks, pure metabolome analyses are too static (Cascante and Marin, 2008). Labeling precursors with stable isotopes, mostly ^{13}C, and then tracing the labeled molecules with various techniques is the predominant method to measure metabolite fluxes (Sauer, 2004; Dauner, 2010). Although this procedure is successfully applied on microbes and plants (Sauer, 2004; Tcherkez et al., 2009), it has several limitations (Wiechert et al., 2007). An exciting alternative in the field of fluxomics was given by the development of fluorescence-based nanosensors. These are genetically modified fluorescent proteins (FP) specifically binding small molecules, which show different fluorescence properties after binding (single FP sensors). Alternatively there are sensors that have altered FRET properties toward another FP after binding (FRET sensors; Frommer at al., 2009). This technology allows cellular imaging of metabolites *in situ*, as shown for glutamate in brain sections (Dulla et al., 2008), as well as *in vivo*, as the results for glucose import in HepG2 cells showed (Takanaga et al., 2008). The development of nanosensors is a technological breakthrough in systems biology because they made possible the identification and quantitation of metabolites on cellular and subcellular levels for the first time.

9.5.3 Trends and Challenges for Changing Tasks in Systems Biology

In the previous chapters, scientific as well as technological trends of high-throughput analysis in the different omics fields and in systems biology are described and also partially interpreted. Particular technological developments and trends were the pacemakers in some fields, and their influence will most likely continue in the next years.

An example of a development with substantial influence on different omics fields as well as on systems biology was undoubtedly the introduction of next-generation sequencing technologies. There are still competing technologies on the market, and the race is still open to determine which of the systems will succeed in the end. Nevertheless, as described above, NGS has revolutionized sequencing and will stay the leading technology in this area for the next decade. This trend will be boosted by the future decrease in prices for DNA sequencers of the second and third generation. With this technology, personalized genome sequencing became possible and will be used for drug discovery and even for clinical applications in the near and middle future (Mukhopadhyay, 2009; Koeppelle, 2010). Technical variations of sequencing increased the influence of sequencing on other fields. As described above, transcriptomics are conquered by RNA-Seq, epigenome analyses can be performed to single base resolution, and microbiomics would be unthinkable without NGS technologies. Therefore the invention of NGS as a fully automated sequencing method with ultrahigh-throughput was one of the major achievements for human-systems biology.

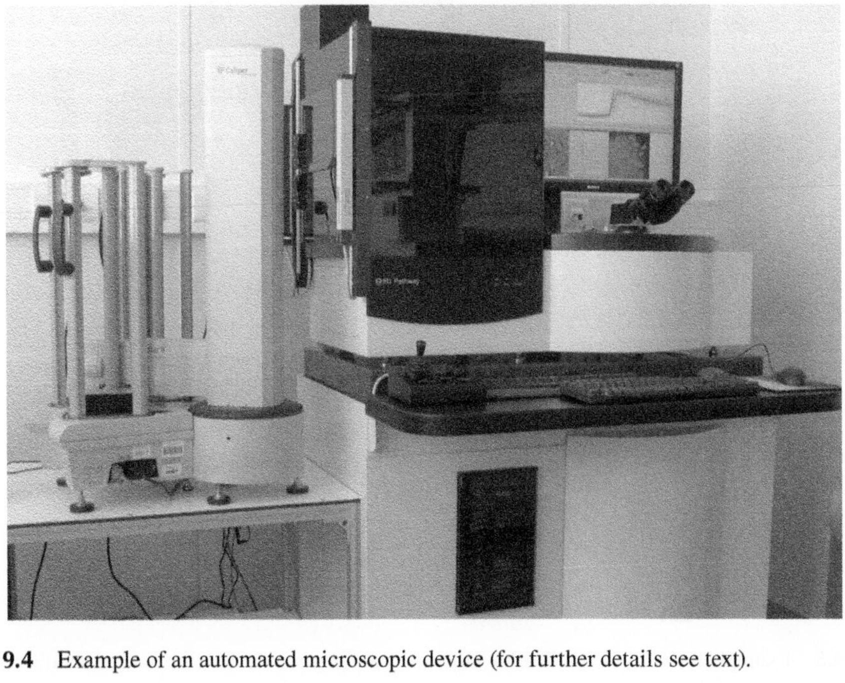

FIGURE 9.4 Example of an automated microscopic device (for further details see text).

Advances have also been made in other areas as diverse as high-throughput antibody production, mass spectrometry, and information technology, which are crucial for the progress of systems biology for biomedicine (Wolkenhauer et al., 2009). With respect to automated high-throughput methods, two areas with high impact on systems biology in the future should be pointed out: automated imaging technologies and flexible laboratory automation.

There is a growing trend in different omics to move away from *in vitro* and HTS assays toward cellular and HCS systems. Namely, MELC in toponomics and the use of nanosensors in fluxomics are examples of the increasing applications of microscopic imaging techniques (see above). Lifting these groundbreaking techniques to the level of high-throughput methods depends on automated imaging equipment. In Figure 9.4, an example of a configuration for automated microscopy of adherent cells is shown. The core unit is a confocal fluorescence microscope with autofocus and

FIGURE 9.5 Example for a small and flexible laboratory automation device (for further details see text).

time-resolved imaging of microtiter plates. Furthermore, it is equipped with a stacker for the delivery of multiple plates and with injectors for automated pipetting procedures in a chamber with a CO_2 atmosphere. The combination of novel imaging techniques with HCS technologies as shown in Figure 9.4 would be a major step toward high throughput of microscopic imaging processes, like the ones mentioned and also other future cellular and subcellular techniques, in systems biology.

The other trend with a substantial impact on the progress of different omics and systems biology will be flexible laboratory automation. In Section 9.3, application and success of laboratory automation in drug discovery was described. Because of the highly repetitive nature of the applied processes and their high-throughput figures and times, the high investment costs were compensated. In systems biology nowadays, there is a growing demand for laboratory automation to master the growing sample numbers as they were described for proteomics, interactomics (see above), and the wide field of small interference RNA manipulations. The costs for purchasing such equipment can only just be covered by large companies or research centers (Wolkenhauer et al., 2009). Unfortunately, without the participation of small and medium enterprises and small research labs, the future tasks of omics and systems biology cannot be accomplished. The development of flexible laboratory automation devices for changing jobs and needs can be a way out of this dilemma. In Figure 9.5, such a solution is illustrated. This development of the Center for Life Science Automation (celisca) and Beckman-Coulter consist of a CO_2 incubator, a liquid handler under a sterile hood, a cell viability counter, and a refrigerator for liquid storage. All of these modules are connected to each other and under control of process management software. The components of the device make the automated processing of most routine works in labs from sample preparation over assay handling to cell culturing possible even for small labs. One advantage of this configuration is its ability and flexibility to fulfill all of these functions. Another is its small footprint and size, when lab capacity for many institutions and small and medium enterprises is limited. The extension of automation not just in major research institutions and pharmaceutical companies in the way described above may lead to major progress in human omics and systems biology.

REFERENCES

Agrestia, J. J., Antipovc, E., Abatea, A. R., et al. (2010). Ultrahigh-throughput screening in drop-based microfluidics for directed evolution. *Proceedings of the National Academy of Sciences USA* 107: 4004–4009.

Alanine, A., Nettekoven, M., Roberts, E., et al. (2003). Lead generation: Enhancing the success of drug discovery by investing in the hit to lead process. *Combinatorial Chemistry & High Throughput Screening* 6: 51–66.

An, H. J., Froehlich, J. W., Lebrilla, C. B. (2009). Determination of glycosylation sites and site-specific heterogeneity in glycoproteins. *Current Opinion in Chemical Biology* 13: 421–426.

Anderson, N. L., Anderson, N. G., Pearson, T. W., et al. (2009). A human proteome detection and quantitation project. *Molecular & Cellular Proteomics* 8: 883–886.

Astle, T. (2000). Recollections of early microplate automation. *Journal of the Association for Laboratory Automation* 5: 30–31.

Bäckhed, F., Crawford, P. A. (2010). Coordinated regulation of the metabolome and lipidome at the host-microbial interface. *Biochimica et Biophysica Acta* 1801: 240–245.

Banks, P. (2009). The microplate market: Past, present and future: Microplates today—the global market. *Drug Discovery World.* Spring 2009: 85–90.

Blanchard, J. E., Elowe, N. H., Huitema, C., et al. (2004). High-throughput screening identifies inhibitors of the SARS coronavirus main protease. *Chemistry & Biology* 11: 1445–1453.

Bonetta, L. (2010). Whole-genome sequencing breaks the cost barrier. *Cell* 141: 917–919.

Borchert, K. M., Galvin, R. J., Frolik, C. A., et al. (2005). High-content screening assay for activators of the Wnt/Fzd pathway in primary human cells. *Assay and Drug Development Technologies* 3: 133–141.

Bougnoux, P., Giraudeau, B., Couet, C. (2006). Diet, cancer, and the lipidome. *Cancer Epidemiology, Biomarkers & Prevention* 15: 416–421.

Boyd, J. C., Felder, R. A., Savory, J. (1996). Robotics and the changing face of the clinical laboratory. *Clinical Chemistry* 42: 1901–1910.

Brown, A. M. (2004). Drugs, hERG and sudden death. *Cell Calcium* 35: 543–547.

Buie, J. (2010). Evolution of microplate technology. *Lab Manager Magazine* 5: 46–47.

Cascante, M., Marin, S. (2008). Metabolomics and fluxomics approaches. *Essays in Biochemistry* 45: 67–81.

Chan, G. K. Y., Richards, G. R. Peters, M., et al. (2005). High content kinetic assays of neuronal signaling implemented on BD™ Pathway HT. *Assay and Drug Development Technologies* 3: 623–636.

Cirulli, E. T., Singh, A., Shianna, K. V., et al. (2010). Screening the *human exome*: A comparison of whole genome and whole transcriptome sequencing. *Genome Biology* 11: R57.

Collins, F. S., Morgan, M., Patrinos, A. (2003). The Human Genome Project: Lessons from large-scale biology. *Science* 300: 286–290.

Craighead, H. (2006). Future lab-on-a-chip technologies for interrogating individual molecules. *Nature* 442: 387–93.

Datson, N. A. (2008). Scaling down SAGE: From miniSAGE to microSAGE. *Current Pharmaceutical Biotechnology* 9: 351–361.

Dauner, M. (2010). From fluxes and isotope labeling patterns towards in silico cells. *Current Opinion in Biotechnology* 21: 55–62.

De Masi, F., Chiarella, P., Wilhelm, H., et al. (2005). High throughput production of mouse monoclonal antibodies using antigen microarrays. *Proteomics* 5: 4070–4081.

Devol, G. C., Jr. (1961). U.S. Patent 2,988,237.

Dittrich, P. S., Manz, A. (2006). Lab-on-a-Chip: Microfluidics in drug discovery. *Nature Reviews Drug Discovery* 5: 210–218.

Dulla, C., Tani, H., Okumoto, S., Frommer, W. B., Reimer, R. J., Huguenard, J. R. (2008). Imaging of glutamate in brain slices using FRET sensors. *Journal of Neuroscience Methods* 168: 306–319.

Ellis, D. I., Dunn, W. B., Griffin, J. L., Allwood, J. W., Goodacre, R. (2007). Metabolic fingerprinting as a diagnostic tool. *Pharmacogenomics* 8: 1243–1266.

Felder, R. A., Sanada, H., Xu, J., et al. (2002). G protein-coupled receptor kinase 4 gene variants in human essential hypertension. *Proceedings of the National Academy of Sciences USA* 99: 3872–3877.

Fertig, N., Blick, R. H., Behrends, J. C. (2002). Whole cell patch clamp recording performed on a planar glass chip. *Biophysical Journal* 82: 3056–3062.

Filmore, D. (2004). It's a GPCR world. *Modern Drug Discovery* 7: 24–28.

Fisler, R., Burke, J. (2004). *High content analysis market outlook*. Cambridge Health Institute, Report No. 36, Needham, MA.

Fleischer, H., Gördes, D., Thurow, K. (2009). High throughput screening applications for enantiomeric excess determination using ESI/MS. *American Laboratory* 41: 21–24.

Fodor, S. P., Read, J. L., Pirrung, M. C., et al. (1991). Light-directed, spatially addressable parallel chemical synthesis. *Science* 251: 767–773.

Fox, S. J. (2006). *A history of high throughput screening for drug discovery: A special report summarizing six comprehensive industry studies in the years 1998–2005*. Moraga, CA: HighTech Business Decisions.

Fox, S., Farr-Jones, S., Sopchak, L., et al. (2006). High-throughput screening: Update on practices and success. *Journal of Biomolecular Screening* 11: 864–869.

Fox, S., Wang, H., Sopchak, L., Farr-Jones, S. (2002). High throughput screening 2002: Moving toward increased success rates. *Journal of Biomolecular Screening* 7: 313–316.

Freeze, H. H. (2006). Genetic defects in the human glycome. *Nature Reviews Genetics* 7: 537–551.

Frommer, W. B., Davidson, M. W., Campbell, R. E. (2009). Genetically encoded biosensors based on engineered fluorescent proteins. *Chemical Society Reviews* 38: 2833–2841.

Gaulton, K. J., Nammo, T., Pasquali, L., et al. (2010). A map of open chromatin in human pancreatic islets. *Nature Genetics* 42: 255–261.

Global Industry Analysts Inc. (2006). *Laboratory automation: A global outlook*. San Jose, CA: Global Industry Analysts Inc.

Global Industry Analysts Inc. (2008). *Drug discovery technologies: A global strategic business report*. San Jose, CA: Global Industry Analysts Inc.

Gnad, F., Ren, S., Cox, J., et al. (2007). PHOSIDA (phosphorylation site database): Management, structural and evolutionary investigation, and prediction of phosphosites. *Genome Biology* 8: R250.

Godolphin, W. (1983). Robots in hospitals: The impact on health care workers. In *Advances in laboratory automation: Robotics*, Edited by Strimaitis, J. R., Hawk, G. L. Hopkinton, MA: Zymark Corp.

Golebiowski, A., Klopfenstein, S. R., Portlock, D. E. (2001). Lead compounds discovered from libraries. *Current Opinion in Chemical Biology* 5: 273–284.

Golebiowski, A., Klopfenstein, S. R., Portlock, D. E. (2003). Lead compounds discovered from libraries: Part 2. *Current Opinion in Chemical Biology* 7: 308–325.

Gribbon, P., Sewing, A. (2005). High throughput drug discovery: What can we expect from HTS? *Drug Discovery Today* 10: 17–22.

Griffiths, W. J., Wang, Y. (2009). Mass spectrometry: From proteomics to metabolomics and lipidomics. *Chemical Society Reviews* 38: 1882–1896.

Gygi, S. P., Corthals, G. L., Zhang, Y., et al. (2000). Evaluation of two-dimensional electrophoresis-based protein analysis technology. *Proceedings of the National Academy of Sciences USA* 97: 9390–9395

Hann M. M., Oprea T. I. (2004). Pursuing the leadlikeness concept in pharmaceutical research. *Current Opinion in Chemical Biology* 8: 255–263.

Henckel, A., Amaud, P. (2010). Genome-wide identification of new imprinted genes. *Briefings in Functional Genomics* 9: 304–314.

HighTech Business Decisions (2002). High-throughput screening 2002: New strategies and technologies. Moraga, CA.

Hoheisel, J. D. (2006). Microarray technology: Beyond transcript profiling and genotype analysis. *Nature Reviews Genetics* 7: 200–210.

Holmes, E., Wilson, I. D., Nicholson, J. K. (2008). Metabolic phenotyping in health and disease. *Cell* 134: 714–717.

Hoofnagle, A. N., Heinecke, J. W. (2009). Lipoproteomics: Using mass spectrometry-based proteomics to explore the assembly, structure, and function of lipoproteins. *Journal of Lipid Research* 50: 1967–1975.

Human Microbiome Jumpstart Reference Strains Consortium (2010). A catalog of reference genomes from the human microbiome. *Science* 328: 994–999.

The Human Proteome Organization (2010). A gene-centric human proteome project. *Molecular & Cellular Proteomics* 9: 427–429.

International Human Genome Sequencing Consortium (2001). Initial sequencing and analysis of the human genome. *Nature* 409: 860–921.

Johnson, S. A., Hunter, T. (2005). Kinomics: Methods for deciphering the kinome. *Nature Methods* 2: 17–25.

Kalorama Information (2008). *The worldwide market for lab automation*. New York: Kalorama Information.

Katagiri, F. (2003). Attacking complex problems with the power of systems biology. *Plant Physiology* 132: 417–419.

Kim, D. W., Nam, S. H., Kim, R. N., et al. (2010). Whole human exome capture for high-throughput sequencing. *Genome* 53: 568–574.

Kitano, H. (2002). Systems biology: A brief overview. *Science* 295: 1662–1664.

Koehn, F. E. (2008). High impact technologies for natural product screening. *Progress in Drug Research* 65: 177–210.

Koeppelle, W. (2010). Pocket money sequencing. *Lab Times* 4: 48–49.

Lee, R. T., Lauc, G., Lee, Y. C. (2005). Glycoproteomics: Protein modifications for versatile functions. *EMBO Reports* 6: 1018–1022.

Lister, R., Pelizzola, M., Dowen, R. H., et al. (2009). Human DNA methylomes at base resolution show widespread epigenomic differences. *Nature* 462: 315–322.

Macarron, R. (2006). Critical review of the role of HTS in drug discovery. *Drug Discovery Today* 11: 270–277.

Macek, B., Mann, M., Olsen, J. V. (2009). Global and site-specific quantitative phosphoproteomics: Principles and applications. *Annual Review of Pharmacology and Toxicology* 49: 199–221.

Malmström, J., Lee, H., Aebersold, R. (2007). Advances in proteomic workflows for systems biology. *Current Opinion in Biotechnology* 18: 378–384.

Manning, G., Whyte, D. B., Martinez, R., et al. (2002). The protein kinase complement of the human genome. *Science* 298: 1912–1934.

Mathivanan, S., Ahmed, M., Ahn, N. G., et al. (2008). Human proteinpedia enables sharing of human protein data. *Nature Biotechnology* 2: 164–167.

Mathivanan, S., Periaswamy, B., Gandhi, T. K. B., et al. (2006). An evaluation of human protein-protein interaction data in the public domain. *BMC Bioinformatics* 7: S19.

Mayr, L., Fuerst, P. (2008). The future of high-throughput screening. *Journal of Biomolecular Screening* 13: 443–448.

Milligan, G. (2003). High-content assays for ligand regulation of G-protein-coupled receptors. *Drug Discovery Today* 8: 579–585.

Morrow, K. J. (2010). Mass spec central to metabolomics. *Genetic Engineering and Biotechnology News* 30: 17–19.

Mounicou, S., Szpunar, J., Lobinski, R. (2009). Metallomics: The concept and methodology. *Chemical Society Reviews* 38: 1119–1138.

Mukhopadhyay, R. (2009). DNA sequencers: The next generation. *Analytical Chemistry* 81: 1736–1740.

Nakao, M., Barrero, R. A., Mukai, Y., et al. (2005). Large-scale analysis of human alternative protein isoforms: Pattern classification and correlation with subcellular localization signals. *Nucleic Acids Research* 33: 2355–2363.
Ng, P. C., Buckingham, K. J., Lee, C., et al. (2010). Exome sequencing identifies the cause of a Mendelian disorder. *Nature Genetics* 42: 30–35.
Ng, P. C., Levy, S., Huang, J., et al. (2008). Genetic variation in an individual human exome. *PLoS Genetics* 4: e1000160.
Nicholson, J. K., Connelly, J., Lindon, J. C., et al. (2002). Metabonomics: A platform for studying drug toxicity and gene function. *Nature Reviews Drug Discovery* 1: 153–162.
Oltvai, Z. N., Barabasi, A.-L. (2002). Life's complexity pyramid. *Science* 298: 763–764.
Ozbal, C. C., LaMarr, W. A., Linton, J. R., et al. (2004). High throughput screening via mass spectrometry: A case study using acetylcholinesterase. *Assay and Drug Development Technologies* 2: 373–381.
Patterson, S. D., Aebersold, R. H. (2003). Proteomics: the first decade and beyond. *Nature Genetics* 33: 311–323.
Pereira, D. A., Williams, J. A. (2007). Origin and evolution of high throughput screening. *British Journal of Pharmacology* 152: 53–61.
Pierre, S., Scholich, K. (2008). Toponomics: Studying protein-protein interactions and protein networks in intact tissue. *Molecular BioSystems* 6: 641–647.
Richards, G. R., Smith, A. J., Parry, F., et al. (2006). A morphology- and kinetic based cascade for human neural cell high content screening, *Assay and Drug Development Technologies* 4: 143–152.
Rothberg, J. M., Leamon, J. H. (2008). The development and impact of 454 sequencing. *Nature Biotechnology* 26: 1117–1124.
Rubenstein, K., Coty, C. (2001). *High throughput screening: Redefining the mission.* International Business Communication.
Rutherford, M. L., Stinger, T. (2001). Recent trends in laboratory automation in the pharmaceutical industry. *Current Opinion in Drug Discovery & Development* 4: 343–346.
Sannes, L. J. (2002). *High content screening: Parallel analysis fuels accelerated discovery and development.* Cambridge: Cambridge Health Institute, Report No. 24.
Sasaki, M., Kageoka, T., Ogura, K., et al. (1998). Total laboratory automation in Japan: Past, present and the future. *Clinica Chimica Acta* 278: 217–227.
Sauer, U. (2004). High-throughput phenomics: Experimental methods for mapping fluxomes. *Current Opinion in Biotechnology* 15: 58–63.
Satomaa, T., Heiskanen, A., Leonardsson, I., et al. (2009). Analysis of the human cancer glycome identifies a novel group of tumor-associated N-acetylglucosamine glycan antigens. *Cancer Research* 69: 5811–5819
Schittmayer, M., Birner-Gruenberger, R. (2009). Functional proteomics in lipid research: Lipases, lipid droplets and lipoproteins. *Journal of Proteomics* 72: 1006–1018.
Schonbrunn, E., Adam, R. A., Steinvurzel, P. E., et al. (2010). High-throughput fluorescence detection using an integrated zone-plate array. *Lab on a Chip* 10: 852–856.
Schreiber, S. L., Nicolaou, K. C., Davies, K. (2002). Diversity-oriented organic synthesis and proteomics: New frontiers for chemistry & biology. *Chemistry & Biology* 9: 1–2.
Schubert, W. (2010). On the origin of cell functions encoded in the toponome. *Journal of Biotechnology*, 149: 252–259.
Schubert, W., Bonnekoh, B., Pommer, A. J., et al. (2006). Analyzing proteome topology and function by automated multidimensional fluorescence microscopy. *Nature Biotechnology* 24: 1270–1278.
Schubert, W., Gieseler, A., Krusche, A., et al. (2009). Toponome mapping in prostate cancer: detection of 2000 cell surface protein clusters in a single tissue section and cell type specific annotation by using a three symbol code. *Journal of Proteome Research* 8: 2696–2707.
Sigworth, F. J., Klemic, K. G. (2002). Patch clamp on a chip. *Biophysical Journal* 82: 2831–2832.
Simpson, P. B. (2006). A model for efficient assay development and screening at a small research site. *Journal of the Association for Laboratory Automation* 11: 100–109.
Snider, J., Kittanakom, S., Damjanovic, D., et al. (2010). Detecting interactions with membrane proteins using a membrane two-hybrid assay in yeast. *Nature Protocols* 5: 1281–1293.
Somani, A.-K., Somani, N. (2008). Toponomics: Visualizing cellular protein networks in health and disease: "A single picture is worth more than a thousand words!" *Journal of Cutaneous Pathology* 35: 791–793.
Stagljar, I., Fields, S. (2002). Analysis of membrane protein interactions using yeast-based technologies. *Trends in Biochemical Sciences* 27: 559–563.
Stamatoyannopoulos, J. A. (2004). The genomics of gene expression. *Genomics* 84: 449–457.

Stelzl, U., Worm, U., Lalowski, M., et al. (2005). A human protein-protein interaction network: A resource for annotating the proteome. *Cell* 122: 957–968.

Sterling, J. (2008). High-throughput screening challenges. *Genetic Engineering & Biotechnology News* 28: 26–27.

Stett, A., Burkhardt, C., Weber, U., et al. (2003). Cytocentering: A novel technique enabling automated cell-by-cell patch clamping with the CytoPatch Chip. *Receptors and Channels* 9: 59–66.

Stoll, N., Thurow, K. (2008). Process management using information systems: Principles and systems. In *Systems Engineering Approach to Medical Automation* (pp. 183–195). Edited by Felder, R., Alwan, M., Zhang, M. Norwood, MA: Artech House Inc.

Stumpf, M. P. H., Thorne, T., de Silva, E., et al. (2008). Estimating the size of the human interactome. *Proceedings of the National Academy of Sciences USA* 105: 6959–6964.

Szpunar, J. (2005). Advances in analytical methodology for bioinorganic speciation analysis: Metallomics, metalloproteomics and heteroatom-tagged proteomics and metabolomics. *The Analyst* 130: 442–465.

Takanaga, H., Chaudhuri, B., Frommer, W. B. (2008). GLUT1 and GLUT9 as major contributors to glucose influx in HepG2 cells identified by a high sensitivity intramolecular FRET glucose sensor. *Biochimica et Biophysica Acta* 1778: 1091–1099.

Talbert, P. B., Henikoff, S. (2010). Histone variants: ancient wrap artists of the epigenome. *Nature Reviews Molecular Cell Biology* 11: 264–275.

Tcherkez, G., Mahé, A., Gauthier, P., et al. (2009). In folio respiratory fluxomics revealed by ^{13}C isotopic labeling and H/D isotope effects highlight the noncyclic nature of the tricarboxylic acid "cycle" in illuminated leaves. *Plant Physiology* 151: 620–630.

Thurow, K., Göde, B., Dingerdissen, U., et al. (2004). Laboratory Information Management Systems for Life Science Applications. *Organic Process Research and Development* 8: 970–982.

Tyers, M., Mann, M. (2003). From genomics to proteomics. *Nature* 422: 493–497.

Uhlen, M., Björling, E., Agaton, C. (2005). A human protein atlas for normal and cancer tissues based on antibody proteomics. *Molecular & Cellular Proteomics* 4: 1920–1932.

Vaisar, T. (2009). Proteomic analysis of lipid-protein complexes. *Journal of Lipid Research* 50: 781–786.

van Meer, G. (2005). Cellular lipidomics. *EMBO Journal* 24: 3159–3165.

Venter, J. C., Adams, M. D., Myers, E. W., et al. (2001). The sequence of the human genome. *Science* 291: 1304–1351.

Via, M., Gignoux, C., Burchard E. G. (2010). The 1000 Genomes project: New opportunities for research and social challenges. *Genome Medicine* 2: 1–3.

Wade, N. (2009). Cost of decoding a gene is lowered. *New York Times* Aug 19: D3.

Wang, Z., Gerstein, M., Snyder, M. (2009). RNA-Seq: A revolutionary tool for transcriptomics. *Nature Reviews Genetics* 10: 57–63.

Weiner, N. (1948). *Cybernetics or communication in the animal and the machine*. Cambridge: MIT Press.

Wheelan, S. J., Murillo, F. M., Boeke, J. D. (2008). The incredible shrinking world of DNA microarrays. *Molecular BioSystems* 4: 726–732.

Wheeler, D. A., Srinivasan, M., Egholm, M. (2008). The complete genome of an individual by massively parallel DNA sequencing. *Nature* 452: 872–877.

Wiechert, W., Schweissgut, O., Takanaga, H., et al. (2007). Fluxomics: Mass spectrometry versus quantitative imaging. *Current Opinion in Plant Biology* 3: 323–330.

Wishart, D. S., Knox, C., Guo, A. C., et al. (2009). HMDB: A knowledgebase for the human metabolome. *Nucleic Acids Research* 37: D603–D610.

Wolkenhauer, O., Fell, D., De Meyts, P., et al. (2009). SysBioMed report: Advancing systems biology for medical applications. *IET Systems Biology* 3: 131–136.

Zhou, M., McDonald, J. F., Fernández, F. M. (2010). Optimization of a direct analysis in real time/time-of-flight mass spectrometry method for rapid serum metabolomic fingerprinting. *Journal of the American Society for Mass Spectrometry* 21: 68–75.

10 Safety in Diagnostic Imaging Techniques Used in Omics

Rachana Visaria
MR Safe Devices LLC
Burnsville, Minnesota

Surya Prakash
Tej Bahadur Saproo Hospital
Allahabad, India

J. Thomas Vaughan and Devashish Shrivastava
University of Minnesota
Minneapolis, Minnesota

CONTENTS

10.1 INTRODUCTION

Ensuring safety while maximizing the imaging efficiency (i.e., clinically useful imaging information over time) is an integral part of diagnostic imaging. Safety standards have been set by regulatory agencies and recommended by additional scientific groups to ensure safety during imaging. This chapter reviews the existing safety standards and their rationale for the commonly used diagnostic imaging modalities in omics of positron emission tomography (PET), single-photon emission computed tomography (SPECT), computed tomography (CT), and magnetic resonance imaging and spectroscopy (MRI and MRS). Current knowledge gaps and research opportunities are discussed to maximize imaging efficiency safely.

10.2 PET AND SPECT

PET and SPECT belong to the family of nuclear imaging techniques that involve intravenous injection of a small amount of the radiopharmaceutical agent fluorodeoxyglucose (FDG or 18F, a glucose-tagged radioactive isotope) is the most widely used compound for PET imaging with a half-life of 110 minutes (Boellaard et al., 2009). Other typically used isotopes with relatively shorter half-lives for PET imaging are carbon-11 (20 min), nitrogen-13 (10 min), and oxygen-15 (2 min). Commonly used isotopes for SPECT imaging are pure gamma emitters like technetium-99m, iodine-123, and indium-111. While travelling through the body, the radioactive-isotope tracer emits positrons as the tagged glucose starts to decay. The positrons collide with the surrounding electrons, emitting gamma rays that are detected by PET cameras. In the case of SPECT, cameras capture photons from the radioactive material. The radiation dose in PET/SPECT imaging is a safety concern because of the high energy of the annihilation photons (511 keV), the abundance (100–2005) of radionuclei, and the desired/undesired presence of radioisotopes in blood vessels and organs. Additionally, the relevant personnel/operator may be exposed to high radiation levels because of external radiation, inhalation of airborne particles, manual handling of radioactive syringes, etc. A significant increase in radiation exposure is experienced when PET or SPECT imaging is combined with CT.

Radiation from PET/SPECT/CT is a safety concern because radiation penetrates into deeper tissues and may cause ionization of atoms and molecules in cells generating free radicals. These free radicals may react with various intracellular organelles, causing oxidation of DNA bases. Radiation-induced oxidation products and hydroxyguanine cause transversion of GC to TA. If DNA is previously injured, coding of DNA is altered, and some actions of DNA are scrambled, leading to irreparable loss of DNA resulting in apoptosis. Cells can repair a single break of DNA strand but not the double break of the strand caused by radiation. DNA damage caused by radiation also causes genetic mutation; thus, spermatogenesis and ova production from gonads are affected. Fetal cells in the uterus are highly sensitive to radiation because they are forming new organs, and any insult in the form of exposure to radiation will give rise to congenital anomalies.

Safety from radiation from a radioisotope tracer has been widely studied. The Food and Drug Administration (FDA)-recommended dose range of FDG for an adult patient is 185–370 MBq (5–10 mCi) for studies of malignancy, cardiology, and epilepsy (FDA, 2005). The European Association of Nuclear Medicine (EANM) recommends an FDG dose of 5 and 2.5 MBq/kg of body weight for two- (2D) and three-dimensional (3D) tumor scans, respectively, assuming fixed scan duration of 5 min and 25% bed overlap (Boellaard et al., 2009). For brain imaging, the EANM recommends a maximum injected FDG activity of 300–600 MBq and 125–250 MBq for 2D and 3D scans, respectively (Varrone et al., 2009). The recommended interval between the FDG administration and the start of acquisition is 40–60 min (FDA, 2005; Boellaard et al., 2009). For obese subjects (>90 kg), an increase of scanning time rather than increase of the FDG activity is recommended to improve image quality. Keeping administered activity below 530 MBq is recommended for patients weighing more than 90 kg (Boellaard et al., 2009; Masuda et al., 2009). For pediatric scanning, lower doses of radiopharmaceutical are recommended if possible without compromising the image quality. Therefore, the EANM dosimetry and pediatrics committee have lowered the recommended minimum activity value for FDG from 70 to 26 MBq for 2D and 14 MBq for 3D scans in pediatric patients (Lassmann et al., 2008a; Stauss et al., 2008). An alternate method of determining FDG dose based on a child's weight is also suggested: 6 and 3 MBq/kg for 2D and 3D scans, respectively (Holm et al., 2007).

In this chapter, the recommended activity levels of only FDG have been presented because FDG is by far the most common nuclide for PET imaging application. To obtain the recommended and/or usual dose range of other radiopharmaceuticals (e.g., Tc-99m, I-131, I-123, and Ga-67) used for different organs, readers can refer to https://www.eanm.org/scientific_info/guidelines. A range of dose is recommended for various radiopharmaceuticals, but no upper safe dose or optimum rate of administration has been established. The EANM's dosage card (Lassmann et al., 2008b) lists minimum activity to be administered for various radiopharmaceuticals for patients weighing 3–68 kg. The final dose needs to be decided by the clinician based on the patient's data including: weight, age, medical condition, PET system (scan duration and overlap), radiopharmaceutical (rate of administration, decay, and pharmacokinetic), particular application, national/regional dose constraints, and guidelines.

Next, according to the United States Nuclear Regulatory Commission (US NRC) (NRC, 1991), the doses of radioactive materials are limited such that the exposure of individual members of the public is less than 10^{-3} Sv in a year to comply with the principle of ALARA (i.e., as low as reasonably

TABLE 10.1
NRC Occupational Dose Limits

Description	Annual Dose Limit (Sv/Year)
Whole body	5 (10^{-2})
Any organ	50 (10^{-2})
Skin	50 (10^{-2})
Extremity	50 (10^{-2})
Lens of eye	15 (10^{-2})
Embryo/fetus of declared pregnant woman	0.5 (10^{-2})
Member of the public	0.1 (10^{-2})

achievable). This excludes dose contributions from background radiation, from any administration received by the individual, from exposure to individuals administered radioactive material, from voluntary participation in medical research, and from the disposal of radioactive material. The annual occupational dose limits for adults is 0.05 Sv, and further limits on individual organs/tissues are listed in Table 10.1. Occupational dose limits for minors are 10% of the annual dose limits specified for adult workers. The dose equivalent to the embryo/fetus during the entire pregnancy, because of the occupational exposure of a declared pregnant woman, should not exceed 0.005 Sv.

For investigations of new radiopharmaceuticals in humans, the Radiation Drug Research Committee limits the radiation dose to the critical organs (effective dose, dose to the lens, and blood-forming organs) to 0.03 Sv from a single dosage and to 0.05 Sv/year during the entire study (Saha, 2010). Only 30 patients over the age of 18 can be studied per protocol, and pregnant women are not allowed to participate in the study.

Regarding future work to maximize safe imaging effectiveness, further research on target-specific radioisotope tracers is desired to minimize radiation exposure. Tissue-specific safe maximum radiation-exposure limit must be developed, beyond which cellular and genetic repairs are adversely affected for appropriate human physiology and pathology models. The effect of tracer radiation itself on the omics must be clearly understood.

10.3 COMPUTED TOMOGRAPHY

CT employs an x-ray-generating tube opposite an x-ray detector in a ring-shaped apparatus rotating around the patient to produce a cross-sectional tomogram. A CT image is acquired in an axial plane; sagittal and coronal plane images are prepared by computer reconstruction. Often radio contrasts are used for enhanced delineation of anatomy of a particular organ. Eight, 16, or 64 detectors may be used by spiral multidetector CT machines to produce images.

Interaction of x-ray radiation with the biological system is a safety concern because of increased risk of cellular mutation in general, cancer, and malignancy. The dose of radiation varies with different diseases in different organs, which may vary from 10 mSv (1000 mrem) to 90 mSv. CT exposure of 10 mSv five times a year is equivalent to the background radiation of three years increasing the cancer risk by 1%. Cancer risk of CT scans performed in year 2007 is one in 270 females and one in 600 males (Smith-Bindman et al., 2009). It is anticipated that the scans will result in about 29,000 future malignancies and 14,500 deaths in the United States alone (Berrington de Gonzalez et al., 2009). Interestingly, about 67% of these cases will be in women because they are more vulnerable to breast cancer after exposure to radiation. At all ages, a women's risk of breast cancer and lung cancer accounts for 80% of total cancers from one CT scan (Einstein et al., 2007). Children are more affected because they are more sensitive to CT-scan exposure and have decades of life ahead. A female child who received CT exposure at the age of 3 years has one in 500 chances of getting cancer in later life, whereas a female who got exposure at the age of 30 years has one in 1,000

TABLE 10.2
ICRP Diagnostic Reference Levels for CT Examinations in Adults

Examination	Weighted Dose Index per Slice or Rotation (mGy)	Dose Length Product per Examination (mGy · cm)
Head	60	1,050
Face and sinuses	35	360
Vertebral trauma	70	460
Chest (general)	30	650
High-resolution CT of lung	35	280
Abdomen	35	780
Liver and spleen	35	900
Pelvis	35	570
Osseous pelvis	25	520

chances, and exposure at the age of 70 years creates a one in 3,333 chance (Berrington de Gonzalez et al., 2009). A CT angiography study suggests a life-time malignancy-risk increase of 50 per 1000 people over the normal natural cancer risk of 212 per 1000 people. CT angiography will result in one in 600 men developing malignancy when done at the age of 40 years; the risk is doubled at the age of 20 years and is halved at the age of 60 years (Smith-Bindman et al., 2009).

There is no upper safe dose range for diagnostic CT applications because of variety of protocols, CT systems, and applications. However, a set of reference values has been recommended by International Commission on Radiological Protection (ICRP) as given in Tables 10.2 and 10.3. In practice, the amount of radiation given to a patient for a specific imaging application is dependent on the clinicians themselves.

The ICRP's recommended dose limits (ICRP, 2006) for members of the public and various tissues of radiation workers are in agreement with that of the US NRC (Table 10.1). However, the annual occupational dose limit recommended by ICRP is 20 mSv (ICRP, 2006) for an average of over 5 years, which significantly deviates from that of NRC's recommended dose limit of 50 mSv. The rationale for the ICRP's recommended values has been developed based on several sources of radiological data including reference anatomical and physiological models of humans, studies at the molecular and cellular level, experimental animal studies, and human epidemiological studies including experimental, cohort, case-control, and ecologic studies (such as the Japanese A-bomb cohort) (SG5, 2006).

Regarding future work to maximize safe imaging effectiveness, as mentioned in the previous section, a tissue-specific safe maximum radiation-exposure limit must be developed, beyond which cellular and genetic repairs are adversely affected for appropriate human physiology and pathology models. A better understanding of omics will help create a better understanding of system physiology in response to CT radiation.

TABLE 10.3
ICRP Diagnostic Reference Levels for CT Examinations in Children

Examination	Weighted Dose Index per Slice or Rotation (mGy) (Years)	Dose Length Product per Examination (mGy · cm) (Years)
Brain	40 (<1), 60 (5), 70 (10)	300 (<1), 600 (5), 750 (10)
Chest	20 (<1), 30 (5), 30 (10)	200 (<1), 400 (5), 600 (10)
Chest (high-resolution CT)	30 (<1), 40 (5), 50 (10)	50 (<1), 75 (5), 100 (10)
Upper abdomen	20 (<1), 25 (5), 30 (10)	330 (<1), 360 (5), 800 (10)
Lower abdomen and pelvis	20 (<1), 25 (5), 30 (10)	170 (<1), 250 (5), 500 (10)

10.4 MAGNETIC RESONANCE IMAGING AND SPECTROSCOPY

MRI and MRS use the interaction of electromagnetic fields with the nuclei of odd atomic weights and/or numbers present in the body to generate signals. Temporally and spatially varying electromagnetic fields are generated using a main static magnet, gradient coils, and radiofrequency (RF) coils to produce images and localized signals. Significantly higher strength of the static magnetic field compared with the earth's magnetic field is a safety concern because it may interfere with or worse, adversely affect the functions of a biological system. Time rate of change of gradient fields is a safety concern because it may produce painful nerve stimulation. Temporally and spatially varying electromagnetic fields produced with the RF coils are a safety concern because it induces heating.

The US FDA limits the strength of the main static magnetic field to 4 tesla (T) in neonates less than 1 month old and 8 T in infants more than 1 month old, children, and adults. Any time rate of change of gradient fields is prohibited that is sufficient to produce peripheral nerve stimulation (PNS). RF power deposited to an imaged tissue per unit tissue weight (i.e., the specific absorption rate [SAR], W/kg) is limited to 3 W/kg for the whole head averaged over any 10 min, 4 W/kg for the whole body averaged over any 15 min, 8 W/kg for a gram of tissue in the head and torso averaged over any 5 min, and 12 W/kg for a gram of tissue in the extremities averaged over any 5 min of the RF power deposition. No limits are provided for the maximum duration of the RF power deposition (i.e., imaging and spectroscopy time) and the maximum temperature induced (CDRH-FDA, 2003).

International Electrotechnical Commission (IEC) concludes no harmful effects from a main static magnetic field up to 7 T. Scanners are restricted to 80 and 100 % of the directly determined PNS threshold level in the normal and first-level operating mode, respectively. The whole-head average SAR from volume coils is limited to 3.2 W/kg in the normal and first-level operating modes averaged over any 6 min. The whole-body average SAR from volume coils is limited to 2.0 W/kg in the normal operating mode and 4.0 W/kg in the first-level operating mode averaged over any 6 min. The whole-body average SAR limit for the first-level operating mode must be decreased to 2 W/kg at the rate of 0.25 W/kg for every degree C rise in room temperature over the room temperature of 25°C. SAR limits for the local transmit coils are significantly higher. The local SAR averaged over any 6 min in the normal operating mode is limited to 10, 10, and 20 W/kg for the head, torso, and extremities, respectively. The local SAR averaged over any 6 min in the first-level operating mode is limited to 20, 20, and 40 W/kg in the head, torso, and extremities, respectively. The short-term SAR limit (i.e., RF power deposition of ≤10 s) is restricted to two times the stated values for the volume and local coils. The long-term RF power deposition is restricted to 240 W·min/kg. The IEC also provides limits for the maximum temperature that can be induced during a scan, acknowledging the fact that thermogenic hazards are more directly related to temperature than a tissue-mass averaged RF power (i.e., the SAR). The maximum local and core body temperature is limited to 39°C, and the rise in core temperature is limited to 0.5°C in the normal operating mode. The maximum local and core temperature is limited to 40°C and the rise in core temperature is limited to 1°C in the first-level operating mode. The IEC claims that the SAR and temperature limits are conservative and would not cause tissue damage. The higher local SAR and temperatures are allowed for specific tissues if no unacceptable risk for the patient occurs (IEC, 2010).

The International Committee on Non-Ionizing Radiation Protection (ICNIRP) limits the strength of the main static magnetic field to 2 T for the head and trunk and 8 T for the extremities. Similar to the IEC guidelines, scanners are restricted to 80 and 100% of the directly determined PNS threshold level in the normal and first-level operating mode, respectively. The whole-head average SAR is limited to 3.0 W/kg in the normal and first-level operating modes averaged over any 6 min. The whole-body average SAR is limited to 2.0 W/kg in the normal operating mode and 4.0 W/kg in the first-level operating mode averaged over any 6 min similar to the IEC guidelines. The local SAR averaged over any 6 min is limited to 10, 10, and 20 W/kg for the head, torso, and extremities, respectively, in the normal as well as first-level operating modes. The short term SAR limit (i.e., RF power deposition of ≤10 s) is restricted to three times the stated values. The ICNIRP limits the maximum local temperature in the two operating modes to 38, 39, and 40°C in the head, trunk, and extremities, respectively. The rise in core temperature is limited to 0.5 and 1°C in the normal and first-level operating modes (ICNIRP, 2004, 2009).

The rationale for the safe usage of main static magnetic field up to 8 T was developed in part based on the human studies conducted by Chakeres et al. (Chakeres et al., 2003a, 2003b). No statistically significant difference was obtained in 25 human subjects when randomly exposed to 0.05- and 8-T static magnetic fields. Evaluations were made based on six standardized neuropsychological exams and auditory reaction times (Chakeres et al., 2003a). In a different study, the effects were studied on the vital signs (i.e., heart rate, respiratory rate, blood pressure, finger-pulse blood-oxygenation level, and sublingual and external auditory-canal core body temperatures) and electrocardiogram of the static magnetic field exposure up to 8 T using 25 human subjects (Chakeres et al., 2003b). A statistically significant increase in the systolic blood pressure of an average of 3.6 mm Hg was observed, which may be clinically irrelevant.

The rationale for the safe limits of time rate of change of gradient fields was developed based on the detailed studies conducted by Bourland et al. (1999) and others (King and Schaefer, 2000; Schaefer et al., 2000; Den Boer et al., 2002; Liu et al., 2003; Zhang et al., 2003; So et al., 2004). The data were fitted to a hyperbolic strength-duration relationship. The relationship was employed to safeguard peripheral nerve stimulation in MR scanners.

The rationale for the safe whole-head or -body SAR limits was developed based on the heating studies conducted by several researchers related to far-field RF radiation, hyperthermia, and MR-RF heating in 1.5 T scanners (Gordon, 1982, 1983, 1987a, 1987b, 1987c, 1992; Gordon and Ferguson, 1984; Gordon et al., 1986a, 1986b; Shellock et al., 1986a, 1986b; Gordon and Ali, 1987; Barber et al., 1990; Adair and Berglund, 1986, 1989, 1992; Adair et al., 1992, 1998, 2003, 2005; Walters et al., 2000; Allen et al., 2003; D'Andrea et al., 2003; Foster and Adair, 2004). The rationale for the local SAR limit of 10 W/kg averaged over any 6 min in a 10-gm tissue of the head and body is based on the maximum allowable safe temperature change of 1°C with no perfusion and conduction in the tissue (SAR = $Cp \cdot dT/dt$ = 3600 Jkg-1K-1. 1 K/(360 s) = 10 W/kg, where Cp = specific heat, dT = temperature change, and dt = time change) (Shrivastava and Vaughan, 2009).

Static magnetic fields effect moving charged particles and thus may interfere with the effectiveness of the functions of cells, tissues, and organs of a mammalian biological system. Regarding future work to maximize safe imaging effectiveness, a detailed study is desired that investigates the short- and long-term, reversible and irreversible effects on the micro- and macroenvironment and related thresholds of relevant mammalian biological systems. Similarly, the effect of the time rate of change of magnetic field and safe thresholds for the functions of cells, tissues, and organs of a relevant mammalian biological system are of interest (ICNIRP, 2004, 2009; IEC, 2010). A relevant mammalian biological system must be identified to determine thresholds because the physical and chemical state of the biological system (i.e., body form, weight, hydration, medication, disease, pregnancy, etc.) may alter the thresholds.

MR-related RF heating and its effect are not well studied at ultrahigh fields (i.e., static magnetic field strength ≥ 3 T) (Vaughan et al., 2001). Studying RF heating in ultrahigh fields is important because more RF power is transferred to an imaged tissue with an increase in the static field strength for a given pulse sequence (Abragam, 1983). Additionally, higher conductivity of biological tissues and frequency-dependent dielectric losses at higher field strengths cause more RF absorption (Gabriel et al., 1996a, 1996b, 1996c). RF power at ultrahigh fields is deposited nonuniformly because of the high gyromagnetic ratio of water proton and nonuniform distribution of tissue electrical properties (Shrivastava et al., 2008, 2009). Nonuniform RF power together with a nonuniform geometry, tissue distribution, and blood flow may cause nonuniform RF heating with the potential of local hot spots.

Reliable and validated means are desired to predict and monitor subject-specific RF heating and local SAR. In principle, RF power, when deposited for a long time, may cause *in vivo* temperature changes greater than 1°C. Thermogenic cellular and systemic hazards are directly related to temperatures and temperature-time history and not to the whole head/body average SAR and local SAR. The effects of nonuniform brain temperature changes in the range of 1–3°C on local and systemic body functions are not well understood and may cause altered thermophysiologic effects (Dewhirst et al., 2003; Shrivastava et al., 2009). Pregnant women and subjects with implanted conductive medical devices are currently contraindicated in MR. RF heating and associated thermophysiological changes must be studied in appropriate animal models to humans to better understand the function and physiology of a relevant human system and ensure safe MR application at the highest fields. The static field-, gradient field-, and RF heating-related studies are imperative to advance omics using MRI and MRS.

10.5 SUMMARY

Safety standards, their rationale, and future research opportunities were reviewed for the four diagnostic imaging techniques of PET, SPECT, CT, and MRI that are commonly used in omics. It has been argued that better understanding of the possible physiological responses activated using the imaging procedures themselves must be developed using appropriate animal and human models to better understand and advance omics using the imaging modalities.

ACKNOWLEDGMENTS

Partial support for this work was provided by the following grants from the National Institute of Health R01-EB007327, R01-EB000895, and BTRR-P41 RR08079 and funds from the Keck Foundation.

REFERENCES

Abragam, A. (1983). Principles of nuclear magnetism. Oxford: Oxford University Press.

Adair, E., Adams, B., Hartman, S. (1992). Physiological interaction processes and radio-frequency energy absorption. *Bioelectromagnetics* 13: 497–512.

Adair, E. R., Berglund, L. G. (1986). On the thermoregulatory consequences of NMR imaging. *Magn. Reson. Imaging* 4: 321–333.

Adair, E. R., Berglund, L. G. (1989). Thermoregulatory consequences of cardiovascular impairment during NMR imaging in warm/humid environments. *Magn. Reson. Imaging* 7: 25–37.

Adair, E. R., Berglund, L. G. (1992). Predicted thermophysiological responses of humans to MRI fields. *Ann. N.Y. Acad. Sci.* 649: 188–200.

Adair, E. R., Blick, D. W., Allen, S. J., Mylacraine, K. S., Ziriax, J. M., Scholl, D. M. (2005). Thermophysiological responses of human volunteers to whole body RF exposure at 220 MHz. *Bioelectromagnetics* 26: 448–461.

Adair, E. R., Kelleher, S. A., Mack, G. W., Morocco, T. S. (1998). Thermophysiological responses of human volunteers during controlled whole-body radio frequency exposure at 450 MHz. *Bioelectromagnetics* 19: 232–245.

Adair, E. R., Mylacraine, K. S., Allen, S. J. (2003). Thermophysiological consequences of whole body resonant RF exposure (100 MHz) in human volunteers. *Bioelectromagnetics* 24: 489–501.

Allen, S. J., Adair, E. R., Mylacraine, K. S., Hurt, W., Ziriax, J. (2003). Empirical and theoretical dosimetry in support of whole body resonant RF exposure (100 MHz) in human volunteers. *Bioelectromagnetics* 24: 502–509.

Barber, B. J., Schaefer, D. J., Gordon, C. J., Zawieja, D. C., Hecker, J. (1990). Thermal effects of MR imaging: Worst-case studies on sheep. *Am. J. Roentgenol.* 155: 1105–1110.

Berrington De Gonzalez, A., Mahesh, M., Kim, K. P., Bhargavan, M., Lewis, R., Mettler, F., Land, C. (2009). Projected cancer risks from computed tomographic scans performed in the United States in 2007. *Arch. Intern. Med.* 169: 2071–2077.

Boellaard, R., O'Doherty, M. J., Weber, W. A., Mottaghy, F. M., Lonsdale, M. N., Stroobants, S. G., Oyen, W. J., Kotzerke, J., Hoekstra, O. S., Pruim, J., et al. (2009). FDG PET and PET/CT: EANM procedure guidelines for tumour PET imaging: Version 1.0. *Eur. J. Nucl. Med. Mol. Imaging* 37: 181–200.

Bourland, J. D., Nyenhuis, J. A., Schaefer, D. J. (1999). Physiologic effects of intense MR imaging gradient fields. *Neuroimaging Clin. N. Am.* 9: 363–377.

CDRH-FDA (2003). Guidance for industry and FDA staff: Criteria for significant risk investigations of magnetic resonance diagnostic devices. U.S. Department of Health and Human Services, FDA, Center for Devices and Radiological Health and Radiological Devices Branch, Division of Reproductive, Abdominal, and Radiological Devices, Office of Device Evaluation, Washington, D.C., http://www.fda.gov/downloads/MedicalDevices/DeviceRegulationandGuidance/GuidanceDocuments/ucm072688.pdf.

Chakeres, D. W., Bornstein, R., Kangarlu, A. (2003a). Randomized comparison of cognitive function in humans at 0 and 8 tesla. *J. Magn. Reson. Imaging* 18: 342–345.

Chakeres, D. W., Kangarlu, A., Boudoulas, H., Young, D. C. (2003b). Effect of static magnetic field exposure of up to 8 tesla on sequential human vital sign measurements. *J. Magn. Reson. Imaging* 18: 346–352.

D'Andrea, J. A., Chou, C. K., Johnston, S. A., Adair, E. R. (2003). Microwave effects on the nervous system. *Bioelectromagnetics* Suppl. 6: S107–S147.

Den Boer, J. A., Bourland, J. D., Nyenhuis, J. A., Ham, C. L., Engels, J. M., Hebrank, F. X., Frese, G., Schaefer, D. J. (2002). Comparison of the threshold for peripheral nerve stimulation during gradient switching in whole body MR systems. *J. Magn. Reson. Imaging* 15: 520–525.

Dewhirst, M. W., Viglianti, B. L., Lora-Michiels, M., Hanson, M., Hoopes, P. J. (2003). Basic principles of thermal dosimetry and thermal thresholds for tissue damage from hyperthermia. *Int. J. Hyperthermia* 19: 267–294.

Einstein, A. J., Henzlova, M. J., Rajagopalan, S. (2007). Estimating risk of cancer associated with radiation exposure from 64-slice computed tomography coronary angiography. *JAMA* 298: 317–323.

FDA (2005). FDG F18 injection, NDA 21-870. http://www.accessdata.fda.gov/drugsatfda_docs/label/2005/021870lbl.pdf.

Foster, K. R., Adair, E. R. (2004). Modeling thermal responses in human subjects following extended exposure to radiofrequency energy. *Biomed. Eng. Online* 3: 4.

Gabriel, C., Gabriel, S., Corthout, E. (1996a). The dielectric properties of biological tissues: I. Literature survey. *Phys. Med. Biol.* 41: 2231–2249.

Gabriel, S., Lau, R. W., Gabriel, C. (1996b). The dielectric properties of biological tissues: II. Measurements in the frequency range 10 Hz to 20 GHz. *Phys. Med. Biol.* 41: 2251–2269.

Gabriel, S., Lau, R. W., Gabriel, C. (1996c). The dielectric properties of biological tissues: III. Parametric models for the dielectric spectrum of tissues. *Phys. Med. Biol.* 41: 2271–2293.

Gordon, C. (1982). Effect of heating rate on evaporative heat loss in the microwave-exposed mouse. *J. Appl. Physiol.* 53: 316–323.

Gordon, C. (1983). Influence of heating rate on control of heat loss from the tail in mice. *Am. J. Physiol.* 244: R778–R784.

Gordon, C. (1992). Local and global thermoregulatory responses to MRI fields. *Ann. N.Y. Acad. Sci.* 649: 273–284.

Gordon, C. J. (1987a). Normalizing the thermal effects of radiofrequency radiation: Body mass versus total body surface area. *Bioelectromagnetics* 8: 111–118.

Gordon, C. J. (1987b). Reduction in metabolic heat production during exposure to radio-frequency radiation in the rat. *J. Appl. Physiol.* 62: 1814–1818.

Gordon, C. J. (1987c). Relationship between preferred ambient temperature and autonomic thermoregulatory function in rat. *Am. J. Physiol.* 252: R1130–R1137.

Gordon, C. J., Ali, J. S. (1987). Comparative thermoregulatory response to passive heat loading by exposure to radiofrequency radiation. *Comp. Biochem. Physiol. A* 88: 107–112.

Gordon, C. J., Ferguson, J. H. (1984). Scaling the physiological effects of exposure to radiofrequency electromagnetic radiation: Consequences of body size. *Int. J. Radiat. Biol. Relat. Stud. Phys. Chem. Med.* 46: 387–397.

Gordon, C. J., Long, M. D., Fehlner, K. S. (1986a). Temperature regulation in the unrestrained rabbit during exposure to 600 MHz radiofrequency radiation. *Int. J. Radiat. Biol. Relat. Stud. Phys. Chem. Med.* 49: 987–997.

Gordon, C. J., Long, M. D., Fehlner, K. S., Stead, A. G. (1986b). Temperature regulation in the mouse and hamster exposed to microwaves in hot environments. *Health Phys.* 50: 781–787.

Holm, S., Borgwardt, L., Loft, A., Graff, J., Law, I., Hojgaard, L. (2007). Paediatric doses—a critical appraisal of the EANM paediatric dosage card. *Eur. J. Nucl. Med. Mol. Imaging* 34: 1713–1718.

ICNIRP (2004). Medical magnetic resonance (MR) procedures: Protection of patients. *Health Physics* 87: 197–216.

ICNIRP (2009). Amendment to the ICNIRP "Statement on medical magnetic resonance (MR) procedures: Protection of patients." *Health Physics* 97: 259–261.

ICRP (2006). Analysis of the criteria used by the International Commission on Radiological Protection to justify the setting of numerical protection level values. *Ann. ICRP* 36: III, 7–76.

IEC (2010). Medical electrical equipment: Part 2-33: Particular requirements for the basic safety and essential performance of magnetic resonance equipment for medical diagnosis. International Electrotechnical Commission, 60601-2-33, http://webstore.iec.ch/webstore/webstore.nsf/Artnum_PK/43851.

King, K. F., Schaefer, D. J. (2000). Spiral scan peripheral nerve stimulation. *J. Magn. Reson. Imaging* 12: 164–170.

Lassmann, M., Biassoni, L., Monsieurs, M., Franzius, C. (2008a). The new EANM paediatric dosage card: Additional notes with respect to F-18. *Eur. J. Nucl. Med. Mol. Imaging* 35: 1666–1668.

Lassmann, M., Biassoni, L., Monsieurs, M., Franzius, C., Jacobs, F. (2008b). The new EANM paediatric dosage card. *Eur. J. Nucl. Med. Mol. Imaging* 35: 1748.

Liu, F., Zhao, H., Crozier, S. (2003). On the induced electric field gradients in the human body for magnetic stimulation by gradient coils in MRI. *IEEE Trans. Biomed. Eng.* 50: 804–815.

Masuda, Y., Kondo, C., Matsuo, Y., Uetani, M., Kusakabe, K. (2009). Comparison of imaging protocols for 18f-FDG PET/CT in overweight patients: Optimizing scan duration versus administered dose. *J. Nucl. Med.* 50: 844–848.

NRC (1991). *Standards for protection against radiation.* Nuclear Regulatory Commission. Final Rule. Fed. Regist. 56: 23360–23474.

Saha, G. B. (2010). *Basics of PET imaging: Physics, chemistry, and regulations.* New York: Springer.

Schaefer, D. J., Bourland, J. D., Nyenhuis, J. A. (2000). Review of patient safety in time-varying gradient fields. *J. Magn. Reson. Imaging* 12: 20–29.

SG5 (2006). Analysis of the criteria used by the International Commission on Radiological Protection to justify the setting of numerical protection level values. *Ann. ICRP* 36.

Shellock, F. G., Gordon, C. J., Schaefer, D. J. (1986a). Thermoregulatory responses to clinical magnetic resonance imaging of the head at 1.5 tesla: Lack of evidence for direct effects on the hypothalamus. *Acta Radiol. Suppl.* 369: 512–513.

Shellock, F. G., Schaefer, D. J., Gordon, C. J. (1986b). Effect of a 1.5 T static magnetic field on body temperature of man. *Magn. Reson. Med.* 3: 644–647.

Shrivastava, D., Hanson, T., Kulesa, J., Delabarre, L., Snyder, C., Vaughan, J. T. (2009). Radio-frequency heating at 9.4 T: In vivo thermoregulatory temperature response in swine. *Magn. Reson. Med.* 62: 888–895.

Shrivastava, D., Hanson, T., Schlentz, R., Gallaghar, W., Snyder, C., Delabarre, L., Prakash, S., Iaizzo, P., Vaughan, J. T. (2008). Radiofrequency heating at 9.4 T: In vivo temperature measurement results in swine. *Magn. Reson. Med.* 59: 73–78.

Shrivastava, D., Vaughan, J. T. (2009). A generic bioheat transfer thermal model for a perfused tissue. *ASME J. Biomech. Eng.* 131: 074506.

Smith-Bindman, R., Lipson, J., Marcus, R., Kim, K. P., Mahesh, M., Gould, R., Berrington De Gonzalez, A., Miglioretti, D. L. (2009). Radiation dose associated with common computed tomography examinations and the associated lifetime attributable risk of cancer. *Arch. Intern. Med.* 169: 2078–2086.

So, P. P., Stuchly, M. A., Nyenhuis, J. A. (2004). Peripheral nerve stimulation by gradient switching fields in magnetic resonance imaging. *IEEE Trans. Biomed. Eng.* 51: 1907–1914.

Stauss, J., Franzius, C., Pfluger, T., Juergens, K. U., Biassoni, L., Begent, J., Kluge, R., Amthauer, H., Voelker, T., Hojgaard, L., et al. (2008). Guidelines for 18f-FDG PET and PET-CT imaging in paediatric oncology. *Eur. J. Nucl. Med. Mol. Imaging* 35: 1581–1588.

Varrone, A., Asenbaum, S., Vander Borght, T., Booij, J., Nobili, F., Nagren, K., Darcourt, J., Kapucu, O. L., Tatsch, K., Bartenstein, P., et al. (2009). EANM procedure guidelines for PET brain imaging using [(18) F]FDG, version 2. *Eur. J. Nucl. Med. Mol. Imaging* 36: 2103–2110.

Vaughan, J. T., Garwood, M., Collins, C. M., Liu, W., Delabarre, L., Adriany, G., Andersen, P., Merkle, H., Goebel, R., Smith, M. B., et al. (2001). 7 T vs. 4 T: RF power, homogeneity, and signal-to-noise comparison in head images. *Magn. Reson. Med.* 46: 24–30.

Walters, T. J., Blick, D. W., Johnson, L. R., Adair, E. R., Foster, K. R. (2000). Heating and pain sensation produced in human skin by millimeter waves: Comparison to a simple thermal model. *Health Phys.* 78: 259–267.

Zhang, B., Yen, Y. F., Chronik, B. A., Mckinnon, G. C., Schaefer, D. J., Rutt, B. K. (2003). Peripheral nerve stimulation properties of head and body gradient coils of various sizes. *Magn. Reson. Med.* 50: 50–58.

Section II

Empirical Research

11 Molecular Genetics of Human Cancers

Modulation of the Tumor Suppressor Function of RB1 Gene Product in Human Vestibular Schwannomas

Mohan L. Gope
Presidency College
Bangalore, India

Rohan Mitra and Rajalakshmi Gope
National Institute of Mental Health and Neurosciences
Bangalore, India

CONTENTS

A tumor arises because of a deregulated proliferation of cells that forms an abnormal mass. Tumorigenesis is a multistep process in which normal controls of cell proliferation and cell-cell interaction are lost. In the human genome, there are various types of genes that control cell growth through very precise mechanisms. Tumorigenesis involves interplay between at least two classes of genes: (1) oncogenes and (2) tumor suppressor genes. Alterations in any one or both of these classes of genes and their pathways could lead to tumor formation. Human primary brain tumors are a diverse group of neoplasms arising from different cells of the central nervous system. The vestibular schwannomas (VS) arise from the vestibular branch of the eighth cranial nerve, the vestibulocochlear nerve. Despite marked improvement in diagnosis, management, and treatment protocols, these tumors continue to cause significant morbidity. If the VS tumor becomes large, it can exert pressure on other cranial nerves, interfere with their functions, and eventually cause mortality. In addition to tumors arising from the vestibulocochlear nerve, patients with VS tumors also develop other secondary malignant tumors such as meningiomas, gliomas, and spinal schwannomas. The *RB1* gene is the first human tumor suppressor gene to be identified. Genetic alterations at the *RB1* gene locus have been reported in human retinoblastomas, which is a human childhood tumor. *RB1* gene alterations have also been reported in other human tumors such as osteosarcoma and soft-tissue sarcoma and cancers of lung, breast, and bladder. The protein product of this gene, pRb, is known to be involved in cell cycle regulation. The functions of pRb are modulated by its phosphorylation and dephosphorylation. There are 16 phosphorable Ser/Thr sites in pRb. Deregulation of phosphorylation or dephosphorylation of pRb is known to affect its function as a tumor suppressor. In this chapter, we will discuss the possible role of the *RB1* gene and its protein product in human VS tumors.

11.1 INTRODUCTION TO HUMAN CANCERS

Cancer is one of the important human diseases with the second highest mortality rate, the first being heart disease. Cancers arise because of either uncontrolled cell proliferation followed by lack of cell differentiation or dedifferentiation of differentiated cells followed by loss of cellular growth control. It is a multistep, complex, genetic disorder where many genes are known to be involved in tumor initiation, progression, and metastasis. Single mutations in genes controlling cell growth and differentiation are usually insufficient to cause cancer. Cancer is known to arise because of accumulation of many genetic alterations, which results in gradual and stepwise effects (Hanahan and Weinberg, 2000; Jacks and Weinberg, 1998; Knudson, 1971; Lane, 1992). The genes that are mainly involved in tumorigenesis can be broadly divided into (i) oncogenes and (ii) tumor suppressor (*TS*) genes. Oncogenes contribute to tumor initiation, progression, and metastasis by promoting cell division and aiding tumor expansion, and *TS* genes are known to inhibit cell growth (Hanahan and Weinberg, 2000; Jacks et al., 1992; Jacks and Weinberg, 1998; Knudson, 1971; Lane, 1992).

11.1.1 Human Tumor Suppressor Genes

The human *TS* genes play crucial role during cell-cycle and cell differentiation. Initially the *TS* genes were identified from hybrid cells generated by fusion of normal cells and tumor cells. These hybrids could not produce tumors in nude mice because some of the genes (possibly *TS* genes) from the normal cells inhibited the tumor-cell growth. The genes responsible for such inhibition are called as anticancer genes, antioncogenes, or tumor suppressor genes. Alterations of oncogene occur sporadically; however, *TS* gene alterations are both familial and sporadic events. The growth inhibitory function of the *TS* genes is reported to be lost in many human tumors (Hanahan and Weinberg, 2000; Jacks et al., 1992; Knudson, 1971; Jacks and Weinberg, 1998; Lane, 1992; Levine, 1997). In 1971, Knudson proposed the *two-hit model* for tumorigenesis, in which he hypothesized that two mutational events inactivating both the alleles of a tumor suppressor gene are necessary for tumor development. This hypothesis originated from studies on the familial childhood eye tumor, retinoblastoma. In familial retinoblastoma, the child is known to inherit a defective retinoblastoma (*RB1*) allele from one of the parents. During somatic cell growth, a second mutation of the normal allele occurs with 90% certainty within susceptible retinoblasts and this second hit causes retinoblastoma. In sporadic cases, the tumor originates upon mutation of both *RB1* alleles within a single cell. Thus, retinoblastoma can occur in both familial and sporadic forms. Most sporadic cases form unilateral tumors, and bilateral tumors are found in familial cases. Similar to the *RB1* gene, the *p53* gene has also been identified as a human tumor suppressor gene. A number of human tumors are known to have mutant p53 protein product. The major difference between the *RB1* and *p53 TS* genes is that both alleles of the *RB1* gene have to be inactivated for tumorigenesis, whereas mutation of a single *p53* allele appears to be sufficient for the same (Hanahan and Weinberg, 2000; Jacks et al., 1992; Jacks and Weinberg, 1998; Knudson, 1971; Lane, 1992; Levine, 1997). The *RB1* gene is one of the widely studied, important human tumor suppressor genes.

11.1.2 The Human Retinoblastoma Gene

The human retinoblastoma gene (*RB1*) is the first *TS* gene to be identified from the human childhood tumor, retinoblastoma (Knudson, 1971). The *RB1* gene spans approximately 200 kb, and it is on human chromosome 13q14. It encodes a 110-kDa protein pRb that is involved in the control of cell-cycle differentiation and apoptosis (Bartek et al., 1997; Classon and Harlow, 2002; Dyson, 1998; Hanahan and Weinberg, 2000). The protein product of the *RB1* gene, pRb, is ubiquitously expressed in a variety of tissue types. It is a nuclear phosphoprotein, and it has a major role in cell-cycle–regulatory pathway during the G_1 to S transition (Figure 11.1). Alterations at the *RB1* gene locus and absence of pRb have been reported for many human tumor types including retinoblastoma, osteosarcoma, soft-tissue sarcoma, small-cell lung cancer, breast cancer, and bladder cancer. pRb is known to arrest cells at the G_1 phase of the cell cycle, and therefore the *RB1* gene is known as the *cellular gatekeeper* gene. This is achieved by repressing transcription of genes required for the G_1 to S phase transition. The function of pRb is regulated by phosphorylation or dephosphorylation of the serine/threonine residues. There are 16 serine/threonine residues in pRb that can be phosphorylated. The pRb is known to be in its active form, having TS function, when it is hypophosphorylated (Lee et al., 1987; Lee et al., 1992; Ma et al., 2003) (Figure 11.1). The active form of pRb binds to various E2F family of transcription factors and inhibits expression of the E2F target genes like c-myc, c-myb, and cdc-2, which are necessary for the DNA synthesis during the S phase (Classon and Harlow, 2002; Dyson, 1998; Lee et al., 1987; Ma et al., 2003). pRb also inhibits components of RNA polymerases I and III that synthesize rRNA and tRNA (Dyson, 1998; Taya et al., 1989). Thus, the *RB1* gene controls cell proliferation not only by arresting the cell cycle but by decreasing the rate of protein synthesis as well. Whenever pRb gets hyperphosphorylated, it becomes inactive in terms of its tumor suppressor function and releases E2F to activate its target genes (Dyson, 1998;

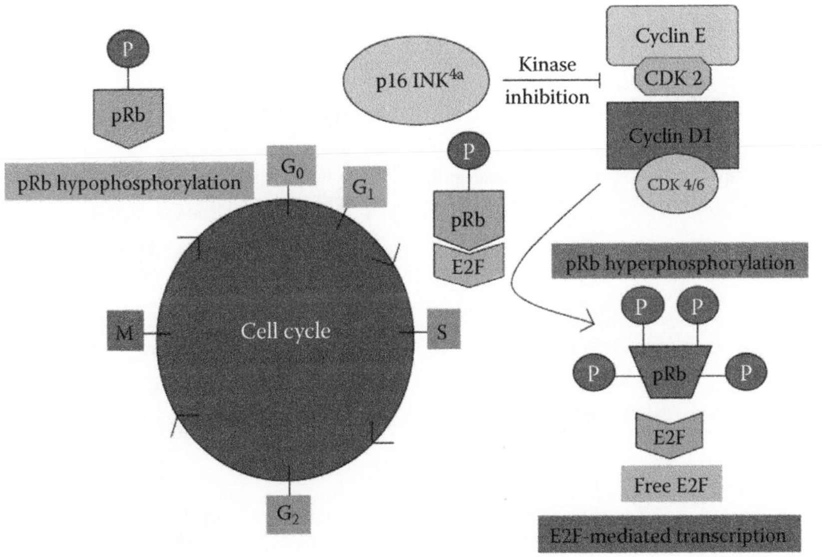

FIGURE 11.1 Function of pRb during cell cycle. G_0/G_1, resting phase; S, synthetic phase; G_2, resting phase 2; M, mitotic phase.

Taya et al., 1989). The E2F family of proteins has the ability to bind to the promoter region of many genes that are necessary for cell-cycle progression, and these genes are repressed by pRb once the cell cycle is completed. The progression of the cell cycle and the transition of cells from the G_1 to the S phase in particular require the phosphorylation of pRb, and it is carried out by cyclin-dependent kinases (CDKs) in association with the corresponding cyclins (Figure 11.1). The pRb protein becomes active through dephosphorylation once the cell exits from the M phase.

This active form of pRb is again able to bind E2F and hold the cell at the G_1/S checkpoint (Figure 11.2). The central pocket domain is essential for the TS function of pRb, and any alteration at this domain could lead to tumor development (Yaya et al., 1989). Therefore, many viral oncoproteins target this pocket domain by binding to pRb, which leads to disruption of its TS function. The sequences in the pocket domain are essential for binding to E2F and many cell-cycle–regulatory proteins. The TS function of pRb has been confirmed by its ability to inhibit tumor-cell growth when the wild-type *RB1* gene is introduced into pRb-deficient tumor cells. In addition to its function in cell-cycle checkpoint, pRb is also known to be involved in many other cellular functions including cell differentiation and apoptosis, and disruption of these pathways also could lead to tumor formation (Hatakeyama and Weinberg, 1995; Jacks and Weinberg, 1998; Lane, 1992; Scherr, 1996). The functions of the *RB1* gene during cell differentiation and apoptosis were identified using transgenic and knockout mice models. The results from these studies show that the *RB1* heterozygous mice do not develop retinoblastomas. However, they are susceptible to other tumors such as pituitary adenomas, and loss of remaining wild-type allele was found in these tumors. Thus, loss of function of both *RB1* alleles appears to be necessary for tumorigenesis in many human cell types. Loss of *RB1* gene function leads to the formation of retinoblastoma only in humans. For example, in mice the *RB1*-null embryos die before the 14th day of gestation. These null embryos also had defective hematopoietic and neuronal development. Analysis of chimeric and *RB1*-null mice have shown that these developmental abnormalities can be rescued by introduction of the wild-type *RB1* gene, thereby directly demonstrating TS function of the *RB1* gene. Results from these studies also showed that the function of the *RB1* gene is important for the neuronal cell survival and differentiation of cells of erythroid lineage (Jacks et al., 1992; Lee et al., 1992; Zacksenhaus et al., 1996). The product of cellular oncogene MDM2 also binds pRb and inhibits its TS function. The tumor virus SV-40 large T antigen is required for virus-mediated tumorigenesis. The SV-40 large T binds to

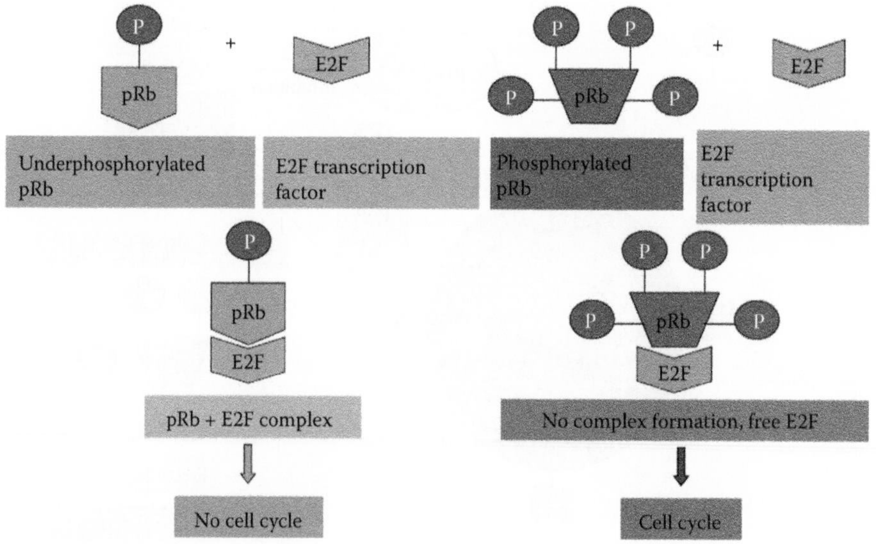

FIGURE 11.2 Function of phosphorylated and underphosphorylated pRb and their ability to form a complex with E2F transcription factor.

pRb and inactivates it, thus releasing the cells from the resting phase. Similarly, other viral proteins with transforming functions such as adenovirus and human papilloma virus, E1a and E7, respectively, are also known to bind to pRb, inactivating its tumor suppressor function by sequestering pRb (Hanahan and Weinberg, 2000; Hatakeyama and Weinberg, 1995; Jacks and Weinberg, 1998; Lane, 1992; Scherr, 1996). Thus, a normal cell can become malignant even in presence of wild-type pRb when its function is lost by mechanisms other than mutations.

Alterations at the *RB1* gene locus have been reported for many tumors (Lie et al., 2005; Loeb, 2001; Nowak et al., 2002). However, many groups have previously reported increased expression of RB1 mRNA and increased phosphorylation of pRb in a few human tumors, such as human colorectal carcinomas, human breast cancers, human malignant melanomas, and human salivary gland acinic cell carcinoma, and in human vestibular schwannomas (Gope et al., 1990, 1991; Gope and Gope, 1991, 1993, 2003; Lie et al., 2005; Roesch et al., 2005; Thomas et al., 2005; Yamamoto et al., 1999). The increased level of *RB1* gene expression and the presence of an increased level of hyperphosphorylated form of pRb are known to have antiapoptotic function (Haas-Kogan et al., 1995; Hofseth et al., 2004), which helps to keep the tumor cells in the cycling phase, resulting in tumor progression (Chatterjee et al., 2004; Cheng et al., 2000; Ma et al., 2003; Roesch et al., 2005; Thomas et al., 2005). It has been hypothesized that the hyperphosphorylation of pRb could be critical for inactivation of the pRb-mediated growth suppression and may play an important role in the progression of these tumors (Tos et al., 2004; Xiao et al., 2005) (Figure 11.2).

11.2 EPIDEMIOLOGY: PREVALENCE

The incidence of VS appears to be greatest in the Far East. They account for 10.6% of primary intracranial neoplasms in India (Dastur et al., 1968) and 10.2% in China (Wen-qing et al., 1982). Occurrence rates of 4.9% and 4% are reported in England (Baker et al., 1976) and the United States, respectively (Kurland et al., 1982). The lowest incidence is found in African blacks, where 2.6% is reported for Kenya (Ruberti and Poppi, 1971), 0.9% is found in Nigeria (Adeloye, 1979), and 0.5% is found in Zimbabwe, which was formerly known as Rhodesia. The last statistic can be contrasted with a 3.7% incidence in the white population of the same country (Levy and Auchterlonie, 1975). A recent study in Denmark reported an incidence rate of 17.4% VS tumors per 1 million inhabitants each year during the years 1996–2001, which is almost double that reported during the years 1990–1995 (Tos et al., 2004). A study conducted over a 26-year period has shown that the number of cases of VS tumors is increasing. However, the size and the average onset of these tumors have been almost unchanged throughout the 26-year period (Stangerup et al., 2004).

11.2.1 Human Vestibular Schwannomas

VS, also referred to as acoustic schwannomas or neuromas, arise from the Schwann cells of the vestibular branch of the eighth cranial nerve, the vestibulocochlear nerve. The majority of these tumors are sporadic and unilateral (Figure 11.3). Approximately 5% of vestibular schwannomas are bilateral, occurring in association with neurofibromatosis type 2 (NF2) (Figure 11.4). NF2, also known as central neurofibromatosis, is a distinct autosomal dominant disorder with a high penetrance (Cohen, 1995; Jacoby et al., 1997; Martuza and Eldridge, 1988; Rouleau et al., 1993). The quickly growing, aggressive form of VS tumor is known as Wishart type, and the slowly growing form is known as Gardner type (Martuza and Eldridge, 1988). In primary cultures, the VS cells appear as round-shaped cells; however, the side view of the cells indicates a kidney-shaped structure (Figure 11.5). Histology of the VS tumor tissues stained with eosin and hematoxylin shows the typical characteristics associated with these tumors. Figure 11.6a shows the lobular growth pattern of the tumor, the Antoni A area containing the densely packed cells, and the Antoni B area with more extracellular clear areas. Figure 11.6b shows the nuclear atypia, which is visible under higher magnification.

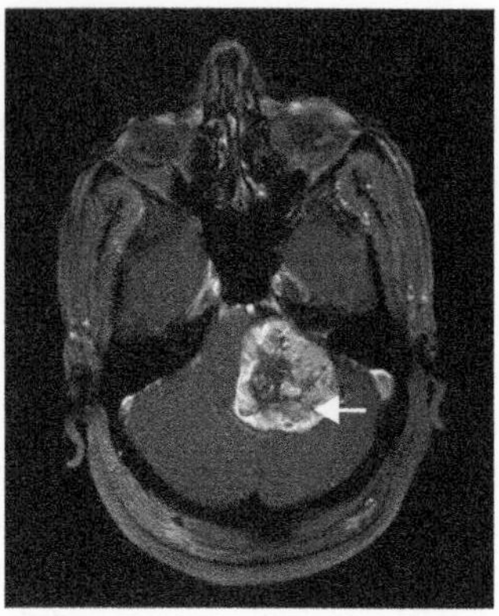

FIGURE 11.3 MRI image. The figure shows the unilateral vestibular schwannoma as indicated by an arrow. (Courtesy of the Department of Neuroimaging and Interventional Radiology, National Institute of Mental Health and Neurosciences, Bangalore, India.)

The anatomical location of these tumors makes management difficult, and patients suffer great morbidity. Hearing loss is usually a symptom in these lesions; other early symptoms include tinnitus and vertigo, and with increasing size there may be facial weakness and numbness. Most of the symptoms associated with vestibular schwannomas are attributed to a mass on the eighth cranial nerve that exerts pressure on both the fifth and seventh cranial nerves. Further enlargement produces cerebellar compression and leads to ataxia, headache, and hydrocephalous (Cohen, 1995; Jacoby et al., 1997; Martuza and Eldridge, 1988; Rouleau et al., 1993).

The NF2 tumors lack a functional NF2 gene and its protein product, merlin (Rouleau et al., 1993). Merlin binds to Mdm2 and inhibits Mdm2-mediated degradation of the tumor suppressor protein p53. Therefore, it is suggested that the merlin protein could be a positive regulator of p53 in terms of tumor suppressor activity (Kim et al., 2004). Merlin is also known to inhibit p21-activated kinase (Kissil et al., 2003). It was not clear whether the *RB1* gene product, pRb, could cooperate with merlin in the normal growth and development of human Schwann cells. If pRb and merlin cooperate in normal cell growth, deregulation of the *RB1* pathway along with functionally altered merlin pathway could contribute to the development of NF2 tumors. It is possible that NF2 mutation is the initiating event in NF2 tumorigenesis, but mutations in other cell-cycle–regulatory genes

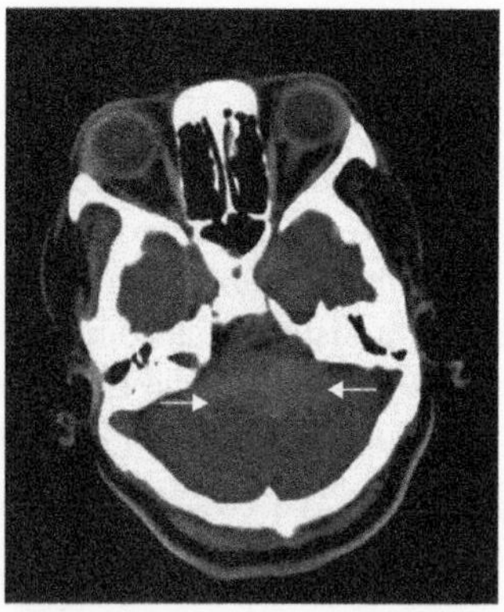

FIGURE 11.4 CT image of bilateral vestibular schwannoma, NF2. The arrows indicate the bilateral tumors. The image shows that the tumors from both the left and right sides have merged because of their large sizes. (Courtesy of the Department of Neuroimaging and Interventional Radiology, National Institute of Mental Health and Neurosciences, Bangalore, India.)

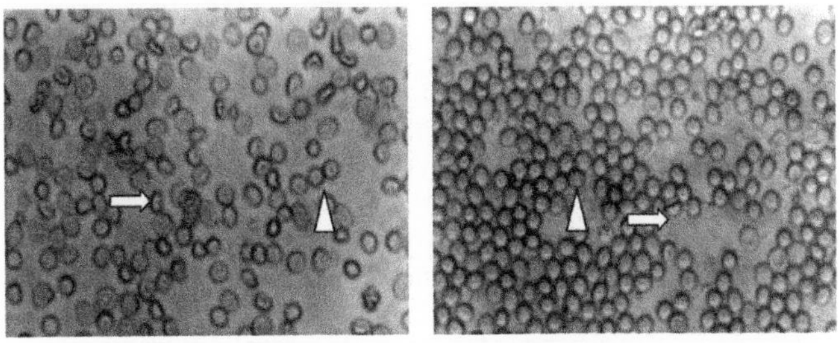

FIGURE 11.5 Phase-contrast photographs of the human VS cells in primary culture. The arrowhead indicates the top view of the cell showing a round shape; the arrow indicates the side view of the cell showing a kidney shape. Magnification, 100×.

may have an additive effect in the progression and can also determine the severity of these tumors (Thomas et al., 2005; Xiao et al., 2005). In order to find answers to these possibilities, the structure and expression of *RB1* in human NF2 tumors and the pRb were analyzed and compared with the matched normal controls (Thomas et al., 2005).

11.2.2 Patient Information

Human VS are generally known to occur in the middle-age group. In certain studies, it has been shown that 60% of the affected cases were in the middle-age group. Earlier reports suggested that men had earlier peak prevalence of VS than women. Approximately 85% of affected males were below 40 years of age, and 52% of females were above 40 years of age. The published peak prevalence appears to be unchanged regardless of the geographical location of the patients. It has been reported that the VS tumors affect more females (56%) than males (44%). Among the sporadic cases alone, 55% were males. However, 80% of confirmed NF2 cases were reported to be females (Table 11.1). NF2 tumors are known to be more common in women and could be related to the female predominance of estrogen binding to these tumors (Martuza and Eldridge, 1988; Thomas et al., 2005).

11.2.3 Loss of Heterozygosity at the *RB1* Gene Locus in Human VS Tumors

Loss of heterozygosity (LOH) in the *RB1* gene was detected in 44% of malignant peripheral–nerve-sheath tumors (Thomas et al., 2005). In VS tumors, LOH at the *RB1* gene locus was reported in 25% informative cases. The observed LOH at the 5' region of the *RB1* gene indicates genetic instability

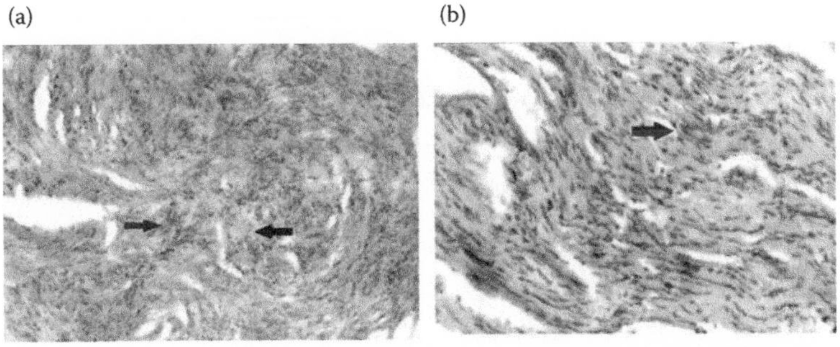

FIGURE 11.6 (See color insert.) Histological features of VS tumor section. Hematoxylin- and eosin-stained VS-tumor tissue sections showing the typical characteristics of these tumors. (a) The tumor tissue showing a typical lobular growth pattern. The Antoni A (left arrow) area with densely packed elongated cells and the Antoni B (right arrow) area with more extracellular clear area can be seen. Magnification, 80×. (b) The nuclear atypia is shown with an arrow. Magnification, 150×.

TABLE 11.1
Patient Data

Parameter	Observation
Average age of patients	26–45 years
Average age of male patients with NF2 tumor	Below 40 years
Average age of female patients with NF2 Tumor	Above 40 years
Average size of tumor	More than 3 cm

at this gene locus that has been reported for other human tumor types (Cohen, 1995; Loeb, 2001; Mawrin et al., 2002; Xu et al., 1993). The LOH at the *RB1* tumor suppressor locus was suggested to be responsible for the development and/or progression of the sporadic NF2 tumors (Table 11.2).

11.2.4 *RB1* Gene Expression in VS Tumors

The *RB1* gene is expressed in a number of human tissues and organs, and the level of expression is similar in all tissues examined (Thomas et al., 2005). Alterations in *RB1* expression have been reported in many human tumor types including lung cancer, osteosarcomas, leukemias, prostate cancer, and bladder cancer (Hanahan and Weinberg, 2000; Jacks and Weinberg, 1998; Scherr, 1996). Increased expression of RB1 mRNA has been reported for many human colon tumor tissues and human colorectal cancer cell lines, breast cancers, and bladder cancers (Botos et al., 2002; Chatterjee et al., 2004; Gope et al., 1990, 1991; Gope and Gope, 1991, 1993, 2003; Thomas et al., 2005; Yamamoto et al., 1999). A recent report indicates variable levels, that is, both up- and down-regulation of RB1 mRNA in human VS (Lasak et al., 2002). A 2–5-fold increase in the RB1 mRNA level was found in large VS tumors of 3–5 cm in diameter (Table 11.2). The observed downregulation of the *RB1* gene expression in two out of eight tumors studied by Lasak et al. (2002) could be due to the size of the tumors used in their study, which were 1.5–3 cm in diameter. In addition, because of the small size of the tumors, Lasak et al. were able to identify the normal vestibular nerve tissues and used them as controls. In patients with large tumors of 3-cm diameter or more, the eighth cranial nerve or the vestibular branch of this nerve was not separately visible. Therefore, the normal diploid cell line WI38 is used as control for Northern and Western blot analysis. Use of WI38 cell line as a control is acceptable when the normal controls are not available (Fitch et al., 2003; Gope et al., 1991; Ma et al., 2003; Nowak et al., 2002). It is important to note that the RB1 mRNA expression in the WI38 cell line was 2-fold higher compared with that of the pooled normal

TABLE 11.2
Status of the *RB1* Gene in NF2 Tumor

Parameter	Percentage of Cases
Informative (for RB1 polymorphism)	75%
LOH at the RB1 locus	25%
Increase in RB1 mRNA level	
1–2-fold	5%
2–3-fold	50%
3–4-fold	30 %
4–5-fold	15%
Increase in pRb level	100%
2–5-fold	
Level of phosphorylated pRb compared with total pRb	50% and above

eighth cranial nerve from autopsy cases. The higher level of expression of the *RB1* gene (Table 11.2) could be correlated to the antiapoptotic function of this gene, which could aid in the progression of these tumors (Gope et al., 1991; Thomas et al., 2005; Yamamoto et al., 1999). Thus, *RB1* gene expression could be used as a proliferative index for NF2 tumor.

11.2.5 Level and Phosphorylation Status of pRb in VS Tumors

Wild-type pRb acts as a survival factor, as is evident from the massive cell death observed in pRb-deficient mice in tissues where pRb is normally highly expressed (Lee et al., 1992). Several lines of evidence suggest that the increased level of pRb may also inhibit cells from undergoing apoptosis (Thomas et al., 2005; Yamamoto et al., 1999). Previous reports from colon tumors, cultured colon tumor cell lines, breast cancer, and bladder cancer indicated a tumor-specific increase in the level of pRb and the phosphorylated form of pRb (Botos et al., 2002; Chatterjee et al., 2004; Gope and Gope, 1991; Yamamoto et al., 1999). A positive role for pRb in malignant transformation is also implied in human colon cancer and bladder cancer (Gope and Gope, 1991; Lane, 1992). pRb has the ability to suppress cell proliferation, and this activity is controlled by its cell-cycle–dependent phosphorylation (Hanahan and Weinberg, 2000; Jacks and Weinberg, 1998; Lane, 1992; Scherr, 1996). There are 16 potential phosphorylation sites in pRb, and each phosphorylation site has a different function with respect to binding to other proteins (Chatterjee et al., 2004; Ma et al., 2003; Scherr, 1996; Taya et al., 1989). The RB1 kinases are responsible for cell-cycle–dependent phosphorylation of pRb (Hanahan and Weinberg, 2000; Jacks and Weinberg, 1998; Lane, 1992; Ma et al., 2003; Scherr, 1996; Taya et al., 1989). It has been found that phosphorylation of sites in the C terminus occurs efficiently in every cell cycle and regulates proliferation. However, high CDK2 activity promotes phosphorylation of Ser^{567} by inducing an intermolecular interaction that leads to release of E2F, degradation of pRb, and susceptibility to apoptosis. Cancer cells can use this as a survival strategy and exploit this dual role of pRb by phosphorylating sites that regulate tumor suppression but avoiding phosphorylation of Ser^{567} and consequent apoptotic stimulus (Gouyer et al., 1994; Haas-Kogan et al., 1995; Ma et al., 2003). Altering the phosphorylation status of pRb can complement the increase in the expression of this gene. In a single study, downregulation of retinoblastoma CDK2 was reported in seven out of eight NF2 cases (Lasak et al., 2002). During the normal cell cycle, CDK2 is involved in phosphorylation of pRb during the transition of cells from the G_1 to the S phase (Hanahan and Weinberg, 2000; Jacks and Weinberg, 1998; Lane, 1992; Scherr, 1996; Taya et al., 1989). Lasak et al. (2002) have also reported that one tumor of 3 × 3 cm, which was the largest used in their study, did not show downregulation of CDK2. Therefore, it appears that the VS smaller than 3 × 3 cm may show CDK2 downregulation, and it may not be a general event for all VS, especially when the tumors are larger than 3 × 3 cm in size. Results from large-size tumors samples appear to be different than that obtained from smaller ones (Lasak et al., 2002) (Table 11.1). Results obtained from larger tumors show a 2.5–5-fold increase in the pRb level and an increase in the percentage of phosphorylated form of pRb in all the VS (Table 11.2). The higher level of the phosphorylated form of pRb in NF2 tumors indicates that most cells in these large tumors are in the cycling phase (Hanahan and Weinberg, 2000; Jacks and Weinberg, 1998; Lane, 1992; Scherr, 1996), and it is also an indication of the presence of cells in an undifferentiated phase (Taya et al., 1989). It has been shown that overexpression of KIP1, which predominantly inhibits CDK2, does not cause growth arrest in colon cancer cells, indicating that CDK2 inhibition is dispensable. In addition, decreased CDK2 levels failed to induce growth arrest in colon cancer cells, cervical cancer, and osteosarcoma cells (Bartek et al., 1997; Chatterjee et al., 2004). A similar mechanism may be involved in human NF2 tumors. It is conceivable that the phosphorylation sites of pRb could be different in VS than those reported for the normal cell cycle (Hanahan and Weinberg, 2000; Jacks and Weinberg, 1998; Lane 1992; Scherr, 1996; Thomas et al., 2005; Yamamoto et al., 1999). In such a case, the pRb could be phosphorylated through pathways other than the pRb-CDK2 pathway in NF2 tumors. A recent report showed that the cells that lack merlin had increased levels of cyclin D1, which is essential to

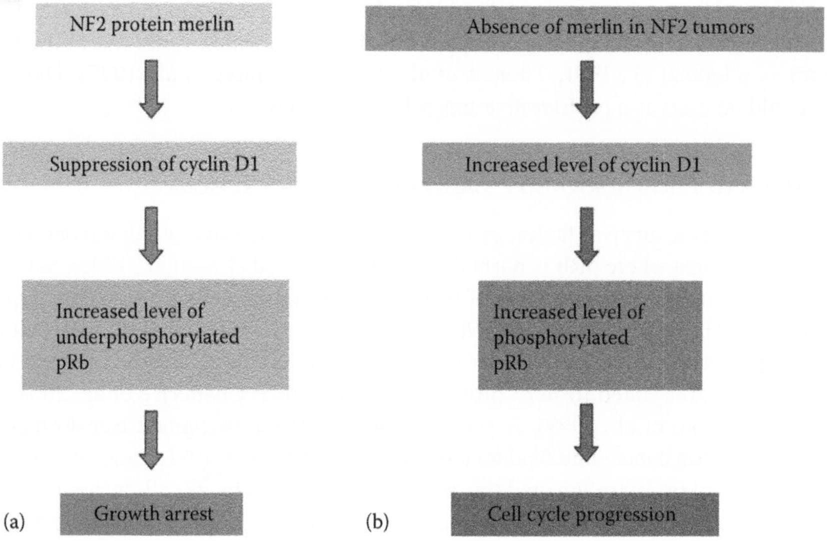

FIGURE 11.7 (a) In the normal cells, merlin inhibits cyclin-D1 expression, which results in an increased level of underphosphorylated pRb and leads to growth arrest. (b) There is no functional merlin protein in NF2 tumors, which results in the increased level of cyclin D1 and increased level of phosphorylated form of pRb. This leads to cell-cycle progression and tumor formation.

phosphorylate pRb. Higher levels of phosphorylated forms of pRb were also present in these cell lines (Xiao et al., 2005). These results are in agreement with the data where higher levels of the phosphorylated form of pRb were reported in human NF2 tumors that lack merlin (Thomas et al., 2005). The exact sites of phosphorylation of pRb in these cell lines and tumors need to be further investigated to understand the possible function of pRb in these tumors and cell lines.

11.3 HYPOTHESIS

Hyperphosphorylation of pRb could inactivate the *RB1*-mediated growth suppression and play an important role in the development of human NF2 tumors (Thomas et al., 2005). This hypothesis was further supported by other reports where merlin has been shown to have an inhibitory role in cell-cycle progression by repressing cyclin-D1 expression. Using cell lines that lack merlin, Xiao et al. (2005) found that cyclin-D1 expression is increased in the absence of merlin. Cyclin D1 is essential for phosphorylation of pRb, and phosphorylated pRb is known to release the cells from a resting phase (G_0) to early cycling phase (G_1) (Bartek et al., 1997; Ma et al., 2003; Taya et al., 1989; Tetsu and McCormick, 2003). Because the NF2 tumors do not have functional merlin (Rouleau et al., 1993), such a condition could lead to an elevated level of cyclin D1 and increased level of hyperphosphorylated pRb (Xiao et al., 2005). In addition to helping cells through cell cycle, hyperphosphorylated pRb is also known to inhibit apoptosis or programmed cell death (Ma et al., 2003). Thus, in the absence of the NF2 protein merlin, the pRb pathway could be deregulated in NF2 tumors, resulting in an increased level of the phosphorylated form of pRb (Figure 11.7). Such an increase in the level of hyperphosphorylated pRb in turn could play a major role in the development of human NF2 tumors, both by advancing cells through the cell cycle and by inhibiting apoptosis (Mawrin et al., 2002; Thomas et al., 2005; Xiao et al., 2005).

11.4 CONCLUSION

The normal cell-cycle mechanism is impaired during tumorigenesis (Classon and Harlow, 2002; Cheng et al., 2000; Lane, 1992; Levine, 1997). Alteration of *RB1* gene structure, gene expression, and phosphorylation of pRb indicate that this gene could play an important role in VS tumors, and its effect could occur at various levels. The observed LOH indicates the genetic instability in the patients with VS. The observed increase in RB1 mRNA and pRb indicates an antiapoptotic function of this gene, which could be involved in the progression of VS. The increased percentage of phosphorylation indicates that pRb could be involved in maintaining the cells in an undifferentiated, proliferating stage that could lead to tumor progression (Thomas et al., 2005; Xiao et al., 2005). Further studies are required to identify alternate kinase pathways, if any, other than CDK2, that are involved in the phosphorylation of pRb and maintenance of a higher percentage of phosphorylated pRb in human VS. It would be interesting to know the pRb phosphorylation sites in the normal Schwann cells of vestibular nerve and in the vestibular schwannomas.

It has been suggested that an increased level of the protein product of the *RB1* gene, pRb, and an increase in its level of phosphorylation could play important roles in the development of human colorectal carcinomas (Gope et al., 1990, 1991; Gope and Gope, 1991, 1993, 2003). It has been suggested that pRb has an important role in the development of human NF2 tumors (Thomas et al., 2005). Many groups have reported hyperphosphorylation of pRb in various human cancers such as breast cancers, malignant melanomas, acinic cell carcinomas, etc. (Loeb, 2001; Mawrin et al., 2002; Roesch et al., 2005; Yamamoto et al., 1999).

Determining the role of the *RB1* gene at a molecular level in human tumors will help us better understand the mechanism of cancer initiation and progression. Many experiments have revealed that reintroduction of the wild-type *TS* genes into the tumor cells can revert the tumorigenesis. Although the inactivation of *TS* gene is known to cause cancer, it is also known that the overexpression of the normal TS protein could also be tumorigenic because of altered posttranslational modifications. Such modification could change TS function of the protein products. The underlying reasons in such a role reversal of *TS* genes are not yet fully understood, and more intensive research is required in these areas. Further understanding of the deregulation pathway of *TS* gene will help us develop new drugs and novel diagnostic and therapeutic procedures against cancer.

INTERNET RESOURCES

http://www.hei.org
http://www.ctf.org

REFERENCES

Adeloye, A. (1979). Neoplasms of the brain in the African. *Surg. Neurol.* 11: 247–255.

Baker, D. J., Weller, R. O., Garfield, J. S. (1976). Epidemiology of primary tumors of the brain and spinal cord: A regional survey in Southern England. *J. Neurol. Neurosurg. Psychiatry* 39: 290–296.

Bartek, J., Bartkova, J., Lukar, J. (1997). The retinoblastoma protein pathway in cell cycle control and cancer. *Exp. Cell Res.* 237: 1–6.

Botos, J., Smith, R., III, Kochevar, D. T. (2002). Retinoblastoma function is a better indicator of cellular phenotype in cultured breast adenocarcinoma cells than retinoblastoma expression. *Exp. Biol. Med.* 227: 354–362.

Chatterjee, S. J., George, B., Goebell, P. J., et al. (2004). Hyperphosphorylation of pRb: A mechanism for RB tumor suppressor pathway inactivation in bladder cancer. *J. Pathol.* 203: 762–770.

Classon, M., Harlow, E. (2002). The retinoblastoma tumor suppressor in development and cancer. *Nat. Rev. Cancer* 2: 910–917.

Cheng, L., Rossi, F., Fang, W., et al. (2000). CDK2-dependent phosphorylation and functional inactivation of the pRb-related protein in pRb(–), p16IINK4A (+) tumor cells. *J. Biol. Chem.* 275: 30317–30325.

Cohen, M. E. (1995). Primary and secondary tumors of the central nervous system. In *Neurology in clinical practice: Vol. II* (pp. 1089–1107). Edited by Bradley, W. G., Daroff, R. B., Fenichel, G. M., et al. Boston: Butterworth-Heinemann.

Dastur, D. K., Lalitha, V. S., Prabhakar, V. (1968). Pathological analysis of intracranial space-occupying lesions in 1000 cases including children: Age, sex and pattern and the tuberculomas. *J. Neurol. Sci.* 6: 575–592.

Dyson, N. (1998). The regulation of E2F by pRB-family proteins. *Genes Dev.* 12: 2245–2262.

Fitch, M. E., Cross, I. V., Turner, S. J., et al. (2003). The DDB2 nucleotide excision repair gene product p48 enhances global genomic repair in p53 deficient human fibroblasts. *DNA Repair* 2: 819–826.

Gope, M. L., Chun, M., Gope, R. (1991). Comparative study of the expression of Rb and p53 genes in human colorectal cancers, colon carcinoma cells lines and synchronized human fibroblasts. *Mol. Cell. Biochem.* 107: 55–63.

Gope, R., Christensen, M. A., Thorson, A., et al. (1990). Increased expression of the retinoblastoma gene in human colorectal carcinomas relative to normal colonic mucosa. *J. Natl. Cancer Inst.* 82: 310–314.

Gope, R., Gope, M. L. (1991). Abundance and state of phosphorylation of retinoblastoma susceptibility gene product in human colon cancer. *Mol. Cell. Biochem.* 110: 123–133.

Gope, R., Gope, M. L. (1993). Effect of sodium butyrate on the expression of retinoblastoma (RB1) and p53 gene and phosphorylation of retinoblastoma protein in human colon tumor cell line HT29. *Cell. Mol. Biol.* 39: 589–597.

Gope, M. L., Gope, R. (2003). Effect of sodium butyrate on the methylation pattern of retinoblastoma (RB1) gene in human colon tumor cell line HT29. *Curr. Sci.* 84: 101–103.

Gouyer, V., Gazzeri, S., Brambilla, E., et al. (1994). Loss of heterozygosity at the RB locus correlates with loss of RB protein in primary malignant neuroendocrine lung carcinomas. *Int. J. Cancer* 58: 811–824.

Haas-Kogan, D. A., Kogan, S., Levi, D., et al. (1995). Inhibition of apoptosis by the retinoblastoma gene product. *EMBO J.* 14: 461–472.

Hanahan, D., Weinberg, R. A. (2000). Hallmarks of cancer. *Cell* 100: 57–70.

Hatakeyama, M., Weinberg, R. A. (1995). The role of RB in cell cycle control. *Prog. Cell Cycle Res.* 1: 9–19.

Hofseth, L. J., Hussain, S. P., Harris, C. C. (2004). p53: 25 years after its discovery. *Trends Pharmacol. Sci.* 25: 177–181.

Jacoby, L. B., Jones, D., Davis, K., et al. (1997). Molecular analysis of the NF2 tumor-suppressor gene in schwannomatosis. *Am. J. Hum. Genet.* 61: 1293–1302.

Jacks, T., Fazeli, A., Schmitt, E. M., et al. (1992). Effects of an Rb mutation in the mouse. *Nature* 359: 295–300.

Jacks, T., Weinberg, R. A. (1998). The expanding role of cell cycle regulators. *Science* 280: 1035–1036.

Kim, H., Kwak, N. J., Lee, J. Y., et al. (2004). Merlin neutralizes the inhibitory effect of Mdm2 on p53. *J. Biol. Chem.* 279: 7812–7818.

Kissil, J. L., Wilker, E. W., Johnson, K. C., et al. (2003). Merlin, the product of the NF2 tumor suppressor gene, is an inhibitor of the p21-activated kinase, Pak1. *Mol. Cell* 12: 841–849.

Knudson, A. G., Jr. (1971). Mutation and cancer: Statistical study of retinoblastoma. *Proc. Natl. Acad. Sci. U.S.A.* 68: 820–823.

Kurland, L. T., Schoenberg, B. S., Annegers, J. F., et al. (1982). The incidence of primary intracranial neoplasms in Rochester, Minnesota 1935–1977. *Ann. N.Y. Acad. Sci.* 381: 6–16.

Lane, D. P. (1992). p53, guardian of the genome. *Nature* 358: 15–16.

Lasak, J. M., Welling, D. B., Akhmametyeva, E. M., et al. (2002). Retinoblastoma-cyclin-dependent kinase pathway deregulation in vestibular schwannomas. *Laryngoscope* 112: 1555–1561.

Lee, E. Y., Chang, C. Y., Hu, N., et al. (1992). Mice deficient for Rb are nonviable and show defects in neurogenesis and haematopoiesis. *Nature* 359: 288–294.

Lee, W. H., Bookstein, R., Hong, F., et al. (1987). Human retinoblastoma susceptibility gene: Cloning, identification, and sequence. *Science* 235: 1394–1399.

Levine, A. J. (1997). p53, the cellular gatekeeper for growth and division. *Cell* 88: 323–331.

Levy, L. F., Auchterlonie, W. C. (1975). Primary cerebral neoplasms in Rhodesia. *Int. Surg.* 60: 286–293.

Lie, T., Zhu, E., Wang, L., et al. (2005). Abnormal expression of Rb pathway-related proteins in salivary gland acinic cell carcinoma. *Hum. Pathol.* 36: 962–970.

Loeb, L. A. (2001). A mutator phenotype in cancer. *Cancer Res.* 61: 3230–3239.

Ma, D., Zhou, P., Harbour, J. W. (2003). Distinct mechanisms regulating the tumor suppressor and antiapoptotic functions of Rb. *J. Biol. Chem.* 278: 19358–19366.

Martuza, R. L., Eldridge, R. (1988). Neurofibromatosis 2 (bilateral acoustic neurofibromatosis). *N. Engl. J. Med.* 318: 684–688.

Mawrin, C., Kirches, E., Boltze, C., et al. (2002). Immunohistochemical and molecular analysis of p53, RB, and PTEN in malignant peripheral nerve sheath tumors. *Virchows Arch.* 6: 610–615.

Nowak, M. A., Komarova, N. L., Sengupta, A., et al. (2002). The role of chromosomal instability in tumor initiation. *Proc. Natl. Acad. Sci. U.S.A.* 99: 16226–16231.

Roesch, A., Becker, B., Meyer, S., et al. (2005). Overexpression and hyperphosphorylation of retinoblastoma protein in the progression of malignant melanoma. *Mod. Pathol.* 18: 565–572.

Rouleau, G. A., Merel, P., Lutchman, M., et al. (1993). Alteration in a new gene encoding a putative membrane-organizing protein causes neuro-fibromatosis type 2. *Nature* 363: 515–521.

Ruberti, R. F., Poppi, M. (1971). Tumors of the central nervous system in the Africans. *East Afr. Med. J.* 48: 576–584.
Scherr, C. J. (1996). Cancer cell cycles. *Science* 274: 1672–1677.
Stangerup, S. E., Tos, M., Caye-Thomasen, P., et al. (2004). Increasing annual incidence of vestibular schwannomas and age at diagnosis. *J. Laryngol. Otol.* 118: 622–627.
Taya, Y., Yasuda, H., Kamijo, M., et al. (1989). In vitro phosphorylation of the tumor suppressor gene RB protein by mitosis-specific histone H1 kinase. *Biochem. Biophys. Res. Commun.* 164: 580–586.
Tetsu, O., McCormick, F. (2003). Proliferation of cancer cells despite CDK2 inhibition. *Cancer Cell* 3: 233–245.
Thomas, R., Prabhu, P. D., Mathivanan J., et al. (2005). Altered structure and expression of RB1 gene and increased phosphorylation of pRb in human vestibular schwannomas. *Mol. Cell. Biochem.* 271: 113–121.
Tos, M., Stangerup, S. E., Caye-Thomasen, P., et al. (2004). What is the real incidence of vestibular schwannoma? *Arch. Otolaryngol. Head Neck Surg.* 130: 216–220.
Wen-qing, H., Shi-ju, Z., Qing-sheng, T., et al. (1982). Statistical analysis of central nervous system tumors in China. *J. Neurosurg.* 56: 555–564.
Xiao, G. H., Gallagher, R., Shetler, J., et al. (2005). The NF2 tumor suppressor gene product, merlin, inhibits cell proliferation and cell cycle progression by repressing cyclin D1 expression. *Mol. Cell. Biol.* 25: 2384–2394.
Xu, H. J., Cairns, P., Hu, S. X., et al. (1993). Loss of RB protein expression in primary bladder cancer correlates with loss of heterozygosity at the RB locus and tumor progression. *Int. J. Cancer* 53: 781–784.
Yamamoto, H., Soh, J.-W., Monden, T., et al. (1999). Paradoxical increase in retinoblastoma protein in colorectal carcinomas may protect cells from apoptosis. *Clin. Cancer Res.* 5: 1805–1815.
Zacksenhaus, E., Jiang, Z., Chung, D., et al. (1996). pRb controls proliferation differentiation, and death of skeletal muscle cells and other lineages during embryogenesis. *Genes Dev.* 10: 3051–3064.

12 Forward Genetics Approach in Genomics

Functional Identification of Unknown Genes

Bishwanath Chatterjee
National Heart, Lung, and Blood Institute/National Institutes of Health
Bethesda, Maryland

CONTENTS

12.1 INTRODUCTION

The first draft of the human genome sequence was published in 2001 (Lander et al., 2001; Venter et al., 2001) and contains 90% of the 3 billion bases, with the completed sequence published in 2004 (International Human Genome Sequencing [IHGS] Consortium, 2004). The most important feature of the sequence was that the number of genes in the human genome was a lot smaller than the earlier estimate of 50,000–140,000. In the recent version of human genome built (NCBI 37.1, http://www.ncbi.nlm.nih.gov/projects/mapview/map_search.cgi?taxid=9606), there are 33,897 transcripts, 21,901 of which are annotated genes (including predicted genes). There are also 3031 nonprotein coding genes with an RNA product. According to estimates, the function of about 40% or more of the transcripts is not known. Besides, there are regulatory elements in the genome, like promoter and enhancer, that coordinate and organize the gene expression (Nobrega et al., 2003; Poulin et al., 2005; Woolfe et al., 2005; Prabhakar et al., 2006; Pennacchio et al., 2006). Efforts are being made to identify such elements and establish their role in gene expression (Visel et al., 2009), the role of these sequences have not been fully explored. Genome-wide association studies dealing with a particular disease or phenotype trait in human patients have often failed to indentify the actual mutation that is responsible for the disease. Also, identification of multiple hits for one disease makes the situation more complicated. One need is to get a comprehensive view of the identified region that includes genes, potential regulatory regions, evolutionary conserved regions, potential transcription factor binding sites, noncoding RNAs, and chromatin structure. It is therefore important to evaluate the role of all of the single-nucleotide polymorphisms in the indentified region to understand the disease process (Coassin et al., 2010). In the post genomic era, the challenge is to understand the function of all of the sequences in the human genome.

12.2 HOW TO MAKE FUNCTIONAL ANNOTATION OF THE MAMMALIAN GENOME

The traditional way to know the function of a gene is to disrupt the function of the gene in the animal model system. To reveal the predisposition of the gene and the role of the gene in the mechanism of disease progression, one needs to create a mutation in the gene of interest in a model system. The organisms of choice are mice; because of their smaller size, they are relatively easy to house and breed. Also, because of the availability of a huge genetic tool kit, which includes comprehensive information on genetically distinct inbred mouse strains (http://www.informatics.jax.org), the mouse is one of the most used organisms for modeling disease. To make such efforts successful worldwide, efforts are being made under the roof of the International Knockout Mouse Consortium (Collins et al., 2007). There are two different approaches to make a knockout; gene targeting and gene trapping. Knockout mice are made either by silencing the gene function by disruption using targeted homologous recombination leading to a phenotype (Thomas and Capecchi, 1986; Doetschman et al., 1988; Austin et al., 2004) or by gene trapping, which is a high-throughput random approach to introduce insertional mutations across the genome in mouse embryonic stem (ES) cells to disrupt gene function (Abuin et al., 2007; Gossler et al., 1989). Presently three major mouse knockout projects are underway guided under the umbrella of the International Gene Trap Consortium (http://www.genetrap.org). A trans-National Institutes of Health initiative called the Knockout Mouse Project (http://www.komp.org) was started in 2007; it aims to generate a comprehensive and public resource comprised of mouse ES cells containing a null mutation in every gene in the mouse genome. This resource will also generate conditional alleles for these known genes. The European Conditional Mouse Mutagenesis Program (http://www.eucomm.org) project is intended to produce 12,000 conditional gene trap mutations and 8000 conditional targeted mutations in mouse ES cells. The third project, the North American Conditional Mouse Mutagenesis Project (http://www.norcomm.org/index.htm) is a Canadian project whose mission is to contribute to the generation of a mouse ES cell resource with characterized mutations in every gene in the genome. Often, such approaches are unsuccessful because of embryonic death when the gene in question is required early in the development of the embryonic mice. To counter this problem, a conditional gene-targeting approach has been used where the particular gene is disrupted in the target tissue at a certain developmental state (Lewandoski, 2001). Although these mice are intended as models for human diseases, a lot of these models miss the actual human diseased locus, given that single-base substitution is largely the cause of most human diseases. For example, see the spectrum of *GATA4* mutations in congenital heart disease (Tomita-Mitchell et al., 2007), the *MKS1* and *MKS3* mutations in Mickel-Grover syndrome (Khaddour et al., 2007), or the *FBN1* mutations in Marfan syndrome (Turner et al., 2009).

12.3 REVERSE VERSUS FORWARD GENETICS APPROACH

Mutants are useful tools to directly understand the function of a gene without understanding its biological or cellular function. Finding mutations in human patients elucidated the function of a number of genes. This approach worked with phenotype-driven screen (disease diagnosis) followed

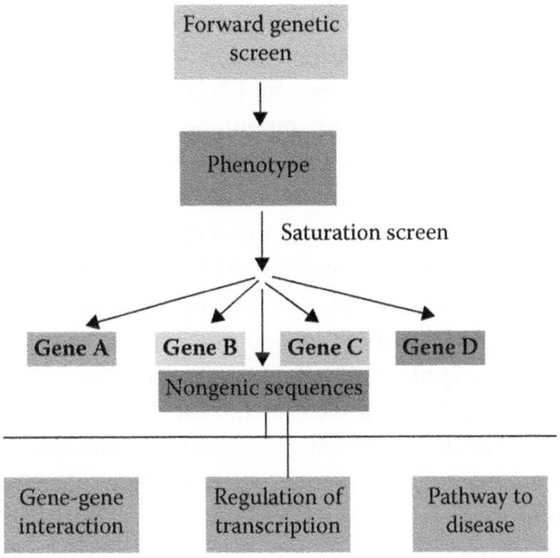

FIGURE 12.1 Saturated screen using forward genetic approach would identify all the genes leading to the disease, given a strict phenotype protocol is used for phenotyping.

by positional cloning. To prove the role of the gene in question, an animal model with the particular gene mutation is the key. A gene-driven mutagenesis approach relies on the availability of information for the particular target. This approach has been successful for researchers who have been studying a particular gene by generating total and/or conditional knockouts. Reverse genetics is the technology of choice for studying the function of a gene. The reverse genetic approach could also be problematical for the multigene families because the other member genes of the family could sometimes take over some of the lost function, partially or fully. In contrast, the forward genetics approach is a gene-independent approach and is open to all parts of the genome at random. This is more like a case of human genetic disease, where one looks for a particular phenotype and then tries to find the gene. The power of such a screen was well established in *Drosophila* during the early 1980s (Nüsslein-Volhard and Wieschaus, 1980). This phenotype-driven screen, however, needs a high-throughput robust phenotyping approach to be completely successful. This approach also gives a unique opportunity for a saturation screen, so that one can identify all of the genes that may be responsible for a particular phenotype. The screen can be continued until no new genes are found to be associated with a particular phenotype. Upon revealing all the genes, one can easily go forward to find out the relationship between all the genes that leads to a phenotype (Figure 12.1). Another advantage of mutation-based forward genetics is the creation of multiple mutations in a gene with similar or different phenotypes, depending upon what region of the gene in question is mutated (see *N*-ethyl-*N*-nitrosourea [ENU]-gene–driven approach later in the chapter).

12.4 *N*-ETHYL-*N*-NITROSOUREA MUTAGEN OF CHOICE

ENU causes random single–base-pair mutations. It acts directly by alkylating the nucleic acid without any metabolic processing required for activation (Justice et al., 1999; Singer and Dosanjh, 1990), which makes it a good reagent for creating mutations in mouse germlines (Russell et al., 1979). ENU can transfer its ethyl group to oxygen or nitrogen radicals in DNA, resulting in mispairing and base-pair substitution, if not repaired (Justice et al., 1999). Most common mutations occur in A-T base pairs. Most of the reported mutations are A-T or T-A transversions or A-T to G-C transitions (Noveroske et al., 2000; Barbaric et al., 2007). Although there are incidences of G-C to C-G transversion and G-C to A-T transitions, they are rarely seen (Augustin, 2005; Yu et al., 2004). ENU being a point mutagen, it can produce several different alleles in the same gene, including loss of function, gain of function, dominant negative, hypomorphs or simple deletion, or addition in protein of interest caused by alternative splicing. Although ENU is a mutagen of choice for random base changes, a study analyzing the actual mutations by ENU approach revealed that the mutations are of higher coding sequence length and higher exon number than the average for the mouse genome. This study also found that ENU-induced mutations were more likely to be found in genes that had a higher G + C content, and neighboring base analysis revealed that the identified ENU mutations were more often directly flanked by G or C nucleotides (Barbaric et al., 2007).

Mutagenesis to produce phenotype goes back to 1923 when x-rays were used to create mutations in mice (Little and Bagg, 1923), which was followed by use of chemical mutagens to introduce the mutations (Muller, 1927). Muller's study and subsequent studies clearly identified ENU as the best chemical mutagen (Ferreiro et al., 1995). To facilitate unbiased gene discovery, ENU became the chemical mutagen of choice, although most of the forward genetics approaches in the case of *Drosophila* used either ethyl methane sulfonate or insertional mutagenesis using transposable elements (FlyBase, http://flybase.org/; Campos et al., 2010; Wang et al., 2010). ENU mutagenesis was successfully tried in *Caenorhabditis elegans* (De Stasio et al., 1997; Flibotte et al., 2010). Use of ENU has been extended to vertebrates that include zebrafish and mice. Zebrafish models for developmental studies are important for optical clarity and external development of the embryos and are suitable for random mutagenesis and mutant screening (Streisinger et al., 1981; Knapik, 2000). A highly efficient protocol for ENU mutagenesis in zebrafish is in place to take advantage of the system (de Bruijn et al., 2009). With the rapid advancement in zebrafish genomics and the availability of genetic information for mapping, large-scale mutagenesis and mapping the mutation have been undertaken in recent past (Chen et al., 2001; Malicki et al., 2002; Rawls et al., 2003; Geisler et al., 2007). Despite the success of zebrafish mutagenesis screening caused by rapid phenotyping and the availability of thousands of mutant embryos in a short span of time, it may not be an ideal organism to truly mimic the human disease model. For example, the zebrafish heart has two chambers, with an atrioventricular valve between the single atrium and the pumping ventricle with a single outflow vessel, the truncus, and it does not have separate circulation for routing deoxygenated and oxygenated blood, making it not a good candidate organism model for human congenital heart disease (Lo et al., 2010). In addition, significant maternal contribution (Dosch et al., 2004; Wagner et al., 2004) and partial tetraploidy (Woods et al., 2000) sometimes make the situation complicated.

12.5 MOUSE AS MODEL SYSTEM FOR ENU MUTAGENESIS

The success of ENU mutagenesis screen in *Drosophila* and zebrafish lead to the mutagenesis efforts in mice. The mouse genome is close to the human genome (Bodenreider et al., 2005). For example, mice have left-right asymmetric four-chambered hearts and separate systemic versus pulmonary circulation—very similar to humans—and therefore are good candidates for modeling congenital heart disease (Lo et al., 2010). The mouse genome has been fully sequenced (Waterston et al., 2002) and fully assembled (Church et al., 2009). The availability of genetic information on several strains of inbred mice makes it an ideal system for doing mutagenesis and mapping of the mutant gene for a well-defined phenotype, even though it is a more complex organism and has relatively slower breeding cycles and small litter sizes, and it is time consuming to map the mutations compared with those of zebrafish. On the other hand, very detailed genetic information has been developed for inbred mice strains, making them ideal for mapping mutations. The availability of a cheaper second generation of sequencing technologies has enhanced the use of the mice in forward genetic research and made it possible to do whole-genome sequencing or sequence captured targeted mapped region sequencing (Summerer, 2009; Tewhey et al., 2009).

The history of mouse ENU mutagenesis tracks to the 1970s, when efforts were made to mutagenize mice (Russell et al., 1979). Subsequently, ENU mutagenesis was established as a tool of mutagenesis in mice because of the success of several groups in mapping and identifying the genes (Bode, 1984; Justice and Bode, 1988; McDonald et al., 1990; Vitaterna et al., 1994; Brunialti et al.,

1996). These studies opened a new era in mouse genetics, including efforts to do a large-scale mutagenesis in several countries including the United Kingdom, Germany, Japan, and the United States. During the past decade, several groups around the world have invested in producing mouse models for human disease that include the large-scale ENU mutagenesis projects.

12.6 ENU MUTAGENESIS APPROACH IN MICE

ENU is administrated as a series of intraperitoneal injections to male adult inbred mice; ENU then acts on spermatogonial stem cells (Brown et al., 2009). Effective dose for carrying out the mutagenesis depends upon genetic background of mice (Hitotsumachi et al., 1985; Justice et al., 2000; Weber et al., 2000). Immediately after the ENU administration, the mice become sterile for 4–8 weeks because of depletion of differentiated spermatogonia. After this period, the male could be bred to produce the offspring carrying the mutation.

ENU has an average high specific locus mutation frequency of 150×10^5 per locus with a maximum dose of 4×100 mg/kg (Hitotsumachi et al., 1985), which can be extrapolated to 30–50 functional mutations in coding sequences. On average, there is one mutation per 1–1.5 Mb, which includes noncoding mutations including mutations in regulatory and noncoding sequences (Brown et al., 2009). It is important that all other mutations except for the one that is responsible for phenotype is eliminated to call it a truly causative mutation for the given phenotype. Hence, the G_1 male has to be bred for at least for 8–10 generations to get rid of most of the mutations except for the mutations that are tightly linked (within a 1–2-Mb region). Often closely linked mutations are hard to separate, resulting in a false alarm, even though sometimes such mutations are in an intergenic region and can be expected not to have any role in phenotype (Zhang et al., 2009).

12.7 ROLE OF PHENOTYPING IN ENU MUTAGENESIS SCREEN

One of the obstacles of the phenotype-driven approach is to generate new mutant resources and then establish the relationship between phenotype and genotype and to discover the function of the human genome by modeling human disease. Once the males are treated with ENU, the relevant phenotype has to be identified, which is followed by identification of the mutation using various genetic approaches. The mice can be bred after recovery from a period of sterility upon ENU treatment. Upon breeding, the offspring of the ENU-treated male produces mice with mutations. To get an effective role of certain mutation in the phenotype of interest, one has to make sure that the other mutations are eliminated. Therefore, it is important that the phenotype is being tracked with each generation of breeding. A fundamental question remains: how do you define the phenotype of interest, and what are the parameters for calling it a phenotype?

Technically, phenotype is comprised of the complex biological output of genetic alleles and the environment (Justice, 2008; Brown et al., 2009). Some defects are easy to recognize visually, although that may sometime require careful necropsy, sectioning, or skeletal preparation, which may be time-consuming for some laboratories. The defects in certain biochemical, metabolic, or physiological parameters/pathways need more rigorous protocol, given the range of values that are basic to the wild-type population. Phenotype can also be measured at molecular or cellular level. Lot of progress has been made in recent past in phenotyping mice. The Mouse Phenome Database (http://phenome.jax.org) was created to coordinate/access data. The phenotypes could be broadly categorized as developmental, metabolic, physiological, cellular behavioral, or suppressor/enhancer phenotype (Stottmann and Beier, 2010). Conversely, phenotyping pipelines can be developed using disease states such as cardiovascular, neurological, cancer, or susceptibility to infectious diseases. In addition for ENU mutagenesis efforts to be successful, a fast phenotyping protocol should be in place. For example, a robust protocol using in utero echocardiography was developed to screen the cardiovascular defects in a large ENU mutagenesis screen (Yu et al., 2004, 2008; Shen et al., 2005; Lo et al., 2010), and robust phenotyping strategies have been developed to assess behavioral phenotypes (Crawley, 2008). The International Mouse Phenotyping Consortium was created to coordinate phenotyping platforms all over the world (Justice, 2008; Abbott, 2010).

12.8 BREEDING SCHEME: IDENTIFICATION OF RECESSIVE VERSUS DOMINANT MUTATION

The beauty of ENU mutagenesis is that it can generate both recessive and dominant mutation. The dominant phenotype could be easily achievable through breeding of the ENU-treated male mouse with wild-type female mice (G_0 generation). The resulting G_1 mice could be easily scanned for the dominant phenotype and further identification of the mutation (Figure 12.2). For recessive phenotype, the G_1 male is again mated with a wild-type female to generate the G_2 males and females. Intercross between the G_2 male and female or backcross of the G_2 female with G_1 male will produce recessive mutant in a Mendelian ratio. It is therefore possible to screen both a dominant and a recessive trait from one breeding scheme (Figure 12.2). A recent study with mutation in the *Kit* gene found distinct gonadal phenotypes for heterozygote and homozygous mutation in accordance with dominant and recessive phenotypes (Wu et al., 2010). The breeding scheme could be modified if someone wants to trap mutations in a particular region using crosses involving balancer chromosomes (Hentges and Justice, 2004; Boles et al., 2009; Probst and Justice, 2010) or make crosses incorporating chromosomal deletion (Rinchik et al., 1990).

12.9 IDENTIFICATION OF THE MUTATED GENE

Once the phenotype is ascertained, one has to the find the actual mutation that may be responsible for the phenotype. The traditional approach involves mapping of the mutation to the chromosomal region, and then another genetic approach can be used to finally locate the gene. Because the ENU mutations are carried out in an inbred mouse strain, a modified breeding scheme would be necessary,

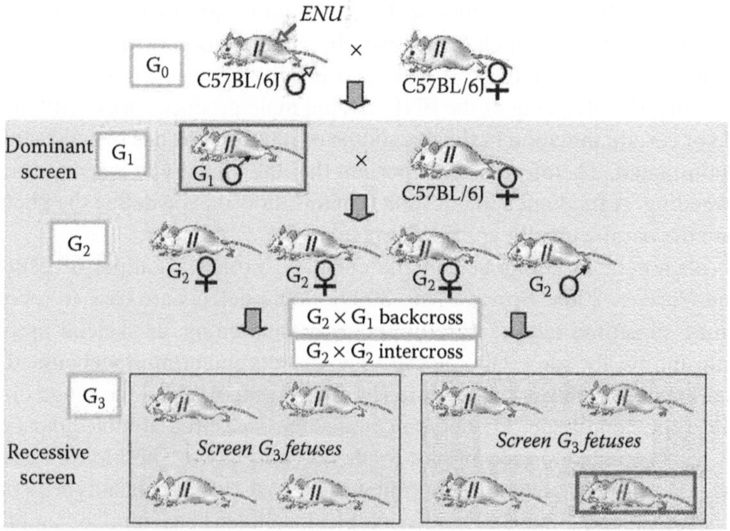

FIGURE 12.2 Breeding scheme for isolation of dominant and recessive traits. An ENU-mutagenized G_0 male was bred with female mice to produce G_1 male offspring. Some of these G_1 offspring have one copy of the mutated gene and can be evaluated for dominant traits. The G_1 males are further bred to produce G_2 females. The G_2 females are then either backcrossed with their G_1 fathers or intercrossed to produce G_3 offspring, some of which will be homozygous for ENU-induced mutations. Those with the phenotype could be evaluated for recessive traits.

involving another inbred strain of mice that could easily be distinguished with genomic resources such as microsatellite or SNP markers. For mapping, the G_2 carrier male (in strain background A) is usually bred with a normal female from another strain background (strain background B), and the resulting G_3 litter is intercrossed to obtain G_4 offspring (Figure 12.3). DNA isolated from G_4 offspring with a phenotype is used for mapping analysis. Whereas this type of breeding could be useful also to remove the unwanted mutation from the specific one that generates the phenotype, an introduction of another strain of mice could complicate the phenotyping assessment of the mutant caused by genetic background effect.

12.9.1 Mapping

The mapping of the mutation is done by tracking segregation of the strain A chromosomal region in G_4 mutants. ENU-induced mutation is expected to lie in a chromosome interval that is consistently homozygous for strain A in all of the affected fetuses. Before the discovery of SNP markers, the mapping process was done through microsatellite markers, which are genetic markers of choice for most populations and are defined as simple sequence repeats that are present as tandem in a genome. Because of differences in the repeat length, these markers could be used to differentiate the two genomes using PCR primers that can amplify the whole repeat region at a particular part of the genome (Iakoubova et al., 2000; Witmer et al., 2003). These markers are usually well suited for mapping distantly related strains. Because of the repetitive nature of such markers, sometimes they are not ideal for every part of the mouse genome. Discovery of SNP markers has changed the scenario of mapping because of the more than 100-fold increase in the number of markers available for these studies, and new markers are being added all the time (http://www.ncbi.nlm.nih.gov/projects/SNP/MouseSNP.cgi). Although these resources are very abundant, researchers can actually find difficulties in finding a marker for a certain cross in a defined region of the chromosome, making the mapping process difficult. Sometimes a combination of both microsatellite and SNP markers is useful in obtaining a better map interval (Zhang et al., 2009).

The mapping and the final identification of a mutation requires step-by-step advancement, depending upon the availability of number of mutants (recombinants) and the number of markers

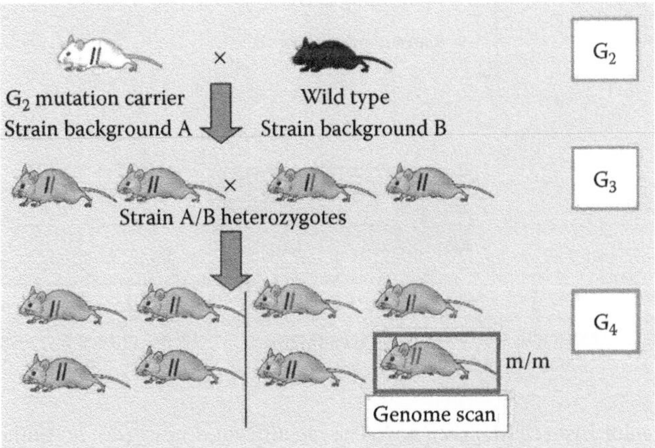

FIGURE 12.3 Breeding scheme for mapping. An ENU-mutagenized G_1 male was bred with female mice in the same background (strain background A, same as G_0 male treated with ENU) to produce G_2 male offspring. The G_2 offspring are further bred with another stain of mice (strain background B) to produce strain A/B heterozygotes G_3. Some of these G_3 will carry one copy of the ENU mutation. G_3 offspring were intercrossed to produce G_4 offspring, some of which will be homozygous for ENU-induced mutations. DNA from those offspring with a phenotype will be analyzed for mapping (genome scan) and further identification of the gene.

(Silver, 1995). A whole-genome mapping panel consisting of 40–60 markers with 10–15 recombinants could easily identify the chromosome with the mutation. In the next step, 10–15 chromosome-specific markers are used to identify the broader area of the mutation in the specific chromosome (50–100 Mb). Recombinants with the break point with a shorter mapped region and new recombinants that could be available will be further analyzed with a high concentration of markers around the break points. The process is repeated until the mapped region goes down to a 5–10 Mb region (Figure 12.4). At this point, bioinformatics is used to analyze the mapped region, such as genes, microRNAs, and other noncoding RNAs, known enhancers, etc. Assimilation of such information with the known phenotypes of the mutated genes in the mapped area could identify the gene using a candidate approach (Figure 12.4).

12.9.2 Finding the Gene

With the mapped interval at less than 5 Mb, a gene-dense area of chromosome may contain 100–200 genes and other RNAs, which could be a tough challenge to address. Several different approaches could give identification of the actual mutation. The candidate approach is the first one that should be tried, given that it can immediately identify the gene even with short mapping or no mapping

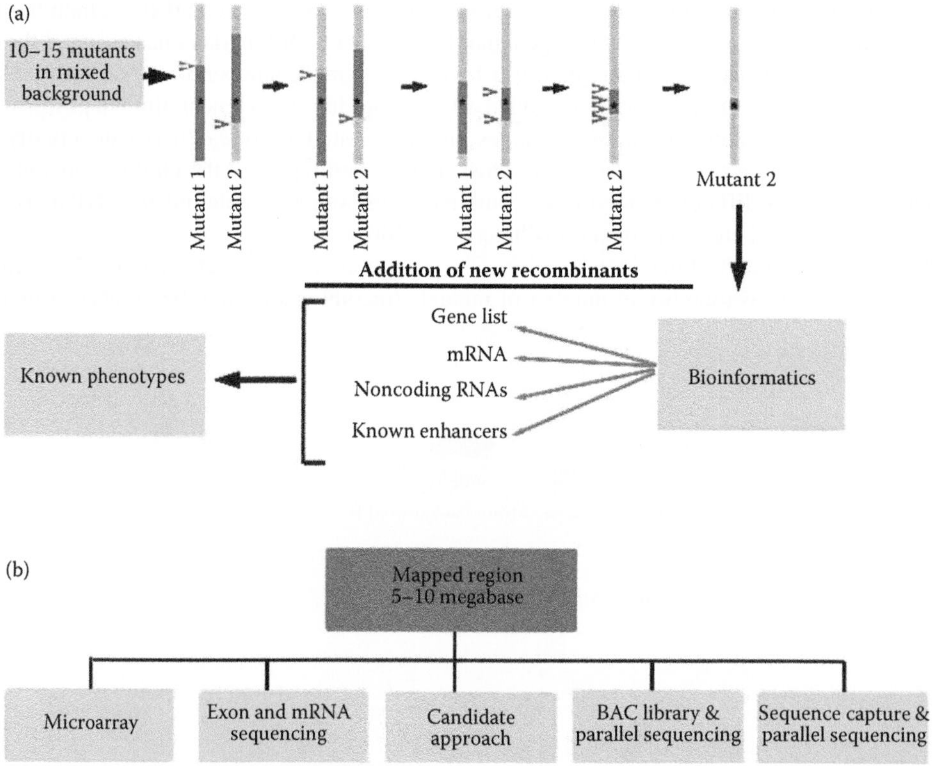

FIGURE 12.4 **(See color insert.)** Mapping and gene identification strategies. (a) Initially 10–15 mutants are used to identify the chromosome where the mutation resides, using 40–60 markers spread all over the genome. Upon successful identification of chromosome, increased concentration of markers is used at the breakpoints (blue arrowheads). The process is repeated with additional mutants and new markers as new breakpoints appear, until the mapped chromosomal regions are in the 5–10-Mb range. The available bioinformatics information is then to find if a candidate gene could be identified. (b) Gene identification methods. With the available bioinformatics tools for the mapped region, one can use several pipelines to identify the gene, depending upon the situation of each phenotype being analyzed.

(Yu et al., 2004). However, there is some chance that there could be another mutation within a short interval of 1 Mb or so that is hard to find or separate from the actual mutation, even after several generations of breeding. The second approach is to do high-throughput exon and cDNA sequencing. cDNA sequencing can identify the alternative splicing, if any, provided proper primers are used to identify them. These approaches are sometimes useful in identification of the gene, but in some cases they can fail to identify the gene because of mutation in promoter and/or enhancers/repressors or because of the presence of an unknown gene in the mapped interval (Zhang et al., 2009). Expression pattern changes caused by mutation in the promoter/enhancer/repressor could be identified by cDNA microarray analysis, leading to the sequencing of the noncoding areas of the gene to identify the actual mutation.

Because of availability of next-generation DNA sequencing (NGS) platforms (Metzker, 2010) and decreases in the cost of sequencing, it has been useful in recent past to identify the mutations associated with the phenotype. However, whole-genome sequencing is still a costlier approach to identify the mutation(s) associated with the phenotype of interest. To use the potential of NGS, approaches like making BAC library from mutant DNA and massively parallel sequencing have been used to identify the mutation (Zhang et al., 2009). Use of selection technologies (Teer et al., 2010) to concentrate the mapped region of interest followed by massive parallel sequencing will be the key to identifying the actual mutation (Figure 12.4).

12.10 PRESERVATION OF MUTATION RESOURCES

In the past two decades, huge resources of mutagenized mice have been produced, and it has been a big challenge to keep up with the growing resources. Many such resources are not analyzed because of a lack of interest or a lack of resources for analysis, funding, space for breeding, or colony management. These resources have to be preserved for future use. Embryo freezing has been used to keep up with this challenge (Pomeroy, 1991). Cryopreservation of oocytes and ovaries was also used as a tool for protecting the resources (Glenister and Thornton, 2000). Freezing embryos requires 500–1000 embryos from 20–50 mated females. The rate of live-born recovery from frozen embryos is often of concern. In mice, this rate can vary according to the method of freezing and the genotype of the embryos being frozen. In freezing mutant and inbred strains of mice, generally low live-born recoveries were observed, in some cases as low as 8 to 10% of the embryos initially frozen (Le, 1981; Mobraaten, 1986). In contrast, each male can give rise to 10–30 million spermatozoa for immediate freezing (Nakagata, 2000). Sperm cryopreservation techniques have been constantly updated since then (Takahashi and Liu, 2010), and this became the method of choice for preservation of the resources.

12.11 ENU GENE-DRIVEN ANALYSIS AND DNA RESOURCES

Parallel archiving of sperm and DNA from ENU mutagenesis screens can be used to get into gene-driven screens of the mutagenized mice. The archived DNA can be used to screen a mutation of choice using various techniques including DNA sequencing or heteroduplex detection methods (Gondo et al., 2009). Once a particular mutation has been identified, a mouse can be reproduced from the frozen sperm archive. This approach was successfully utilized in a ENU mouse mutagenesis program in the United Kingdom, with 2230 DNA samples for the identification of mutation in the *Cx26* (*Gjb2*) gene (Coghill et al., 2002). The authors further tested 6000 more individual mutants for 15 genes (whole gene or selected regions) and identified 15 new alleles in nine genes (Quwailid et al., 2004). These studies found the rate of functional mutation close to the predicted level (Brown et al., 2009). A gene-driven mutagenesis approach has been established in other laboratories (Augustin et al., 2005; Michaud et al., 2005; Sakuraba, et al., 2005; Takahasi et al., 2007). From these studies, a total of 40,000 G_1 mouse lines have been cryopreserved, carrying about 1.2×10^8 base substitutions. On average, the archive should contain 80 point mutations per gene,

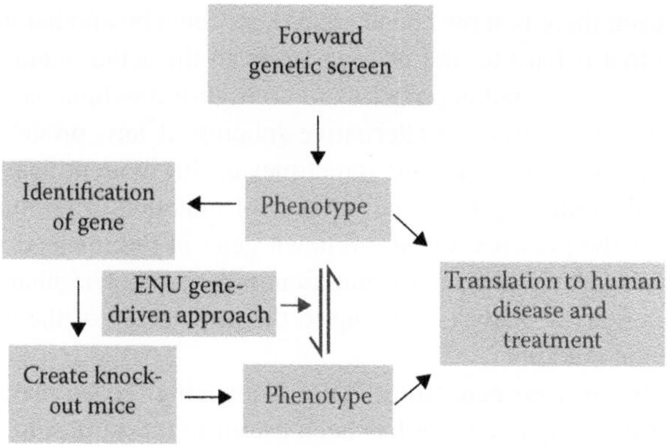

FIGURE 12.5 Advantage of forward genetic screen using ENU mutagenesis. Once a phenotype and the associated gene have been identified, a reverse genetic approach using knockout or an ENU-gene–driven approach can be used to fully dissect the function of the gene in question. Together, this phenotype-genotype information could be translated to human disease and treatment.

which could be large enough given the average gene size of 2 kb (Gondo, 2008). Although ENU as a mutagen may be biased in targeting certain bases depending upon structure and length of the gene (Barbaric et al., 2007), all of the allelic series of mutations in a particular gene may be hard to obtain, whereas some of the bases could be frequently found to be mutated with increases in the number of mutagenized G_1 male DNA.

12.12 CONCLUSION

In conclusion, ENU mutagenesis is a powerful tool in mice not only for elucidating function(s) of new gene(s) or other sequences in the genome, but also for gene-targeting studies with the availability of large archives of ENU-mutagenized G_1 male collections worldwide. This, together with the knockout and conditional knockout model approach, will dissect the function of all of the genes and nongenic regions in the genome (Figure 12.5). Using the combination of available resources on any new gene and recovery of serial array of mutation on the same gene can be used to dissect the function of the gene in detail. A large-scale phenotyping pipeline will also be useful to connect other genes that may not otherwise be known to interact with a gene of choice.

REFERENCES

Abbott, A. (2010). Mouse project to find each gene's role. *Nature* 465: 410.

Abuin, A., Hansen, G., Zambrowicz, B. (2007). Gene trap mutagenesis. *Handb. Exp. Pharmacol.* 178: 129–147.

Augustin, M., Sedlmeier, R., Peters, T., et al. (2005). Efficient and fast targeted production of murine models based on ENU mutagenesis. *Mamm. Genome* 16: 405–413.

Austin, C., Battey, J., Bradley, A., et al. (2004). The knockout mouse project. *Nat. Genet.* 36: 921–924.

Barbaric, I., Wells, S., Russ, A., et al. (2007). Spectrum of ENU-induced mutations in phenotype-driven and gene-driven screens in the mouse. *Environ. Mol. Mutagen.* 48: 124–142.

Bode, V. (1984). Ethylnitrosourea mutagenesis and the isolation of mutant alleles for specific genes located in the T region of mouse chromosome 17. *Genetics* 108: 457–470.

Bodenreider, O., Hayamizu, T., Ringwald, M., et al. (2005). Of mice and men: Aligning mouse and human anatomies. *AMIA Annu. Symp. Proc.* 2005: 61–65.

Boles, M., Wilkinson, B., Maxwell, A., et al. (2009). A mouse chromosome 4 balancer ENU-mutagenesis screen isolates eleven lethal lines. *BMC Genet.* 10: 1–12.

Brown, S., Wurst, W., Kühn, R., et al. (2009). The functional annotation of mammalian genomes: The challenge of phenotyping. *Annu. Rev. Genet.* 43: 305–333.

Brunialti, A., Harding, C., Wolff, J., et al. (1996). The mouse mutation sarcosinemia (sar) maps to chromosome 2 in a region homologous to human 9q33-q34. *Genomics* 36: 182–184.

Campos, I., Geiger, J., Santos, A., et al. (2010). Genetic screen in *Drosophila melanogaster* uncovers a novel set of genes required for embryonic epithelial repair. *Genetics* 184: 129–140.

Chen, J., van Bebber, F., Goldstein, A., et al. (2001). Genetic steps to organ laterality in zebrafish. *Comp. Funct. Genomics* 2: 60–68.

Church, D., Goodstadt, L., Hillier, L., et al. (2009). Lineage-specific biology revealed by a finished genome assembly of the mouse. *PLoS Biol.* 7: e1000112.

Coassin, S., Brandstätter, A., Kronenberg, F. (2010). Lost in the space of bioinformatic tools: A constantly updated survival guide for genetic epidemiology: The GenEpi Toolbox. *Atherosclerosis* 209: 321–335.

Coghill, E., Hugill, A., Parkinson, N. (2002). A gene-driven approach to the identification of ENU mutants in the mouse. *Nat. Genet.* 30: 255–256.

Collins, F., Rossant, J., Wurst, W. (2007). International Knockout Mouse Consortium: A mouse for all reasons. *Cell* 128: 9–13.

IHGS Consortium (2004). Finishing the euchromatic sequence of the human genome. *Nature* 431: 931–945.

Crawley, J. (2008). Behavioral phenotyping strategies for mutant mice. *Neuron* 57: 809–818.

de Bruijn, E., Cuppen, E., Feitsma, H. (2009). Highly efficient ENU mutagenesis in zebrafish. *Methods Mol. Biol.* 546: 3–12.

De Stasio, E., Lephoto, C., Azuma, L., et al. (1997). Characterization of revertants of unc-93(e1500) in *Caenorhabditis elegans* induced by N-ethyl-N-nitrosourea. *Genetics* 147: 597–608.

Doetschman, T., Maeda, N., Smithies, O. (1988). Targeted mutation of the Hprt gene in mouse embryonic stem cells. *Proc. Natl. Acad. Sci. U.S.A.* 85: 8583–8587.

Dosch, R., Wagner, D., Mintzer, K., et al. (2004). Maternal control of vertebrate development before the midblastula transition: Mutants from the zebrafish I. *Dev. Cell* 6: 771–780.

Ferreiro, J., Sierra, L., Comendador, M. (1995). Methodological aspects of the white-ivory assay of *Drosophila melanogaster*. *Mutat. Res.* 335: 151–1261.

Flibotte, S., Edgley, M., Chaudhry, I., et al. (2010). Whole-genome profiling of mutagenesis in *Caenorhabditis elegans*. *Genetics* 185: 431–441.

Geisler, R., Rauch, G., Geiger-Rudolph, S. (2007). Large-scale mapping of mutations affecting zebrafish development. *BMC Genomics* 8: 11.

Glenister, P., Thornton, C. (2000). Cryoconservation: Archiving for the future. *Mamm. Genome* 11: 565–571.

Gondo, Y. (2008). Trends in large-scale mouse mutagenesis: From genetics to functional genomics. *Nat. Rev. Genet.* 9: 803–810.

Gondo, Y., Fukumura, R., Murata, T., et al. (2009). Next-generation gene targeting in the mouse for functional genomics. *BMB Rep.* 42: 315–323.

Gossler, A., Joyner, A., Rossant, J., et al. (1989). Mouse embryonic stem cells and reporter constructs to detect developmentally regulated genes. *Science* 244: 463–465.

Hentges, K., Justice, M. (2004). Checks and balancers: Balancer chromosomes to facilitate genome annotation. *Trends Genet.* 20: 252–259.

Hitotsumachi, S., Carpenter, D., Russell, W. (1985). Dose-repetition increases the mutagenic effectiveness of N-ethyl-N-nitrosourea in mouse spermatogonia. *Proc. Natl. Acad. Sci. U.S.A.* 82: 6619–6621.

Iakoubova, O., Olsson, C., Dains, K. (2000). Microsatellite marker panels for use in high-throughput genotyping of mouse crosses. *Physiol. Genomics* 3: 145–148.

Justice, M. (2008). Removing the cloak of invisibility: Phenotyping the mouse. *Dis. Model. Mech.* 1: 109–112.

Justice, M., Bode, V. (1988). Genetic analysis of mouse t haplotypes using mutations induced by ethylnitrosourea mutagenesis: The order of T and qk is inverted in t mutants. *Genetics* 120: 533–543.

Justice, M., Carpenter, D., Favor, J. (2000). Effects of ENU dosage on mouse strains. *Mamm. Genome* 11: 484–488.

Justice, M., Noveroske, J., Weber, J., et al. (1999). Mouse ENU mutagenesis. *Hum. Mol. Genet.* 8: 1955–1963.

Khaddour, R., Smith, U., Baala, L. (2007). Spectrum of MKS1 and MKS3 mutations in Meckel syndrome: A genotype-phenotype correlation. *Hum. Mutat.* 28: 523–524.

Knapik, E. (2000). ENU mutagenesis in zebrafish: From genes to complex diseases. *Mamm. Genome* 11: 511–519.

Lander, E., Linton, L., Birren, B., et al. (2001). Initial sequencing and analysis of the human genome. *Nature* 409: 860–921.

Le, M. (1981). *The Jackson Laboratory genetics stocks resources redipository: Frozen stocks of laboratory animals*. New York: Gustav Fischer Verlag.

Lewandoski, M. (2001). Conditional control of gene expression in the mouse. *Nat. Rev. Genet.* 2: 743–755.

Little, C. C., Bagg, H. J. (1923). The occurrence of two heritable types of abnormality among the descendents of x-rayed mice. *Am. J. Roentgenol. Radiat. Therap.* 10: 975–989.

Lo, C., Yu, Q., Shen, Y. (2010). *Exploring genetic basis for congenital heart disease with mouse ENU mutagenesis: Heart development and regeneration.* Orlando, FL: Academic Press.

Malicki, J., Pujic, Z., Thisse, C., et al. (2002). Forward and reverse genetic approaches to the analysis of eye development in zebrafish. *Vision Res.* 42: 527–533.

McDonald, J., Bode, V., Dove, W., et al. (1990). The use of N-ethyl-N-nitrosourea to produce mouse models for human phenylketonuria and hyperphenylalaninemia. *Prog. Clin. Biol. Res.* 340C: 407–413.

Metzker, M. (2010). Sequencing technologies: The next generation. *Nat. Rev. Genet.* 11: 31–46.

Michaud, E., Culiat, C., Klebig, M. (2005). Efficient gene-driven germ-line point mutagenesis of C57BL/6J mice. *BMC Genomics* 6: 164.

Mobraaten, L. (1986). Mouse embryo cryobanking. *J. In Vitro Fert. Embryo Transf.* 3: 28–32.

Muller, H. (1927). Artificial transmutation of the gene. *Science* 66: 84–87.

Nakagata, N. (2000). Cryopreservation of mouse spermatozoa. *Mamm. Genome* 11: 572–576.

Nobrega, M., Ovcharenko, I., Afzal, V., et al. (2003). Scanning human gene deserts for long-range enhancers. *Science* 302: 413.

Noveroske, J., Weber, J., Justice, M. (2000). The mutagenic action of N-ethyl-N-nitrosourea in the mouse. *Mamm. Genome* 11: 478–483.

Nüsslein-Volhard, C., Wieschaus, E. (1980). Mutations affecting segment number and polarity in *Drosophila. Nature* 287: 795–801.

Pennacchio, L., Ahituv, N., Moses, A., et al. (2006). In vivo enhancer analysis of human conserved non-coding sequences. *Nature* 444: 499–502.

Pomeroy, K. (1991). Cryopreservation of transgenic mice. *Genet. Anal. Tech. Appl.* 8: 95–101.

Poulin, F., Nobrega, M., Plajzer-Frick, I., et al. (2005). In vivo characterization of a vertebrate ultraconserved enhancer. *Genomics* 85: 774–781.

Prabhakar, S., Poulin, F., Shoukry, M., et al. (2006). Close sequence comparisons are sufficient to identify human cis-regulatory elements. *Genome Res.* 16: 855–863.

Probst, F., Justice, M. (2010). Mouse mutagenesis with the chemical supermutagen ENU. *Methods Enzymol.* 477: 297–312.

Quwailid, M., Hugill, A., Dear, N. (2004). A gene-driven ENU-based approach to generating an allelic series in any gene. *Mamm. Genome* 15: 585–591.

Rawls, J., Frieda, M., McAdow, A. (2003). Coupled mutagenesis screens and genetic mapping in zebrafish. *Genetics* 163: 997–1009.

Rinchik, E., Carpenter, D., Selby, P. (1990). A strategy for fine-structure functional analysis of a 6- to 11-centimorgan region of mouse chromosome 7 by high-efficiency mutagenesis. *Proc. Natl. Acad. Sci. U.S.A.* 87: 896–900.

Russell, W., Kelly, E., Hunsicker, P., et al. (1979). Specific-locus test shows ethylnitrosourea to be the most potent mutagen in the mouse. *Proc. Natl. Acad. Sci. U.S.A.* 76: 5818–5819.

Sakuraba, Y., Sezutsu, H., Takahasi, K., et al. (2005). Molecular characterization of ENU mouse mutagenesis and archives. *Biochem. Biophys. Res. Commun.* 336: 609–616.

Shen, Y., Leatherbury, L., Rosenthal, J., et al. (2005). Cardiovascular phenotyping of fetal mice by noninvasive high-frequency ultrasound facilitates recovery of ENU-induced mutations causing congenital cardiac and extracardiac defects. *Physiol. Genomics* 24: 23–36.

Silver, L. (1995). *Mouse genetics: Concept and application*, New York: Oxford University Press.

Singer, B., Dosanjh, M. (1990). Site-directed mutagenesis for quantitation of base-base interactions at defined sites. *Mutat. Res.* 233: 45–51.

Stottmann, R., Beier, D. (2010). Using ENU mutagenesis for phenotype-driven analysis of the mouse. *Methods Enzymol.* 477, 329–348.

Streisinger, G., Walker, C., Dower, N., et al. (1981). Production of clones of homozygous diploid zebra fish (*Brachydanio rerio*). *Nature* 291: 293–296.

Summerer, D. (2009). Enabling technologies of genomic-scale sequence enrichment for targeted high-throughput sequencing. *Genomics* 94: 363–368.

Takahashi, H., Liu, C. (2010). Archiving and distributing mouse lines by sperm cryopreservation, IVF, and embryo transfer. *Methods Enzymol.* 476: 53–69.

Takahasi, K., Sakuraba, Y., Gondo, Y. (2007). Mutational pattern and frequency of induced nucleotide changes in mouse ENU mutagenesis. *BMC Mol. Biol.* 8: 52.

Teer, J., Bonnycastle, L., Chines, P., et al. (2010). Systematic comparison of three genomic enrichment methods for massively parallel DNA sequencing. *Genome Res.* 20: 1420–1431.

Tewhey, R., Nakano, M., Wang, X., et al. (2009). Enrichment of sequencing targets from the human genome by solution hybridization. *Genome Biol.* 10: R116.

Thomas, K., Capecchi, M. (1986). Introduction of homologous DNA sequences into mammalian cells induces mutations in the cognate gene. *Nature* 324: 34–38.

Tomita-Mitchell, A., Maslen, C., Morris, C., et al. (2007). GATA4 sequence variants in patients with congenital heart disease. *J. Med. Genet.* 44: 779–783.

Turner, C., Emery, H., Collins, A., et al. (2009). Detection of 53 FBN1 mutations (41 novel and 12 recurrent) and genotype-phenotype correlations in 113 unrelated probands referred with Marfan syndrome, or a related fibrillinopathy. *Am. J. Med. Genet. A* 149A: 161–170.

Venter, J., Adams, M., Myers, E., et al. (2001). The sequence of the human genome. *Science* 291: 1304–1351.

Visel, A., Rubin, E., Pennacchio, L. (2009). Genomic views of distant-acting enhancers. *Nature* 461: 199–205.

Vitaterna, M., King, D., Chang, A., et al. (1994). Mutagenesis and mapping of a mouse gene, Clock, essential for circadian behavior. *Science* 264: 719–725.

Wagner, D., Dosch, R., Mintzer, K., et al. (2004). Maternal control of development at the midblastula transition and beyond: Mutants from the zebrafish II. *Dev. Cell* 6: 781–790.

Wang, H., Chattopadhyay, A., Li, Z., et al. (2010). Rapid identification of heterozygous mutations in *Drosophila melanogaster* using genomic capture sequencing. *Genome Res.* 20: 981–988.

Waterston, R., Lindblad-Toh, K., Birney, E., et al. (2002). Initial sequencing and comparative analysis of the mouse genome. *Nature* 420: 520–562.

Weber, J., Salinger, A., Justice, M. (2000). Optimal N-ethyl-N-nitrosourea (ENU) doses for inbred mouse strains. *Genesis* 26: 230–233.

Witmer, P., Doheny, K., Adams, M., et al. (2003). The development of a highly informative mouse simple sequence length polymorphism (SSLP) marker set and construction of a mouse family tree using parsimony analysis. *Genome Res.* 13: 485–491.

Woods, I., Kelly, P., Chu, F., et al. (2000). A comparative map of the zebrafish genome. *Genome Res.* 10: 1903–1914.

Woolfe, A., Goodson, M., Goode, D., et al. (2005). Highly conserved non-coding sequences are associated with vertebrate development. *PLoS Biol.* 3: e7.

Wu, B., Yin, L., Yin, H., et al. (2010). A mutation in the Kit gene leads to novel gonadal phenotypes in both heterozygous and homozygous mice. *Hereditas* 147: 62–69.

Yu, Q., Leatherbury, L., Tian, X., et al. (2008). Cardiovascular assessment of fetal mice by in utero echocardiography. *Ultrasound Med. Biol.* 34: 741–752.

Yu, Q., Shen, Y., Chatterjee, B., et al. (2004). ENU induced mutations causing congenital cardiovascular anomalies. *Development* 131: 6211–6223.

Zhang, Z., Alpert, D., Francis, R., et al. (2009). Massively parallel sequencing identifies the gene Megf8 with ENU-induced mutation causing heterotaxy. *Proc. Natl. Acad. Sci. U.S.A.* 106: 3219–3224.

13 Proteomics of Phagosomal Pathogens

Lessons from Listeria monocytogens *and New Tools in Immunology*

Estela Rodriguez-Del Rio, Carlos Carranza-Cereceda, Lorena Fernandez-Prieto, José Ramós-Vivas, and Carmen Alvarez-Dominguez
Servicio de Inmunología, Fundación Marqués de Valdecilla-IFIMAV and Hospital Santa Cruz de Liencres
Liencres, Spain

CONTENTS

13.1 INTRODUCTION

Phagocytes are immune cells with the commitment to destroy pathogens. In this regard, macrophages usually perform this task intracellularly, while other phagocytes are specialized in the destruction of extracellular pathogens.

Phagosomes are the intracellular vesicles where pathogens are first confined in compartments to await their destruction. The microbicidal potential of phagosomes depends on oxidative and nonoxidative reagents. Although the multienzymatic complex of the nicotinamide adenine dinucleotide phosphate (NADPH) oxidase, also known as *phox*, regulates the oxidative phagosomal mechanisms, nonoxidative mechanisms are less well characterized. In this regard, nonoxidative mechanisms include the interactions of phagosomes with endosomes and lysosomes that transport hydrolytic enzymes. These enzymes are able to degrade pathogens and, with the bactericidal potential of the activity of the H^+-ATPase complex, are responsible for phagosomes low pH. These microbicidal abilities of phagosomes are modulated by proinflammatory cytokines that increase significantly both oxidative and nonoxidative mechanisms.

In this chapter, we will review our group's studies of the features of phagosomes containing the facultative intracellular bacteria and human pathogen *Listeria monocytogenes* (LM). This pathogen resides in macrophage phagosomes for a relative short time, varying between 30 and 90 min. Thereafter, LM lyses the phagosomal membrane, escapes to the cytosol, and replicates. However, the activation of macrophages by proinflammatory cytokines avoids LM lysis of the phagosomal membrane and causes LM degradation within this phagosomal environment. The focus of our research implies an understanding of the differences between nonbactericidal and bactericidal LM phagosomes. Our initial proteomics studies involved cell fractionation and one-dimensional gel electrophoresis of highly purified LM phagosomes from different situations. We combined these initial proteomic studies with functional analysis of LM viability in *in vitro* fusion assays with other organelles, biochemical studies, and studies of the immunological abilities of LM phagosomes. This overall analysis has given us a global picture of those putative proteins involved in phagosomal degradation and LM immune responses. We have completed our studies with endosomal-lysosomal protein-deficient mice that confirmed the role of several lysosomal proteins in LM phagosomal degradation and innate immunity.

Our research is currently focused on setting up the conditions for advanced proteomic analysis of LM phagosomes, fractionated into endophagosomes, lysophagosomes, and autophagosomes by applying differential two-dimensional gel electrophoresis (DIGE) for proteomic comparison of all of these phagosome types. The immune analysis of these different phagosomes will reveal important knowledge about how to use phagosomes as immune tools in vaccination protocols.

13.2 PATHOGENS RESIDING IN THE PHAGOSOMAL COMPARTMENT: AN OVERVIEW

In cell biology, phagosomes are just vacuoles formed around a particle absorbed by phagocytosis. However, in immunology, phagosomes are cellular compartments able to kill and digest pathogenic microorganisms, which bear the specialized immune machinery for major histocompatibility complex (MHC) class I and class II antigen presentation (Jutras et al., 2008; Ramachandra et al., 1999) (Figure 13.1).

A surprising number of pathogens have developed strategies to use the phagosomes for their own advantage. Nowadays, more than 20 taxons of pathogens are able to enter cells and survive within phagosomes. Some of them, such as *Coxiella* or *Helicobacter pylori*, replicate inside phagolysosomes or, like *Rickettsia*, escape to the cytoplasm before the phagosome fuses with lysosomes. In general, pathogens manipulate the host vacuolar trafficking to prevent phagosome maturation and avoid contact with the products of the NADPH oxidase or the lysosomal enzymes. Incomplete maturation of phagosomes creates a favorable environment for the pathogens, allows them to avoid the bactericidal mechanisms and the direct exposure to antibodies, and allows them to acquire appropriate the nutrients and signaling pathways essential for their survival (Alvarez-Dominguez et al., 2004).

Each organism has its own adaptation for survival within the phagosomes, and the range of possible interactions between pathogens and host cells is extremely high. In this chapter, we have only focused to the phagosomal pathway and classify the strategies of pathogens into two main groups: pathogens residing within phagosomes and pathogens escaping from phagosomes. In both groups, pathogens manipulate the phagosomes independently and reside short or long term.

13.2.1 PATHOGENS INHABITING WITHIN PHAGOSOMES

This first group is the largest one and includes those pathogens that inhabit in vacuoles and present at least some interactions with the endosomal or lysosomal compartments, while avoiding complete fusion with lysosomes. The majority of this group needs a low pH (4.5–5) to grow and survive

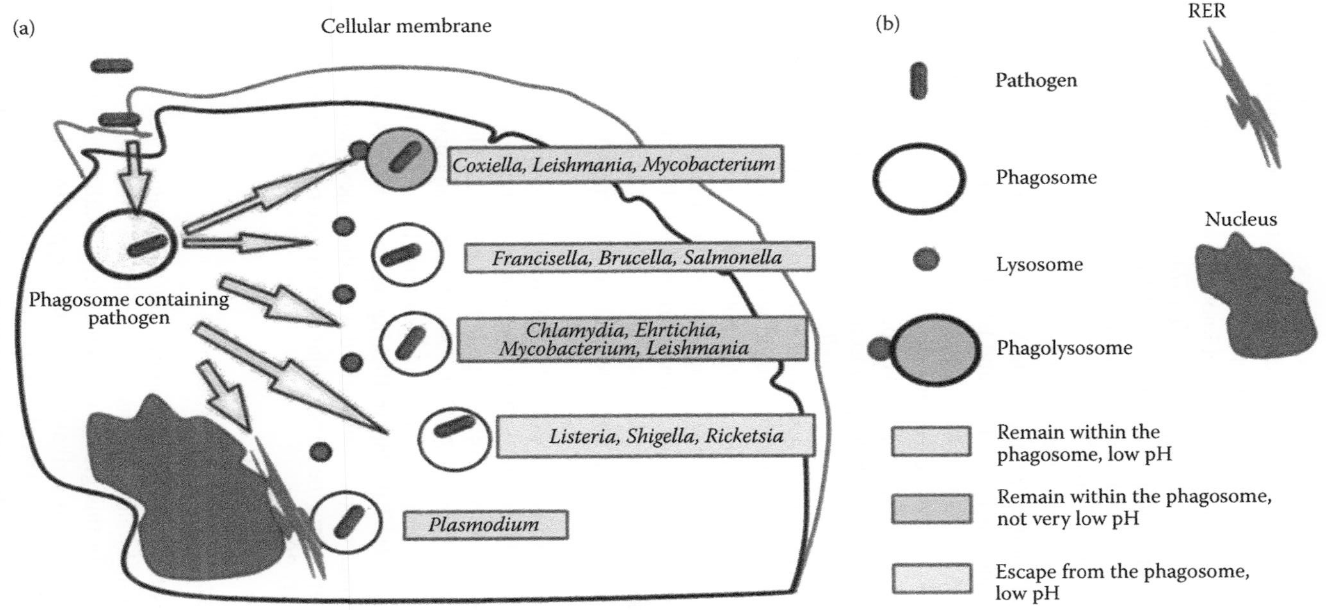

FIGURE 13.1 **(See color insert.)** Trafficking pathways of phagosomes containing pathogens in a macrophage (a) and legend of symbols used in the drawing (b). Some of the pathogens appear in more than one box because they can develop different pathways, such as *Mycobacterium* or *Leishmania*.

within phagosomes; they include the following bacteria: *Brucella spp.*, *Coxiella*, *Francisella*, *Helicobacter*, and *Salmonella* and the protozoa *Leishmania spp.* However, some of them, such as *Coxiella*, *Helicobacter*, and the amastigote form of *Leishmania spp.*, are able to show some interactions with lysosomes or late endosomes. The rest of the pathogens forming this group, such as *Chlamydia spp.*, *Erlichia*, and *Legionella*, reside in phagosomes with a pH value not lower than 6.0 and avoid fusion with lysosomes. There are also special cases, such as *Erlichia*, that induce a morulae (a large membrane-bound phagosome that binds to mitochondria) and *Plasmodium spp.* phagosomes that associate with the rough endoplasmic reticulum.

The molecular strategies of pathogens are not known in all cases; however, some infection pathways have been characterized in the past 10 years. *H. pylori* phagosomal interference seems targeted to the function of Rab7 in late transport events. However, these phagosomes show typical lysosomal markers such as lysosome-associated transmembrane protein 1 (LAMP-1) or Rab7 and lack the CI-M6PR. The bacterial secreted toxin VacA is responsible for this interference. Although less well characterized, *Salmonella*'s intracellular strategy is beginning to be elucidated. This pathogen is able to bypass the normal endocytic route while residing in a compartment with low pH caused by early endocytic markers such as Rab5, TfR, and EEA1. It also has some lysosomal markers, such as Rab7, LAMP-1, and LAMP-2, whereas it lacks others, such as cathepsin-D and M6PR. In macrophages, the phagosomal compartment seems slightly different. Fusion with lysosomes is avoided, whereas interactions with endosomes are allowed because of the accumulation of Rab5-NSF and the lack of Rab7. Two bacterial proteins belonging to the type III secretion system seems to be responsible for the trafficking interference of *Salmonella*: SpiC, a protein secreted into the macrophage cytosol, seems to inhibit phagosome-lysosome fusion, and SopE, also secreted into the macrophage cytosol, appears to bind to Rab5, promoting fusion with endosomes. *Brucella* phagosomes are also unique compartments because at first they interact with endosomes acquiring the endosomal marker EEA1 but later on they transform into autophagosomes containing the autophagic marker MDC, the ER marker sec61β, and the late endosomal markers LAMP-1 and LAMP-2. This specialized phagosomal compartment diverts from the phagocytic pathway to a safer compartment in the ER with autophagocytic features, where it replicates (Steele-Mortimer et al., 1999; Birmingham et al., 2006; Mallo et al., 2008).

Legionella pneumophila is another example of a well-characterized intracellular strategy. The phagosomal location of this bacterium is autophagosomes that avoid fusion with endosomes and lysosomes but allow interactions with other organelles. *Legionella* phagosomes lack early markers such as TfR and late markers such as LAMP-1, LAMP-2, and cathepsin-D but contained 5'-nucleotidase and CR3 complement receptors. These phagosomes lack Rab5, the early endosomal regulator, but show recruitment of Rab7, the late trafficking regulator. Rab7 is lost once these phagosomes are transformed into autophagosomes and studded with ribosomes. Two bacterial genes encoded by a type IV secretion system control the intracellular interference of *Legionella*: the *dot* and *icm* genes (Alvarez-Dominguez et al., 2004).

13.2.2 Pathogens Escaping from Phagosomes

A second group includes the pathogens escaping from the phagosomes to reside free in the cytoplasm. They require low pH values because the machinery that allows them to escape is active only at low pH conditions. In this group we find bacterial pathogens such as *L. monocytogenes*, *Shigella flexneri*, and *Ricketsia*, as well as the protozoa *Trypanosoma cruzi*. Also depending on the infection process, bacteria such as *Mycobacterium spp.* might belong to this group because they lyse the phagosomal membrane after long infection times, increasing the host cell toxicity. All of these pathogens show toxins able to permeate the phagosomal membranes and liberate them in the cytosol. *Shigella* is able to escape from the phagosomes using a gene region called SER. Once in the cytosol, *Shigella* is able to escape autophagy using the type III secretion factor IcsB that binds to the bacterial factor VirG, a protein required for actin-based cell motility with the ability to bind to the autophagic factor Aptg5 (Huang and Brumell, 2009; Allaoui et al., 2010). *L. monocytogenes* uses a cytolysin called listeriolysin O (LLO) that with the help of phosphatidylinositol-specific phospholipase C (PI-PLC), a phospholipase secreted by the bacterium, escapes from the phagosome. Once in the cytosol in certain conditions, LM is able to induce autophagy, also requiring this pore-forming toxin, LLO (Russell, 2001). *Mycobacterium spp.* shows a very slow phagocytic pattern; however, after 2–3 days postinfection, it is also able to escape from phagosomes using a protein called ESX-1 that belongs to the type VII secretion system. The phagosomal features of *Mycobacterium spp.* imply the removal of the H^+-ATPase and prevention of this environment acidification. The lack of phagosomal maturation is observed by several features: an interference with the trafficking machinery involving a blockage between the function of Rab5 and Rab7; a low percentage of interactions with lysosomes reflected by a low presence of late endocytic markers such as CD63, LAMP-1, and cathepsin-D; and finally, the retention of TACO, a coat protein with an unknown function. Nevertheless, because *Mycobacterium* resides for long times within phagosomes, it is possible that the lysis of the phagosomal membrane has been subverted in certain cases (Kayal and Charbit, 2001; Vergne et al., 2004; Seto et al., 2009).

Some steps of *T. cruzi* strategy within phagosomes have been recently discovered. These pathogen phagosomes recruit very early lysosomes but also show interactions with endosomes regulating the activation of Rab5 and EEA1 and colocalizing both markers in the phagosomes. This colocalization has the purpose of promoting phagosome-endosome fusion and activating the phosphatidylinositol 3-kinase (PI3K), which it is indispensable for recruiting lysosomes to the site of entry. The products of the PI3K activate EEA1 and recruit Rab5 in its active form. Once the phagosomes fuse with lysosomes, this pathogen uses the low pH of lysosomes to activate its pore-forming protein TcTox, which degrades the lysosomal membrane and allows the pathogen to escape to the cytosol (Maganto-Garcia et al., 2008).

The phagosomal features and specific interactions with the intracellular organelles indicate that each pathogen diverges from others in critical steps along the phagocytic route, including the entry process, the lysis or maintenance of the phagosomal membrane integrity, the compartment for replication, and the exploitation of the actin-based machinery of the cell for movement.

13.3 DIFFERENT PHAGOSOMAL COMPARTMENTS IN THE CASE OF *L. MONOCYTOGENES*

L. monocytogenes is a Gram-positive bacterium able to interfere with the phagosomal trafficking in resting macrophages that show a low bactericidal potential. However, after macrophage activation caused by contact with proinflammatory cytokines, the overall LM interference is overcome because of the high bactericidal potential of these phagocytes. In this regard, phagosomes from resting macrophages are considered nonlistericidal compartments, whereas phagosomes from activated macrophages are considered listericidal organelles (Prada-Delgado et al., 2001). Our initial proteomic analysis of both compartments in macrophages and other epithelial cells has indicated that LM phagosomes show features of early and late endosomes, lysosomes, or even autophagosomes, depending on the phagocyte conditions. Here, we will describe the specific characteristics of these varied phagosomes and the predominance of one type or the other depending on the activation state of these cells.

13.3.1 Phagosomes with Endocytic Features

Endosomes are organelles whose trafficking is regulated by two Rab families of GTPases, part of the ras superfamily of genes, regulated through binding of GTP and hydrolysis of bound GTP to GDP acting in sequence: Rab5a and Rab7. Although Rab5a controls the early endocytic events, Rab7 seems to participate in late endocytic events and interactions with lysosomes. The function of these small GTPases requires a GTP-active state that is achieved after two different regulators acted onto them. First, cytosolic GDP forms of these GTPases are recruited to membranes by guanosine nucleotide dissocation inhibitor (GDI), and GDP membranes forms are activated by GDP/GTP exchange that is exerted by a GDP/GTP-specific exchange factor. The GDP forms exchanged to GTP forms perform their specific functions in trafficking, and a GTPase protein acts to hydrolyze GTP to GDP. Next, GDI factors are able to remove GDP forms back to the cytosol and restart the whole cycle (Bucci et al., 1992).

The features of LM phagosomes in resting macrophages and epithelial cells are similar and show mainly endosomal characteristics. These LM phagosomes contain early and late endocytic markers such as TfR, Rab5a, Rab5c, Rab7, and ManR, and they lack typical lysosomal markers such as LAMP-1, LIMP-2, CD-M6PR, and cathepsin-D as detected by biochemical and initial proteomic analysis. Our proteomic studies also detected regulators of the interactions with endosomes, such as the fusogenic factors NSF and SNAP or the activator of Rab5a function, Rabex-5, a GDP/GTP exchange factor. Surprisingly, we could not detect enough of the Rab5a-GTP activator EEA1 in LM phagosomes, suggesting that Rab5a-GTP-active forms were not detected in LM phagosomes. All of these features suggested that LM phagosomes might be considered between early and late endosomal-like compartments. Once we observed the electron microscopic (EM) appearance of these compartments, we observed very spacious vesicles lacking the presence of electrodense material, a characteristic of lysosomes. However, we also observed some multimeric structures within these compartments that may correspond to the so-called multivesicular bodies characteristic of late endocytic compartments. These studies were confirmed with biochemical-proteomic studies and indicated that LM phagosomes were able to interact with early and late endosomes but lack the ability to interact with lysosomes. We proved this hypothesis isolating phagosomes of different ages and establishing *in vitro* fusion assays with different organelles: early and late endosomes and lysosomes. These assays confirm that LM phagosomes were able to interact at early time points with early and late endosomes to decrease the level of fusion rates upon the time. However, they showed no significant interactions with lysosomes at any time point, either early or late. Moreover, a combination of the proteomic isolation of LM phagosomes with biochemical and functional assays indicated that LM phagosomes bear Rab5a-GDP inactive forms and lack the presence of Rab5a-GTP active forms (Alvarez-Dominguez et al., 1996, 1997; Alvarez-Dominguez and Stahl, 1999; Prada-Delgado et al., 2001, 2005). All of these data strongly suggested an interference of the pathogen with the function of the early endosomal regulator Rab5a at the level of GDP/GTP exchange. In fact, LM phagosomal interference was based on the recruitment and retention of Rab5a in the phagosomes as detected with the proteomic analysis of isolated phagosomes and a delay of Rab5a GDP/GTP exchange activity. Because heat-killed bacteria do not show this phagosomal interference, we suggested that a putative heat-sensitive inhibitory factor from this pathogen was responsible for this strategy. This analysis also revealed that the putative factor demonstrated a proteical and enzymatic nature using a LM crude extract that was able to mimic both Rab5a recruitment to phagosomes and the delay in Rab5a GDP/GTP exchange. Therefore, we designed a detailed biochemical-proteomic approach to identify the Rab5a inhibitory bacterial factor using Rab5a-affinity columns loaded with the LM crude extract. The major bacterial protein eluted from these Rab5a-affinity columns, a p40 *Listeria* protein, was sequenced and identified by mass spectrometry as the glyceraldehyde-3-phosphate dehydrogenase (GAPDH-LM) (Uniprot ID: Q8Y4I1; Locus tag: Lmo2459; GeneID: 987377; gi: 16804497). Two additional bands were also eluted and observed by Coomassie: p56 and p130. However, p56 and p130 sequencing failed because p130 was below the detection sensitivity, and p56 gave no significant LM-specific peptide map. These proteomic approaches allowed us to perform structural studies and homology alignments that indicated the close relationship of GAPDH-LM with GAPDH from *Streptococcus pyogenes* (GAPDH-SP) at the N terminus, showing NAD-binding domains and probably sharing the enzymatic ability of GAPDH-SP of ADP-ribosylation (Huynh et al., 2007). Therefore, the proteomic and structural analysis revealed interesting functional data of this bacterial LM factor that were confirmed *in vitro* with recombinant Rab5a as well as *in vivo* with isolated phagosomes. GAPDH-LM was able to ADP-ribosylate Rab5a, causing its inactivation as it decreases the binding ability of Rab5a-GDP with its specific exchange factor, Rabex-5.

Also ADP-ribosylation of Rab5a by GAPDH-LM retained Rab5a in the phagosomal membranes and explained the interference strategy of LM phagosomes with the endosomal compartment. All of the enzymatic and inhibitory ability of GAPDH-LM was located at the N terminus in a 22-mer peptide that shows a special tridimensional structure able to adopt an amphipatic helix conformation because of its richness in arginine residues that might endow peptides with cationic charges and hydrogen bonding properties for interaction with the anionic components of the phagosomal membranes. All of these data indicated that this 22-mer N-terminal peptide of GAPDH-LM might interact with the anionic components of the phagosomal membranes and encounter Rab5a-GDP to bind to. Proteomic studies of isolated LM phagosomes detected this bacterial factor into the phagosomal membranes as well as Rab5a-GDP. The combination of these results with immunofluorescence colocalization analysis of GAPDH-LM and Rab5a into LM phagosomal membranes and coprecipitation studies allowed us to propose the following model of LM interference (Alvarez-Dominguez et al., 2008). The GAPDH-LM N-terminal peptide might access LM phagosomal membranes and bind to the inactive Rab5a-GDP forms, causing ADP-ribosylation of this GTPase. This covalent modification of Rab5a-GDP would decrease the interaction with its GDP/GTP-specific exchange factor and delay the GDP/GTP exchange activity. Moreover, Rab5a-GDP and ADP-ribosylated by GAPDH-LM would also avoid the possibility that the specific factor, GDI, could be responsible for removing Rab5a-GDP from membranes, might not access this conformation, might retain longer Rab5a-GDP in the phagosomal membranes, and might affect the interactions with endosomes. In fact, we confirmed this model using *in vitro* fusion assays and observed that ADP-ribosylation of Rab5a by GAPDH-LM decreases the ability of phagosomes to interact with endosomes, suggesting a delay in phagosomal maturation. All of this delay in LM phagosome maturation occurs before LLO action takes place. In fact, LLO shows two sequential actions in LM phagosomes: first preventing the fusion with lysosomes and next, with the aid of PI-PLC, permeating the phagosomal membranes and allowing the pathogen to escape to the cytosol (Portnoy et al., 1992). Therefore, nonbactericidal LM phagosomes appear as endosomal-like compartments showing TfR, ManR, Rab5a-GDP, and Rab7 that cannot progress to matured lysosomes because of the action of two LM virulence factors: GAPDH-LM and LLO. The former virulence factor blocks Rab5a function and allows some interactions with endosomes, and the latter avoids the fusion with lysosomes (Shaughnessy et al., 2006). The different proteomic approaches with isolated phagosomes and LM crude extract combined with functional assays have indicated the power of proteomics to reveal the intracellular strategies of pathogens and identify at the molecular level the components on both sides: the bacterial factors causing the interference and the cellular regulators inhibited. Moreover, proteomics allowed us to characterize the features of nonbactericidal phagosomes that would be essential to compare with other pathogens and establish general rules of action.

13.3.2 Phagosomes with Lysosomal Features

LM phagosomes in activated macrophages showed mainly lysosomal features different than phagosomes from resting cells. Activation of macrophages with the proinflammatory cytokine, interferon γ (IFN-γ), caused the intracellular destruction of this pathogen at much higher levels than 60–70% of LM killing observed in resting macrophages. Moreover, EM images have proven that these IFN-γ-activated macrophages were not bacteria-free in the cytosol (Portnoy et al., 1989). Because IFN-γ signaling induces bactericidal mechanisms both in phagosomes and in the cytosol, we initiated our proteomic studies with the purpose of discovering the contribution of bactericidal mechanisms acting within phagosomes to the overall destruction of this pathogen in macrophages. Therefore, we first isolated phagosomes from resting and IFN-γ-activated macrophages and checked LM viability. We verified with this analysis that resting phagosomes were nonlistericidal compartments, whereas phagosomes from IFN-γ-activated macrophages were listericidal organelles. Next, we performed a comparative proteomic analysis of both types of LM phagosomes using biochemical approaches and one-dimensional sodium dodecyl sulfate-polyacrylamide gel electrophoresis (SDS-PAGE) analysis.

This proteomic approach allowed us to characterize in detail IFN-γ listericidal phagosomes and revealed the molecular strategy used to bypass LM interference within the phagosomal environment (Prada-Delgado et al., 2001). In brief, IFN-γ listericidal phagosomes showed lysosomal features as compared with the nonlistericidal phagosomes detailed in Section 13.3.1. Our proteomic analysis and EM studies revealed that IFN-γ listericidal phagosomes contained damaged bacteria and electrodense material. They also showed high levels of the lysosomal markers LAMP-1, LIMP-2, and cathepsin-D active forms but unusual high levels of the early endosomal regulator, Rab5a. Other endosomal markers such as Rab5c or Rab7 were detected at similar ranges in listericidal and nonlistericidal phagosomes. These proteomic results clearly suggested that IFN-γ listericidal phagosomes showed different fusogenic abilities with other organelles than nonlistericidal phagosomes. In this regard, nonlistericidal phagosomes showed interactions with early and late endosomes until phagosomes reached an age of 15 min and very low percentage of fusion with lysosomes at any phagosomal age tested (Alvarez-Dominguez et al., 1996, 1997). However, IFN-γ listericidal phagosomes presented high percentages of interactions with late endosomes and lysosomes and similar rates of fusion with early endosomes than nonlistericidal phagosomes (Prada-Delgado et al., 2001). Altogether, these results suggested that the delay in phagosome maturation and the lack of fusion with lysosomes was bypassed in IFN-γ listericidal phagosomes. A detailed molecular analysis of Rab5a function in IFN-γ listericidal phagosomes revealed that in contrast to the retention of Rab5a-GDP inactive forms in nonlistericidal phagosomes, IFN-γ listercidal phagosomes showed high levels of Rab5a-GTP active forms with an accelerated rate of GDP/GTP exchange. Therefore, the action of GAPDH-LM virulence factor to inhibit Rab5a and block phagosome maturation was not observed in IFN-γ listericidal phagosomes. These data explained the high levels of fusion with endosomes that requires the continuous presence of Rab5a-GTP active forms in phagosomes and endosomes (Prada-Delgado et al., 2005; Del Cerro-Vadillo et al., 2006). The high percentage of fusion of IFN-γ listericidal phagosomes with lysosomes also explained the high levels of active forms of cathepsin-D compared with nonlistericidal phagosomes (Prada-Delgado et al., 2001). Because the lack of fusion with lysosomes was assigned to LLO action (Shaughnessy et al., 2006), these results also suggested that IFN-γ listericidal phagosomes should present an inactive LLO on their phagosomal membranes. In this regard, proteomic studies of IFN-γ listericidal phagosomes showed the presence of LLO degraded forms exclusively, whereas nonlistericidal phagosomes presented mainly LLO intact forms (Carrasco-Marin et al., 2009). We could explain the molecular mechanism of LLO inactivation in IFN-γ listericidal phagosomes using cathepsin-D genetically deficient mice or a protocol that reduced the murine expression of this lysosomal protease *in vivo* and a similar proteomic approach as before. We isolated nonlistericidal and IFN-γ listericidal phagosomes from cells deficient in cathepsin-D and observed that the listericidal ability of these phagosomes depended on the cathepsin-D enzymatic activity. In fact, cathepsin-D enzymatic action was responsible for LLO inactivation in the phagosomes, decreasing the bacterial viability within the phagosomal lumen and avoiding LM escape to the cytosol. We used a proteomic approach with highly pure and active cathepsin-D and recombinant LLO to reveal the molecular action of cathepsin-D onto LLO. This proteomic analysis and mass spectrometry studies of the LLO peptides after cathepsin-D digestion revealed that the cathepsin-D cleavage site was localized within the undecapeptide-conserved region of this pore-forming toxin. The cathepsin-D cleavage site was mapped between the two sequential residues ^{491}WW^{492} of the WWEWR domain recognized by a monoclonal antibody raised against *Streptococcus* pneumolysin, PLY (anti-PLY5). The confirmation of these results was performed after a proteomic approach of isolated LM phagosomes combined with biochemical analysis using this anti-PLY5 antibody (Del Cerro-Vadillo et al., 2006; Carrasco-Marin et al., 2009). In fact, we could conclude that the two sequential residues ^{491}WW^{492} were responsible for LLO anchoring into the phagosomal membrane and that this enzymatic cleavage affected LLO binding to the membranes and avoided pore formation of this toxin. Moreover, using LM mutants on each of these sites, ΔLLO/W491 and ΔLLO/W492, we could reveal that the W^{491} site was the relevant LLO site linked to cathepsin-D action and this lysosomal protease role in innate immunity against LM.

In summary, IFN-γ signaling in macrophages affected not only the enzymatic activity of cathepsin-D but also the function of Rab5a inducing this GTPase synthesis, processing, and activation to active GTP forms (Alvarez-Dominguez and Stahl, 1998) and bypassing the action of the virulence factor, GAPDH-LM (Prada-Delgado et al., 2001; Del Cerro-Vadillo et al., 2006; Carrasco-Marin et al., 2009). Moreover, down-regulation of Rab5a expression in macrophages affected the fusion of phagosomes with late endosomes and lysosomes, whereas not with early endosomes, indicating that Rab5a function in phagocytosis was not restricted to regulation of the early endosomal trafficking but also to control the late endosomal trafficking (Alvarez-Dominguez and Stahl, 1999). In this regard, macrophages with low expression of Rab5a showed an impaired transport of cathepsin-D active forms to LM phagosomes that was not reverted by IFN-γ signaling (Prada-Delgado et al., 2001). Therefore, IFN-γ listericidal phagosomes bypassed the action of both bacterial virulence factors, GAPDH-LM and LLO, that inhibited Rab5a and the transport of lysosomal enzymes to LM phagosomes, respectively.

The combination of proteomic, biochemical, and functional approaches using isolated phagosomes from genetically deficient mice allowed us to reveal the involvement of the lysosomal enzyme cathepsin-D in this pathogen immunity. In fact, cathepsin-D appeared as a relevant component of the phagosomal stage of LM innate immune response and validated the use of proteomic analysis as immunological tools. Therefore, nonoxidative mechanisms, as in the case of lysosomal proteases action, such as cathepsin-D, were effective listericidal agents playing relevant roles within the phagosomes.

IFN-γ listericidal phagosomes showed not only high levels of active forms of cathepsin-D but also high levels of two members of the family of lysosomal membrane proteins, LAMP-1 and LIMP-2 (Prada-Delgado et al., 2001). LAMP are type I transmembrane glycoproteins associated with the membrane, containing a short cytoplasmic domain bearing a tyrosine-based motif to guide to the lysosomes, trans-Golgi network (TGN), and plasma membranes. LIMP are type III transmembrane glycoproteins inserted into the membrane that contain a short cytoplasmic domain with a dileucine motif to guide to lysosomes or TGN membranes (Eskelinen et al., 2003). We used mice with genetic deficiencies in LAMP-1 and LIMP-2 lysosomal proteins (Huynh et al., 2007) and isolated nonlistericidal phagosomes and IFN-γ listericidal phagosomes to decipher LAMP-1 and LIMP-2 role as phagosomal bactericidal mechanisms. These proteomic studies indicated that Rab5a-GTP active forms acted in concert with LIMP-2, but not with LAMP-1, to modulate the number of bacteria able to escape to the cytosol. Therefore, the specific phagosomal role of LIMP-2 was linked to the fusion with late endocytic compartments and not directly with LM intraphagosomal viability. We performed a detailed analysis of isolated LM phagosomes from cathepsin-D and LIMP-2 genetically deficient mice to reveal the sequential phagosomal action of both lysosomal proteins on LM destruction. We concluded that cathepsin-D action affected directly LM viability within the phagosomes, acting in first place. Later on, LIMP-2 participated in controlling the number of bacteria escaping to the cytosol by regulating the fusion events with late endocytic compartments, including MHC class II antigen-processing compartment vesicles that belong to the antigen-presentation MHC class II compartments. The purpose of this tight control on the number of LM escaping to the cytosol was the induction of early acute-phase proinflammatory cytokines (i.e., monocyte chemotactic protein-1, interleukin-6, and tumor necrosis factor-α). Only low numbers of cytosolic LM are able to trigger these immune signaling pathways, whereas huge numbers of cytosolic LM aborted this signaling. Therefore, LIMP-2 acted as a linker molecule between phagosomal trafficking and the immune signaling to activate macrophages. In summary, LIMP-2 participated also in LM innate immunity at the phagosomal stage, but acting in sequence after the phagosomal action of cathepsin-D (Carrasco-Marin et al., 2010).

Altogether, these results indicate that proteomic analysis of isolated phagosomes combined with genetically deficient mice, functional and biochemical studies, and immune assays may constitute powerful techniques that allow dissertation of the molecular mechanisms of pathogen interactions with the immune cells such as macrophages.

13.3.3 Autophagosomes

The bulk degradation of cytoplasmic proteins and organelles is largely mediated by autophagy. As with phagocytosis, autophagy has been exploited in eukaryotes as a host defense mechanism, targeting intracellular pathogens for destruction. The autophagic process starts with the formation of autophagosomes. These double-membrane–layered vesicles are normally filled with cytosolic material. These autophagosomes fuse with lysosomes to form autolysosomes where their contents are then degraded by acidic lysosomal hydrolases. Two of the most important lysosomal components are LAMP-1 and LAMP-2. LAMPs were believed to function in the maintenance of the integrity of the lysosomal membrane by protecting it from the luminal environment. LAMP-2 is needed for maturation of phagosomes and autophagosomes, with or without engulfed bacteria (Saftig et al., 2008). The origin of autophagosomal membranes is still a matter for research, but some lines of evidence suggest that in mammalian cells they are formed from parts of the endoplasmic reticulum. Actually, specific and quantitative methods such as electron microscopy, chemical inhibitors, fluorescent light chain 3 (LC3) colocalizations, and proteomic and gene targeting approaches help us to study this mechanism. A standard autophagosome marker is microtubule-associated protein 1-LC3 (Kabeya et al., 2000). This protein is covalently conjugated to phosphatidylethanolamine and is localized to autophagosomes during their formation, transport to, and fusion with lysosomes. Visualization of bacteria-containing autophagosomes is possible with the use of GFP-LC3 constructions by means of transient or stable transfections in cells (Birmingham and Brumell, 2009; Ogawa et al., 2009). Autophagy contributes to immunity against pathogens mainly in two ways: targeting free cytosolic bacteria (i.e., *Listeria* and *Salmonella*) and targeting phagosomes to aid in the clearance of their bacterial contents (*i.e.,* bacterial toxins or whole microorganisms) (Figure 13.2).

In epithelial cells and macrophages, the autophagosomal marker LC3 colocalizes with intracellular *Salmonella enterica* and *L. monocytogenes*, respectively (Birmingham and Brumell, 2009; Meyer-Morse et al., 2010). In *S. enterica*, most bacteria having LC3-labeling retain colocalization with *Salmonella*-containing vacuole (SCV) membrane proteins as LAMP-1, which suggests that damaged SCVs are rapidly targeted by the autophagic system. However, only a low percentage of the SCV is surrounded by LC3 in infected epithelial cells displaying maximal autophagic activity. Therefore, it remains to be demonstrated whether bacteria enclosed in the SCV of epithelial cells subvert autophagy or not. In macrophages, *S. enterica* induces autophagy and cell death by injecting SipB, an effector protein secreted by *Salmonella* pathogenicity island-1 typhimurium type III secretion system (Birmingham and Brumell, 2006).

In the case of *L. monocytogenes*, this pathogen induces a large number of autophagic vesicles as observed EM and immunofluorescence studies (Birmingham et al., 2007). The main bulk of internalized LM are degraded within the phagosomes with early-late endocytic features (described in Section 13.3.1). Those bacteria escaping from the phagosomes because of the concerted action of LLO and PI-PLC reached the cytosol, nucleate actin filaments because of the action of ActA virulence factor and used the actin-based motility of the cell to move and infect the next cell, a process known as cell-to-cell spread. However, a low percentage of cytosolic bacteria might be recognized by the host autophagy system to be delivered to autophagolysosomes or evade growth restriction by autophagy. Those bacteria reaching the autophagic vesicles depended on the expression of the virulence factor, LLO for maximal efficiency (Birmingham et al., 2007). Interestingly, in the case of *L. monocytogenes*, cytosolic *S. flexneri*, they efficiently evade autophagy (Ogawa et al., 2009). The importance of the autophagic machinery in immunity against bacterial pathogens is evident by the development of pathogen countermeasures to evade destruction in the autophagosomes. Pathogens such as *Brucella abortus*, *L. pneumophila*, *Porphyromonas gingivalis*, and *Coxiella burnetti* subvert autophagy by blocking the fusion of lysosomes with the autophagosome (Sanjuan and Green, 2008). Autophagy also modulates the survival of *Mycobacterium tuberculosis* inside macrophages by releasing the maturation arrest imposed by *M. tuberculosis* on its vacuole (Gutierrez et al., 2004). Intravacuolar survival of *M. tuberculosis* is also controlled by IFN-γ. Thus,

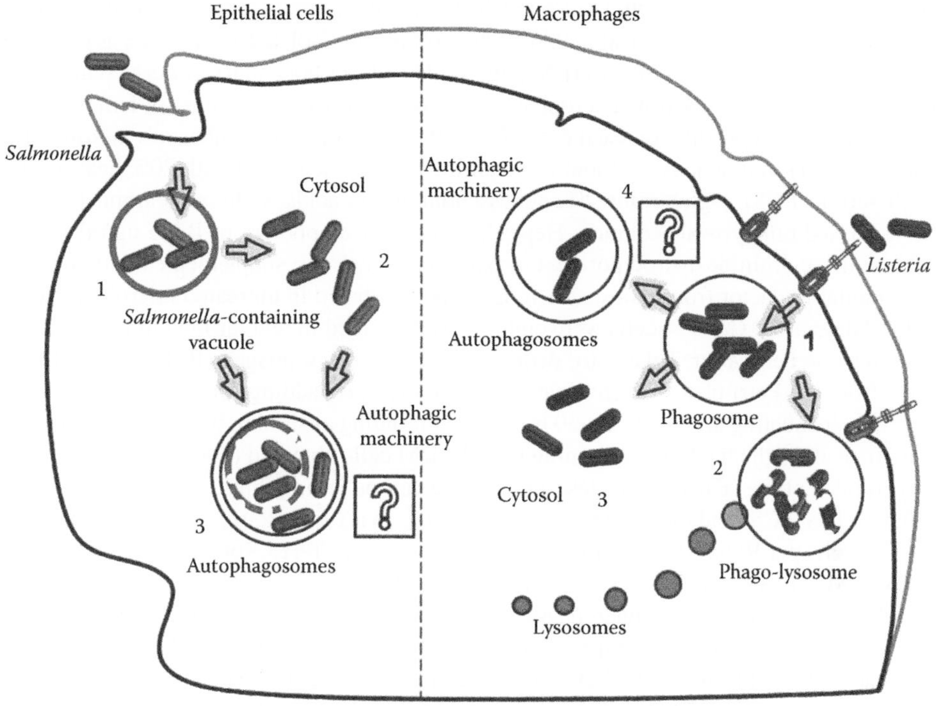

FIGURE 13.2 **(See color insert.)** Autophagy interactions with intracellular pathogens: the *Salmonella* and *Listeria* models. In epithelial cells, *Salmonella* initiate the infection enclosed in the SCV (left panel). This compartment is further modeled to sustain bacterial proliferation (1). *Salmonella* replicates, however, at higher rates in the cytosol of epithelial cells (2). A proportion of damaged SCV and cytosolic bacteria is proposed to be targeted by the autophagic system and ubiquitinated proteins, in epithelial cells, and also in fibroblast (3). In the *Listeria* model (right panel), the pathogen interacts with receptors in the macrophage and the bacteria are internalized in phagosomes (1). The intracellular destruction of some bacteria is due to the interaction of phagosomes with lysosomes (2). *Listeria* can escape from the phagosomes (3), and some cytosolic bacteria may interact with the autophagic machinery (4).

immunity-related GTPases activated upon IFN-γ stimulation promote antimycobacterial effect by inducing autophagy (Singh et al., 2006).

In the case of LM, no evidence of the role of autophagy in IFN-γ-treated macrophages has been performed. Moreover, the features of LM autophagosomes are barely known: only EM and immunofluorescence images of colocalization with LC3 marker. However, other autophagic markers such as LAMP-2 have not been studied in LM phagosomes. We propose that proteomic studies such as the ones we have performed with other types of phagosomes would be required to make significant conclusions about the role of autophagy for phagosomal LM destruction either within nonlistericidal phagosomes or IFN-γ listericidal phagosomes. Once again, proteomics in combination with biochemical functional studies and immune assays using genetically deficient mice would be helpful in the study of autophagy and LM infection.

13.4 ISOLATION AND PURIFICATION OF LISTERIA PHAGOSOMES: INITIAL PROTEOMICS

The main protocol to establish all of the initial proteomics with isolated phagosomes containing live LM is well established in our laboratory (Prada-Delgado et al., 2001; 2005; Del Cerro-Vadillo et al., 2006; Alvarez-Dominguez et al., 2008; Carrasco-Marin et al., 2009, 2010). Moreover, this methodology is the first one of its kind published with live microorganisms and has had a great relevance (Alvarez-Dominguez et al., 1996). Phagosomes are isolated from either a macrophage-cell line able to be fully activated with IFN-γ, the J-774 cells, or bone marrow derived macrophages (BM-DM). In brief, bone marrow macrophages are extracted from the femurs of female mice with different genotypes depending on each case. The cells are cultured in Dulbecco's modified Eagle's medium (DMEM)-high glucose without L-glutamine and supplemented with 20% heat-inactivated fetal calf serum (FCS), L-glutamine 2%, nonessential amino acids, sodium bicarbonate (NAHCO 3, 7.5% w/v), sodium pyruvate solution, Hepes buffer (1 M), amphotericin-B (25 μg/mL), gentamicin (30 μg/mL), L-arginine hydrochloride, L-asparagine anhydrous, supplemented with macrophage colony-stimulating factor from mouse (10 ng/mL) and incubated in untreated microbiological Petri plates for 7 days at 37°C. J-774 cells were cultured as described in regular medium with 10% FCS.

Bone-marrow cells after 7 days are differentiated into macrophages (BM-DM), and they are detached from the Petri plates by eliminating all the medium, adding filtered phosphate-buffered saline (PBS), incubating at 4°C for 20–30 min, and continuing aspiration with a pipette. Thereafter, we centrifuge all of the medium containing the BM-DM cells obtained (700 × *g*, 10 min) in a conical polypropilene tube (Falcon). We resuspended the cell pellet in 30 mL of D10 (DMEM supplemented with 10% FCS) and incubate it overnight at 37°C in a T-125 Roux, a disposable flask for cell culture (10^7 cells per Roux). J-744 cells are cultured directly in T-125 Roux in D10 medium similar to the BM-DM cells overnight at 37°C at same quantity before infection with LM.

The following day, we washed macrophages (both BM-DM and J-774 cells) by removing the medium from the T-125 flasks and adding 25 mL of prewarmed, filtered DMEM (37°C, 1 h). Concurrently, we prepared the bacterial inoculum (10^8 bc) by adding 1 mL of filtered DMEM on an Eppendorf filter and centrifugation of bacteria (2500 × *g*, 5 min, 4°C) and resuspended in 50 μL of filtered DMEM adding 20 mL of extra DMEM. Then we decanted the DMEM from the T-125 flasks containing the macrophage cells and added the 20 mL of filtered DMEM containing the bacteria for infection (15–20 min, slow shaking, 37° C). Once incubation was finished, we decanted the medium containing extracellular bacteria and added 10 mL of Hanks' balanced salt solution at room temperature three times for washing. The following step consists of detaching the macrophage cells carefully with a scrapper (a procedure we repeated three times, checking under the optical microscope for cell viability and detaching of cells from flasks). The three washes for detaching the cells were combined and centrifuged (720 × *g*, 5 min, 4°C), the supernatants were decanted, and the pellets were resuspended with 3 mL of homogenization buffer (HBE; 20 mm Hepes-KOH, 250 mM sucrose, 0.5 mM EGTA, pH 7.2) and centrifuged in same conditions three times. The supernatants were decanted, and 300 μl of HBE was added; the cells were resuspended carefully and incubated for 15 min on ice in an Eppendorf microcentrifuge. Then the cells were gently homogenized in a 1-mL syringe loaded with a 19-gauge needle after 13 passages through the syringe on ice. Homogenized cells were centrifuged (1200 × *g*, 7 min, 4°C) to pellet nuclei and unbroken cells, and we collected the postnuclear supernatants (PNS) containing all intracellular organelles including the phagosomes and froze them into liquid nitrogen for storage at −80°C.

We performed several controls to check the percentage of phagosomes and the viability the bacteria in PNS by plating solubilized PNS into blood agar plates for colony-forming unit (CFU) analysis. We also pelleted all intracellular membranes (ultracentrifugation of PNS at 50,000 × *g*, 10 min, 4°C) and resuspended the pellets in solubilized buffer (PBS with 0.1% Triton X-100) to plate them as above.

We next purified the phagosomes from the PNS with the following procedure: PNS were thawed on ice and diluted with 1 mL of HBE and spun down (1 min, 37,000 × *g*, 4°C), and supernatants were centrifuged to obtain a phagosomal-enriched fraction (50,000 × *g*, 10 min, 4°C) (Figure 13.3). Next, the phagosomal enriched fraction (8.5% sucrose solution) was loaded into the following discontinuous sucrose cushion (5 mL of 40% sucrose, 5 mL of 20% sucrose, 0.5-mL sample) and centrifuged (1700 × *g*, 45 min, 4°C). Phagosomes were recovered from the 8.5–20% interphase and upper 20%. Phagosomes were next spun down (50,000 × *g*, 10 min, 4°C). The pellets were resusupended into 50 μL of HBE, and one-tenth of this purified phagosome preparation was used for different controls

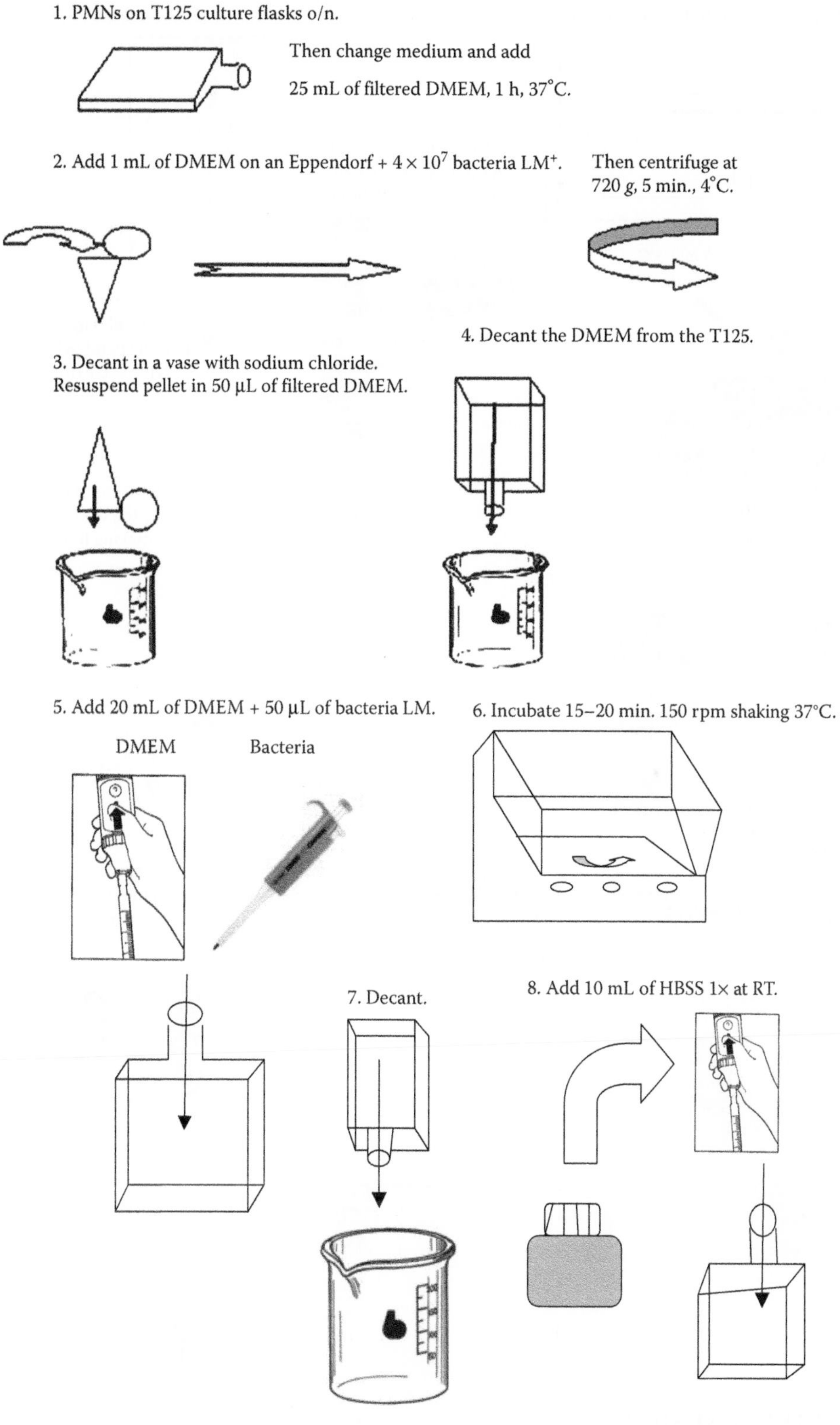
1. PMNs on T125 culture flasks o/n.
Then change medium and add
25 mL of filtered DMEM, 1 h, 37°C.
2. Add 1 mL of DMEM on an Eppendorf + 4 × 10⁷ bacteria LM⁺.
Then centrifuge at
720 g, 5 min., 4°C.
4. Decant the DMEM from the T125.
3. Decant in a vase with sodium chloride.
Resuspend pellet in 50 μL of filtered DMEM.
5. Add 20 mL of DMEM + 50 μL of bacteria LM.
6. Incubate 15–20 min. 150 rpm shaking 37°C.
DMEM
Bacteria
7. Decant.
8. Add 10 mL of HBSS 1× at RT.

FIGURE 13.3 Detailed proteomics protocol steps for isolation of *L. monocytogenes* phagosomes.

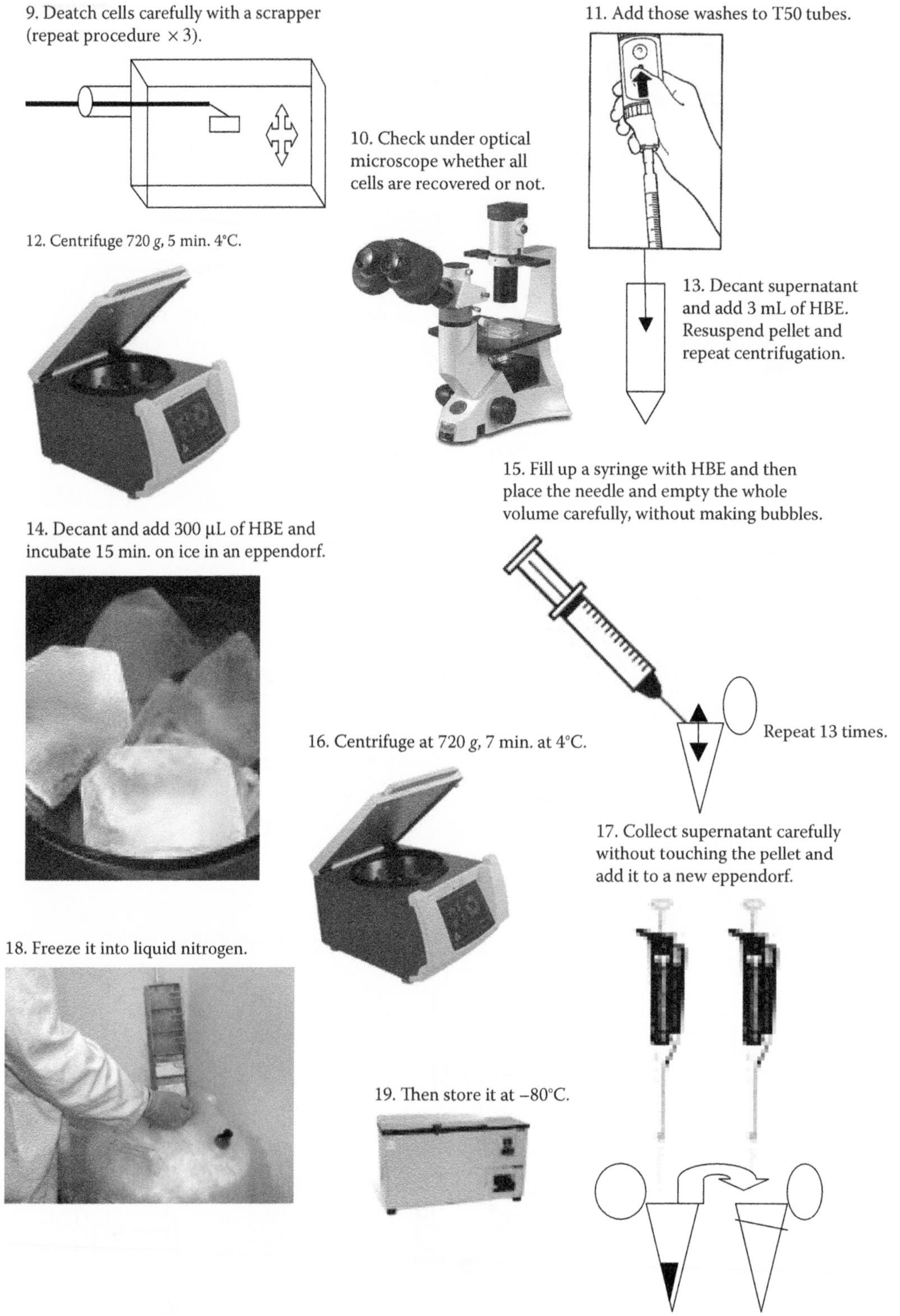
9. Deatch cells carefully with a scrapper (repeat procedure × 3).
10. Check under optical microscope whether all cells are recovered or not.
11. Add those washes to T50 tubes.
12. Centrifuge 720 g, 5 min. 4°C.
13. Decant supernatant and add 3 mL of HBE. Resuspend pellet and repeat centrifugation.
14. Decant and add 300 µL of HBE and incubate 15 min. on ice in an eppendorf.
15. Fill up a syringe with HBE and then place the needle and empty the whole volume carefully, without making bubbles.
Repeat 13 times.
16. Centrifuge at 720 g, 7 min. at 4°C.
17. Collect supernatant carefully without touching the pellet and add it to a new eppendorf.
18. Freeze it into liquid nitrogen.
19. Then store it at −80°C.

FIGURE 13.3 (CONTINUED)

such as the checking of LM viability and phagosomal percentage calculations by plating onto blood agar plates or the purity of the organelles. The purity of the phagosomes was monitored by biochemical analysis to check contamination with other cell components. Plasma membrane contamination was assayed after internalization of LM and surface labeling of cells with horseradish peroxidase (HRP) (500 mg/mL) for 30 min at 4°C. Macrophages bound and internalized HRP by two mechanisms: pinocytosis and receptor-mediated endocytosis through mannose receptor because of mannan-inhibited HRP binding and uptake. The cells were washed with PBS and homogenized, and phagosome isolation was performed as above. HRP was next measured in the phagosomal preparation as a proof of the plasma membrane contamination. The galactosyl transferase activity was checked as a marker for Golgi contamination as described previously. Endosome contamination was recorded by mixing an aliquot of PNS after bacteria uptake and an aliquot of a different set of PNS after 5 min of HRP uptake. Phagosomes were isolated as above, and endosome contamination was measured as the HRP recovered in the phagosomes compared with the total activity present in PNS containing the endosomal probe. Recovery of 0.10% HRP from a total recovery of 70% of the bacteria indicated a low enrichment of endosomes in the preparation. Endosomal and phagosomal fractions were centrifuged (150,000 × *g*, 10 min) to obtain endosomal or phagosomal membranes, respectively. Phagosomal membranes were applied onto 10% SDS-PAGE for biochemical analysis of different markers using specific antibodies. Moreover, phagosomal membranes were also applied to two-dimensional gels for proteomic analysis procedures and DIGE comparison of nonlistericidal phagosomes with listericidal phagosomes, selecting those differential spots for sequencing by matrix-assisted laser desorption ionization time-of-flight (performed by A. Paradela and J. A. Albar at the Proteomic Service of the Center of Biotechnology in Madrid, Spain). This whole procedure is considered as initial proteomics of LM phagosomes. The results obtained by the combination of the biochemical and proteomic analysis were presented in Sections 13.3.1 and 13.3.2. Actually, we are setting up the conditions for a procedure using fluorochrome-labeled samples of phagosomal membranes and DIGE analysis that we called *advanced proteomics*, and we plan to apply for autophagosomes and the full characterization of all phagosomal components, including the bacterial virulence factors contained in the phagosomal membranes.

13.5 UTILITIES OF PHAGOSOMAL PROTEOMICS: TOOLS FOR IMMUNE ANALYSIS

The isolation of phagosomes containing live LM is not only relevant for proteomic purposes and the identification of all components of nonlistericidal and listericidal phagosomes, but they might also be used as a powerful tool in vaccination procedures. We have already shown in Section 13.3 that phagosomal proteomics combined with genetically deficient mice and immune assays might reveal important components of the phagosomal innate immune response against LM in macrophages. However, there are two important proteomic features of phagosomes that implied that these vesicles could induce a good immune and specific response relevant for memory T-cell induction. The presence of MHC class II molecules in these phagosomes was detected at high levels in listericidal phagosomes compared with nonlistericidal phagosomes containing loaded peptides, and these phagosomes were able to degrade one of the main virulence factors of LM, LLO. In this regard, LLO is the main antigen that elicits a good T-cell memory, a requirement for efficient vaccines against LM (Pamer, 2004; Lauvau and Glaichenhaus, 2004; Brockstedt and Dubensky, 2008). Therefore, we have initially set up the conditions for using these vesicles to immunize mice and check the ability of this vaccination protocols to protect against a LM infection. The proteomic protocol we have used implies the inoculation of mice with 30–50 μg of purified phagosomal membranes from nonlistericidal and listericidal phagosomes intraperitoneally for 7 days. Later on, we inoculated the mice with another similar phagosomal dose for 7 days before infection of mice with a sublethal doses of LM (5×10^3 bc per mouse). Next, the mice were bled to get sera that are stored at −80°C, and the mice are sacrificed to obtain the spleens and livers; after homogenization, the homogenates are plated onto blood agar plates for CFU analysis. Sera are used for analysis of TH1 or TH2 cytokine production that indicates the ability of the specific immune response elicited against LM. Similarly, another set of mice are infected with a lethal doses of LM (5×10^4 bc per mouse). Then we check the number of dead mice in each condition: mice without vaccination, mice vaccinated with nonlistericidal phagosomes, mice vaccinated with listericidal phagosomes, and mice vaccinated with endosomes as controls. Our results indicate that nonlistericidal phagosomes were able to protect 20% of the mice from death after 15 days postinfection compared with nonvaccinated mice or mice vaccinated with endosomes that showed an average of 1% of protection. However, listericidal phagosomes were able to protect 80–85% of the mice from death. The vaccination protocols with the sublethal doses of LM indicated that the listericidal phagosomes elicited mainly TH1 proinflammatory cytokines, whereas the amount of TH1 proinflammatory cytokines produced by nonlistericidal phagosomes was limited, explaining the limited efficiency of these vesicles as vaccination vectors. Our current research is focused on the analysis of T-cell populations relevant for a good vaccination protocol as well as the combination of genetically deficient mice in each of the above studied lysosomal proteins to decipher their role in specific immune responses against LM and their role in vaccine designs. We will combine these studies with the proteomics of all components of phagosomes in order to set up diagnostic tools relevant for the preparation of efficient vaccines against LM.

REFERENCES

Alvarez-Dominguez, C., et al. (1996). Phagocytosed live *Listeria monocytogenes* influences Rab5-regulated in vitro phagosome-endosome fusion. *J. Biol. Chem.* 271: 13834.

Alvarez-Dominguez, C., Roberts, R., Stahl, P. D. (1997). Internalized *Listeria monocytogenes* modulates intracellular trafficking and delays maturation of the phagosome. *J. Cell Sci.* 110: 731.

Alvarez-Dominguez, C., Stahl, P. D. (1998). Interferon-g selectively induces Rab5a synthesis and processing in mononuclear cells. *J. Biol. Chem.* 273: 33901.

Alvarez-Dominguez, C., Stahl, P. D. (1999). Increased expression of Rab5a correlates directly with accelerated maturation of *Listeria monocytogenes* phagosomes. *J. Biol. Chem.* 274: 11459.

Alvarez-Dominguez, C., Peña-Macarro, C., Prada-Delgado, A. (2004). Endosome-phagosome interactions in pathogenesis. In *Intracellular pathogen in membrane interactions and vacuole biogenesis* (pp. 51–64). Edited by Golvel, J. P. Austin, TX: Landes Biosciences.

Alvarez-Dominguez, C., et al. (2008). Characterization of a *Listeria monocytogenes* protein interfering with Rab5a. *Traffic* 9: 325.

Birmingham, C. L., Brumell J. H. (2006). Autophagy recognizes intracellular *Salmonella enterica* serovar *typhimurium* in damaged vacuoles. *Autophagy* 2: 156.

Birmingham, C. L., Brumell, J. H. (2009). Methods to monitor autophagy of *Salmonella enterica* serovar *typhimurium*. *Methods Enzymol.* 452: 325.

Birmingham, C. L., et al. (2006). Autophagy controls *Salmonella* infection in response to damage to the *Salmonella*-containing vacuole. *J. Biol. Chem.* 281: 11374.

Birmingham, C. L., et al. (2007). *Listeria monocytogenes* evades killing by autophagy during colonization of host cells. *Autophagy* 3: 442.

Brockstedt, D. G., Dubensky, T. W. (2008). Promises and challenges for the development of *Listeria monocytogenes*-based immunotherapies. *Expert Rev. Vaccines* 7: 1069.

Bucci, C., et al. (1992). The small GTPase Rab5 functions as a regulatory factor in the early endocytic pathway. *Cell* 70: 715.

Carrasco-Marin, E., et al. (2009). The innate immunity role of cathepsin-D is linked to Trp-491 and Trp-492 residues of listeriolysin O. *Mol. Microbiol.* 72: 668.

Carrasco-Marin, E., et al. (2010). LIMP-II links late phagosomal trafficking with the onset of the innate immune response to *Listeria monocytogenes*: A role in macrophage activation. *J. Biol. Chem.* 286: 3332.

Del Cerro-Vadillo, E., et al. (2006). Cutting edge: A novel nonoxidative phagosomal mechanism exerted by cathepsin-D controls *Listeria monocytogenes* intracellular growth. *J. Immunol.* 176: 1321.

Eskelinen, E. L., Tanaka, Y., Saftig, P. (2003). At the acidic edge: Emerging functions for lysosomal membrane proteins. *Trends Cell Biol.* 13: 137.

Gutierrez, M. G., et al. (2004). Autophagy is a defense mechanism inhibiting BCG and *Mycobacterium tuberculosis* survival in infected macrophages. *Cell* 119: 753.

Huang, J., Brumell, J. H. (2009). Autophagy in immunity against intracellular bacteria. *Curr. Top. Microbiol. Immunol.* 335: 189.

Huynh, K. K., et al. (2007). LAMP proteins are required for fusion of lysosomes with phagosomes. *EMBO J.* 16: 313.

Jutras, I., et al. (2008). Modulation of the phagosome proteome by interferon-g. *Mol. Cell Proteomics* 7: 697.

Kabeya, Y., et al. (2000). LC3, a mammalian homologue of yeast Apg8p, is localized in autophagosome membranes after processing. *EMBO J.* 19: 5720.

Kayal, S., Charbit, A. (2006). Listeriolysin O: A key protein of *Listeria monocytogenes* with multiple functions. *FEMS Microbiol. Rev.* 30: 514.

Kayath, C. A., et al. (2010). Escape of intracellular *Shigella* from autophagy requires binding to cholesterol through the type III receptor, IcsB. *Microbes Infect.* 12: 956.

Lauvau, G., Glaichenhaus, N. (2004). Presentation of pathogen-derived antigens in vivo. *Eur. J. Immunol.* 34: 913.

Maganto-Garcia, E., et al. (2008). Rab5 activation by Toll-like receptor 2 is required for *Trypanosoma cruzi* internalization and replication in macrophages. *Traffic* 9: 1299.

Mallo, G. V., et al. (2008). SopB promotes phosphatidylinositol 3-phosphate formation on *Salmonella* vacuoles by recruiting Rab5 and Vps34. *J. Cell Biol.* 182: 741.

Meyer-Morse, N., et al. (2010). Listeriolysin O is necessary and sufficient to induce autophagy during *Listeria monocytogenes* infection. *PLoS One* 5: e8610.

Ogawa, M., et al. (2009). *Streptococcus-*, *Shigella-*, and *Listeria*-induced autophagy. *Methods Enzymol.* 452: 363.

Pamer, E. G. (2004). Immune responses to *Listeria monocytogenes*. *Nat. Rev.* 4: 812.

Portnoy, D. A., et al. (1989). Gamma interferon limits access of *Listeria monocytogenes* to the macrophage cytoplasm. *J. Exp. Med.* 170: 2141.

Portnoy, D. A., et al. (1992). Molecular determinants of *Listeria monocytogenes* pathogenesis. *Infect. Immun.* 60: 1263.

Prada-Delgado, A., et al. (2001). Interferon-g listericidal action is mediated by novel Rab5a functions at the phagosomal environment. *J. Biol. Chem.* 276: 19059.

Prada-Delgado, A., et al. (2005). Inhibition of Rab5a Exchange activity is a key step for *Listeria monocytogenes* survival. *Traffic* 6: 252.

Ramachandra, L., Song, R., Harding, C. V. (1999) Phagosomes are fully competent antigen-processing organelles that mediate the formation of peptide:class II MHC complexes. *J. Immunol.* 162: 3263.

Russell, D. G. (2001). *Mycobacterium tuberculosis*: Here today, and here tomorrow. *Nat. Rev. Mol. Cell. Biol.* 2: 569.

Saftig, P., Beertsen, W., Eskelinen, E. L. (2008) LAMP-2: A control step for phagosome and autophagosome maturation. *Autophagy* 16: 510.

Sanjuan, M. A., Green, D. R. (2008). Eating for good health: Linking autophagy and phagocytosis in host defense. *Autophagy* 4: 607.

Seto, et al. (2009). Dissection of Rab7 localization on *Mycobacterium tuberculosis* phagosome. *Biochem. Biophys. Res. Commun.* 387: 272.

Shaughnessy, L. M., et al. (2006). Membrane perforations inhibit lysosome fusion by altering pH and calcium in *Listeria monocytogenes* vacuoles. *Cell Microbiol.* 8: 781.

Singh, S. B., et al. (2006). Human IRGM induces autophagy to eliminate intracellular mycobacteria. *Science* 313: 1438.

Steele-Mortimer, O., et al. (1999). Biogenesis of *Salmonella typhimurium*-containing vacuoles in epithelial cells involves interactions with early endocytic pathway. *Cell Microbiol.* 1: 133.

Vergne, I., et al. (2004). Cell biology of *Mycobacterium tuberculosis* phagosome. *Annu. Rev. Cell. Dev. Biol.* 20: 367.

Section III

Computational and Systems Biology

14 In a Quest to Uncover Governing Principles of Cellular Networks

A Systems Biology Perspective

Kumar Selvarajoo and Masa Tsuchiya
Keio University
Tsuruoka, Japan

CONTENTS

14.1 INTRODUCTION

Recent years have seen high-throughput experimental technologies generating large sets of biological data that are helping to unravel the detailed molecular composition and complexity of living organisms. However, the real challenge is in the understanding of how such information is incorporated to yield biological properties such as aging, growth, metabolism, and biological diversity.

Each cell is packed with a large number of molecules constituting of DNA, RNA, proteins, and metabolites. These molecules interplay to execute numerous inter- and intracellular processes in response to environmental changes or perturbations that allow them to survive under diverse conditions. It is now evident that the molecular interactions create a large degree of interconnectivity that controls and regulates vital complex processes such as the immune responses, cell cycle, cell division, and apoptosis.

A natural direction in the postgenomic era has been in the systemic high-throughput cataloguing of genes, proteins, and metabolites and identifying all possible interactions between them (Ideker et al., 2001). Today, advances made in the integration of traditional wet-bench biology with high-throughput technologies and theoretical methodologies, the so-called systems biology, suggest that universal laws govern cellular networks (Albert, 2005; Barabási and Oltvai, 2004). This recent change of mindset to a multidisciplinary approach, utilizing the law of causality, will revolutionize our view of modern biology and disease pathologies.

14.2 ORGANIZATION OF BIOLOGICAL NETWORKS

Various studies have demonstrated that biological networks are robust to random errors (Barabási and Oltvai, 2004; Kitano, 2004). Typically, random deletion, mutations, or duplication of genes have been shown not to affect the overall network behavior or phenotypic outcome of living systems, revealing the persistence of stable and robust behavior under diverse perturbations (Bennett and Hasty, 2008; Ishii et al., 2007). On the other hand, biological networks were also shown to be stable over a wide range of biochemical parameter variation (Barkai and Leibler, 1997; Alon et al., 1999).

Despite being robust and stable, biological systems have been shown to be fragile on a number of occasions and are vulnerable to attacks (Rao and Arkin, 2001; Stromberg and Carlson, 2006; Lemoine et al., 2005; Carlson and Doyle, 2002). The removal of a relatively few highly linked molecules (molecules that interacts with many other molecules) can lead to system failure, a property found in scale-free networks (Albert and Barabasi, 2000). This postulation suggests that biological networks are not connected randomly but center around a small proportion of *hub* and *connector* elements (Barabási and Oltvai, 2004; Albert and Barabasi, 2000; Yu et al., 2007).

Although uncovering the general organization of biological networks, robust to perturbation, and fragile to targeted attacks obeying the scale-free or power law, the comprehensive understanding of dynamic cellular behaviors and their control mechanisms still remains poor. Thus, there is an increasing need for the development of integrated theoretical and experimental approaches to model and understand the topological and dynamic properties of the various networks that, in quantifiable terms, control the behavior of the living cell (Barabási and Oltvai, 2004).

14.3 MODELING COMPLEX BIOLOGICAL NETWORKS

The molecular environment within a cell is highly inhomogenous, consisting of specialized compartments, e.g., mitochondria within the cytoplasm, to carry out fundamental survival and regulatory functions such as energy metabolism and signal transduction pathways (Alberts et al., 2008). Landmark experimental and imaging works have provided evidence for the complexity in spatiotemporal organization of signal transduction pathways (Fink et al., 1999, 2000) from the activation of receptor at the membrane surface inducing intracellular cascades such as phosphorylation, ubiquitylation, and sumoylation. These effectors can be gene-regulated transcription factor proteins, metabolic components, or parts of a cytoskeleton for cell shape or movement (Alberts et al., 2008). The results from these efforts have prompted recent investigations to focus on understanding biochemical reaction networks with transport mechanisms using deterministic and stochastic spatiotemporal partial differential equations (PDEs) (Eungdamrong and Iyengar, 2004; Costa et al., 2009; Kholodenko et al., 2010).

Although these efforts have grown significantly in recent years, the studies have mainly focused on the progress made in the computational methodological aspects, rather than the actual understanding of cytosolic signal transduction or nuclear gene expressions. The main reasons for this lie in the difficulties of conceptualization of the spatiotemporal events of the biochemical reactions, the cells ever-changing shape and environment, the determination of detailed molecular reaction principles that govern each interaction, and the experimental determination of reaction-diffusion parameter values and their sensitivity caused by the resultant nonlinear PDEs. Moreover, very specialized experimental *in vivo* techniques for cellular systems to measure species concentration with high accuracy and fine time resolution measurements of such variables are far from accomplished (Eungdamrong and Iyengar, 2004; Kholodenko et al., 2010; Dehmelt and Bastiaens, 2010). Therefore, in the current situation, the ability to use deterministic and stochastic spatiotemporal PDEs to model complex biological processes is limited.

On the other hand, remarkably, many studies using simpler, nonparametric approaches have yielded notable successes for the understanding of the average cell response of biological networks. These approaches can be broadly classified into qualitative and quantitative. The qualitative approaches consist of stoichiometric methods, which deal with reaction stoichiometry to identify reaction mechanisms (Tsuchiya and Ross, 2001; Schilling and Palsson, 2008); singular-value decomposition methods and statistical clustering for identifying coordinated behavior from high-throughput data (Yeung et al., 2002; Alter et al., 2000); and Bayesian network approaches for constructing network connectivity using probabilistic graphs (Friedman et al., 2000; Friedman, 2004). Examples of quantitative approaches include perturbation-response experiments, which infer network connectives using perturbation and activation dynamics of reaction species (Bruggemann et al., 2002; Morán et al., 2007; Arkin et al., 1997) and Boolean sequential dynamical approaches, which decipher causal networks by mapping input and output data (Yeo, 2007; Brazma and Schlitt, 2003).

In this chapter, we review the progress made in the understanding of network biology, through quantitative perturbation-response approaches, especially focusing on immune cell signaling dynamics using the universal law of conservation in conjunction with simple rules.

14.4 PERTURBATION-RESPONSE EXPERIMENTS REVEAL GOVERNING PRINCIPLES IN BIOLOGICAL NETWORKS

14.4.1 The Use of First-Order Mass-Action Response Equations to Detect Network Connectivities

Perturbation-response studies involve the controlled perturbation of a certain (input) reaction species of a system kept at steady state and monitor average responses of concentration/activation levels of other species (output) of the system. Ross and colleagues (Vance et al., 2002; Ross, 2008) applied pulse perturbation to deterministic kinetic evolution equation and constructed rules for detecting causal connectivities of species in a reaction network. By studying the effects of giving small perturbation to the concentration of one or more species and using first-order mass-action response equations to monitor the response profiles of other species within a mass-conserved system, Ross and colleagues proposed the analysis of time to reach peak, and the actual peak levels for all species can be used to determine the causal relationships or connectivities between them.

To illustrate, consider a linear chain of reactions ($X_1 \rightarrow X_2 \rightarrow X_3 \rightarrow \ldots$) at steady-state conditions (Vance et al., 2002). If the concentration of X_1 is pulse perturbed, concentrations of X_2, X_3, X_4, etc., will increase, go through a maximum, and then decrease back to steady-state values in sequential order (stable perturbation-response phenomenon) (Figure 14.1). Thus, the temporal order of the responses can yield the causal (direct) connectivities of the species in the reaction mechanism.

The basic principle of Ross et al. is that simple linear rules might exist for the propagation of small perturbation (response wave) within a biochemical system. That is, the response rate of species in a mass-conserved system at initial steady state can be approximated by the first-order mass-action response equation, given small perturbation to one or more species.

The experiments, based on the law of conservation, preserve the total input and output fluxes. Under such conditions, a linear superposition of propagation response wave connects the species between input and output fluxes. Thus, simple linear rules can be derived for the system, notably:

1. The time to reach peak values (u_1, u_2, etc. in Figure 14.1) increases, and its amplitude decreases as we move down the reaction network, unless there are other features such as feedback reactions.

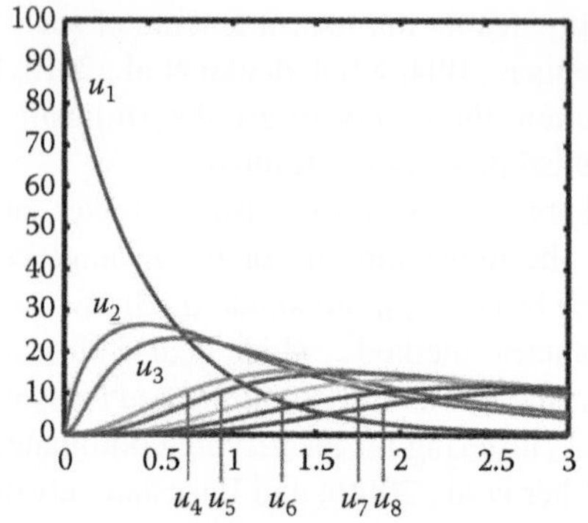

FIGURE 14.1 Temporal response profile of concentration in linear chain of first-order mass-action reaction network for pulse perturbation of specie X_1 (see main text). The units are arbitrary, scaled by rate coefficients. (Adapted from Vance, W., Arkin, A., and Ross, J., *Proceedings of the National Academies of Sciences USA*, 99, 5816–5821, 2002.)

2. The initial response gradient can be used to determine the location of a reactant species in a network, i.e., the steepest gradient is closest to the perturbed reactant species and the lowest gradient is farthest.
3. Reactant species that are not connected to the system do not show any response profile.
4. Because the law of mass conservation is used for pulse propagation, the sum of all species' deviations from steady state (weighted by stoichiometric coefficients) is constant. Therefore, it can help determine correct stoichiometric coefficients.

Using the simple linear rules, Torralba et al. (2003) investigated the glycolysis pathway, a well-appreciated network module involving energy metabolism. They confirmed that simple linear rules can be used to infer again the causal connectivity of the glycolysis pathway. In other words, the perturbation-response study clearly provides evidence that complex biological networks could be governed by first-order mass-action response equations, resulting in the linear superposition of propagation response waves.

14.4.2 Detecting Feedback Control of Transcription Factor NF-κB Using Mass-Action Response Equations

The transcription factor nuclear factor-κB (NF-κB) is a well-known key regulator of several cellular functions such as the immune response, cell growth, survival, and apoptosis (Vallabhapurapu and Karin, 2009; Baud and Karin, 2009). Hence, numerous investigations have been devoted to the elucidation of its regulatory properties in health and disease. Furthermore, the prospect of developing therapeutic target of NF-κB for proinflammatory diseases and cancer looks worthwhile (Baud and Karin, 2009; Karin et al., 2004).

To understand the dynamics of the IκB kinase (IKK)-IκB-NF-κB signaling module and to elucidate the distinct gene expression programs for different stimuli in murine embryonic fibroblasts, Werner et al. (2005) developed a computational model based on simple mass-action response equations with IKK perturbation. Because temporal control of NF-κB activity can lead to selective gene expressions (Hoffmann et al., 2002), stimulus-specific temporal control of IKK activity might allow for distinct biological responses if signal processing within the IKK-IκB-NF-κB signaling module resulted in distinct NF-κB activity profiles.

Compared with experimental data for both tumor-necrosis factor (TNF) and lipopolysaccharide (LPS) stimulation, Werner et al. (2005) simulations suggested that the mechanism for the rapid termination of IKK activity in TNF stimulation is due to negative feedback control by postinduction of A20, whereas the prolonged activation of IKK observed in LPS stimulation is due to positive feedback of autocrine TNF signaling. Therefore, from a simple mass-action response model, the switching nature of IKK activity, by regulating certain key molecules such as A20, can produce a distinct cellular response that could be harnessed for therapeutic benefits.

14.4.3 The Power of First-Order Mass-Action Response in Elucidating Novel Innate Immune Signaling Features

In our works, we investigated the well characterized innate immune response to invading pathogens based on the Toll-like receptor (TLR) signaling pathways. Using the perturbation-response approach, we developed *Response Rules* (see Section 14.4.3.3) that infer novel pathway features as of the TLR3 and 4 signaling:

1. Signaling intermediates (Selvarajoo, 2006; Helmy et al., 2009)
2. Crosstalk mechanisms (Helmy et al., 2009; Selvarajoo, 2007)
3. Signaling flux redistribution (SFR) (Selvarajoo et al., 2008b)

To illustrate our approach, we perturb a stable biological network consisting of n species from the reference steady states. In general, the resultant changes in the concentration of species are governed by the kinetic evolution equation:

$$\frac{\partial X_i}{\partial t} = F_i(X_1, X_2, \ldots, X_n), i = 1, .., n \tag{14.1}$$

where the corresponding vector form of Equation 14.1 is $\frac{\partial \boldsymbol{X}}{\partial t} = F(\boldsymbol{X})$. $\boldsymbol{F}$ is a vector of any nonlinear function including diffusion and reaction of the species vector $\boldsymbol{X} = (X_1, X_2, \ldots X_n)$, which represents activated concentration levels of signaling molecules, for example, through phosphorylation and binding concentration of transcription factors to promoter regions. The response to perturbation can be written by $\boldsymbol{X} = \boldsymbol{X}_0 + \delta\boldsymbol{X}$, where $\boldsymbol{X}_0$ is the reference steady-state vector, and $\delta\boldsymbol{X}$ is the relative response from steady states ($\delta\boldsymbol{X}_{t=0} = 0$).

We consider small perturbation around steady state in which higher-order terms become negligible in Equation 14.1, resulting in the approximation of the first-order term. In vector form

$$\frac{d\delta\boldsymbol{X}}{dt} \cong \left.\frac{\partial F(\boldsymbol{X})}{\partial \boldsymbol{X}}\right|_{X=X_0} \delta\boldsymbol{X} \tag{14.2}$$

(note the change from partial derivative to total derivative of time), where the zeroth-order term $\boldsymbol{F}(\boldsymbol{X}_0) = 0$ at the steady-state $\boldsymbol{X}_0$ and $\boldsymbol{J}$ is the *Jacobian* matrix or *linear stability matrix*. The elements of $\boldsymbol{J}$ are chosen by fitting $\delta\boldsymbol{X}$ with corresponding experimental profiles and knowing the activation topology. Note that $\boldsymbol{J}$ elements (response coefficients) can include not only reaction information but also spatial information such as diffusion and transport mechanisms (see below).

14.4.3.1 TLR4 Signaling

14.4.3.1.1 Predicting Missing Molecules/Processes

Briefly, in TLR4 signaling, upon LPS recognition, myeloid differentiation primary response gene 88 (MyD88) and Toll/IL-1 receptor-domain–containing adaptor-inducing interferon-β (TRIF) bind to TLR4 and trigger the MyD88-dependent and TRIF-dependent pathways, respectively (Figure 14.2a) (Akira et al., 2001). We developed a wild-type model based on first-order response equations and

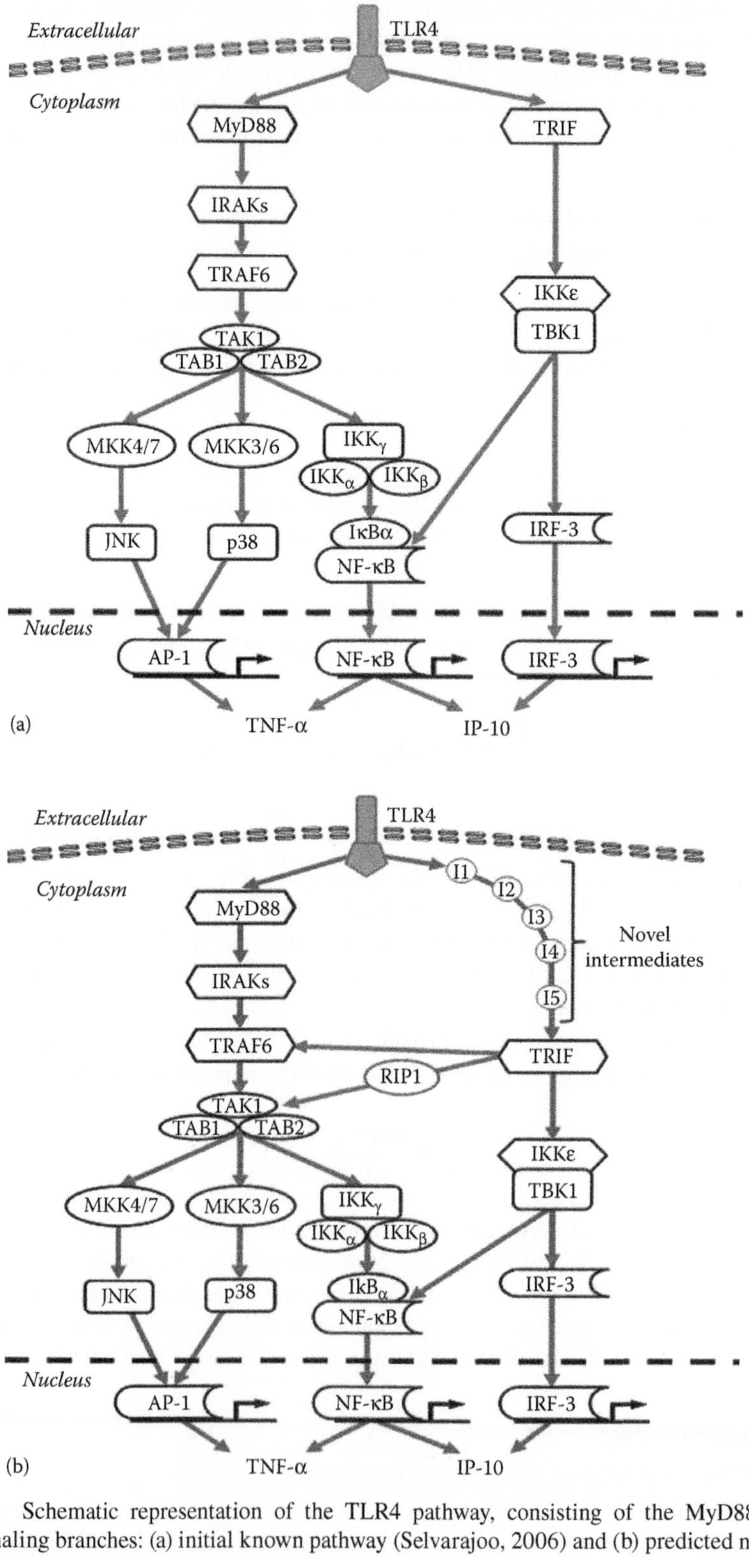

FIGURE 14.2 Schematic representation of the TLR4 pathway, consisting of the MyD88- and TRIF-dependent signaling branches: (a) initial known pathway (Selvarajoo, 2006) and (b) predicted novel signaling intermediates (41) and crosstalk mechanisms (Selvarajoo, 2007, 2008a).

tested the *in silico* MyD88 knock-out (KO) (Selvarajoo, 2006). Our initial wild-type model simulation could not reproduce the delayed experimental time-series activation profile of downstream NF-κB observed in the MyD88 KO (Selvarajoo, 2006).

Time-delay processes can occur for several reasons such as directed transport machinery, complex formation, missing reactions, etc. The Jacobian matrix $\boldsymbol{J}$ can represent such information through additional terms. For example, to simulate the desired 20-minute–delayed experimental activation of NF-κB in MyD88 KO compared with wild-type, we added five novel mass-action response terms (intermediates) along the TRIF-dependent pathway (Figure 14.2b) to reproduce the delayed activation; here the additional terms can be treated as a chain of dynamical events that may include diffusion process (also approximated to follow mass-action response) and reaction terms independently (see Section 14.4.3.3). It is important to note that to maintain a peak activation level, reducing response coefficients alone cannot provide a time-delay activation profile.

Although the five additional hypothetical terms may not exactly represent five actual intermediate molecules, as mentioned above, these terms broadly represent missing biological molecules/processes including spatial effects. Nevertheless, our predicted novel terms (Selvarajoo, 2006) were subsequently confirmed experimentally to be:

1. The phosphorylation of toll-IL-1R domain-containing adaptor-inducing IFN-β-related adaptor molecule (TRAM) by protein kinase Cε upstream of TRIF (Mcgettrick et al., 2006)
2. Binding of TNF receptor-associated factor 3 (TRAF3) to TRIF (Hacker et al., 2006)
3. Upstream sequential events (endocytosis of TLR4) leading to TRAM activation (Kagan et al., 2008)

Using this approach further on the TLR4 signaling, analyzing both NF-κB and mitogen-activated protein (MAP) kinases in wild-type, MyD88 KO, TRAF6 KO, TRIF KO, TAK1 KO, and MyD88/TRAF6 double KO, we were able to propose two additional crosstalk mechanisms between TRIF and TRAF6 and between TRIF to RIP1, which are necessary to explain the induction of MyD88-dependent genes *Tnf* and *Ilb* in MyD88 KO (Selvarajoo, 2007) (Figure 14.2b).

14.4.3.1.2 Discovering SFR at Pathway Junctions

Again in TLR4 signaling, we focused on the behavior of the alternative TRIF (TRAM)-dependent pathway in wild-type and MyD88 KO condition and showed that *in silico* removal of MyD88 resulted in the increased activation of the TRAM-dependent pathway (Figure 14.3), through signaling flux (the rate of signaling molecules' activation) conservation. That is, the total signaling flux propagation through the network, from receptor activation through downstream gene activation, is conserved. The concept of the conservation derives from the idea that the removal and addition of a molecule at a signaling pathway junction enhances and impairs, respectively, the entire alternative pathway through SFR (Selvarajoo et al., 2008).

To illustrate this, the removal of MyD88 solely abolishes the propagation of signal transduction from TLR4 to MyD88. As a result, interaction between TLR4 and TRAM increases because of the law of mass conservation derived from first-order response:

$$-\frac{d(TLR4)}{dt}+\sum\frac{d(MyD88\,pathway)}{dt}+\sum\frac{d(TRAMpathway)}{dt}=0 \tag{14.3}$$

where $\frac{d(TLR4)}{dt}$ is the rate of TLR4 activation by LPS perturbation, and $\sum\frac{d(MyD88\,pathway)}{dt}$ and $\sum\frac{d(TRAMpathway)}{dt}$ are signaling flux of MyD88 and TRAM pathways, respectively. In *in silico* MyD88 KO, $\sum\frac{d(MyD88\,pathway)}{dt}=0$ and, therefore, $\sum\frac{d(TRAMpathway)}{dt}$ increase, resulting in enhanced activation of the entire TRAM-dependent pathway, i.e., SFR.

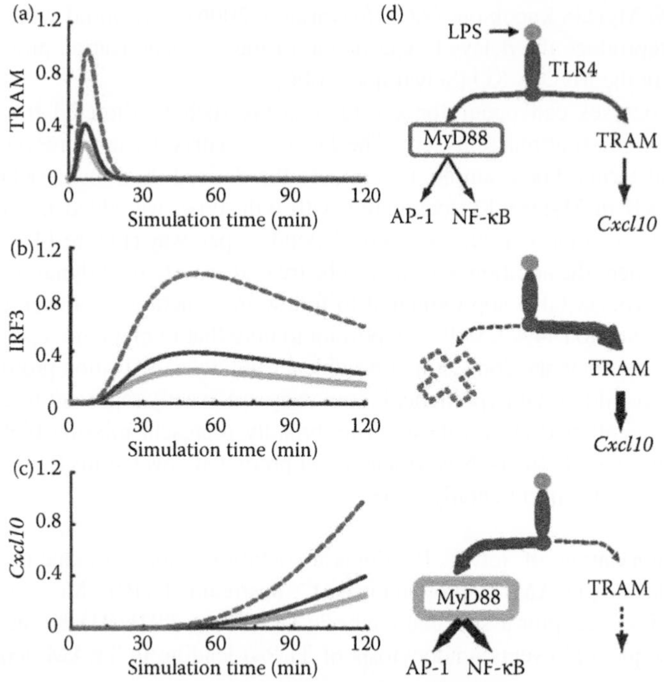

FIGURE 14.3 Enhancement of TRAM-dependent pathway in MyD88 KO due to SFR. (a–c) Simulation profiles (arbitrary units) of TRAM activation (a), IRF3 activation (b), and *Cxcl10* induction (c) in wild-type (dark gray), MyD88 KO (dotted), and two-fold overexpression (light gray) of MyD88; (d) schematic of SFR. (d, top) Wild-type; fluxes propagate through both the MyD88-dependent and TRAM-dependent pathways. (d, middle) MyD88 KO; more fluxes propagate or overflows through the TRAM-dependent pathway resulting in increased *Cxcl10* induction; (d, bottom) MyD88 overexpression by two-fold. (Adapted from Selvarajoo, K., Takada, Y., Gohda, J., et al., *PLoS One*, 3, e3430, 2008.)

Furthermore, overexpressing MyD88 reduced activation of TRAM-dependent molecules. That is, when molecules at the pathway junction are decreased (or increased), the activation of alternative pathways enhances (or reduces) through SFR. These results were validated experimentally at two pathway junctions of TLR4 (Selvarajoo et al., 2008b). Although we demonstrated SFR for molecules with a common binding domain, SFR might also occur between molecules with a different binding domain at pathway junctions from the law of mass flow conservation (Figure 14.4). Thus, SFR, a simple linear rule, illustrates a novel mechanism for enhanced activation of alternative pathways when molecules at pathway junctions are removed.

14.4.3.2 Predicting Novel Features of TLR3 Signaling

Similarly to the TLR4 signaling, we investigated the mammalian innate immune response against viral attacks by recognizing double-stranded RNA or its synthetic analog polyinosinic-polycytidylic acid (poly(I:C)) by the TLR3. This leads to the activation of MAP kinases and NF-κB, which results in the induction of type I interferons and proinflammatory cytokines to combat the viral infection. Here, using the first-order response equations and signaling flux conservation on the activation dynamics of the NF-κB and MAP kinases in wild-type, TRAF6 KO, and TRADD KO condition, we inferred:

1. The existence of missing intermediary steps between extracellular poly(I:C) stimulation and intracellular TLR3 binding
2. The presence of a novel pathway that is essential for JNK and p38, but not NF-κB, activation (Helmy et al., 2009)

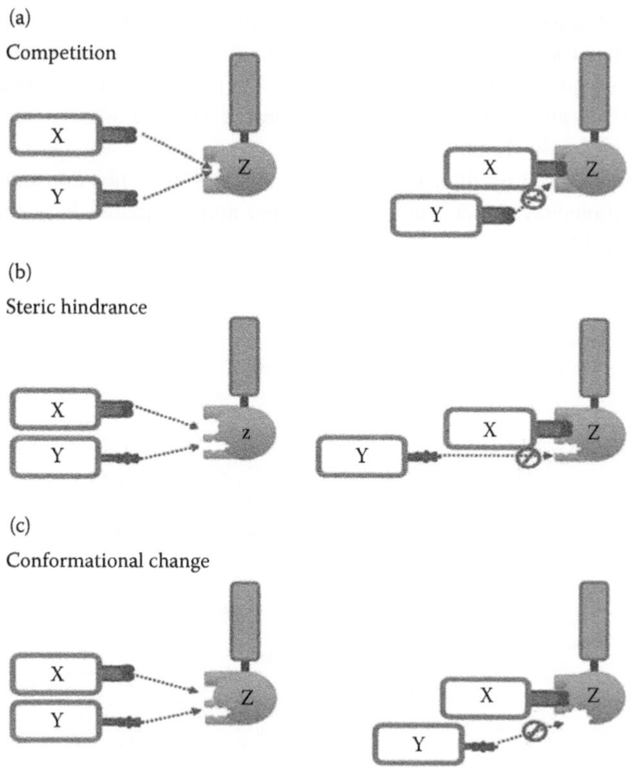

FIGURE 14.4 Possible mechanisms of action for SFR (but not limited). (a) Competition: Molecules X and Y compete to bind with molecule Z. X and Y share binding sites at Z. (b) Steric hindrance: When X binds to Z, the complex prevents the binding of Y to another binding site at Z. (c) Conformational change: when X binds to Z, structural changes to Z lowers the affinity of Y binding to Z. (Adapted from Selvarajoo, K., Takada, Y., Gohda, J., et al., *PLoS One*, 3, e3430, 2008.)

14.4.3.3 Setting Response Rules for Revealing Novel Network Features

From our TLR3 and TLR4 investigations, the law of signaling flux conservation and the first-order response derive four Response Rules for analyzing biological topology with experimental data alone:

- Rule 1 – time delay: Any time delay between initial (e.g., receptor) and target signaling molecule (e.g., NF-κB) activations represents missing cellular processes such as directed transport machinery, protein complex formation, and novel molecular interactions. For example, three mass conserved intermediate terms, X_i, X_{i+1}, X_{i+2}, can be linearly added to a network $\xrightarrow{K_{i-1}} X_i \xrightarrow{K_i} X_{i+1} \xrightarrow{K_{i+1}} X_{i+2} \xrightarrow{K_{i+2}}$ to form time delay (see Figure 14.5a and Section 14.3).
- Rule 2 – SFR at pathway junction: A special case where the removal or addition of a molecule at a pathway junction results in the enhancement or repression of the entire alternative pathway (Figures 14.3 and 14.5b).
- Rule 3 – balancing signaling flux: When the removal of an upstream molecule does not completely abolish the activation of a directly connected downstream molecule, the presence of novel bypass pathway(s) is uncovered, with or without intermediates(s) if no SFR is observed (Figure 14.5c), and with intermediates if SFR is observed (Figure 14.5d).

- Rule 4 – no SFR at pathway junction: When the removal of a molecule at pathway junction does not enhance the alternative pathway, intermediate(s) between the removed molecule and the pathway junction is proposed (Figure 14.5e), and crosstalk between the removed molecule and the alternative pathway is suggested (Figure 14.5f). Because no SFR is observed, the crosstalk should be dominant.
- Rule 5 – crucial signaling flux: Crucial pathways and branches are identified when target signaling molecule activations are quantified and compared between wild-type and mutants (Figure 14.5g).

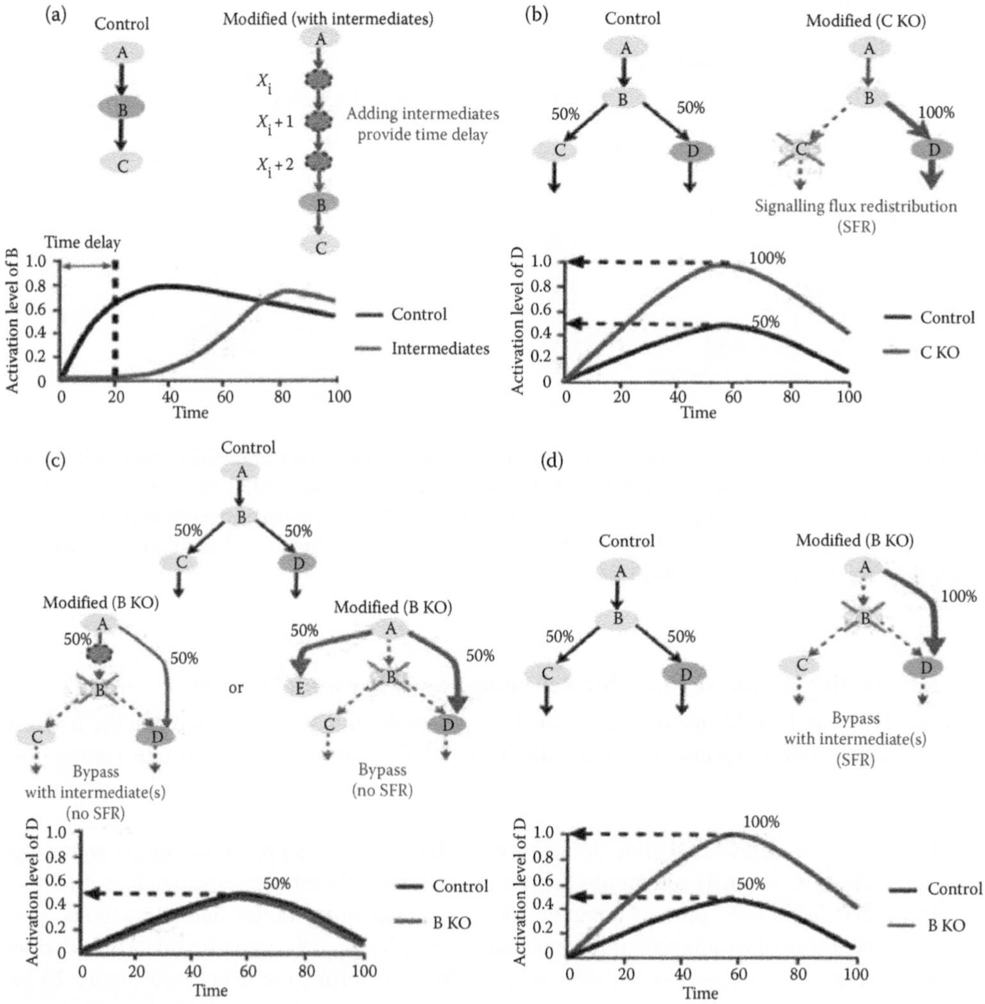

FIGURE 14.5 Response rules for revealing novel network features. (a) Rule 1 – time-delay mechanisms: X_i, X_{i+1}, X_{i+2} are added between molecules A and B to achieve delayed activation. (b) Rule 2 – signaling flux redistribution. (c and d) Rule 3 – balancing signaling flux: novel bypass pathway with or without intermediates if no SFR observed (c) and without intermediate(s) if SFR observed (d). (e and f) Rule 4 – no SFR at pathway junction: novel intermediate(s) (e) and crosstalk at pathway junction (f). (g) Rule 5 – crucial signaling flux: key pathway. See text for details. Note that the schematic illustrates the strength of simple rules to decipher novel network features and are not meant to show exhaustively all possible network connectivities. The x-axis represents time, and the y-axis represents activated concentration, both in arbitrary units.

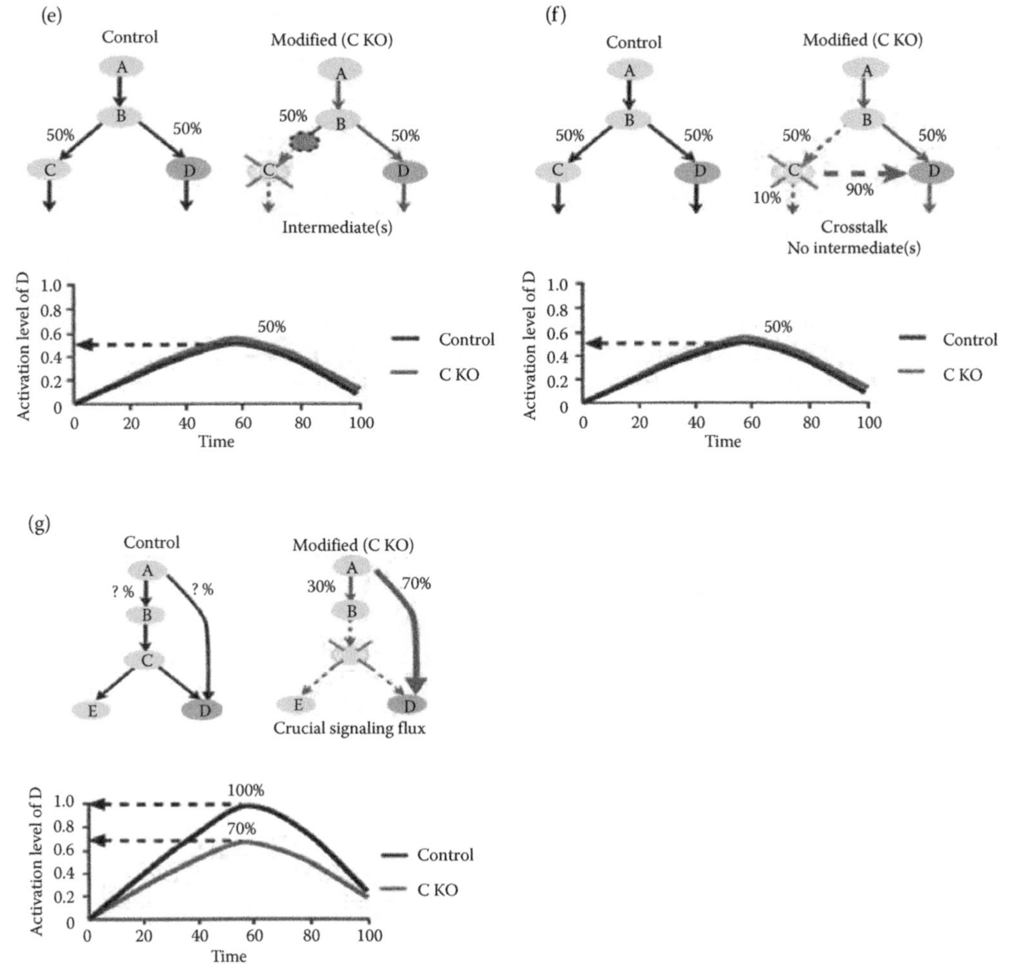

FIGURE 14.5 (CONTINUED)

14.4.4 Reverse Engineering to Decipher Local Biological Network Connectivities: The Nonintegral Connectivity Method

So far, we have shown that the complex cellular processes can be understood by simple linear rules derived from the signaling flux conservation and the first-order response equations. Using this approach to analyze cellular network is forward engineering, where the initial network is first obtained from the current knowledge and subsequently modified to fit experimental outcome. In this section, we show a reversed engineering approach, where *a priori* knowledge of a network is not required to analyze its biological response. We review the nonintegral connectivity method (NICM) (Selvarajoo and Tsuchiya, 2007; Selvarajoo et al., 2009), which uses the first-order response solutions of δX (sum of exponentials) and delineates them into formation and depletion propagation response waves. Such delineation of response waves can be used for the construction of local biological network connectivities.

For pulse perturbation, we can linearize Equation 14.1 (Section 14.4.3) to solve δX by linear transformation, $\delta \boldsymbol{X} = \boldsymbol{P}\delta \boldsymbol{Y}$ and diagonalizing $\boldsymbol{J}$, such that $\frac{d\delta Y}{dt} = P^{-1}\boldsymbol{J}\boldsymbol{P}\delta \boldsymbol{Y}$. The diagonal elements of $\boldsymbol{P}^{-1}\boldsymbol{J}\boldsymbol{P}$ are eigenvalues of the Jacobian matrix, when matrix $\boldsymbol{P} = (\hat{e}_1, \hat{e}_2,.., \hat{e}_n)$ is given by column unit eigenvector $\hat{\mathbf{e}}_i$ of $\boldsymbol{J}$ (i.e., $\boldsymbol{J}\hat{\boldsymbol{e}}_i = \lambda_i \hat{\boldsymbol{e}}_i$ with eigenvalue λ_i, $i = 1, 2, ...n$). Then the solution vector, $\delta\boldsymbol{X} = (\delta X_1, \delta X_2, ... \delta X_n)$, becomes the summation of exponential terms:

$$\delta\boldsymbol{X} = \left(\sum_{i=1}^{n} a_{1,j} e^{-\lambda_j t}, ..., \sum_{i=1}^{n} a_{n,j} e^{-\lambda_j t} \right) \tag{14.4}$$

where a_{ij} is the jth perturbation coefficient of the ith molecule, X_i. Theoretical analysis shows that the eigenvalues λ_i are nonpositives, because otherwise the output signal will be unbounded. The eigenvalues are usually simple (distinct) and real; however, complex eigenvalues might exist. If complex values exist, then they occur in conjugated pairs that produce trigonometric functions of time in Equation 14.4, modulated by exponentially decaying amplitudes. Note that in some cases such linearization can lead to error compensation or amplification (Vlad et al., 2004).

To illustrate NICM, let us consider an example of pulse perturbation (α, 0) at $t = 0$, given to a simple two-intermediate first-order mass-action model, $\boldsymbol{X} = (X_1, X_2)$: $X_1 \xrightarrow{k_1} X_2 \xrightarrow{k_2}$.

The perturbation wave $\delta\boldsymbol{X} = \begin{pmatrix} \delta X_1 \\ \delta X_2 \end{pmatrix}$ applied to the system with rate constants k_1 and k_2 for X_1 and X_2, where X_1 has k_1 as depletion, and X_2 has k_1 as formation and k_2 as depletion, can be represented by:

$$\frac{d\delta\boldsymbol{X}}{dt} = \begin{pmatrix} -k_1 & 0 \\ k_1 & -k_2 \end{pmatrix} \delta\boldsymbol{X} \tag{14.5}$$

With initial conditions, $\boldsymbol{X}_0$, the temporal profiles of intermediates are $\boldsymbol{X}_0 + \delta\boldsymbol{X}$ and yield sum-of-exponential solutions:

$$\delta\boldsymbol{X}_1 = \alpha e^{-k_1 t} \tag{14.6}$$

$$\delta X_2 = \frac{k_1}{k_2 - k_1}(e^{-k_1 t} - e^{-k_2 t}) \tag{14.7}$$

Factorize Equation 14.7 with respect to $e^{-k_1 t}$ if $k_2 > k_1$ (or $e^{-k_2 t}$ if $k_1 > k_2$) and obtain:

$$\delta X_2 = \frac{k_1}{k_2 - k_1}(1 - e^{-(k_2 - k_1)t})e^{-k_1 t} \tag{14.8}$$

In NICM, we use Equation 14.8 as a foundation to set up a generalized expression directly for the concentration or activated concentration level of an intermediate X from steady state, acting along a pathway, to constitute of both the formation and depletion wave terms:

$$\delta X = \alpha \times \underline{\left(1 - e^{-p_1 t}\right)} \times e^{-p_2 t} \tag{14.9}$$

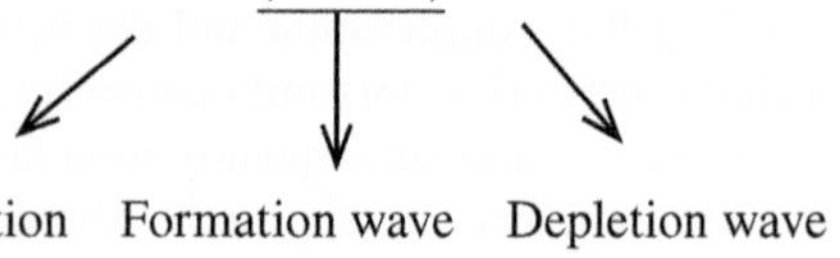

where α represents the amount of perturbation, and p_1 and p_2 represent the measure of formation and depletion response propagation waves, respectively. In general, p_1 and p_2 are not equal to k_1 and k_2, respectively, and are determined by fitting with experimental data.

Symbolically the formation and depletion wave terms are defined by:

Formation wave term: $\delta X = (1 - e^{-p_1 t}) \xrightarrow{p1} \boxed{X}$

Depletion wave term: $\delta X = e^{-p_2 t} \boxed{X} \xrightarrow{p2}$

Combining the two terms with unit perturbation, we obtain: $\delta X = (1 - e^{-p_1 t})e^{-p_2 t} \xrightarrow{p1} \boxed{X} \xrightarrow{p2}$. When a perturbation of α (where α can be a constant or function of time) is given to X_1, it induces change to X_1 (i.e., δX_1), which subsequently induces change to X_2 (δX_2) and so on, until the perturbation a is propagated throughout the network.

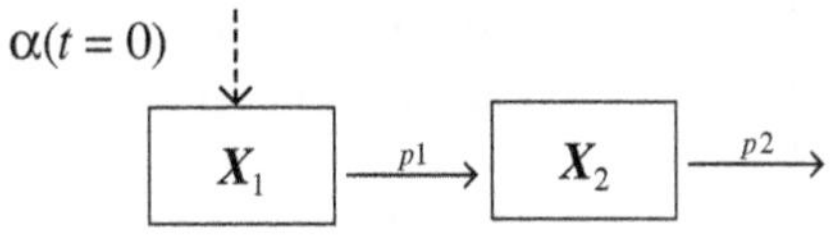

Mathematically:

$$\delta X_1 = \alpha e^{-p_1 t} \tag{14.10}$$

$$\delta X_2 = \alpha(1 - e^{-p_1 t})e^{-p_2 t} \tag{14.11}$$

For mass conservation, the depletion coefficient of X_1 (p_1) is equal to the formation coefficient of X_2. The expansion of the linear-chain motif shows that time delay can be easily observed by NICM representation; for instance, consider adding three intermediates to the chain reaction between X_1 and X_5. From Table 14.1 (linear-chain motif), the activation profile of δX_5 becomes $\alpha(1 - e^{-p_1 t}) \ldots (1 - e^{-p_5 t})e^{-p_6 t}$, where each of the three intermediate terms represents a formation term $(1 - e^{-p_i t}) < 1$ ($i = 2, 3, 4$), thereby their three times multiplication reveals a time-delay process.

Basically the time events of the formation and depletion wave terms for each reactant species obey the law of superposition of propagation waves, which is the essence of NICM. This can be used to construct theoretical (NICM) motifs of commonly found subnetwork structures such as linear-chain, merging, branching, reversible, feedback loop (FBL), and feed-forward loop (FFL) motifs (Table 14.1). By fitting experimental concentration profiles with commonly found network motifs constructed using NICM, we are able to detect the local connectivity of reactant species. In time-dependent perturbation or regulation such as FBL and FFL, p_1 and p_2 may not be constant values. For illustration, the construction of FBL motif is an extension of linear-chain motif with nonconstant power of exponential:

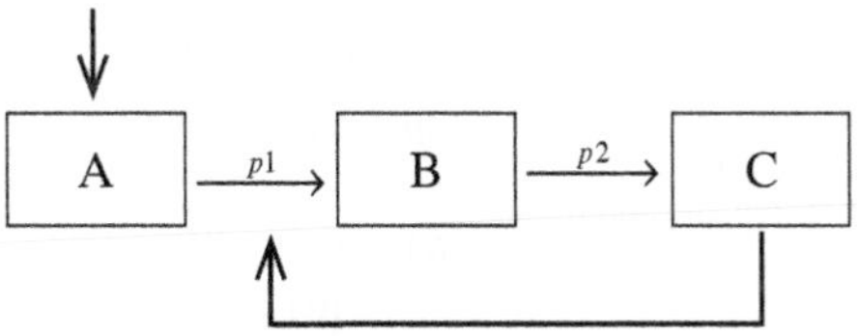

(14.12)

$$\delta A = A_1 e^{-p_1(C)t} \tag{14.13}$$

$$\delta B = A_1(1 - e^{-p_1(C)t})e^{-p_2 t} \tag{14.14}$$

$$\delta C = A_1(1 - e^{-p_1(C)t})(1 - e^{-p_2 t}) \tag{14.15}$$

where $p_1(C)$ is function of species C concentration/activation level. For positive FBL, p_1 is a function such that it increases in line with C. For negative FBL, p_1 is a function that decreases as C increases. A similar approach can be taken for FFL. In order to determine the exact function of formation or depletion terms in FBL and FFL, we search for the best fitting experimental data.

To show how NICM works and prevents false positive connectivities, we created a three-chain linear motif using first-order mass-action kinetics with predefined rate constants and pulse

TABLE 14.1
NICM Motifs and Expressions

Motif	NICM Expressions Based on Response Propagation Rules
Linear-chain motif 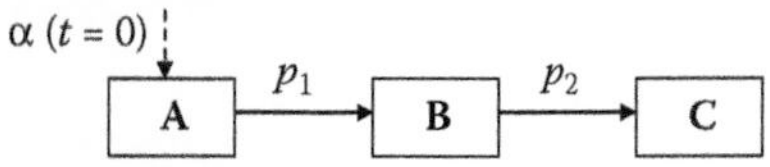	$\delta A=\alpha e^{-p_1 t}$ $\delta B=\alpha(1-e^{-p_1 t})e^{-p_2 t}$ $\delta C=\alpha(1-e^{-p_1 t})(1-e^{-p_2 t})$
Merging motif 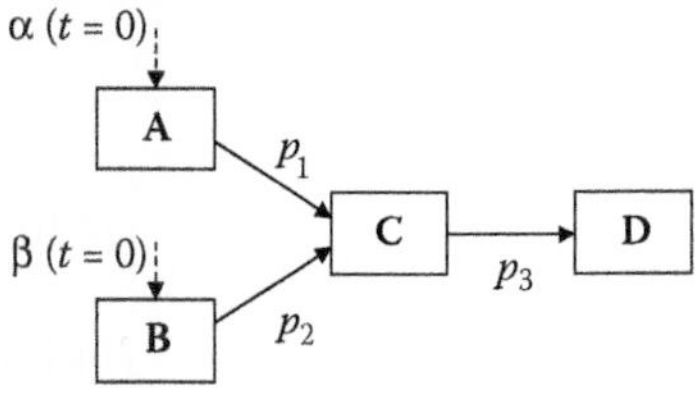	$\delta A=\alpha e^{-p_1 t}$ $\delta B=\beta e^{-p_2 t}$ $\delta C=[\alpha(1-e^{-p_1 t})+\beta(1-e^{-p_2 t})]e^{-p_3 t}$ $\delta D=[\alpha(1-e^{-p_1 t})+\beta(1-e^{-p_2 t})](1-e^{-p_3 t})$
Diverging motif 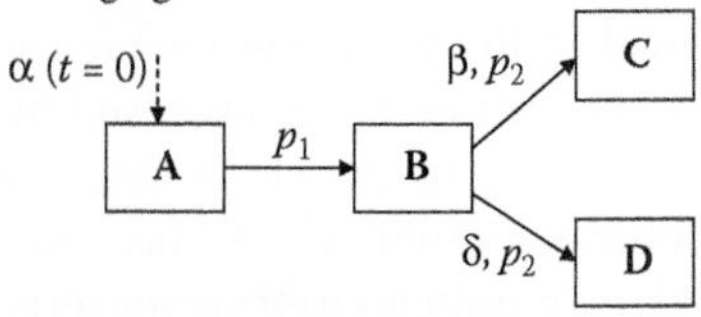	$\delta A=\alpha e^{-p_1 t}$ $\delta B=\alpha(1-e^{-p_1 t})e^{-p_2 t}$ $\delta C=\beta(1-e^{-p_1 t})(1-e^{-p_2 t})$ $\delta D=\delta(1-e^{-p_1 t})(1-e^{-p_2 t})$ Note: $\beta+\delta=1$ for mass-conserved system
(i) Linear-chain motif with reversible step 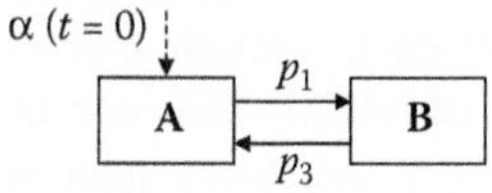	$\delta A(i)=\alpha e^{-p_1 t}$ $\delta A(ii)=B^*(1-e^{-p_3 t})=\beta(1-e^{-p_1 t})(1-e^{-p_3 t})$
For A: 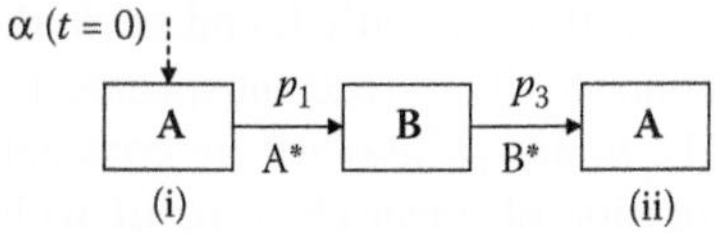	$\delta A=\delta A(i)+\delta A(ii)$ $\delta A=\alpha e^{-p_1 t}+\beta(1-e^{-p_1 t})(1-e^{-p_3 t})$
For B: 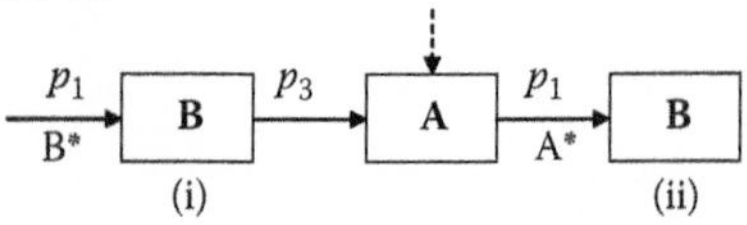	$\delta B(i)=B^* e^{-p_3 t}=\beta(1-e^{-p_1 t})e^{-p_3 t}$ $\delta B(ii)=A^*=(\alpha-\beta)(1-e^{-p_1 t})$ $\delta B=\delta B(i)+\delta B(ii)$ $\delta B=\beta(1-e^{-p_1 t})e^{-p_3 t}+(\alpha-\beta)(1-e^{-p_1 t})$
(ii) Linear-chain motif with reversible step and additional reactant 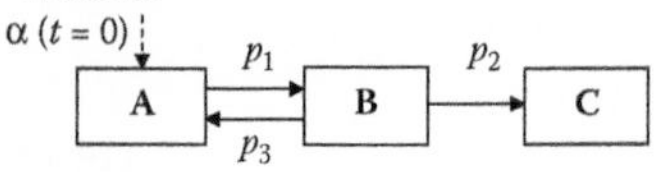	$\delta A=\alpha e^{-p_1 t}+\beta(1-e^{-p_1 t})(1-e^{-p_3 t})e^{-p_2 t}$ $\delta B=\begin{bmatrix}\beta(1-e^{-p_1 t})e^{-p_3 t}\\ +(\alpha-\beta)(1-e^{-p_1 t})\end{bmatrix}e^{-p_2 t}$ $\delta C=\alpha(1-e^{-p_1 t})(1-e^{-p_2 t})$
For A: 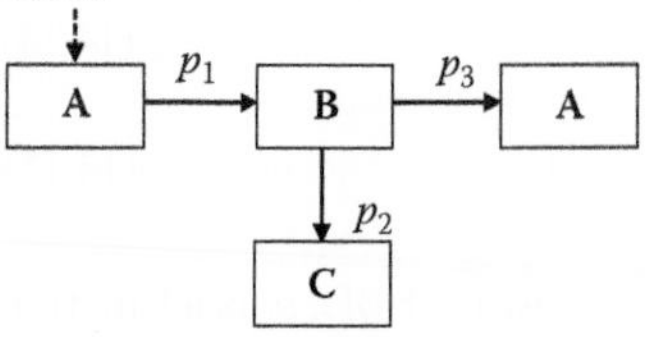	

TABLE 14.1 (CONTINUED)
NICM Motifs and Expressions

Motif	NICM Expressions Based on Response Propagation Rules
For B: 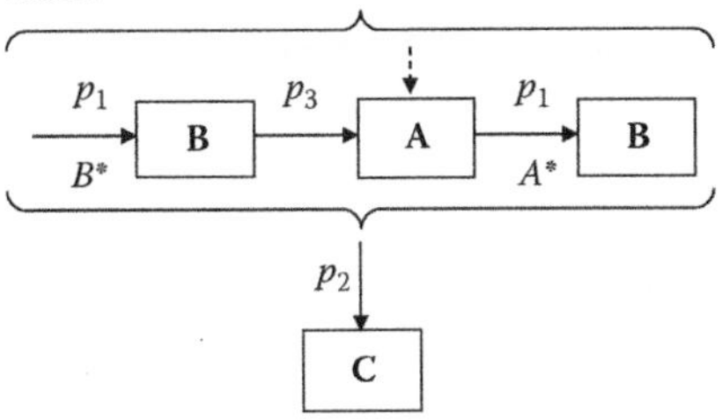	
Feedback motif 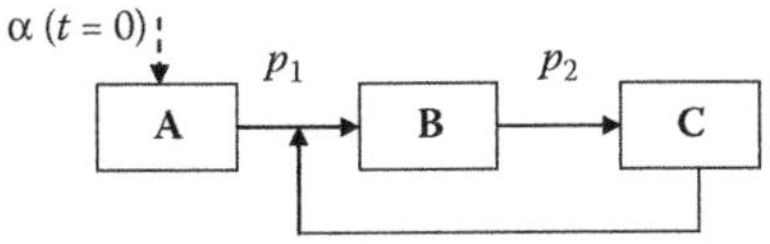	$\delta A = \alpha e^{-p_1(C)t}$ $\delta B = \alpha(1-e^{-p_1(C)t})e^{-p_2 t}$ $\delta C = \alpha(1-e^{-p_1(C)t})(1-e^{-p_2 t})$
Feed-forward motif 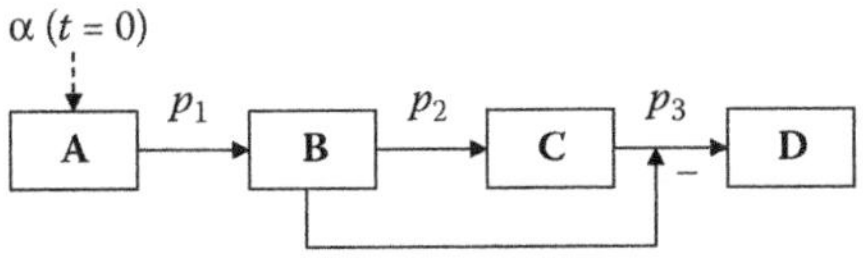	$\delta A = \alpha e^{-p_1 t}$ $\delta B = \alpha(1-e^{-p_1 t})e^{-p_2 t}$ $\delta C = \alpha(1-e^{-p_1 t})(1-e^{-p_2 t})e^{-\frac{p_3}{B}t}$ $\delta D = \alpha(1-e^{-p_1 t})(1-e^{-p_2 t})(1-e^{-\frac{p_3}{B}t})$ for $B > 0$

perturbation. By using various NICM network motifs (Table 14.1) without bias, we analyzed the time-series plots of the three concentration profiles and found that only the correct motif yielded the best fitness score (Figure 14.6 shows for two motifs only).

Interestingly, further experimental evidence of the superposition of formation and depletion propagation waves, which reveals first-order mass-action response, can be visibly seen in the temporal activation profiles of phosphoproteomics data generated for epidermal growth factor signaling (see Figure 3A and B from Blagoev et al., 2004). Although the NICM approach, utilizing simple linear rules, has only been applied to infer novel regulation in yeast glycolysis (Selvarajoo et al., 2009), we believe that it will eventually become a powerful approach in the future when more complete quantitative temporal data of molecular species become readily available.

14.5 DISCUSSION

Complex living systems have shown remarkably well-orchestrated, self-organized, robust, and stable behaviors under a wide range of diverse perturbations. However, even despite the recent generation of high-throughput experimental data sets, basic cellular properties such as cell division, differentiation, and apoptosis still remain elusive. One of the key reasons is the lack of understanding of the governing principles in complex living systems. Here we reviewed the success of perturbation-response experiments, where without the requirement of detailed *in vivo* biochemical parameters, the analysis of temporal concentrations or activation profiles (responses) using simple linear rules reveals biological network structures such as causal relationships of reactant species, regulatory motifs, and novel network intermediates and interactions.

The successes of linear-response approaches to interpret the dynamics of complex biological networks may suggest that the spatial transport mechanisms described by the diffusion terms in PDE models can be represented by additional intermediate mass-action terms. That is, the spatio-transport of inactivated molecules and the temporal-eventual activation of them can be treated as a

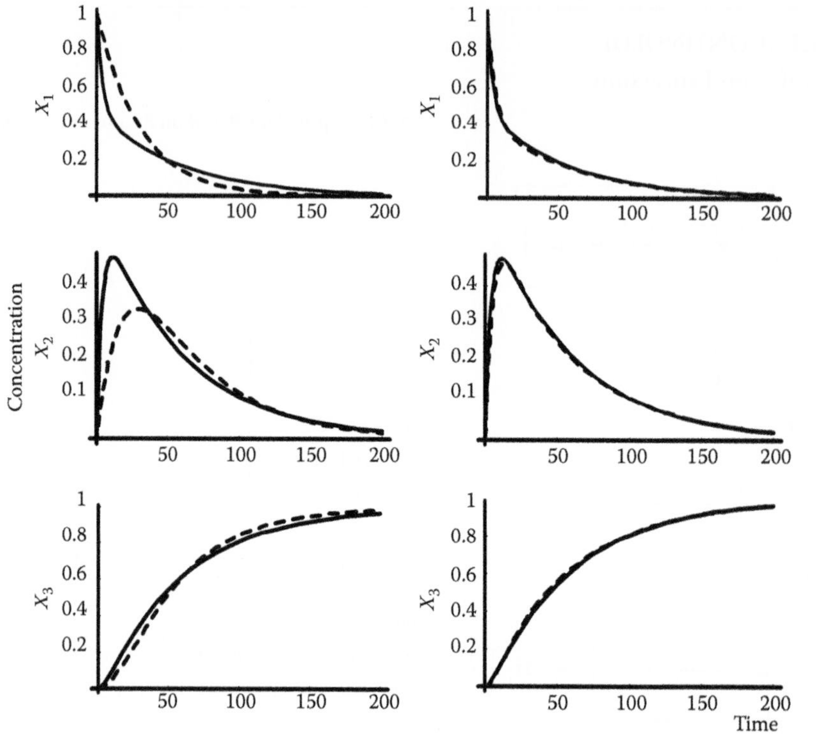

FIGURE 14.6 Generation of time-series plots for linear-chain motif with reversible step using first-order mass-action response equations with pulse perturbation and comparison with best fitted NICM motifs using genetic algorithm, compared with three linear-chain (X_1, X_2, and X_3) NICM motif (a) and three linear-chain NICM motif with reversible step (b) (Table 14.1). The maximum error for (a) is 42%; for (b) it is 4% (see Selvarajoo and Tsuchiya, 2007). The x-axis represents time, and the y-axis represents concentration—both in arbitrary units. For all of the subsequent plots, we adopt the same representation of the axes (x: time; y: concentration of reactant). The solid lines indicate the first-order mass-action kinetics with pulse perturbation, and the dotted lines indicate the results of NICM motifs.

chain of temporal events where both diffusion and reaction processes are considered independently, and thus signaling profiles can be analyzed by simple linear rules. Although we have proposed this, we do acknowledge that more research is required along these lines to confirm the validity of this approximation to model spatial effects in signaling networks.

Nevertheless, it is interesting to know why the simplifications of signaling processes by simple rules could hold in a heterogeneous environment where diffusion processes, molecular crowding, and stochastic processes are expected to make the response highly complex and nonlinear, which are in fact observed for single cell response (Elowitz et al., 2002; Chiam et al., 2006; Ellis and Minton, 2003). The answer may perhaps lie in the averaging effect of a single cell response when ensembles of them form a cell population.

It is still intriguing how a specific deterministic path can be chosen by a cell having enormous number of molecules, among the uncountable number of possibilities that can arise, through the complex multimolecular interactions during cellular process such as differentiation. However, at the same time, these molecular networks are shown to be robust, highly predictable, and stable under drastic environment perturbations (Albert, 2005; Barabási and Oltvai, 2004). The basis for such determinism may stem from the averaging effect caused by the reduction of response fluctuations, which reduces according to the law of large numbers demonstrating by two of our recent investigations on microarray data set. In both the innate immune response of macrophages to LPS (Tsuchiya et al., 2009) and the neutrophil differentiation process (Tsuchiya et al., 2010), we showed that correlation fluctuations reduced when genes were grouped, thereby deciphering the hidden collective genome-wide expression dynamics.

For LPS response, the ensemble property, in wild-type and mutant conditions, uncovered local and global effects of LPS; local being the well-known proinflammatory response of a small number of highly expressed genes (about 100–200), and global being the novel collective activation of diverse processes comprising the rest of majority number of the lowly expressed genes. The global property of the immune response emerged as a result of the transition from large scatter in expression distributions for single gene to smooth linear lines for grouped genes that can be linearly superposed to decipher the global gene regulatory differential control principle of the transcriptional and mRNA decay machineries between wild-type and mutant genomes.

For neutrophil differentiation, collective motion of fractal-like genome ensembles, notably consisting of mostly low and moderate variable genes, acted as *genome vehicle* to guide cell-fate decision, deciphering the significance of low expressed genes that are often considered noisy and insignificant. Thus, there must be organizational principles to explain such highly coordinated average responses of biological processes, which indicate the existence of simple rules that control and regulate complex biological processes.

It remains an open question as to which direction future investigations of cellular networks should follow: the highly fluctuating nature of single cell dynamics or the robust character of cell populations that represent the average response of an entire tissue, organ, or living system. Nevertheless, without the discovery of organizing principles in biology, it will be a daunting task to analyze and interpret large-scale data set and elucidate the mystery of life.

ACKNOWLEDGMENTS

We thank Kentaro Hayashi and Vincent Piras of Keio University for technical support. This work was supported by Japan Science and Technology Agency/Core Research for Evolutional Science and Technology; the Ministry of Education, Culture, Sports, Science and Technology of Japan; Yamagata Prefecture; and Tsuruoka City, Japan.

REFERENCES

Akira, S., Takeda, K., Kaisho, T. (2001). Toll-like receptors: Critical proteins linking innate and acquired immunity. *Nat. Immunol.* 2: 675–680.

Albert, R. (2005). Scale-free networks in cell biology. *J. Cell Sci.* 118: 4947–4957.

Albert, J. H., Barabasi, A. L. (2000). Error and attack tolerance of complex networks. *Nature* 406: 378–382.

Alberts, B., Johnson, A., Lewis, J., et al. (Eds.) (2008). *Molecular biology of the cell.* New York: Garland Science.

Alon, U., Surette, M. G., Barkai, N., et al. (1999). Robustness in bacterial chemotaxis. *Nature* 397: 168–171.

Alter, O., Brown, P. O., Botstein, D. (2000). Singular value decomposition for genome-wide expression data processing and modeling. *Proc. Natl. Acad. Sci. U.S.A.* 97: 10101–10106.

Arkin, A., Shen, P., Ross, J. (1997). A test case of correlation metric construction of a reaction pathway from measurements. *Science* 277: 1275–1279.

Barabási, A. L., Oltvai, Z. N. (2004). Network biology: Understanding the cell's functional organization. *Nat. Rev. Genet.* 5: 101–113.

Barkai, N., Leibler, S. (1997). Robustness in simple biochemical networks. *Nature* 387: 913–917.

Baud, V., Karin, M. (2009). Is NF-κB a good target for cancer therapy? Hopes and pitfalls. *Nat. Rev. Drug Discov.* 8: 33–40.

Bennett, M. R., Hasty, J. (2008). Systems biology: Genome rewired. *Nature* 452: 824–825.

Blagoev, B., Ong, S. E., Kratchmarova, I., et al. (2004). Temporal analysis of phosphotyrosine-dependent signaling networks by quantitative proteomics. *Nat. Biotechnol.* 22: 1139–1145.

Brazma, A., Schlitt, T. (2003). Reverse engineering of gene regulatory networks: A finite state linear model. *Genome Biol.* 4: 1–31.

Bruggemann, F. J., Westerhoff, H. V., Hoeck, J. B., et al. (2002). Modular response analysis of cellular regulatory networks. *J. Theor. Biol.* 218: 507–520.

Carlson, J. M., Doyle, J. (2002). Complexity and robustness. *Proc. Natl. Acad. Sci. U.S.A.* 99: 2538–2545.

Chiam, K. H., Tan, C. M., Bhargava, V., et al. (2006). Hybrid simulations of stochastic reaction-diffusion processes for modeling intracellular signaling pathways. *Phys. Rev. E* 74: 51910–51913.

Costa, M. N., Radhakrishnan, K., Wilson, B. S., et al. (2009). Coupled stochastic spatial and non-spatial simulations of ErbB1 signaling pathways demonstrate the importance of spatial organization in signal transduction. *PLoS One* 4: e6316.

Dehmelt, L., Bastiaens, P. I. H. (2010). Spatial organization of intracellular communication: Insights from imaging. *Nat. Rev. Mol. Cell Biol.* 11: 440–452.

Ellis, R. J., Minton, A. P. (2003). Cell biology: Join the crowd. *Nature* 425: 27–28.

Elowitz, M. B., Levine, A. J., Siggia, E. D., et al. (2002). Stochastic gene expression in a single cell. *Science* 297: 1183–1186.

Eungdamrong, N. J., Iyengar, R. (2004). Modeling cell signaling networks. *Biol. Cell* 96: 355–362.

Fink, C. C., Slepchenko, B., Moraru, I. I., et al. (1999). Morphological control of inositol-1,4,5-trisphosphate-dependent signals. *J. Cell Biol.* 147: 929–936.

Fink, C. C., Slepchenko, B., Moraru, I. I., et al. (2000). An image-based model of calcium waves in differentiated neuroblastoma cells. *Biophys. J.* 79: 163–183.

Friedman, N. (2004). Inferring cellular networks using probabilistic graphical models. *Science* 303: 799–805.

Friedman, N., Linial, M., Nachman, I., et al. (2000). Using Bayesian network to analyze expression data. *J. Comp. Biol.* 7: 601–620.

Hacker, H., Redecke, V., Blagoev, B., et al. (2006). Specificity in Toll-like receptor signalling through distinct effector functions of TRAF3 and TRAF6. *Nature* 439: 204–207.

Helmy, M., Gohda, J., Inoue, J., et al. (2009). Predicting novel features of Toll-like receptor 3 signaling in macrophages. *PLoS One* 4: e4661.

Hoffmann, A., Levchenko, A., Scott, M. L., et al. (2002). The IκB-NF-κB signaling module: Temporal control and selective gene activation. *Science* 298: 1241–1245.

Ideker, T., Thorsson, V., Ranish, J. A. (2001). Integrated genomic and proteomic analyses of a systematically perturbed metabolic network. *Science* 292: 929–934.

Ishii, N. K., Nakahigashi, K. T., Baba, T., et al. (2007). Multiple high-throughput analyses monitor the response of *E. coli* to perturbations. *Science* 316: 593–597.

Kagan, J. C., Su, T., Horng, T., Chow, A., et al. (2008). TRAM couples endocytosis of Toll-like receptor 4 to the induction of interferon-beta. *Nat. Immunol.* 9: 361–368.

Karin, M., Yamamoto, Y., Wang, Q. M. (2004). The IKK NF-κB system: A treasure trove for drug development. *Nat. Rev. Drug Discov.* 3: 17–26.

Kholodenko, B. N., Hancock, J. F., Kolch, W. (2010). Signalling ballet in space and time. *Nat. Rev. Mol. Cell Biol.* 11: 414–426.

Kitano, H. (2004). Biological robustness. *Nat. Rev. Genet.* 5: 826–837.

Lemoine, F. J., Degtyareva, N. P., Lobachev, K., et al. (2005). Chromosomal translocations in yeast induced by low levels of DNA polymerase a model for chromosome fragile sites. *Cell* 120: 587–598.

Mcgettrick, A. F., Brint, E. K., Palsson-Mcdermott, E. M., et al. (2006). TRIF-related adapter molecule is phosphorylated by PKCε during Toll-like receptor 4 signaling. *Proc. Natl. Acad. Sci. U.S.A.* 103: 9196–9201.

Morán, F., Vlad, M. O., Buestos, M., et al. (2007). Species connectivities and reaction mechanisms from neutral response experiments. *J. Phys. Chem. A* 111: 1844–1851.

Rao, C. V., Arkin, A. P. (2001). Control motifs for intracellular regulatory networks. *Annu. Rev. Biomed. Eng.* 3: 391–419.

Ross, J. (2008). From the determination of complex reaction mechanisms to systems biology. *Annu. Rev. Biochem.* 77: 479–494.

Schilling, C. H., Palsson, B. O. (1998). The underlying pathway structure of biochemical reaction networks. *Proc. Natl. Acad. Sci. U.S.A.* 95: 4193–4198.

Selvarajoo, K. (2006). Discovering differential activation machinery of the toll-like receptor 4 signaling pathways in MyD88 knockouts. *FEBS Lett.* 580: 1457–1464.

Selvarajoo, K. (2007). Decoding the signaling mechanism of Toll-like receptor 4 pathways in wild type and knockouts. In Ghosh, S., Tomita, M., Arjunan, S. N. V., Dhar, P. K. (Eds.), *E-Cell system: Basic concepts and applications.* Austin, TX: Kluwer Academic Publishers-Landes Bioscience.

Selvarajoo, K., Tsuchiya, M. (2007). *Systematic determination of biological network topology: Nointegral connectivity method (NICM).* In Choi, S (Ed.), *Introduction to systems biology.* Totowa, NJ: Humana Press.

Selvarajoo, K., Helmy, M., Tomita, M., et al. (2008a). Inferring the mechanistic basis for the dynamic response of the MyD88-dependent and -independent pathways. *Proc. 10th Int. Conf. Mol. Systems Biol.* 110–114.

Selvarajoo, K., Takada, Y., Gohda, J., et al. (2008b). Signaling flux redistribution at Toll-like receptor pathway junctions. *PLoS One* 3: e3430.

Selvarajoo, K., Tomita, M., Tsuchiya, M. (2009). Can complex cellular processes be understood by simple linear rules? *J. Bioinformatics Comp. Biol.* 7: 243–268.

Stromberg, S. P., Carlson, J. (2006). Robustness and fragility in immunosenescence. *PLoS Comput. Biol.* 2: e160.

Torralba, A. S., Yu, K., Shen, P., Oefner, P. J., et al. (2003). Experimental test of a method for determining causal connectivities of species in reactions. *Proc. Natl. Acad. Sci. U.S.A.* 100: 1494–1498.

Tsuchiya, M., Ross, J. (2001). Application of genetic algorithm to chemical kinetics: Systematic determination of reaction mechanism and rate coefficients for a complex reaction network. *J. Phys. Chem. A* 105: 4052–4058.

Tsuchiya, M., Piras, V., Choi, S., et al. (2009). Emergent genome-wide control in wild-type and genetically mutated lipopolysaccarides-stimulated macrophages. *PLoS One* 4: e4905.

Tsuchiya, M., Piras, V., Giuliani, A., et al. (2010). Collective dynamics of specific gene ensembles crucial for neutrophil differentiation: The existence of genome vehicles revealed. *PLoS One* 5: e12116.

Vallabhapurapu, S., Karin, M. (2009). Regulation and function of NF-κB transcription factors in the immune system. *Annu. Rev. Immunol.* 27: 693–733.

Vance, W., Arkin, A., Ross, J. (2002). Determination of causal connectivities of species in reaction networks. *Proc. Natl. Acad. Sci. U.S.A.* 99: 5816–5821.

Vlad, M. O., Arkin, A., Ross, J. (2004). Response experiments for nonlinear systems with application to reaction kinetics and genetics. *Proc. Natl. Acad. Sci. U.S.A.* 101: 7223–7238.

Werner, S. L., Barken, D., Hoffmann, A. (2005). Stimulus specificity of gene expression programs determined by temporal control of IKK activity. *Science* 309: 1857–1861.

Yeo, Z., Wong, S. T., Arjunan, S. N. V., et al. (2007). Sequential logic model deciphers dynamic transcriptional control of gene expressions. *PLoS One* 2: e776.

Yeung, M. K. S., Tegner, J., Collins, J. J. (2002). Reverse engineering gene networks using singular value decomposition and robust regression. *Proc. Natl. Acad. Sci. U.S.A.* 99: 6163–6168.

Yu, H., Kim, P. M., Sprecher, E., et al. (2007). The importance of bottlenecks in protein networks: Correlation with gene essentiality and expression dynamics. *PloS Comput. Biol* 3: e59.

15 Intermolecular Interaction in Biological Systems

Yoshifumi Fukunishi
National Institute of Advanced Industrial Science and Technology
Tokyo, Japan

CONTENTS

15.1 INTERMOLECULAR INTERACTION IN SOLVENT

15.1.1 Hydrophobic Interaction and Hydrophilic Interaction

Intermolecular interactions derive almost entirely from van der Waals and electrostatic forces. The quantum chemical effect is relatively small. Even in hydrogen bonding, 90% of the energy is contributed by electrostatic interaction, and only 10% is contributed by the quantum chemical effect. For convenience, intermolecular interactions in a solvent are classified into hydrophobic and hydrophilic interactions.

15.1.2 Scaled Particle Theory

The intermolecular interaction in a solvent can be estimated based on the solvation free energy. Solvation free energy represents the probability of insertion of a molecule into the solvent. Thus, the solvation free energy can be calculated based on the density of the solvent, the size of the inserted molecule, and the attractive interaction between the solvent and the molecule. Scaled particle theory gives a simple estimation of solvation free energy (Pierotti, 1976). Namely, the solvation free energy is proportional to the area of the molecular surface.

Figure 15.1 shows the thermodynamic cycle. The energy depends only on the state, and the energy difference between the two states does not depend on the pathway of the state change. The molecular interaction in the solvent (*E0*) is equal to *E1* − *E2* + *E3*. Molecular interaction in the vacuum (*E1*) is easily calculated by van der Waals (vdW) and electrostatic potential functions. The transfer free energy from vacuum to solvent (solvation free energy) is in question (*E2* and *E3*). The solvation process is divided into two steps for analysis (Figure 15.2). The first step is the cavity

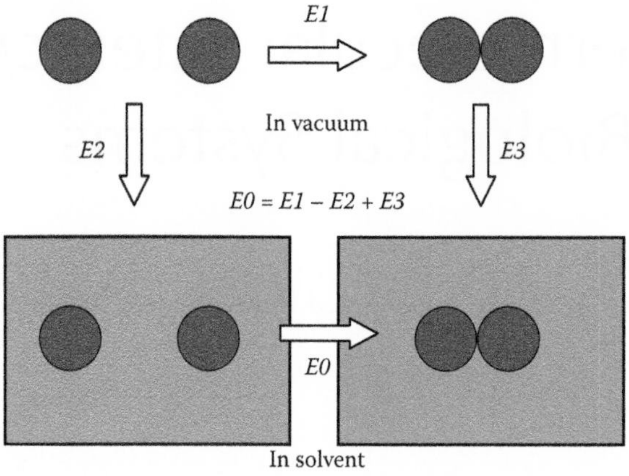

FIGURE 15.1 Thermodynamic cycle. The filled circles and gray squares represent solute molecules and solvent, respectively.

formation process, in which a molecular-size vacuum cavity is formed. The second process is the insertion of the molecule into the cavity, where the molecule can interact with the surrounding solvent molecules. Thus, the solvation free energy (ΔG) is given by:

$$\Delta G = \Delta G_{\text{cav}} + \Delta G_{\text{int}} \tag{15.1}$$

where ΔG_{cav} and ΔG_{int} are the cavity formation energy and the solute-solvent interaction energy.

Free energy represents probability. P_{cav} is the probability of finding a cavity of molecular size, and then the cavity formation energy ΔG_{cav} is $-k_BT \log P_{\text{cav}}$, where k_B and T are the Boltzmann constant and temperature, respectively. If the radius of a molecule is zero (the molecule is just a point), $P_{\text{cav}} = 1 - \rho$, where ρ is the occupancy (filling fraction) of the solvent molecules. For water at room temperature, ρ is approximately 0.7. In the scaled particle theory (SPT), the ΔG_{cav} of a spherical cavity of radius R is given by the following polynomial:

$$\Delta G_{\text{cav}} = c_0 + c_1 R + c_2 R^2 + c_3 R^3 + c_4 R^4 + \ldots \tag{15.2}$$

When $R = 0$, $P_{\text{cav}} = 1 - \rho$, then $c_0 = -k_BT \ln(1-\rho)$. The R^3 term corresponds to the volume. Thus:

$$c_3 R^3 = PV \tag{15.3}$$

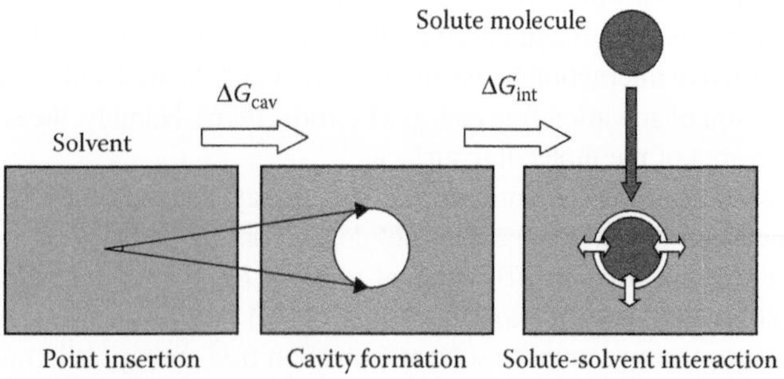

FIGURE 15.2 Solvation process. The filled circle, open circle, and gray square represent solute molecule, cavity, and solvent, respectively.

where P is pressure and V is volume. The coefficients c_n ($n > 3$) should be zero. In the SPT, c_1 and c_2 are determined based on the analytic continuation:

$$\Delta G_{cav} = -k_B T \left\{ \ln(1-\rho) + \left(\frac{3\rho}{1-\rho} \right) R + \left[\frac{3\rho}{1-\rho} + \frac{9}{2} \left(\frac{\rho}{1-\rho} \right)^2 \right] R^2 \right\} + PV \quad (15.4)$$

It has been known that the R^2 term (surface area) gives the major contribution to ΔG_{cav} and that ΔG_{cav} is almost proportional to the accessible surface area (ASA). The accessible surface is the surface of contact between the solvent molecule and solute molecule.

ΔG_{int} is the sum of the solute-solvent vdW interaction and electrostatic interaction. The solvent molecules in the first hydration shell (solvent molecules contacting with the solute) contribute the most to the ΔG_{int} value. Thus, both ΔG_{cav} and ΔG_{int} are proportional to the accessible surface area, and in general, the ΔG could be approximated as:

$$\Delta G = \sum_{i=1}^{n} \sigma_i \cdot A_i \quad (15.5)$$

where σ_i and A_i are the so-called atomic solvation parameter of the ith atom and the accessible surface area of the ith atom. Atomic solvation parameters were determined based on the experimental ΔG values of many small compounds.

The ASA and its gradient (force acting on the atom) are usually calculated by the analytical method proposed by Richmond (1984). The gradient of the ASA of the general shape can be calculated by the following numerical method. cij, n_i, and n_j are the boundary of the accessible surfaces of atoms i and j, the normal vector of the ith atom surface, and that of the jth atom surface, respectively (Figure 15.3). The force acting on the ith atom (F_i) is:

$$\vec{F}_i = \oint_{cij} \frac{\vec{n}_j \sigma_i - \vec{n}_i \sigma_j}{\left| \vec{n}_i \times \vec{n}_j \right|} ds \quad (15.6)$$

15.1.3 Atomic Solvation Parameter

The SPT showed that the solvation free energy is given by the atomic solvation parameter (σ_i) × molecular surface (A_i). This simple model is very useful and widely used, but the accuracy is poor. This simple model could be improved by consideration of the hydration shells.

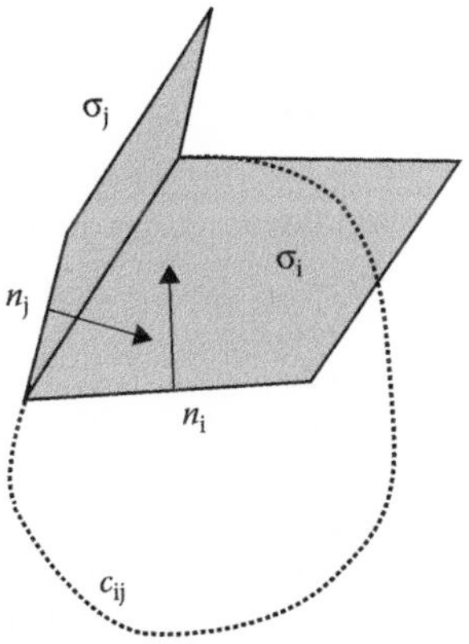

FIGURE 15.3 Schematic representation of boundary of solvent-accessible surfaces (dashed line) between two atoms i and j. The gray plates represent solvent-accessible surfaces.

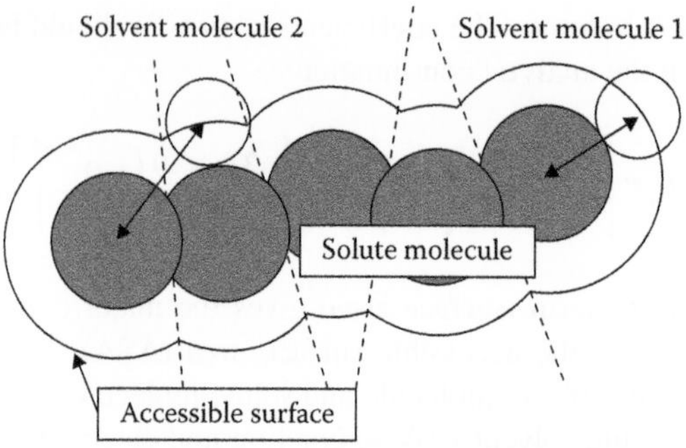

FIGURE 15.4 Solvent-accessible surface of solute and its Voronoi fraction. The gray circles and small open circles represent solute atoms and solvent molecules (probe), respectively. The dashed lines represent Voronoi boundaries.

The idea of the atomic solvation parameter is somewhat strange, and this quantity is quite different from mass and charge. We can divide a material into pieces, and each mass fraction/charge fraction satisfies a physical law. On the contrary, if we get a surface element from the accessible surface, the solvent molecules approach both sides of the surface element, and the interface disappears. The surface element is not an interface anymore. The atomic solvation parameters are determined based on the experimental ΔG values. Theoretically, the atomic solvation parameter is derived from the energy contribution of the Voronoi fraction around the solute molecule (Figure 15.4).

Figure 15.4 suggests that the solvent molecule around the middle of the solvent molecule can interact with many solute atoms and that the solvent molecules around the edge of the solute molecule can interact with a small number of solute atoms. The ΔG_{int} value in the middle of the solvent molecule should be lower than that at the edge of the solvent molecule. This means that a CH_3 group is more hydrophobic than a CH_2/CH group. Considering this effect, the prediction accuracy of ΔG is drastically improved as shown in Table 15.1. Obviously, the experimental ΔG value is neither proportional to the number of carbons nor the ASA. The ΔG values of hydrocarbons are summarized in Table 15.1, and the data are sorted by experimental ΔG values (ΔG_{exptl}). In model 1, only one σ (8.2 cal/mol/Å^2) value was applied to all carbon atoms, and the hydrogen atoms were ignored. In model 2, a σ of 11.7 cal/mol/Å^2 was applied to CH_3 groups, and a σ of 5.1 cal/mol/Å^2 was applied to other carbon atoms, and the hydrogen atoms were ignored. The correlation coefficient between the ΔG_{exptl} value and the ΔG values obtained by model 1 was only 0.24; this means that there was almost no correlation. On the contrary, the coefficient between

TABLE 15.1
Solvation Free Energies of Hydrocarbons

Compound	Number of Carbons	ASA (Å^2)	ΔG_{exptl} (kcal/mol)	ΔG (Model 1) (kcal/mol)	ΔG (Model 2) (kcal/mol)
Cyclopropane	3	195.69	0.79	1.61	1.00
Cyclopentane	5	235.56	1.21	1.93	1.20
Cyclohexane	6	255.00	1.23	2.09	1.30
Ethane	2	177.93	1.83	1.46	2.08
Propane	3	207.65	1.96	1.70	2.13
Methane	1	145.27	2.00	1.19	1.70
Butane	4	237.15	2.08	1.95	2.28
Pentane	5	258.52	2.33	2.12	2.27
Hexane	6	280.17	2.49	2.30	2.30
Isobutane	4	233.17	2.68	1.91	2.62

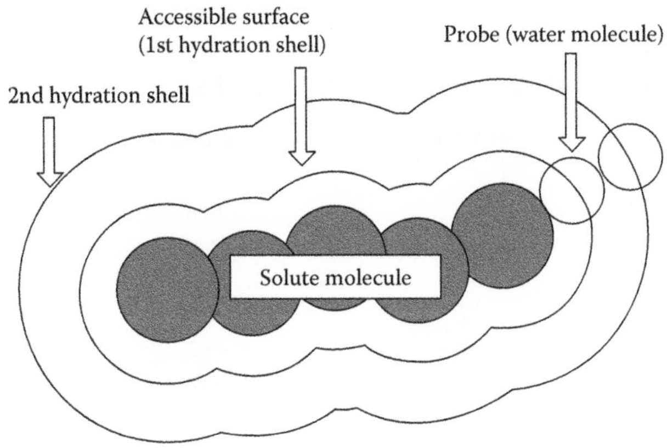

FIGURE 15.5 First hydration shell (solvent-accessible surface) and second shell (extended solvent-accessible surface). The gray circles and small open circles represent solute atoms and solvent molecules (probe), respectively.

the $\Delta G_{\mathrm{exptl}}$ value and the ΔG values obtained by model 2 was 0.95; this indicates a very strong correlation. A single parameter extension drastically improved the ΔG estimation.

In surface chemistry, the atomic solvation parameter corresponds to the surface tension. We must note that the accessible surface is on the atomic scale and that the interface in surface chemistry is a macroscopic smooth or flat surface. Because the accessible surface area is larger than the area of macroscopic interface, the atomic solvation parameter is slightly different from the surface tension.

In the SPT, ΔG_{int} is proportional to the accessible surface are. Because the electrostatic interaction is a long-range interaction, the polar or charged solute atoms can interact with the solvent molecules in the second and third hydration shells. Thus, the equation is simply extended as:

$$\Delta G = \sum_{i=1}^{n} \sigma_i \cdot A_i + \sum_{i=1}^{n} \sigma_i^2 \cdot A_i^2 \tag{15.7}$$

where σ_i^2 and A_i^2 are the solvation parameter and the surface area of the second hydration shell, respectively (Figure 15.5). Using this equation, the parameters were determined based on the ΔG values of about 30 compounds. The results are summarized in Table 15.2. For nonpolar atoms (CH_n carbons), only the first hydration shell was considered, and all of the hydrogen atoms were ignored. The σ^2 values are all negative. This means that the interaction between the solute and the second hydration shell is all attractive. The prediction accuracy of ΔG is drastically improved; namely, the average error using only one hydration shell was 0.57 kcal/mol, and that using Equation 15.7 was 0.15 kcal/mol.

15.1.4 Hofmeister Effect

The electrostatic interaction depends on the density of salt in a solution, because the salt density changes the dielectric constant. Hydrophobic interactions are intensified by the salt. This phenomenon has been known as the Hofmeister effect since 1888. Based on the SPT, the increase of hydrophobic interaction is proportional to the density of the salt.

Figure 15.6 shows the potential of mean force (free energy profile) of two methane molecules in water at room temperature obtained by a molecular dynamics simulation. This figure suggests that the interaction between two methane molecules, which are totally hydrophobic, is increased by the salt. Based on the SPT,

$$\Delta G_{\mathrm{cav}} = -k_B T \ln P_{\mathrm{cav}} \tag{15.8}$$

TABLE 15.2
Atomic Solvation Parameters for C, N, O, and S

	vdW Radius (Å)	σ (cal/mol/Å²)	σ² (cal/mol/Å²)
Carbon			
CH_3	2	10.6	—
CH_2	2	4.2	—
>CH–	2	1.2	—
>C<	2	4.2	—
CH (aromatic)	1.7	−1.2	−0.2
C (aromatic)	1.7	−57	−15.2
CH_2= (alkene)	1.7	7	—
–CH= (alkene)	1.7	−2.5	—
>C= (alkene)	1.7	−131	—
Nitrogen			
NH_2	1.5	−9	−2.3
NH	1.5	−33	−2.3
N	1.5	−33	−2.3
NH_2 (amide)	1.5	−6	−1.3
NH/N (amide)	1.5	−30	−2.0
Oxygen			
OH	1.5	−8	−2.6
O=CH	1.5	0	−2.0
O=CH	1.5	0	−2.4
–O–	1.5	10	−1.6
O (amide)	1.5	−30	−5.0
Sulfar			
SH	2	5	−1.0
S	2	5	−1.4

For nonpolar atoms, the second hydration shell is not necessary; —, no data.

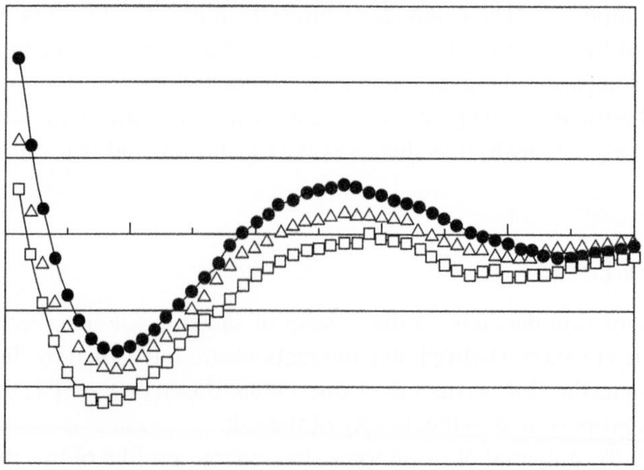

FIGURE 15.6 Potential of mean force (pmf: free energy profile) of two methane molecules in water obtained by a molecular dynamics simulation at room temperature. The two methane molecules were located in a water cube of 18.3 Å × 18.3 Å × 18.3 Å, and the periodic boundary condition was applied to this system. The filled circles, open triangles, and open squares represent the results of a pure water system, a water system consisting of two Na^+Cl^-, and a water system consisting of four Na^+Cl^-, respectively.

salt ions attract water molecules rather than the hydrophobic molecules by electrostatic interaction. ρ_{ion} and V_{ion} are the density of salt and the volume around the salt that the hydrophobic solute molecule cannot penetrate. Then, Equation 15.8 becomes

$$\Delta G_{\text{cav}} = -k_B T \ln(P_{\text{cav}} - \rho_{\text{ion}} V_{\text{ion}}) \tag{15.9}$$

When the density of salt is low,

$$\begin{aligned} \Delta G_{\text{cav}} &= -k_B T \ln(P_{\text{cav}} - \rho_{\text{ion}} V_{\text{ion}}) = -k_B T \ln\left[P_{cav}\left(1 - \frac{\rho_{\text{ion}} V_{\text{ion}}}{P_{\text{cav}}}\right)\right] \\ &\cong -k_B T \ln P_{\text{cav}} + \frac{\rho_{\text{ion}} V_{\text{ion}}}{P_{\text{cav}}} \end{aligned} \tag{15.10}$$

ΔG_{int} does not depend on the ρ_{ion} so much. Thus, the salt dependence of the hydrophobic interaction is proportional to the salt density.

15.2 APPLICATIONS

15.2.1 Protein-Compound Docking

Protein-compound docking software predicts the protein-compound complex structure and its binding energy based on the three-dimensional (3D) structures of the target protein and the compound. There have been dozens of protein-compound docking programs reported in the last three decades (Kuntz et al., 1982; Rarey et al., 1996; Jones et al., 1997; Fukunishi et al., 2005). These programs estimate the protein-compound interaction energy using the so-called scoring function. The scoring function is based on the SPT; however, these methods do not consider the salt density.

The computational procedure of the protein-compound docking program Sievgene is as follows:

1. Sievgene reads the atomic coordinates and atomic charges of the target protein and the compound.
2. Sievgene generates multiconformers of the compound by rotating the chemical bonds of the compound.
3. Sievgene puts the conformers on the protein surface.
4. The protein-compound interaction energy is calculated by a scoring function.
5. The 3rd and 4th steps are performed many times (usually 100,000 times in a few seconds). The most stable structure is selected among many protein-compound complex structures generated in the 3rd and 4th steps.

The scoring function roughly estimates the protein-compound interaction. The scoring function of Sievgene is as follows: the receptor-ligand interactions that Sievgene accounts for include vdW, Coulomb, and hydrophobic interactions and hydrogen bonds:

$$\Delta G_{\text{calc}} = c_{\text{rot}} \cdot N_{\text{rot}} + c_{\text{AV}} \cdot (E_{\text{ASA}} + E_{\text{vdW}}) + c_{\text{ele}} \cdot E_{\text{ele}} + c_{\text{hyd}} \cdot E_{\text{hyd}} + c_{\text{intra-vdw}} \cdot E_{\text{intra-vdW}} \tag{15.11}$$

where N_{rot}, E_{ASA}, E_{vdW}, E_{ele}, E_{hyd}, and $E_{\text{intra-vdW}}$ represent the number of rotatable bonds, the hydrophobic energy from the accessible surface area, the vdW energy, the protein-ligand electrostatic interaction, the hydrogen bond energy, and the intramolecular vdW energy of the ligand, respectively. The optimized coefficients for each energy term are c_{rot}, c_{AV}, c_{ele}, c_{hyd}, and $c_{\text{intra-vdW}}$. The parameters were determined based on the experimental binding energies of many protein-compound complexes.

The accuracy of the scoring function is not so good compared to the solvation free energy (ΔG) estimation. Namely, the average error is about 2.5 kcal/mol, and the prediction accuracy of the protein-compound complex structure is not so good. About 20–30% of protein-compound complex structures are reproduced with a coordinate error of 2 Å. In 70–80% cases, the prediction fails.

15.2.2 Molecular Dynamics Simulation

The intermolecular interaction in solvent can be calculated by a molecular dynamics (MD) simulation. MD simulations reveal how complicated molecular interactions are. The molecular interaction strongly depends on the water model.

MD simulations simulate the molecular system at an atomic scale and are used to generate a canonical ensemble for free energy calculation based on statistical physics. The atomic coordinates are calculated by a differential equation

$$\sum_{n=1}^{N} m_i^n \frac{d^n x_i}{dt^n} = F_i + \eta_i \tag{15.12}$$

where m_i^n, x_i, t, F_i, and η_i are the ith mass parameter, coordinates of the ith atom, time, the force acting on the ith atom and random force acting on the ith atom, respectively. Usually, the Newtonian-like equation

$$m_i \frac{d^2 x_i}{dt^2} + \mu_i \frac{dx_i}{dt} = F_i \tag{15.13}$$

is used, where m_i and μ_i are the mass of the ith atom and a control variable, which controls the temperature of the system, respectively.

The force F is the gradient of the potential function. The potential function used in the MD is comprised of the so-called 1-2, 1-3, 1-4, and 1-5 interactions: namely, the chemical bond, angle, torsion, and intermolecular interactions. Chemical bonds in the MD are treated as springs with suitable strength. The intermolecular interaction consists of the vdW and electrostatic interactions. If the water molecules are ignored, the ASA model or the generalized-Born surface area model is applied to a biological system. If water molecules are explicitly considered, the hydrophobic and hydrophilic interactions are automatically calculated from the vdW and electrostatic interactions. Therefore, we can discount the SPT.

We can obtain a canonical ensemble by solving the equation of motion (Equation 15.13), but we do not know why it is possible. The canonical ensemble is an ensemble of many systems consisting of the same molecules. The MD simulation generates a trajectory of only one system. There is a theoretical gap between the ensemble of many systems and the trajectory of only one system. Nonetheless, the MD simulation can be used to understand the molecular interaction in the solvent, the binding free energy between the protein and ligand, the fluctuation of protein and so on.

The number N of Equation 15.12 can be a noninteger. The derivation of the exponential function is:

$$\frac{d^n e^{at}}{dt^n} = a^n e^{at} \tag{15.14}$$

Even if the number n is noninteger, the exponential function can satisfy this equation. Because the Fourier transform shows that any kind of function can be a sum of exponential functions, we can apply the noninteger order differential equation. But only $n = 1$ or $n = 2$ is used in the MD simulation, because the actual calculation for noninteger n is much too complicated, and it is difficult to find the physical meaning of a differential equation of a noninteger order.

The MD simulation is exhaustive in many meanings. We can see the behavior of all atoms of the system. Because the time step of the MD simulation is 10^{-15} second, we can see the vibration of chemical bonds (the vibration period is about 10^{-14} second) and the rotation and migration of atoms. The MD simulation is very time consuming. Even the fastest specialized super computer can simulate just 10^{-3} second of one protein. Most of the simulated times reported in the literature are on the order of 10^{-9} seconds.

However, we must note that our solvent model is quite old fashioned. The most widely used water model is the TIP3P model. The TIP3P model was developed in 1985, and only oxygen atom has a vdW parameter; the radius of the hydrogen atom is set to zero. The intermolecular interaction in solvent strongly depends on the size and shape of the solvent. Three models of water molecule were tried. The hydrophobic interaction between the two methane molecules strongly depends on the water model as suggested by the SPT. The TIP3P model is well established, and many researchers have supported its reliability. However, it is not perfect.

At the distance of 5–6 Å, there is an energy barrier in the free energy surface (Figure 15.6). The reason is that water molecules cannot exist/penetrate between two methanes even if the two methanes do not contact each other (Figure 15.7), and the freedom of water molecules is decreased. If the solute molecules are charged, the energy barrier should become even greater. Between the two solvent molecules, a void volume (vacuum), in which water cannot exist, is created (Figure 15.7) (Fukunishi and Suzuki, 1996, 1997). In this void, the dielectric constant is 1. The electrostatic energy E_{ele} is:

$$E_{\text{ele}} = \int \frac{1}{2\varepsilon} \vec{E}^2 \, dxdydz \tag{15.15}$$

where E is the electric field and ε is the dielectric constant. For small values of ε, the E_{ele} value becomes large. This effect is not considered in the SPT.

15.2.3 *In Silico* Drug Screening

Currently, several million compounds are commercially available, at an average cost of about US$100. The first step of a drug-development project is finding active (hit or seed) compounds that bind the target protein from the compound library. We do not design a drug in the initial step. Computer-aided drug design (CADD) is unreliable, and synthesis of the designed compound is too burdensome a task. Usually, the CADD is performed to modify active compounds, which are found by screening experiments *a priori*.

There are two approaches to *in silico* drug screening. One is structure-based drug screening, which is based on the 3D structure of the target protein, and the other is ligand-based drug

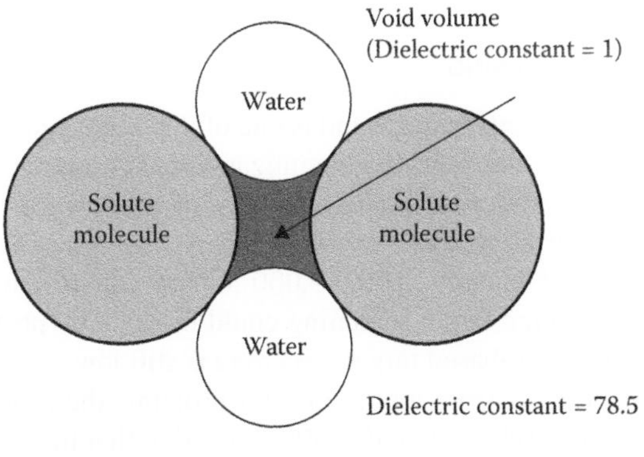

FIGURE 15.7 Void volume between two solute molecules.

screening, which is based on known active compounds (Leach and Andrew, 2001; Fukunishi, 2009). In structure-based drug screening, the protein-compound binding energy is estimated by a protein-compound docking program, and compounds that show a strong affinity (binding energy), are selected as the active compounds (Kuntz et al., 1982; Rarey et al., 1996; Jones et al., 1997; Fukunishi et al., 2005). In the ligand-based drug screening, compounds that are similar to the known active compound are selected as candidate active compounds.

Structure-based *in silico* drug screening is not equal to protein-compound docking simulation. The calculated binding energy by the protein-compound docking is unreliable, so that postprocessing of docking information is necessary (Warren et al., 2006). We get more reliable binding energy of a compound based on many calculated binding energies of the compound than a binding energy obtained by a single docking simulation. There have been many methods proposed. The multiple target screening (MTS) method calculates the relative binding energy of each compound for many proteins (Fukunishi et al., 2006). The consensus score is the average binding energy of the many predicted energies by many computational methods. Combinations of the machine-learning approach and MTS/consensus scoring have also been proposed.

The prediction accuracy (hit rate) of the *in silico* drug screening is about 1–10% when it is successful, whereas the hit rate of a random screening experiment is 0.01%. This means that a screening experiment guided by *in silico* screening is several hundred times more efficient than a random screening without bioinformatics. *In silico* drug screening is widely used in many drug-development projects. The hit rate depends on how many compounds are purchased. The accuracy of screening is quantitatively evaluated by the database enrichment curve and the area under the curve (AUC). The result of AUC of ~100% is an ideal result, and that of AUC of ~0% is poor. Random screening gives an AUC of 50%. The screening result strongly depends on the 3D coordinates of the target protein. The protein is flexible, and many screening results come from many structural snapshots of the same protein. This is also problematic in structure-based drug screening.

In addition to the screening method, the diversity of chemical compounds is also an important problem (Tanaka et al., 2009). The several million commercially available compounds are generated from a small number of building blocks. When two similar compounds are connected in a line, the chemical compounds form a small-world network. The randomness of the compound library is important, because the number of examined compounds is limited. Many researchers are working to enlarge this chemical space.

In structure-based drug screening, the molecular interaction is roughly estimated by a so-called scoring function. Only a 1% increase in the error of energy estimation causes a drastic decrease of the prediction accuracy. The screening results were obtained by the MTS method for 14 target proteins with changing the error of Sievgene score. The average AUC values were 79 and 73% for the additional 0% error of scores and the additional 1% error of scores, respectively. Thus, the estimation of molecular interaction is essential. Many kinds of improvements have been proposed, and more will be proposed in the future.

15.2.4 *In Silico* Target Screening

Finding a target protein for a bioactive compound is one of the important topics in medicinal chemistry and is known as *target profiling*. Selective optimization of side activities is a method to develop a new drug by finding a new target of the side activity of a known drug. This target profiling becomes possible by computer simulation. *In silico* target screening is performed based on the similarity of compounds. As of January 2010, 63,000 3D protein structures have been registered in the Protein Data Bank. *In silico* target screening could be done by protein-compound docking; however, the accuracy of structure-based target screening is still low.

If we select the 5% of proteins from a pool of total proteins, the probability of finding a true target protein for a given bioactive compound is 30%, meaning that this computer prediction is six times better than a random target profiling (Fukunishi et al., 2006). Because the total number of proteins in humans is about 100,000, this prediction accuracy is not good enough to perform actual experiments.

For practical use, the similarity of compounds is used (Keiser et al., 2009; Okuno et al., 2008). The concept is that similar proteins bind similar compounds. For a given bioactive compound, a protein that is similar to a protein whose ligand is similar to the given bioactive compound could be the target protein. Protein-compound interaction forms a various protein-protein and protein-compound networks, and several methods to predict the network have been developed. These methods and data will play an important role in systems biology in the near future.

REFERENCES

Fukunishi, Y. (2009). Structure-based drug screening and ligand-based drug screening with machine learning. *Comb. Chem. High Throughput Screen.* 12: 397–408.

Fukunishi, Y., Suzuki, M. (1996). Reproduction of potential of mean force by a modified solvent-accessible surface method. *J. Phys. Chem.* 100: 5634–5636.

Fukunishi, Y., Suzuki, M. (1997). Potential of mean forces of calculation of solute molecules in water by modified solvent-accessible surface method. *J. Compt. Chem.* 18: 1656–1663.

Fukunishi, Y., Kubota, S., Nakamura, H. (2006). Noise reduction method for molecular interaction energy: Application to in silico drug screening and in silico target protein screening. *J. Chem. Inf. Model.* 46: 2071–2084.

Fukunishi, Y., Mikami, Y., Nakamura, H. (2005). Similarities among receptor pockets and among compounds: Analysis and application to in silico ligand screening. *J. Mol. Graph. Model.* 24: 34–45.

Jones, G., Willet, P., Glen, R. C., et al. (1997). Development and validation of a genetic algorithm for flexible docking. *J. Mol. Biol.* 267: 727–748.

Keiser, M. J., Setola, V., Irwin, J. J., et al. (2009). Predicting new molecular targets for known drugs. *Nature.* 462: 175–182.

Kuntz, I. D. Blaney, J. M. Oatley, S. J., et al. (1982). A geometric approach to macromolecule-ligand interactions. *J. Mol. Biol.* 161: 269–288.

Leach, A. R. (2001). *Molecular modeling: Principles and applications.* 2nd Edition. Edinburgh Gate, UK: Pearson Education Ltd.

Okuno, Y., Tamon, A., Yabuuchi, H., et al. (2008). GLIDA: GPCR-ligand database for chemical genomics drug discovery: Database and tools update. *Nucleic Acids Res.* 36: D907–D912.

Pierotti, R. A. (1976). A scaled particle theory of aqueous and nonaqueous solutions. *Chem. Rev.* 76: 717–726.

Rarey, M., Kramer, B., Lengauer, T., et al. (1996). A fast flexible docking method using an incremental construction algorithm. *J. Mol. Biol.* 261: 470–489.

Richmond, T. J. (1984). Solvent accessible surface area and excluded volume in proteins. Analytical equations for overlapping spheres and implications for the hydrophobic effect. *J. Mol. Biol.* 178: 63–89.

Tanaka, N., Ohno, K., Niimi, T., et al. (2009). Small-world phenomena in chemical library networks: Application to fragment-based drug discovery. *J. Chem. Inf. Model.* 49: 2677–2686.

Warren, G. L., Webster Andrews, C., Capelli, A. M., et al. (2006). A critical assessment of docking programs and scoring functions. *J. Med. Chem.* 49: 5912–5931.

16 Implanted Brain–Machine Interfaces in Rats

A Modern Application of Neuromics

Weidong Chen, Jianhua Dai, Ting Zhao, Shaomin Zhang, and Xiaoxiang Zheng
Zhejiang University
Hangzhou, China

Xiaoling Hu
Hong Kong Polytechnic University
Hung Hom, China

CONTENTS

16.1 INTRODUCTION

With the ability of establishing communication pathways between a brain and external devices, brain-machine interfaces (BMIs) have been considered as a revolutionary way of treating or compensating for neurological dysfunction (Donoghue, 2008; Hatsopoulos and Donoghue, 2009; Lebedev and Nicolelis, 2006; Nicolelis and Lebedev, 2009; Wolpaw et al., 2002). The field of BMIs has grown at a tremendous rate since 1990, thanks to the progress made in neuromics and its interaction with some other disciplines, including computer science, material science, medical science, and electronic engineering. More and more types of BMIs became available, along with the development of systems that are driven by motor-cortex activities or cognitive processes. The attraction of BMIs for researchers with different backgrounds has been continuing for over two decades, according to the rapidly increased production of the related resources and literature in the interdisciplinary fields of biomedical engineering and neuroengineering.

There are two major categories of BMI systems, implanted BMIs and nonimplanted BMIs, depending on how activities of a brain are extracted (Lebedev and Nicolelis, 2006). In an implanted system, arrays of tens to hundreds of electrodes are inserted into the brain or laid on the surface of the brain for recording the firing patterns of a population of neurons. In contrast, nonimplanted systems interact with a brain through a noninvasive recording method, which measures neuron signal transmitted outside of the brain. The most common noninvasive recording method is electroencephalography (EEG), which records changes in EEG state by emplacing multielectrode arrays on the skull. Other recording methods used in nonimplanted systems include magnetoencephalography, thermography, functional magnetic resonance imagery interpretation, and analysis of near-infrared spectrum activity.

It is too early to conclude whether implanted BMIs or nonimplanted BMIs will have better results and patient compliance in the long run. Most of the nonimplanted BMIs systems to date are based on EEG signal recording at the scalp of a person because it is relatively easy and safe. However, EEG systems have significant drawbacks, such as limited bandwidth (loss of higher frequencies caused by scalp filtering), significant noise and environmental artifact (muscle contamination), the need for help attaching sensors to the scalp, variability in sensor contact over time, tethering to instruments, and appearance issues.

Implanted BMIs have better decoding fidelity and information transfer rate, although their duration of functioning needs to be improved. There are two main types of electrical potentials recorded from brain: action potentials (spike) and field potentials. They carry information about what is going on in nervous systems (Waldert et al., 2009). It is widely accepted that neural information is mainly encoded by spike rate, the number of spikes in a specific interval, or a related function. The vast majority of implanted systems use information coding at the level of spikes from neurons. It is not surprising that spike signals in most implanted BMI systems have better spatiotemporal resolution and noise immunity than EEG signals (Nicolelis, 2007).

16.1.1 Research on Implanted BMIs

Implanting multielectrode arrays in rodent and primate animals, many groups have successfully collected the activities of a large number of cortical neurons and decoded the activities of the behaviors of subjects. The first modern BMI study was carried out by Chapin and his colleagues (1999) in rats. In such studies, rats were trained to press a lever to move a water dripper arm and receive a water reward. The activities of the motor cortex and thalamus were recorded by dozens of microwires. The position of the lever was synchronously sampled. Various algorithms, such as population rate, principal component analysis, and recurrent neural networks, were then used to decode the lever position from the neural recordings (Linderman et al., 2008). After training, the decoder took over the control of the dripper arm. Serruya et al. (2002) and Taylor et al. (2002) applied similar ideas on monkeys and showed that the monkeys were able to control the position of a two-dimensional cursor on screen or a three-dimensional cursor in virtual reality through brain signal.

Recently, more fine motor prostheses like those supporting dexterous hand and finger movements have been incorporated into BMIs (Acharya et al., 2008; Aggarwal et al., 2008). To improve the stability of motor prostheses, sensory feedback from the robot was included to contain tactile and proprioceptive information feedback (O'Doherty et al., 2009). Furthermore, some basic research about fundamental mechanisms such as cortical plasticity, neuroscience, and neurophysiology is involved in BMIs (Ganguly and Carmena, 2009; Nicolelis and Lebedev, 2009). Several reviews detailed the state of the art in this field (Donoghue, 2008; Hatsopoulos and Donoghue, 2009; Lebedev and Nicolelis, 2006; Patil and Turner, 2008; Scherberger, 2009).

Instead of reviewing every aspect of implanted BMIs, the goal of this chapter is to introduce some implanted BMIs studies on rats. The first rat BMI research can be traced back to the end of last century. A seminal study done by Chapin demonstrated that neuronal population functions derived from the neural firing patterns in the primary motor cortex and ventrolateral thalamus of a rat produced an accurate prediction of the corresponding lever trajectories. Furthermore, the neural activities could be converted into real-time commands for robot-arm control. Soon after this first demonstration, some similar studies were reported to decode neural signals extracted from pre- or supplementary motor cortex. With the aid of sophisticated nonlinear classification algorithms, the ensemble activities were used to predict an output supervisory command to potentially control a vehicle. These studies focused on decoding neural activity from forearm areas of cortex to operate a robotic arm or perform other manipulation tasks. In contrast, a recent study demonstrated that the kinematics of trunk and hindlimbs could also be decoded simultaneously during locomotion (Fitzsimmons et al., 2009; Song et al., 2009). Instead of recording electrical potential from the brain to read out ongoing neural activities, another form of implanted BMIs relies on applying electrical stimulation to some specific part of the brain to restore or modulate the activities in a nervous system. The most successful case to date is the roborat shown by Chapin and his colleagues (Talwar et al., 2002). Electrodes were implanted in the medial forebrain bundle (MFB) and sensorimotor cortex of a rat brain. The rat received remote stimulation via its backpack containing a radio receiver and electrical stimulator. The virtual cues and rewards caused by electrical stimulus remotely direct the rat to move left, right, and forward. These efforts not only advanced the knowledge necessary to build BMIs but also made fundamental contributions to understanding nervous systems.

16.2 BMIs FOR RATS

16.2.1 Bidirectional Telemetry BMI Systems for Freely Behaving Rats

With the development of neural ensemble recording and brain stimulation technologies, BMIs have become a new research field for patients to restore their sensory and motor function lost because of injury or disease (Lebedev and Nicolelis, 2006; Mussa-Ivaldi and Miller, 2003;

Nicolelis, 2001; Wolpaw et al., 2002). Based on the directions of information transmission, BMIs can be divided into two types: afferent BMIs and efferent BMIs. An afferent BMI sends electrical signals to the brain to restore sensory functions (visual, auditory, tactile, etc.). An example of this type is the cochlear implant, which provides a sense of sound to the disabled using a series of electrodes inserted into cochlear nucleus to send microelectrical stimulation. Instead of sending signals to a brain, an efferent BMI uses brain signals to control external devices such as switches, computers, or prostheses. They have been successfully built on nonhuman primates and paralyzed patients. Microelectrical stimulation and neuronal activity recording are the essential techniques for afferent and efferent BMIs, respectively; thus, in this section, we will focus on some issues in these two areas.

16.2.1.1 Telemetry Recording System

The development of neural ensemble recording methods makes various types of efferent BMI systems available (Nicolelis, 2007). In the mid-1950s, the first multielectrode recording experiment in monkeys was reported (Lilly, 1956), but the modern methods for recording extracellular activity of large populations of individual neurons in behaving animals did not emerge until the early 1980s (Nicolelis and Lebedev, 2009). During the following decades, the rapid development of multichannel neural activity recording approaches has allowed the establishment of direct real-time BMIs.

A number of commercial instruments can be used in most of the efferent BMI systems. Examples are Plexon (Dallas, TX) and Blackrock Microsystems (Salt Lake City, UT). In general, commercial hardware for recording neural activity consists of a headstage, an analog front end, and a digital board. The headstage is an impedance transformer circuit, which provides an interface between electrodes and a preamplifier. The headstage and the preamp are usually connected by a shielded cable. The headstage ensures that the impedance seen by the cable at the source has uniform electrical characteristics including low impedance. Low impedance is important because it reduces the interference caused by cable flexion (cable artifact). There are two designs for headstage, including conventional voltage follower circuit and decoupled headstage amplifier with gain. The analog board, e.g., Plexon multichannel acquisition processer, applies switchable gains ranging 40 from 116 dB and band-pass filtering (250–8 kHz). Analog signals can be passed to a digital board for sampling (generally, the frequency over 30 kHz, 12 bit A/D resolution) and processing (including spike detection and spike sorting). Previous studies have shown that these commercial instruments are stable for real-time BMIs, but the use of cables limits the subject's freedom of movement.

A practical BMI system should not be built with large instruments and many cables. Thus, some wireless commercial instruments, such as Alpha Omega (Nazareth, Israel) and Triangle BioSystems (Durham, NC), were developed for neuroscience research and BMI applications. Some research labs have also developed some telemetry systems in recent years (Ativanichayaphong et al., 2008; Chien and Jaw, 2005; Grohrock et al., 1997; Hawley et al., 2002; Jurgens and Hage, 2006; Obeid et al., 2004; Schregardus et al., 2006; Vyssotski et al., 2006). The wireless recording system reported by Hawley et al. (2002) is the first system that transmits location-specific activities of hippocampal cells from untethered rats. It is highly reliable, and its signal can be detected over the distance of 20 m. Since then, many multichannel telemetry systems have been developed. A typical multichannel telemetry system was successfully used to record neural activities from awake, chronically implanted macaque and owl monkeys (Obeid et al., 2004). Another interesting report was a lightweight telemetry system for recording neuronal activities in zebra finches (Schregardus et al., 2006), and other multichannel recording telemetry systems have been also described (Jurgens and Hage, 2006; Nieder, 2000). Although there are lots of wireless systems for recording neural signals, none of them is ready for clinical use. One main reason is that these systems are too large to be fully implantable. Therefore, some groups are interested in making miniaturized electronic circuits for the system (Chestek et al., 2009; Harrison et al., 2007; Olsson and Wise, 2005). A fully implantable 96-channel neural data acquisition system was the first demonstration in the literature of the full implantation of a high channel-count neural data acquisition system for BMIs (Rizk et al., 2009).

There are two kinds of wireless communication modes for the telemetry systems: analog mode and digital mode. The analog communication mode is widely used for transferring electrical physiology signals from a small animal body to a recording device because it has the advantages of high-speed communication, small size, light weight, and low power consumption. However, signal transmission in an analog system is vulnerable to noise, which may complicate neuronal spike sorting. Another problem of the analog mode is the difficulty of adding a stimulation module. Digital mode for wireless neural signal transmission, such as Bluetooth, wireless LAN et al., has the advantages of less susceptible to noise and easy to add complex transmission protocol.

In summary, the neural ensemble recording system is a key component of a clinically viable brain-machine interface. The next generation recording systems are expected to have the features of miniaturization, wireless, and full implantation.

16.2.1.2 Brain Stimulation

Historical highlights and numerous conclusions of brain stimulation are described in Tehovnik's review (1996). Here, we will mainly introduce the potential role of electrical stimulation for a better efferent BMI control system.

Real-time performance is important for BMIs because feedback has the corrective mechanisms of on-line correction of errors and gradual adaption of motor commands across trials (Mussa-Ivaldi et al., 2007). Almost all BMI systems to date provide the animal subjects sensory information from the environment through visual feedback, and it has been shown that vision is important for planning and makes predictions of motor parameters more stable (Lebedev and Nicolelis, 2006). Although visual feedback is considered as the most important approach, other feedback forms are worth investigation too. Natural control in efferent BMIs requires interplay between the transmission of control commands and the reception of various sensory feedback such as visual, tactile, and proprioceptive, so visual feedback alone is not enough, especially for those who lost proprioceptive sense.

Recently, a few experiments have been done to investigate the perceptual effects of microelectrical stimulation in the somatosensory cortex. Although it is not yet known whether brain stimulation could provide better feedback to guide moments in BMIs, several research groups are interested in providing this sensory feedback in efferent BMIs through microelectrical stimulation of brain. The main idea of this technology is not only to extract neural signals to control external device but also to let the user feel the device like a real part of the body. Romo et al. (2000) successfully demonstrated that monkeys could distinguish different stimulation frequencies, whether artificial (microelectrical stimulation) or natural (peripheral vibration). This research indicated that tactile feedback might be provided in BMIs through direct cortical stimulation. In another demonstration, electrical stimulation of the somatosensory cortex (SI) and MFB was used as cues and rewards to condition distant rats to execute controlled turns remotely (Talwar et al., 2002). Xu et al. (2004) designed this typical stimulation system that could be used for controlling freely roaming animals. More recently, the Nicolelis lab at Duke University showed that owl monkeys could do a choice task (reaching left or right) by multichannel spatiotemporal critical microstimulation (Fitzsimmons et al., 2007; O'Doherty et al., 2009). They developed two systems for this experiment: one was based on National Instruments data acquisition cards, and the second was used an embedded microprocessor, which can autonomously produce the pulsed stimulation train. Another team led by Miller at Northwestern University has studied the effectiveness of electrical stimulation in the proprioceptive area of the primary somatosensory cortex as a potential means to deliver an artificial sense of proprioception to a behaving monkey (London et al., 2008). In the future, this electrical stimulation feedback will be developed as "encoders" into many BMIs paradigms for users to receive information about the state (position and movement) of the virtual or external devices.

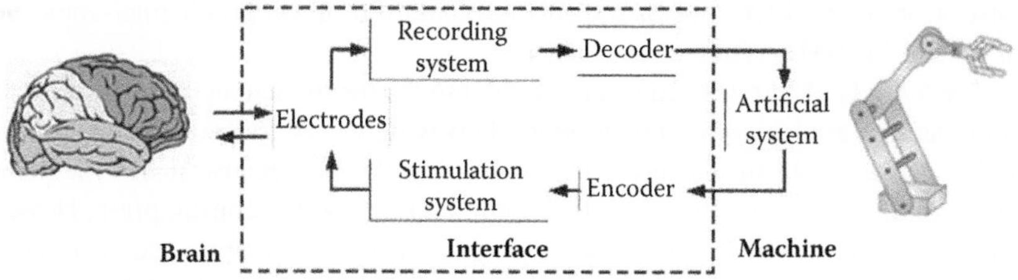

FIGURE 16.1 An ideal bidirectional BMI.

16.2.1.3 Bidirectional BMIs

Both afferent and efferent BMIs are one-way BMIs. In a one way BMI, a machine either accepts commands from a brain or sends signal to it, but doing both tasks in the same system is not possible. Bidirectional BMIs have no such limitation. The bidirectional communication module (Figure 16.1) of such a BMI allows a brain and a device to exchange information with each other. This can significantly enhance the performance and capability of BMIs because the brain can receive immediate feedback from its counterpart machine. For example, a motor prosthesis can send direct sensory feedback to its master brain and make the controlling procedure more natural; thus more accurate commands can be generated from the brain. An interesting experiment has been done to build bidirectional BMIs at Northwestern University (Kositsky et al., 2009). The researchers were able to control a mobile robot using neural signals recorded from a motor area of the lamprey brain stem, while providing feedback by electrically stimulating a sensory portion of its brain stem. The stimuli were derived from light sensors on the robot. Although this experiment only demonstrated a successful bidirectional BMIs between an *in vitro* lamprey CNS preparation and a small wheeled robot, it is strong evidence that bidirectional BMIs may become an effective paradigm for neural control of motor prostheses (Figure 16.1).

Here, we will describe a new telemetry system developed in our lab for both brain stimulation and neuronal activity recording in freely behaving small animals. It could not only deliver brief trains of electrical stimulation to four different brain locations, each implanted with a pair of electrodes, but could also acquire neuronal action potentials (APs) and local field potentials (LFPs) from two different places of the brain simultaneously. The wireless communication mode based on Bluetooth to the USB interface of a portable personal digital assistant (PDA) is used as uniform mode both for transmission and reception (Figure 16.2).

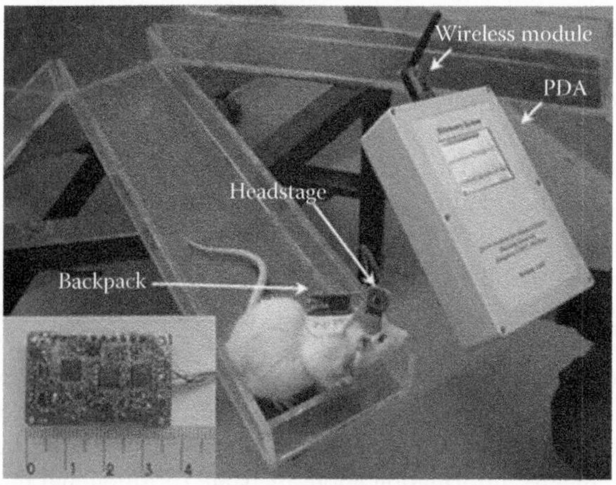

FIGURE 16.2 Bidirection telemetry system for freely behaving rats.

As shown in Figure 16.2, the system consists of three main components: headstage, portable PDA, and backpack. The headstage preamplifiers on a tiny board are directly connected to the recording electrode array on the head of the rat. The direct connection minimizes artifacts caused by environmental interference when the signals are transferred through the cables between the head of a rat and the main board, which is placed on the back of the rat. The gain of the first amplification is about 50, and the gain of the second one is about 2. The input filter network, which is composed of resistance/capacitance components with high common mode rejection ratio, serves to attenuate the DC offset. A low-pass filter (0–4,500 Hz) is used for high-frequency noise removal. A right leg driver circuit is used to make the analog ground follow the animal ground in order to reduce the common mode inference. Low-noise and low-voltage amplifiers TLV2264 (Texas Instruments Inc.) are employed for their low power consumption (500 A maximum).

The PDA of the system consists of five components:

1. A core board (ARMSYS2410-CORE, Liyutai Elec. Inc., Hangzhou, China) based on a 266MHz ARM920T processor
2. A 320 × 240 TFT LCD and touch screen module (TS35ND2502, TopSun Optronics Inc.)
3. A Bluetooth module (BTS2502C1H, Jinoux Inc.), connected to the USB port of the processor
4. A base board, which connects the Bluetooth, LCD, and touch screen module to the core board
5. A 1-GB USB mass storage device

The operation system of the PDA is GNU/Linux 2.6.15, which contains a USB-serial driver. The program, written in C++, configures the virtual USB-serial port and outputs different commands. The stimulation parameters, including pulse width, pulse number, duration, amplitude, and constant current/voltage mode, can be set as needed by the PDA. It also receives the data of neuronal signals from the backpack and then recovers the data, saves them in the USB mass storage device, and shows them on the LCD.

The backpack executes the commands to generate electrical stimuli and acquire and send neuronal signals to the portable PDA. C8051F411 is used as the microcontroller of the main board. The details of stimulation circuit design are similar to that mentioned in the navigation system. A more important function of the main board is to amplify the neural signals from the headstage once more. There are four-cascade conditioning circuits for each channel. The first cascade has two different gains for the signals of LFPs and neuronal APs, about one for the bandwidth of 1–40 Hz, and 6.1 for the bandwidth of 300–4,500 Hz. The second cascade is a second-order low-pass filter with 3-dB attenuation at the cutoff frequency of 4,500 Hz. The amplification gain of the third cascade is about 14.7. The amplification gain of the headstage is about 100; thus the total amplification gains of the backpack for the neuronal APs and LFPs are about 8,960 and 1,470, respectively. The wireless module based on Bluetooth can transfer data in the range of 100 m with the communication speed up to 70 kb/s, which is sufficient for two-channel neuronal signal acquisition and some applications with more than two channels. The 460,800 baud is chosen to satisfy the requirement of power consumption.

Experiments were made to prove the feasibility of our system. We first tested the stimulation function by a navigation experiment (Feng et al., 2007) and then tested the recording function by acquiring neuron signals from the primary motor area (M1) of a freely moving rat. At last, we tested both stimulation and recording function synchronously in real time in the same rat by the experiment of amygdaloid-kindled seizures. The results indicate that our design establishes a closed loop system between a specific region of the nervous system and the artificial device.

16.2.1.4 Challenges of Bidirectional Telemetry System

Although there are a number of wireless technologies for transmitting neural signals, it is still a problem to transmit a large amount of neural signals. Power consumption, bandwidth, and transmission distance of wireless technologies are the main bottlenecks. Currently, some solutions have been proposed for transmitting as much neural information as possible under limited bandwidth. One of these solutions is not to transmit the original neural signals but only to transmit preprocessed data. However, this method may throw away some useful information. For example, Bossetti et al. (2004) made their system only transmit the time at which a spike appears. In such a system, they lost the waveform of each spike, which is necessary for spike sorting. Therefore, a new approach based on vector-quantization data compression has been proposed to overcome the limited bandwidth of wireless BMI systems (Cho et al., 2007).

Using electrical stimulation as sensory feedback and reducing electrical stimulus artifact are other key issues of closed-loop BMIs. Stimulus artifact may make operational amplifiers saturation and affect the recording of neural signals. Software-based (filter, template subtraction, or sample interpolate) and hardware-based (filter or sample hold) methods have been developed to minimize signal distortion caused by stimulus artifact (Heffer and Fallon, 2008), but how to remove stimulus artifact online in closed-loop BMIs is still a problem. Current methods of electrical stimulation to provide sensory input from the environment are still limited. BMIs for motor prostheses should incorporate this approach for optimal performance. If various sensory feedback could be sent to the nervous system to build bidirectional BMIs, the performance of motor prostheses would be augmented.

16.2.2 Rat Navigation under the Control of the Cortical Stimulating System in Complicated Environments

Electrical activation of a nervous system is not only a technique to study the form and function of the nervous system but also a method to restore function to persons with neurological disorders. The cochlear implant has been shown to restore audition in hearing impaired patients, and the deep-brain stimulator is an effectively therapeutic tool for relieving the symptoms of Parkinson's disease and dystonia (Mussa-Ivaldi et al., 2007).

16.2.2.1 Intracranial Microstimulation and BMIs

Directly inputting external information into the biological brain with the aid of implanted BMIs technologies has only recently been realized thanks to the advances in electronics, neuroscience, and medicine. In the early 2000s, Romo et al. (2000) successfully demonstrated that peripherally provided tactile information could be mimicked through direct cortical stimulation. In 2002, Chapin and his fellow researchers successfully trained rats to navigate through a complex, three-dimensional terrain, following control cues delivered by cortical microstimulation (Talwar et al., 2002). This study implies that the animals with electrodes implanted in brain can be remotely guided by applying virtual cue or rewards stimuli to perform real-world applications and tasks, such as searching for victims of earthquakes or bombings. Because of their advantages in motor function and power supply, this type of hybrid robotics can overcome the limitations of traditional robotics in flexibility, maneuverability, mobility, and adaptability. These findings not only hinted that cortical microstimulation had potential for directly delivering the information but also represent an extension of intracranial electrical stimulation into useful real-world applications.

In contrast to the operant paradigm used by Chapin, a brain can be stimulated in proper regions to generate specific locomotor movement commands without training. This has been shown in cockroaches (Holzer and Shimoyama, 1997). The authors measured locomotory reactions of a cockroach (*Periplaneta americana*) evoked by various electrical stimuli given to the antennae of the insects. They then built a simple mathematical model of antennal stimulation and locomotion responses. Based on that model, limited directional locomotion control could be achieved through an electronic backpack, which serves to stimulate the afferent nerve fibers of the antennae.

Recent research on such training-free stimulation has been extended to different locomotions in different species. Sato et al. (2009) demonstrated the remote control of insects in free flight via an implantable radio-equipped miniature neural stimulating system. The pronotum mounted system consisted of neural stimulators, muscular stimulators, a radio transceiver-equipped microcontroller, and a microbattery. Flight initiation, cessation, and elevation control were accomplished through neural stimulus of the brain, which elicited, suppressed, or modulated wing oscillation. Turns were triggered through the direct muscular stimulus of either of the basalar muscles. Kobayashi et al. (2009) confirmed the midbrain nuclei as the swimming center for goldfish. Stimulation of sites on the midline induced forward movement, whereas that of sites off midline induced turning toward the stimulated side. Using a wirelessly controlled two-channel microstimulator, the locomotion of goldfish in the horizontal plane could be controlled directly by stimulating medial longitudinal fasciculus (Nflm) for movements involving the trunk and tail. Forward and turning movements could be arbitrarily induced in goldfish equipped with the stimulation device and electrodes implanted in or near the right and left Nflm.

16.2.2.2 Rat Navigation in Complicated Environments

Although many studies have been done researching how to effectively realize cortical microstimulation in brain and how it can be used for functional purposes, there is still much more to be learned. This section described a remote control training system made in our lab for rat navigation in complicated environments (Feng et al., 2007).

All of the procedures were carried out in accordance with the Guide for the Care and Use of Laboratory Animals (Ministry of Health, China). Adult Sprague-Dawley rats (230–380 g) were anesthetized with pentobarbital sodium (40 mg/kg, intraperitoneally) and placed in a stereotaxic apparatus (Stoelting Co.). The scalp was incised at the midline. 1.5-mm diameter holes were drilled in the skull for inserting four stimulating electrodes into the brain and for anchoring stainless steel screws.

Stimulating electrodes were made from pairs of insulated nichrome wires (65 μm in diameter) with a 0.3–0.5-mm vertical tip separation. One bipolar stimulating electrode was placed in right MFB (AP: −3.6, ML: −1.8, DV: +8.2) (Paxinos and Watson, 2005). Two stimulating electrodes were implanted symmetrically in the whisker barrel fields of left and right somatosensory cortices (SI) (AP: −1.8, ML: ±5.0, DV: +2.3). The fourth bipolar stimulating electrode was implanted in the left dorsolateral periaqueductal gray matter (dPAG) (AP: −8.0, ML: +0.6, DV: +4.5). Dental acrylic was used to fix the electrodes and screws to the skull. At least 7 days were allowed for postoperative recovery (Figure 16.3).

The behavioral effectiveness of MFB stimulations were evaluated in an operant chamber equipped with a lever. The chamber was connected with the computer through a COM port. Each lever-pressing action produced a negative pulse that was delivered to computer by the COM port. When the computer detected the pulse signal, it counted the number of the lever pressing. At the same time, the computer sent a command string to trigger the remote control stimulator. Then a

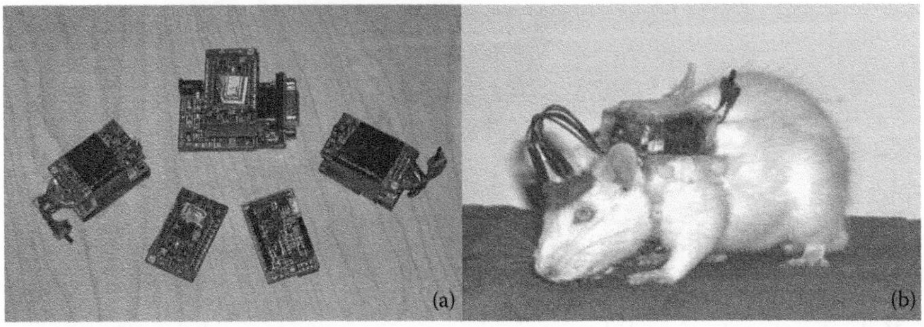

FIGURE 16.3 (a) The Bluetooth modules used for the transmitter, the wireless receiver, and microstimulator. (b) A rat with backpack.

train of biphasic pulses to the MFB followed each lever press. Each stimulation train was set at 10 biphasic pulses at 100 Hz with a pulse duration of 1 ms and a peak-to-peak pulse amplitude of 8 V.

When the rats were placed in the operant chamber, they occasionally touched the lever mounted on one side of the wall. MFB stimulation reward was immediately given through the stimulator on the backpack for the lever press. Under this reinforcement schedule, all of the subjects could press the lever continuously to obtain the MFB stimulation reward, reaching pressing rates as high as 120/min for over 5 minutes.

Immobile behavioral responses evoked by dPAG stimulation were recorded in an open field. The first step of the procedure, 1.5 seconds of training with pulses of 40 Hz, with a pulse duration of 1.0 ms, was applied in the simulation, whereas stimulation amplitude was increased in stepwise increments of 2 V up to 8 V. Immobile behavior was scored through the freezing time. Freezing was operationally defined as the total absence of movement of the body and vibrissae, accompanied by at least two of the following responses: arched back, retraction of the ears, piloerection, or exophthalmos (Kobayashi et al., 2009). Most subjects showed immobile behaviors and held freezing for 3–5 seconds with 6-V amplitude. When the amplitude exceeded 8 V, the escape behaviors were recorded from all rats. In the next step of the procedure, the dPAG stimulation was followed by one or two trains of MFB stimulation. No matter whether the rats were freezing or not, they immediately began to run forward after the termination of MFB stimulation. It might suggest the virtual rewards of MFB play a key role in switching immobile behavior to motion initialization.

An eight-arm radial maze was made for steering training of animals' locomotion. The subjects were motivated to keep running forward in the eight-arm maze by receiving MFB rewarding stimulations delivered at a speed around 0.5–1 Hz. Electrical stimulation delivered to one side of SI was used as turning instruction when they moved into the central area of the maze, which would give the subject a virtual sense of object touch in the contralateral side. By using this cue, rats could be trained to turn correctly on their navigation route by associating MFB rewarding stimulation with each correct turning. After 5–7 days of training, the average correct rate of turns reached more than 90% in 20 test runs per rat. In each test run the rats were given 10 left and 10 right turning instructions during their navigations in the maze. The instruction stimulations were 1–2 trains with 10 pulses at 100 Hz, with pulse duration of 1 ms and amplitude of 6–8 V.

In the end, the animals were placed in an open environment with several obstacles consisting of bottles, an elevated runway, ledges, and shelters. The rats could be easily guided around bottles, over the bridge and across the runway and be made to pause in a shelter. They could even be instructed to climb or jump if sufficient rewards were offered (Figure 16.4).

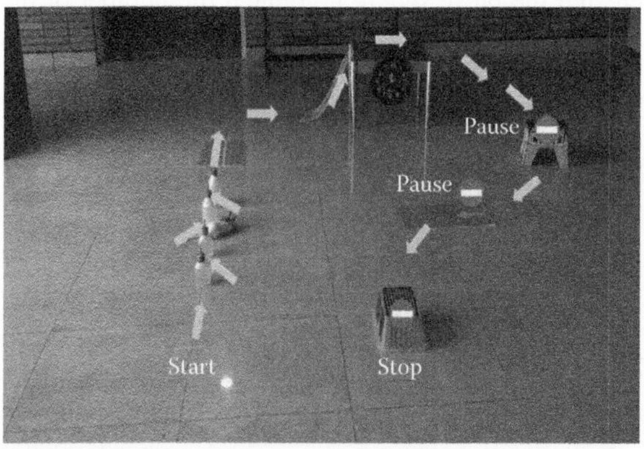

FIGURE 16.4 An example of guided rat navigation using brain microstimulation. The picture was constructed from digitized video recordings. The arrows indicate the movement tracks. The circles denote the places where the rat should pause with immobility.

In this study, operant conditioning paradigm was used to train rats to perform turning task. Meanwhile, some brain regions were stimulated to evoke direct specific locomotor movements. Both of these methods showed that the external information could be converted into electrical signals and input into the biological brain with the aid of implanted electrodes. Although little is known about the responses that temporally patterned microstimulation produces in neural areas and the mechanisms by which the brain can use these responses to guide behavior, intracranial stimulation could be sufficient for a variety of BMIs applications.

16.3 DECODING NEURAL INFORMATION IN RAT MOTOR CORTEX FROM NEURAL ENSEMBLE RECORDING

16.3.1 Neural Decoding in BMIs

Decoding cortical neural activities into the corresponding animal behaviors is a core part of every BMI system. Especially, the decoding of movement information according to cortex neural activities has been a focus of the research on BMIs (Lebedev and Nicolelis, 2006; Ye et al., 2008). Although traditional recording technology of single electrode could record the electrical activity of a single neuron, it hardly demonstrates the synergic characteristic of neural firing in population level. In recent years, the development of techniques for simultaneous population-level neural recording enables the research of distributed coding in the motor cortex.

Since the early 1980s, many methods have been developed to decode motor-cortex activity, including population vector algorithm, optimal linear estimation, artificial neural networks, and Bayesian methods (Georgopoulos et al., 1988; Salinas and Abbott, 1994; Sanchez et al., 2003; Taylor et al., 2002; Zhang et al., 1998). The population vector algorithm has been a successful neural ensemble decoding method (Georgopoulos et al., 1988; Taylor et al., 2002), in which the activity of each neuron is characterized by the firing rate at its preferred direction. Population vector is an optimal linear method when the preferred directions are uniformly distributed. Optimal linear estimation (OLE) is used when the preferred directions are not uniformly distributed (Salinas and Abbott, 1994). Therefore, OLE is a more general method. Population vector and OLE hardly consider the dynamic relation between neural activity and movement, because they assume that the firing rates are linearly related to the underlying movement parameters. Artificial neural networks (ANN) have also been proven successful in neural ensemble decoding in motor cortex, but the results were not significantly different than those from population vector or OLE. This is not surprising because artificial neural networks actually encode a linear model as the population vector method (Sanchez et al., 2003). Later on, the Kalman filter was adopted with the assumption that movement is a linear map of neural activity. It presents a Bayesian estimation problem that can be solved by an efficient recursive algorithm. The Kalman filter provides a clear observation model as a tuning function and also considers dynamic information of the temporal kinematics modeling. In the study, neural decoding based on probabilistic neural network (PNN) is also introduced, which has no linear assumption on the relation between neural firing and movement (Specht, 1990).

In this section, we tested several neural decoding methods on our own data from rats. In our experiment, a rat performed prelimb movement, and we recorded both neural ensemble firing and prelimb movement of the rat. Kalman filter-, OLE-, and PNN-based neural decoding are detailed in this section as examples of neural decoding methods.

16.3.2 Decoding Methods

16.3.2.1 Kalman Filter

A Kalman filter provides an efficient recursive way to optimally estimate the posterior probability where the state and observation models are linear, and the prior and posterior are Gaussian (Wu et al., 2002). Therefore, in our example, we assume that the relationship between the prelimb motion and the neural activity was linear, and the noise in the neural ensemble firing obeyed Gaussian distribution.

Rat's forelimb pressure was defined as $p(n)$ in the decoding process at time $t_n = n\Delta t$, where Δt was bin size. The neural ensemble activities of N neurons were defined as $f_i(n)$, which represents the firing number of the ith neuron in nth Δt, and N represents the total number of recorded neurons. Hence, the state $x(n)$ was defined as the prelimb pressure $P(n)$.

The Kalman filter contained two parts in our work. First, it obtained the linear relationship between the pressure state to the neural activity from training data. Second, it decoded pressure state from concurrent neural ensemble firing by the obtained linear relation.

In the encoding part, the pressure states were assumed to propagate in time according to the following linear dynamic equation, namely the system model:

$$x(n+1) = A_n x(n) + q_n \tag{16.1}$$

demonstrating that the pressure state at time $n + 1$ was linearly related to the pressure state at time n, where A_n was a $N \times N$ square matrix and q_n denoted the zero-mean noise term with covariance $Q_n(q_n \sim N(0, Q_n))$. The pressure state output could be obtained from generative model:

$$y(n) = H_n x(n) + v_n \tag{16.2}$$

where H_n was a matrix that linearly relates the pressure to the neural firing, v_n was the zero mean measurement noise with covariance V_n, and $y(n)$ was a vector representing the neuron ensemble activities. In the model, A_n, H_n, Q_n, and V_n would change across the time step n. However, in our model, we assumed them as constants for simplification. We could compute parameters in a Kalman filter from training data, which contain both neural ensemble activity and pressure state, by the least squares estimation below:

$$A = \arg\min_{A} \sum_{n=1}^{L-1} \|x(n+1) - Ax(n)\|^2 \tag{16.3}$$

$$H = \arg\min_{H} \sum_{n=1}^{L-1} \|y(n) - Hx(n)\|^2 \tag{16.4}$$

Matrix A and Q could be estimated by training data as follows:

$$A = X_1 X_0^{\mathrm{T}} \cdot \left(X_1 X_1^{\mathrm{T}}\right)^{-1} \tag{16.5}$$

$$Q = \frac{(X_1 - AX_0)(X_1 - AX_0)^{\mathrm{T}}}{L-1} \tag{16.6}$$

where X_1 was a train of states, and X_0 was another train with same length and after X_1 by one time step. Estimation of matrix H is similar as matrix A.

In the decoding part, the following equations were used to estimate pressure state step by step:

$$P'(n+1) = AP(n)A^{\mathrm{T}} + Q \tag{16.7}$$

$$K(n+1) = P'(n+1)H^{\mathrm{T}}(HP'(n+1)H^{\mathrm{T}})^{-1} \tag{16.8}$$

$$\tilde{x}(n+1) = A\tilde{x}(n) + K(n+1)(Y(n+1) - HA\tilde{x}(n)) \tag{16.9}$$

$$P(n+1) = (I - K(n+1)H)P'(n+1) \tag{16.10}$$

where $K(n)$ represents observer gain, and $P(n)$ represents error variance. Equation 16.8 was used to update the Kalman gain, which produced a state estimate that minimizes the mean squared error of the reconstruction. Equations 16.7 and 16.10 were used to update the state error covariance in each step. Using the current state estimate $\tilde{x}(n)$ and neural activity in the next bin $Y(n+1)$, the next state value was estimated by Equation 16.9.

16.3.2.2 Optimal Linear Estimation

OLE was also adopted here as a comparison with other methods. It was estimated by

$$P(n) = y(n)\beta \tag{16.11}$$

where $p(n)$ represented the estimated pressure value by neural ensemble activity $y(n)$ in time n. The coefficient matrix can be obtained by least-square estimation:

$$\beta = (Y^{\mathrm{T}}Y)^{-1}Y^{\mathrm{T}}P \tag{16.12}$$

where Y and P were neural ensemble activity and pressure signal of a train of training data.

16.3.2.3 Probabilistic Neural Network

PNN is a typical nonlinear classifier that uses minimum Bayesian risk criterion (Specht, 1990). We discretized the pressure value into several levels as the decision attributes of PNN. The number of discretization levels could be changed for different demands. Only two to five levels are needed for judging whether the rat has pressed the lever. For more precise decoding, the pressure value should be discretized into much more levels, i.e., 10–100 levels. Compared with a traditional BP neural network, PNN takes less time in training and does not need to train again when more training data are added or reduced. Moreover, it can always obtain the optimal solution in Bayesian criteria when the training data are sufficient, regardless of the complexity of the classification problem.

In our case, a set of N-dimensional data are to be classified into $D(D \geq 2)$ classes, $C_1 \ldots C_i, \ldots C_D$, corresponding to levels of pressure values. The Bayes' decision C_i is made by the following rule: a data point X belongs to C_i if $H_i L_i f_i(X) > H_j L_j f_j(X)$ for all $j = 1, 2, \ldots D$ and $j \neq i$, where H_i denotes the prior probability that X belongs to C_i, L_i is the loss factor when a mistake occurs for each X_i, and $f_i(X)$ is the Parzen probability density function estimator. Using the spherical Gaussian radial basis function as the weight function of $f_i(X)$, we have

$$f_i(X) = \frac{1}{(2\pi)^{N/2}\sigma^N}\frac{1}{M_i}\sum_{k=1}^{M_i}\exp\left[\frac{-(X - X_{ik})^{\mathrm{T}}(X - X_{ik})}{2\sigma^2}\right] \tag{16.13}$$

where N is the dimension of X, M_i is the number of training vector in class C_i, X_{ik} is the kth training vector in class C_i, and σ was the smooth factor or bandwidth. All of the data vectors are normalized to have the same length before classification. So we have

$$(X - X_{ik})^{\mathrm{T}}(X - X_{ik}) = -2(X^{\mathrm{T}}X_{ik} - 1) \tag{16.14}$$

Combining Equations 16.14 and 16.15, we get a simplified classification rule: X belongs to C_i if

$$L_i\sum_{k=1}^{M_i}\exp\left[\frac{X^{\mathrm{T}}X_{ik} - 1}{\sigma^2}\right] > L_j\sum_{k=1}^{M_j}\exp\left[\frac{X^{\mathrm{T}}X_{jk} - 1}{\sigma^2}\right] \tag{16.15}$$

for all $j = 1, 2, \ldots D$ and $j \neq i$. In the view of a neural network, the PNN network is a structure composed of four layers: input, pattern, summation, and decision layers. The input layer receives and

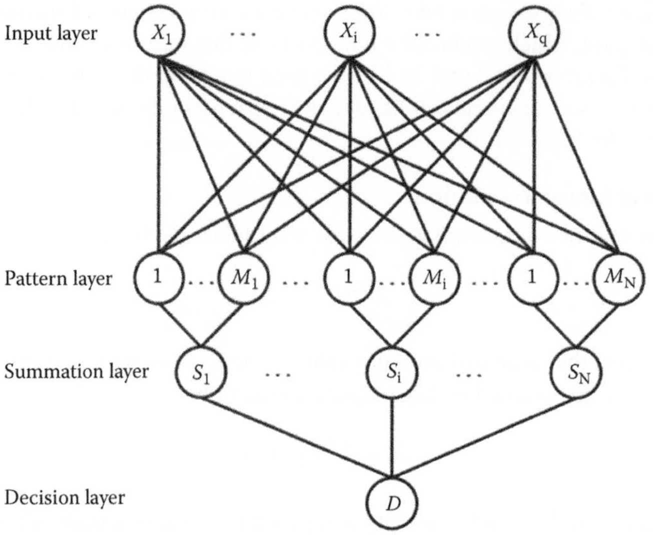

FIGURE 16.5 Probabilistic neural network.

normalizes input vector; each unit in the pattern layer represents a training vector with the response function $\exp\left[(X^{T}X_{ik}-1)/\sigma^{2}\right]$. The summation layer computes the summation of each pattern and multiplies the loss factor. The decision layer selects the largest one in the summation layer as the classification result. More prior knowledge expands the PNN horizontally and may increase its classification capacity (Figure 16.5).

Two kinds of PNN decoders were defined here with different input vectors as Equations 16.16 and 16.17.

$$X(n)=[f_1(n)\,f_2(n)\ldots f_N(n)] \tag{16.16}$$

$$X(n)=[f_1(n)\,f_2(n)\ldots f_N(n)\,\tilde{p}(n-1)] \tag{16.17}$$

where $f_i(n)$ denoted firing rate of ith neuron in nth bin, and $\tilde{p}(n-1)$ denoted the previous estimated pressure. The decoder with input vector Equation 16.16 was named as a PNN decoder, and the decoder with input vector Equation 16.17 was called a Modified PNN (MPNN) decoder here. The decoding process of the PNN decoder is as follows. Firstly, the structure of PNN was trained by a set of training data, whose decision attributes were $p(n)$, and in the testing step, we estimated $\tilde{p}(n)$ by classifying the class of $X(n)$ with the PNN structure. The process of MPNN decoder is similar, but in the training process, the $\tilde{p}(n-1)$ in Equation 16.6 was replaced by the actually recorded pressure value $p(n-1)$.

16.3.3 Experiments

Sprague-Dawley rats (approximately 280 g) were trained to perform a conditional operant task. When a rat pressed the lever by its forelimb with the pressure value above the threshold, the animal was rewarded with a drop of water. All of the rats were trained to achieve a success rate over 75% before surgery. After training, a 2 × 8-channel chronic microwire electrode array (35 μm in diameter, 300 μm for a space between rows, and 200 μm for a space within rows; California Fine Wire) was implanted into the forepaw region of the forelimb primary motor cortex of each rat. The electrode tips were positioned in layer *V*, and the approximate depth in layer *V* was in the range of

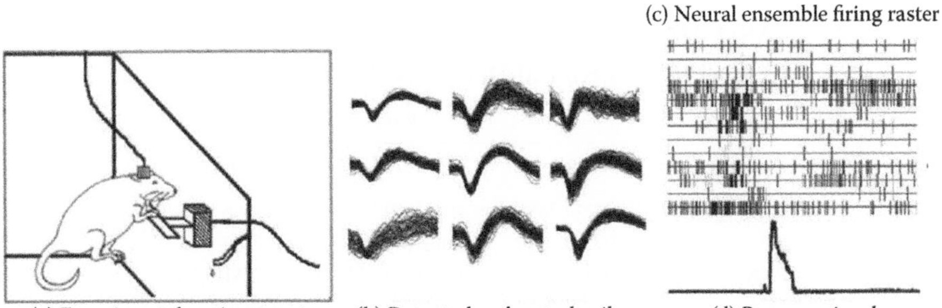

FIGURE 16.6 Experimental environment and simultaneously recorded neural signal and pressure signal. (a) Experimental environment. Water was rewarded when rat pressed the lever. (b) Part of extracted spikes in rat by 2 × 8 microelectrode array. (c) Raster of neural ensemble firing in 5 seconds, each vertical short line denoted a spike firing. (d) Pressure value of the lever from pressure sensor. An abrupt change of the pressure value indicated a pressing event of rat.

1.1–1.8 mm beneath the pia in the cerebral cortex and contained neurons that projected directly to the spinal cord to activate peripheral motor neurons. To decrease the common mode noise, one of the 16 electrodes was used as the reference channel. Before being recorded at the 30-kHz sampling rate, all 16 channels of the analog neural signals were filtered and amplified using Cerebus 128™ (Cyberkinetics Inc.) (Figure 16.6):

1. Experimental environment: Water was rewarded when the subject rat pressed the lever.
2. Part of extracted spikes in rat by 2 × 8 microelectrode array.
3. Raster of neural ensemble firing in 5 seconds, each vertical short line denoted a spike firing.
4. Pressure value of the lever from pressure sensor. An abrupt change of the pressure value indicated a pressing event of rat.

The reference channel was not considered for later decoding. Neural spiking activities were extracted from the other 15 channels of neural signals by a thresholding method. After that, spikes from each channel were sorted into one to three types by principal component analysis and K-Means clustering, and each type was supposed to represent one neuron. A total of 22–58 neurons were found in all 15 channels per rat. To compute neural spiking frequency for each neuron, firing number in a time bin ($\Delta t = 100$ ms) was counted. Meanwhile, the pressure signal was recorded at 500 Hz and averaged into bin size ($\Delta t = 100$ ms). Discretization was implemented to transfer the pressure value into discrete decision attributes of PNN.

16.3.4 Decoding Results

To evaluate the performance of a decoding method, the mean square error and the correlation coefficient (CC) between the recorded pressure and the corresponding predication were calculated for the testing data. The results of decoders were computed in Table 16.1, where the bin size was 100 ms, and the discretization for PNN and MPNN level was 100. The decoding outputs of rat S9-03 were showed in Figure 16.7, where the dotted line denotes the pressure signal recorded by pressure sensor, and the solid line denotes the decoding outputs. The result of the MPNN decoder was significantly better than others with CC = 0.8657. There were cleavages in yellow line peaks in PNN decoder output, which seemed to be high-frequency noise. The dashed line denotes target pressure value, and the solid lines denote estimated value by each method.

TABLE 16.1
CC and MSE of Neural Decoders (Bin Size = 100 ms)

	Rats		
Decoders	**S9-03**	**S9-04**	**S9-12**
OLE	0.7778·0.2487	0.8817·0.1200	0.7066·0.4448
KF	0.7973·0.4047	0.8018·0.3956	0.7583·0.4825
PNN	0.7061·0.5866	0.5782·0.8420	0.6151·0.7682
MPNN	0.8657·0.2563	0.9322·0.1749	0.8840·0.2248

KF, Kalman filter.

The output of OLE was similar with those from the other decoding methods. It predicted the pressure values in a reasonable accuracy (CC = 0.7778), but the result contained a lot of high-frequency artifacts. The Kalman filter produced smoother prediction than OLE did and estimated the pressing events better (CC = 0.7973). These linear methods did not perform as accurately as MPNN decoder, according to the CC values. The results of the other two rats, S9-04 and S9-12, were also computed in Table 16.1.

The decoding result of MPNN was smoother and had higher CC than that of PNN. Although the structure of PNN decoder was also nonlinear in Bayesian criterion, it provided no estimation of uncertainty and was difficult to analyze complex temporal movement. Moreover, we found that the result of a Kalman filter was not as good as MPNN decoder here, and the output fluctuates between pressing events. A Kalman filter is a Bayesian filter with the linear and Gaussian assumption. Here, the pressing events in our experiment appeared to happen accidentally, and the pressure value seemed changed abruptly; these changes are beyond a linear process. Therefore, under the least-square rule, the training of evolving matrix was compromised for the abruptly changed pressure value, resulting disorder between pressing events.

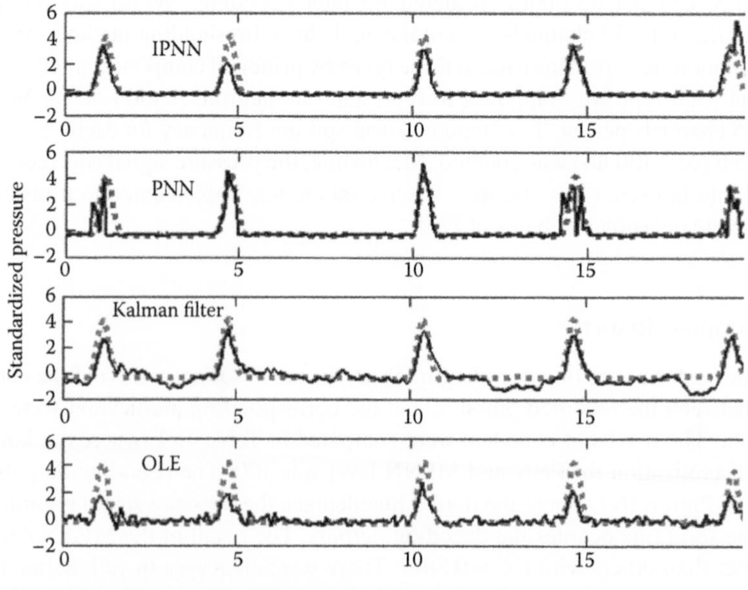

FIGURE 16.7 Examples of pressure decoding results. The dashed lines denote target pressure value, and the solid lines denote estimated value by each method.

16.4 BUILDING REAL-TIME BMI SYSTEMS ON RATS

Although the correlation between arm movements and firing rates in M1 were observed several decades ago (Evarts, 1968; Fetz, 1969; Humphrey et al., 1970), the first implanted BMI was not reported until about 30 years later when Chapin et al. (1999) published their work. They showed how rats learned to control a simple robot arm for water reward using multichannel neural activities recorded in motor cortex. This paradigm was then demonstrated in primates for different tasks, such as controlling an elaborate robotic arm (Wessberg et al., 2000), reaching and grasping an object (Carmena et al., 2003), moving a cursor in two dimensions (Serruya et al., 2002), or moving a ball in three dimensions (Taylor et al., 2002) and even activating paralyzed muscles (Moritz et al., 2008; Pohlmeyer et al., 2009).

The most impressive application was developed by Schwartz's group in which two monkeys were implanted with intracortical microelectrode arrays in their primary motor cortices, and each monkey used the signals to control a robotic arm to feed itself (Velliste et al., 2008). The recorded signal was also used by the subject to open and close the gripper as it grasped and moved the food to the mouth. The endpoint velocity and gripper command were extracted from the instantaneous firing rates of simultaneously recorded units. The algorithm used to extract control signal from the real-time stream of neural data was a version of the population vector algorithm, which is a linear algorithm relying on the cosine directional tuning function and preferred direction of each unit (Georgopoulos et al., 1982). In this study, monkeys were for the first time trained to continuously control a robotic arm and hand for self-feeding. It has expanded the capabilities of prosthetic devices through the use of observation-based training and closed-loop cortical control.

16.4.1 Real-Time Control in BMIs on Rats

In this section, we will present an integrated real-time brain-machine interface control paradigm in rats. This representative system includes microelectrode, surgery, behavior training, data acquisition, neural decoding algorithms, and the final results. Additionally, we implemented the decoding algorithm in field-programmable gate array (FPGA) platform for portable and real-time BMIs applications. The results indicate that the performance of current FPGAs is competent for portable BMI applications.

In each setup of our experiment, a rat was trained to perform a lever pressing task for water rewards in a box. Before the training, the rat was kept on water restriction and maintained at 12 ml of water per day in the training process. The motor cortical neural signals from primary motor cortex were recorded with a Cerebus™ 128-channel data acquisition system (Cyberkinetics Inc.), and the pressure signal of the lever was recorded simultaneously. After training, the system was switched from mechanical control to neural control. The spikes were sorted in real time and sent to the decoding algorithm. The output of the decoding was then sent to a water supplying system to reward the rat with some water.

The decoding algorithm used here is PNNs, a kind of radial basis artificial neural network that is usually used for classification problems. PNNs have the advantages of much faster training process and more accurate results using the minimum Bayesian risk criterion compared with other networks such as multilayer perception networks. In this study, we used this neural network for decoding neural ensemble activities. The pressure data were quantized in several levels, which correspond to the output classes. The PNN model is used to decide which class the current neural activity belongs to.

PNNs were first described by Specht (1990) in detail. There are four layers in the structure, including an input layer, a pattern layer, a summation layer, and a decision layer. When an input is presented, the pattern layer calculates the distances from the input vector to each training vector and produces a vector whose elements indicate how close the input is to each training vector. The active function used in this pattern layer is an exponential function. The third (summation) layer sums these contributions for each class of inputs to produce its net output as a vector of probabilities.

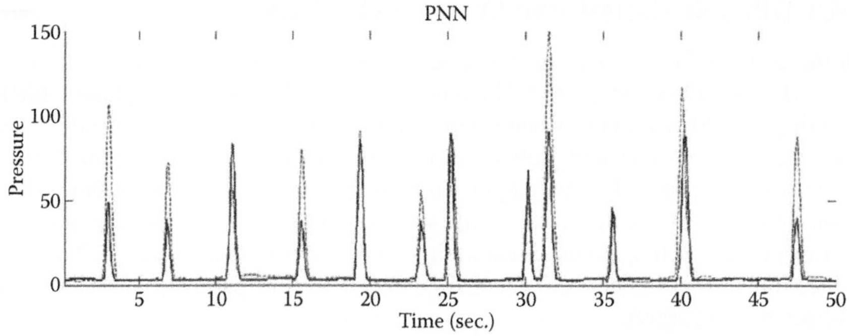

FIGURE 16.8 Test of MATLAB®-based PNN model for neural decoding. The result of the MATLAB algorithm (solid line) shows a high correlation with the actual pressure value (dashed line).

Finally, a competitive transfer function on the output of the third layer picks the maximum of these probabilities, and produces a 1 for that class and a 0 for the other classes.

We implemented the PNN algorithm in C++ and tested it offline on real data. In the experiment of one rat, a total of 31 units were sorted from 16-channel neural recordings and the number of spikes in a time bin (100 ms) was calculated for each neuron. The vector of 31 spike firing rate of units was defined as the input of the PNN. A periods of 200 seconds of neural recordings and pressure signals, containing 2,000 neural bins and 33 pressing events, were taken as training data, and another 300 seconds worth of data were used to test the model. The results are illustrated in Figure 16.8, where the solid line indicates PNNs outputs, and the dashed line indicates actual pressure data. The prediction was very accurate. Nearly every pressing event is predicted, and only few mistakes are made.

Considering the feasibility of this algorithm, a real-time online neural control system was developed. In this system, the mechanical lever was removed, and rats could drink water only when the decoding algorithm output exceeded a predefined threshold. After a period of time of learning, most rats became skillful in the neural control, which indicated that the system was feasible.

16.4.2 FPGA Implementation of PNN

FPGA, which has a parallel computer architecture, is well-suited for implementing ANN models (Omondi and Rajapakse, 2006). Many studies have proposed FPGA implementation of various neural networks for different applications (Grossi and Pedersini, 2008; Lin and Lee, 2009). An FPGA-based neural network can be much faster than a PC-based one. Therefore we implemented the PNN algorithm in FPGA to meet the real-time requirement of a portable BMI.

Specifically, Xilinx Virtex-5 SXT FPGA is chosen because of its rich resources and high-speed computation. The FPGA implementation is developed in the Xilinx ISE 10.1 Foundation Design Software Environment using Verilog hardware description language. The system architecture and data flow of the overall design for the PNN model are shown in Figure 16.9. In the PNN algorithm flowchart, there are four computation modules, which are the distance calculation module, the exponential module, the summation module, and the compare module. Random access memory (RAM) blocks are used to restore the training data at the beginning of computation. The controller is responsible for FPGA's state logic and its communication with an external master. The following will give some more detailed implementation about the modules.

Before computing pressure values from neural data, the training data must be fed into FPGA. These data are stored in the RAM blocks. The distance between an input vector and each training vector has to be computed. The number of distances to compute for the input vector is the same as the number of the training vectors, but the computation can be done efficiently with the parallel usage of three-stage pipeline floating-point computing units. All of the distances

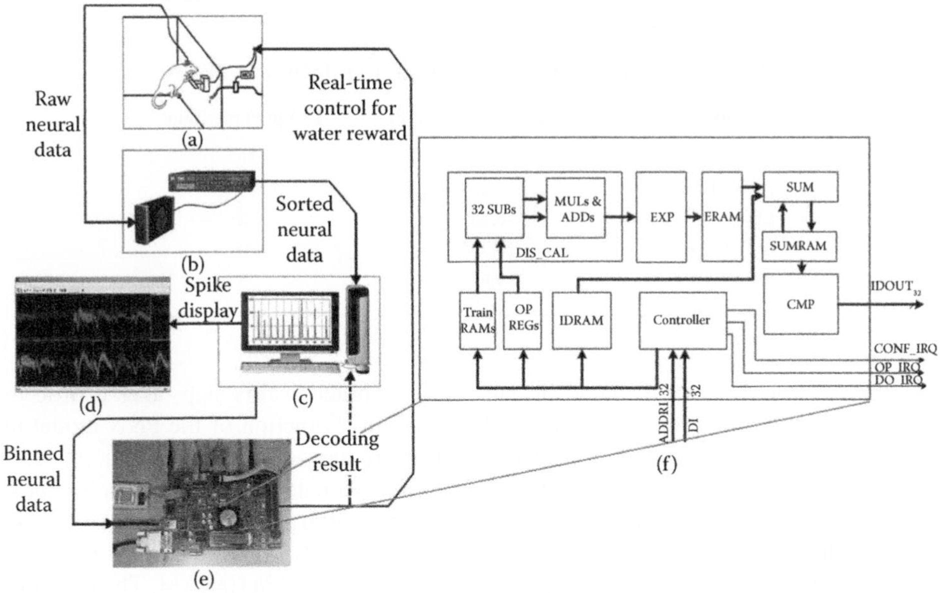

FIGURE 16.9 System setup for FPGA implementation of spike trains decoding in rats. (a) Experimental paradigm: rat performing a lever pressing task for water rewards. (b) Neural data acquisition system (Cyberkinetics Inc., Salt Lake City, UT). (c) PC workstation. (d) Sixteen-channel spike trains recorded in primary motor cortex (M1). (e) FPGA board (Virtex-5, ML506 board; Xilinx Inc.) realizing the PNN mapping algorithm. (f) Data flow diagram for FPGA implementation of PNN model. Some calculation modules and interfaces are contained in this implementation.

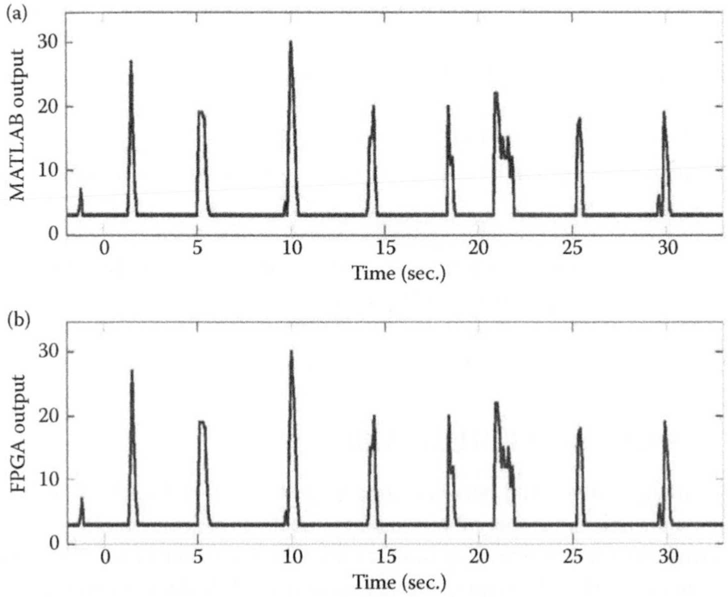

FIGURE 16.10 (a) Prediction results of the PNN output using MATLAB implementation. (b) Prediction results of the PNN output using FPGA implementation.

TABLE 16.2
Resource Utilization of PNN Implementation in FPGA

Resources	Usage Amount (kB)	Usage Percentage
Blocks of RAMs	140	26
DSP48E	234	81
LUT	26,855	82
LUT-FF pairs	10,735	30
Slice registers	19,423	59

for the input vector can be calculated in one clock. In practice this step takes a little longer time because of the time propagation delay. The activation function of the PNN model in our design is the exponential function. Because the hardware has no built-in exponential function, we used a Taylor series and a look-up table (LUT) to approximate it. A seven-order Taylor series was chosen for the hardware implementation of the exponential function. The results of exp($-x$) were temporally stored in the ERAM block (shown in Figure 16.9) before all 1,000 results were accomplished. The class label of each exponential output is stored in IDRAM. The third layer of the PNN adds up the exponential functions of each class and stores the result in the SUMRAM block. After the summation for each class, the maximum value was found, and its corresponding class label will be the output of the current neural activity. All of these mathematical operations are generated by ISE10.1 Floating-point V4.0, which is implemented using FPGA on-chip DSP48E Slices.

PNN implemented in FPGA involves 31 inputs, 1,000 neurons in the pattern layer, 100 neurons in the summation layer, and one output. For each session, the network receives a vector containing 31 neural binned data and then gives out the result indicating which class the current neural state maps to. Figure 16.10 shows the FPGA simulation result together with the MATLAB® result. As shown in the figure, the two results have no perceptible difference.

Table 16.2 summarizes the resources used in the implementation. Block RAMs are mainly used for storing training data (TrainData and IDRAM, $33 \times 4 \times 1$ kB) and some intermediate variables such as the results of exponential operation (ERAM, $1 \times 4 \times 1$ kB) and summation (SUMRAM, $1 \times 4 \times 1$ kB). A total amount of 140-kB RAM is consumed. All of the arithmetic such as addition, subtraction, and multiplication are realized using 232 on-chip DSP48E slices, each of which consumes two DSP48Es. The number of LUT, LUT-FF pairs, and slice registers involved depends on the precision and the word length in FPGA. LUT is also used for exponential function computation. The details are listed in Table 16.2.

In our implementation, each session requires a total of 8,589 clocks for generating an output. It means that if we adopt a reference clock frequency of 100 MHz as the global clock signal, each session generates a result in 85.89 μs. This is 37.9 times as fast as a personal computer (PC) did. This showed that the FPGA implementation of a probabilistic neural network model could improve the performance of BMI systems significantly and thus add possibilities for portable BMI applications.

16.5 CONCLUSIONS AND FUTURE WORK

The breadth of challenges that stand between concept generation and realization of practical BMIs systems calls for synergistic development of the many facets of neurotechnology. Here we have provided several examples of how to develop bidirectional telemetry interfaces and their applications in rat navigation, motor cortex decoding, and neurocontrolled devices. However, implanted BMIs have only been demonstrated in labs, and to bring them into real applications, several problems have to be solved.

16.5.1 Existing Problems

16.5.1.1 Stability of Long-Term Recording

Implanted BMIs provide an innovative tool to understand physiological processes at the cellular level and to help restore functions in an abnormal nervous system. Recording directly from neurons requires more sophisticated signal processing and computationally intensive algorithms to interpret neural activities. Electrodes must remain stable for a long period. The ideal implant should have a very small cross-section to displace or damage as little tissue as possible during insertion (Cogan, 2008). However, no neural recording electrodes have currently been fully verified to stably and reliably record action potentials from multiple single units in a complicated situation for a long time (i.e., many years) (Hatsopoulos and Donoghue, 2009). Reliable chronic recording has been a big concern, although both cone electrodes and microelectrode arrays can record for many months in rodents, primates, and humans (Kennedy et al., 1992). Especially, more efforts should be spent on reducing the tissue encapsulating around the electrode and degradation of electrode materials.

16.5.1.2 Dimensions of Neural Decoding

BMI technologies are expected to control a prosthesis with numerous degrees of freedom (DOFs). For example, an anthropomorphic prosthetic hand might have more than 16 DOFs. However, the state-of-the-art BMI can only recover four DOFs (three spatial dimensions plus grasp) (Matrone et al., 2010). Better neural signal recording and decoding is required to fill the gap.

16.5.2 Future Research Directions

16.5.2.1 Fully Implantable BMIs

In most of the reported BMI systems, implanted multielectrode arrays using bundles of fine wires were used to record neural signals. Electronic devices for amplification and recording connect to a recording electrode array through a skull-mounted connector. This presents two major barriers to the development of practical BMI devices:

1. The transcutaneous connector makes the wounded area vulnerable for infection.
2. Neural signals may by subject to the contamination of too much noise and unexpected signal (Neihart and Harrison, 2005).

Fully implantable BMIs break the barriers by transmitting neural signals to remote devices wirelessly. Early wireless neural recording systems were built from discrete modules, whereas more recent wireless systems have utilized an integrated circuit for amplification and several off-chip components for power rectification (Harrison et al., 2007). Future work on fully implantable BMI systems for neuroprosthetic applications might focus on power dissipation and data reduction.

16.5.2.2 Close-Loop Control

A promising way to improve the performance of BMIs for motor prostheses is to provide real-time feedback from an artificial or natural actuator. With feedback provided, a brain can not only correct the errors during the execution of a movement but also learn to perform better because of its plasticity (Mussa-Ivaldi and Miller, 2003). Visual and auditory feedback has been applied in current BMI systems. Researchers are also trying to use less natural forms of feedback, such as mechanical stimulation of peripheral tissues and electrical stimulation of nervous systems (Chatterjee et al., 2007). For this kind of feedback, there is a map between the state of the prosthesis and the features of the stimulation. Learning is required for a brain to establish such a map. It is not yet known how to provide natural or artificial modalities for neuroprosthetic applications.

16.5.2.3 Real-Time Neural Decoding

Current decoding algorithms in BMIs are time-consuming. Real-time neural decoding is one of the key issues in BMIs. One of the solutions is to reduce the computational complexity of the decoding algorithms; however, this has not been successful because of the complexity of brains. The other solution is to improve computational performance of the hardware platform for a BMI. Large-scale cyber-workstation (Digiovanna et al., 2010) and parallel-computation FPGA (Zhou et al., 2009) are two promising candidates. They provide rich and scalable computing resources to enable more sophisticated computational neuroscience investigation and allow scientists to develop new models and integrate them with existing models. Meanwhile, it also facilitates the development of portable BMIs, which have been requested in some clinical applications.

REFERENCES

Acharya, S., Tenore, F., Aggarwal, V., et al. (2008). Decoding individuated finger movements using volume-constrained neuronal ensembles in the M1 hand area. *IEEE Transactions on Neural Systems and Rehabilitation Engineering* 16: 15–23.

Aggarwal, V., Acharya, S., Tenore, F., et al. (2008). Asynchronous decoding of dexterous finger movements using M1 neurons. *IEEE Transactions on Neural Systems and Rehabilitation Engineering* 16: 3–14.

Ativanichayaphong, T., He, J. W., Hagains, C. E., et al. (2008). A combined wireless neural stimulating and recording system for study of pain processing. *Journal of Neuroscience Methods* 170: 25–34.

Bossetti, C. A., Carmena, J. M., Nicolelis, M. A., et al. (2004). Transmission latencies in a telemetry-linked brain-machine interface. *IEEE Trans. Biomed. Eng.* 51: 919–924.

Carmena, J. M., Lebedev, M. A., Crist, R. E. (2003). Learning to control a brain-machine interface for reaching and grasping by primates. *PLoS Biology* 1: 193–208.

Chapin, J. K., Moxon, K. A., Markowitz, R. S., et al. (1999). Real-time control of a robot arm using simultaneously recorded neurons in the motor cortex. *Nature Neuroscience* 2: 664–670.

Chatterjee, A., Aggarwal, V., Ramos, A., et al. (2007). A brain-computer interface with vibrotactile biofeedback for haptic information. *Journal of Neuroengineering and Rehabilitation* 4.

Chestek, C. A., Gilja, V., Nuyujukian, P., et al. (2009). HermesC: Low-power wireless neural recording system for freely moving primates. *Neural systems and rehabilitation engineering: IEEE Transactions* 17: 330–338.

Chien, C.-N., Jaw, F.-S. (2005). Miniature telemetry system for the recording of action and field potentials. *Journal of Neuroscience Methods* 147: 68–73.

Cho, J., Paiva, A. R. C., Kim, S.-P., et al. (2007). Self-organizing maps with dynamic learning for signal reconstruction. *Neural Networks* 20: 274–284.

Cogan, S. F. (2008). Neural stimulation and recording electrodes. *Annual Review of Biomedical Engineering* 10: 275–309.

Digiovanna, J., Rattanatamrong, P., Zhao, M., et al. (2010). Cyber-workstation for computational neuroscience. *Frontiers in Neuroengineering* 2: 17.

Donoghue, J. P. (2008). Bridging the brain to the world: A perspective on neural interface systems. *Neuron* 60: 511–521.

Evarts, E. V. (1968). Relation of pyramidal tract activity to force exerted during voluntary movement. *Journal of Neurophysiology* 31: 14–27.

Feng, Z. Y., Chen, W. D., Ye, X. S., et al. (2007). A remote control training system for rat navigation in complicated environment. *Journal of Zhejiang University-Science A* 8: 323–330.

Fetz, E. E. (1969). Operant conditioning of cortical unit activity. *Science* 163: 955–958.

Fitzsimmons, N. A., Drake, W., Hanson, T. L., et al. (2007). Primate reaching cued by multichannel spatiotemporal cortical microstimulation. *J. Neurosci.* 27: 5593–5602.

Fitzsimmons, N. A., Lebedev, M. A., Peikon, I. D., et al. (2009). Extracting kinematic parameters for monkey bipedal walking from cortical neuronal ensemble activity. *Front. Integr. Neurosci.* 3: 3.

Ganguly, K., Carmena, J. M. (2009). Emergence of a stable cortical map for neuroprosthetic control. *PLoS Biology* 7: e1000153.

Georgopoulos, A. P., Kalaska, J. F., Caminiti, R. (1982). On the relations between the direction of two-dimensional arm movements and cell discharge in primate motor cortex. *Journal of Neuroscience* 2: 1527–1537.

Georgopoulos, A. P., Kettner, R. E., Schwartz, A. B. (1988). Primate motor cortex and free arm movements to visual targets in three-dimensional space. II. Coding of the direction of movement by a neuronal population. *Journal of Neuroscience* 8: 2928–2937.

Grohrock, P., Haussler, U., Jurgens, U. (1997). Dual-channel telemetry system for recording vocalization-correlated neuronal activity in freely moving squirrel monkeys. *Journal of Neuroscience Methods* 76: 7–13.

Grossi, G., Pedersini, F. (2008). FPGA implementation of a stochastic neural network for monotonic pseudo-Boolean optimization. *Neural Networks* 21: 872–879.

Harrison, R. R., Watkins, P. T., Kier, R. J., et al. (2007). A low-power integrated circuit for a wireless 100-electrode neural recording system. *IEEE Journal of Solid State Circuits* 42: 123.

Hatsopoulos, N. G., Donoghue, J. P. (2009). The science of neural interface systems. *Annual Review of Neuroscience* 32: 249–266.

Hawley, E. S., Hargreaves, E. L., Kubie, J. L., et al. (2002). Telemetry system for reliable recording of action potentials from freely moving rats. *Hippocampus* 12: 505–513.

Heffer, L. F., Fallon, J. B. (2008). A novel stimulus artifact removal technique for high-rate electrical stimulation. *Journal of Neuroscience Methods* 170: 277–284.

Holzer, R., Shimoyama, I. (1997). Locomotion control of a bio-robotic system via electric stimulation. Paper presented at the 1997 IEEE/RSJ International Conference on Intelligent Robots and Systems (IROS '97), Grenoble, France, September 1997.

Jurgens, U., Hage, S. R. (2006). Telemetric recording of neuronal activity. *Methods* 38: 195–201.

Kennedy, P. R., Mirra, S. S., Bakay, R. A. E. (1992). The cone electrode: Ultrastructural studies following long-term recording in rat and monkey cortex. *Neuroscience Letters* 142: 89–94.

Kobayashi, N., Yoshida, M., Matsumoto, N., et al. (2009). Artificial control of swimming in goldfish by brain stimulation: Confirmation of the midbrain nuclei as the swimming center. *Neuroscience Letters* 452: 42–46.

Kositsky, M., Chiappalone, M., Alford, S. T., et al. (2009). Brain-machine interactions for assessing the dynamics of neural systems. *Frontiers in Neurorobotics* 3: 1.

Lebedev, M. A., Nicolelis, M. A. (2006). Brain-machine interfaces: Past, present and future. *Trends Neurosci.* 29: 536–546.

Lilly, J. C. (1956). Distribution of 'motor' functions in the cerebral cortex in the conscious, intact monkey. *Science* 124: 937.

Lin, C. J., Lee, C. Y. (2009). FPGA implementation of a recurrent neural fuzzy network with on-chip learning for prediction and identification applications. *Journal of Information Science and Engineering* 25: 575–589.

Linderman, M. D., Santhanam, G., Kemere, C. T., et al. (2008). Signal processing challenges for neural prostheses. *IEEE Signal Processing Magazine* 25: 18–28.

London, B. M., Jordan, L. R., Jackson, C. R., et al. (2008). Electrical stimulation of the proprioceptive cortex (area 3a) used to instruct a behaving monkey. *Neural systems and rehabilitation engineering: IEEE Transactions* 16: 32–36.

Matrone, G. C., Cipriani, C., Secco, E. L., et al. (2010). Principal components analysis based control of a multi-DOF underactuated prosthetic hand. *Journal of Neuroengineering and Rehabilitation* 7: 16.

Moritz, C. T., Perlmutter, S. I., Fetz, E. E. (2008). Direct control of paralysed muscles by cortical neurons. *Nature* 456: 639–642.

Mussa-Ivaldi, F. A., Miller, L. E. (2003). Brain-machine interfaces: Computational demands and clinical needs meet basic neuroscience. *Trends in Neurosciences* 26: 329–334.

Mussa-Ivaldi, F. A., Miller, L. E., Rymer, W. Z., et al. (2007). Neural Engineering. In Parasuraman, R. and Rizzo, M. (Eds.), *Neuroergonomics: The brain at work.* New York: Oxford University Press.

Neihart, N. M., Harrison, R. R. (2005). Micropower circuits for bidirectional wireless telemetry in neural recording applications. *IEEE Transactions on Biomedical Engineering* 52: 1950–1959.

Nicolelis, M. A. (2001). Actions from thoughts. *Nature* 409: 403–407.

Nicolelis, M. A. L. (2007). *Methods for neural ensemble recordings (2nd ed.).* Boca Raton, FL: CRC Press.

Nicolelis, M., Lebedev, M. (2009). Principles of neural ensemble physiology underlying the operation of brain-machine interfaces. *Nature Reviews Neuroscience* 10: 530–540.

Nieder, A. (2000). Miniature stereo radio transmitter for simultaneous recording of multiple single-neuron signals from behaving owls. *Journal of Neuroscience Methods* 101: 157–164.

O'Doherty, J. E., Lebedev, M., Hanson, T. L., et al. (2009). A brain-machine interface instructed by direct intracortical microstimulation. *Frontiers in Integrative Neuroscience* 3: 20.

Obeid, I., Nicolelis, M. A. L., Wolf, P. D. (2004). A multichannel telemetry system for single unit neural recordings. *Journal of Neuroscience Methods* 133: 33–38.

Olsson, R. H., III, Wise, K. D. (2005). A three-dimensional neural recording microsystem with implantable data compression circuitry. *IEEE Journal of Solid-State Circuits* 40: 2796–2804.

Omondi, A. R., Rajapakse, J. C. (Eds.) (2006). *FPGA implementations of neural networks*. Dordrecht, the Netherlands: Springer.

Patil, P. G., Turner, D. A. (2008). The development of brain-machine interface neuroprosthetic devices. *Neurotherapeutics* 5: 137–146.

Paxinos, G., Watson, C. (2005). *The rat brain in stereotaxic coordinates*. San Diego: Academic Press.

Pohlmeyer, E. A., Oby, E. R., Perreault, E. J., et al. (2009). Toward the restoration of hand use to a paralyzed monkey: Brain-controlled functional electrical stimulation of forearm muscles. *PLoS ONE* 4: e5924.

Rizk, M., Bossetti, C. A., Jochum, T. A., et al. (2009). A fully implantable 96-channel neural data acquisition system. *Journal of Neural Engineering* 6: 14.

Romo, R., Hernández, A., Zainos, A., et al. (2000). Sensing without touching: Psychophysical performance based on cortical microstimulation. *Neuron* 26: 273–278.

Salinas, E., Abbott, L. F. (1994). Vector reconstruction from firing rates. *Journal of Computational Neuroscience* 1: 89–107.

Sanchez, J., Erdogmus, D., Rao, Y., et al. (2003). Learning the contributions of motor, premotor, and posterior parietal cortices for hand trajectory reconstruction in a brain machine interface. *Proceedings of the 1st International IEEE/EMBS Conference on Neural Engineering*, Capri, Italy, March 2003 (pp. 59–62).

Sato, H., Berry, C. W., Peeri, Y., et al. (2009). Remote radio control of insect flight. *Front. Integr. Neurosci.* 3: 24.

Scherberger, H. (2009). Neural control of motor prostheses. *Current Opinion in Neurobiology* 19: 629–633.

Schregardus, D. S., Pieneman, A. W., Ter Maat, A., et al. (2006). A lightweight telemetry system for recording neuronal activity in freely behaving small animals. *Journal of Neuroscience Methods* 155: 62–71.

Serruya, M. D., Hatsopoulos, N. G., Paninski, L., et al. (2002). Instant neural control of a movement signal. *Nature* 416: 141–142.

Song, W. G., Ramakrishnan, A., Udoekwere, U. I., et al. (2009). Multiple types of movement-related information encoded in hindlimb/trunk cortex in rats and potentially available for brain-machine interface controls. *IEEE Transactions on Biomedical Engineering* 56: 2712–2716.

Specht, D. F. (1990). Probabilistic neural networks. *Neural Networks* 3: 109–118.

Talwar, S. K., Xu, S., Hawley, E. S., et al. (2002). Rat navigation guided by remote control. *Nature* 417: 37–38.

Taylor, D. M., Tillery, S. I., Schwartz, A. B. (2002). Direct cortical control of 3d neuroprosthetic devices. *Science* 296: 1829–1832.

Tehovnik, E. J. (1996). Electrical stimulation of neural tissue to evoke behavioral responses. *Journal of Neuroscience Methods* 65: 1–17.

Velliste, M., Perel, S., Spalding, M. C., et al. (2008). Cortical control of a prosthetic arm for self-feeding. *Nature* 453: 1098–1101.

Vyssotski, A. L., Serkov, A. N., Itskov, P. M., et al. (2006). Miniature neurologgers for flying pigeons: Multichannel EEG and action and field potentials in combination with GPS recording. *Journal of Neurophysiology* 95: 1263–1273.

Waldert, S., Pistohl, T., Braun, C., et al. (2009). A review on directional information in neural signals for brain-machine interfaces. *Journal of Physiology–Paris* 103: 244–254.

Wessberg, J., Stambaugh, C. R., Kralik, J. D., et al. (2000). Real-time prediction of hand trajectory by ensembles of cortical neurons in primates. *Nature* 408: 361–365.

Wolpaw, J. R., Birbaumer, N., McFarland, D. J., et al. (2002). Brain-computer interfaces for communication and control. *Clinical Neurophysiology* 113: 767–791.

Wu, W., Black, M. J., Gao, Y., et al. (2002). Inferring hand motion from multi-cell recordings in motor cortex using a Kalman filter. *SAB'02-Workshop on Motor Control in Humans and Robots: On the Interplay of Real Brains and Artificial Devices*, Edinburgh, Scotland, August 2002 (pp. 66–73).

Xu, S., Talwar, S. K., Hawley, E. S., et al. (2004). A multi-channel telemetry system for brain microstimulation in freely roaming animals. *Journal of Neuroscience Methods* 133: 57–63.

Ye, X. S., Wang, P., Liu, J., et al. (2008). A portable telemetry system for brain stimulation and neuronal activity recording in freely behaving small animals. *J. Neurosci. Methods* 174: 186–193.

Zhang, K., Ginzburg, I., McNaughton, B. L., et al. (1998). Interpreting neuronal population activity by reconstruction: Unified framework with application to hippocampal place cells. *Journal of Neurophysiology* 79: 1017–1044.

Zhou, F., Liu, J., Yu, Y., et al. (2009). Field-programmable gate array implementation of a probabilistic neural network for motor cortical decoding in rats. *Journal of Neuroscience Methods* 185: 299–306.

Section IV

Therapeutics

17 Pharmacogenomics in Development of Disease-Specific Therapeutic Strategy

Sanjeev Sharma
Apollo Health City
Hyderabad, India

Anjana Munshi
King Saud University
Riyadh, Saudi Arabia

CONTENTS

17.1 INTRODUCTION

The sequencing of the human genome has become the foundation for one of the most significant scientific contributions to mankind; the idea that although all human individuals are genetically similar, each retains a unique genetic identity. The publication of the human blueprint has triggered an explosion in pharmaceutical research to utilize this knowledge in the prescription of drugs to be tailored according to the genetic makeup of susceptible individuals or in other words *personalized medicine*. In fact, half century ago, well before the Human Genome Project, scientists had realized that inheritance was an important factor responsible for individual variation in drug response (Kalow, 1962; Venter et al., 2001). This led to the birth of the term pharmacogenetics. Pharmacogenetics is the study of the role of inheritance in interindividual variation in drug response. Although human beings are 99.9% similar in their genetic makeup, this 0.1% variability in terms of single-nucleotide polymorphisms is significantly accountable for an individual's susceptibility to diseases and inter- and intraindividual variation of drug response (Brooks, 1999). In recent years, the convergence of advances in pharmacogenetics with the genomic revolution has led to the evolution of pharmacogenetics into pharmacogenomics. Pharmacogenomics refers to the entire spectrum of genes responsible for pharmacokinetic and pharmacodynamic variation of drug response (Dubey et al., 2008). Therefore the traditional classic approach of clinical therapy, trial and error or one dose fits for all, is to be reconsidered in the case of drugs with narrow therapeutic range that result in drug toxicity and treatment failure (Spear et al., 2001).

The effectiveness of pharmacotherapy depends upon a sufficient amount of drug concentration reaching the target site or in systemic circulation (Dubey et al., 2008). For most of the drugs, a considerable amount of variation in therapeutic efficacy has been observed among patients. In addition to this, life-threatening adverse drug effects have also been observed. Variability in drug response can be due to many factors such as gender, dietary habits, concurrent disease, concomitant medications, environmental factors, and genetic factors (Spear et al., 2001).

A recent survey has shown that efficacy rates of drug therapy in most of the diseases vary from 25 to 80%, and among these, approximately 20% of patients are nonresponders, and a significant number of patients develop adverse drug reactions (Spear et al., 2001). The phenotype of a drug is highly complicated and is an outcome of gene-drug interactions. The mutations in the genetic code may influence the metabolism (via mutation in metabolizing enzymes such as CYP450 isoforms), transport proteins (uranyl glycoproteins), or target binding site of a drug (adenosine triphosphate [ATP] cassette-binding protein). Clinically variable drug responses are classified as inefficacy, efficacy, resistance, and toxicity (Xie et al., 2005). A patient receiving the drug treatment in the therapeutic range but without any benefit from the treatment results in inefficacy. Efficacy can be defined in terms of drug reaching in systemic circulation in sufficient amount and exerting its expected therapeutic benefit. Resistance is defined as the patient on drug treatment without any benefit of the drug. In fact, most of the approved drugs result in adverse effects in a subpopulation of patients, and more than 2 million patients have been found to be affected with severe adverse drug reactions annually, along with 100,000 deaths in the United States (Lazarou et al., 1998).

The main goal of pharmacogenomics is to identify the specific biomarkers to treat the cause of disease and to guide to healthcare professionals to prescribe the right drug at the right dose for the right diseases. It will also guide the pharmaceutical companies to develop novel therapeutic agents designed to bind to specific target sites (based on identification of individual single-nucleotide polymorphism analysis) (Dubey et al., 2008).

A number of success stories of pharmacogenomics-based therapeutic emergence have emerged with the help of genome-wide association studies in cases of complex diseases such as cancer, asthma, inflammatory diseases of gastrointestinal tracts, cardiovascular diseases, anticoagulant therapy, diabetes, hyperlipidemia, obesity, epilepsy, osteoporosis, and other disorders of metabolic origin (Grant and Haknorson, 2007). The integration of pharmacogenomic knowledge into clinical medicine would be very helpful in developing the therapeutic strategies for the best possible management of diseases. The chapter provides an overview of the clinical significance of pharmacogenomics with examples of therapeutic variation of drug response based on genetic polymorphism.

17.2 CARDIOVASCULAR DISEASES

A number of factors like age, race, concomitant diseases, medicines, and renal and hepatic functions are taken into account when selecting drug therapy for a patient with cardiovascular diseases.

In addition lot of data from expert panels and large clinical trials and consensus guidelines are available that help to steer drug therapy decisions. Despite all of these efforts, there is no guarantee that a given treatment will be effective and well tolerated in a given patient.

Recently, genetic variants have been found to be associated with therapeutic variation in the case of cardiovascular drug treatment. Pharmacogenetics and pharmacogenomics involves the search for these genetic variants that influence responses to drug therapy. A number of studies have dictated that pharmacogenomic tailored cardiovascular drug therapy is more effective as compared to a traditional classical approach (Johnson and Cavallari, 2005). The following drugs are generally prescribed for various cardiovascular diseases.

17.2.1 Anticoagulant: Warfarin

The anticoagulants such as coumarin/warfarin are widely prescribed in the prophylaxis and treatment of thromboembolic disorders and stroke (Martin, 2009). The major problem concerned with warfarin treatment is a wide range of variation in dosage from individual to individual. The effectiveness of warfarin critically depends on prothrombin time measured as international normalized ratio (INR). INR is supratherapeutic more than one-third of the time during the first month of treatment (Beyth et al., 2000). Treatment is not effective if INR is very low, and a higher range may lead to an increased risk of bleeding. The pharmacokinetics of warfarin is highly influenced by the cytochrome P450 2C9 gene and vitamin K2, 3 epoxide reductase gene (*VKORC1*) (Sconce et al., 2005; Rieder et al., 2005). Warfarin is metabolized by *CYP2C9* and inhibits the activation of *VKORC1*, which further inhibits the activation of clotting factors II, VII, IX, and X. Recent studies have shown that at least 30 genes influence the metabolism of warfarin. However, the point mutations in *CYP2C9* and *VKORC1* have been found to be strongly associated with variation in dosage of warfarin among patients (Wadelius and Pirmohamed, 2007).

A number of studies have demonstrated that association of genetic variants of *CYP2C9* with substantially lower warfarin dose, prolonged time in dose stabilization, and increased risk of bleeding (Aithal et al., 1999; Daly and King, 2003). In addition to genetic variants, other factors such as food interaction and environmental factors also contribute towards variation in drug response from individual to individual. In 2007, the Food and Drug Administration (FDA) approved the addition of pharmacogenomic information on the product label of warfarin. Therefore, pharmacogeneomic/pharmacogenetic information of warfarin along with consideration of nongenetic factors and critical observation of INR may improve the therapeutic benefits.

17.2.2 Anti-Inflammatory and Antiplatelet Agents

Major anti-inflammatory agents, which are widely prescribed consists of nonsteroidal anti-inflammatory agents (NSAIDs) like ibuprofen, diclofenac, lornoxicam, and piroxicam. The major enzymes involved in the metabolism of NSAIDs are CYP2C8 and CYP2C9. The higher blood concentration of flubiprofen, diclofenac (minor influence), ibuprofen, and piroxicam have been found in patients with *CYP2C9*3* variant alleles (Rollason et al., 2008). NSAID-induced side effects, such as gut bleeding, has been found to be more in patients with *CYP2C9*3* mutant allele (Pilotto et al., 2008). Acetylsalicylic acid (aspirin) is one of the NSAIDs used in a number of cardiovascular events including primary and secondary prevention of myocardial infarction and stroke (Goodman et al., 2007). Aspirin acts by irreversibly inhibiting the cyclooxygenase (COX) enzyme and peripheral arterial occlusion, which converts arachidonic acid to prostaglandins and further thromboxane synthase, which is a potent vasoconstrictor. Despite the benefits of aspirin treatment, many patients continue to suffer the cardiovascular diseases because of therapeutic failure (Kaur et al., 2006).

Aspirin resistance is defined as the failure to produce an expected therapeutic effect such as inhibition of platelet aggregation, suppression of thromboxane A2 production, or prolongation of the bleeding time (Wang et al., 2006). Cellular factors that influence the aspirin's therapeutic efficacy include inadequate suppression of platelet COX-1. Additionally, it has been found that aspirin resistance is attributed to COX-2 mRNA overexpression by platelets and endothelial cells (Weber et al., 1999; Zimmermann et al., 2003). The generation of 8-iso prostaglandins by peroxidation of arachidonate may also lead to aspirin resistance by binding to thromboxane receptors (Davi et al., 2002). Acetylation of COX-2 by aspirin generates resolvins (a family of bioactive omega-3 fatty acid metabolites), which mediate inflammatory response. Deficiency of these products could also influence therapeutic failure (Chen et al., 2004). Although no formal definition of aspirin resistance exists, it may involve clinical failure of therapeutic dose of aspirin (75–150 mg for at least 5 days) to protect individuals from arterial thrombotic events or laboratory methods, indicating the failure of aspirin to inhibit platelet activity.

Aspirin resistance has been reported to be associated with a 10-fold increase in the risk of recurrent vascular events patients who had a history of stroke (Kaur et al., 2006). Increased incidence of recurrent cerebral ischemic attacks in patients resistant on aspirin has been confirmed by various studies. A number of studies have demonstrated that genetic variants of genes like the *COX1* gene (*840G/C50T*), glycoprotein receptor gene, and collagen receptor gene significantly contribute towards development of aspirin resistance in patients on aspirin therapy (Undas et al., 1999; Szczeklik et al., 2000; Lepantalo et al., 2006; Maree et al., 2005).

17.2.3 Clopidogrel

Clopidogrel is a prodrug that is metabolized in the liver by cytochrome P450 3A4 (CYP3A4) into its active metabolite. It inhibits platelet aggregation by irreversibly binding to the platelet receptor (P2Y12) on the platelet surface (Wang et al., 2006). A number of studies have revealed that patients on clopidogrel treatment develop resistance, and about 4–30% do not respond well to conventional doses of clopidogrel. The clinical implications of clopidogrel are still unknown. The mechanism of development of clopidogrel resistance might be due to polymorphism in the *P2Y12* ADP receptor gene (Serebruany et al., 2005). It has been reported that carriers of the *P2Y12* gene (SNP 34C>T) do not respond well to clopidogrel treatment and had a 4-fold higher risk of developing adverse neurological events (Ziegler et al., 2005). Carriers of CYP2C19 (with *2 or *4) have been found to be responsible for reduced platelet inhibition, even in loading doses of clopidogrel (Gladding et al., 2008). Therefore genotyping of the relevant gene polymorphism may help to optimize the drug treatment.

17.2.4 Antihypertensives

Hypertension, a major public health problem, is associated with a high rate of morbidity and mortality (Padmanabhan et al., 2010). A combination of drugs and lifestyle interventions are the cornerstones of hypertension treatment. Despite many safe and effective antihypertensive drugs, blood pressure is controlled in <50% of patients. The high prevalence of disease, treatment resistance, and complications leading to increased cardiovascular risk offers hypertension as a potential candidate disease to study the pharmacogenomics of antihypertensive agents. The candidate genes studied in response to antihypertensive agents are genes belonging to the renin-angiotensin-aldosterone system (RAAS), the adrenergic system, and genes involved in sodium transport in the kidneys (Padmanabhan et al., 2008). The RAAS plays a major role in the development and progression of cardiovascular diseases by promoting vasoconstriction, sodium reabsorption, cardiac remodeling, norepinephrine release, and other potentially detrimental effects. Evidence has suggested that there is substantial variability in individual responses to antihypertensive agents such as angiotensin-converting–enzyme (ACE) inhibitors and angiotensin II type 1 receptor blockers (Materson et al., 1995). Less than 50% of hypertensive patients achieve adequate blood-pressure control with ACE inhibitor monotherapy. Single-nucleotide polymorphisms (SNPs) in angiotensinogen (T1198C), apolipoprotein B (APOB G10108A), and adrenoreceptor α2A (A1817G) have been found to be associated significantly with alteration in the left ventricular mass during treatment with antihypertensive agents (Liljedahl et al., 2004).

The interindividual variation in response to hydrochlorothiazide has been found to be associated with polymorphism of genes in renal sodium-transport systems like With No Lysine protein kinases gene (*WNK1*), β2 adrenergic receptor gene (*ADRB2*), and sodium-channel subunit gene (*SCNN1G*) (Turner et al., 2005). The promise of antihypertensive pharmacogenomics is that it may provide a more effective way of identifying the responders and nonresponders, thus explaining the heterogeneity of drugs prescribed for hypertensives (Padamnabhan et al., 2010). A number of studies have shown a genetic association between antihypertensive drugs like diuretics and single-nucleotide polymorphism in the G protein β3 gene and α-adducin gene (Cusi et al., 1997; Turner et al., 2005).

17.2.5 Beta Blocker Drugs

Approximately 60% of hypertensive patients do not achieve adequate blood-pressure lowering from monotherapy with beta blockers (Materson et al., 1995). It may be due to genetic variation in β-adrenergic receptor genes. Polymorphism in drug-metabolizing enzymes result in absent or nonfunctional protein; for example, metabolism of the beta blocker metoprolol is controlled by the cytochrome P450 2D6 (CYP2D6) enzyme, and patients with a mutation in the gene encoding for CYP2D6 have no functional protein present. These patients are poor metabolizers (nonresponders), resulting in more than five times metoprolol plasma concentration compared with the responders (Lennard et al., 1983; McGourty et al., 1985).

17.2.6 Statins

Elevated cholesterol is one of the most common disorders in individuals above 40 years of age. Statins are widely prescribed first-line drugs in the treatment of primary and secondary coronary artery disease, atherosclerosis, and stroke. Statins are the most effective cholesterol-lowering agents and act by inhibiting hydroxymethylglutaryl coenzyme A reductase enzyme activity (Sacks et al., 1996; Downs et al., 1998; Shepherd et al., 1995) Therapeutic variability has been observed in patients on statin treatment. Genetic variation in the apolipoprotein E (APOE) gene has been found to be associated with altered plasma concentrations of the lipoprotein (Zannis et al., 1981). *APOE2* carriers have been reported to be more responsive to lipid-lowering therapies (Ordovas et al., 1995). A differential response to statin medications has been reported in relation to a patient's genotype status of cholesteryl ester transfer gene, β-fibrinogen gene, and lipoprotein lipase gene (Kuivenhoven et al., 1998).

17.3 CANCER

The current armamentarium of cancer therapies has been found to be hindered by drug resistance and drug-induced toxicities. The tumor genomic factors, as well as heritable genetic variants, have been reported to affect interindividual responses to anticancer therapy (Sing-Huang et al., 2008). The urgent need is to adopt the pharmacogenomic/pharmacogenetic-based therapy, which would be able to predict the individual's response to cancer therapy and to determine which tumors are likely to recur following the therapy (Grant and Haknorson, 2007). Clinical application of pretreatment pharmacogenetic testing to determine drug response and toxicity is still very limited in oncology, with the only availability of CYP2D6 testing.

Most of the anticancer drugs are based on the mechanism of target-specific molecular abnormalities of tumor cells (Sing-Huang et al., 2008). Tumor responses to the inhibitors of oncogenic tyrosine kinases have been found to be associated with the presence of activating mutations within the genes encoding for target kinases. For example, activation by somatic mutations or overexpressed increased tyrosine kinase activity of epidermal growth factor receptor (EGFR)/ErB1 plays an important role in the proliferation and metastasis of tumor cells. Currently available EGFR-targeted antitumor drugs are designed to inhibit the selective critical molecules and specific signaling pathways involved in tumor growth (Haung-Gaung et al., 2005).

17.3.1 Drugs Used in Breast Cancer

The human epidermal growth factor receptor gene (*HER2*) (*ErB2*) serves as a target for anti-HER2 antibody trastuzumab (herceptin), a humanized monoclonal antibody. The antiangiogenesis agents bevacizumab and trastuzumab are now approved treatments for breast cancer. The HER-2 drug is ineffective in almost two-thirds of patients who do not overexpress the drug's target (Slamon et al., 2001). *HER2* testing has become an integral part of the optimal management of breast cancer patients from past several years. A prospective study among Asian breast cancer patients treated with doxorubicin has revealed that SNPs in metabolizing enzymes such as carbonyl reductases exert major effects on drug metabolism. Two common variants in the *CRB* gene have been found to affect the pharmacokinetic and pharmacodynamic profile of doxorubicin. The CRB 311G>A polymorphism was found to be associated with reduced metabolism of the doxorubicin along with hematologic toxicities. On the other hand, the variant CRB 3730G>A was reported to increase conversion of doxorubicin to doxorubicinol without any haematologic toxicities (Sing-Huang et al., 2008). Similarly, in studies including genetic variants of ATP cassette-binding protein (*ABCB1*) (the gene encoding for P-glycoprotein), the variants have been found to be associated with drug resistance. Kafka et al. (2003) have reported significant association of ABCB1 variant 3435T with a better drug response of anthracyclins with or without taxanes in breast cancer patients.

17.3.2 Drugs Used in Chronic Myeloid Leukemia

Imatinib used for the treatment of chronic myeloid leukemia (CML) is a competitive inhibitor of ATP cassette-binding protein, which binds to a nonreceptor tyrosine kinase (ABL kinase), thus inhibiting the constitutively activated breakpoint cluster region-abelson fusion gene (*BCR-ABL*) and produces rapid and durable clinical response in CML patients with minimum toxic effects. The United States FDA has already approved that imatinib can be used to treat *BCR-ABL*-positive CML and gastrointestinal stromal tumors associated with activating mutations (Haung-Gaung et al., 2005).

17.3.3 Drugs Used in Pancreatic Cancer

Pancreatic cancer is a genetic disease associated with various other forms of cancers. It is one of the solid tumors with chemoresistant biology (Kang and Saif, 2008). Chemotherapy in the treatment of pancreatic cancer has produced unsatisfactory therapeutic outcome. Gemcitabine is the standard drug used for the treatment of pancreatic cancer. Treatment with gemcitabine is effective in 20–30% of patients. Equilibrative concentrative nucleoside transporters (hCNT) are the transport proteins of gemcitabine. More than 58 single-nucleotide polymorphisms have been found in equilibrative hCNT having a functional impact on gemcitabine transport. The high expression of cytidine kinase enzyme (polymorphism 208G>A) has been found to be associated with altered pharmacokinetic of gemcitabine and its resistance (Kang and Saif, 2008). Similarly, cetuximab, a recombinant humanized monoclonal antibody used in the treatment of cancer, showed resistance in patients with mutation in EGFR. Pharmacogenomic-based therapy in pancreatic cancer offers a great potential toward effectiveness of chemotherapeutic agents such as 5-fluorouracil and cisplatin.

17.4 RHEUMATOID ARTHRITIS

Genetic mapping studies have revealed that most of the rheumatic diseases and idiopathic systemic lupus arthritis are polygenic in nature. The polymorphisms in drug-metabolizing genes are one of major problems associated with ineffectiveness of rheumatic disease therapy (Gaffney et al., 2002). For instance, azathiopurine, prescribed for systemic lupus arthritis and inflammatory myopathies and less commonly in rheumatoid arthritis (RA), is converted into 6- mercatopurine and then activated to thioguanine nucleotides, which are incorporated in DNA, thereby blocking the DNA replication. 6-Mercatopurine is further inactivated by thiopurine methyltransferase (TPMT).

Pharmacogenetic studies on azathiopurine have shown that genetic polymorphism in the *TPMT* gene (*TPMT*2*, *TPMT*3A*, and *TPMT*3C*) contributes toward impaired metabolism of azathiopurine and 6-mercatopurine and further leads to toxic hematological effects (Krynetski and Evans, 2000; McLeod et al., 2000). The alteration in TPMT activity has been observed in most patients. Eight alleles have been identified in the *TPMT* gene encoding for the *TPMT* enzyme. It has also been observed that patients with deficient TPMT activity develop fatal bone marrow toxicity in response to higher doses of azathiopurine (Black et al., 1998; Evans et al., 2001). Because there is a significant relation between *TPMT* genotype and TPMT activity measured in peripheral erythrocyte, individualization of drug dosing could be improved by determining genotype (Xie et al., 2005).

Methotrexate (MTX) is another drug used in the treatment of RA. The variation in effectiveness and toxicity induced by MTX has offered it as a candidate to study the pharmacogenomic/genetic aspects. The methotraxate has been reported to be discontinued because of its adverse effects in individuals with mutation in the methylene tetrahydrofolate reductase (*MTHFR C677T*) gene. The nonresponsiveness of MTX has also been observed in individuals with genetic variants of transport protein (ATP cassette-binding protein) (Wessels et al., 2006).

17.5 EPILEPSY

Epilepsy is a brain disorder with high unpredictability of therapeutic response in patients. Antiepileptic drugs such as phenytoin and carbamazepine are commonly prescribed as first-line drugs in the treatment of epilepsy. Accumulating evidence suggests that patients on phenytoin and carbamazepine therapy show a wide range of inter- and intraindividual dosage variation caused by nonlinear pharmacokinetics of drugs (Xie et al., 2007). Adverse drug reactions and antiepileptic drug toxicity are the most common problems associated with epilepsy treatment. The influence of genetic variants on dose variation has been evident in a number of association studies (Depondt, 2008).

The efficacy of antiepileptic drugs is influenced by a number of factors such as concomitant medications, environmental factors, lifestyle factors, and genetic variants. It has been found that single-nucleotide polymorphism in drug-metabolizing enzymes cytochrome P450 such as CYP2C9, CYP2C19, and CYP3A4 is mainly involved in inter- and intraindividual variation among epileptic patients (Daly, 2003). It has been reported that patients with the homozygous mutant for the variant allele of *CYP2C9*3* and *CYP2C9*6* experience severe phenytoin toxicity with a usual therapeutic dose (Brandolese et al., 2001; Kidd et al., 2001). Single-nucleotide polymorphism in gene encoding the α subunit of voltage-gated neuronal sodium channel (SCN1 rs 3812718) has also been reported to be significantly associated with a variable dosage of phenytoin in patients on phenytoin treatment.

Recently, the human leukocyte antigen (HLA B*1502) has been found to be strongly associated with carbamazepine-induced hypersensitivity reactions (severe cutaneous reactions) in the Han Chinese population (Uetrecht, 2003). The FDA has already approved human leukocyte antigen (HLAB*1502) as a biomarker for hypersensitive patients in case of carbamazepine therapy. The identification of gene variants associated with antiepileptic drug response may guide for the development of more efficacious antiepileptic drug treatment.

17.6 OSTEOPOROSIS

Osteoporosis is characterized by decreased bone mineral density (BMD) and increased susceptibility to fractures. Both men and women can develop osteoporosis, but postmenopausal women are more susceptible to develop chronic form of osteoporosis (Melton et al., 1992). Calcium and vitamin D supplements along with antiresorptive agents such as biphosphonates (alendronate, etidronate, ibandornate, and risedronate) are mainly prescribed for osteoporosis management. Previous studies have reported a strong association between variation in genetic factors and osteoporosis (Heaney et al., 2000; Mora et al., 2003; Krall et al., 1993). It has been found that female children of osteoporotic women have reduced BMD (Seeman et al., 1989).

Several studies have demonstrated that vitamin D receptor (VDR) polymorphism significantly contributes toward a small variation in BMD (Morrison et al., 2005). Vitamin D receptor polymorphism affects BMD (increased rate) in postmenopausal women (with b allele) on alendronate and hormonal replacement therapy. In other studies, polymorphism in the estrogen receptor gene (*ESR1*), *PvuII*, and *XbaI* has been found to influence bone mass and response to estrogen therapy in pre- and postmenopausal women (Greene et al., 2009). Studies have shown that BMD change was significantly associated with the interaction between VDR polymorphisms and various antiresorptive drug therapies (Greene et al., 2009).

17.7 OBESITY

Obesity, a common multifactorial disorder, affects many people worldwide. It predisposes to type 2 diabetes, coronary heart disease, and hypertension and is a major cause of morbidity and mortality. Leptin protein has been found to play an important role in the organization of energy control and obesity. It has been reported that 40–70% of the variation in obesity-related phenotypes, such as body-mass index, skin thickness, fat mass, and leptin levels, is heritable (Allison et al., 1996). Leptin therapy has been found to be effective only in patients with a mutation in the leptin gene.

Studies have shown that a viscerally obese male heterozygous for apolipoprotein B-EcoRI polymorphism is prone to develop the dense low-density lipoprotein phenotype, and an increased risk of coronary disease also occurs with some lipoprotein lipase gene variants. Genetic variations in the glucocorticoid receptors (bcl/l restriction fragment polymorphism) and fatty acid-binding protein 2 (*Ala54Thr FABP2*) gene were associated with visceral adiposity in lean men (Buemann et al., 1997).

The response of the centrally acting noradrenaline and serotonin reuptake inhibitor sibutramine has been examined in a study to reduce weight and was attributed to C825T polymorphism in the guanine-nucleotide-binding protein β-3 (*GNB3*) gene. Therefore, genotyping of the GNB3C825T polymorphism may help in predicting the obese individuals who may benefit from sibutramine therapy (Grant and Haknorson, 2007).

The genes that are mainly implicated in obesity include agouti-gene–related transcript, neuropeptide Y (NPY) and its receptors (NPY5R and Y6R), proopiomelanocortin, uncoupling protein 2 (UCP2), and the melanocortin-4 receptor. NPY is one of the most potent appetite stimulators in animals, and it also appears to be one of the mediators of the ob gene in the brain. A number of studies suggest that body fat is controlled by a lipostat mechanism in which leptin is the afferent signal, and the hypothalamus serves as an integrator and activates an output loop that further modulates feeding behavior, energy expenditure, and fat and glucose metabolism (Friedman, 1997). The next generation medicine may target the leptin pathway. A drug or multiple therapeutic agents that suppress appetite, increasing metabolic rate and reducing the amount of body fat, should be designed specifically based on genetic information. The new approach of developing centrally acting agents such as NPY Y1,Y5 antagonists remains an attractive possibility in obesity treatment.

17.8 DIABETES MELLITUS TYPE 2

A number of drugs such as biguanides, sulfonylureas, thiazolidinediones (TZD), meglitinides, α-glucosidase inhibitors, amylin mimetics, glucagon-like peptide-1 mimetics, dipeptidyl peptidase-4 inhibitors, and insulin are mainly used as therapeutic agents in the treatment of diabetes mellitus (Wolford et al., 2004; Reitman and Schadt, 2007). Recent pharmacogenomic studies have revealed a marked variation in antidiabetic therapeutic efficacy caused by polymorphism at receptor sites, effector proteins, and transporter proteins and in metabolizing enzymes (Avery et al., 2009).

Metformin is a first-line drug that is widely prescribed in the treatment of type 2 diabetes. The pharmacological effect of metformin is exerted via activation of adenosine monophosphate protein kinase (AMPK). AMPK is activated by serine threonine kinase (LKB1), which suppresses hepatic gluconeogenesis. Despite being a first-line drug with a good safety profile, metformin still fails to reach glycemic goals (Avery et al., 2009).

A number of studies have suggested that interpatient variability of metaformin could be the result of polymorphism in organic cation transporter genes (*SLC221* and *SLC22A2*) and multidrug or toxin extrusion gene (*MATE*). The single-nucleotide polymorphism at organic cation transporter gene (*SLC47A1*) has been found to reduce the glucose lowering effect of metformin. An alternative to metformin is sulfonylureas, which acts on potassium channels (K_{ATP}). Polymorphism in genes encoding for potassium channel further reduces insulin secretion. Also it has been reported that TZD target proteins, effector and receptor protein, significantly contribute toward variability in therapeutic response of TZD (Aquilante, 2007).

Maturity onset diabetes of the young (MODY) is the autosomal dominantly inherited form of diabetes without insulin dependency, characterized by b cell dysfunction and is diagnosed at a relatively young age (<25 years) (Tattersall, 1974; Tattersall et al., 1975). Of the seven MODY genes identified so far, the most common forms present are as a consequence of mutations in the genes encoding the glycolytic enzyme, glucokinase, and the transcription factor-1-α (HNF1 α) (Froguel et al.,1993; Frayling et al., 1997). MODY patients with mutation in HNF1 α are highly sensitive to hypoglycemic effects of sulfonylureas representing evidence of pharmacogenetic relation (Heiervang et al., 1989; Pearson et al., 2000). Recently, a new hypoglycemic therapy acting as a glucokinase activator has been developed for type 2 diabetes mellitus (Grant and Haknorson, 2007).

17.9 INFLAMMATORY BOWEL DISEASE

Inflammatory-bowel disease (IBD) is a common inflammatory disorder affecting the gastrointestinal tract that is resistant to most of the available therapies (Grant and Haknorson, 2007). The common forms of IBD, Crohn's disease (CD) and ulcerative colitis (UC), have been reported to be due to genetic variants in the caspase activator recruitment domain containing protein 15 gene (*CARD 15*) on chromosome 16q12 (Oostenbrug et al., 2006).

Mesalamine is used in the treatment of both UC and CD and is metabolized by the enzyme arylamine *N*-acetyltransferase (NAT). It has been suggested that polymorphism in the NAT gene is responsible for variation in mesalamine response (Egan et al., 2006). Thiopurine methyl transferase and azathiopurine are mainly used as maintenance treatment to prevent clinical relapse in patients with IBD. The polymorphism in the *TPMT* gene has been found to be significantly associated with altered enzyme activity. Therefore, genetic analysis in patients is very important before prescribing these drugs.

The therapy of IBD has been improved, with the introduction of tumor necrosis factor-α (TNF-α) inhibitor drugs. The major problems associated with anti-TNF therapy are high treatment cost and interindividual variation in drug-response rate (Grant and Haknorson, 2007). Predicting the responders and nonresponders in response to IBD treatment would be beneficial for designing effective therapeutic measures. A number of routes of TNF inhibition have been investigated, and among the most extensively evaluated is the use of monoclonal antibodies against TNF-α (infliximab). A number of control trials have indicated that infliximab has a role in treating patients with moderate to severely active CD.

17.10 ASTHMA

Current therapeutic agents for asthma have encountered problems such as interindividual variability and drug adverse effects. A number of patients do not respond well to leukotriene inhibitors. β-Agonists are the most commonly prescribed medications in the treatment of asthma. Short-acting β-agonists and long acting β-agonists act by binding to the adrenergic receptor (ADRB2). Receptor binding results in activation of adenylyl cyclase through stimulatory G proteins that activate protein kinase A and further phosphorylate several proteins, leading to a reduced level of intracellular calcium and causing smooth muscle relaxation in the respiratory tract. A polymorphism at the 16th position in ADRB2 identifies the patients with adverse response to regular use of short-acting β-agonists (Kazani et al., 2010). Most of the clinical studies in relation to the use of both short- and long-term acting β-agonists have focused on the Gly16Arg polymorphism.

SNPs in arachidonate-5 lipoxygenase enzyme (ALOX5AP) have been found to be associated with variation in responses to leukotriene modifiers such as zileuton. Klotsman et al. (2007) identified two SNPs in ALOX5 to be associated with an 18–25% improvement in forced expiratory volume (FEV1) when compared with an 8–10% improvement in those bearing the wild-type alleles treated with montelukast (antiasthmatic drug). Lima et al. (2006) also found an association between a different SNP in ALOX5 (rs2115189) and differential FEV1 responses to montelukast. A polymorphism in the ATP cassette-binding protein also has been associated with differential changes in FEV1 in the LTRA pathway analysis (Kazani et al., 2010). A candidate gene study in three populations has suggested an association of SNPs in the corticotrophin-releasing hormone receptor 1 (*CRHR1*) gene affecting the response to inhaled corticosteroids.

17.11 CONCLUSION/CLINICAL PERSPECTIVES

The inherited variations in the genetic makeup of an individual may lead to considerable variation in drug phenotype. By determining the variation in genes of an individual, it becomes easier for clinicians to select an appropriate drug at an appropriate dose. Use of a simple and inexpensive genetic testing system, based on high throughput DNA microarrays and microfluidic devices, would allow patients to be screened for relevant polymorphisms before initiating drug treatment.

The main goal of disease-specific pharmacogenomics is to select the best drug for the best candidate (patient's requirement), according to genotypic analysis. This will further result in a safe and effective therapeutic outcome, reduced hospital visits, and reduced drug-associated adverse effects. Pharmacogenomic/pharmacogenetic data can be utilized in clinical trials for screening of certain drugs in patients, and this will further reduce the cost of drug development. Better, more stable vaccines can be developed to provide more effective treatment.

Regulatory agencies such as the FDA in 2003 have recognized the potential benefits of pharmacogenomics and issued a draft of guidelines for pharmacogenomic data submission for pharmaceutical industries. The emerging genomic technologies would enable the search for relevant candidate genes and their mutants to be determined for personalized medicine. This could identify the genetic variants responsible for disease susceptibility. The major problem of translating pharmacogenomic/pharmacogenetic analysis into current clinical treatment is the feasibility of the system in terms of cost effectiveness and individual acceptance of the same.

Although a number of molecular mechanisms affecting the drug response have been discovered, the journey toward personalized medicine remains uncertain. The effective translation of pharmacogenomics into clinical practice depends on a number of social, ethical, and economical considerations. The patients have to be educated to understand the importance of genetic testing in response to drug treatment. Physicians/clinicians should be well trained in genomic medicine so that they can easily understand data obtained from genetic analysis in order to prepare an expert system that can be used in prescribing or tailor-making medicines. Additionally, social and ethical issues should be addressed so that patients and clinicians can embrace the pharmacogenetics/pharmacogenomics enthusiastically.

The first and foremost need is to protect patient confidentiality and enhance public confidence to avoid any discrimination or stratification of individuals. The patients should be informed about the genetic tests and give their consent. Political acceptance is also important to avoid any discrimination against individuals receiving the benefits of genomic medicine (Figure 17.1).

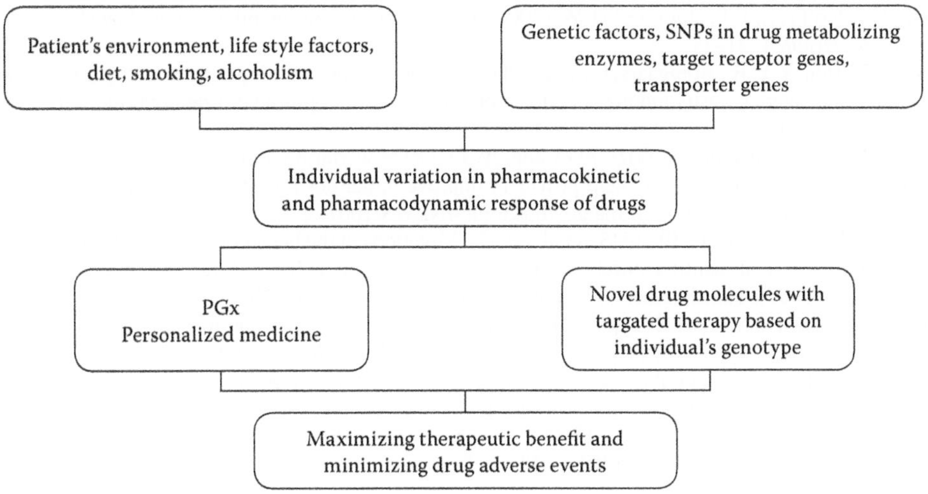

FIGURE 17.1 Pharmacogenetics: a tool for safe and effective pharmacotherapy.

REFERENCES

Aithal, G. P., Day, C. P., Kesteven, P. J., et al. (1999). Association of polymorphisms in the cytochrome P450 CYP2C9 with warfarin dose requirement and risk of bleeding complications. *Lancet* 353: 717–719.

Allison, D. B., Kaprio, J., Korkelia, M., et al. (1996). The heritability of body mass index among an international sample of monozygotic twins reared apart. *Int. J. Obesity Relat. Metab. Disord.* 20: 501–506.

Aquilante, C. L. (2007). Pharmacogenetics of thiazolidinediones therapy. *Pharmacogenomics.* 8: 917–931.

Avery, P. Mousa, S. S. Mousa, S. A. (2009). Pharmacogenomics in type II diabetes mellitus management: Steps toward personalized medicine. *Pharmacogenomics Personalized Med.* 2: 79–81.

Beyth, R. J., Quinn, L., Landefeld, C. S. (2000). A multicomponent intervention to prevent major bleeding complications in older patients receiving warferin: A randomized, controlled trial. *Ann. Intern. Med.* 133: 687–695.

Black, A. J., McLeod, H. L., Capell, H. A., et al. (1998). Thiopurine methyltranferase deficiency and heterozygosity among S-methyltransferase genotype predicts therapy-limiting severe toxicity from azathiopurine. *Ann. Intern. Med.* 129: 716–718.

Brandolese, R., Scordo, M. G., Spina, E., et al. (2001). Severe phenytoin intoxication in a subject homozygous for CYP2C9*3. *Clin. Pharmacol. Ther.* 70: 391–394.

Brooks, A. J. (1999). The essence of SNPs. *Gene* 234: 177–186.

Buemann, B., Vohl, M. C., Chagnon, M., et al. (1997). Abdominal visceral fat is associated with a bcli restriction fragment length polymorphism at the glucocorticoid receptor gene locus. *Obesity Res.* 5: 186–192.

Chen, W. H., Lee, P. Y., Ng, W., et al. (2004). Aspirin resistance is associated with a high incidence of myonecrosis, after nonurgent precutaneous coronary intervention despite clopidogril pretreatment *J. Am .Coll. Cardiol.* 41: 37–42.

Cusi, D., Barlassina, C., Azzani, T., et al. (1997). Polymorphisms of alpha adducin and salt-sensitivity in patients with essential hypertension. *Lancet* 349: 1353–1357.

Daly, A. K. (2003). Pharmacogenetics of the major polymorphic metabolizing enzymes. *Fund. Clin. Pharmacol.* 17: 27–41.

Daly, A. K., King, B. P. (2003). Pharmacogenetics of oral anticoagulants. *Pharmacogenetics* 13: 247–252.

Davi, G., Guagnano, M. T., Ciabattoni, G., et al. (2002). Platelet activation in obese women: Role of inflammation and oxidant stress. *JAMA* 288: 2008–2014.

Depondt, C. (2008). Pharmacogenetics in epilepsy treatment: Sense or nonsense? *Personalized Med.* 5: 123–131.

Downs, J. R., Clearfield, M., Weis, S., et al. (1998). Primary prevention of acute coronary events with lovastatin in men and women with average cholesterol levels: Results of AFCAPS/TexCAPS. (Air Force/Texas Coronary Atherosclerosis Prevention Study). *JAMA* 279: 1615–1622.

Dubey, A. K., Subish, P. P., Shankar, R. P., et al. (2008). Understanding the essentials of pharmacogenomics: The potential implications for the future pharmacotherapy. *J. Clin. Diag. Res.* 2: 681–689.

Egan, L. G., Derijks, L. J. J., Hommes, D. W. (2006). Pharmacogenomics in inflammatory bowel disease. *Clin. Gastro. Hepat.* 4: 21–28.

Evans, W. E., Hon, Y. Y. B., Bomgaars, L., et al. (2001). Preponderance of thiopurine S-methyltransferase deficiency and heterozygosity among patients intolerant to mercaptopurine or azathiopurine. *J. Clin. Oncol.* 19: 2293–2301.

Food and Drug Administration. (2007). FDA approves updated warfarin (coumarin) prescribing information: New genetic information may help providers improve initial dosing estimates of the anticoagulant for individual patients. http://www.fda.gov/bbs/topics/NEWS/2007.

Frayling, T. M., Bulamn, M. P., Ellard, S., et al. (1997). Mutations in the hepatocyte nuclear factor-1α gene are a common cause of maturity-onset diabetes of the young in the U.K. *Diabetes* 46: 720–725.

Friedman, J. M. (1997). The alphabet of weight control. *Nature* 385: 119–120.

Froguel, P., Zouali, H., Vionnet, N., et al. (1993). Familial hyperglycemia due to mutations in glucokinase: Definition of a subtype of diabetes mellitus. *N. Engl. J. Med.* 328: 697–702.

Gaffney, P. M., Moser, K. L., Graham, R. R., et al. (2002). Recent advances in the genetics of systemic lupus erythematosus. *Rheum. Dis. Clin. North Am.* 28: 111–126.

Gladding, P., Webster, M., Zeng, I., et al. (2008). The pharmacogenetics and pharmacodynamics of clopidogril response: An analysis from the PRINC (Plavix Response in Coronary Intervention) Trial. *J. Cardiovas. Interven.* 1: 620–627.

Goodman, T., Sharma, P., Ferro, A. (2007). The genetics of Aspirin. *Int. J. Clin. Practice* 61: 826–834.

Grant, S. F., Haknorson, A. (2007). Recent development in pharmacogenomics: From candidate genes to genome wide association studies. *Expert. Rev. Mol. Diagn.* 4: 371–393.

Greene, R. Mousa, S. S., Ardawiz, M., et al. (2009). Pharmacogenomics in osteoporosis: Steps toward personalized medicine. *Pharmacogenomics Personalized Med.* 2: 69–78

Haung-Gaung, X., Freuh, F. W. (2005). Pharmacogenomics steps toward personalized medicine. *Personalized Med.* 2: 325–337.

Heaney, R. P., Abrams, S., Dawson- Hughes, B., et al. (2000). Peak bone mass. *Osteoporos. Int.* 11: 985–1009.

Heiervang, E., Folling, I., Sovik, O., et al. (1989). Maturity-onset diabetes of the young: Studies in a Norwegian family. *Acta Paediatr. Scand.* 78: 74–80.

Johnson, J. A., Cavallari, L. H. (2005). Cardiovascular pharmacogenomics. *Exp Physiol.* 90: 283–289.

Kafka, A., Sauer, G., Jaeger, C., et al. (2003). Polymorphism C3435T of the MDR1 gene predicts response to preoperative chemotherapy in locally advanced breast cancer. *Int. J. Oncol.* 22: 1117–1121.

Kalow, W. (1962). *Pharmacogenetics: Heredity and the response to drugs.* Philadelphia: W.B. Saunders.

Kang, S. P., Saif, M. W. (2008). Pharmacogenomics and pancreatic cancer treatment: Optimizing current therapy and individualizing future therapy. *J. Pancreas* 9: 251–266.

Kaur, D., Tandon, V. R., Kapoor, B., et al. (2006). Aspirin resistance. *New Horizons* 8: 116–117.

Kazani, S., Wechsler, M. E., Israel, E. (2010). The role of pharmacogenomics in improving the management of asthma. *Mechan. Allergic Dis.*125: 295–302.

Kidd, R. S., Curry, T. B., Gallagher, S., et al. (2001). Identification of a null allele of CYP2C9 in a African-American exhibiting toxicity to phenytoin. *Pharmagenetics* 11: 803–808.

Klotsman, M., York, T. P., Pillai, S. G., et al. (2007). Pharmacogenetics of the 5-lipoxygenase biosynthetic pathway and variable clinical response to montelukast. *Pharmacogenet. Genomics* 207: 189–196.

Krall, E. A., Dawson-Hughes, B. (1993). Heritable and life style determinants of bone mineral density. *J. Bone Miner. Res.* 8: 1–9.

Krynetski, E. Y., Evans, W. E. (2000). Genetic polymorphisms of thiopurine S-methyltransferase: Molecular mechanisms and clinical importance. *Pharmacology* 61: 136–146.

Kuivenhoven, J. A., Jukema, J. W., Zwinderman, A. H., et al. (1998). The role of a common variant of the cholesteryl ester transfer protein gene in the progression of coronary atherosclerosis: The Regression Growth Evaluation Statin Study Group. *N. Engl. J. Med.* 338: 86–93.

Lazarou, J., Pomeranz, B. H., Corey, P. N. (1998). Incidence of adverse drug reactions in hospitalized patients: A meta analysis of prospective studies. *JAMA* 279: 1200–1205.

Lennard, M. S., Tucker, G. T., Silas, J. H., et al. (1983). Differential stereoselective metabolism of metaprolol in extensive and poor debrisoquine metabolizers. *Clin. Pharmacol. Ther.* 34: 732–737.

Lepantalo, A., Mikkelsson, J., Resendiz, J. C., et al. (2006). Polymorphisms of COX-1 and GPVI associate with the antiplatelet effect of aspirin in coronary artery disease patients. *Thromb. Haemost.* 95: 253–259.

Liljedahl, U., Kahan, T., Malmqvist, K., et al. (2004). Single nucleotide polymorphisms predict the changes in left ventricular mass in response to antihypertensive treatment. *J. Hypertens.* 22: 2321–2328.

Lima, J. J., Zhang, S., Grant, A., et al. (2006). Influence of leukotriene pathway polymorphisms on response to montelukast in asthma. *Am. J. Respir. Crit. Care Med.* 173: 379–385.

Maree, A. O., Curtin, R. J., Chubb, A., et al. (2005). Cyclooxygenase-1 haplotype modulates platelet response to aspirin. *J. Thromb. Haemost.* 3: 2340–2345.

Martin, J. H. (2009). Pharmacogenetics of warfarin: Is testing clinically indicated. *Australian Prescriber* 32: 76–80.

Materson, B. J., Reda, D. J., Cushman, W. C. (1995). Department of Veterans Affairs single-drug therapy of hypertension study: Revised figures and new data. Department of Veterans Affairs cooperative study group on antihypertensive agents. *Am. J. Hypertens.* 8: 189–192.

McGourty, J. C., Silas, J. H., Lennard, M. S., et al. (1985). Metaprolol metabolism and debrisoquine oxidation polymorphism-population and family studies. *Br. J. Clin. Pharmacol.* 20: 555–566.

Mcleod, H. L., Krynetski, E. Y., Relling, M. V., et al. (2000). Genetic polymorphisms of thiopurine methyltransferase and its clinical relevance for childhood lymphoblastic leukemia. *Leukemia* 14: 567–572.

Melton, L. J., Chrischillies, E. A., Cooper, C., et al. (1992). Perspective: How many women have osteoporosis? *J. Bone Miner. Res.* 7: 1005–1010.

Mora, S., Gilsanz, V. (2003). Establishment of peak bone mass. *Endocrinol. Metab. Clin. North. Am.* 32: 39–63.

Morrison, N. A., George, P. M., Vaughan, T., et al. (2005). Vitamin D receptor genotypes influence the success of calcitriol therapy for recurrent vertebral fracture in osteoporosis. *Pharmacogenet. Genomics* 15: 127–135.

Oostenbrug, L. E., Nolte, I. M., Oosterom, E., et al. (2006). CARD15 in inflammatory bowel disease and Crohn's disease phenotypes: An association study and pooled analysis. *Dig. Liv. Dis.* 38: 834–845.

Ordovas, J. M., Lopez-Miranda, J., Perez-Jimenez, F., et al. (1995). Effect of apolipoprotein E and A-IV phenotypes on the low density lipoprotein response to HMG CoA reductase inhibitor therapy. *Atherosclerosis* 113: 157–166.

Padmanabhan, S., Melander, O., Hastie, C., et al. (2008). Hypertension and genome-wide association studies: Combining high fidelity phenotyping and hypercontrols. *J. Hypertens.* 26: 1275–1281.

Padamnabhan, S., Paul, L., Dominczak, A. F. (2010). The pharmacogenomics of antihypertensive therapy. *Pharmaceuticals* 3: 1779–1791.

Pearson, E. R., Liddell, W. G., Shepherd, M., et al. (2000). Sensitivity to sulphonylureas in patients with hepatocyte nuclear factor-1α gene mutations: Evidence for pharmacogenetics in diabetes. *Diabet. Med.* 17: 543–545.

Pilotto, A., Seripa, D., Francesch, I. M., et al. (2008). Genetic susceptibility to nonsteroidal anti-inflammatory drug-related gastroduodenal bleeding: Role of cytochrome P450 2C9 polymorphisms *Gastroenterology* 133: 465–471.

Reitman, M. L., Schadt, E. E. (2007). Pharmacogenetics of metformin response: A step in the path toward personalized medicine. *J. Clin. Invest.* 117: 1226–1229.

Rieder, M. J., Reiner, A. P., Gage, B. F., et al. (2005). Effect of VKORC1 haplotypes on transcriptional regulation and warfarin dose. *N. Engl. J. Med.* 352: 2285–2293.

Rollason, V., Samer, C., Piguet, V., et al. (2008). Pharmacogenetics of analgesics: Toward the individualization of prescription. *Pharmacogenomics*. 9: 905–933.

Sacks, F. M., Pfeffer, M. A., Moye, L. A., et al. (1996). The effect of pravastatin on coronary events after myocardial infarction in patients with average cholesterol levels: Cholesterol and Recurrent Events Trial Investigators. *N. Engl. J. Med.* 335: 1001–1009.

Sconce, E. A., Khan, T. I., Wynne, H. A., et al. (2005). The impact of CYP 2C9 and VKORC1 genetics polymorphism and patient characteristics upon warfarin dose requirements: Proposal for a new dosing regimen. *Blood* 106: 2329–2333.

Seeman, E., Hopper, J. L., Bach, L. A., et al. (1989). Reduced bone mass in daughters of women with osteoporosis. *N. Engl. J. Med.* 320: 554–558.

Serebruany, V. L., Steinhubl, S. R., Berger, P. B., et al. (2005). Variability in platelet responsiveness to clopidogril among 544 individuals. *J. Am. Coll. Cardiol.* 45: 246–251.

Shepherd, J., Cobbe, S. M., Ford, I., et al. (1995). Prevention of coronary heart disease with pravastatin in men with hypercholesterolemia. West of Scotland Coronary Prevention Study Group. *N. Engl. J. Med.* 333: 1301–1307.

Sing-Huang, T., Soo-Chin, L., Boon-Cher, G., et al. (2008). Pharmacogenetics of breast cancer therapy. *Clin. Cancer Res.* 14: 8027–8041.

Slamon, D. J., Leyland-Jones, B., Shak, S., et al. (2001). Use of chemotherapy plus a monoclonal cancer that overexpresses HER2 for metastatic breast cancer that overexpresses HER2. *N. Engl. J. Med.* 344: 783–792.

Spear, B. B., Heath-Chiozzi, M., Huff, J. (2001). Clinical application of pharmacogenetics. *Trends Mol. Med.* 7: 201–204.

Szczeklik, A., Undas, A., Sanak, M., et al. (2000). Relationship between bleeding time, aspirin and the PlA1/A2 polymorphism of platelet glycoprotein IIIa. *Br. J. Haematol.* 110: 965–967.

Tattersall, R. B. (1974). Mild familial diabetes with dominant inheritance. *Q. J. Med.* 43: 339–357.

Tattersal, R. B., Fajans, S. S. (1975). Prevalence of diabetes and glucose intolerance in 199 offspring of thirty-seven conjugal diabetic parents. *Diabetes* 24: 452–462.

Turner, S. T., Schwartz, G. L., Chapman, A. B., et al. (2005). WNK1 kinase polymorphism and blood pressure response to a thiazide diuretic. *Hypertension* 46: 758–765.

Uetrecht, J. (2003). Screening for the potential of a drug candidate to cause idiosyncratic drug reactions. *Drug Discov. Today* 8: 832–837.

Undas, A., Sanak, M., Musial, J., et al. (1999). Platelet glycoprotein IIIa polymorphism, aspirin, and thrombin generation. *Lancet* 353: 982–983.

Venter, J. C., Adams, M. D., Myers, E. W., et al. (2001). The sequence of the human genome. *Science* 291: 1304–1351.

Wadelius, M., Pirmohamed, M. (2007). Pharmacogenetics of warfarin: Current status and future challenges. *Pharmacogenomics J.* 7: 99–111.

Wang, T. H., Bhatt, D. L., Topol, E. J. (2006). Aspirin and clopidogril resistance: An emerging clinical entity. *Eur. Heart J.* 27: 647–654.

Weber, A. A., Zimmerman, K. C., Meyer-Kirchrath, et al. (1999). Cyclooxygenase-2 in human platelets as a possible factor in aspirin resistance. *Lancet* 353: 900.

Wessels, J. A., de Vries-Bouwstra, J. K., Heijmans, B. T., et al. (2006). Efficacy and toxicity of methotrexate in early rheumatoid arthritis are associated with single nucleotide polymorphisms in genes coding for folate pathway enzymes. *Arthritis Rheum.* 54: 1087–1095.

Wolford, J. K., Vozarova de Courten, B. (2004). Genetic basis of type 2 diabetes mellitus: Implications for therapy. *Treat. Endocrinol.* 3: 257–267.

Xie, H., Frueh, F. W. (2005). Pharmacogenomics steps towards personalized medicine. *Per. Med.* 2: 325–337.

Zannis, V. I., Just, P. W., Breslow, J. L. (1981). Human apolipoprotein E isoprotein subclasses are genetically determined. *Am. J. Hum. Genet.* 33: 11–24.

Ziegler, S., Schillinger, M., Funk, M., et al. (2005). Association of a functional polymorphism in the clopidogrel target receptor gene, P2Y12, and the risk for ischemic cerebrovascular events in patients with peripheral artery disease. *Stroke* 36: 1394–1399.

Zimmermann, N., Wenk, A., Kim, U., et al. (2003). Functional and biochemical evaluation of platelet aspirin resistance after coronary artery bypass surgery. *Circulation* 108: 542–547.

18 Omics Approaches in Cancer Biomarker and Targeted Anticancer Drug Discovery

Dipali Dhawan and Harish Padh
B. V. Patel Pharmaceutical Education and Research Development Centre
Ahmedabad, India

CONTENTS

18.1 INTRODUCTION

The term omics embodies the study of biological processes, physiologic functions, and structures as systems and facilitates the analysis of interactions between the various complex networks and pathways involved in the biological system (Keusch, 2006). Since the completion of the Human Genome Project, the field of omics has grown tremendously and enabled researchers to explore new areas of diagnosis and treatment of most diseases, including cancer (Keusch, 2006; Nicholson, 2006; Finn, 2007; Hamacher et al., 2008). Omics has proved to be a revolutionary tool in our efforts to treat cancer. This has proved to be a remarkable process.

The conventional drug-development approach has courted more failure than it has success. It has been observed that the attrition rate for drugs in clinical development is high: the percentage of tested products entering phase I trials that eventually gain regulatory approval has been estimated at a trivial 8% (O'Connell and Roblin, 2006). Many of these failures occur late in clinical trials after large amounts of money have been wasted in the drug-development process. Only a few drugs can make it out of the clinical research pipeline. There is global agreement that biomarkers are useful evaluative tools for improving clinical research and will shape the future of clinical drug development. The Food and Drug Administration (FDA) *Critical Path Opportunities List* (2006) mentions biomarkers for their potential to speed the development and approval of medical products.

Recent drug-development failures emphasize the need for biomarkers to guide clinical research. At least 34 drugs were withdrawn from the market between 1995 and 2005 because of hepatotoxicity or cardiotoxicity (Ingelman-Sundberg, 2008); one in that list was the failure of the high-profile drug monoclonal antibody TGN1412, which is against human CD28 antigen, intended for the treatment of B cell chronic lymphocytic leukemia. The various omics approaches in cancer-biomarker and targeted anticancer-drug–discovery research are evaluated in this chapter.

18.2 OVERVIEW OF OMICS TECHNOLOGIES

Several omics technologies have been utilized to develop biomarker assays and to identify new drug targets, especially in the field of cancer research. This section gives an overview of the various omics technologies and is schematically represented in Figure 18.1.

18.2.1 Genomics

Genomics can be used to identify the deregulated pathways that are responsible for cancer occurrence and progression. Furthermore, it can also enable the identification of the different probable targets for cancer therapy. Figure 18.2 depicts the different types of mutations that are usually studied for discovering potential biomarkers. The field of personalized medicine has grown remarkably over the past few years with the advancement of newer techniques available for the analysis of the genomes of individuals. With the lowered cost of next generation sequencing, gradually it has become affordable

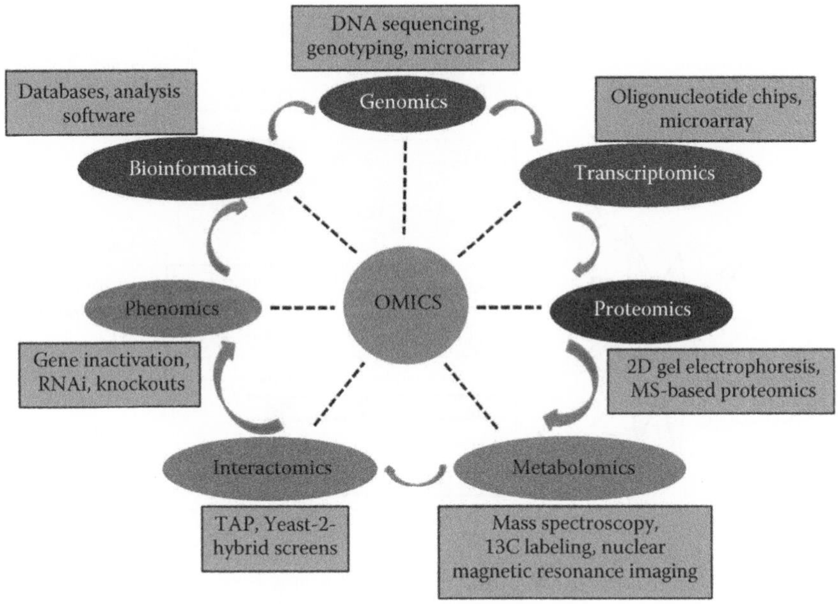

FIGURE 18.1 Various omics technologies.

to sequence the genomes of individuals to discover newer single nucleotide polymorphisms and structural variations and understand their effect on the phenotype (Dhawan and Padh, 2009).

18.2.2 Epigenomics

The field of epigenomics is fairly new, with more knowledge on the regulation of gene expression leading to the current concepts. It is well known that genetic variants along with somatic epigenetic modifications lead to cancer. Hence it becomes necessary to analyze epigenomic targets to better understand the cause of alteration in gene expression. Also with the newer technologies available, it has become easier to analyze epigenomes.

18.2.3 Transcriptomics

The field of transcriptomics studies the transcriptome, which is a collection of all the messenger RNA (mRNA) molecules in a population of cells. A significant fraction of the mammalian genome can be transcribed into mRNA, which can be either coding or noncoding. Transcription is the first step in gene regulation, and information about the transcript levels is needed for understanding gene regulatory networks. Transcriptomics is being widely used for cancer diagnosis and prognosis on the basis of gene expression profiling of mRNA (He, 2006). MicroRNAs, single-stranded small noncoding RNA molecules, are increasingly being studied because they have an important epigenetic regulatory function (Negrini et al., 2007). The widely used techniques for studying gene expression include cDNA microarrays and oligoarrays, SAGE, and cDNA-amplified fragment length polymorphism.

18.2.4 Proteomics

Considerable advances have been achieved in the field of clinical proteomics because of the availability of newer technologies (Rosenblatt et al., 2004). Differential analysis of biological samples has been made possible by tools such as two-dimensional difference gel electrophoresis, protein microarray, mass spectrometry (MS) platforms including matrix-assisted laser desorption/

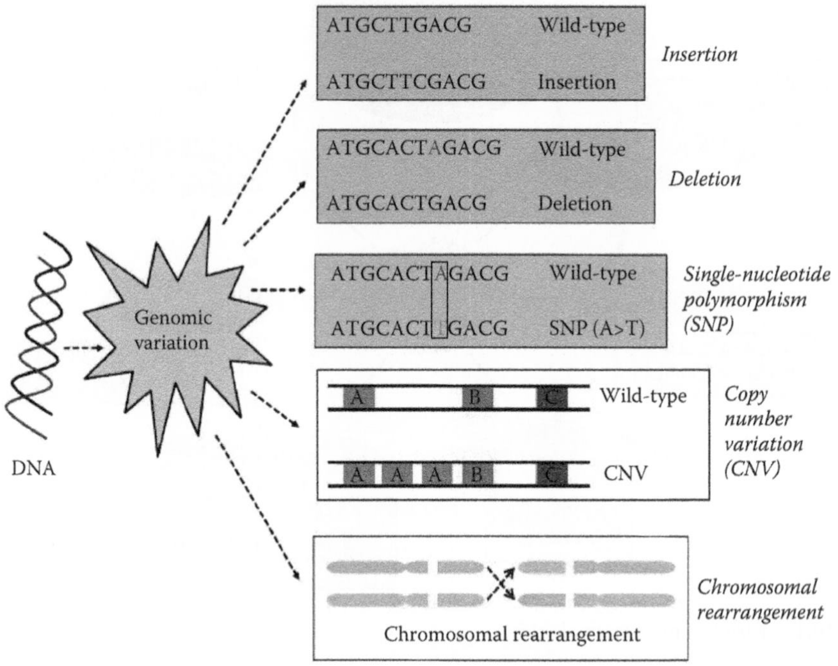

FIGURE 18.2 Types of genomic variations studied for identifying potential biomarkers.

ionization, electrospray ionization, surface-enhanced laser desorption/ionization, isotope-coded affinity tag, isobaric tags for relative and absolute quantification, as well as multidimensional protein identification technology (Cho and Cheng, 2007; van der Merwe et al., 2007). Oncoproteomics has the potential to revolutionize clinical practice, including early cancer diagnosis and screening based on proteomic portraits, personalized medicine, assessment of therapeutic efficacy and toxicity, and rational modulation of therapy based on changes in the cancer protein network associated with prognosis and drug resistance (Cho, 2007a).

18.2.5 Metabolomics

Metabolomics is a newly emerging field of omics research concerned with the comprehensive characterization of the small molecule metabolites in the biological systems. It has been reported by a number of scientists that metabolome is the most predictive of the phenotype (Fiehn, 2002; Weckwerth, 2003). The study of the metabolome relies primarily on nuclear magnetic resonance (NMR) or mass spectrometry coupled to chromatography. The Human Metabolome Project attempts to identify and catalogue all of the metabolites found in the human body. Its goal is to complete a metabolite inventory for human beings, thereby generating resources that can facilitate metabolomics research across many different disciplines. It also provides detailed information about the linkage between human metabolites and the genes, proteins, and pathways in which they are involved (Wishart et al., 2006; Wishart, 2007). As more quantitative metabolomic databases evolve, we can integrate them with data sets from the other omics technologies to enhance their data value and provide greater biological insight than any one omics technique alone can offer.

18.2.6 Bioinformatics

Bioinformatics is the application of information technology to the field of molecular biology. It uses computers and statistics to perform extensive omics-related research by searching biological databases and comparing gene sequences and proteins on a vast scale to identify sequences or proteins that differ between diseased and healthy tissues or between different phenotypes of the same disease. Bioinformatics provides a connecting link between information and modeling in cancer (Stransky et al., 2007).

18.3 OMICS IN CANCER BIOMARKERS

18.3.1 Diagnostic and Prognostic Biomarkers

During the progression of cancer, it has been observed that the levels of certain proteins are elevated. These abnormally increased protein molecules can be used as biomarkers for gaining insight into the course of the disease. A number of studies have identified some specific proteins that are highly expressed in certain cancer types (Agaoglu et al., 2004; Okano et al., 2004; Zhou et al., 2004). These biomarkers in the body fluids can aid in the early detection of cancer and help in monitoring cancer progression. Prognostic markers can be defined as factors that can predict an outcome in the absence of systemic therapy or predict an outcome different from patients who are devoid of the marker, despite empiric therapy (Sargent et al., 2005). Hence, prognostic markers can be utilized to classify patients into appropriate groups for treatment (Sargent et al., 2005). Although a lot of effort has been put into research looking for and studying new biological diagnostic and prognostic markers, only a few out of several hundred have progressed to clinical use. With the use of newer omics technologies, more markers can be validated and would be available to reach the market.

18.3.1.1 Diagnostic Screening

Some of the serum tumor markers like carcinoembryonic antigen and cancer antigens CA15.3 and CA27.29 are not confirmed to be sensitive for early detection; however, the levels of these markers do reflect disease progression and recurrence (Hayes, 1996). Mammoglobin (Zehenter and Carter, 2004) and mammary serine, the protease inhibitor (MASPIN), have been shown to be markers of early breast-cancer detection (O'Brien et al., 2002; Maass, 2002). The early detection of circulating breast cancer cells by morphologic methods is currently being challenged by ultrasensitive proteomic (Li et al., 2002) and PCR-based methods often enhanced by immunomagnetic bead-based cell capture (Kvalheim, 1996; Hu and Chow, 2000). The role of MASPIN in diagnosis of head and neck carcinomas is being explored. It has been reported that low or absent MASPIN cytoplasmic expression was frequently observed in oral carcinomas with lymph node metastasis (Marioni et al., 2009).

18.3.1.2 Messenger RNA

The detection of cancer-specific mRNA in saliva can be used as a diagnostic marker for head and neck squamous cell carcinoma (HNSCC). It has been seen that there was a 3.5-fold increase in the mRNA levels of seven biomarkers in oral SCC patients compared to healthy individuals. However, this test needs to be validated before it can be used as a diagnostic marker for HNSCC (Li et al., 2004).

18.3.1.3 Microsatellite Instability and DNA Repair

Defects in the mismatch repair genes such as *MLH1*, *MSH2*, or *MSH6* or methylation of the MLH1 promoter lead to erroneous replication of segments of simple nucleotide repeats, leading to microsatellite instability (MSI), and increase the risk of cancer occurrence. However, MSI is uncommon in cancers of the breast as compared with some other cancers like colorectal carcinoma (Ozer et al., 2002). It has been observed that prognosis in MSI-positive breast cancer patients is worse than that of patients with MSI-negative tumors (Paulson et al., 1996; Tomita et al., 1999). About 30% of HNSCC patients have MSI (El-Naggar et al., 1996). A number of studies have evaluated the potential of detecting these markers in tumor samples as well as in saliva from HNSCC patients (Spafford et al., 2001).

Stage II and III colorectal cancer (CRC) patients with high microsatellite instability have been reported to show improved survival and better relapse-free survival as compared with microsatellite stable patients (Gryfe and Gallinger, 2001). Also other studies have shown better prognosis with MSI (Lim et al., 2004; Popat et al., 2005).

18.3.1.4 DNA Methylation

Methylation changes in the precarcinoma stage can be used as markers for diagnosis in cancer if these epigenetic changes are not present in normal cells. With the development of techniques like sodium bisulfite modification and methylation-specific PCR, it has become easier to detect methylation changes qualitatively as well as quantitatively. A number of genes have been studied for methylation changes including the tumor suppressor gene, RAS-associated domain family protein 1A (*RASSF1A*) (Dammann et al., 2001) and adenomatous polyposis coli (Esteller et al., 2000). Both of these genes show hypermethylation in breast cancers but are unmethylated in normal cells. Gene promoter hypermethylation in death-associated protein kinase 1 (*DAPK1*) (Sanchez-Cespedes et al., 2000), *p16* (Van der Riet et al., 1994), and *RASSF1A* (Hasegawa et al., 2002) can be used as biomarkers for detection of HNSCCs (Cho, 2007b).

In breast cancer, DNA methylation has shown promise as a potential marker for early detection, therapy monitoring, and assessment of prognosis or prediction of therapeutic response. DNA methylation markers have been reported to show the worst outcome in colorectal cancer patients (Ward et al., 2003). Methylation studies in genes like *p16*, myogenic differentiation 1 (*MYOD1*), and inhibitor of DNA binding 4 (*ID4*) have shown association with unfavorable prognosis in the patients affected with colorectal cancer (Hiranuma et al., 2004; Maeda et al., 2003; Umetani et al., 2004).

18.3.1.5 Circulating Tumor Cells

Many scientists have determined the prognostic value of determining the number of circulating tumor cells (CTCs) in patients with breast cancer (Cristofanilli et al., 2004, 2005; Budd et al., 2006; Beveridge, 2007). It has also been reported that CTC levels have greater prognostic value than other conventionally used markers (Cristofanilli et al., 2004).

18.3.1.6 Viral Markers

Human papilloma virus (HPV) 16 has been linked to HPV-positive HNSCCs (Gillison et al., 2000; Schiffman et al., 2005). About 20% of HNSCCs are HPV-positive, whereas in the case of oropharyngeal SCCs, about half of the patients test positive for HPV (Gillison and Shah, 2003; Herrero et al., 2003). There is a biological and clinical difference in HPV-positive and HPV-negative HNSCC patients. Nasopharyngeal carcinoma can be detected by identifying the Epstein-Barr virus (EBV) by *in situ* hybridization for EBV-encoded RNAs or by immunohistochemical analysis (Cho, 2007b). There is an overexpression of p16 in HPV-positive tumors, which serves as a surrogate marker for high-risk HPVs. In a number of studies, it has been observed that HPV-positive tonsillar cancers have a much better prognosis than HPV-negative cancers (Li et al., 2003; Schwartz et al., 2001; Weinberger et al., 2006).

18.3.1.7 Oncogenes and Tumor Suppressor Genes

The *c-myc* proto-oncogene is well known to be amplified in about 16% of breast cancer cases and is associated with decreased patient survival (Deming et al., 2000; Mizukami et al., 1995). Stathmin levels have been seen to correlate with prognosis in nasopharyngeal carcinomas (Cho, 2007b). *p53*, a tumor-suppressor gene, has a lower mutation rate in breast cancer and is associated with aggressive disease and worse overall survival (Borresen-Dale, 2003). However, conflicting results have been observed by some scientists regarding the role of *p53* in the prognosis of breast cancer.

18.3.1.8 Cytogenetics

Techniques like comparative genomic hybridization have identified complex genetic variants associated with adverse prognosis in breast cancer (Monni et al., 2001; Isola et al., 1995). Cytogenetics or the study of chromosomes is also used as a prognostic tool in head and neck cancers (Pandey and Mishra, 2007). In the case of HNSCCs, loss of heterozygosity on distal arm of 18q has been reported to be associated with poorer survival (Pearlstein et al., 1998). Rearrangements affecting 11q13 are also correlated to reduced survival in HNSCC patients (Akervall et al., 1995). CRC patients with chromosome 18q loss show worse disease-free and overall survival (Lanza et al., 1998; Popat and Houlston, 2005). Contradictory results have been observed by scientists studying the effect of ploidy and S-phase status on disease-free and overall survival (Ross, 1996). Although this marker is in clinical use by some institutions, it is not a generally accepted prognostic marker.

18.3.1.9 Cell Cycle Markers

Markers like Ki-67 staining detecting cell proliferation have been observed to be significantly correlated with breast cancer outcome (Mohsenifar et al., 2007; de Azambuja et al., 2007). One-fifth of breast-cancer patients demonstrate an amplification or overexpression of cyclin D1 (*PRAD1* or bcl-1) (Wolman et al., 1992), which is responsible for the conversion from *in situ* to invasive ductal breast cancer (Weinstat-Saslow et al., 1995). However, the role of cyclin D1 in breast cancer prognosis is less clear (Arnold and Papanikolaou, 2005). A number of conflicting results have been reported for association of altered expression of *p21* with breast cancer outcome (Oh et al., 2001; Gohring et al., 2001; Lau et al., 2001) and low *p27* expression with poor prognosis (Barbareschi, 1999; Barbareschi et al., 2000; Leivonen et al., 2001; Nohara et al., 2001).

Cyclin D1 is reported to be overexpressed in about 30% of HNSCC (Callender et al., 1994). Several studies have shown the association of overexpression of cyclin D1 with clinical outcome (Bellacosa et al., 1996; Akervall et al., 1997; Bova et al., 1999; Kyomoto et al., 1997; Michalides et al., 1997).

18.3.1.10 Cell Adhesion Molecules

The epithelial cell adhesion molecule is the most widely studied adhesion molecule in breast cancer and is associated with survival (Gastl et al., 2000). In the case of patients with HNSCC, reduced or aberrant expression of E-cadherin protein was associated with the presence of cervical metastases (Tanaka et al., 2003; Bosch et al., 2005; Andrews et al., 1997). Reduced E-cadherin expression has also been studied as a prognostic factor in cases of primary HNSCC (Bosch et al., 2005).

18.3.1.11 Proteases and Proteins Involved in Invasion

Elevated levels of cathepsin D, an estrogen-regulated lysosomal aspartyl protease, have been shown to be a predictor of survival in breast cancer (Rochefort et al., 2001). The detection was by an immunoassay, which has limitations, and hence has not been widely used for prognostic assessment. Recently, scientists have tried to assess the levels of cathepsin D by immunohistochemistry (IHC) (Barthell et al., 2007). The two main groups of enzymes playing a vital role in promoting invasion and metastasis include serine proteases and the matrix metalloproteases (MMPs). Breast cancer invasion studies have been focused on urokinase plasminogen activator (uPA) and its receptor and plasminogen activator inhibitor (PAI)-1. uPA converts plasminogen to plasmin, degrades the extracellular matrix (ECM), and is inhibited by PAI-1. uPA and its inhibitor PAI-1 have shown prognostic value for survival in breast-cancer patients (Visscher et al., 1993). High uPA and PAI-1 levels in fresh tissue extracts and tumor cytosol have been associated with disease recurrence and patient survival (Mokbel and Elkak, 2001; Harbeck et al., 2002a, 2002b; Duffy, 2002). MMPs, a group of about 19 zinc metalloenzymes including collagenases, gelatinases, stromelysins, and membrane-type MMPs, are involved in breast-cancer initiation, invasion, and metastasis (Egeblad and Werb, 2002). Studies have shown an association of high levels of MMP-2, MMP-9, and MMP-11 with poor disease outcome in breast cancer (Egeblad and Werb, 2002; Brinckerhoff and Matrisian, 2002; McCawley and Matrisian, 2000; Benaud et al., 1998).

Thomas et al. (1999) have shown that an increase in the levels of MMPs correlated with metastasis and with tumor behavior and prognosis. Many other studies have confirmed the role of MMPs-2 and -9 in oral squamous-cell carcinomas with poor prognosis (Hong et al., 2000; Miyajima et al.,

1995). The level of fibronectin, an adhesive glycoprotein of the ECM, has been correlated to poor prognosis of head and neck cancers (Martins et al., 2003). Cathepsin D and annexin 1 may have a prognostic role in head and neck cancers (Cho, 2007b). uPA and PAI-1 have shown association with survival in rectal cancer patients but not in colon cancer patients (Langenskiöld et al., 2009).

18.3.1.12 Estrogen and Progesterone Receptors

The prognostic role of estrogen receptor (ER) and progesterone receptor (PR) has been well established in breast cancer. ERα and ERβ are the two intracellular receptors of the ER pathway mediating the functions of estrogen (Sommer and Fuqua, 2001; Kuiper et al., 1996). Overexpression of ERα has shown an association with prognosis and prediction in breast cancer, whereas the role of ERβ is not well defined (Speirs and Kerin, 2000; Dotzlaw et al., 1999; Fuqua et al., 1999; Su et al., 2000). Immunohistochemistry is the standard method used for analyzing ER and PR status.

18.3.1.13 Growth Factors and Receptors

Overexpression of the *HER-2* gene is observed in about 60% of ductal carcinomas *in situ* and in 20–30% of infiltrating breast carcinomas (King et al., 1985; Slamon et al., 1987) and is important in its pathogenesis (Di-Fiore et al., 1987; Guy et al., 1992; Esteva-Lorenzo et al., 1998). Overexpression of *HER-2* gene can be elucidated using IHC or fluorescence *in situ* hybridization (FISH). It has been shown that in patients with axillary node-positive breast cancer, *HER-2* amplification is correlated with poor disease-free survival (Slamon et al., 1987; Esteva et al., 2000; Borg et al., 1990). The transforming growth factor-α is an activating ligand for epidermal growth factor receptor (EGFR) and is seen to be correlated with recurrence of breast cancer and adverse prognosis (Castellani et al., 1994; Umekita et al., 2000). Vascular endothelial growth factor (VEGF) and its receptors have been well studied in breast cancer and have shown conflicting results as prognostic markers for the disease (Kinoshita et al., 2001; Linderholm et al., 2001; Foekens et al., 2001; Manders et al., 2002; Coradini et al., 2001; De Paola et al., 2002; MacConmara et al., 2002).

EGFR is overexpressed in about 80–100% of HNSCC tumors (Grandis and Tweardy, 1993; Grandis et al., 1996). *EGFR* overexpression has been widely studied by a number of scientists and has been correlated with poor prognosis and decreased overall survival (Mayer et al., 1993; Bartlett et al., 1996; Klijn et al., 1992; Volm et al., 1998). A number of different techniques are available for detection at DNA, RNA, and protein level; however, each method has advantages as well as limitations. *EGFR* gene copy number is increased in HNSCC patients and has been correlated with worse progression-free survival and overall survival (Chung et al., 2006).

18.3.1.14 Telomerase

Telomerase, known to maintain the ends of chromosomes, is upregulated in about 90% of breast cancers but not in normal tissues (Herbert et al., 2001). There have been conflicting reports of prognostic significance of telomerase expression in breast cancer (Carey et al., 1999; Mokbel et al., 1999; Mueller et al., 2002; Kimura et al., 2003). The catalytic subunit of telomerase, human telomerase reverse-transcriptase (hTERT), and an internal RNA component (hTR) have shown significant association with disease outcome (Bieche et al., 2000; Poremba et al., 2002).

18.3.2 Therapeutic Biomarkers

Biological markers that can predict therapeutic outcome facilitate the choice of treatment (Duffy, 2005). Depending on the levels of the markers, it could be estimated whether the patient would respond to a particular therapy or not; hence, it would help in deciding if the patient should undergo that therapy or opt for alternatives. With the availability of such predictive biological markers, treatment has become more efficient and cost-effective. Cancer-therapy predictive markers are more difficult to evaluate as compared with prognostic markers especially for adjuvant therapy because there is no measurable disease. It has been known that anti-cancer drugs cause adverse drug reactions

(ADRs) in patients undergoing therapy (Hassett et al., 2006). Hence it would be desirable to identify patients who are likely to develop ADRs when undergoing any particular chemotherapy. This would save the patient from life-threatening toxicity, hospital charges, and time of treatment. Some of the well studied and validated markers for predicting response and toxicity will be discussed in the following sections.

18.3.2.1 Therapy Response and Drug-Resistance Markers in Breast Cancer

In case of serine proteases, plasminogen protease levels have been effectively used as predictors of chemotherapy response (Harbeck et al., 2002b). *HER-2* gene overexpression has a role as a prognostic marker and is also correlated with improved response to doxorubicin-based chemotherapy (Wood et al., 1994; Muss et al., 1994; Thor et al., 1998; Paik et al., 1998; Ravdin et al., 1998). Trastuzumab, a monoclonal immunoglobulin G1 class humanized murine antibody, has become one of the major therapeutic options for patients with HER-2/neu-positive breast cancer. It is being widely used in breast cancer patients as a second-line treatment and in neoadjuvant treatment regime (McKeage and Perry, 2002; Shawver et al., 2002; Ligibel and Winer, 2002). However, it should be noted that there are conflicting results about *HER-2* overexpression and response to hormonal therapy (Ravdin et al., 1998; De Placido et al., 2003; Elledge et al., 1998). Overexpression of thymidylate synthase (TS) has been seen to be correlated with resistance to 5-fluorouracil (5-FU) therapy (Nishimura et al., 1999). Another example of a proven therapeutic predictive marker is the ER test for response to hormonal therapy. It has been seen that ER/HER-2/neu-positives may show resistance to tamoxifen treatment but respond to an aromatase inhibitor (Ellis et al., 2001; Dowsett et al., 2001).

The very well-studied multiple-drug resistance gene, *MDR1*, that encodes an integral transmembrane protein, the P-glycoprotein (Pgp), has been associated with resistance to chemotherapeutics in breast-cancer patients (Decker et al., 1993). Drugs that are affected by Pgp expression include anthracyclines, vinca alkaloids and taxanes. It is seen that the glutathione *S*-transferase (GST)-π gene is associated with Pgp expression with the resultant increased intracellular drug detoxification and multidrug resistance in breast cancer. GST-π expression is being used as a marker for drug resistance to alkylating agents in breast cancer patients (Batiste et al., 1986; Satta et al., 1992).

18.3.2.2 Therapy Response and Drug-Resistance Markers in Head and Neck Cancer

Platinum agents are increasingly being used as chemotherapeutics in carcinomas of the head and neck. The cytotoxic effect of these drugs is by covalent binding to the DNA molecule. Resistance to platinum agents is seen in tumors that have increased DNA repair capabilities. The excision-repair cross-complementation group 1 (ERCC1) is a rate-limiting enzyme in the nucleotide excision repair pathway and plays a role in removing platinum-containing DNA adducts. High levels of ERCC1 have been associated with resistance to platinum-based chemotherapy and poor survival in patients with non-small–cell lung cancer (NSCLC) (Olaussen et al., 2006). Also ERCC1-positive HNSCC patients had worse survival and early progression after cisplatin-based chemoradiotherapy (Handra-Luca et al., 2007). The anti-EGFR antibody cetuximab is being used in the management of HNSCC. Other EGFR tyrosine-kinase inhibitors being used in NSCLC include gefitinib and erlotinib. Many studies have been conducted for analyzing potential predictive markers of response to EGFR inhibitors. It has been seen that activating mutations in the EGFR kinase domain and increased *EGFR* gene copy numbers detected by FISH in NSCLC patients are associated with better response and survival when treated with gefitinib (Tsao et al., 2005; Cappuzzo et al., 2005). Also higher *EGFR* copy number determined by FISH is associated with a better response to erlotinib (Agulnik et al., 2007).

Cisplatin is a widely used chemotherapeutic in head and neck cancer patients. However, the effect of the drug is limited by the resistance observed in these patients; and it has been seen that hypermethylation in some of the genes responsible for cytotoxicity causes the resistance, for example, the *S100P* gene (Chang et al., 2010).

18.3.2.3 Therapy Response and Drug-Resistance Markers in Colorectal Cancer

The current chemotherapy options for patients with metastatic CRC include 5-FU, irinotecan, oxaliplatin, and capecitabine, bevacizumab (the anti-VEGF monoclonal antibody), and cetuximab and panitumumab (both anti-EGFR monoclonal antibodies), either as monotherapy or as combination therapy (National Comprehensive Cancer Network, 2008). A number of studies have shown a significant association of intratumoral *TS* levels with response to fluoropyrimidine-based therapy (Johnston et al., 1995; Lenz et al., 1998; Salonga et al., 2000). Further, it has been shown that patients with low levels of *TS*, thymidine phosphorylase, and dihydropyrimidine dehydrogenase (*DPD*) showed response compared to patients with elevated levels of any one of these genes (Salonga et al., 2000). *DPD* has been shown to be a marker of toxicity, but its role as a response-predicting marker needs to be further explored (van Kuilenberg, 2004). A variant in the isoform 1A1 of the uridine diphosphate glucuronyltransferase (UGT) enzyme leading to an additional TA repeat in the TATA sequence (*UGT1A1*28*) results in less detoxification of irinotecan and higher toxicity (Iyer et al., 1999, 2002). The new irinotecan label has been modified to educate patients homozygous for the *UGT1A1*28* against the toxic effects of the drug (Innocenti et al., 2006).

It has been seen that the presence of gain-of-function mutations in patients with CRC led to resistance to monotherapy with EGFR-targeted monoclonal antibodies and was associated with poorer prognosis (Benvenuti et al., 2007; Lievre et al., 2006, 2008; Khambata-Ford et al., 2007; Di Fiore et al., 2007; De Roock et al., 2008; Freeman et al., 2007; Amado et al., 2008; Tejpar et al., 2008). The RAS G-protein activation of the mitogen-activated protein kinase signalling cascade downstream of EGFR is thought to be the causal factor for the tumor resistance to EGFR inhibitors (Benvenuti et al., 2007; Barh et al., 2011).

18.4 OMICS IN DRUG TARGET DISCOVERY AND VALIDATION

The process of discovering new medicines in the pharmaceutical industry is long (10–15 years) and expensive ($0.8–1.5 billion) with uncertain success (Austin and Babiss, 2006). To improve the drug-discovery process, appropriate biomarkers need to be developed and validated. The utilization of biomarkers at each stage of drug development would help in decreasing the costs of late-phase testing of an ineffective product. The development and validation process of biomarkers also requires a considerable amount of time, which is depicted in Figure 18.3.

18.4.1 Genomics in Drug Target Discovery and Validation

There have been enormous technological advances in the field of genomics in the past few decades. The field of omics is defined as the elucidation and functional understanding of the genes encoded by the human genome. With the completion of the human genome sequencing, it has been possible to analyze the number of potentially interesting drug targets encoded by the genes. Hopkins and Groom (2002) have provided a summary of the number of human genes in the various drug-friendly gene families. About 3,000 genes are part of the drug target families. When it was analyzed whether pharmaceutical compounds actually bind to the presumed target, it was revealed that the frequently quoted number of druggable genes is in fact much lower and possibly only about 25% of the 483 targets originally identified (Drews, 1996, 2000).

The erbB oncogene family is an important example of how we can translate advances from basic research into clinically relevant markers that may also serve as new therapeutic targets (Jardines et al., 1993). A number of tools are being used nowadays for genomic analysis including mutation analysis, expression arrays, microRNA arrays, array CGH, ChIP-on-chip, methylation arrays, genome-wide association studies, and integrated functional genomics analysis. Genomic studies will likely identify driver mutations or genes that affect the growth and metastasis. These genes can be used to identify drugs that are more efficient and less toxic for cancer treatment. Specific

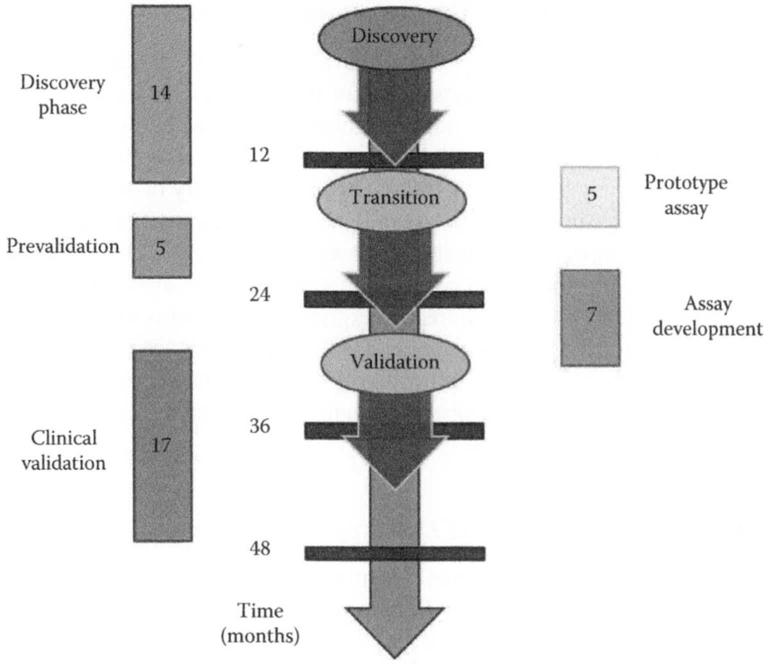

FIGURE 18.3 **(See color insert.)** Different phases and estimated timelines for the discovery and validation of biomarkers.

genetic signatures can be used to predict drug responsiveness. Hence these biomarkers will enable an increase in the power and efficiency of clinical trials by selecting the suitable patient group and may lead to successful clinical drug development.

18.4.2 Transcriptomics in Drug Target Discovery and Validation

Transcriptomics allows the profiling of the subset of genes transcribed in a given organism and provides a dynamic link between the genome, proteome, and cellular phenotype (Hu et al., 2005). This area of omics been widely used in drug discovery and development to discern genes that are associated with specific diseases to identify drug targets and annotate the functions of the different genes. Transcriptomics offers a tremendous growth in target discovery by profiling the expression patterns of thousands of genes from complex biological mixtures. Nowadays, DNA microarray is the most extensively used tool for transcriptomics because it allows measurement of the expression level of thousands of genes, or even entire genomes, simultaneously. An association of the novel genes with the disease can be established by comparing the signature of the clustering profiles with future transcriptome assessments.

Many new approaches are being utilized for target discovery in oncology by using a panel of biomarkers that have been identified from transcriptomic arrays (Liotta and Petricoin, 2000; Ono et al., 2000; Sallinen et al., 2000; Finlin et al., 2001; Gruvberger et al., 2001; Sorlie et al., 2001; West et al., 2001; van't Veer et al., 2002; van de Vijver et al., 2002; Sotiriou et al., 2003; Rhodes et al., 2004). Gene expression profiling in normal and pathological tissues has been reported for identification and validation of potential biomarkers and also identification of new molecular targets (Zhang, 2007; Zhang et al., 2007). These microarrays can also enable the profiling of pharmacological effects of lead compounds, further facilitating in the identification of molecular mechanism of drug action and of genes and expression patterns related to toxicity, drug sensitivity, or resistance.

18.4.3 Proteomics in Drug Target Discovery and Validation

Proteins involved in signal transduction pathways have being widely studied in many pathologies and have emerged as candidate drug targets. Some of the leading drug candidates are regulatory enzymes that catalyze protein posttranslational modifications. These proteins provide further candidates for drug therapeutics and potential cellular markers of disease. The clarity of requirements for drug specificity and selectivity has been facilitated by the availability of structural information on proteins, particularly the ones in complex with small molecular ligands. The advent of new technologies has accelerated protein structural analysis and methods for high-throughput protein production and crystallization. Two-dimensional gel electrophoresis and MS are the two most widely used techniques for proteome analysis. The differential expression of proteins in normal as compared to cancerous tissues can be used to identify the biological pathways involved in the pathogenesis, and molecules from that pathway can then be utilized as drug targets. It has been reported that specific protein changes in response to drug administration in humans can have a significant role in clinical research (Lee et al., 2005; Patil et al., 2007).

18.4.4 Metabolomics in Drug Target Discovery and Validation

Metabolomics studies take into account genetic regulation, altered kinetic activity of enzymes, and changes in metabolic reactions (Griffin and Shockcor, 2004; Mendes et al., 1996, 1992). Usually, NMR spectroscopy and MS are the two major spectroscopic techniques used in metabolic analysis. The metabolome is now increasingly being analyzed for identification of new targets. Metabolomics is being used in developing therapeutics, examples being tyrosine kinase inhibitors, proapoptotic agents, and heat shock protein inhibitors (Gottschalk et al., 2004; Serkova and Boros, 2005; Hasmann and Schemainda, 2003; Muruganandham et al., 2005; Lyng et al., 2007; Blankenberg et al., 1997; Chung et al., 2003).

Metabolomics can be utilized for identification of multivariate biomarkers, including fingerprints, profiles, or signatures, which would enable the characterization of the state of cancer. Metabolomics is being used for assessing the pharmacodynamic parameters of novel agents and for the characterization of toxic effects. Further development and application of metabolomics will depend on several factors including the establishment of spectral databases of metabolites and associated biochemical identities along with cross-validation of NMR- or MS-obtained metabolites and correlation with other assays.

18.4.5 Systems Biology in Drug Target Discovery and Validation

Systems biology comprises the global, integrated analysis of large-scale data sets that encode different levels of biological information (Hood et al., 2004). There has been considerable progress in the field of systems biology because of the availability of large-scale experimental and computational tools that help in generating, analyzing, and integrating different types of omics data (Hood et al., 2004; Auffray et al., 2009). The application of systems approach to biomarker discovery is determined by the fact that biological processes underlying health and disease responses are complex and interconnected. The system-based biomarker discovery approach functions by inferring and analyzing the molecular networks underlying the disease. This approach provides the functional, contextual information about how genes and proteins are implicated in disease, which might not be possible by using only genomic or proteomic approaches. It will help researchers to attain enhanced knowledge about the occurrence of disease and response to therapy. Some obstacles in the development of a systems-driven approach to biomarker discovery include technological, educational, and policy challenges (Lemberger, 2007). These limitations can be overcome by making use of cost-effective and more advanced technologies and instruments for analysis.

18.5 REGULATORY REQUIREMENTS FOR BIOMARKER ASSAY VALIDATION

The number of molecular biomarkers for drug applications is expected to increase in the future because most of the pharmaceutical companies have some biomarker programs being developed. Not only discovering potential molecular biomarkers for use in drug development but also validation and building up a robust evidence base for regulatory review is a great challenge being faced by drug companies. Because the FDA has recognized a significant increase in the use of genomic biomarkers in drug development and applications, it has issued guidance for the industry (Goodsaid and Frueh, 2007; FDA, 2010) to enable the sponsors to submit pharmacogenomic data through the Voluntary Genomic Data Submissions (VGDS) program (Orr et al., 2007; Goodsaid et al., 2010). The VGDS has been renamed the Voluntary Exploratory Data Submissions (VXDS) because of the diversity of the biomarker data received. The genomic data submitted through VXDS are exploratory research data, and hence it is not obligatory to include them in an investigational new drug or a new drug application. VXDS provides FDA reviewers with this research data and increases communication with the industry for the further development of the genomic biomarkers to enable personalized medicine.

Biomarker qualification is a graded, fit-for-purpose evidence-based process linking a biomarker with biological processes and clinical end points, dependent on the intended application. A qualified biomarker is defined as the one that is measured in an analytical test system with well-established performance characteristic and for which there is an established scientific framework or body of evidence that elucidates the physiologic, toxicologic, pharmacologic, or clinical significance of the test results. The use of qualified biomarkers in drug discovery, development, and postapproval is believed to be necessary because it has the potential to facilitate the development of safer and more effective medicines, to guide dose selection, and to enhance the benefit-risk profile of approved drug products. The FDA has established a biomarker qualification pilot process to provide a framework for the consideration of qualification requests from consortia, partnerships, and other submitters (Hong et al., 2010).

18.6 ISSUES IN BIOMARKER ASSAY RESEARCH AND VALIDATION

Biomarker assays utilize both low-end technologies, like immunohistochemistry, electrophoresis, reverse transcription-PCR, and ELISA, and high-end technologies, including platforms for genomics (gene chips), proteomics (surface-enhanced laser desorption/ionization time-of-flight), and multiplex ligand-binding assays (Jones et al., 2003; Johann et al., 2004; Seligson, 2005; Scholler et al., 2006). Biomarker analytical methods are labor-intensive, are prone to variability, and frequently lack sensitivity. The analysis also depends on the integrity of reagents such as antibodies that have problems of supply, quality control, and stability because they are derived from biologic sources. The aim of biomarker assay development and validation should be to develop a valid assay rather than to validate a developed assay (Smith and Sittampalam, 1998). It is better to define the biologic end point of the assay before the biomarker assay validation is done.

The key issues with biomarker discovery research are measurement feasibility or difficulty. The difficulties include mixture complexity, analyte polarity, sensitivity, and disease specificity. Other issues include method validation and method suitability. Sometimes the measurements are difficult to reproduce. Some of the major issues with biomarker discovery and development are summarized in Table 18.1.

18.6.1 BIAS

Bias has been defined as "the systematic erroneous association of some characteristic with a group in a way that distorts a comparison with another group" (Ransohoff, 2005). In cancer research, bias has been increasingly recognized as a problem in cases of disputed results or irreproducible results. It is difficult to address the problem of bias because it is often unintentional. Bias cannot be minimized by increasing the sample size; it requires careful control and planning of experiments. Sample blinding and randomization can help in controlling bias.

TABLE 18.1
Major Issues in Biomarker Discovery and Development

Stage	Major Issues
Biomarker discovery	1. Molecular profiling • *Experimental design* • *Data interpretation* 2. Imaging biology • *Throughput* • *Cost* 3. Biomarker validation • *Epidemiology studies*
Prototype development	1. Assay development • *Platforms* • *Reagents* • *Protocols* 2. Analytic validation • *Sensitivity/specificity* • *Alpha-site testing*
Biomarker test development	1. Production • *Process standardization* • *Quality assurance* 2. Clinical validation • *Clinical trials* • *Beta-site testing* 3. Marketing • *Registration*

18.6.2 Statistical

It is important to have adequate replicates in both the measurement process and biological replicates. It has been observed that biological replicates are more important than methodological or technical replicates (Rocke, 2004). There are three important factors that can contribute to a false-positive yet statistically significant observation (Sterne and Davey Smith, 2001). First is the absolute magnitude of the p value; second is the statistical power of the method; and third is "the fraction of tested hypotheses that is true."

18.6.3 Methodological

One of the aspects worth considering is the variation of biomarkers in different parts (example: individual cells, subpopulations of cells, tissues, organs, and organisms). There might be discrepancy caused by a lack of understanding of the baseline or normal value for biological variation and for variation within different parts. At times small individual omics differences have importance in disease states and have to be measured but might be superimposed on a variation not associated with the target disease (Table 18.1).

18.7 OMICS IN CANCER CLINICAL PRACTICE

Nowadays a number of biomarkers are being successfully used on a routine basis in the clinic. Reports suggest that a combination of genes measured concurrently imparts better information about the clinical effect than a single gene. Figure 18.4 depicts the stepwise selection of therapies based on the oncogenic variations. Oncotype DX™ is the first multigene predicting test of prognosis for breast-cancer patients receiving anti-estrogen therapy (Genomic Health). The FDA has also

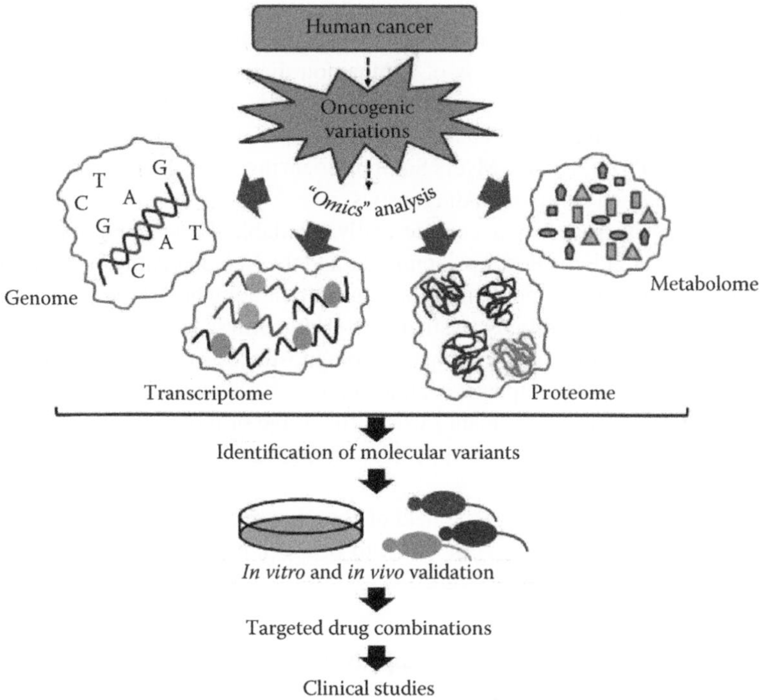

FIGURE 18.4 Stepwise selection of anticancer therapies based on oncogenic variations.

approved the test MammaPrint, which predicts relapses in breast cancer patients by analyzing the activity of 70 genes. In case of prostate cancer, the prostate-specific antigen (PSA) test has been approved. PSA levels help in detecting prostate cancer and also predict a recurrence in patients suffering from the disease. a-Fetoprotein has been approved for the diagnosis and monitoring of patients with nonseminoma testicular cancer. The levels of this protein are higher in almost all yolk sac tumors and malignant liver tumors. Further, another test approved by the FDA is the cancer antigen CA-125, which detects patients with ovarian cancer. In case of medullary-thyroid carcinoma, which is a rare cancer originating in the parafollicular C cells in the thyroid gland, calcitonin levels are used to detect this cancer. However, a major challenge remains to validate a marker and

TABLE 18.2
Biomarkers Approved by the FDA for Cancer Staging and Prognosis

Biomarker	Mechanism/Role	Cancer Type	Clinical Use
α-Fetoprotein	Glycoprotein	Nonseminomatous testicular cancer	Staging
CA125	Mucin glycoprotein	Ovarian cancer	Monitoring
Carcinoembryonic antigen	Cell-surface adhesion molecule	CRC	Monitoring
CA27.29	Cancer antigen, elevated in breast cancer	Breast cancer	Monitoring
CA15.3	Breast-associated mucin, elevated in breast cancer	Breast cancer	Monitoring
Cytokeratins	Intermediate filament proteins	Breast cancer	Prognosis
CA19.9	Carbohydrate antigen	Pancreatic cancer	Monitoring
PSA	Protein secreted by prostate gland	Prostate cancer	Screening and monitoring

prove its clinical benefit rather than just showing an association at the research level (Pusztai, 2004). Table 18.2 gives a list of markers approved for cancer staging and prognosis.

One of the recently developed areas is of companion diagnostics where a diagnostic is developed along with a drug to screen for patients for predictable clinical outcome. The combination of diagnostic test and the drug can be codeveloped and comarketed. A number of pharmaceutical companies like Abbott, Pfizer, Bristol-Myers Squibb, Boehringer Ingelheim, GlaxoSmithKline, and AstraZeneca have developed companion diagnostics.

The U.S. FDA has approved two commercially available HER-2/neu IHC kits, the Dako Herceptest™ (DakoCytomation, Glostrup, Denmark) and the Ventana Pathway™ (Ventana Medical Systems Inc., Oro Valley, AZ) for predicting the eligibility among breast-cancer patients to receive the antiHER-2/neu therapeutic antibody trastuzumab (Herceptin®; Genentech, San Francisco, CA). There are several other examples where companion diagnostic tests have been approved for predicting the outcome of a particular therapeutic regime before subjecting the patient to that treatment. In the last decade there has been a significant raise in the usage of molecular biomarkers in labels of FDA-approved drugs. Some examples of molecular biomarkers in current labels of FDA-approved drugs are listed in Table 18.3.

DxS Diagnostics is another leader in the field of companion diagnostics. The company has partnered with some big pharmaceutical companies to codevelop biomarker tests. An incentive for

TABLE 18.3
Molecular Biomarkers in Labels of Drugs Approved by the FDA

Biomarker	Drug	Date	Section in Label
CYP2C19	Plavix	May 2009	Clinical pharmacology, precautions, dosage, and administration
	Vfend	March 2008	Clinical pharmacology
	Effient	July 2009	Use in specific populations, clinical pharmacology, and clinical studies
	Celebrex	July 2005	Clinical pharmacology
CYP2C9	Celebrex	July 2005	Clinical pharmacology
	Effient	July 2009	Use in specific populations, clinical pharmacology, and clinical studies
	Coumadin	August 2007	Clinical pharmacology and precautions
CYP3A4	Celebrex	July 2005	Clinical pharmacology
	Codeine sulfate	July 2009	Drug interactions and clinical pharmacology
CYP3A5	Effient	July 2009	Use in specific populations, clinical pharmacology, and clinical studies
CYP2B6	Effient	July 2009	Use in specific populations, clinical pharmacology, and clinical studies
CYP2D6	Strattera	July 2008	Dosage and administration, warnings and precautions, drug interactions, and clinical pharmacology
	Prozac	September 2006	Clinical pharmacology and precautions
	Codeine-sulfate tablets	July 2009	Warnings in precautions, drug interactions, use in specific populations, and clinical pharmacology
VKORC1	Coumadin	August 2007	Clinical pharmacology and precautions
UGT1A1	Camptosar	July 2005	Clinical pharmacology, warnings, dosage, and administration
	Tasigna	October 2007	Drug interactions and clinical pharmacology
*HLA-B**1502	Tegretol	December 2007	Warnings and precautions
*HLA-B**5701	Ziagen	July 2008	Warnings and precautions
Deletion of 5q	Revlimid	January 2009	Hematologic toxicity, clinical studies, precautions, and adverse reactions

See Hong et al. (2010) for more information.

pharmaceutical companies is to develop a diagnostic test not only for preventing drugs from failing after reaching the market but also because the drug approval process might become easier.

There are several diagnostic companies in the market place involved in developing biomarker-based tests, for example, DxS Diagnostics, Almac Diagnostics, and Genentech. At times these smaller players are taken over by or licensed out to larger pharmaceutical companies who have interest in codevelopment of drug-diagnostic test (DxRx). For example, Pfizer has collaborations with many smaller companies who develop the biomarkers. Larger diagnostic companies develop biomarkers to monitor the activity of their therapeutics and seek large market opportunities and low risk through their well established manpower and infrastructure. Smaller companies can gain help from the research funding of the inventor's lab and then gradually take the test to the market. Sometimes the larger companies help in bringing the biomarkers of the smaller companies to the market.

18.8 FUTURE CHALLENGES: DISCOVERING NEW BIOMARKERS

At present omics technologies are not entirely being used in the clinical setup as diagnostic tools; however, by further development in the field, far better, simple, straightforward, and cost-effective assays can be made available for routine use of the patients (Zhang et al., 2007). There is an increasingly growing list of potential biomarkers that can be used at various stages by the cancer patients; however, thorough validation and issues of sensitivity, specificity, reproducibility, and accuracy must be dealt with. A major challenge lies in the integration of various omics data and their functional elucidation along with clinical results. Molecular studies of disease genes have paved the way for the revelation of significant biochemical and physiological pathways that can enable identification of newer potential targets for therapeutic intervention. A lot of information is being gathered from around the globe with researchers sharing their data on variations in nucleotide or protein sequences and compiling huge databases for general use by the public.

A tremendous amount of work still remains to be done for the elucidation of potential biomarkers, and with the advancement in the omics technologies, it will become easier to achieve these goals. For example, the G-protein–coupled receptors (GPCRs) superfamily is one of the largest families of proteins in the mammalian genome (Venter et al., 2001; Lander et al., 2001). It is known that about half of the modern drugs are targeted at these receptors (Flower, 1999). However, these drugs have been targeted to a very small number of GPCRs, and there is immense opportunity in this area for potential target discovery.

Finally, validation is the key step in biomarker development process; hence it is necessary to perform validation in a large number of samples. In addition, taking into consideration the diversity among different populations; studies should be conducted on different populations to confirm the effects in each of those.

18.9 CONCLUSION

Until now a number of biomarkers have been identified and validated for being routinely used in a number of diseases in cancer. Also, some of these biomarkers have been approved by the FDA as essential tests before prescription of certain drugs. Hence, it is evident how development of omics technologies have enabled research and its translation to bedside for the common man and will continue to do so in the future. It is, however, important that before these biomarkers are sent to the market, careful validation is completed.

Omics analyses in combination with the clinical development of numerous targeted drugs as well as those already approved lend hope for a more effective and individualized anticancer therapy. However, for more advancement in the field, it is essential that people from academia, pharmaceutical companies, and regulatory agencies work hand in hand.

Ultimately, the aim of omics approaches is to acquire a comprehensive, integrated understanding of biology by studying all biological processes to identify the various genes, RNA, proteins, and metabolites rather than each of those individually. Omics approaches currently attempt to address specific biological questions, often without the need for prior understanding of a biological basis. With the advancement in technologies, omics research might aim to explain more complex systemic questions and become a tool in diagnostics and drug development.

REFERENCES

Agaoglu, F. Y., Dizdar, Y., Dogan, O., et al. (2004). P53 overexpression in nasopharyngeal carcinoma. *In Vivo* 18: 555–560.

Agulnik, M., da Cunha Santos, G., Hedley, D., et al. (2007). Predictive and pharmacodynamic biomarker studies in and skin tissue samples of patients with recurrent or metastatic squamous cell carcinoma of the head and neck treated with erlotinib. *J. Clin. Oncol.* 25: 2184–2190.

Akervall, J., Jin, Y., Wennerberg, J., et al. (1995). Chromosomal abnormalities involving 11q13 are associated with poor prognosis in squamous cell carcinoma of the head and neck. *Cancer* 76: 853–859.

Akervall, J. A., Michalides, R. J., Mineta, H., et al. (1997). Amplification of cyclin D1 in squamous cell carcinoma of the head and neck and the prognostic value of chromosomal abnormalities and cyclin D1 overexpression. *Cancer* 79: 380–389.

Amado, R. G., Wolf, M., Peters, M., et al. (2008). Wild-type KRAS is required for panitumumab efficacy in patients with metastatic colorectal cancer. *J. Clin. Oncol.* 26: 1626–1634.

Andrews, N. A., Jones, A. S., Helliwell, T. R., et al. (1997). Expression of the E-cadherin-catenin cell adhesion complex in primary squamous cell carcinomas of the head and neck and their nodal metastases. *Br. J. Cancer* 75: 1474–1480.

Arnold, A., Papanikolaou, A. (2005). Cyclin D1 in breast cancer pathogenesis. *J. Clin. Oncol.* 23: 4215–4224.

Auffray, C., Chen, Z., Hood, L. (2009). Systems medicine: The future of medical genomics and healthcare. *Genome Med.* 1: 2.

Austin, M. J. F., Babiss, L. (2006). Commentary: Where and how could biomarkers be used in (2016). *AAPS J.* 8: E185–E189

Barbareschi, M. (1999). p27 expression: A cyclin-dependent kinase inhibitor in breast carcinoma. *Adv. Clin. Path.* 3: 119–127.

Barbareschi, M., van Tinteren, H., Mauri, F. A., et al. (2000). P27 (kip1) expression in breast carcinomas: An immunohistochemical study on 512 patients with long-term follow-up. *Int. J. Cancer* 89: 236–241.

Barh, D., Agate, V., Dhawan, D., et al. (2011). Cancer biomarkers for diagnosis, prognosis and therapy. In Whitehouse D., Rapley R. (Eds.) *Cellular and molecular therapeutics*. New York: John Wiley & Sons, Inc.

Bartlett, J. M., Langdon, S. P., Simpson, B. J., et al. (1996). The prognostic value of epidermal growth factor receptor mRNA expression in primary ovarian cancer. *Br. J. Cancer* 73: 301–306.

Barthell, E., Mylonas, I., Shabani, N., et al. (2007). Immunohistochemical visualisation of cathepsin-D expression in breast cancer. *Anticancer Res.* 27: 2035–2039.

Batiste, G., Tulpule, A., Shinha, B. K., et al. (1986). Overexpression of a novel and an ionic glutathionic transferase in multi-drug-resistant human breast cancer cells. *J. Biol. Chem.* 261: 15549–15554.

Bellacosa, A., Almadori, G., Cavallo, S., et al. (1996). Cyclin D1 gene amplification in human laryngeal squamous cell carcinomas: Prognostic significance and clinical implications. *Clin. Cancer Res.* 2: 175–180.

Benaud, C., Dickson, R. B., Thompson, E. W. (1998). Roles of the matrix metalloproteinases in mammary gland development and cancer. *Breast Cancer Res. Treat.* 50: 97–116.

Benvenuti, S., Sartore-Bianchi, A., Di Nicolantonio, F., et al. (2007). Oncogenic activation of the RAS/RAF signaling pathway impairs the response of metastatic colorectal cancers to anti-epidermal growth factor receptor antibody therapies. *Cancer Res.* 67: 2643–2648.

Beveridge, R. (2007). Circulating tumor cells in the management of metastatic breast cancer. *Commun. Oncol.* 4: 79–82.

Bieche, I., Nogues, C., Paradis, V., et al. (2000). Quantitation of hTERT gene expression in sporadic breasts with a real-time reverse transcription-polymerase chain reaction assay. *Clin. Cancer Res.* 6: 452–459.

Blankenberg, F. G., Katsikis, P. D., Storrs, R. W., et al. (1997). Quantitative analysis of apoptotic cell death using proton nuclear magnetic resonance spectroscopy. *Blood* 89: 3778–3786.

Borg, A., Tandon, A. K., Sigurdsson, H., et al. (1990). HER-2/neu amplification predicts poor survival in node-positive breast cancer. *Cancer Res.* 50: 4332–4337.

Borresen-Dale, A. L. (2003). TP53 and breast cancer. *Hum. Mutat.* 21: 292–300.

Bosch, F. X., Andl, C., Abel, U., et al. (2005). E-cadherin is a selective and strongly dominant prognostic factor in squamous cell carcinoma: A comparison of E-cadherin with desmosomal components. *Int. J. Cancer* 114: 779–790.

Bova, R. J., Quinn, D. I., Nankervis, J. S., et al. (1999). Cyclin D1 and p16INK4A expression predict reduced survival in carcinoma of the anterior tongue. *Clin. Cancer Res.* 5: 2810–2819.

Brinckerhoff, C. E., Matrisian, L. M. (2002). Matrix metalloproteinases: A tail of a frog that became a prince. *Nat. Rev. Mol. Cell Biol.* 3: 207–214.

Budd, G. T., Cristofanilli, M., Ellis, M. J., et al. (2006). Circulating cells versus imaging- predicting overall survival in metastatic breast cancer. *Clin. Cancer Res.* 12: 6403–6409.

Callender, T., el-Naggar, A. K., Lee, M. S., et al. (1994). PRAD-1 (CCND1)/cyclin D1 oncogene amplification in primary head and neck squamous cell carcinoma. *Cancer* 74: 152–158.

Cappuzzo, F., Hirsch, F. R., Rossi, E., et al. (2005). Epidermal growth factor receptor gene and protein and gefitinib sensitivity in non-small-cell lung cancer. *J. Natl. Cancer Inst.* 97: 643–655.

Carey, L. A., Kim, N. W., Goodman, S., et al. (1999). Telomerase activity and prognosis in primary breast cancers. *J. Clin. Oncol.* 17: 3075–3081.

Castellani, R., Visscher, E. W., Wykes, S., et al. (1994). Interaction of transforming growth factor-α and epidermal growth factor receptor in breast carcinoma. *Cancer* 73: 344–349.

Chang, X., Monitto, C. L., Demokan, S., et al. (2010). Identification of hypermethylated genes associated with cisplatin resistance in human cancers. *Cancer Res.* 70: 2870–2879.

Cho, W. C. (2007a). Proteomic approaches to cancer target identification. *Drug Discov. Today Ther. Strateg.* 4: 245–250.

Cho, W. C. (2007b). Nasopharyngeal carcinoma: Molecular biomarker discovery and progress. *Mol. Cancer* 6: 1.

Cho, W. C., Cheng, C. H. (2007). Oncoproteomics: Current trends and future perspectives. *Expert Rev. Proteomics* 4: 401–410.

Chung, C. H., Ely, K., McGavran, L., et al. (2006). Increased epidermal growth factor receptor gene copy number is associated with poor prognosis in head and neck squamous cell carcinomas. *J. Clin. Oncol.* 24: 4170–4176.

Chung, Y. L., Troy, H., Banerji, U., et al. (2003). Magnetic resonance spectroscopic pharmacodynamic markers of the heat shock protein 90 inhibitor 17-allylamino,17-demethoxygeldanamycin (17AAG) in human colon cancer models. *J. Natl. Cancer Inst.* 95: 1624–1633.

Coradini, D., Boracchi, P., Daidone, M. G., et al. (2001). Contribution of vascular endothelial growth factor to the Nottingham Prognostic Index in node-negative breast cancer. *Br. J. Cancer* 85: 795–797.

Cristofanilli, M., Budd, G. T., Ellis, M. J., et al. (2004). Circulating cells, disease progression and survival in metastatic breast cancer. *N. Engl. J. Med.* 351: 781–791.

Cristofanilli, M., Hayes, D. F., Budd, G. T., et al. (2005). Circulating cells: A novel prognostic factor for newly diagnosed metastatic breast cancer. *J. Clin. Oncol.* 23: 1420–1430.

Dammann, R., Yang, G., Pfeifer, G. P. (2001). Hypermethylation of the CpG island of Ras association domain family 1A (RASSF1A), a putative suppressor gene from the 3p21.3 locus, occurs in a large percentage of human breast cancers. *Cancer Res.* 61: 3105–3109.

de Azambuja, E., Cardoso, F., de Castro, G., Jr., et al. (2007). Ki-67 as a prognostic marker in early breast cancer: A meta-analysis of published studies involving 12155 patients. *Br. J. Cancer* 96: 1504–1513.

De Paola, F., Granato, A. M., Scarpi, E., et al. (2002). Vascular endothelial growth factor and prognosis in patients with node-negative breast cancer. *Int. J. Cancer* 98: 228–233.

De Placido, S., De Laurentiis, M., Carlomagno, C., et al. (2003). Twenty-year results of the Naples GUN randomized trial: Predictive factors of adjuvant tamoxifen efficacy in early breast cancer. *Clin. Cancer Res.* 9: 1039–1046.

De Roock, W., Piessevaux, H., De, S. J., et al. (2008). KRAS wild-type state predicts survival and is associated to early radiological response in metastatic colorectal cancer treated with cetuximab. *Ann. Oncol.* 19: 508–515.

Decker, D. A., Morris, L. W., Levine, A. J., et al. (1993). Multi-drug resistance phenotype: A potential marker of chemotherapy resistance in breast cancer. *Lab. Med.* 24: 574–578.

Deming, S. L., Nass, S. J., Dickson, R. B., et al. (2000). C-myc amplification in breast cancer: A meta-analysis of its occurrence and prognostic relevance. *Br. J. Cancer* 83: 1688–1695.

Dhawan, D., Padh, H. (2009). Pharmacogenetics: Technologies to detect copy number variations. *Curr. Opin. Mol. Ther.* 11: 670–680.

Di Fiore, F., Blanchard, F., Charbonnier, F., et al. (2007). Clinical relevance of KRAS mutation detection in metastatic colorectal cancer treated by cetuximab plus chemotherapy. *Br. J. Cancer* 96: 1166–1169.

Di-Fiore, P. P., Pierce, J. H., Kraus, M. H., et al. (1987). ErbB-2 is a potent oncogene when overexpressed in NIH/3T3 cells. *Science* 237: 178–182.

Dotzlaw, H., Leygue, E., Watson, P. H., et al. (1999). Estrogen receptor-beta messenger RNA expression in human breast biopsies: Relationship to steroid receptor status and regulation by progestins. *Cancer Res.* 59: 529–532.

Dowsett, M., Harper-Wynne, C., Boeddinghaus, I., et al. (2001). HER-2 amplification impedes the antiproliferative effects of hormone therapy in estrogen receptor: Positive primary breast cancer. *Cancer Res.* 61: 8452–8458.

Drews, J. (1996). Genomic sciences and the medicine of tomorrow. *Nat. Biotechnol.* 14: 1516–1518.

Drews, J. (2000). Drug discovery: A historical perspective. *Science* 287: 1960–1964.

Duffy, M. J. (2002). Urokinase plasminogen activator and its inhibitor, PAI-1, as prognostic markers in breast cancer: From pilot to level 1 evidence studies. *Clin. Chem.* 48: 1194–1197.

Duffy, M. J. (2005). Predictive markers in breast and other cancers: A review. *Clin. Chem.* 51: 494–503.

Egeblad, M., Werb, Z. (2002). New functions for the matrix metalloproteinases in cancer progression. *Nat. Rev. Cancer* 2: 161–174.

Elledge, R. M., Green, S., Ciocca, D., et al. (1998). HER-2 expression and response to tamoxifen in estrogen receptor-positive breast cancer: A Southwest Oncology Group Study. *Clin. Cancer Res.* 4: 7–12.

Ellis, M. J., Coop, A., Singh, B., et al. (2001). Letrozole is more effective neoadjuvant endocrine therapy than tamoxifen for ErbB-1- and/or ErbB-2-positive, estrogen receptor-positive primary breast cancer: Evidence from a Phase III randomized trial. *J. Clin. Oncol.* 19: 3808–3816.

El-Naggar, A. K., Hurr, K., Huff, V., et al. (1996). Microsatellite instability in preinvasive and invasive head and neck squamous carcinoma. *Am. J. Pathol.* 148: 2067–2072.

Esteller, M., Sparks, A., Toyota, M., et al (2000). Analysis of adenomatous polyposis coli promoter hypermethylation in human cancer. *Cancer Res.* 60: 4366–4371.

Esteva, F. J., Pusztai, L., Symmans, W. F., et al. (2000). Clinical relevance of HER-2 amplification and overexpression in human cancers. *Ref. Gynecol. Obst.* 7: 267–276.

Esteva-Lorenzo, F. J., Sastry, L., King, C. R. (1998). The *erbB-2* gene: From research to application. In Dickson, R. B., Saloman, D. S. (Eds.), *Hormones and Growth Factors in Development and Neoplasia* (pp. 421–444). New York: John Wiley & Sons.

Fiehn, O. (2002). Metabolomics: The link between genotypes and phenotypes. *Plant Mol. Biol.* 48: 155–171.

Finlin, B. S., Gau, C. L., Murphy, G. A., et al. (2001). RERG is a novel ras-related, estrogen-regulated and growth-inhibitory gene in breast cancer. *J. Biol. Chem.* 276: 42259–42267.

Finn, W. G. (2007). Diagnostic pathology and laboratory medicine in the age of "omics": A paper from the 2006 William Beaumont Hospital Symposium on Molecular Pathology. *J. Mol. Diagn.* 9: 431–436.

Flower, D. R. (1999). Modelling G-protein-coupled receptors for drug design. *Biochim. Biophys. Acta* 1422: 207–234.

Foekens, J. A., Peters, H. A., Grebenchtchikov, N., et al. (2001). High levels of vascular endothelial growth factor predict poor response to systemic therapy in advanced breast cancer. *Cancer Res.* 61: 5407–5414.

Food and Drug Administration (2006). *FDA Critical Path Opportunities List.* http://www.fda.gov/downloads/scienceResearch/SpecialTopics/CriticalPathinitiative/CriticalPathOpportunitiesReports/UCM077258.pdf.

Food and Drug Administration (2010) *Systems Toxicology.* http://www.fda.gov/AboutFDA/CentersOffices/NCTR/WhatWeDo/ResearchDivisions/ucm079059.htm.

Food and Drug Administration (2011). *Critical Path Initiative.* http://www.fda.gov/oc/initiatives/criticalpath/reports/opp_list.pdf.

Freeman, D., Juan, T., Meropol, N. J., et al. (2007). Association of somatic KRAS gene mutations and clinical outcome in patients (pts) with metastatic colorectal cancer (mCRC) receiving panitumumab monotherapy. *Eur. J. Cancer Suppl.* 5: 239 (abstr O3014).

Fuqua, S. A., Schiff, R., Parra, I., et al. (1999). Expression of wild-type estrogen receptor beta and variant forms in human breast cancer. *Cancer Res.* 59: 5425–5428.

Gastl, G., Spizzo, G., Obrist, P., et al. (2000). Ep-CAM overexpression in breast cancer as a predictor of survival. *Lancet* 356: 1981–1982.

Gillison, M. L., Shah, K. V. (2003). Role of mucosal human papillomavirus in nongenital cancers. *J. Natl. Cancer. Inst. Monogr.* 31: 57–65.

Gillison, M. L., Koch, W. M., Capone, R. B., et al. (2000). Evidence for a causal association between human papillomavirus and a subset of head and neck cancers. *J. Natl. Cancer Inst.* 92: 709–720.

Gohring, U. J., Bersch, A., Becker, M., et al. (2001). P21 (waf) correlates with DNA replication but not with prognosis in invasive breast cancer. *J. Clin. Pathol.* 54: 866–870.

Goodsaid, F., Armur, S., Aubrecht, J., et al. (2010). Voluntary exploratory data submissions to the US FDA and the EMA: Experience and impact. *Nat. Rev. Drug Discov.* 9: 435–445.

Goodsaid, F., Frueh, F. W. (2007). Implementing the U.S. FDA guidance on pharmacogenomic data submissions. *Environ. Mol. Mutagen.* 48: 354–358.

Gottschalk, S., Anderson, N., Hainz, C., et al. (2004). Imatinib (STI571)-mediated changes in glucose metabolism in human leukemis BCR-ABL-positive cells. *Clin. Cancer Res.* 10: 6661–6668.

Grandis, J., Melhem, M., Barnes, E., et al. (1996). Quantitative immunohistochemical analysis of transforming growth factor-alpha and epidermal growth factor receptor in patients with squamous cell carcinoma of the head and neck. *Cancer* 78: 1284–1292.

Grandis, J. R., Tweardy, D. J. (1993). TGF-alpha and EGFR in head and neck cancer. *J. Cell Biochem.* 17F (suppl.): 188–191.

Griffin, J. L., Shockcor, J. P. (2004). Metabolic profiles of cancer cells. *Nat. Rev. Cancer* 4: 551–561.

Gruvberger, S., Ringnér, M., Chen, Y., et al. (2001). Estrogen receptor status in breast cancer is associated with remarkably distinct gene expression patterns. *Cancer Res.* 61: 5979–5984.

Gryfe, R., Gallinger, S. (2001). Microsatellite instability, mismatch repair deficiency, and colorectal cancer. *Surgery* 130: 17–20.

Guy, C. T., Webster, M. A., Schaller, M., et al. (1992). Expression of the neu protooncogene in the mammary epithelium of transgenic mice induces metastatic disease. *Proc. Natl. Acad. Sci. U.S.A.* 89: 10578–10582.

Hamacher, M., Herberg, F., Ueffing, M., et al. (2008). Seven successful years of omics research: The Human Brain Proteome Project within the National German Research Network (NGFN). *Proteomics* 8: 1116–1117.

Handra-Luca, A., Hernandez, J., Mountzios, G., et al. (2007). Excision repair cross complementation group 1 immunohistochemical expression predicts objective response and cancer-specific survival in patients treated by Cisplatin-based induction chemotherapy for locally advanced head and neck squamous cell carcinoma. *Clin. Cancer Res.* 13: 3855–3859.

Harbeck, N., Kates, R. E., Schmitt, M. (2002a). Clinical relevance of invasion factors urokinase-type plasminogen activator and plasminogen activator inhibitor type 1 for individualized therapy decisions in primary breast cancer is greatest when used in combination. *J. Clin. Oncol.* 20: 1000–1007.

Harbeck, N., Schmitt, M., Kates, R. E., et al. (2002b). Clinical utility of urokinase-type plasminogen activator and plasminogen activator inhibitor-1 determination in primary breast cancer tissues for individualized therapy concepts. *Clin. Breast Cancer* 3: 196–200.

Hasegawa, M., Nelson, H. H., Peters, E., et al. (2002). Patterns of gene promoter methylation in squamous cell cancer of the head and neck. *Oncogene* 21: 4231–4236.

Hasmann, M., Schemainda, I. (2003). FK866, a highly specific noncompetitive inhibitor of nicotinamide phosphoribosyltransferase, represents a novel mechanism for induction of cell apoptosis. *Cancer Res.* 63: 7436–7442.

Hassett, M. J., O'Malley, A. J., Pakes, J. R., et al. (2006). Frequency and cost of chemotherapy-related serious adverse effects in a population sample of women with breast cancer. *J. Natl. Cancer Inst.* 98: 1108–1117.

Hayes, D. F. (1996). Serum (circulating) markers for breast cancer. *Recent Results Cancer Res.* 140: 101–113.

He, Y. D. (2006). Genomic approach to biomarker identification and its recent applications. *Cancer Biomark* 2: 103–113.

Herbert, B. S., Wright, W. E., Shay, J. W. (2001). Telomerase and breast cancer. *Breast Cancer Res.* 3: 146–149.

Herrero, R., Castellsague, X., Pawlita, M., et al. (2003). Human papillomavirus and oral cancer: The International Agency for Research on Cancer multicenter study. *J. Natl. Cancer Inst.* 95: 1772–1783.

Hiranuma, C., Kawakami, K., Oyama, K., et al. (2004). Hypermethylation of the MYOD1 gene is a novel prognostic factor in patients with colorectal cancer. *Int. J. Mol. Med.* 13: 413–417.

Hong, H., Goodsaid, F., Shi, L., Tong, W. (2010). Molecular biomarkers: A US FDA effort. *Biomarkers Med.* 4: 215–225.

Hong, S. D., Hong, S. P., Lee, J. I., Lim, C. Y. (2000). Expression of matrix metalloproteinase-2 and -9 in oral squamous cell carcinomas with regard to the metastatic potential. *Oral Oncol.* 36: 207–213.

Hood, L., Heath, J. R., Phelps, M. E., et al (2004). Systems biology and new technologies enable predictive and preventative medicine. *Science* 306: 640–643.

Hopkins A. L., Groom C. R. (2002). The druggable genome. *Nat. Rev. Drug Discov.* 1: 727–730.

Hu, X. C., Chow, L. W. (2000). Detection of circulating breast cancer cells by reverse transcriptase-polymerase chain reaction (RT-PCR). *Eur. J. Surg. Oncol.* 26: 530–535.

Hu, Y., Kaplow, J., He, Y. (2005). From trasitional biomarkers to transcriptome analysis in drug development. *Curr. Mol. Med.* 5: 29–38.

Ingelman-Sundberg M. (2008). Pharmacogenomic biomarkers for prediction of severe adverse drug reactions. *N. Engl. J. Med.* 358: 637–639.

Innocenti, F., Vokes, E. E., Ratain, M. J. (2006). Irinogenetics: What is the right star? *J. Clin. Oncol.* 24: 2221–2224.

Isola, J. J., Kallioniemi, O.-P., Chu, L. W., et al. (1995). Genetic aberrations detected by comparative genomic hybridization predict outcome in node-negative breast cancer. *Am. J. Pathol.* 147: 905–911.

Iyer, L., Das, S., Janisch, L., et al. (2002). UGT1A1*28 polymorphism as a determinant of irinotecan disposition and toxicity. *Pharmacogenomics J.* 2: 43–47.

Iyer, L., Hall, D., Das, S., et al. (1999). Phenotype-genotype correlation of in vitro SN-38 (active metabolite of irrinotecan) and bilirubin glucuronidation in human liver tissue with UGT1A1 promoter polymorphism. *Clin. Pharmacol. Ther.* 65: 576–582.

Jardines, L., Weiss, M., Fowble, B., et al. (1993). Neu (c-erbB-2/HER2) and the epidermal growth factor receptor (EGFR) in breast cancer. *Pathobiology* 61: 268–282.

Johann, D. J., Jr., McGuigan, M. D., Patel, A. R., et al. (2004). Clinical proteomics and biomarker discovery. *Ann. N.Y. Acad. Sci.* 1022: 295–305.

Johnston, P. G., Lenz, H. J., Leichman, C. G., et al. (1995). Thymidylate synthase gene and protein expression correlate and are associated with response to 5-fluorouracil in human colorectal and gastric tumors. *Cancer Res.* 55: 1407–1412.

Jones, C. D., Yeung, C., Zehnder, J. L. (2003). Comprehensive validation of a real-time quantitative bcr-abl assay for clinical laboratory use. *Am. J. Clin. Pathol.* 120: 42–48.

Keusch, G. T. (2006). What do -omics mean for the science and policy of the nutritional sciences? *Am. J. Clin. Nutr.* 83: S520–S522.

Khambata-Ford, S., Garrett, C. R., Meropol, N. J., et al. (2007). Expression of epiregulin and amphiregulin and K-ras mutation status predict disease control in metastatic colorectal cancer patients treated with cetuximab. *J. Clin. Oncol.* 25: 3230–3237.

Kimura, M., Koida, T., Yanagita, Y., et al. (2003). A study on telomerase activity and prognosis in breast cancer. *Medical Oncol.* 20: 117–126.

King, C. R., Kraus, M. H., Williams, L. T., et al. (1985). Human cell lines with EGF receptor gene amplification in the absence of aberrant sized mRNAs. *Nucleic Acids Res.* 13: 8477–8486.

Kinoshita, J., Kitamura, K., Kabashima, A., et al. (2001). Clinical significance of vascular endothelial growth factor-C (VEGF-C) in breast cancer. *Breast Cancer Res. Treat.* 66: 159–164.

Klijn, J. G., Berns, P. M., Schmitz, P. I., et al. (1992). The clinical significance of epidermal growth factor receptor (EGF-R) in human breast cancer: A review on 5232 patients. *Endocr. Rev.* 12: 3–17.

Kuiper, G. G., Enmark, E., Pelto-Huikko, M., et al. (1996). Cloning of a novel receptor expressed in rat prostate and ovary. *Proc. Natl. Acad. Sci. U.S.A.* 93: 5925–5930.

Kvalheim, G. (1996). Detection of occult tumor cells in bone marrow and blood in breast cancer patients: Methods and clinical significance. *Acta Oncol.* 35 (Suppl. 8): 13–18.

Kyomoto, R., Kumazawa, H., Toda, Y., et al. (1997). Cyclin-D1-gene amplification is a more potent prognostic factor than its protein over-expression in human head-and-neck squamous-cell carcinoma. *Int. J. Cancer* 74: 576–581.

Lander, E. S., Linton, L. M., Birren, B., et al. (2001). The International Human Genome Sequencing Consortium: Initial sequencing and analysis of the human genome. *Nature* 409: 860–921.

Langenskiöld, M., Holmdahl, L., Angenete, E., et al. (2009). Differential prognostic impact of uPA and PAI-1 in colon and rectal cancer. *Biology* 30: 210–220.

Lanza, G., Matteuzzi, M., Gafa, R., et al. (1998). Chromosome 18q allelic loss and prognosis in stage II and III colon cancer. *Int. J. Cancer* 79: 390–395.

Lau, R., Grimson, R., Sansome, C., et al. (2001). Low levels of cell cycle inhibitor p27kip1 combined with high levels of Ki-67 predict shortened disease-free survival in T1 and T2 invasive breast carcinomas. *Int. J. Oncol.* 18: 17–23.

Lee, K. H., Yim, E. K., Kim, C. J., et al. (2005). Proteomic analysis of anti-cancer effects by paclitaxel treatment in cervical cancer cells. *Gynecol. Oncol.* 98: 45–53.

Leivonen, M., Nordling, S., Lundin, J., et al. (2001). p27 expression correlates with short-term but not with long-term prognosis in breast cancer. *Breast Cancer Res. Treat.* 6: 15–22.

Lemberger, T. (2007). Systems biology in human health and disease. *Mol. Syst. Biol.* 3: 136.

Lenz, H. J., Hayashi, K., Salonga, D., et al. (1998). p53 point mutations and thymidylate synthase messenger RNA levels in disseminated colorectal cancer: An analysis of response and survival. *Clin. Cancer Res.* 4: 1243–1250.

Li, J., Zhang, Z., Rosenzweig, J., et al. (2002). Proteomics and bioinformatics approaches for identification of serum biomarkers to detect breast cancer. *Clin. Chem.* 48: 1296–1304.

Li, W., Thompson, C. H., O'Brien, C. J., et al. (2003). Human papillomavirus positivity predicts favourable outcome for squamous cell carcinoma of the tonsil. *Int. J. Cancer* 106: 553–558.

Li, Y., St John, M. A., Zhou, X., et al. (2004). Salivary transcriptome diagnostics for oral cancer detection. *Clin. Cancer Res.* 10: 8442–8450.

Lievre, A., Bachet, J. B., Boige, V., et al. (2008). KRAS mutations as an independent prognostic factor in patients with advanced colorectal cancer treated with cetuximab. *J. Clin. Oncol.* 26: 374–379.

Lievre, A., Bachet, J. B., Le Corre, D., et al. (2006). KRAS mutation status is predictive of response to cetuximab therapy in colorectal cancer. *Cancer Res.* 66: 3992–3995.

Ligibel, J. A., Winer, E. P. (2002). Trastuzumab/chemotherapy combinations in metastatic breast cancer. *Semin. Oncol.* 29: 38–43.

Lim, S. B., Jeong, S. Y., Lee, M. R., et al. (2004). Prognostic significance of microsatellite instability in sporadic colorectal cancer. *Int. J. Colorectal Dis.* 19: 533–537.

Linderholm, B. K., Lindahl, T., Holmberg, L., et al. (2001). The expression of vascular endothelial growth factor correlates with mutant p53 and poor prognosis in human breast cancer. *Cancer Res.* 61: 2256–2260.

Liotta, L., Petricoin, E. (2000). Molecular profiling of human cancer. *Nat. Rev. Genet.* 1: 48–56.

Lyng, H., Sitter, B., Bathen, T. F., et al. (2007). Metabolic mapping by use of high-resolution magic angle spinning ^{1}H NMR spectroscopy for assessment of apoptosis in cervical carcinomas. *BMC Cancer* 7: 11.

Maass, N., Nagasaki, K., Ziebart, M., et al. (2002). Expression and regulation of suppressor gene maspin in breast cancer. *Clin. Breast Cancer* 3: 281–287.

MacConmara, M., O'Hanlon, D. M., Kiely, M. J., et al. (2002). An evaluation of the prognostic significance of vascular endothelial growth factor in node-positive primary breast carcinoma. *Int. J. Oncol.* 20: 717–721.

Maeda, K., Kawakami, K., Ishida, Y., et al. (2003). Hypermethylation of the CDKN2A gene in colorectal cancer is associated with shorter survival. *Oncol. Rep.* 10: 935–938.

Manders, P., Beex, L. V., Tjan-Heijnen, V. C., et al. (2002). The prognostic value of vascular endothelial growth factor in 574 node-negative breast cancer patients who did not receive adjuvant systemic therapy. *Br. J. Cancer* 87: 772–778.

Marioni, G., Staffieri, C., Staffieri, A., et al. (2009). MASPIN tumor-suppressing activity in head and neck squamous cell carcinoma: emerging evidence and therapeutic perspectives. *Acta Otolaryngol.* 129: 476–480.

Martins, G. B., Reis, S. R., Silva, T. M. (2003). Expressão do colágeno I em carcinomas epidermóides da cavidade oral. *Pesqui Odontol. Bras.* 17: 82–88.

Mayer, A., Takimoto, M., Fritz, E., et al. (1993). The prognostic significance of proliferating cell nuclear antigen, epidermal growth factor receptor, and mdr gene expression in colorectal cancer. *Cancer* 71: 2454–2460.

McCawley, L. J., Matrisian, L. M. (2000). Matrix metalloproteinases: Multifunctional contributors to progression. *Mol. Med. Today* 6: 149–156.

McKeage, K., Perry, C. M. (2002). Trastuzumab: A review of its use in the treatment of metastatic breast cancer overexpressing HER2. *Drugs* 62: 209–243.

Mendes, P., Kell, D. B., Westerhoff, H. V. (1992). Channelling can decrease pool size. *Eur. J. Biochem.* 204: 257–266.

Mendes, P., Kell, D. B., Westerhoff, H. V. (1996). Why and when channeling can decrease pool size at constant net flux in a simple dynamic channel. *Biochim. Biophys. Acta* 1289: 175–186.

Michalides, R. J., van Veelen, N. M., Kristel, P. M., et al. (1997). Overexpression of cyclin D1 indicates a poor prognosis in squamous cell carcinoma of the head and neck. *Arch. Otolaryngol. Head Neck Surg.* 123: 497–502.

Miyajima, Y., Nakano, R., Morimatsu, M. (1995). Analysis of expression of matrix metalloproteinases-2 and -9 in hypopharingeal squamous cell carcinoma by in situ hybridization. *Ann. Oto. Rhinol. Laryngol.* 104: 678–684.

Mizukami, Y., Nonomura, A., Takizawa, T., et al. (1995). N-myc protein expression in human breast carcinoma: Prognostic implications. *Anticancer Res.* 15: 2899–2905.

Mohsenifar, J., Almassi-Aghdam, M., Mohammed-Taheri, Z., et al. (2007). Prognostic values of proliferative markers ki-67 and repp86 in breast cancer. *Arch. Iranian Med.* 10: 27–31.

Mokbel, K., Elkak, A. (2001). Recent advances in breast cancer (the 37th ASCO meeting, May 2001). *Curr. Med. Res. Opin.* 17: 116–122.

Mokbel, K., Parris, C. N., Radbourne, R., et al. (1999). Telomerase activity and prognosis in breast cancer. *Eur. J. Surg. Oncol.* 25: 269–272.

Monni, O., Hyman, E., Mousses, S., et al. (2001). From chromosomal alterations to target genes for therapy: Integrating cytogenetic and functional genomic views of the breast cancer genome. *Semin. Cancer Biol.* 11: 395–401.

Mueller, C., Riese, U., Kosmehl, H., et al. (2002). Telomerase activity in microdissected human breast cancer tissues: Association with p53, p21 and outcome. *Int. J. Oncol.* 20: 385–390.

Muruganandham, M., Alfieri, A. A., Matei, C., et al. (2005). Metabolic signatures associated with a NAD synthesis inhibitor-induced apoptosis identified by 1H-decoupled-31P magnetic resonance spectroscopy. *Clin. Cancer Res.* 11: 3503–3513.

Muss, H. B., Thor, A. D., Berry, D. A., et al. (1994). c-erbB-2 expression and response to adjuvant therapy in women with node-positive early breast cancer. *N. Engl. J. Med.* 330: 1260–1266.

National Comprehensive Cancer Network (2008). *Clinical Practice Guidelines in Oncology: Colon Cancer. V1.2008*. http://www.nccn.org.

Negrini, M., Ferracin, M., Sabbioni, S., et al. (2007). MicroRNAs in human cancer: From research to therapy. *J. Cell Sci.* 120: 1833–1840.

Nicholson, J. K. (2006). Reviewers peering from under a pile of omics data. *Nature* 440: 992.

Nishimura, R., Nagao, K., Miyayama, H., et al. (1999). Thymidylate synthase levels as therapeutic and prognostic predictor in breast cancer. *Anticancer Res.* 19: 5621–5626.

Nohara, T., Ryo, T., Iwamoto, S., et al. (2001). Expression of cell-cycle regulator p27 is correlated to the prognosis and ER expression in breast carcinoma patients. *Oncology* 60: 94–100.

O'Brien, N., Maguire, T. M., O'Donovan, N., et al. (2002). Mammoglobin: A promising marker for breast cancer. *Clin. Chem.* 48: 1362–1364.

O'Connell, D., Roblin, D. (2006). Translational research in the pharmaceutical industry: From bench to bedside. *Drug Discov. Today* 11: 833–838.

Oh, Y. L., Choi, J. S., Song, S. Y., et al. (2001). Expression of p21Waf1, p27Kip1 and cyclin D1 proteins in breast ductal carcinoma *in situ*: Relation with clinicopathologic characteristics and with p53 expression and estrogen receptor status. *Pathol. Int.* 51: 94–99.

Okano, H., Shinohara, H., Miyamoto, A., et al. (2004). Concomitant overexpression of cyclooxygenase-2 in HER-2 positive on Smad 4-reduced human gastric carcinomas is associated with a poor patient outcome. *Clin. Cancer Res.* 10: 6938–6945.

Olaussen, K. A., Dunant, A., Fouret, P., et al. (2006). DNA repair by ERCC1 in non-small-cell lung cancer and cisplatin-based adjuvant chemotherapy. *N. Engl. J. Med.* 355: 983–991.

Ono, K., Tanaka, T., Tsunoda, T., et al. (2000). *Cancer Res.* 60: 5007–5011.

Orr, M. S., Goodsaid, F., Amur, S., et al. (2007). The experience with voluntary genomic data submissions at the FDA and a vision for the future of the voluntary data submission program. *Clin. Pharm. Ther.* 81: 294–297.

Ozer, E., Yuksel, E., Kizildag, S., et al. (2002). Microsatellite instability in early-onset breast cancer. *Pathol. Res. Pract.* 198: 525–530.

Paik, S., Bryant, J., Park, C., et al. (1998). erbB-2 and response to doxorubicin in patients with axillary lymph node-positive, hormone receptor-negative breast cancer. *J. Natl. Cancer Inst.* 90: 1361–1370.

Pandey, A., Mishra, A. (2007). Cytogenetics in head and neck cancer. *Indian J. Otolaryngol. Head Neck Surg.* 59: 317–321.

Patil, S. T., Higgs, R. E., Brandt, J. E., et al. (2007). Identifying pharmacodynamic protein markers of centrally active drugs in humans: a pilot study in a novel clinical model. *J. Proteome Res.* 3: 955–966.

Paulson, T. G., Wright, F. A., Parker, B. A., et al. (1996). Microsatellite instability correlates with reduced survival and poor disease prognosis in breast cancer. *Cancer Res.* 56: 4021–4026.

Pearlstein, R. P., Benninger, M. S., Carey, T. E., et al. (1998). Loss of 18q predicts poor survival in patients with squamous cell carcinoma of the head and neck. *Genes Chromosomes Cancer* 21: 333–339.

Popat, S., Houlston, R. S. (2005). A systematic review and meta-analysis of the relationship between chromosome 18q genotype, DCC status and colorectal cancer prognosis. *Eur. J. Cancer* 41: 2060–2070.

Popat, S., Hubner, R., Houlston, R. S. (2005). Systematic review of microsatellite instability and colorectal cancer prognosis. *J. Clin. Oncol.* 23: 609–618.

Poremba, C., Heine, B., Diallo, R., et al. (2002). Telomerase as a prognostic marker in breast cancer: High-throughput tissue microarray analysis of hTERT and hTR. *J. Pathol.* 198: 181–189.

Pusztai, L. (2004). Perspectives and challenges of clinical pharmacogenomics in cancer. *Pharmacogenomics* 5: 451–454.

Ransohoff, D. F. (2005). Bias as a threat to the validity of cancer molecular-marker research. *Nat. Rev. Cancer* 5: 142–149.

Ravdin, P. M., Green, S., Albain, K. S., et al. (1998). Initial report of the SWOG biological correlative study of C-erbB-2 expression as a predictor of outcome in a trial comparing adjuvant CAF T with tamoxifen (T) alone. *Proc. Am. Soc. Clin. Oncol.* 17: A374 (abstract).

Rhodes, D. R., Yu, J., Shanker, K., et al. (2004). Large-scale meta-analysis of cancer microarray data identifies common transcriptional profiles of neoplastic transformation and progression. *Proc. Natl. Acad. Sci. U.S.A.* 101: 9309–9314.

Rochefort, H., Chalbos, D., Cunat, S., et al. (2001). Estrogen regulated proteases and antiproteases in ovarian and breast cancer cells. *J. Steroid Biochem. Mol. Biol.* 76: 119–124.

Rocke, D. M. (2004). Design and analysis of experiments with high throughput biological assay data. *Semin. Cell Dev. Biol.* 15: 703–713.

Rosenblatt, K. P., Bryant-Greenwood, P., Killian, J. K., et al. (2004). Serum proteomics in cancer diagnosis and management. *Annu. Rev. Med.* 55: 97–112.

Ross, J. S. (1996). *DNA ploidy and cell cycle analysis in pathology* (pp. 54–55). New York: Igaku-Shoin Publishing.

Sallinen, S. L., Sallinen, P. K., Haapasalo, H. K., et al. (2000). Identification of differentially expressed genes in human gliomas by DNA microarray and tissue chip techniques. *Cancer Res.* 60: 6617–6622.

Salonga, D., Danenberg, K. D., Johnson, M., et al. (2000). Colorectal tumors responding to 5-fluorouracil have low gene expression levels of dihydropyrimidine dehydrogenase, thymidylate synthase, and thymidine phosphorylase. *Clin. Cancer Res.* 6: 1322–1327.

Sanchez-Cespedes, M., Esteller, M., Wu, L., et al. (2000). Gene promoter hypermethylation in tumors and serum of head and neck cancer patients. *Cancer Res.* 60: 892–895.

Sargent, D. J., Conley, B. A., Allegra, C., et al. (2005). Clinical trial designs for predictive marker validation in cancer treatment trials. *J. Clin. Oncol.* 23: 2020–2027.

Satta, T., Isobe, K., Yamauchi, M., et al. (1992). Expression of MDR1 and glutathione S-transferase genes and chemosensitivities in human gastrointestinal cancer. *Cancer* 69: 941–946.

Schiffman, M., Khan, M. J., Solomon, D., et al. (2005). A study of the impact of adding HPV types to cervical cancer screening and triage tests. *J. Natl. Cancer Inst.* 97: 147–150.

Scholler, N., Crawford, M., Sato, A., et al. (2006). Bead-based ELISA for validation of ovarian cancer early detection markers. *Clin. Cancer Res.* 12: 2117–2124.

Schwartz, S. R., Yueh, B., McDougall, J. K., et al. (2001). Human papillomavirus infection and survival in oral squamous cell cancer: A population based study. *Otolaryngol. Head Neck Surg.* 125: 1–9.

Seligson, D. B. (2005). The tissue micro-array as a translational research tool for biomarker profiling and validation. *Biomarkers* 10 (Suppl. 1): S77–S82.

Serkova, N., Boros, L. G. (2005). Detection of resistance to imatinib by metabolic profiling: Clinical and drug development implications. *Am. J. Pharmacogenomics* 5: 293–302.

Shawver, L. K., Slamon, D., Ullrich, A. (2002). Smart drugs: Tyrosine kinase inhibitors in cancer therapy. *Cancer Cell* 1: 117–123.

Slamon, D. J., Clark, G. M., Wong, S. G., et al. (1987). Human breast cancer: Correlation of relapse and survival with amplification of the HER-2/neu oncogene. *Science* 235: 177–182.

Smith, W. C., Sittampalam, G. S. (1998). Conceptual and statistical issues in the validation of analytic dilution assays for pharmaceutical applications. *J. Biopharm. Stat.* 8: 509–532.

Sommer, S., Fuqua, S. A. (2001). Estrogen receptor and breast cancer. *Semin. Cancer Biol.* 11: 339–352.

Sorlie, T., Perou CM., Tibshirani, R., et al. (2001). Gene expression patterns of breast carcinomas distinguish subclasses with clinical implications. *Proc. Natl. Acad. Sci. U.S.A.* 98: 10869–10874.

Sotiriou, C., Neo, S. Y., McShane, L. M., et al. (2003). Breast cancer classification and prognosis based on gene expression profiles from a population-based study. *Proc. Natl. Acad. Sci. U.S.A.* 100: 10393–10398.

Spafford, M. F., Koch, W. M., Reed, A. L., et al. (2001). Detection of head and neck squamous cell carcinoma among exfoliated oral mucosal cells by microsatellite analysis. *Clin. Cancer Res.* 7: 607–612.

Speirs, V., Kerin, M. J. (2000). Prognostic significance of oestrogen receptor beta in breast cancer. *Br. J. Surg.* 87: 405–409.

Sterne, J. A., Davey Smith, G. (2001). Sifting the evidence: What's wrong with significance tests? *Br. Med. J.* 322: 226–231.

Stransky, B., Barrera, J., Ohno-Machado, L., et al. (2007). Modeling cancer: Integration of "omics" information in dynamic systems. *J. Bioinform. Comput. Biol.* 5: 977–986.

Su, J. L., McKee, D. D., Ellis, B., et al. (2000). Production and characterization of an estrogen receptor beta subtype-specific mouse monoclonal antibody. *Hybridoma* 19: 481–487.

Tanaka, N., Odajima, T., Ogi, K., et al. (2003). Expression of E-cadherin, alpha-catenin, and beta-catenin in the process of lymph node metastasis in oral squamous cell carcinoma. *Br. J. Cancer* 89: 557–563.

Tejpar, S., De Roock, W., Biesmans, B., et al. (2008). High aphiregulin and epiregulin expression in KRAS wild-type colorectal primaries predicts response and survival benefit after treatment with cetuximab and irrinotecan for metastatic disease. *ASCO Gastrointestinal Cancers Symposium* (abstr. 411).

Thomas, G. T., Lewis, M. P., Speight, P. M. (1999). Matrix metalloproteinases and oral cancer. *Oral Oncol.* 35: 227–233.
Thor, A. D., Berry, D. A., Budman, D. R., et al. (1998). erbB-2, p53, and efficacy of adjuvant therapy in lymph node-positive breast cancer. *J. Natl. Cancer Inst.* 90: 1361–1370.
Tomita, S., Deguchi, S., Miyaguni, T., et al. (1999). Analyses of microsatellite instability and the transforming growth factor-β receptor Type II gene mutation in sporadic breast cancer and their correlation with clinicopathological features. *Breast Cancer Res. Treat.* 53: 33–39.
Tsao, M. S., Sakurada, A., Cutz, J. C., et al. (2005). Erlotinib in lung cancer: Molecular and clinical predictors of outcome. *N. Engl. J. Med.* 353: 133–144.
Umekita, Y., Ohi, Y., Sagara, Y., et al. (2000). Co-expression of epidermal growth factor receptor and transforming growth factor-α predicts worse prognosis in breast cancer patients. *Int. J. Cancer* 89: 484–487.
Umetani, N., Takeuchi, H., Fujimoto, A., et al. (2004). Epigenetic inactivation of ID4 in colorectal carcinomas correlates with poor differentiation and unfavourable prognosis. *Clin. Cancer Res.* 10: 7475–7483.
van de Vijver, M. J., He, Y. D., van't Veer, L. J., et al. (2002). A gene-expression signature as a predictor of survival in breast cancer. *N. Engl. J. Med.* 347: 1999–2009.
van der Merwe, D. E., Oikonomopoulou, K., Marshall, J., et al. (2007). Mass spectrometry: Uncovering the cancer proteome for diagnostics. *Adv. Cancer Res.* 96: 23–50.
Van der Riet, P., Nawroz, H., Hruban, R. H., et al. (1994). Frequent loss of chromosome 9p21–22 early in head and neck cancer progression. *Cancer Res.* 54: 1156–1158.
van Kuilenberg, A. B. (2004). Dihydropyrimidine dehydrogenase and the efficacy and toxicity of 5-fluorouracil. *Eur. J. Cancer* 40: 939–950.
van't Veer, L. J., Dai, H., van de Vijver, M. J., et al. (2002). Gene expression profiling predicts clinical outcome of breast cancer. *Nature* 415: 530–536.
Venter, J. C., Adams, M. D., Myers, E. W., et al. (2001). The sequence of the human genome. *Science* 291: 1304–1351.
Visscher, D. W., Sarkar, F., LoRusso, P., et al. (1993). Immunohistologic evaluation on invasion-associated proteases in breast carcinoma. *Mod. Pathol.* 6: 302–306.
Volm, M., Rittgen, W., Drings, P. (1998). Prognostic value of ERBB-1, VEGF, cyclin A, FOS, JUN and MYC in patients with squamous cell lung carcinomas. *Br. J. Cancer* 77: 663–669.
Ward, R. L., Cheong, K., Ku, S. L., et al. (2003). Adverse prognostic effect of methylation in colorectal cancer is reversed by microsatellite instability. *J. Clin. Oncol.* 21: 3729–3736.
Weckwerth, W. (2003). Metabolomics in systems biology. *Annu. Rev. Plant Biol.* 54: 669–689.
Weinberger, P. M., Yu, Z., Haffty, B. G., et al. (2006). Molecular classification identifies a subset of human papillomavirus-associated oropharyngeal cancers with favorable prognosis. *J. Clin. Oncol.* 24: 736–747.
Weinstat-Saslow, D., Merino, M. J., Manrow, R. E., et al. (1995). Overexpression of cyclin D mRNA distinguishes invasive and in situ breast carcinomas from non-malignant lesions. *Nat. Med.* 1: 1257–1260.
West, M., Blanchette, C., Dressman, H., et al. (2001). Predicting the clinical status of human breast cancer by using gene expression profiles. *Proc. Natl. Acad. Sci. U.S.A.* 98: 11462–11467.
Wishart, D. S. (2007). Proteomics and the human metabolome project. *Expert Rev. Proteomics* 4: 333–335.
Wishart, D. S., Knox, C., Guo, A., et al. (2006). DrugBank: A comprehensive resource for *in silico* drug discovery and exploration. *Nucleic Acids Res.* 34: D668–D672.
Wolman, S. R., Pauley, R. J., Mohamed, A. N., et al. (1992). Genetic markers as prognostic indicators in breast cancer. *Cancer* 70: 1765–1774.
Wood, W. C., Budman, D. R., Korzun, A. H., et al. (1994). Dose and dose intensity of adjuvant chemotherapy for stage II, node-positive breast carcinoma. *N. Engl. J. Med.* 330: 1253–1259.
Zehenter, B. K., Carter, D. (2004). Mammaglobin: A candidate diagnostic marker for breast cancer. *Clin. Biochem.* 37: 249–257.
Zhang, X. W. (2007). Biomarker validation: Movement towards personalized medicine. *Expert Rev. Mol. Diagn.* 7: 469–471.
Zhang, X. W., Li, L., Wei, D., et al. (2007). Moving cancer diagnostics from bench to bedside. *Trends Biotechnol.* 25: 166–173.
Zhou, X., Tan, M., Stone Hawthorne, V., et al. (2004). Activation of the Akt/mammalian target of rapamycin/4E-BP1 pathway by ErbB2 overexpression predicts progression in breast cancers. *Clin. Cancer Res.* 10: 6779–6788.

19 Recent Advances in MicroRNA Expression Profiling towards a Molecular Anatomy of Tumorigenesis and Applications for Diagnosis, Prognosis, and Therapeutics

Hiroaki Ohdaira and Kenichi Yoshida
Meiji University
Kawasaki, Japan

CONTENTS

19.1 INTRODUCTION

The number of mature microRNAs (miRNAs) in the human genome is estimated to be around several hundred. This small group of molecules has a major impact on gene regulatory networks because they can target a thousand overlapping messenger RNAs (mRNAs), or roughly one-third of all human mRNAs. The degradation or translational suppression of mRNAs is induced based on the degree of complementarity between an miRNA and its target mRNAs (Table 19.1). Research on miRNAs, which compose small noncoding fragments of RNA, has focused on the unique roles of these molecules in tumorigenesis in humans, although the importance of miRNAs,

TABLE 19.1
A Wide Variety of RNAs

Classical RNAs	mRNA, tRNA, rRNA
Small noncoding RNAs	snRNA, snoRNA, miRNA, etc.

Classical RNAs are functionally involved in the translational step. Noncoding RNAs (ncRNAs) are not translated in proteins. In this context, tRNA and rRNA are characterized as ncRNAs. Among ncRNAs, small nuclear RNA (snRNA), small nucleolor RNA (snoRNA), and miRNA are specially termed as small ncRNAs based on their sizes. miRNAs uniquely form complementary base pairs with target mRNAs and suppress translation from target mRNAs.

specifically *lin-4* and *let-7* encoding 22- and 21-nucleotide RNAs, respectively, was originally implicated in the timing of stem cell division and differentiation in the nematode *Caenorhabditis elegans* (Lagos-Quintana et al., 2001; Lau et al., 2001; Lee and Ambros, 2001). Subsequently, *let-7* has been shown to act as a tumor suppressor, because the down-regulation of *let-7* resulted in a tumor-prone phenotype, a less-differentiated state of cancer cells, and a poor prognosis of tumor patients (Büssing et al., 2008; Jérôme et al., 2007; Shell et al., 2007). The finding that the *mir-17-92* cluster is upregulated in B cell lymphoma, compared with normal tissue, strongly suggesting that miRNA may be a potential human oncogene (He et al., 2005). A relationship between miRNA and cancer has also been suggested by the fact that about half of the miRNA genes are located in cancer-associated genomic regions or in fragile sites of the genome, and several miRNAs located in deleted regions have been shown to be down-regulated in cancer samples (Calin et al., 2004). Accumulating evidence supports the notion that miRNAs have critical roles that contribute to tumor formation including onset, progression, invasion, migration, and metastasis as oncogenes or tumor-suppressor genes (Table 19.2), and miRNAs are expected to be useful as biomarkers for the diagnosis, prognosis, and treatment of cancer in the near future (Ahmed, 2007; Bandres et al., 2007; Baranwal and Alahari, 2010; Barbarotto et al., 2008; Bartels and Tsongalis, 2009; Blenkiron and Miska, 2007; Calin and Croce, 2006a, 2006b; Croce, 2009; Cummins and Velculescu, 2006; Deng et al., 2008; Dillhoff et al., 2009; Drakaki and Iliopoulos, 2009; Esquela-Kerscher and Slack, 2006; Fabbri et al., 2008; Garzon et al., 2006, 2009; Gusev and Brackett, 2007; Hagan and Croce, 2007; Hammond, 2006; Hernando, 2007; Hwang and Mendell, 2007; Iorio and Croce, 2009; Jay et al., 2007; Kent and Mendell, 2006; Lee and Dutta, 2006, 2009; Lotterman et al., 2008; Lu et al.,

TABLE 19.2
Representative miRNAs as Oncogene and Tumor-Suppressor Gene

Oncogene	*mir-17-92*	Overexpressed in cancers, including B-cell lymphoma (He et al., 2005; Olive et al., 2009), acute leukemia (Mi et al., 2010), renal-cell carcinoma (Chow et al., 2010), hepatocellular carcinoma (Connolly et al., 2008), and lung cancer (Hayashita et al., 2005)	1. Functionally associated with c-Myc (He et al., 2005) 2. Downregulation by p53 (Yan et al., 2009) 3. Autofeedback loop with activator E2Fs (Sylvestre et al., 2007; Woods et al., 2007)
Tumor-suppressor gene	*let-7*	Underexpressed in cancers, including retinoblastoma (Mu et al., 2010) and lung cancer (Takamizawa et al., 2004)	*let-7* targets HMG2A, Ras, Myc, E2F2, and cyclin D (Dong et al., 2010; Johnson et al., 2005; Lee and Dutta, 2007; Sampson et al., 2007)

2005; Ma and Weinberg, 2008; Medina and Slack, 2008; Metias et al., 2009; Mirnezami et al., 2009; Mocellin et al., 2009; Negrini et al., 2007, 2009; Nelson and Weiss, 2008; Osaki et al., 2008; Rossi et al., 2008; Ruan et al., 2009; Sassen et al., 2008; Schmittgen, 2008; Shenouda and Alahari, 2009; Shi et al., 2008b; Sioud and Cekaite, 2008; Slack and Weidhaas, 2006; Stahlhut Espinosa and Slack, 2006; Stefani, 2007; Tili et al., 2007; Trang et al., 2008; Tong and Nemunaitis, 2008; Vandenboom et al., 2008; Ventura and Jacks, 2009; Visone and Croce, 2009; Voorhoeve, 2010; Voorhoeve and Agami, 2007; Waldman and Terzic, 2009; Wang and Lee, 2009; Wang and Wu, 2009; Wiemer, 2007; Wijnhoven et al., 2007; Winter et al., 2009; Wu et al., 2007; Yu et al., 2007, Zhang et al., 2006, 2007).

Until recently, miRNA-expression profiling has lagged behind mRNA expression profiling. However, evaluations of the importance of miRNA signatures as molecular classifiers in various human tumors have become a timely issue. In general, tumors often display remarkable heterogeneity in their clinical behavior, ranging from spontaneous regression to rapid progression and resistance to therapy. The clinical behavior of these tumors is associated with many factors, including patient age, histopathology, and genetic abnormalities. Therefore, understanding the nature of cancer by linking such data from clinical specimens with miRNA signatures is an intriguing approach. Importantly, the etiology and clinical significance of miRNA alterations remain questionable. In the present chapter, the miRNA-expression profiles of clinical tumors will be introduced, with particular focus on the miRNA microarray-based approach and not the real-time reverse transcription-polymerase chain reaction (RT-PCR) candidate approach that dominated research at the beginning of the study of miRNA expressional regulation; this information will be used to search for specifically regulated miRNAs that may exist in clinical tumor samples. Later, the current status of miRNA for potential applications in the early diagnosis of cancers and pharmaceutical drug development will be discussed. To prove a functional relationship between miRNA and tumorigenesis, the study of cell lines or mouse models will be critical for establishing solid evidence required for the further development of miRNA-based therapeutics.

19.2 miRNA SIGNATURES IN TUMORS

The identification of activated and/or suppressed miRNAs in various tumors is a priority for current cancer research. Expressional changes of mature and/or precursor miRNA transcripts (Table 19.3) found in tumors are thought to be caused by mutations, deletions, or amplifications, the methylation of miRNAs and the deregulation of upstream transcription factors, based on the literature mentioned in the above section. Working towards a comprehensive understanding of the status of miRNAs in tumors, knowledge of the miRNA-expression profiles of various tumor specimens has been rapidly expanding and is more popular than ever. The miRNome, which is defined as the full complement of miRNAs in a genome, has enabled us to design miRNA microarrays for

TABLE 19.3
miRNA Biogenesis

Primary miRNA (pri-miRNA)
Long nucleotide sequences containing hairpin-shaped stem-loop secondary structure transcribed by RNA polymerase II or III and processed by Drosha and its cofactor DGCR8/Pasha
Precursor miRNA (pre-miRNA)
Approximately 60–70-nucleotide length of stem-loop structure, transported into cytoplasm by Exportin-5 and processed by Dicer into short double-stranded RNA
Mature miRNA
Approximately 20–25-nucleotide length of single-stranded RNA unwounded by helicase and incorporated into RNA-induced silencing complex

predicting miRNA signatures in tumorigenesis. Systematic evaluations of the evidence provided by the miRNA-expression profiles of clinically derived tumor specimens and outlooks on the importance of miRNAs as early diagnostic biomarkers are now timely topics. The dismal outcome of cancer patients is mainly due to a lack of reliable prognostic markers. Therefore, miRNA signatures are also expected to be of great use as reliable prognostic markers. In addition, miRNA signatures can potentially be applied to personalized medicine, such as the monitoring of anticancer drug sensitivity, and miRNA-based biomarkers are very attractive for this purpose because samples containing small RNA can be easily obtained from a tiny biopsy, and a similar method can be used to deliver small RNA into living individuals.

19.2.1 Colorectal and Gastric Cancers

Every year, more than one million people in the world suffer from colorectal cancer, and this disease represents the second leading cause of death from cancer in the United States. The traditional staging system for colorectal cancer is not perfect. Thus, the effectiveness of gene-expression profiles as predictors for aiding therapeutic decision making has been tested. mRNA-expression profiles are expected to be useful for the prognosis of patients with stage II colorectal cancer, but further studies are needed to identify and validate specific gene signatures to maximize the effectiveness of the prognosis of a poor outcome in stage II colorectal cancer patients (Lu et al., 2009). To add further value to gene-expression profiles, miRNA signatures are expected to play an important role, if they can be coupled with mRNA signatures. Indeed, the miRNA expression analysis of colorectal cancer is expected to present novel therapeutic targets and to predict the prognosis and support the diagnosis of poorly differentiated tumors (Aslam et al., 2009; Faber et al., 2009; Monzo et al., 2008; Slaby et al., 2009; Tang and Fang, 2009; Valeri et al., 2009; Yang et al., 2009). miRNA signatures are also predicted to serve as potent biomarkers for gastrointestinal tumors (Bhatti et al., 2009). Based on miRNA profiling, the deregulated expression of numerous miRNAs in colorectal cancer has been linked to a functional effect on tumor cell malignancy (Schepeler et al., 2008). miRNA expression patterns are coordinately altered in colon adenocarcinomas, and the upregulation of *miR-21* has been shown to be associated with a poor survival and poor therapeutic outcome (Schetter et al., 2008). A comparison between colonic cancer tissues without lymph-node metastasis and neighboring noncancerous areas revealed the upregulation of *miR-106b*, *miR-135b*, *miR-18a*, *miR-18b*, *miR-196b*, *miR-19a*, *miR-224*, *miR-335*, *miR-424*, *miR-20a**, *miR-301b*, and *miR-374a* and the downregulation of *miR-378* and *miR-378** (Wang et al., 2010). When colon tumors were compared with normal colon tissues, the colon tumors exhibited the differential expression of miRNAs depending on the mismatch repair status of the cancer specimens (Sarver et al., 2009). Based on other experiments, *miR-31*, *miR-183*, *miR-17-5p*, *miR-18a*, *miR-20a*, and *miR-92* were shown to be upregulated, and *miR-143* and *miR-145* were shown to be downregulated in cancer tissues other than normal tissues (Motoyama et al., 2009). miRNA expression analysis revealed that *miR-92* is significantly elevated in the plasma of colorectal cancer patients and thus may be a potential noninvasive molecular marker for colorectal cancer screening (Ng et al., 2009).

Gastric cancer is one of the top five most common cancers, and infection with *Helicobacter pylori* is highly suspected of being involved in the pathogenesis of stomach cancer. Comparisons between nontumor mucosa and histological subtypes of gastric cancer have revealed that miRNA signatures can even distinguish histological subtypes of gastric cancer, such as diffuse- and intestinal-type cancer. *miR-125b*, *miR-199a*, and *miR-100* have been identified as the most important miRNAs involved in the progression-related signature (Ueda et al., 2009). The first report on an miRNA expression analysis in gastric cancer showed that *miR-139-5p*, *miR-497*, and *miR-768-3p* were highly expressed in nontumor tissues, whereas *miR-340**, *miR-421*, and *miR-658* were highly expressed in tumors (Guo et al., 2009). *miR-421* is a promising early diagnosis marker of gastric cancer; however, the expression level of *miR-421* is not perfectly correlated with clinicopathological features (Jiang et al., 2010). On the other hand, *miR-9* and *miR-433* are downregulated in gastric carcinoma (Luo et al., 2009). Taken together, the miRNA signatures in colorectal and gastric tumors strongly support the idea that tumors are too heterogeneous to select commonly regulated miRNAs with the exception of a few miRNAs that are commonly regulated in different clinical samples. Thus, the unique miRNA signature of each specimen might be applicable to the personalization of prognosis and treatment options in the near future.

19.2.2 Hepatocellular Carcinomas

Mainly because of the endemic status of chronic hepatitis B and C viruses, the prevalence of hepatocellular carcinoma (HCC) is higher in Asia than in other regions of the world. HCC is the third most common cause of cancer-related death worldwide. Viral infection causes chronic injury to the liver, with subsequent progression to severe fibrosis and cirrhosis and, ultimately, the development of an increased risk of HCC. HCC is an aggressive cancer with a dismal outcome largely because of metastasis and postsurgical recurrence; it is also a target cancer for which miRNAs are expected to be useful as diagnostic and therapeutic molecules (Gramantieri et al., 2008; Varnholt, 2008). *miR-222* has been reported to be a commonly upregulated miRNA in HCC, compared with adjacent cirrhotic liver tissue (Wong et al., 2010). The miRNA profiling of full-fledged HCC in both normal and cirrhotic tissue identified a set of miRNA signatures in which *miR-221* and *miR-222* were the most strongly upregulated miRNAs in the tumor samples (Pineau et al., 2010). A comparison of HCC and adjacent nontumor tissue using miRNA profiling identified the downregulation of *miR-338* in association with clinical aggressiveness (Huang et al., 2009). The differential expressional levels for miRNAs, including *miR-214*, *miR-199a*, *miR-150*, *miR-125a*, and *miR-148a*, have been identified by comparing hepatoblastoma with HCC (Magrelli et al., 2009).

19.2.3 Lung Cancers

Lung cancer is still one of the most common causes of cancer-related death among middle-aged men and women worldwide. It is the leading cause of cancer deaths in the United States, and smoking practices and gender are well-known risk factors for lung carcinoma (Egleston et al., 2009; Harichand-Herdt and Ramalingam, 2009; Planchard et al., 2009). The importance of miRNAs in lung cancer development and progression and their use as therapeutic and diagnostic targets has been recognized (Du and Pertsemlidis, 2010; Ortholan et al., 2009; Wang, Q. Z. et al., 2009; Wu et al., 2009). The differentiation of histological subtypes and the prediction of survival in patients with non-small cell lung cancer (NSCLC) are urgently needed. For this purpose, molecular diagnostic marker such as miRNA are highly anticipated. NSCLC mainly consists of adenocarcinoma and squamous-cell carcinoma (SCC). All the members of the *let-7* family were downregulated in SCC; thus, the miRNA-expression profile is potentially applicable for the differentiation of adenocarcinoma from SCC, particularly during the early stages (Landi et al., 2010). The miRNA-expression profiles in SCC versus normal lung tissue revealed that *miR-146b* showed the best prediction accuracy and that miRNA signatures were superior to mRNA signatures for predicting the prognosis of patients with SCC (Raponi et al., 2009). An expression analysis of miRNA revealed that a specific set of miRNAs can be used as a potential marker for the diagnostic classification of stage I NSCLC cases with or without recurrence (Patnaik et al., 2010).

19.2.4 Breast Cancers

It is estimated that more than one million women are diagnosed as having breast cancer every year. Mammographic screening for breast cancer is likely to reduce cancer mortality (Gøtzsche and Nielsen, 2009). miRNAs are highly expected to be useful for individualized breast-cancer treatment and have been considered to have potential clinical applications as novel biomarkers for breast cancer diagnosis and prognosis (Adams et al., 2008; Heneghan et al., 2009; Iorio et al., 2008; Khoshnaw et al., 2009; Lowery et al., 2008; Negrini and Calin, 2008; Shi and Guo, 2009; Verghese et al., 2008; Zoon et al., 2009). A microarray-based comparison between normal breast tissue and breast cancer has enabled the deregulation of miRNAs such as *miR-125b*, *miR-145*, *miR-21*, and *miR-155* to be distinguished (Iorio et al., 2005). Unlike in other types of tumors, *miR-205* has been shown to be uniquely downregulated in breast tumors, compared with matched normal breast tissues (Wu and Mo, 2009). A recent miRNA signature approach comparing breast tumors containing breast-cancer stem cells and nontumorigenic cancer cells revealed that *miR-200c* seems to be implicated in the clonal expansion of stem cells (Shimono et al., 2009). Circulating miRNAs are thought to be related to breast cancer, presenting the possibility of blood-based miRNA profiling for breast cancer prediction (Heneghan et al., 2010).

19.2.5 Brain Tumors

The utilization of specific miRNA for the treatment of brain tumors including glial tumor, glioblastomas, medulloblastomas, and neuroblastoma, all of which are aggressive malignancies with a high incidence and are devastating solid tumors in childhood, has been discussed, and specific miRNA subsets are expected to be of diagnostic and prognostic value for some brain tumors (Ferretti et al., 2009; Lawler and Chiocca, 2009; Mathupala et al., 2007; Nicoloso and Calin, 2008; Pang et al., 2009; Schulte et al., 2009; Shai et al., 2008; Silber et al., 2009; Stallings, 2009). Based on the miRNA signatures of stage 4 neuroblastoma patients, miRNAs could be applicable for the prognosis of high-risk neuroblastoma (Scaruffi et al., 2009). The overexpression of *miR-517c* and *miR-520g* embedded in the miRNA amplicon region has been shown to be linked to aggressive primitive neuroectodermal brain tumors but to be very rare in other brain tumors (Li et al., 2009). Together, these findings indicate that miRNA signatures could be useful for characterizing the cancer phenotypes of brain tumors.

19.2.6 Hematological Tumors

miRNAs are expected to be useful for the diagnosis and prognosis of leukemia including acute myeloid leukemia (AML), chronic myeloid leukemia, myeloblastic leukemia, acute lymphoblastic leukemia (ALL), chronic lymphoid leukemia (CLL), diffuse large B-cell lymphoma, follicular lymphoma, and primary effusion lymphoma (Barbarotto and Calin, 2008; Calin and Croce, 2007; Chen et al., 2010; Codony et al., 2009; Havelange et al., 2009; Lawrie, 2007a, 2007b; Marcucci et al., 2009; Mills, 2008; Mraz et al., 2009; Mrózek et al., 2009; Nicoloso et al., 2007; Wang and Zhang, 2008; Yong and Melo, 2009). A unique miRNA signature has been shown to be associated with a prognostic factor and disease progression in CLL based on its ability to distinguish normal B cells from malignant B cells in patients with CLL (Calin et al., 2005). miRNA-expression profiling identified that *miR-128a* and *miR-128b* are significantly upregulated and that *let-7b* and *miR-223* are significantly downregulated in ALL, compared with AML (Mi et al., 2007). ALL is an immunophenotypically heterogeneous disease. The discrimination of T-lineage versus B-lineage can be achieved using a specific set of miRNAs including *miR-148*, *miR-151*, and *miR-424* (Fulci et al., 2009). CLL can be characterized based on accompanying chromosomal abnormalities (Montserrat and Moreno, 2009). miRNA-based classifications of CLL patients could discriminate disease progression in CLL harboring specific karyotype cytogenetic subgroups (Visone et al., 2009). The specific overexpressions of *let-7e*, *miR-125a-5p*, and *miR-99b* have been identified as specific miRNA signatures in multiple myeloma (Lionetti et al., 2009). A high expression level of *miR-191* and *miR-199a* was associated with a significantly worse outcome in AML patients, compared with a low expression level in AML patients (Garzon et al., 2008). Finally, an miRNA microarray of plasma from leukemia patients revealed that the ratio of *miR-92a/miR-638* in plasma could be an excellent biomarker for the detection of leukemia (Tanaka et al., 2009).

19.2.7 Urinary Tumors

At present, prostate cancer is the second leading cause of cancer deaths in American men. Androgen ablation therapy is the first choice for prostate cancer; however, the development of androgen-independent cancer cells is problematic. As part of a basic screening program for prostate cancer, the measurement of the prostate-specific antigen (PSA) level in combination with a digital rectal examination are the most commonly used screening tests; however, some pitfalls, including the absence of a normal PSA level, exist, and a risk of unnecessary prostate biopsy has appeared (Catalona and Loeb, 2010; Snow and Klein, 2010). The accurate diagnosis of clinically significant prostate cancer remains a challenge. When microarray technology first began to be used, prostate carcinoma tumors were shown to be characterized by their miRNA signatures according to whether they were androgen dependent or independent (Porkka et al., 2007). To date, the miRNA profiling of prostate cancer remains at an early stage, and the data obtained have been largely inconsistent; nevertheless, such information is expected to be useful for identifying diagnostic and therapeutic targets as it has been for other malignant tumors (Gandellini et al., 2009; Shi et al., 2008a). miRNA microarrays have shown that *miR-28*, *miR-185*, *miR-27*, and *let-7f-2* are upregulated in renal cell carcinomas, compared with in normal kidneys, while *miR-223*, *miR-26b*, *miR-221*, *miR-103-1*, *miR-185*, *miR-23b*, *miR-203*, *miR-17-5p*, *miR-23a*, and *miR-205* are upregulated in bladder cancers, compared with in normal bladder mucosa (Gottardo et al., 2007). *miR-96*, *miR-182*, *miR-183*, and *miR-375* are upregulated, and *miR-16*, *miR-31*, *miR-125b*, *miR-145*, *miR-149*, *miR-181b*, *miR-184*, *miR-205*, *miR-221*, and *miR-222* are downregulated in prostate cancer, in parallel with the clinicopathologic data, suggesting that mRNA signatures can be defined in prostate cancer (Schaefer et al., 2010). In addition, *miR-221* and *miR-141* have also been identified as molecular biomarkers (Sørensen and Ørntoft, 2010). Coupling mRNA and miRNA signatures revealed that cell cycle-related genes are deeply implicated in the etiology and progression of prostate cancer, especially for aggressive phenotypes versus nonaggressive phenotypes (Wang et al., 2009). Interestingly, mRNA and miRNA expression analyses revealed that miRNA and its host genes are simultaneously deregulated; the MCM7 and *miR-106b-25* cluster, which maps to intron 13 of MCM7, and the C9orf5 and *miR-32* cluster, which maps to intron 14 of C9orf5, showed higher levels of expression in prostate tumors than in nontumor prostate tissue (Ambs et al., 2008). In addition to prostate cancer, the miRNA profiling of human bladder cancer and matched normal urothelial epithelium controls revealed that miRNAs are differentially expressed in bladder cancer tissues, and *miR-143* may function as a tumor suppressor in bladder cancer (Lin et al., 2009). In renal cell carcinoma, specifically clear cell carcinomas, *miR-141* and *miR-200c* were identified as the most significantly downregulated miRNAs, compared with the levels in normal kidney tissue (Nakada et al., 2008). In urothelial carcinoma of the bladder, the alteration of miRNA expression was observed in a tumor phenotype-specific manner, and these alterations were seen early during tumor onset. High-grade urothelial carcinoma was characterized by the upregulation of miRNA, including *miR-21* (Catto et al., 2009).

19.2.8 Genital Tumors

The miRNA-expression profiles of ovarian tumors, which are highly lethal tumors and a leading cause of gynecologic cancer-related deaths, have been reported, and miRNAs are recognized as potential biomarkers during ovarian cancer development (Dahiya and Morin, 2010; Toloubeydokhti et al., 2008). So far, several ovarian cancer markers have been reported as circulating autoantibodies to tumor-associated antigens such as p53, homeobox proteins (HOXA7, HOXB7), etc., and these markers are believed to be early diagnostic markers of ovarian epithelial carcinoma; however, the specificity of these biomarkers for ovarian cancers is unknown (Piura and Piura, 2009). An miRNA microarray was used to identify potential tumor suppressor miRNAs and their marked downregulation in ovarian tumors, which was partially accounted for by genomic copy number loss and epigenetic silencing, based on the results of array-based comparative genomic hybridization, a complementary DNA microarray, and a tissue array (Zhang et al., 2008). The downregulation of *miR-200a*, *miR-200b*, and *miR-429* may be predictive of a poor survival outcome in ovarian cancer patients (Hu et al., 2009). miRNA signatures have also been useful for serous ovarian carcinoma (Nam et al., 2008). All of the patients who received platinum-based chemotherapy before surgical treatment for ovarian cancer showed a unique set of miRNAs that was closely correlated with the patient outcome (Eitan et al., 2009). An miRNA array for ovarian tumor revealed that *miR-221* stands out as a highly elevated miRNA, whereas *miR-21* and members of the *let-7* family were downregulated (Dahiya et al., 2008).

19.2.9 Endocrine Tumors

Pancreatic adenocarcinoma is the most deadly of the solid tumors and is associated with an extremely poor prognosis; thus, an effective screening method is urgently required. miRNA profiling has been shown to be capable of characterizing pancreatic ductal adenocarcinoma (Seux et al., 2009). The miRNA expression pattern in miRNA microarrays has been shown to differentiate pancreatic cancer from normal pancreas and chronic pancreatitis (Bloomston et al., 2007). miRNA signatures are also promising as diagnostic markers for distinguishing benign from malignant thyroid carcinomas (Menon and Khan, 2009; Nikiforova et al., 2009; Vriens et al., 2009). Similar results have been reported for the usefulness of miRNAs in malignant melanoma and also in pancreatic ductal adenocarcinoma (Mardin and Mees, 2009; Mueller and Bosserhoff, 2009).

19.3 IMPORTANT REMINDER

When evaluating miRNA signatures, one must remember that although a large number of miRNAs are transcribed, not all of them are processed to mature miRNA (Lee et al., 2008) (Table 19.3). In addition to miRNAs, small nucleolar RNAs and antisense RNA, double-stranded RNA, and long RNA species have also been found as noncoding RNAs, functioning as regulators of mRNAs and proteins (Mallardo et al., 2008) (Table 19.1). Therefore, the overall RNA signature might need to be considered. Epigenetic modifications of miRNAs have been recognized as important regulators influencing miRNA involvement in tumorigenesis. The epigenetic silencing of miRNAs, especially of tumor suppressors, by CpG island hypermethylation and histone modifications has come to be realized as a common hallmark of human tumors (Fabbri, 2008; Lujambio and Esteller, 2007; Rouhi et al., 2008; Yang et al., 2008a). Indeed, the DNA methylation of certain miRNAs is critically linked to the conversion of normal cells to tumor cells (Vrba et al., 2010). *miR-148a*, *miR-34b/c*, and *miR-9* were found to undergo specific hypermethylation-associated silencing in cancer cells, compared with in normal tissues, based on the results of a microarray analysis (Lujambio et al., 2008).

According to the published reports, the miRNA expression level has been given too much importance; however, miRNA itself has a comparatively weak effect on the suppression of target mRNA translation but a rather potent effect on modulations of the transcriptional network. Although miRNAs are causally linked to the onset and development of cancer by acting as oncogenes or tumor suppressors, careful attention should be paid to the functional aspects of miRNAs, such as the exact biological significance of mRNAs regulated by miRNA, with regard to the development of therapeutics for tumors. Notably, in proliferating cells, mRNAs can escape from miRNA-regulated translational suppression by shortening their 3' UTRs through mRNAs terminating at upstream polyadenylation sites (Sandberg et al., 2008).

19.4 PERSPECTIVES

Vogelstein's model of multistage carcinogenesis for colorectal carcinoma clearly showed that the concomitant deregulation of specific genes, characterized as oncogenes and tumor-suppressor genes, is critical for the transformation of normal cells into benign and malignant cancerous cells (Cho and Vogelstein, 1992; Williams et al., 1993). The deregulation of p53 is frequently detected, because its gene function is lost in up to half of all tumor specimens. Among miRNAs, the p53-dependent regulation of *miR-34a* and *miR-34b/c* has been shown to be involved in the regulation of apoptosis (He, L., et al., 2007a; Hermeking, 2010; Raver-Shapira et al., 2007; Tarasov et al., 2007). Indeed, p53 has been shown to act as a crucial transcriptional regulator of certain miRNAs (He, L., et al., 2007b; He, X., et al., 2007; Hermeking, 2007; Raver-Shapira and Oren, 2007). Moreover, p53 can be implicated in the biogenesis of certain miRNAs (Suzuki et al., 2009). Notably, a regulatory circuit including auto-regulation or a positive/negative feedback loop between gene products and miRNAs has been widely recognized (Carthew, 2006; Shalgi et al., 2007; Tsang et al., 2007) (Table 19.4) and even implicated in tumorigenesis. Examples of such pathways include *miR-326*/Notch-1 feedback in gliomas (Kefas et al., 2009), *miR-29*/NF-κB/YY1 in rhabdomyosarcomas (Wang et al., 2008), *miR-106b-25* cluster/E2F1/TGFb in gastric cancer (Petrocca et al., 2008), *miR-449a/b* and pRb/E2F1 (Yang et al., 2009), *miR-34a*/SIRT1/p53 (Yamakuchi et al., 2008; Yamakuchi and Lowenstein, 2009), *miR-17-92*/E2F/Myc (Aguda et al., 2008), c-Myc/*miR-17-92* cluster/E2F1 (O'Donnell et al., 2005), *miR-17-5p*/*miR-20a*/cyclin D1 (Yu et al., 2008), and *miR-20a* of *miR-17-92* cluster/E2F1~E2F3 (Sylvestre et al., 2007). The prototype of feedback regulation can be found in developmental regulation, such as the larval-to-adult transition that occurs in *C. elegans* (Hammell et al., 2009; Roush and Slack, 2009) and the juvenile-to-adult phases in *Arabidopsis* (Wu et al., 2009). These findings suggest that miRNA signatures are strongly correlated with classical oncogenes and tumor suppressors such as p53 via the architecture of a complex regulatory network. Thus, miRNA-expression profiles are critical to understanding the perturbed transcriptional networks underlying the molecular mechanisms of tumorigenesis.

The coupling of miRNA signatures with histological examinations may enable accurate diagnosis and therapeutic approaches if such combinations can predict the behavior of cancer cells more precisely. Indeed, methodologies employing miRNA microarrays of paraffin-embedded and fresh-frozen samples originating from different tumor tissues and metastases have established that miRNA signatures can be used to trace the tissue of origin of cancers of unknown primary origin (Rosenfeld et al., 2008). Moreover, the identification of miRNA expression patterns might be useful

TABLE 19.4
Regulatory Circuits between Pivotal mRNAs/Proteins and miRNAs

Upstream Regulator	miRNA (Regulation)	Downstream Target	Reference
Notch	*miR-326* (down)	Notch	Kefas et al. (2009)
NF-κB, YY1	*miR-29* (down)	YY1	Wang et al. (2008)
E2F1	*miR-106b-25* (up)	TGFβ	Petrocca et al. (2008)
E2F1	*miR-449a/b* (up)	CDK6 and CDC25A	Yang et al. (2009)
E2F1–E2F3	*miR-17-92* (up)	E2F2 and E2F3	Sylvestre et al. (2007)
E2F3	*miR-17-92* (up)	E2F1	Woods et al. (2007)
c-Myc	*miR-17-92* (up)	E2F1	O'Donnell et al. (2005)
p53	*miR-145* (up)	c-Myc	Sachdeva et al. (2009)
p53	*miR-34a* (up)	SIRT1 and E2F	Tazawa et al. (2007) Yamakuchi et al. (2008)

for predicting recurrence after surgical resection. The relationship between miRNAs and drug sensitivity is another important issue that needs to be investigated (Garofalo et al., 2008). miRNA microarrays have led to the identification of the downregulation of *let-7i* expression in chemotherapy-resistant patients (Yang et al., 2008b). Even 1 ml of human serum from patients with cancer, including prostate, colon, ovarian, breast, and lung cancer, is sufficient for use in a miRNA microarray analysis (Lodes et al., 2009) because of the protective property of miRNAs against endogenous RNase activity present in human plasma. miRNA levels not only in the serum but also in other body fluids have been evaluated and used for the early diagnosis of clinical conditions (Gilad et al., 2008; Mitchell et al., 2008). Gene therapy can be challenged by regulating miRNA expression levels in tumors by applying synthetic pre-miRNA molecules or antisense oligonucleotides (Casalini and Iorio, 2009). For the suppression of oncogenic miRNAs, the introduction of miRNA-antagonizing oligonucleotides into tumors is a fascinating concept (Stenvang and Kauppinen, 2008). Together, miRNA signatures could be applied for tailored medicine specifically targeting the pathogenesis and development of individual cancers. The potential importance of miRNAs as novel minimally invasive diagnostic, prognostic, invasion and metastasis, and therapeutic biomarkers of tumors is strongly indicated.

19.5 CONCLUSIONS

Expression profiling of miRNAs is an important milestone on the road to understanding the full spectrum of the molecular etiology of tumorigenesis, because the deregulation of miRNAs is now considered a hallmark of a variety of human tumors. Molecular biological and genetic approaches employing cancer cell lines and model animals will be accelerated once a platform of miRNA signatures of various tumors has been established using miRNA-expression profiling. As a powerful tool for analyzing miRNA global expression changes in clinical samples, the miRNA signatures obtained using miRNA microarrays should be recognized as the gold standard, but once knowledge of a specific set of miRNAs that are certainly involved in tumorigenesis as oncogenes or tumor suppressors has been obtained, real-time RT-PCR could be useful as a substitute for miRNA microarrays. The recent advent of miRNA microarrays is superior to real-time RT-PCR with regard to its quantitative capabilities. However, in the near future, once a specific set of oncogenic and anti-oncogenic miRNAs has been determined in a specific cell type, quantitative RT-PCR detection might become a major means of investigation because of its qualitative and economic merits. Recently, high-throughput PCR using a 384-well format has been established (Schmittgen et al., 2008). Arguments regarding the experimental evidence of the biological impacts of individual miRNAs on apoptosis, cell differentiation and proliferation, and development via specific mRNA targets in cultured cell and animal models will be described elsewhere. Specific miRNAs have received much attention because their functional importance in tumorigenesis has been experimentally established. The present chapter has not focused on such miRNAs but has rather emphasized global expression analyses of miRNA signatures in clinical tumors (Table 19.5). Collectively, a comprehensive understanding of miRNAs based on expression signatures and the biological importance of target mRNAs will be essential for understanding the molecular etiology of tumorigenesis.

TABLE 19.5
Selected miRNAs from miRNA Array Analyses as Potential Biomarkers

Overexpressed miRNAs	Tumor Types
miR-18a, *miR-21*, miR-92	Colorectal cancers (Motoyama et al., 2009; Ng et al. 2009; Schetter et al., 2008; Wang et al., 2010)
miR-222	Hepatocellular carcinomas (Pineau et al., 2010)

ACKNOWLEDGMENTS

This study was supported by a grant from the Vehicle Racing Commemorative Foundation.

REFERENCES

Adams, B. D., Guttilla, I. K., White, B. A. (2008). Involvement of microRNAs in breast cancer. *Semin. Reprod. Med.* 26: 522–536.

Aguda, B. D., Kim, Y., Piper-Hunter, M. G., et al. (2008). MicroRNA regulation of a cancer network: Consequences of the feedback loops involving miR-17-92, E2F, and Myc. *Proc. Natl. Acad. Sci. U.S.A.* 105: 19678–19683.

Ahmed, F. E. (2007). Role of miRNA in carcinogenesis and biomarker selection: A methodological view. *Expert Rev. Mol. Diagn.* 7: 569–603.

Ambs, S., Prueitt, R. L., Yi, M., et al. (2008). Genomic profiling of microRNA and messenger RNA reveals deregulated microRNA expression in prostate cancer. *Cancer Res.* 68: 6162–6170.

Aslam, M. I., Taylor, K., Pringle, J. H., et al. (2009). MicroRNAs are novel biomarkers of colorectal cancer. *Br. J. Surg.* 96: 702–710.

Büssing, I., Slack, F. J., Grosshans, H. (2008). Let-7 microRNAs in development, stem cells and cancer. *Trends Mol. Med.* 14: 400–409.

Bandres, E., Agirre, X., Ramirez, N., et al. (2007). MicroRNAs as cancer players: Potential clinical and biological effects. *DNA Cell Biol.* 26: 273–282.

Baranwal, S., Alahari, S. K. (2010). miRNA control of tumor cell invasion and metastasis. *Int. J. Cancer* 126: 1283–1290.

Barbarotto, E., Calin, G. A. (2008). Potential therapeutic applications of miRNA-based technology in hematological malignancies. *Curr. Pharm. Des.* 14: 2040–2050.

Barbarotto, E., Schmittgen. T. D., Calin, G. A. (2008). MicroRNAs and cancer: Profile, profile, profile. *Int. J. Cancer* 122: 969–977.

Bartels, C. L., Tsongalis, G. J. (2009). MicroRNAs: Novel biomarkers for human cancer. *Clin. Chem.* 55: 623–631.

Bhatti, I., Lee, A., Lund, J., et al. (2009). Small RNA: A large contributor to carcinogenesis? *J. Gastrointest. Surg.* 13: 1379–1388.

Blenkiron, C., Miska, E. A. (2007). miRNAs in cancer: Approaches, aetiology, diagnostics and therapy. *Hum. Mol. Genet.* 16: R106–R113.

Bloomston, M., Frankel, W. L., Petrocca, F., et al. (2007). MicroRNA expression patterns to differentiate pancreatic adenocarcinoma from normal pancreas and chronic pancreatitis. *JAMA* 297: 1901–1908.

Calin, G. A., Croce, C. M. (2006a). MicroRNA-cancer connection: The beginning of a new tale. *Cancer Res.* 66: 7390–7394.

Calin, G. A., Croce. C. M. (2006b). MicroRNA signatures in human cancers. *Nat. Rev. Cancer* 6: 857–866.

Calin, G. A., Croce, C. M. (2007). Investigation of microRNA alterations in leukemias and lymphomas. *Methods Enzymol.* 427: 193–213.

Calin, G. A., Sevignani, C., Dumitru, C. D., et al. (2004). Human microRNA genes are frequently located at fragile sites and genomic regions involved in cancers. *Proc. Natl. Acad. Sci. U.S.A.* 101: 2999–3004.

Calin, G. A., Ferracin, M., Cimmino, A., et al. (2005). A microRNA signature associated with prognosis and progression in chronic lymphocytic leukemia. *N. Engl. J. Med.* 353: 1793–1801.

Carthew, R. W. (2006). Gene regulation by microRNAs. *Curr. Opin. Genet. Dev.* 16: 203–208.

Casalini, P., Iorio, M. V. (2009). MicroRNAs and future therapeutic applications in cancer. *J. BUON* 14: S17–S22.

Catalona, W. J., Loeb, S. (2010). Prostate cancer screening and determining the appropriate prostate-specific antigen cutoff values. *J. Natl. Compr. Canc. Netw.* 8: 265–270.

Catto, J. W., Miah, S., Owen, H. C., et al. (2009). Distinct microRNA alterations characterize high- and low-grade bladder cancer. *Cancer Res.* 69: 8472–8481.

Chen, J., Odenike, O., Rowley, J. D. (2010). Leukaemogenesis: More than mutant genes. *Nat. Rev. Cancer* 10: 23–36.

Cho, K. R., Vogelstein, B. (1992). Genetic alterations in the adenoma: Carcinoma sequence. *Cancer* 70: 1727–1731.

Chow, T. F., Mankaruos, M., Scorilas, A., et al. (2010). The miR-17-92 cluster is over expressed in and has an oncogenic effect on renal cell carcinoma. *J. Urol.* 183: 743–751.

Codony, C., Crespo, M., Abrisqueta, P., et al. (2009). Gene expression profiling in chronic lymphocytic leukaemia. *Best Pract. Res. Clin. Haematol.* 22: 211–222.

Connolly, E., Melegari, M., Landgraf, P., et al. (2008). Elevated expression of the miR-17-92 polycistron and miR-21 in hepadnavirus-associated hepatocellular carcinoma contributes to the malignant phenotype. *Am. J. Pathol.* 173: 856–864.

Croce, C. M. (2009). Causes and consequences of microRNA dysregulation in cancer. *Nat. Rev. Genet.* 10: 704–714.

Cummins, J. M., Velculescu, V. E. (2006). Implications of micro-RNA profiling for cancer diagnosis. *Oncogene* 25: 6220–6227.

Dahiya, N., Morin, P. J. (2010). MicroRNAs in ovarian carcinomas. *Endocr. Relat. Cancer* 17: F77–F89.

Dahiya, N., Sherman-Baust, C. A., Wang, T. L., et al. (2008). MicroRNA expression and identification of putative miRNA targets in ovarian cancer. *PLoS One* 3: e2436.

Deng, S., Calin, G. A., Croce, C. M., et al. (2008). Mechanisms of microRNA deregulation in human cancer. *Cell Cycle* 7: 2643–2646.

Dillhoff, M., Wojcik, S. E., Bloomston, M. (2009). MicroRNAs in solid tumors. *J. Surg. Res.* 154: 349–354.

Dong, Q., Meng, P., Wang, T., et al. (2010). MicroRNA let-7a inhibits proliferation of human prostate cancer cells in vitro and in vivo by targeting E2F2 and CCND2. *PLoS One* 5: e10147.

Drakaki, A., Iliopoulos, D. (2009). MicroRNA gene networks in oncogenesis. *Curr. Genomics* 10: 35–41.

Du, L., Pertsemlidis, A. (2010). MicroRNAs and lung cancer: Tumors and 22-mers. *Cancer Metastasis Rev.* 29: 109–122.

Egleston, B. L., Meireles, S. I., Flieder, D. B., et al. (2009). Population-based trends in lung cancer incidence in women. *Semin. Oncol.* 36: 506–515.

Eitan, R., Kushnir, M., Lithwick-Yanai, G., et al. (2009). Tumor microRNA expression patterns associated with resistance to platinum based chemotherapy and survival in ovarian cancer patients. *Gynecol. Oncol.* 114: 253–259.

Esquela-Kerscher, A., Slack, F. J. (2006). Oncomirs: MicroRNAs with a role in cancer. *Nat. Rev. Cancer* 6: 259–269.

Fabbri, M. (2008). MicroRNAs and cancer epigenetics. *Curr. Opin. Investig. Drugs* 9: 583–590.

Fabbri, M., Croce, C. M., Calin, G. A. (2008). MicroRNAs. *Cancer J.* 14: 1–6.

Faber, C., Kirchner, T., Hlubek, F. (2009). The impact of microRNAs on colorectal cancer. *Virchows Arch.* 454: 359–367.

Ferretti, E., De Smaele, E., Po, A., et al. (2009). MicroRNA profiling in human medulloblastoma. *Int. J. Cancer* 124: 568–577.

Fulci, V., Colombo, T., Chiaretti, S., et al. (2009). Characterization of B- and T-lineage acute lymphoblastic leukemia by integrated analysis of MicroRNA and mRNA expression profiles. *Genes Chromosomes Cancer* 48: 1069–1082.

Gandellini, P., Folini, M., Zaffaroni, N. (2009). Towards the definition of prostate cancer-related microRNAs: Where are we now? *Trends Mol. Med.* 15: 381–390.

Garofalo, M., Condorelli, G., Croce, C. M. (2008). MicroRNAs in diseases and drug response. *Curr. Opin. Pharmacol.* 8: 661–667.

Garzon, R., Calin, G. A., Croce, C. M. (2009). MicroRNAs in Cancer. *Annu. Rev. Med.* 60: 167–179.

Garzon, R., Fabbri, M., Cimmino, A., et al. (2006). MicroRNA expression and function in cancer. *Trends Mol. Med.* 12: 580–587.

Garzon, R., Volinia, S., Liu, C. G., et al. (2008). MicroRNA signatures associated with cytogenetics and prognosis in acute myeloid leukemia. *Blood* 111: 3183–3189.

Gilad, S., Meiri, E., Yogev, Y., et al. (2008). Serum microRNAs are promising novel biomarkers. *PLoS One* 3: e3148.

Gottardo, F., Liu, C. G., Ferracin, M., et al. (2007). Micro-RNA profiling in kidney and bladder cancers. *Urol. Oncol.* 25: 387–392.

Gøtzsche, P. C., Nielsen, M. (2009). Screening for breast cancer with mammography. *Cochrane Database Syst Rev.* CD001877.

Gramantieri, L., Fornari, F., Callegari, E., et al. (2008). MicroRNA involvement in hepatocellular carcinoma. *J. Cell Mol. Med.* 12: 2189–2204.

Guo, J., Miao, Y., Xiao, B., et al. (2009). Differential expression of microRNA species in human gastric cancer versus non-tumorous tissues. *J. Gastroenterol. Hepatol.* 24: 652–657.

Gusev, Y., Brackett, D. J. (2007). MicroRNA expression profiling in cancer from a bioinformatics prospective. *Expert Rev. Mol. Diagn.* 7: 787–792.

Hagan, J. P., Croce, C. M. (2007). MicroRNAs in carcinogenesis. *Cytogenet. Genome Res.* 118: 252–259.

Hammell, C. M., Karp, X., Ambros, V. (2009). A feedback circuit involving let-7-family miRNAs and DAF-12 integrates environmental signals and developmental timing in *Caenorhabditis elegans*. *Proc. Natl. Acad. Sci. U.S.A.* 106: 18668–18673.

Hammond, S. M. (2006). MicroRNAs as oncogenes. *Curr. Opin. Genet. Dev.* 16: 4–9.

Harichand-Herdt, S., Ramalingam, S. S. (2009). Gender-associated differences in lung cancer: Clinical characteristics and treatment outcomes in women. *Semin. Oncol.* 36: 572–580.

Havelange, V., Garzon, R., Croce, C. M. (2009). MicroRNAs: New players in acute myeloid leukaemia. *Br. J. Cancer* 101: 743–748.

Hayashita, Y., Osada, H., Tatematsu, Y., et al. (2005). A polycistronic microRNA cluster, miR-17-92, is overexpressed in human lung cancers and enhances cell proliferation. *Cancer Res.* 65: 9628–9632.

He, L., He, X., Lim, L. P., et al. (2007a). A microRNA component of the p53 tumour suppressor network. *Nature* 447: 1130–1134.

He, L., He, X., Lowe, S. W., et al. (2007b). MicroRNAs join the p53 network: Another piece in the tumour-suppression puzzle. *Nat. Rev. Cancer* 7: 819–822.

He, L., Thomson, J. M., Hemann, M. T., et al. (2005). A microRNA polycistron as a potential human oncogene. *Nature* 435: 828–833.

He, X., He, L., Hannon, G. J. (2007). The guardian's little helper: MicroRNAs in the p53 tumor suppressor network. *Cancer Res.* 67: 11099–11101.

Heneghan, H. M., Miller, N., Lowery, A. J., et al. (2009). MicroRNAs as novel biomarkers for breast cancer. *J. Oncol.* 2009: 950201.

Heneghan, H. M., Miller, N., Lowery, A. J., et al. (2010). Circulating microRNAs as novel minimally invasive biomarkers for breast cancer. *Ann. Surg.* 251: 499–505.

Hermeking, H. (2007). p53 enters the microRNA world. *Cancer Cell* 12: 414–418.

Hermeking, H. (2010). The miR-34 family in cancer and apoptosis. *Cell Death Differ.* 17: 193–199.

Hernando, E. (2007). MicroRNAs and cancer: Role in tumorigenesis, patient classification and therapy. *Clin. Transl. Oncol.* 9: 155–160.

Hu, X., Macdonald, D. M., Huettner, P. C., et al. (2009). A miR-200 microRNA cluster as prognostic marker in advanced ovarian cancer. *Gynecol. Oncol.* 114: 457–464.

Huang, X. H., Wang, Q., Chen, J. S., et al. (2009). Bead-based microarray analysis of microRNA expression in hepatocellular carcinoma: miR-338 is downregulated. *Hepatol. Res.* 39: 786–794.

Hwang, H. W., Mendell, J. T. (2007). MicroRNAs in cell proliferation, cell death, and tumorigenesis. *Br. J. Cancer* 96: R40–R44.

Iorio, M. V., Croce, C. M. (2009). MicroRNAs in cancer: Small molecules with a huge impact. *J. Clin. Oncol.* 27: 5848–5856.

Iorio, M. V., Casalini, P., Tagliabue, E., et al. (2008). MicroRNA profiling as a tool to understand prognosis, therapy response and resistance in breast cancer. *Eur. J. Cancer* 44: 2753–2759.

Iorio, M. V., Ferracin, M., Liu, C. G., et al. (2005). MicroRNA gene expression deregulation in human breast cancer. *Cancer Res.* 65: 7065–7070.

Jay, C., Nemunaitis, J., Chen, P., et al. (2007). miRNA profiling for diagnosis and prognosis of human cancer. *DNA Cell Biol.* 26: 293–300.

Jérôme, T., Laurie, P., Louis, B., et al. (2007). Enjoy the silence: The story of let-7 microRNA and cancer. *Curr. Genomics* 8: 229–233.

Jiang, Z., Guo, J., Xiao, B., et al. (2010). Increased expression of miR-421 in human gastric carcinoma and its clinical association. *J. Gastroenterol.* 45: 17–23.

Johnson, S. M., Grosshans, H., Shingara, J., et al. (2005). RAS is regulated by the let-7 microRNA family. *Cell* 120: 635–647.

Kefas, B., Comeau, L., Floyd, D. H., et al. (2009). The neuronal microRNA miR-326 acts in a feedback loop with notch and has therapeutic potential against brain tumors. *J. Neurosci.* 29: 15161–15168.

Kent, O. A., Mendell, J. T. (2006). A small piece in the cancer puzzle: MicroRNAs as tumor suppressors and oncogenes. *Oncogene* 25: 6188–6196.

Khoshnaw, S. M., Green, A. R., Powe, D. G., et al. (2009). MicroRNA involvement in the pathogenesis and management of breast cancer. *J. Clin. Pathol.* 62: 422–428.

Lagos-Quintana, M., Rauhut, R., Lendeckel, W., et al. (2001). Identification of novel genes coding for small expressed RNAs. *Science* 294: 853–858.

Landi, M. T., Zhao, Y., Rotunno, M., et al. (2010). MicroRNA expression differentiates histology and predicts survival of lung cancer. *Clin. Cancer Res.* 16: 430–441.

Lau, N. C., Lim, L. P., Weinstein, E. G., et al. (2001). An abundant class of tiny RNAs with probable regulatory roles in *Caenorhabditis elegans*. *Science* 294: 858–862.

Lawler, S., Chiocca, E. A. (2009). Emerging functions of microRNAs in glioblastoma. *J. Neurooncol.* 92: 297–306.
Lawrie, C. H. (2007a). MicroRNAs and haematology: Small molecules, big function. *Br. J. Haematol.* 137: 503–512.
Lawrie, C. H. (2007b). MicroRNA expression in lymphoma. *Expert. Opin. Biol. Ther.* 7: 1363–1374.
Lee, E. J., Baek, M., Gusev, Y., et al. (2008). Systematic evaluation of microRNA processing patterns in tissues, cell lines, and tumors. *RNA* 14: 35–42.
Lee, R. C., Ambros, V. (2001). An extensive class of small RNAs in *Caenorhabditis elegans*. *Science* 294: 862–864.
Lee, Y. S., Dutta, A. (2006). MicroRNAs: Small but potent oncogenes or tumor suppressors. *Curr. Opin. Investig. Drugs* 7: 560–564.
Lee, Y. S., Dutta, A. (2007). The tumor suppressor microRNA let-7 represses the HMGA2 oncogene. *Genes Dev.* 21: 1025–1030.
Lee, Y. S., Dutta, A. (2009). MicroRNAs in cancer. *Annu. Rev. Pathol.* 4: 199–227.
Li, M., Lee, K. F., Lu, Y., et al. (2009). Frequent amplification of a chr19q13.41 microRNA polycistron in aggressive primitive neuroectodermal brain tumors. *Cancer Cell* 16: 533–546.
Lin, T., Dong, W., Huang, J., et al. (2009). MicroRNA-143 as a tumor suppressor for bladder cancer. *J. Urol.* 181: 1372–1380.
Lionetti, M., Biasiolo, M., Agnelli, L., et al. (2009). Identification of microRNA expression patterns and definition of a microRNA/mRNA regulatory network in distinct molecular groups of multiple myeloma. *Blood* 114: e20–e26.
Lodes, M. J., Caraballo, M., Suciu, D., et al. (2009). Detection of cancer with serum miRNAs on an oligonucleotide microarray. *PLoS One* 4: e6229.
Lotterman, C. D., Kent, O. A., Mendell, J. T. (2008). Functional integration of microRNAs into oncogenic and tumor suppressor pathways. *Cell Cycle* 7: 2493–2499.
Lowery, A. J., Miller, N., McNeill, R. E., et al. (2008). MicroRNAs as prognostic indicators and therapeutic targets: Potential effect on breast cancer management. *Clin. Cancer Res.* 14: 360–365.
Lu, A. T., Salpeter, S. R., Reeve, A. E., et al. (2009). Gene expression profiles as predictors of poor outcomes in stage II colorectal cancer: A systematic review and meta-analysis. *Clin. Colorectal Cancer* 8: 207–214.
Lu, J., Getz, G., Miska, E. A., et al. (2005). MicroRNA expression profiles classify human cancers. *Nature* 435: 834–838.
Lujambio, A., Esteller, M. (2007). CpG island hypermethylation of tumor suppressor microRNAs in human cancer. *Cell Cycle* 6: 1455–1459.
Lujambio, A., Calin, G. A., Villanueva, A., et al. (2008). A microRNA DNA methylation signature for human cancer metastasis. *Proc. Natl. Acad. Sci. U.S.A.* 105: 13556–13561.
Luo, H., Zhang, H., Zhang, Z., et al. (2009). Down-regulated miR-9 and miR-433 in human gastric carcinoma. *J. Exp. Clin. Cancer Res.* 28: 82.
Ma, L., Weinberg, R. A. (2008). Micromanagers of malignancy: Role of microRNAs in regulating metastasis. *Trends Genet.* 24: 448–456.
Magrelli, A., Azzalin, G., Salvatore, M., et al. (2009). Altered microRNA expression patterns in hepatoblastoma patients. *Transl. Oncol.* 2: 157–163.
Mallardo, M., Poltronieri, P., D'Urso, O. F. (2008). Non-protein coding RNA biomarkers and differential expression in cancers: A review. *J. Exp. Clin. Cancer Res.* 27: 19.
Marcucci, G., Mrózek, K., Radmacher, M. D., et al. (2009). MicroRNA expression profiling in acute myeloid and chronic lymphocytic leukaemias. *Best Pract. Res. Clin. Haematol.* 22: 239–248.
Mardin, W. A., Mees, S. T. (2009). MicroRNAs: Novel diagnostic and therapeutic tools for pancreatic ductal adenocarcinoma? *Ann. Surg. Oncol.* 16: 3183–3189.
Mathupala, S. P., Mittal, S., Guthikonda, M., et al. (2007). MicroRNA and brain tumors: A cause and a cure? *DNA Cell Biol.* 26: 301–310.
Medina, P. P., Slack, F. J. (2008). MicroRNAs and cancer: An overview. *Cell Cycle* 7: 2485–2492.
Menon, M. P., Khan, A. (2009). Micro-RNAs in thyroid neoplasms: molecular, diagnostic and therapeutic implications. *J. Clin. Pathol.* 62: 978–985.
Metias, S. M., Lianidou, E., Yousef, G. M. (2009). MicroRNAs in clinical oncology: At the crossroads between promises and problems. *J. Clin. Pathol.* 62: 771–776.
Mi, S., Lu, J., Sun, M., et al. (2007). MicroRNA expression signatures accurately discriminate acute lymphoblastic leukemia from acute myeloid leukemia. *Proc. Natl. Acad. Sci. U.S.A.* 104: 19971–19976.
Mi, S., Li, Z., Chen, P., et al. (2010). Aberrant overexpression and function of the miR-17-92 cluster in MLL-rearranged acute leukemia. *Proc. Natl. Acad. Sci. U.S.A.* 107: 3710–3715.

Mills, K. (2008). Gene expression profiling for the diagnosis and prognosis of acute myeloid leukaemia. *Front. Biosci.* 13: 4605–4616.

Mirnezami, A. H., Pickard, K., Zhang, L., et al. (2009). MicroRNAs: Key players in carcinogenesis and novel therapeutic targets. *Eur. J. Surg. Oncol.* 35: 339–347.

Mitchell, P. S., Parkin, R. K., Kroh, E. M., et al. (2008). Circulating microRNAs as stable blood-based markers for cancer detection. *Proc. Natl. Acad. Sci. U.S.A.* 105: 10513–10518.

Mocellin, S., Pasquali, S., Pilati, P. (2009). Oncomirs: From tumor biology to molecularly targeted anticancer strategies. *Mini. Rev. Med. Chem.* 9: 70–80.

Montserrat, E., Moreno, C. (2009). Genetic lesions in chronic lymphocytic leukemia: Clinical implications. *Curr. Opin. Oncol.* 21: 609–614.

Monzo, M., Navarro, A., Bandres, E., et al. (2008). Overlapping expression of microRNAs in human embryonic colon and colorectal cancer. *Cell Res.* 18: 823–833.

Motoyama, K., Inoue, H., Takatsuno, Y., et al. (2009). Over- and under-expressed microRNAs in human colorectal cancer. *Int. J. Oncol.* 34: 1069–1075.

Mraz, M., Pospisilova, S., Malinova, K., et al. (2009). MicroRNAs in chronic lymphocytic leukemia pathogenesis and disease subtypes. *Leuk. Lymphoma* 50: 506–509.

Mrózek, K., Radmacher, M. D., Bloomfield, C. D., et al. (2009). Molecular signatures in acute myeloid leukemia. *Curr. Opin. Hematol.* 16: 64–69.

Mu, G., Liu, H., Zhou, F., et al. (2010). Correlation of overexpression of HMGA1 and HMGA2 with poor tumor differentiation, invasion, and proliferation associated with let-7 down-regulation in retinoblastomas. *Hum. Pathol.* 41: 493–502.

Mueller, D. W., Bosserhoff, A. K. (2009). Role of miRNAs in the progression of malignant melanoma. *Br. J. Cancer* 101: 551–556.

Nakada, C., Matsuura, K., Tsukamoto, Y., et al. (2008). Genome-wide microRNA expression profiling in renal cell carcinoma: Significant down-regulation of miR-141 and miR-200c. *J. Pathol.* 216: 418–427.

Nam, E. J., Yoon, H., Kim, S. W., et al. (2008). MicroRNA expression profiles in serous ovarian carcinoma. *Clin. Cancer Res.* 14: 2690–2695.

Negrini, M., Calin, G. A. (2008). Breast cancer metastasis: A microRNA story. *Breast Cancer Res.* 10: 203.

Negrini, M., Nicoloso, M. S., Calin, G. A. (2009). MicroRNAs and cancer: New paradigms in molecular oncology. *Curr. Opin. Cell Biol.* 21: 470–479.

Negrini, M., Ferracin, M., Sabbioni, S., et al. (2007). MicroRNAs in human cancer: From research to therapy. *J. Cell Sci.* 120: 1833–1840.

Nelson, K. M., Weiss, G. J. (2008). MicroRNAs and cancer: Past, present, and potential future. *Mol. Cancer Ther.* 7: 3655–3660.

Ng, E. K., Chong, W. W., Jin, H., et al. (2009). Differential expression of microRNAs in plasma of patients with colorectal cancer: A potential marker for colorectal cancer screening. *Gut* 58: 1375–1381.

Nicoloso, M. S., Calin, G. A. (2008). MicroRNA involvement in brain tumors: From bench to bedside. *Brain Pathol.* 18: 122–129.

Nicoloso, M. S., Kipps, T. J., Croce, C. M., et al. (2007). MicroRNAs in the pathogeny of chronic lymphocytic leukaemia. *Br. J. Haematol.* 139: 709–716.

Nikiforova, M. N., Chiosea, S. I., Nikiforov, Y. E. (2009). MicroRNA expression profiles in thyroid tumors. *Endocr. Pathol.* 20: 85–91.

O'Donnell, K. A., Wentzel, E. A., Zeller, K. I., et al. (2005). c-Myc-regulated microRNAs modulate E2F1 expression. *Nature* 435: 839–843.

Olive, V., Bennett, M. J., Walker, J. C., et al. (2009). miR-19 is a key oncogenic component of mir-17-92. *Genes Dev.* 23: 2839–2849.

Ortholan, C., Puissegur, M. P., Ilie, M., et al. (2009). MicroRNAs and lung cancer: New oncogenes and tumor suppressors, new prognostic factors and potential therapeutic targets. *Curr. Med. Chem.* 16: 1047–1061.

Osaki, M., Takeshita, F., Ochiya, T. (2008). MicroRNAs as biomarkers and therapeutic drugs in human cancer. *Biomarkers* 13: 658–670.

Pang, J. C., Kwok, W. K., Chen, Z., et al. (2009). Oncogenic role of microRNAs in brain tumors. *Acta Neuropathol.* 117: 599–611.

Patnaik, S. K., Kannisto, E., Knudsen, S., et al. (2010). Evaluation of microRNA expression profiles that may predict recurrence of localized stage I non-small cell lung cancer after surgical resection. *Cancer Res.* 70: 36–45.

Petrocca, F., Visone, R., Onelli, M. R., et al. (2008). E2F1-regulated microRNAs impair TGFbeta-dependent cell-cycle arrest and apoptosis in gastric cancer. *Cancer Cell* 13: 272–286.

Pineau, P., Volinia, S., McJunkin, K., et al. (2010). miR-221 overexpression contributes to liver tumorigenesis. *Proc. Natl. Acad. Sci. U.S.A.* 107: 264–229.

Piura, B., Piura, E. (2009). Autoantibodies to tumor-associated antigens in epithelial ovarian carcinoma. *J. Oncol.* 2009: 581939.

Planchard, D., Loriot, Y., Goubar, A., et al. (2009). Differential expression of biomarkers in men and women. *Semin. Oncol.* 36: 553–565.

Porkka, K. P., Pfeiffer, M. J., Waltering, K. K., et al. (2007). MicroRNA expression profiling in prostate cancer. *Cancer Res.* 67: 6130–6135.

Raponi, M., Dossey, L., Jatkoe, T., et al. (2009). MicroRNA classifiers for predicting prognosis of squamous cell lung cancer. *Cancer Res.* 69: 5776–5783.

Raver-Shapira, N., Oren, M. (2007). Tiny actors, great roles: MicroRNAs in p53's service. *Cell Cycle* 6: 2656–2661.

Raver-Shapira, N., Marciano, E., Meiri, E., et al. (2007). Transcriptional activation of miR-34a contributes to p53-mediated apoptosis. *Mol. Cell* 26: 731–743.

Rosenfeld, N., Aharonov, R., Meiri, E., et al. (2008). MicroRNAs accurately identify cancer tissue origin. *Nat. Biotechnol.* 26: 462–469.

Rossi, S., Sevignani, C., Nnadi, S. C., et al. (2008). Cancer-associated genomic regions (CAGRs) and noncoding RNAs: Bioinformatics and therapeutic implications. *Mamm. Genome* 19: 526–540.

Rouhi, A., Mager, D. L., Humphries, R. K., et al. (2008). miRNAs, epigenetics, and cancer. *Mamm. Genome* 19: 517–525.

Roush, S. F., Slack, F. J. (2009). Transcription of the *C. elegans* let-7 microRNA is temporally regulated by one of its targets, hbl-1. *Dev. Biol.* 334: 523–534.

Ruan, K., Fang, X., Ouyang, G. (2009). MicroRNAs: Novel regulators in the hallmarks of human cancer. *Cancer Lett.* 285: 116–126.

Sachdeva, M., Zhu, S., Wu, F., et al. (2009). p53 represses c-Myc through induction of the tumor suppressor miR-145. *Proc. Natl. Acad. Sci. U.S.A.* 106: 3207–3212.

Sampson, V. B., Rong, N. H., Han, J., et al. (2007). MicroRNA let-7a down-regulates MYC and reverts MYC-induced growth in Burkitt lymphoma cells. *Cancer Res.* 67: 9762–9770.

Sandberg, R., Neilson, J. R., Sarma, A., et al. (2008). Proliferating cells express mRNAs with shortened 3' untranslated regions and fewer microRNA target sites. *Science* 320: 1643–1647.

Sarver, A. L., French, A. J., Borralho, P. M., et al. (2009). Human colon cancer profiles show differential microRNA expression depending on mismatch repair status and are characteristic of undifferentiated proliferative states. *BMC Cancer* 9: 401.

Sassen, S., Miska, E. A., Caldas, C. (2008). MicroRNA: Implications for cancer. *Virchows Arch.* 452: 1–10.

Scaruffi, P., Stigliani, S., Moretti, S., et al. (2009). Transcribed-ultra conserved region expression is associated with outcome in high-risk neuroblastoma. *BMC Cancer* 9: 441.

Schaefer, A., Jung, M., Mollenkopf, H. J., et al. (2010). Diagnostic and prognostic implications of microRNA profiling in prostate carcinoma. *Int. J. Cancer* 126: 1166–1176.

Schepeler, T., Reinert, J. T., Ostenfeld, M. S., et al. (2008). Diagnostic and prognostic microRNAs in stage II colon cancer. *Cancer Res.* 68: 6416–6424.

Schetter, A. J., Leung, S. Y., Sohn, J. J., et al. (2008). MicroRNA expression profiles associated with prognosis and therapeutic outcome in colon adenocarcinoma. *JAMA* 299: 425–436.

Schmittgen, T. D. (2008). Regulation of microRNA processing in development, differentiation and cancer. *J. Cell. Mol. Med.* 12: 1811–1819.

Schmittgen, T. D., Lee, E. J., Jiang, J. (2008). High-throughput real-time PCR. *Methods Mol. Biol.* 429: 89–98.

Schulte, J. H., Horn, S., Schlierf, S., et al. (2009). MicroRNAs in the pathogenesis of neuroblastoma. *Cancer Lett.* 274: 10–15.

Seux, M., Iovanna, J., Dagorn, J. C., et al. (2009). MicroRNAs in pancreatic ductal adenocarcinoma: New diagnostic and therapeutic clues. *Pancreatology* 9: 66–72.

Shai, R. M., Reichardt, J. K., Chen, T. C. (2008). Pharmacogenomics of brain cancer and personalized medicine in malignant gliomas. *Future Oncol.* 4: 525–534.

Shalgi, R., Lieber, D., Oren, M., et al. (2007). Global and local architecture of the mammalian microRNA-transcription factor regulatory network. *PLoS Comput. Biol.* 3: e131.

Shell, S., Park, S. M., Radjabi, A. R., et al. (2007). Let-7 expression defines two differentiation stages of cancer. *Proc. Natl. Acad. Sci. U.S.A.* 104: 11400–11405.

Shenouda, S. K., Alahari, S. K. (2009). MicroRNA function in cancer: Oncogene or a tumor suppressor? *Cancer Metastasis Rev.* 28: 369–378.

Shi, M., Guo, N. (2009). MicroRNA expression and its implications for the diagnosis and therapeutic strategies of breast cancer. *Cancer Treat. Rev.* 35: 328–334.
Shi, X. B., Tepper, C. G., White, R. W. (2008a). MicroRNAs and prostate cancer. *J. Cell. Mol. Med.* 12: 1456–1465.
Shi, X. B., Tepper, C. G., deVere White, R. W. (2008b). Cancerous miRNAs and their regulation. *Cell Cycle* 7: 1529–1538.
Shimono, Y., Zabala, M., Cho, R. W., et al. (2009). Downregulation of miRNA-200c links breast cancer stem cells with normal stem cells. *Cell* 138: 592–603.
Silber, J., James, C. D., Hodgson, J. G. (2009). MicroRNAs in gliomas: Small regulators of a big problem. *Neuromol. Med.* 11: 208–222.
Sioud, M., Cekaite, L. (2008). Expression profiling of microRNAs in cancer cells: Technical considerations. *Methods Mol. Biol.* 439: 179–190.
Slaby, O., Svoboda, M., Michalek, J., et al. (2009). MicroRNAs in colorectal cancer: Translation of molecular biology into clinical application. *Mol. Cancer* 8: 102.
Slack, F. J., Weidhaas, J. B. (2006). MicroRNAs as a potential magic bullet in cancer. *Future Oncol.* 2: 73–82.
Snow, D. C., Klein, E. A. (2010). Use of nomograms for early detection in prostate cancer. *J. Natl. Compr. Cancer Netw.* 8: 271–276.
Sørensen, K. D., Ørntoft, T. F. (2010). Discovery of prostate cancer biomarkers by microarray gene expression profiling. *Expert Rev. Mol. Diagn.* 10: 49–64.
Stahlhut Espinosa, C. E., Slack, F. J. (2006). The role of microRNAs in cancer. *Yale J. Biol. Med.* 79: 131–140.
Stallings, R. L. (2009). MicroRNA involvement in the pathogenesis of neuroblastoma: Potential for microRNA mediated therapeutics. *Curr. Pharm. Des.* 15: 456–462.
Stefani, G. (2007). Roles of microRNAs and their targets in cancer. *Expert Opin. Biol. Ther.* 7: 1833–1840.
Stenvang, J., Kauppinen, S. (2008). MicroRNAs as targets for antisense-based therapeutics. *Expert Opin. Biol. Ther.* 8: 59–81.
Suzuki, H. I., Yamagata, K., Sugimoto, K., et al. (2009). Modulation of microRNA processing by p53. *Nature* 460: 529–533.
Sylvestre, Y., De Guire, V., Querido, E., et al. (2007). An E2F/miR-20a autoregulatory feedback loop. *J. Biol. Chem.* 282: 2135–2143.
Takamizawa, J., Konishi, H., Yanagisawa, K., et al. (2004). Reduced expression of the let-7 microRNAs in human lung cancers in association with shortened postoperative survival. *Cancer Res.* 64: 3753–3756.
Tanaka, M., Oikawa, K., Takanashi, M., et al. (2009). Down-regulation of miR-92 in human plasma is a novel marker for acute leukemia patients. *PLoS One* 4: e5532.
Tang, J. T., Fang, J. Y. (2009). MicroRNA regulatory network in human colorectal cancer. *Mini. Rev. Med. Chem.* 9: 921–926.
Tarasov, V., Jung, P., Verdoodt, B., et al. (2007). Differential regulation of microRNAs by p53 revealed by massively parallel sequencing: miR-34a is a p53 target that induces apoptosis and G1-arrest. *Cell Cycle* 6: 1586–1593.
Tazawa, H., Tsuchiya, N., Izumiya, M., et al. (2007). Tumor-suppressive miR-34a induces senescence-like growth arrest through modulation of the E2F pathway in human colon cancer cells. *Proc. Natl. Acad. Sci. U.S.A.* 104: 15472–15477.
Tili, E., Michaille, J. J., Gandhi, V., et al. (2007). miRNAs and their potential for use against cancer and other diseases. *Future Oncol.* 3: 521–537.
Toloubeydokhti, T., Bukulmez, O., Chegini, N. (2008). Potential regulatory functions of microRNAs in the ovary. *Semin. Reprod. Med.* 26: 469–478.
Tong, A. W., Nemunaitis, J. (2008). Modulation of miRNA activity in human cancer: A new paradigm for cancer gene therapy? *Cancer Gene Ther.* 15: 341–355.
Trang, P., Weidhaas, J. B., Slack, F. J. (2008). MicroRNAs as potential cancer therapeutics. *Oncogene* 27: 52–57.
Tsang, J., Zhu, J., van Oudenaarden, A. (2007). MicroRNA-mediated feedback and feedforward loops are recurrent network motifs in mammals. *Mol. Cell* 26: 753–767.
Ueda, T., Volinia, S., Okumura, H., et al. (2009). Relation between microRNA expression and progression and prognosis of gastric cancer: A microRNA expression analysis. *Lancet Oncol.* 11: 136–146.
Valeri, N., Croce, C. M., Fabbri, M. (2009). Pathogenetic and clinical relevance of microRNAs in colorectal cancer. *Cancer Genomics Proteomics* 6: 195–204.
Vandenboom, T. G., II, Li, Y., Philip, P. A., et al. (2008). MicroRNA and cancer: Tiny molecules with major implications. *Curr. Genomics* 9: 97–109.
Varnholt, H. (2008). The role of microRNAs in primary liver cancer. *Ann. Hepatol.* 7: 104–113.

Ventura, A., Jacks, T. (2009). MicroRNAs and cancer: Short RNAs go a long way. *Cell* 136: 586–591.
Verghese, E. T., Hanby, A. M., Speirs, V., et al. (2008). Small is beautiful: MicroRNAs and breast cancer: Where are we now? *J. Pathol.* 215: 214–221.
Visone, R., Croce, C. M. (2009). miRNAs and cancer. *Am. J. Pathol.* 174: 1131–1138.
Visone, R., Rassenti, L. Z., Veronese, A., et al. (2009). Karyotype-specific microRNA signature in chronic lymphocytic leukemia. *Blood* 114: 3872–3879.
Voorhoeve, P. M. (2010). MicroRNAs: Oncogenes, tumor suppressors or master regulators of cancer heterogeneity? *Biochim. Biophys. Acta* 1805: 72–86.
Voorhoeve, P. M., Agami, R. (2007). Classifying microRNAs in cancer: The good, the bad and the ugly. *Biochim. Biophys. Acta* 1775: 274–282.
Vrba, L., Jensen, T. J., Garbe, J. C., et al. (2010). Role for DNA methylation in the regulation of miR-200c and miR-141 expression in normal and cancer cells. *PLoS One* 5: e8697.
Vriens, M. R., Schreinemakers, J. M., Suh, I., et al. (2009). Diagnostic markers and prognostic factors in thyroid cancer. *Future Oncol.* 5: 1283–1293.
Waldman, S. A., Terzic, A. (2009). A study of microRNAs in silico and in vivo: Diagnostic and therapeutic applications in cancer. *FEBS J.* 276: 2157–2164.
Wang, H., Garzon, R., Sun, H., et al. (2008). NF-kappaB-YY1-miR-29 regulatory circuitry in skeletal myogenesis and rhabdomyosarcoma. *Cancer Cell* 14: 369–381.
Wang, L., Tang, H., Thayanithy, V., et al. (2009). Gene networks and microRNAs implicated in aggressive prostate cancer. *Cancer Res.* 69: 9490–9497.
Wang, Q. Z., Xu, W., Habib, N., et al. (2009). Potential uses of microRNA in lung cancer diagnosis, prognosis, and therapy. *Curr. Cancer Drug Targets* 9: 572–594.
Wang, V., Wu, W. (2009). MicroRNA-based therapeutics for cancer. *BioDrugs* 23: 15–23.
Wang, X. S., Zhang, J. W. (2008). The microRNAs involved in human myeloid differentiation and myelogenous/myeloblastic leukemia. *J. Cell. Mol. Med.* 12: 1445–1455.
Wang, Y., Lee, C. G. (2009). MicroRNA and cancer: Focus on apoptosis. *J. Cell. Mol. Med.* 13: 12–23.
Wang, Y. X., Zhang, X. Y., Zhang, B. F., et al. (2010). Initial study of microRNA expression profiles of colonic cancer without lymph node metastasis. *J. Dig. Dis.* 11: 50–54.
Wiemer, E. A. (2007). The role of microRNAs in cancer: No small matter. *Eur. J. Cancer* 43: 1529–1544.
Wijnhoven, B. P., Michael, M. Z., Watson, D. I. (2007). MicroRNAs and cancer. *Br. J. Surg.* 94: 23–30.
Williams, A. C., Browne, S. J., Manning, A. M., et al. (1993). Biological consequences of the genetic changes which occur during human colorectal carcinogenesis. *Semin. Cancer Biol.* 4: 153–159.
Winter, J., Jung, S., Keller, S., et al. (2009). Many roads to maturity: MicroRNA biogenesis pathways and their regulation. *Nat. Cell Biol.* 11: 228–234.
Wong, Q. W., Ching, A. K., Chan, A. W., et al. (2010). miR-222 overexpression confers cell migratory advantages in hepatocellular carcinoma through enhancing AKT signaling. *Clin. Cancer Res.* 16: 867–875.
Woods, K., Thomson, J. M., Hammond, S. M. (2007). Direct regulation of an oncogenic microRNA cluster by E2F transcription factors. *J. Biol. Chem.* 282: 2130–2134.
Wu, G., Park, M. Y., Conway, S. R., et al. (2009). The sequential action of miR156 and miR172 regulates developmental timing in *Arabidopsis*. *Cell* 138: 750–759.
Wu, H., Mo, Y. Y. (2009). Targeting miR-205 in breast cancer. *Expert Opin. Ther. Targets* 13: 1439–1448.
Wu, W., Sun, M., Zou, G. M., et al. (2007). MicroRNA and cancer: Current status and prospective. *Int. J. Cancer* 120: 953–960.
Wu, X., Piper-Hunter, M. G., Crawford, M., et al. (2009). MicroRNAs in the pathogenesis of lung cancer. *J. Thorac. Oncol.* 4: 1028–1034.
Yamakuchi, M., Lowenstein, C. J. (2009). MiR-34, SIRT1 and p53: The feedback loop. *Cell Cycle* 8: 712–715.
Yamakuchi, M., Ferlito, M., Lowenstein, C. J. (2008). miR-34a repression of SIRT1 regulates apoptosis. *Proc. Natl. Acad. Sci. U.S.A.* 105: 13421–13426.
Yan, H. L., Xue, G., Mei, Q., et al. (2009). Repression of the miR-17-92 cluster by p53 has an important function in hypoxia-induced apoptosis. *EMBO J.* 28: 2719–2732.
Yang, L., Belaguli, N., Berger, D. H. (2009). MicroRNA and colorectal cancer. *World J. Surg.* 33: 638–646.
Yang, N., Coukos, G., Zhang, L. (2008a). MicroRNA epigenetic alterations in human cancer: One step forward in diagnosis and treatment. *Int. J. Cancer* 122: 963–968.
Yang, N., Kaur, S., Volinia, S., et al. (2008b). MicroRNA microarray identifies Let-7i as a novel biomarker and therapeutic target in human epithelial ovarian cancer. *Cancer Res.* 68: 10307–10314.
Yang, X., Feng, M., Jiang, X., et al. (2009). miR-449a and miR-449b are direct transcriptional targets of E2F1 and negatively regulate pRb-E2F1 activity through a feedback loop by targeting CDK6 and CDC25A. *Genes Dev.* 23: 2388–2393.

Yong, A. S., Melo, J. V. (2009). The impact of gene profiling in chronic myeloid leukaemia. *Best Pract. Res. Clin. Haematol.* 22: 181–190.

Yu, S. L., Chen, H. Y., Yang, P. C., et al. (2007). Unique microRNA signature and clinical outcome of cancers. *DNA Cell Biol.* 26: 283–292.

Yu, Z., Wang, C., Wang, M., et al. (2008). A cyclin D1/microRNA 17/20 regulatory feedback loop in control of breast cancer cell proliferation. *J. Cell Biol.* 182: 509–517.

Zhang, B., Pan, X., Cobb, G. P., et al. (2006). MicroRNAs as oncogenes and tumor suppressors. *Dev. Biol.* 302: 1–12.

Zhang, L., Volinia, S., Bonome, T., et al. (2008). Genomic and epigenetic alterations deregulate microRNA expression in human epithelial ovarian cancer. *Proc. Natl. Acad. Sci. U.S.A.* 105: 7004–7009.

Zhang, W., Dahlberg, J. E., Tam, W. (2007). MicroRNAs in tumorigenesis: A primer. *Am. J. Pathol.* 171: 728–738.

Zoon, C. K., Starker, E. Q., Wilson, A. M., et al. (2009). Current molecular diagnostics of breast cancer and the potential incorporation of microRNA. *Expert Rev. Mol. Diagn.* 9: 455–467.

20 Marine Metabolomics in Cancer Chemotherapy

David M. Pereira, Georgina Correia-da-Silva, Patrícia Valentão, Natércia Teixeira, and Paula B. Andrade
University of Porto
Porto, Portugal

CONTENTS

20.1 METABOLOMICS: AN OVERVIEW

Metabolomics is the study of the complete collection of all metabolites within an organism, which today remains an unachievable goal (Sumner et al., 2003). This arises as a consequence of the high number of different small molecules present in organisms and the low concentration at which they occur. In addition, presently there are no analytical techniques or combination of techniques that allow for the identification of all of these metabolites. Small molecules are very diverse in their physical and chemical properties and occur in a wide concentration range. For example, within lipids alone, both high abundant compounds, such as fatty acids, triglycerides, or phospholipids, and trace level components with important regulatory effects, like eicosanoids derived from arachidonic acid, are encountered (Dettmer et al., 2007; Pereira et al., 2010). Several different approaches can be used depending on the objective of the work and the type of information needed (Dunn et al., 2005).

In metabolite profiling, compounds related through similar chemistry or metabolic pathways are studied. An example of this would be the study of lipids (lipidomics) or the choline pathway. In this kind of approach, selective extraction procedures are usually employed, which allows the removal of other metabolites that are regarded as interferences, thus improving the analysis of the target molecules. From an instrumental point of view, liquid chromatography (LC)-mass spectrometry (MS), gas chromatography (GC)-MS, and LC-nuclear magnetic resonance (NMR) are the most used options (Hollywood et al., 2006). Whenever there is a hypothesis regarding the target molecules, other analytical techniques that provide less structural information, such as high-performance liquid chromatography (HPLC)-diode array detection, can be used when standards are available.

In some cases, however, the identities and the quantification of all of the metabolites studied are not required. Instead, a metabolic fingerprint is generated, in which a profile of several peaks is obtained. This approach can be used, for instance, in the search of disease biomarkers by comparing the profile of healthy subjects and patients or in quality control for the identification of adulterations. After statistical treatment, the peaks that represent major differences between samples are identified, further studied, and structurally elucidated (Pereira et al., 2010). In this metabolomic strategy, the sample preparation is reduced, once the objective is a global, rapid, and high-throughput analysis of crude samples (Dunn et al., 2005).

The widespread use of cell cultures in several research areas gave way to a new approach called *metabolite footprinting*. In this technique, the analysis is performed on the culture media. By using this method, compounds secreted as a consequence of cellular metabolism can be studied without the need of cell disruption and lysis.

Target metabolite analysis is another approach in which there is complete knowledge concerning the metabolite to be found. Therefore, all other metabolites can be ignored, and the extraction, sample preparation, and analysis techniques can be adapted to the target metabolite (Pereira et al., 2010).

20.2 METABOLOME PROFILE OF PLANTS, ANIMALS, AND OTHER ORGANISMS AND THEIR IMPACT ON HUMAN HEALTH

Metabolomics is a powerful tool in diagnosis and prognosis and even in the evaluation of the response to therapy. Nevertheless, there are other applications; for instance, when applied to vegetable and animal species, metabolomics can be a precious tool for discovering molecules to be used in human therapeutics. Natural compounds have already proved to be an amazing source of chemical diversity and, as a consequence, possess interesting pharmacological properties. This is easily represented if we consider that nearly 60% of all drugs introduced into therapy between 1981 and 2006 were first identified as natural products. This number increases to 75% if we consider only cancer drugs (Newman and Cragg, 2007). When compared with synthetic compounds, natural molecules usually have well-defined three-dimensional structures and fit biological targets that are conserved across species. Thus, metabolomics is nowadays a powerful tool in drug discovery, and many molecules from natural products are currently used either in therapeutics or in clinical trials.

20.3 METABOLOMICS: APPLICATIONS IN DIAGNOSIS AND PROGNOSIS

In recent years, the application of metabolomics in diagnosis, prognosis, and therapy has experienced an exponential growth, partly because of the exponential growth that metabolomics itself has experienced (Figure 20.1). The increasing understanding of how the human metabolome is affected in health and disease has marked a new era in which several metabolites could be used to identify pathological conditions, sometimes even before the onset of symptoms.

Without a doubt, one of the most exciting promises of metabolomics is the potential discovery of new disease biomarkers. This involves the metabolite profile of healthy subjects compared with that of patients. A first approach would be to use techniques that provide an overview of the metabolome, such as NMR to monitor the individual's metabolic fingerprint. In this level, no definitive identification of the metabolites is performed. Instead, statistical tools are employed in order to identify the peaks that are exclusive to patients, which could be possible candidates for new biomarkers. The last step consists in identifying the selected metabolites.

In the case of cancer, tumor cell metabolism is distinct from normal cells. These differences are explored in metabolomic studies that try to find new biomarkers for cancer onset, evolution, and phenotype. In breast cancer, for instance, *in vitro* studies with cells with distinct estrogen receptor status and metastatic potential indicated that the levels of phosphodiesters and uridine diphosphosugar metabolites were significantly higher in estrogen receptor-positive and low metastatic potential cell lines (Sterin et al., 2001). Glycolysis is also impaired in cancer conditions (Warburg effect), with most tumors displaying increased glucose uptake. A positive correlation between glucose uptake and tumor aggression has been reported (Gatenby and Gillies, 2004).

Diagnosis of prostate cancer is another area that could benefit from metabolomics studies. Although prostate cancer is characterized by high mortality and a long-term asymptomatic disease course, current laboratory diagnosis of this pathology relies on prostate-specific antigen (PSA) and suffers from poor accuracy (Lokhov et al., 2009). The same authors used a MS-based

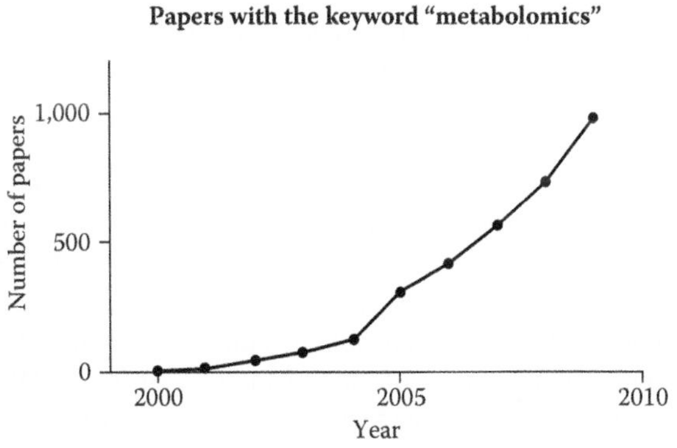

FIGURE 20.1 Evolution of the number of scientific papers including the keyword "metabolomics" between 2000 and 2009 (search performed using "Scopus" on May 5, 2010).

metabolite fingerprinting approach to analyze blood plasma for the diagnosis of prostate cancer. Overall, this new technique displayed sensitivity, specificity, and accuracy of 95%, 96.7%, and 95.7%, respectively, which was superior to enzyme-linked immunosorbent assay PSA test, which exhibited 35%, 83%, and 52% accuracy, respectively. This topic will be discussed in detail in other chapters.

20.4 ANALYTICAL TECHNOLOGIES IN METABOLOMICS

For obvious reasons, sample preparation is of major importance in the field of metabolomics, being far more critical for quantification than for identification. Sample preparation largely depends on the metabolomic strategies (metabolic fingerprinting and target metabolite analysis, among others). The selection of a suitable procedure for extraction is highly dependent on the type of metabolites to be determined and the nature of the sample (its physical state, polarity, and the quantity of metabolites, among others). For an efficient extraction, the solvent must be able to solubilize the target molecules while leaving the sample matrix intact, and its polarity should closely match that of the target compounds. Thus, mixing solvents of different polarities is a strategy that can be used to extract a wide range of chemical classes.

The conditions employed should be as mild as possible in order to avoid the chemical artifacts arising from hydrolysis, oxidation, and isomerization. Some classes of compounds are known to be highly affected by extraction procedures, such as carotenoids, that readily photoisomerize (Ferreres et al., 2010a). There should also be certainty that compounds are not covalently bound to the matrix, which can happen, for instance, for some low-molecular-weight acids. Usually, a digestion step can overcome this difficulty. Detailed discussion of extraction protocols is beyond the scope of this chapter.

Several techniques with high sensitivity and reproducibility are currently available for metabolomics studies. Some of the current technologies for determining the metabolome require separation techniques, such as HPLC and GC, and detection techniques like MS and NMR (Hollywood et al., 2006). However, for other metabolomic strategies (metabolic fingerprinting and footprinting), it is possible to do direct injection of the sample onto detectors without chromatographic separation (Hollywood et al., 2006).

Nowadays, NMR and MS hyphenated techniques constitute the most powerful and useful tools in metabolomics. Mass spectrometry is a technique that relies on the fragmentation of target molecules as a result of exposure to high energy and can provide important information about the target compounds, such as the molecular weight and functional groups. In addition to its high sensitivity and selectivity, MS can detect NMR-invisible moieties, such as sulfates (Dettmer et al., 2007). For some chemical classes, the characteristic fragmentation pattern can allow the characterization of the unknown compound's structure. However, in most cases MS does not allow differentiation of isomers, which are particularly important in a biological context, because the bioactivity of some molecules is highly dependent on the isomer and conformation. This is mostly important in the case of ligand-receptor interactions. For this reason, chemical identification of a compound usually requires the use of NMR at some point.

In metabolomics, the isolated use of MS can be found in very few situations (the already mentioned metabolic fingerprinting and footprinting, for instance). In the majority of cases, this technique is coupled to a chromatographic separation step, usually GC or LC. Each technique has its own applications, strengths, and weaknesses. In the case of GC, compounds with high to medium volatility are ideal candidates. Analysis of nonvolatile compounds is possible, although derivatization steps are required that result in a time-consuming analysis. Still, GC-MS offers excellent reproducibility and low detection limits. Given the fact that several databases of compound identification exist, a tentative identification in the absence of standards is possible.

LC-MS is a robust technique that has proved to be a useful method for metabolomics. MS is based on the production of ions from the analyte. The fragmentation products are impelled and focused through a magnetic mass analyzer to be further collected, and each selected ion is measured in a detector (Ferreres et al., 2010b). The choice of the ionization method to be employed is dependent on the physical-chemical properties of the analyte(s) of interest (volatility, molecular weight, thermolability, and the complexity of the matrix in which the analyte is contained, among others). Regarding ion sources, the most popular include electron ionization, chemical ionization, electrospray ionization, atmospheric pressure chemical ionization, atmospheric pressure photoionization, and matrix-assisted laser desorption/ionization.

Analysis by UV-visible and MS, when in total ion count mode, has detection limits in the range of 10 ng. When single-ion monitoring (SIM) mode is used, MS analysis provides better detection limits, usually below 1 ng (Cuyckens and Claeys, 2004). SIM mode, however, causes a loss of valuable information concerning the fragmentation pattern, which is very important for the identification of many compounds (Pereira et al., 2010).

Although it is less sensitive than MS, NMR can provide strong structural information, particularly in the case of isomers. Sensitivity is perhaps the most important requirement for metabolomics. Here, ^{1}H NMR, with a detection threshold of around approximately 5 nmol, is several orders of magnitude less sensitive than MS, which has a detection threshold of 10^{-12} mol. However, in this field there has been increasing success in improving NMR sensitivity, for instance by the use of cryoprobes, in which the sensitivity is increased by cooling the detection system, whereas the sample remains at room temperature. This limits the noise voltage associated with signal detection, and when compared with regular probes, the signal-to-noise ratio is ameliorated by a factor of 3–4 (Krishnan et al., 2005).

NMR coupled to LC is an analytical tool even more powerful, but its prohibitive price has prevented its widespread use. When working in on-line flow NMR, the acquisition time is limited by the short presence of the sample in the detection coil as a consequence of the flow rates commonly used, thus resulting in poor signal-to-noise ratio values. Also, when the flow rate is reduced by a factor of 3–10, a better signal/noise can be registered, followed by an increase in experimental acquisition time, which may lead to diffusion processes that can influence the separation of peaks eluted from the LC column. In order to surpass this problem, accumulation of peaks into storage loops for off-line NMR at a later stage is possible (Wolfender et al., 2003). In natural products chemistry, the use of standard compounds is not possible in most cases, because of the novelty of the compounds, and for this reason, a multi-instrumental approach is usually required.

20.5 METABOLOME PROFILE OF MARINE ORGANISMS AND THEIR IMPACT ON HUMAN DISEASE AND CHEMOTHERAPY

For years, nature has been a source of molecules with marked biological activities. Among natural products, plants were the main origin of bioactive compounds. Recently, marine natural products have proved to be an amazing source of chemical diversity, and consequently, many marine metabolites were shown to exhibit a number of different pharmacological activities. These include pharmacological treatments that range from cancer to inflammation and allergy, among others. This remarkable diversity arises from the fact that many marine organisms, having only primitive immune systems and soft bodies, produce secondary metabolites that can be an effective defense against predation. In fact, organisms with lower physical defenses, such as sponges and mollusks, are usually the ones with the most bioactive molecules.

Their marked biological activity results from the diversity of these small molecules often found in marine chemistry. Several cellular targets have been described for these compounds, and some of them constitute very promising options to counter the frequent problem of chemotherapy resistance. Because of the high chemical diversity, a chemical-based classification was used to describe the cellular effects of these compounds. An overview of the compounds' cellular targets is presented in Figure 20.2. It should be noted that the majority of these compounds have mitochondria

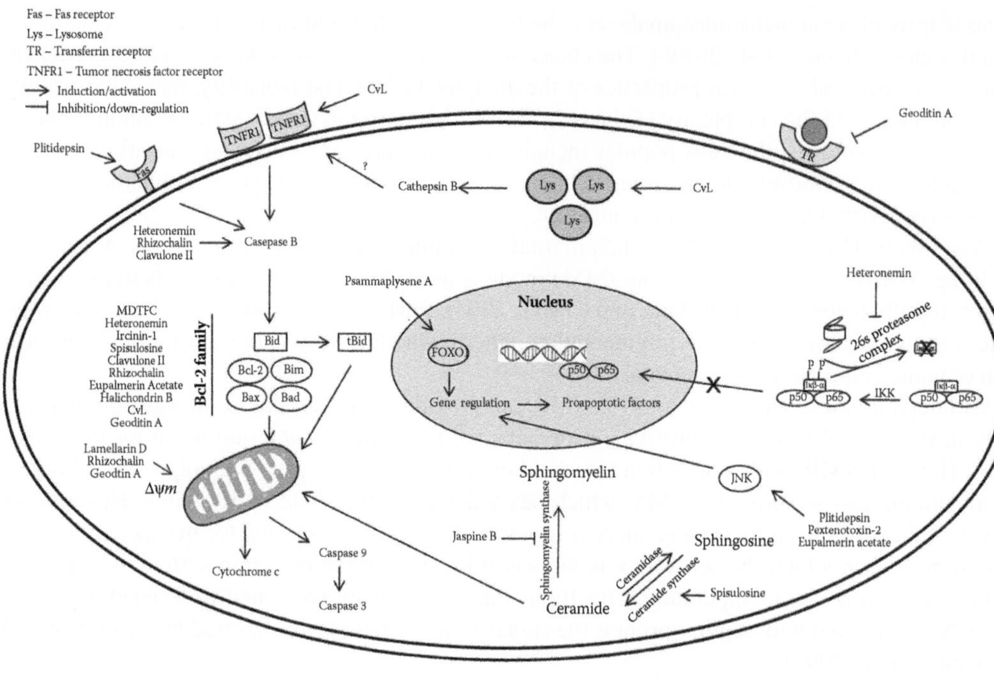

FIGURE 20.2 Major targets of cell death of some of the compounds discussed in this chapter.

as cellular targets, either directly or indirectly. The cell lines used in the cited studies can be found in Table 20.1.

20.5.1 Terpenes

20.5.1.1 MDTFC

Arepalli et al. (2009) conducted a study on the soft coral *Sinularia kavarittiensis*. Dichloromethane and methanol were used to extract the material, and following ethyl acetate extraction of the residue, a silica-gel column chromatography allowed the recovery of five fractions. Further purification led to the isolation of the furano-sesquiterpene methyl 5-[(1*E*,5*E*)-2,6-dimethyl-octa-1,5,7-trienyl] furan-3-carboxylate (MDTFC) (Figure 20.3), identified by LC-MS and ^{1}H and ^{13}C NMR (Arepalli et al., 2009).

Several cell lines were exposed to the compound, namely U937, THP-1, Jurkat, Raji, Daudi, HL-60, HeLa, Mia-Pa-Ca-2, and A375 (Table 20.1). Among these, THP-1 cell line was the most susceptible to MDTFC cytotoxicity, with an IC_{50} of 29.59 μM. After 3 hours of incubation, low concentrations of the compound (5–30 μg/mL) induced the appearance of apoptotic cells, whereas a concentration of 10 μg/mL caused cell arrest in the sub-G_1 phase in approximately 60% at 24 hours.

The authors also found several traits of apoptosis, including membrane blebbing, chromatin condensation, DNA fragmentation, and a decrease in the levels of procaspase-3 and -9, as well as an increase in the ratio Bcl-2 associated protein x (Bax)/B-cell lymphoma 2 (Bcl-2). Members of the Bcl-2 family can act either as antiapoptotic or proapoptotic factors. Proteins with antiapoptotic activity contain four Bcl-2 homology domains (BH1-4), and their majority is found within outer mitochondrial membrane (OMM), although some exist in the cytoplasm and endoplasmatic reticulum. These molecules directly inhibit proapoptotic Bcl-2 proteins, thus protecting OMM integrity. Major members of this class are Bcl-2 long isoform (BCL-xL), Bcl-2-related gene A1 (A1), and myeloid cell leukemia 1 (Chipuk et al., 2010). One further division concerning proapoptotic

TABLE 20.1
Cell Lines Used in the Works Covered by This Review

Cell Line	Designation
786-O	Human renal carcinoma
A375	Human amelanotic melanoma
A431	Human squamous carcinoma
A549	Human lung carcinoma
AsPC-1	Human pancreas adenocarcinoma
B16	Murine melanoma
BT474	Human breast ductal carcinoma
Daudi	Human Burkitt's lymphoma
ECC1	Human endometrial adenocarcinoma
HCT116	Human colon carcinoma
HeLa	Human cervix epitheloid carcinoma
Hep3B	Human hepatocellular carcinoma
HepG2	Human hepatocellular carcinoma
HL-60	Human promyelocytic leukemia
HT29	Human colon adenocarcinoma
Ishikawa	Human endometrial adenocarcinoma
Jurkat	Human T cell leukemia
K562	Human chronic myelogenous leukemia
H292	Human lung mucoepidermoid carcinoma
LNCaP	Human $p53^+$, AR^+ prostate cancer
MCF7	Human breast adenocarcinoma
Mia-Pa-Ca-2	Human pancreatic carcinoma
N2a	Mouse neuroblastoma
NIH-3T3	Mouse embryonic fibroblasts
NCI-H460	Human large cell lung carcinoma
P388	Murine leukaemia
PC3	Human $p53^-$, AR^- prostate adenocarcinoma
Raji	Human Burkitt's lymphoma
RH-7777	Rat hepatoma
SK-BR-3	Human breast adenocarcinoma carcinoma
SK-hep-1	Human liver adenocarcinoma
SK-MEL-28	Human melanoma
SW1573	Human non-small cell lung cancer
THP-1	Human monocytoid leukemia
UACC-257	Human melanotic melanoma
U251	Human glioblastoma cells
U87-MG	Human malignant glioma (wild-type p53)
U373-MG	Human malignant glioma (mutant p53)
U937	Human histiocytic lymphoma

proteins includes effector proteins (Bcl-2 antagonist killer 1 [Bak] and Bax) and BH3-only ones (Bid, Bim, and Bad).

The treatment of the cells with caspase-3 and -9 inhibitors moderately prevented DNA fragmentation, thus showing that the intrinsic or mitochondrial pathway was activated (Figure 20.2). In this pathway, permeabilization of outer mitochondrial membrane results in the leakage of proapoptotic molecules, such as cytochrome *c*, Smac/DIABLO, HtrA2/Omi, apoptosis inducing factor, and endonuclease G. Once released, cytochrome *c* initiates the assembly of apoptotic protease-activating factor 1 and procaspase-9 into a holoenzyme complex called the *apoptosome*, which in turn activates the initiator caspase-9 (Dirsch et al., 2003) (Figure 20.2).

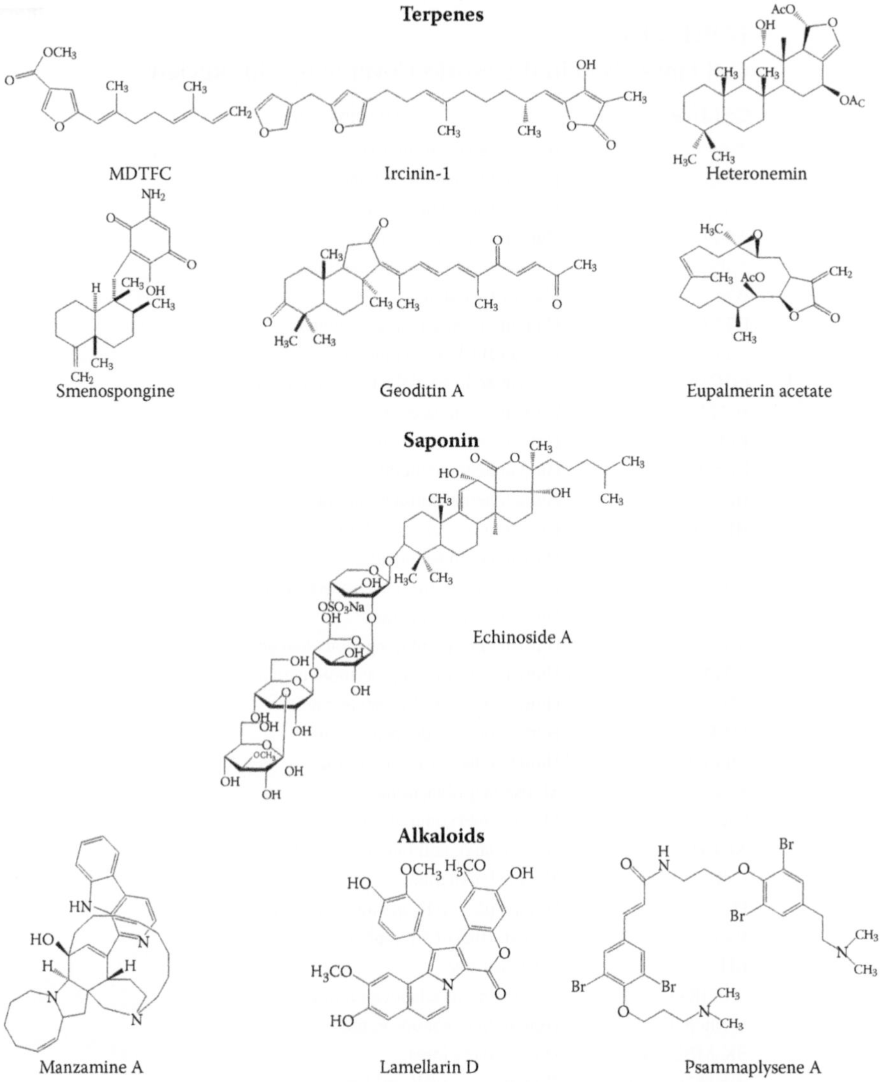

FIGURE 20.3 Structures of terpenes, saponin, and alkaloids discussed in this chapter. Ac, acetyl.

The release of cytochrome *c* to the cytosol indicated that MDTFC-mediated toxicity involved mitochondrial dysfunction, which was further confirmed by the study of the effect of this molecule on mitochondrial membrane potential ($\Delta\psi m$). In fact, MDTFC proved to cause depolarization of the mitochondrial membrane within 6 hours of exposition. This action could be a consequence of the above referred modulation of the level of the proapoptotic protein Bax and of the antiapoptotic protein Bcl-2, which play important roles in apoptotic cell death (Chipuk et al., 2010).

20.5.1.2 Heteronemin

Heteronemin (Figure 20.3) is a scalarene sesterterpene first described in 1974 (Kazlauskas et al., 1976), isolated from the sponge *Heteronema erecta*. In the following year, its ^{13}C NMR spectrum and stereochemistry were fully described (Kashman and Rudi, 1977). Recent works have demonstrated the ability of heteronemin to modulate nuclear factor κB (NF-κB) activity. This molecule is an inducible transcription factor that promotes prosurvival pathways and has a ubiquitous cell distribution. It consists of a dimer of proteins belonging to the Rel family, which include RelA (p65), RelB, c-Rel, p50, and p52 (Keutgens et al., 2006), with the p50/p65 heterodimer being the most common form.

In a basal situation, NF-κB can be found in cytoplasm in an inactive form because of the binding of the inhibitor IκB. Upon stimulation, the IκB kinase complex (IKK) is activated and results in IκB phosphorylation, which marks the protein for degradation by the 26S proteasome. Once free from IκB, NF-κB translocates to the nucleus, where it exerts a number of actions (Evans, 2005) (Figure 20.2).

The causal relation between NF-κB activation and the onset of some types of cancer is well reported (Coussens and Werb, 2002). Schumacher et al. (2010) showed that heteronemin modulates tumor necrosis factor α (TNFα)-induced NF-κB activation on the chronic myelogenous leukemia cell line K562 (Table 20.2), as well as cell-cycle regulation. In concentrations starting at 4 μM, heteronemin was able to reduce the activating effect of TNFα on NF-κB. This result was confirmed by studying the ability of the compound to prevent DNA binding of NF-κB, which showed that, in the same concentration, complete inhibition was achieved. The same behavior was found when the Jurkat cell line was used. This is a very interesting target for heteronemin given the fact that, to our knowledge, few molecules from marine sources are able to target DNA binding of NF-κB (Bremner and Heinrich, 2002).

As described before, one of the first steps to activate NF-κB is IκB degradation, which is followed by translocation of p50/p65 to the nucleus. In the case of heteronemin, the authors showed that the compound completely prevented IκB degradation via attenuation of phosphorylation (Schumacher et al., 2010). Given the fact that IκB degradation can be affected either by IKK or proteasome activities, these authors showed that, whereas no inhibition of IκB activity took place, 0.5 μM heteronemin inhibited proteasome activity (55% of chymotrypsin- and trypsin-like proteolytic activity and 30% of caspase-like activity).

Additional work showed that heteronemin affected cell viability in a dose-dependent manner, reaching 80% of viability loss with a concentration of 5.6 μM. Further studies showed an increase in activated caspase-3, -8, and -9, as well as truncated Bid, which suggested that a cross-talk between the intrinsic and extrinsic pathways of apoptosis was taking place.

20.5.1.3 Ircinin-1

Ircinin-1 is a sesterterpene lactone (Figure 20.3) isolated from the sponge *Sarcotragus* sp. When SK-ML-28 cells were exposed to 25 μM ircinin, accumulation of cells in the G_1 phase was observed, whereas treatments with higher concentrations induced a decrease in the G_1 phase population (Choi et al., 2005). G_1 phase arrest was both dose and time dependent. In addition to the effect in cell cycle, morphological changes compatible with apoptosis were also described. The apoptotic pathway was further confirmed by the change in the Bax/Bcl-2 ratio, cleavage of poly(ADP-ribose) polymerase (PARP), cytochrome *c* release, and activation of caspase-3 and -9 (Figure 20.2 and Table 20.2).

Because of the effect of the compound in cell-cycle populations, these authors studied the expression of several cell-cycle regulating proteins, by Western blot and reverse transcription (RT)-polymerase chain reaction (PCR). Overall, ircinin-1 was shown to significantly reduce the expression of cyclin D, as well as of Cdk4 and Cdk6. The levels of the casein kinase 1 $p21^{WAF1/CIP1}$, a regulator of cell-cycle progression, were also markedly increased (Table 20.2).

20.5.1.4 Geoditin A

Isomalabaricane triterpene geoditin A (Figure 20.3 and Table 20.2), a compound that has been described in the sponge *Geodia japonica*, induced a dose-dependent toxic effect, with an IC_{50} of 20 μg/mL in the HT29 colon cancer cell line (Table 20.2). Human dermal fibroblasts showed an IC_{50} value three times higher. In addition, geoditin A also downregulated transferrin receptor (Cheung et al., 2010) (Figure 20.2). This ability is particularly important considering that, although iron is required for normal cell metabolism, transferrin receptor is upregulated in cancer cells. Iron

TABLE 20.2
Overview of the Compounds Discussed in This Chapter: Chemical Class, Organism of Origin, Cellular Target, and Cell Line Used

	Compound	Class	Organism	Cellular Target	Cell Line	Reference
Terpenes	MDTFC	Furano-sesquiterpene	*Sinularia kavarittiensis*	OMM; Caspase-3, -9: Bax/Bcl2	THP-1; U937; Jurkat; Raji; Daudi HL-60; HeLa; Mia-Pa-Ca-2; A375	Arepalli et al. (2009)
	Heteronemin	Scalarene sesterterpene	*Hyrtios erecta*	IκB; 26s proteasome; Caspase3,- 8 and -9;	K562	Schumacher et al. (2010)
	Ircinin-1	Sesterterpene lactone	*Sarcotragus* sp.	Bax/Bcl2; Caspase-3, -9; Cyclins D, Cdk4, Cdk6, p21	SK-MEL-2	Choi et al. (2005)
	Geoditin A	Cembrenolide triterpene	*Geodia japonica*	Transferrin receptor; induction of oxidative stress	HT29	Cheung et al. (2010)
	Eupalmerin acetate	Diterpene	*Eunicea succinea*	Bax; JNK pathway	U87-MG; U373-MG	Iwamaru et al. (2007)
	Smenospongine	Sesquiterpene Aminoquinone	*Dactylospongia elegans*	P21-Rb pathaway	HL-60; K562	Kong et al. (2008)
Saponin	Echinoside A	Triterpenic saponin	*Holothuria nobilis*	Topoisomerase 2α	See Li et al., 2010	Li et al. (2010)
Alkaloids	Manzamine A (HB-071)	3-Alkylpyridine alkaloid	*Haliclona* sp./ *Xestospongia* sp./ *Pellina* sp.	GSK3β; Cyclin D1; p65	AsPC-1	Guzman et al. (2010)
	Lamellarin D	Dihydroisoquinoline pyrrole alkaloid	*Lamellaria* sp.	Δψm; calcium uptake; mitochondrial respiration; ATP synthesis	Jurkat; P388	Ballot et al. (2010)
	Psammaplysene A	Phenylethylamine alkaloid	*Psammaplysilla* sp.	FOXO1	Ishikawa; ECC1	Berry et al. (2009)

Lipids	Spisulosine	Sphingoid-type base	*Spisula polynema*	*De novo* synthesis of ceramide; PKCζ; caspase-3, -9	PC-3; LNCaP; N2a; NIH-3T3; RH7777	Sanchez et al. (2008); Salcedo et al. (2007)
	Jaspine B	Anhydrophytosphingosine	*Pachastrissa* sp.; *Jaspis* sp.	Sphingomyelin synthase	SK-MEL-28; B16	Salma et al. (2009)
	Rhizachalin	Two-headed sphingolipid	*Rhizochalina incrustata*	Caspase-3, -8, -9; Δψm	HL-60	Jin et al. (2009)
	Clavulone II	Cyclopentenone prostaglandin	*Clavularia viridis*	Cyclin D1; caspase-3, -8, -9; Bcl-2	HL-60	Huang et al. (2005)
Peptides	Kahalalide F	Depsipeptide	*Elysia rufescens*	ErbB3; PI3K-Akt signalling	SKBR3; BT474; MCF-7; A549; A431; NCI-H292 SW1573; SKhep1; HepG2; Hep3B; HT29	Janmaat et al. (2005)
	Plitidepsin	Depsipeptide	*Aplidium albicans*	Rac1/JNK pathway	SK-MEL-28; UACC-257	Munoz-Alonso et al. (2008)
Macrolides	Iejimalide B	Macrolide	*Eudistoma rigida*	Androgen receptor mRNAR; cyclinA, D1, and E mRNA; p21 mRNA	LNCaP; PC3	Wang et al. (2008)
	Candidaspongiolide	Macrolide	*Candidaspongia* sp.	eIF2; protein synthesis; caspase-12	U251; HCT116	Trisciuoglio et al. (2008)
	Halichondrin B	Polyether macrolide	*Aninella* sp.	GADD45 α, β, and γ proteins; Bad; caspase-3	A549	Catassi et al. (2006)
	Pextenotoxin-2	Macrolide	Dinoflagellates; sponges: shellfish	JNK pathway; actin	U937; HL-60; THP-1; K562	Moon et al. (2008)
Miscellaneous	CvL	Lectin	*Cliona varians*	Cathepsin B; TNFR1; p65; p21; pRb	K562; Jurkat	Queiroz et al. (2009)

depletion either by iron chelators or by decreased expression of transferrin or transferrin receptor has proved to be an effective cancer treatment (Daniels et al., 2006).

Regular cell metabolism gives rise to a number of oxidant species that can have severe deleterious effects on proteins and metabolic machinery. For this reason, a balance of oxidants and antioxidants is required to keep cellular homeostasis, which can be perturbed either by an increase of the former or a decrease of the latter. In order to elucidate the mechanism of action of geoditin A, prior to exposure to geoditin A, cells were incubated with diphenyleneidonium and rotenone, which are NADPH oxidase and electron transport complex I inhibitors, respectively. Given the fact that iron can also give rise to reactive oxygen species via the Fenton reaction (Halliwell and Gutteridge, 1995), an iron chelant, salicylaldehyde isonicotinoyl hydrazone, was also used. Neither of these compounds could prevent nor minimize geoditin A-induced toxicity. However, pretreatment of cells with *N*-acetylcysteine was able to reduce apoptosis by 60%. *N*-Acetylcysteine is a very important molecule in cell antioxidant defense because it is the precursor of glutathione, one of the most powerful antioxidant molecules. Overall, the authors proved that the mechanism of action of this isomalabaricane triterpene was related with its ability to induce oxidative stress.

In another study, geoditin A was evaluated for its apoptotic activity in the HL-60 cell line (Liu et al., 2005). Other isomalabaricane triterpenes were also evaluated, with geoditin A being the most potent and displaying a concentration-dependent activity (IC_{50} of 3 μg/mL). In addition to the morphological changes compatible with apoptotic death, found by microscopy, the authors also reported dissipation of mitochondrial membrane potentials and caspase-3 activation.

20.5.1.5 Eupalmerin Acetate

There have been increasing reports of marine molecules targeting c-Jun NH_2-terminal kinase pathway. Such is the case of eupalmerin acetate (Figure 20.2 and Table 20.2), a cembrenolide diterpene isolated from gorgonian octocoral *Eunicea succinea* (Iwamaru et al., 2007). This compound was assayed for its cytotoxicity against the human malignant glioma cells U87-MG (wild-type p53) and U373-MG (mutant p53) (Table 20.2).

Flow cytometry studies showed that after exposure to eupalmerin acetate in a concentration of 20 μM, the population of cells in sub-G_1 phase rose up to 91 and 57% in U87-MG and U373-MG cell lines, respectively. However, when eupalmerin acetate was used in a lower concentration (10 μM), G_2/M arrest was also detected.

Further studies revealed that eupalmerin acetate caused apoptosis via the mitochondrial pathway, in particular by inducing Bax translocation to the mitochondria. Also, the authors showed that the c-Jun NH_2-terminal kinase (JNK) pathway was the upstream signal pathway activated by eupalmerin acetate (Figure 20.2 and Table 20.2). These *in vitro* results were confirmed using mice with xenografts derived from U87-MG cells and treated with eupalmerin acetate at 50 mg/kg. After 19 days of treatment, the animal tumors were 80% smaller, thus showing the potential of this molecule in the chemotherapy of malignant glioma conditions (Iwamaru et al., 2007).

20.5.1.6 Smenospongine

Smenospongine (Figure 20.3 and Table 20.2) is a sesquiterpene aminoquinone isolated from the sponge *Dactylospongia elegans* (Kong et al., 2008). This compound proved to induce G_1 phase arrest in the K562 cell line, an effect that was not found in HL-60 and U937 human cell lines after incubation with the compound for 24 hours. Instead, HL60 and U937 cells presented a sub-G1 accumulation, indicating the presence of apoptotic cells. In K562 cells, this molecule proved to upregulate p21, a very important protein in cell-cycle regulation. In addition to this effect, exposure to smenospongine at 15 μM inhibited phosphorylation of Rb in K562 cells after a 48-hour incubation, which resulted in increased levels of underphosphorylated Rb protein (Kong et al., 2008). Rb is located downstream of p21 and has been proven to regulate G_1/S progression.

20.5.2 Saponin

20.5.2.1 Echinoside A

Echinoside A (Figure 20.3 and Table 20.2), a triterpenic saponin from sea cucumber *Holothuria nobilis*, was evaluated for its effect on the growth of 24 human cell lines (Li et al., 2010). The mechanism of action of this saponin was proved to be related to its nonintercalative inhibition of the catalytic activity of topoisomerase 2α (Top2α), an enzyme that unwinds and winds DNA to facilitate replication. Subsequent studies by the same group showed that echinoside A reduced the noncovalent binding of Top2α to DNA via competition for the DNA-binding site. Interference with pre- and poststrand passage DNA cleavage/religation equilibrium in the catalytic site of Top2α was also reported to be important to the activity of echinoside A and constitutes a new mechanism of action not described before.

These *in vitro* studies were further confirmed with *in vivo* studies using tumors in mouse models and human prostate xenografts in nude mouse models. In mouse S-180 sarcoma model, echinoside A at 3 mg/kg/day revealed a growth inhibition value of 80%. In the mouse H22 hepatoma, inhibition reached 66.1%.

20.5.3 Alkaloids

20.5.3.1 Manzamine A

Manzamine A (Figure 20.3 and Table 20.2) an 3-alkylpyridine alkaloid isolated from several organisms, including sponges of the genera *Haliclona* (Edrada et al., 1996), *Xestospongia* (Watanabe et al., 1998), and *Pellina* (Ichiba et al., 1994). This alkaloid has proved to display several pharmacological activities, including antitumor, antibacterial, antimalarial, and anti-inflammatory activity (Guzman et al., 2010).

Recently, Guzman et al. (2010) studied the effect of this compound in the pancreatic cancer cells AsPC-1. A cytotoxicity assay, as determined by the 3-(4,5-dimethylthiazol-2-yl)-2,5-diphenyltetrazolium bromide (MTT) assay, showed that manzamine A IC_{50} was 4.2 μM, after a 72-hour period of exposure.

Pancreatic cancer is a highly metastatic and lethal condition in which cells are usually resistant to apoptosis. Migration studies were performed in a collagen matrix, revealing the ability of this compound to prevent migration of AsPC-1 at a concentration of 10 μM. The ability of pancreatic cancer cells to migrate is one of the factors that contribute to its aggressive phenotype. In addition, manzamine A was able to induce apoptosis as demonstrated by annexin V binding assay.

20.5.3.2 Lamellarin D

Lamellarin D is a dihydroisoquinoline pyrrole alkaloid (Figure 20.3 and Table 20.2) that has shown cytotoxicity via induction of apoptosis in several cancer cell lines, even some that are multidrug-resistant because of its enhanced P-glycoprotein-mediated drug efflux (Vanhuyse et al., 2005). Its effect has equally been demonstrated in mutated p53 cells (Kluza et al., 2006). Kluza et al. (2006) showed the importance of mitochondrion to lamellarin D-mediated apoptosis, although the precise mechanism remained unknown. Recently, other authors (Ballot et al., 2010) demonstrated that induction of apoptosis by lamellarin D in the Jurkat and P388 cell lines was a result of reduction of $\Delta\psi m$, mitochondria swelling, and cytochrome *c* release. Calcium uptake by mitochondrion was also decreased by exposition to lamellarin D. Likewise, mitochondrial respiration and ATP synthesis was compromised (Ballot et al., 2010).

20.5.3.3 Psammaplysene A

Psammaplysene A is a phenylethylamine alkaloid from *Psammaplysilla* sp. Berry et al. (2009) studied the contribution of FOXO1 to induction of apoptosis in the endometrial cancer cell lines Ishikawa and ECC-1 by psammaplysene A (Figure 20.3 and Table 20.2).

Psammaplysene A was able to induce G_2/M in both cell lines at a concentration of 1 μM and an increase in cleaved PARP, thus showing that apoptosis was taking place. In addition, immunofluorescent staining studies revealed that psammaplysene A, at a concentration of 1 μM, increased nuclear FOXO1 protein levels (Berry et al., 2009).

The transcription factor FOXO1 plays an important role in several critical cellular processes that include cell survival and cell-cycle progression, among others (Burgering and Medema, 2003). When translocated to the nucleus, this transcription factor is also known to induce apoptosis by activating the expression of death receptor ligands or of Bcl-2 family members.

20.5.4 Lipids

20.5.4.1 Spisulosine

The effect of spisulosine (Figure 20.4 and Table 20.2), a sphingoid-type base compound isolated from the mollusk *Spisula polynema*, was studied in prostate tumor PC-3 and LNCaP cell lines (Sanchez et al., 2008). The compound revealed a mechanism of action that interfered with ceramide metabolism, as confirmed by the protection provided by fumonisi N1, a ceramide synthase enzyme inhibitor. Nowadays, three pathways are known to modulate ceramide levels in cells: biosynthesis of sphingomyelin through sphingomyelin synthase, biosynthesis of cerebrosides through glycosyl-transferases, and production of sphingosine and sphingosine-1-phosphate via ceramidase and sphingosine kinase, respectively (Hannun and Obeid, 2008). Subsequent studies showed that spisulosine-mediated toxicity (1 μM) was a consequence of an increase of *de novo* synthesis of ceramide (Figure 20.2). In particular, the authors showed that spisulosine could activate protein kinase C (PKC) ζ, an atypical isoform of PKC that has been shown to directly bind ceramide (Lozano et al., 1994).

In another study (Salcedo et al., 2007), the compound was evaluated for its cytotoxicity in neuroblastoma cell lines. Spisulosine showed that it was capable of activating caspase-3 and -12 and

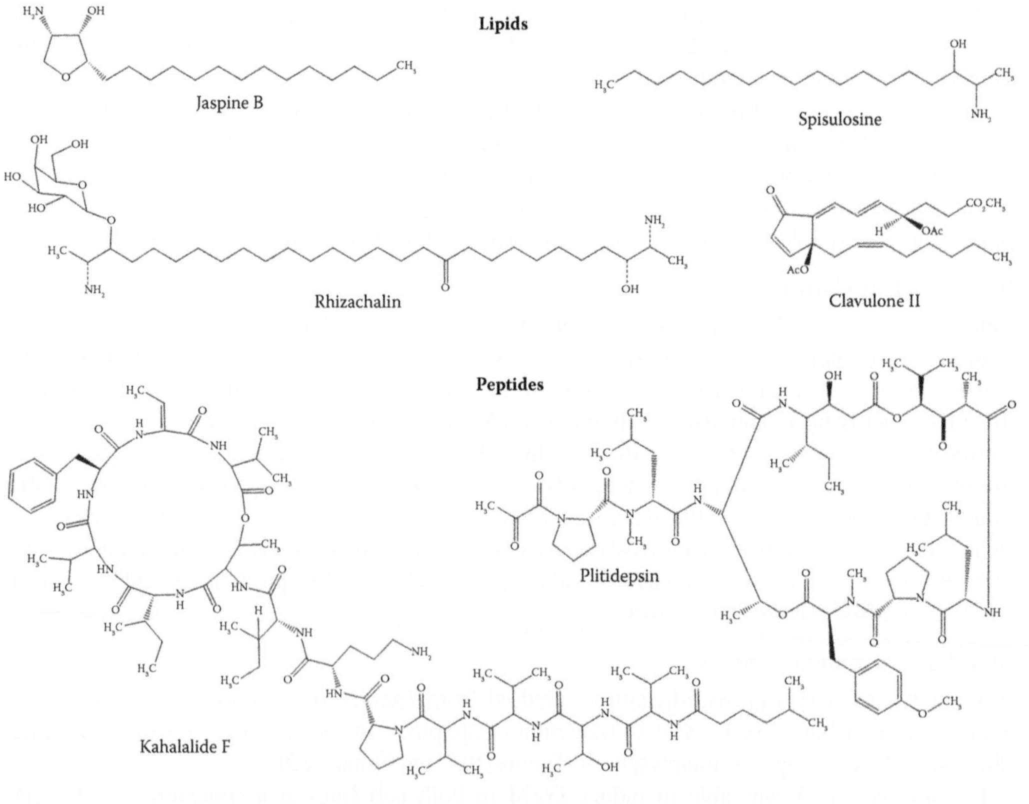

FIGURE 20.4 Structures of the lipids and peptides discussed in this chapter.

that it modified the phosphorylation of p53. In contrast, other pathways widely implicated in regulating cell survival/apoptosis, such as JNK, extracellular signal-regulated kinases, or Akt, were not affected.

20.5.4.2 Jaspine B

Interference with ceramide levels is also the mechanism of action of jaspine B (Figure 20.4 and Table 20.2), an anhydrophytosphingosine isolated from extracts of the sponge *Jaspis* sp. on the melanoma cell lines SK-Mel-28 and B16 (Salma et al., 2009). The target of jaspine B proved to be sphingomyelin synthase (Figure 20.2), which was further confirmed by the fact that sphingomyelin synthase small interference RNA-silenced cells were more sensitive to the compound's toxicity than their control counterparts. However, it should be highlighted that specific studies concerning sphingomyelin synthase were conducted on the murine cell line B16. Confirmation of the results in human cell lines is still required.

20.5.4.3 Rhizochalin

Rhizochalin (Figure 20.4 and Table 20.2) is a two-headed sphingolipid that was first reported in 1989 as a secondary metabolite of the sponge *Rhizochalina incrustata* (Makarieva et al., 1989). Because of the rare α,ω-bifunctionalized structures of these compounds, this class is highly promising when it comes to pharmacological properties.

Jin et al. (2009) studied the effect of rhizochalin, its aglycone, and other derivatives on the HL-60 leukemia cell line. Neutral red assay revealed a marked cytotoxic effect at a concentration of 10 μM, with the aglycon being more potent than the parent compound (Jin et al., 2009).

The apoptotic effect of the compounds was first evaluated by annexin V/propidium iodide staining. Rhizochalin showed dose-dependent activity, with 10 and 25 μM causing 33 and 86% of apoptotic cells, respectively, and the aglycone displaying higher activity, with 5 μM causing 93% of apoptotic cells. Further studies showed that the levels of procaspase-3, -8, and -9 were decreased after exposition to rhizochalin, with simultaneous rise in the levels of caspase-3, -8, and -9 (Figure 20.2). Cleavage of PARP was also found, and disruption of $\Delta\psi m$ was also observed following incubation with the compounds.

20.5.4.4 Clavulone II

Clavulone II is a cyclopentenone prostaglandin (Figure 20.4 and Table 20.2) isolated from the soft coral *Clavularia viridis*, whose structure includes a α-β carbonyl group, thus distinguishing this compound from regular prostaglandins. This molecule was evaluated for its antiproliferative and apoptotic effects in human acute promyelocytic leukemia cell line, HL-60 (Huang et al., 2005).

Initially, the MTT assay was used to evaluate clavulone II cell growth inhibition, with an IC_{50} of 1.5 μM being determined after a 24-hour period of incubation. Exposure to clavulone II also caused a significant increase in a sub-G_1 population of cells, with a concomitant decrease in G_0/G_1, S, and G_2/M cells. Cyclin D1 downregulation clavulone II was also proved.

Further studies on the mechanism of cell death showed that depolarization of mitochondrial membrane was taking place, and the involvement of caspases in the process of cell death was confirmed, mainly because of the effect of caspase-3, -8, and -9 and Bcl-2 family proteins. The activation of caspase-8 suggested a cross-talk interaction between extrinsic and intrinsic pathways, which was further confirmed by the finding of cleaved Bid (Figure 20.2).

20.5.5 Peptides

20.5.5.1 Kahalalide F

Most therapeutic agents that are used in cancer chemotherapy are inducers of apoptosis or negative modulators of cell cycle and division, although some exceptions are known. Such is the case with kahalalide F, a cyclic toxin (Figure 20.4 and Table 20.2) isolated from the marine mollusk *Elysia rufescens* (Janmaat et al., 2005).

Janmaat et al. (2005) studied the effects of kahalalide F in a number of different cell lines, such as breast cells (SK-BR-3, BT474, and MCF7), vulval (A431), non-small-cell lung cancer (H460, A549, SW1573, and H292), hepatic (Skhepl, HepG2, and Hep3B), and colon carcinoma (HT29). These authors showed that no phosphatidylserine externalization took place, suggesting that no apoptotic death was taking place, whereas the hypodiploid cell population reached nearly 100% after a 72-hour period, as revealed by cell-cycle analysis.

Pretreatment of cells with broad-range caspase inhibitors also could not protect cells from kahalalide F cytotoxicity, thus discarding the involvement of caspases in cell death. The authors also studied the involvement of cathepsins in the cytotoxic effect, but cathepsin inhibitors could not protect cells from death.

Subsequent studies revealed that ErbB3 (HER3) protein levels were positively correlated with cell line sensitivity to kahalalide F-mediated toxicity. In addition, the authors reported that after incubation with kahalalide F, ErbB3 protein expression was downregulated. However, although this target seems to be very important for this compound toxicity, other targets must exist, because the cell line A549, with low ErbB3 levels, was also highly sensitive to kahalalide F toxicity. Inhibition of phosphatidylinositol 3-kinase-Akt signalling pathway was also found to be one of the targets causing the necrotic-like cell death promoted by kahalalide F.

20.5.5.2 Plitidepsin

SK-MEL-28 (nonpigmented) and UACC-257 (melanotic) human melanoma cell lines were incubated for 48 hours with plitidepsin, a depsipeptide from the tunicate *Aplidium albicans* (Figure 20.4 and Table 20.2). This depsipeptide activity proved to be equivalent in both cell lines, with an IC_{50} of 12–14 nM being determined (Munoz-Alonso et al., 2008). This compound displayed a dose-dependent effect, with low doses (15–45 nM) causing G_1 phase arrest and high doses (150–450 nM) inducing PARP cleavage and sub-G_1 accumulation, which are classic markers of apoptosis. The above referred JNK pathway was found to be the target site of plitidepsin (Figure 20.2), which demonstrated cell-cycle inhibition and induced apoptosis via the Racl/c-Jun NH_2-terminal kinase pathway in these human melanoma cells.

In another study, Mitsiades et al. (2008) showed the antimyeloma activity of this compound against a panel of 19 human multiple myeloma cell lines. As reported in the work of Munoz-Alonso et al. (2008), the JNK pathway was found to be one of the targets of plitidepsin, as well as p38 activation, translocation of Fas/CD95 to lipid rafts, and caspase activation. In addition to this, Mitsiades et al. (2008) could also show that these molecules caused suppression of proliferative/antiapoptotic genes (e.g., *myc*, *mybl2*, *bubl*, *mcm2*, *mcm4*, *mcm5*, and survivin) and upregulation of several potential regulators of apoptosis (including c-Jun, TRAIL, CASP9, and Smac).

20.5.6 Macrolides

20.5.6.1 Iejimalide B

The tunicate *Eudistoma rigida* is the source of iejimalide B, a macrolide (macrocyclic lactone) composed of a 24-member ring with two methoxy groups and an *N*-formyl-L-serine side chain (Figure 20.5 and Table 20.2). As happens with many molecules from this chemical class, this compound showed cytotoxic activity against a wide range of cell lines, at nanomolar concentration levels (Kobayashi et al., 1988). Despite the recognized biological activities of this molecule, the mechanisms underlying such activities are not yet fully elucidated.

Wang et al. (2008) studied the effect of iejimalide B in prostate cancer cells, with special attention on its effects on cell growth and apoptosis. The LNCap and PC-3 cell lines were used, representing the earlier and later stages of prostate cancer, respectively. When exposed to the compound, LNCap cells exhibited a G_0/G_1 arrest followed by apoptosis at 48 hours. In contrast,

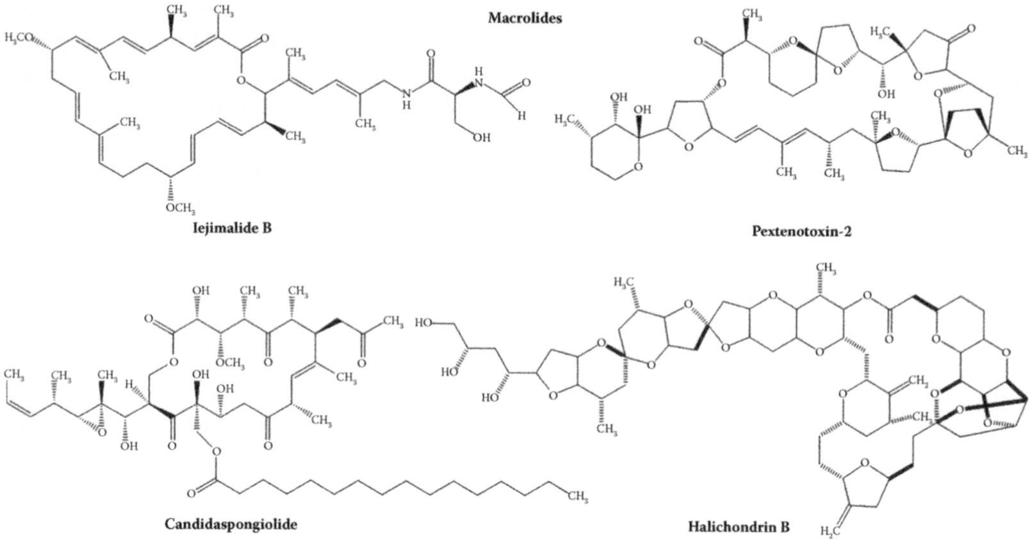

FIGURE 20.5 Structures of the macrolides discussed in this chapter.

although the PC-3 cell line suffered G_0/G_1 arrest followed by S phase arrest, apoptosis was not induced.

Quantitative PCR studies revealed that iejimalide B was responsible for a dose-dependent decrease in the levels of the androgen receptor mRNA. The levels of cyclin D1 and E mRNA were equally decreased, a trend that did not take place in PC-3 cells. The most expressive effect of iejimalide B on major cyclin genes was in cyclin B mRNA, which was downregulated more than 20-fold in LNCap cells and, to a lower extent, in PC-3 cells (Wang et al., 2008). However, PC-3 cell line upregulation of p21 and downregulation of cyclin A, as a consequence of iejimalide B exposure, was observed, thus showing that this molecule exerts its activity by distinct mechanisms in these two cancer cell lines.

20.5.6.2 Pextenotoxin-2

Pextenotoxin-2 (Figure 20.5 and Table 20.2) is a polyether macrolide compound isolated from dinoflagellates, sponges, and shellfish (Jung et al., 1995). This compound displayed significant cytotoxicity against human leukemia cell lines U937, HL-60, THP-1, and K562 (Moon et al., 2008). Cells incubated with pextenotoxin-2 10 ng/mL for 72 hours failed to complete mitotic division and remained in the one-cell stage, but with multiple nuclei, as a consequence of cytokinesis inhibition caused by this compound. In addition, the authors reported the marked G_2/S arrest following pextonotoxin-2 exposure, as well as endoreplication and apoptosis through the JNK pathway via actin depolymerization (Figure 20.2) (Moon et al., 2008).

20.5.6.3 Candidaspongiolide

Candidaspongiolide is a tedanolide macrolide (Figure 20.5 and Table 20.2) isolated from *Candidaspongia* sp. At low concentrations, the compound (10 nM) reversibly inhibited protein synthesis in the U251 glioblastoma and HCT116 colorectal cancer cells. An irreversible effect was obtained with higher concentrations (50 nM) (Trisciuoglio et al., 2008).

Among the several targets that were studied, eukaryotic initiation factor (eIF2), an important regulatory protein for mRNA translation eIF2, proved to be the most important site for candidaspongiolide activity. Phosphorylation of the α subunit of eIF2 is a described mechanism for inhibition of protein synthesis (Trisciuoglio et al., 2008). Candidaspongiolide was able to induce eIF2α phosphorylation in a dose-dependent manner, as well as an increase in the phosphorylation of the elongation factor eEF2 at Thr-56, a phosphorylation site associated with inhibition of its activity (Trisciuoglio et al., 2008; Pause et al., 1994). In addition, this macrolide could trigger apoptosis in the human cancer cells used in this study. Deeper insights in the mechanism of action revealed that caspase-12 was a cellular target for candidaspongiolide. This conclusion arose from the use of several caspase inhibitors, such as Ac-ZATAD-fmk, a caspase-12 inhibitor that substantially prevented candidaspongiolide toxicity. The role of caspase-12 in human programmed cell death is still controversial, although it has been associated with the endoplasmatic reticulum stress-dependent apoptosis (Szegezdi et al., 2003).

20.5.6.4 Halichondrin B

Catassi et al. (2006) conducted a study in which several marine compounds, such as halichondrin B, a macrolide from *Aninella* sp. (Figure 20.5 and Table 20.2), were evaluated for their ability to influence cell cycle and induce cell death in the A549 lung cancer cell line. After a 72-hour exposition period, halichondrin B showed a marked cytotoxic effect, evaluated by the MTT assay, with an IC_{50} of 3.2 nM. The authors proved that apoptosis was taking place 24 hours after treatment with halichondrin B, in a time-dependent manner.

As a result of exposure to cytotoxic agents, in many situations a cell-surviving pathway is triggered, with NF-κB being usually activated, with all its deleterious effect at long term. In fact, many of the therapeutic agents in use also activate this transcription factor. By using an electrophoretic mobility shift assay, the authors proved that this marine-derived compound did not activate NF-κB, which can constitute a therapeutical advantage.

Given the fact that this compound is described as tubulin-interacting molecule, the authors studied its effects on the expression of GADD45 α, β, and γ. In recent years, GADD45 proteins have been considered stress sensors, being mediated by a complex interplay of physical interactions with other cellular proteins that are implicated in cell-cycle regulation and in the response of cells to stress (Liebermann and Hoffman, 2008) (Figure 20.2). The authors found an upregulation of GADD45 α and γ and downregulation of β. The role of p53 in modulating GADD45 was discharged.

Bad is a member of the BH3-only protein, a subfamily whose action is highly affected by its phosphorylation state. Phosphorylated Bad is localized in the cytoplasm in an inactive form where dephosphorylated Bad binds to Bcl-2 and Bcl-XL, thus preventing their antiapoptotic function. However, when phosphorylation takes place (serines 112 and 136), the molecule migrates to cytosol where it binds to 14-3-3 proteins, a form in which it is no longer able to induce cell death (Catassi et al., 2006) (Figure 20.2). Halichondrin B was found to inhibit phosphorylation of Bad at serine 136, thus increasing the levels of Bad localized in the mitochondria and bound to the antiapoptotic factors, Bcl-2 and Bcl-xL.

Release of cytochrome *c* to cytoplasm and activation of caspase-3 was also proved to be a consequence of treatment for all compounds, thus showing that these molecules triggered cell death via the mitochondrial apoptosis pathway as a consequence of Bad dephosphorylation.

20.5.7 Miscellaneous

20.5.7.1 CvL

Queiroz et al. (2009) studied the effect of a lectin named CvL (Table 20.2), isolated from the sponge *Cliona varians* in Jurkat and K562 cell lines. The compound showed toxicity against both cell lines, with K562 being more sensitive, and no inhibitory effects were noticed against normal proliferating lymphocytes. No cytotoxic effect was found against the adherent tumors B16, 786-O, and PC3, suggesting the selectivity of this compound against leukemia cells.

Cells treated with CvL exhibited several features of apoptotic cells, such as condensed chromatin and nuclear fragmentation, which was further confirmed by annexin V binding assay. CvL was not able to activate caspase-mediated cell death. Laser scanning confocal microscopy showed the translocation of cathepsin B from the lysosomes to the cytoplasm as a consequence of exposure to CvL (Queiroz et al., 2009). The involvement of cathepsin B on the cellular response to CvL was confirmed using E-64, a broad spectrum cathepsin inhibitor, which completely prevented CvL-mediated toxicity. In fact, cathepsin B has been increasingly associated with the death receptor pathway of apoptosis, as well as the mitochondrial pathway (Foghsgaard et al., 2001; Turk and Stoka, 2007), although the mechanism of this cathepsin-mediated death is still poorly understood. This led the authors to study the expression of tumor necrosis factor receptor 1 (TNFR1) and NF-κB, a transcription factor associated with TNFR1 signalling. The results showed an upregulation of TNFR1. Western blot analysis revealed that the expression of the NF-κB p65 subunit was decreased, an event accompanied by increased amounts of Bcl-2 and Bax, which suggests that multiple pathways are involved in CvL-mediated toxicity (Figure 20.2). Upregulation of p21 protein expression and downregulation of pRb, compatible with cell-cycle arrest, was also reported.

20.6 INTERNET RESOURCES

Because of the high amounts of data and the diversity of metabolites involved in metabolomics, a number of databases that congregate several molecules are available offering free access. Some of these databases provide detailed information regarding specific metabolites, such as NMR and MS profiles, chemical structure, and physical properties, among others. Several examples could be enunciated, such as the Human Metabolite Database (http://www.metabolibrary.ca), the Human Metabolome Project (http://www.metabolomics.ca), the Lipid Library (http://www.lipidlibrary.co.uk), or the European Carbohydrate Databases (http://www.eurocarbdb.org).

20.7 FUTURE PROSPECTIVE

It is undeniable that the area of metabolomics has experienced exponential growth in the last few years. However, it should be highlighted that a remarkable improvement of the analytical techniques is the major contributor to this fulfilment of the definition of metabolomics: the characterization of all metabolites in a matrix. In fact, although chemical analysis has been performed for over an hundred years, only now are the instrumental techniques sensible and powerful enough to allow a glimpse of the whole metabolome. Thus, it is to be expected that the major developments in the area of metabolomics will result from the advent of new and more sophisticated instrumental techniques.

In the area of applied metabolomics, the application of the knowledge we have acquired so far will facilitate the finding of new and better disease biomarkers. Likewise, the application of metabolomics to drug discovery will allow the characterization of new molecules that will, without a doubt, positively affect human therapeutics. After a period in which no new natural molecules were introduced into therapeutics, acknowledgment of the huge potential that the marine environment encompasses has triggered a return to nature as source of bioactive molecules.

REFERENCES

Arepalli, S. K., Sridhar, V., Rao, J. V., et al. (2009). Furano-sesquiterpene from soft coral, *Sinularia kavarittiensis*, induces apoptosis via the mitochondrial-mediated caspase-dependent pathway in THP-1, leukemia cell line. *Apoptosis* 14: 729–740.

Ballot, C., Kluza, J., Lancel, S., et al. (2010). Inhibition of mitochondrial respiration mediates apoptosis induced by the anti-tumoral alkaloid lamellarin D. *Apoptosis* 15: 769–781.

Berry, E., Hardt, J. L., Clardy, J., et al. (2009). Induction of apoptosis in endometrial cancer cells by psammaplysene A involves FOXO1. *Gynecologic Oncology* 112: 331–336.

Bremner, P., Heinrich, M. (2002). Natural products as targeted modulators of the nuclear factor-kappa B pathway. *Journal of Pharmacy and Pharmacology* 54: 453–472.

Burgering, B. M. T., Medema, R. H. (2003). Decisions on life and death: FOXO Forkhead transcription factors are in command when PKB/Akt is off duty. *Journal of Leukocyte Biology* 73: 689–701.

Catassi, A., Cesario, A., Arzani, D., et al. (2006). Characterization of apoptosis induced by marine natural products in non small cell lung cancer A549 cells. *Cellular and Molecular Life Sciences* 63: 2377–2386.

Cheung, F. W., Li, C., Che, C. T., et al. (2010). Geoditin A induces oxidative stress and apoptosis on human colon HT29 cells. *Marine Drugs* 8: 80–90.

Chipuk, J. E., Moldoveanu, T., Llambi, F., et al. (2010). The BCL-2 family reunion. *Molecular Cell* 37: 299–310.

Choi, H. J., Choi, Y. H., Yee, S. B., et al. (2005). Ircinin-1 induces cell cycle arrest and apoptosis in SK-MEL-2 human melanoma cells. *Molecular Carcinogenesis* 44: 162–173.

Coussens, L. M., Werb, Z. (2002). Inflammation and cancer. *Nature* 420: 860–867.

Cuyckens, F., Claeys, M. (2004). Mass spectrometry in the structural analysis of flavonoids. *Journal of Mass Spectrometry* 39: 1–15.

Daniels, T. R., Delgado, T., Rodriguez, J. A., et al. (2006). The transferrin receptor part I: Biology and targeting with cytotoxic antibodies for the treatment of cancer. *Clinical Immunology* 121: 144–158.

Dettmer, K., Aronov, P. A., Hammock, B. D. (2007). Mass spectrometry-based metabolomics. *Mass Spectrometry Reviews* 26: 51–78.

Dirsch, V. M., Muller, I. M., Eichhorst, S. T., et al. (2003). Cephalostatin 1 selectively triggers the release of Smac/DIABLO and subsequent apoptosis that is characterized by an increased density of the mitochondrial matrix. *Cancer Research* 63: 8869–8876.

Dunn, W. B., Bailey, N. J., Johnson, H. E. (2005). Measuring the metabolome: Current analytical technologies. *Analyst* 130: 606–625.

Edrada, R. A., Proksch, P., Wray, V., et al. (1996). Four new bioactive manzamine-type alkaloids from the Philippine marine sponge *Xestospongia ashmorica. Journal of Natural Products* 59: 1056–1060.

Evans, P. C. (2005). Regulation of pro-inflammatory signalling networks by ubiquitin: Identification of novel targets for anti-inflammatory drugs. *Expert Reviews in Molecular Medicine* 7: 1–19.

Ferreres, F., Pereira, D. M., Gil-Izquierdo, A., et al. (2010a). HPLC-PAD-atmospheric pressure chemical ionization-MS metabolite profiling of cytotoxic carotenoids from the echinoderm *Marthasterias glacialis* (spiny sea-star). *Journal of Separation Science* 33: 2250–2257.

Ferreres, F., Pereira, D. M., Valentao, P., et al. (2010b). First report of non-coloured flavonoids in *Echium plantagineum* bee pollen: Differentiation of isomers by liquid chromatography/ion trap mass spectrometry. *Rapid Communications in Mass Spectrometry* 24: 801–806.

Foghsgaard, L., Wissing, D., Mauch, D., et al. (2001). Cathepsin B acts as a dominant execution protease in tumor cell apoptosis induced by tumor necrosis factor. *Journal of Cell Biology* 153: 999–1009.

Gatenby, R. A., Gillies, R. J. (2004). Why do cancers have high aerobic glycolysis? *Nature Reviews Cancer* 4: 891–899.

Guzman, E. A., Johnson, J. D., Linley, P. A., et al. (2010). A novel activity from an old compound: Manzamine A reduces the metastatic potential of AsPC-1 pancreatic cancer cells and sensitizes them to TRAIL-induced apoptosis. *Investigational New Drugs* 2010: 1–9.

Halliwell, B., Gutteridge, J. M. (1995). The definition and measurement of antioxidants in biological systems. *Free Radical Biology & Medicine* 18: 125–126.

Hannun, Y. A., Obeid, L. M. (2008). Principles of bioactive lipid signalling: Lessons from sphingolipids. *Nature Reviews Molecular Cell Biology* 9: 139–150.

Hollywood, K., Brison, D. R., Goodacre, R. (2006). Metabolomics: Current technologies and future trends. *Proteomics* 6: 4716–4723.

Huang, Y. C., Guh, J. H., Shen, Y. C., et al. (2005). Investigation of anticancer mechanism of clavulone II, a coral cyclopentenone prostaglandin analog, in human acute promyelocytic leukemia. *Journal of Biomedical Science* 12: 335–345.

Ichiba, T., Corgiat, J. M., Scheuer, P. J., et al. (1994). 8-Hydroxymanzamine-a, a beta-carboline alkaloid from a sponge, *Pachypellina* sp. *Journal of Natural Products* 57: 168–170.

Iwamaru, A., Iwado, E., Kondo, S., et al. (2007). Eupalmerin acetate, a novel anticancer agent from Caribbean gorgonian octocorals, induces apoptosis in malignant glioma cells via the c-Jun NH_2-terminal kinase pathway. *Molecular Cancer Therapeutics* 6: 184–192.

Janmaat, M. L., Rodriguez, J. A., Jimeno, J., et al. (2005). Kahalalide F induces necrosis-like cell death that involves depletion of ErbB3 and inhibition of Akt signaling. *Molecular Pharmacology* 68: 502–510.

Jin, J. O., Shastina, V., Park, J. I., et al. (2009). Differential induction of apoptosis of leukemic cells by rhizochalin, two headed sphingolipids from sponge and its derivatives. *Biological & Pharmaceutical Bulletin* 32: 955–962.

Jung, J. H., Sim, C. J., Lee, C. O. (1995). Cytotoxic compounds from a two-sponge association. *Journal of Natural Products* 58: 1722–1726.

Kashman, Y., Rudi, A. (1977). The ^{13}C-NMR spectrum and stereochemistry of heteronemin. *Tetrahedron* 33: 2997–2998.

Kazlauskas, R., Murphy, P. T., Quinn, R. J., et al. (1976). Heteronemin, a new scalarin type sesterterpene from sponge heteronema-erecta. *Tetrahedron Letters* 2631–2634.

Keutgens, A., Robert, I., Viatour, P., et al. (2006). Deregulated NF-kappaB activity in haematological malignancies. *Biochemical Pharmacology* 72: 1069–1080.

Kluza, J., Gallego, M. A., Loyens, A., et al. (2006). Cancer cell mitochondria are direct proapoptotic targets for the marine antitumor drug lamellarin D. *Cancer Research* 66: 3177–3187.

Kobayashi, J., Cheng, J. F., Ohta, T., et al. (1988). Iejimalide-A and iejimalide-B, novel 24-membered macrolides with potent antileukemic activity from the Okinawan tunicate *Eudistoma cf rigida. Journal of Organic Chemistry* 53: 6147–6150.

Kong, D., Aoki, S., Sowa, Y., et al. (2008). Smenospongine, a sesquiterpene aminoquinone from a marine sponge, induces G1 arrest or apoptosis in different leukemia cells. *Marine Drugs* 6: 480–488.

Krishnan, P., Kruger, N. J., Ratcliffe, R. G. (2005). Metabolite fingerprinting and profiling in plants using NMR. *Journal of Experimental Botany* 56: 255–265.

Li, M., Miao, Z. H., Chen, Z., et al. (2010). Echinoside A, a new marine-derived anticancer saponin, targets topoisomerase 2 alpha by unique interference with its DNA binding and catalytic cycle. *Annals of Oncology* 21: 597–607.

Liebermann, D. A., Hoffman, B. (2008). Gadd45 in stress signaling. *Journal of Molecular Signaling* 3: 15.

Liu, W. K., Ho, J. C., Che, C. T. (2005). Apoptotic activity of isomalabaricane triterpenes on human promyelocytic leukemia HL60 cells. *Cancer Letters* 230: 102–110.

Lokhov, P. G., Dashtiev, M. I., Bondartsov, L. V., et al. (2009). Metabolic fingerprinting of blood plasma for patients with prostate cancer. *Biomeditŝinskaiâ Khimiiâ* 55: 247–254.

Lozano, J., Berra, E., Municio, M. M., et al. (1994). Protein-kinase-Cz isoform is critical for kB-dependent promoter activation by sphingomyelinase. *Journal of Biological Chemistry* 269: 19200–19202.

Makarieva, T. N., Denisenko, V. A., Stonik, V. A., et al. (1989). Rhizochalin, a novel secondary metabolite of mixed biosynthesis from the sponge *Rhizochalina incrustata. Tetrahedron Letters* 30: 6581–6584.

Mitsiades, C. S., Ocio, E. M., Pandiella, A., et al. (2008). Aplidin, a marine organism-derived compound with potent antimyeloma activity in vitro and in vivo. *Cancer Research* 68: 5216–5225.

Moon, D. O., Kim, M. O., Kang, S. H., et al. (2008). Induction of G(2)/M arrest, endoreduplication, and apoptosis by actin depolymerization agent pextenotoxin-2 in human leukemia cells, involving activation of ERK and JNK. *Biochemical Pharmacology* 76: 312–321.

Munoz-Alonso, M. J., Gonzalez-Santiago, L., Zarich, N., et al. (2008). Plitidepsin has a dual effect inhibiting cell cycle and inducing apoptosis via Rac1/c-Jun NH_2-terminal kinase activation in human melanoma cells. *Journal of Pharmacology and Experimental Therapeutics* 324: 1093–1101.

Newman, D. J., Cragg, G. M. (2007). Natural products as sources of new drugs over the last 25 years. *Journal of Natural Products* 70: 461–477.

Pause, A., Belsham, G. J., Gingras, A. C., et al. (1994). Insulin-dependent stimulation of protein-synthesis by phosphorylation of a regulator of 5'-cap function. *Nature* 371: 762–767.

Pereira, D. M., Valentão, P., Ferreres, F., et al. (2010). Metabolomic analysis of natural products. In *Reviews in Pharmaceutical and Biomedical Analysis.* Oak Park, IL: Bentham.

Queiroz, A. F. S., Silva, R. A., Moura, R. M., et al. (2009). Growth inhibitory activity of a novel lectin from *Cliona varians* against K562 human erythroleukemia cells. *Cancer Chemotherapy and Pharmacology* 63: 1023–1033.

Salcedo, M., Cuevas, C., Alonso, J. L., et al. (2007). The marine sphingolipid-derived compound ES 285 triggers an atypical cell death pathway. *Apoptosis* 12: 395–409.

Salma, Y., Lafont, E., Therville, N. (2009). The natural marine anhydrophytosphingosine, Jaspine B, induces apoptosis in melanoma cells by interfering with ceramide metabolism. *Biochemical Pharmacology* 78: 477–485.

Sanchez, A. M., Malagarie-Cazenave, S., Olea, N., et al. (2008). Spisulosine (ES-285) induces prostate tumor PC-3 and LNCaP cell death by de novo synthesis of ceramide and PKC zeta activation. *European Journal of Pharmacology* 584: 237–245.

Schumacher, M., Cerella, C., Eifes, S., et al. (2010). Heteronemin, a spongean sesterterpene, inhibits TNF alpha-induced NF-kappa B activation through proteasome inhibition and induces apoptotic cell death. *Biochemical Pharmacology* 79, 610–622.

Sterin, M., Cohen, J. S., Mardor, Y., et al. (2001). Levels of phospholipid metabolites in breast cancer cells treated with antimitotic drugs: A P-31-magnetic resonance spectroscopy study. *Cancer Research* 61: 7536–7543.

Sumner, L. W., Mendes, P., Dixon, R. A. (2003). Plant metabolomics: Large-scale phytochemistry in the functional genomics era. *Phytochemistry* 62: 817–836.

Szegezdi, E., Fitzgerald, U., Samali, A. (2003). Caspase-12 and ER-stress-mediated apoptosis: The story so far. *Annals of the New York Academy of Sciences* 1010: 186–194.

Trisciuoglio, D., Uranchimeg, B., Cardellina, J. H., et al. (2008). Induction of apoptosis in human cancer cells by candidaspongiolide, a novel sponge polyketide. *Journal of the National Cancer Institute* 100: 1233–1246.

Turk, B., Stoka, V. (2007). Protease signalling in cell death: Caspases versus cysteine cathepsins. *FEBS Letters* 581: 2761–2767.

Vanhuyse, M., Kluza, J., Tardy, C., et al. (2005). Lamellarin D: A novel pro-apoptotic agent from marine origin insensitive to P-glycoprotein-mediated drug efflux. *Cancer Letters* 221: 165–175.

Wang, W. L. W., Mchenry, P., Jeffrey, R., et al. (2008). Effects of iejimalide B, a marine macrolide, on growth and apoptosis in prostate cancer cell lines. *Journal of Cellular Biochemistry* 105: 998–1007.

Watanabe, D., Tsuda, M., Kobayashi, J. (1998). Three new manzamine congeners from *Amphimedon* sponge. *Journal of Natural Products* 61: 689–692.

Wolfender, J. L., Ndjoko, K., Hostettmann, K. (2003). Liquid chromatography with ultraviolet absorbance-mass spectrometric detection and with nuclear magnetic resonance spectroscopy: A powerful combination for the on-line structural investigation of plant metabolites. *Journal of Chromatography A* 1000: 437–455.

21 Genome-Wide Association Studies of Type 2 Diabetes
Current Status, Open Challenges, and Future Perspectives

Huixiao Hong, Lei Xu, Donna L. Mendrick, and Weida Tong
U. S. Food and Drug Administration
Jefferson, Arkansas

CONTENTS

21.1 TYPE 2 DIABETES

21.1.1 Symptoms and Epidemiology of Type 2 Diabetes

The term diabetes, often referred to as diabetes mellitus, is a disorder characterized by high blood glucose levels, because of metabolic defects in either the body's ability to produce or appropriately respond to insulin. Type 2 diabetes (T2D) is the most common form of diabetes.

The symptoms of T2D develop gradually, including fatigue, frequent urination, increased thirst and hunger, weight loss, blurred vision, and slow healing of wounds or sores (Cooke and Plotnick, 2008).

Diabetes occurs throughout the world but is more common in the more developed countries. According to the World Health Organization, at least 171 million people worldwide suffer from diabetes, or 2.8% of the population. It is estimated that by 2030, this number will almost double (Wild et al., 2004). Incidence of diabetes in North America has been increasing substantially. The Centers for Disease Control reports that there are approximately 24 million people with diabetes in the United States (17.9 million diagnosed with T2D and 5.7 million undiagnosed). In addition, 57 million people are estimated to have prediabetes (http://www.cdc.gov/Features/diabetesfactsheet). The risk of T2D increases with aging (Wild et al., 2004). Of great concern is the recent increase in T2D in children (Bloomgarden, 2004).

21.1.2 Diagnosis and Mechanism of T2D

Type 1 diabetes results from destruction of the islet cells that produce insulin. T2D, the most prevalent form, is caused by cells' failure to use insulin properly and accounts for 90–95% of people with diabetes. A diagnosis of T2D is made if a fasting plasma glucose concentration is ≥7.0 mmol/L (≥126 mg/dL) or when plasma levels of glucose, 2 hours after a standard glucose challenge, is ≥11.1 mmol/L (≥200 mg/dL) (World Health Organization, 2006).

Insulin is the main hormone that regulates uptake of glucose from the blood into most cells, primarily muscle and fat cells, but not cells of the central nervous system. It also controls the conversion of glucose to glycogen for internal storage in liver and muscle cells (Figure 21.1). When glucose levels fall, controlled by the hormone glucagon, the release of insulin from the β-cells is reduced, and the reverse conversion of glycogen to glucose is triggered.

21.1.3 Genetics and T2D

T2D is caused by a combination of genetic and environmental risk factors. More than four decades ago Neel (1962) recognized the input of genetics and environment on the prevalence of T2D and hypothesized that this disease once comprised a *thrifty genotype*, which had a selective advantage (e.g., during famine) as this genotype is very efficient in the utilization of food. The major risk factors for T2D are obesity (≥120% ideal body weight or a body mass index ≥30 k/m^2) and a sedentary lifestyle (van Dam, 2003; Shaw and Chisholm, 2003). It has been estimated that approximately 80% of all new T2D cases are due to obesity (Lean, 2000). The other major T2D risk factor is physical inactivity. Physical activity, such as daily walking or cycling for more than 30 minutes, has been shown to significantly reduce the risk of T2D (Hu, 2003). In recent years, because most populations have experienced a continuous supply of calorie-dense processed foods and a decrease in physical activity, this has led to a worldwide rise in T2D prevalence.

It has long been known that T2D is, in part, inherited. Family studies have revealed that first-degree relatives of individuals with T2D are about three-fold more likely to develop the disease than individuals without a positive family history of T2D (Florez et al., 2003; Gloyn, 2003; Hansen, 2003). One approach used to identify T2D susceptibility genes is based on the identification of candidate genes. More than 50 genes involved in pancreatic β-cell function, insulin action/glucose metabolism, or other metabolic conditions that increase T2D risk have been selected as candidate genes and studied in various populations worldwide. However, many of the candidate genes have proven to be controversial, and this might be due to small sample sizes, differences in T2D susceptibility across ethnic groups, variation in environmental exposures, and gene-environmental interactions.

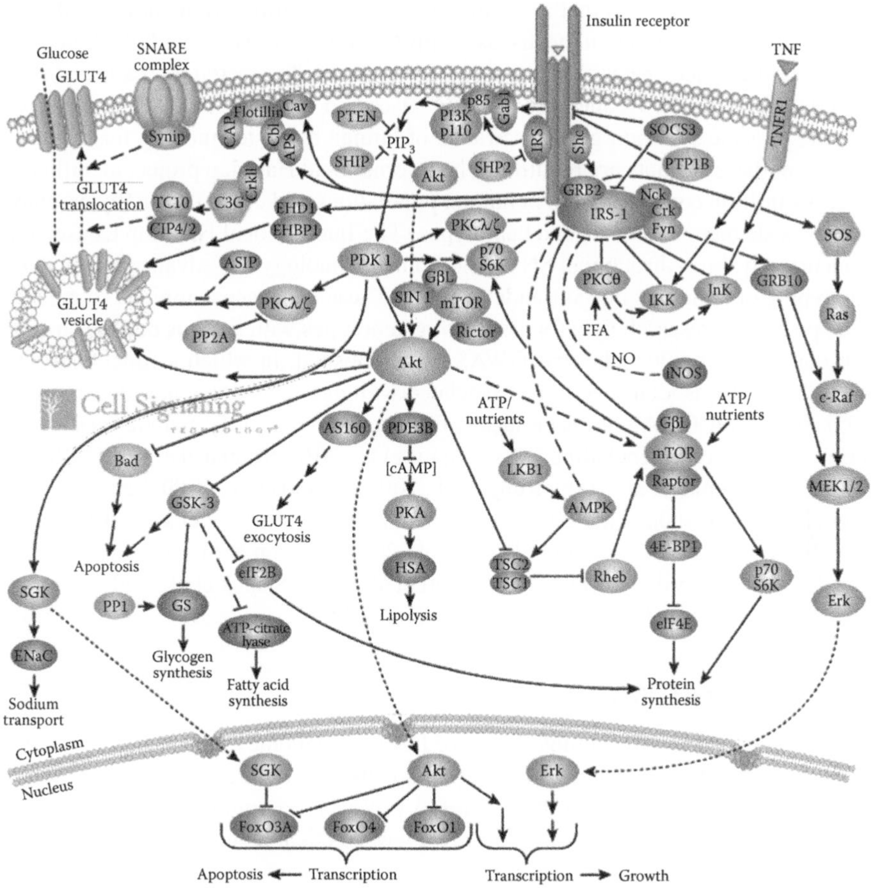

FIGURE 21.1 **(See color insert.)** Signal transduction by the insulin receptor. Insulin activates the insulin receptor (IR) tyrosine kinase, which phosphorylates and recruits different substrate adaptors including the IRS family of proteins. Tyrosine-phosphorylated IRS then displays binding sites for numerous signaling partners. Among them, phosphatidylinositol 3-kinase (PI3K) has a major role in insulin function, mainly via the activation of the Akt/protein kinase B (PKB) and the protein kinase Cζ (PKCζ) cascades. Activated Akt induces glycogen synthesis through inhibition of GSK-3, protein synthesis via mTOR and downstream elements, and cell survival through inhibition of several pro-apoptotic agents (Bad, Forkhead family transcription factors, GSK-3). Insulin stimulates glucose uptake in muscle and adipocytes via translocation of GLUT4 vesicles to the plasma membrane. GLUT4 translocation involves the PI3K/Akt pathway and IR-mediated phosphorylation of CAP and formation of the CAP:Cbl:CrkII complex. Insulin signaling also has growth and mitogenic effects that are mostly mediated by the Akt cascade as well as by activation of the Ras/MAPK pathway. A negative feedback signal emanating from Akt/PKB, PKCζ, p70 S6K, and the MAPK cascades results in serine phosphorylation and inactivation of IRS signaling. (Reproduced from Cell Signaling Technology, Inc., http://www.cellsignal.com.)

21.2 GENOME-WIDE ASSOCIATION STUDY

21.2.1 Background

Genome-wide association studies (GWAS) aim to identify genetic variants, single nucleotide polymorphisms (SNPs), across the entire human genome that are associated with phenotypic traits, such as disease status and/or drug response. In 1996, Lander proposed the common disease, common variant model (Lander, 1996). Based on both theoretical arguments and examples of heterogeneity of disease-associated alleles, this model hypothesized that the genetic profile of widespread diseases is determined by genetic variants that are common in the population (frequency > 0.01 at least) and have, individually, a small effect on the disease. This fits well in the scope of GWAS because, if this model is correct, the genetic basis of universal diseases can be discovered by searching for common variants with different allele frequencies between cases and controls. To conduct a GWAS, both a map of all the possible common genetic variants of human and a technology for massive parallel measurements of these variants are required. The International HapMap project identified and catalogued over 3.1 million common SNPs in human populations and computationally assembled them into a genome-wide map of SNP-tagged haplotypes (The International HapMap Consortium, 2005, 2007). Concurrently, high-throughput SNP genotyping technology has advanced to enable simultaneous genotyping of hundreds of thousands of SNPs. These advances combined to make GWAS a feasible and promising research field for associating genotypes with various disease susceptibilities and health outcomes. In 2005, the first GWAS was published, in which a functional SNP in the complement factor H was identified to be associated with age-related macular degeneration (Klein et al., 2005). Since then, GWAS has been successfully applied to identify genetic variants associated with a variety of phenotypes, including T2D (Dupuis et al., 2010; Saxena et al., 2007, 2010; Bouatia-Naji et al., 2009; Lyssenko et al., 2009; Rung et al., 2009; Sparsø et al., 2009; Takeuchi et al., 2009; Timpson et al., 2009; Yasuda et al., 2008; Unoki et al., 2008; Vaxillaire et al., 2008; Zeggini et al., 2007, 2008; Florez et al., 2007; Gudmundsson et al., 2007; Hanson et al., 2007; Hayes et al., 2007; Rampersaud et al., 2007; Salonen et al., 2007; Scott et al., 2007; Sladek et al., 2007; Steinthorsdottir et al., 2007; Wellcome Trust Case Control Consortium, 2007).

21.2.2 Workflow of GWAS

In a GWAS, markers across the complete sets of DNA, or genomes, of many people are scanned to find genetic variations associated with a particular disease, such as T2D, for developing better strategies to detect, treat, and prevent the disease. A GWAS is a complicated process depicted by a simplified carton shown in Figure 21.2. There are a few excellent review articles that describe each

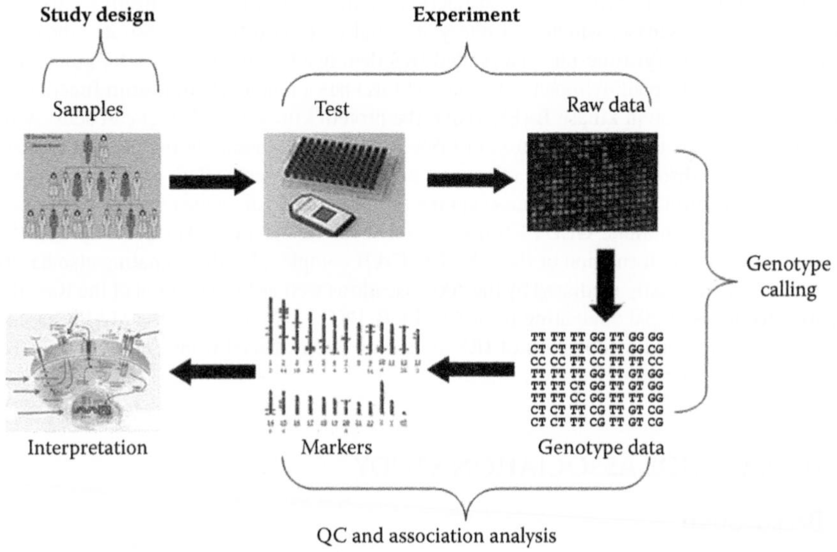

FIGURE 21.2 Overview of workflow for a GWAS. First, cases and controls are selected. Then the samples from cases and controls are subjected to genotyping. The raw data from the genotyping experiment are used to determine genotypes for SNPs of the samples. Downstream association analyses are conducted using the genotype data to identify significantly associated SNPs for the trait in study. Finally, interpretation of the possible biological functions of significant SNPs may aid understanding of the trait.

step in GWAS in great detail (Zondervan and Cardon, 2007; McCarthy et al., 2008; Pearson and Manolio, 2008).

To carry out a GWAS, a case-control design is typically used: people with the disease being studied and similar people without the disease. A DNA sample from each participant is obtained (usually from blood), purified, and placed on tiny chips and scanned on automated laboratory machines. In this way, each participant's genome is interrogated for strategically selected markers of genetic variation, SNPs. Thereafter, the raw data are used for determining genotypes of SNPs using a genotype calling algorithm. By comparing the genotypes, the SNPs that are found to be significantly more frequent in one group than the other, case or control, are said to be associated with the disease in study. However, the associated SNPs themselves may not directly cause the disease. Often, additional steps are needed to identify the exact genetic change involved in the disease.

21.3 GWAS OF T2D

21.3.1 Background

Until 2006, hypothesis-free genome-wide linkage mapping in families with multiply affected subjects or association studies within candidate genes using case-control samples or parent-offspring trios were the main approaches used to narrow down the genetic variants associated with the risk of T2D (Prokopenko et al., 2008). Linkage analysis suffered from being seriously underpowered for susceptibility models, because linkage is best placed to detect genetic variants of high penetrance. However, genetic variants of high penetrance do not contribute substantially to the risk of T2D, and few if any robust signals have emerged from such efforts (McCarthy, 2003). The candidate gene-association approach has difficulties associated with selection of candidate genes and sample size required to offer confident detection of genetic variants for the small effect sizes. However, GWAS can efficiently uncover genetic variants that influence T2D at a much larger scale than the candidate gene-association approach.

21.3.2 Initial GWAS of T2D

As of this writing, there have been 14 initial GWAS of T2D published in top scientific journals (Table 21.1), most of them during the first wave of GWAS in 2007. A large number of genetic variants from all chromosomes were identified as associated with T2D, and the papers in which they were reported are cited (Table 21.2). These studies utilized different sizes of cases and control sets, in some instances different selection criteria, populations, etc. A brief overview of each study and their findings is provided below.

In the Wellcome Trust Case Control Consortium GWAS study of T2D (Wellcome Trust Case Control Consortium 2007), there were 37 genetic variants identified to a significant degree (defined here as with a p value less than 0.0001) in association with the risk of developing T2D. These included some genetic markers in or near the three widely replicated genes, *PPARG* (encoding the peroxisomal proliferative activated receptor γ), *KCNJ11* (the inwardly-rectifying Kir6.2 component of the pancreatic β-cell KATP channel), and *TCF7L2* (transcription factor 7-like 2). They were detected with effect sizes consistent with previous reports.

In the Diabetes Genetics Initiative (Saxena et al., 2007), a GWAS of 1464 patients with T2D and 1467 matched controls was conducted and identified 17 genetic variants to be significantly associated with T2D. Among these were three loci, a noncoding region near gene *CDKN2A* (cyclin-dependent kinase inhibitor 2A) and *CDKN2B* (cyclin-dependent kinase inhibitor 2B), an intron in gene *IGF2BP2* (insulin-like growth factor 2 mRNA binding protein 2), and an intron of gene *CDKAL1* (CDK5 regulatory subunit associated protein 1-like 1), confirmed by another GWAS.

In a large-scale GWAS of T2D (Steinthorsdottir et al., 2007), 1,399 Icelandic cases and 5,275 Icelandic controls were genotyped for the association study. Three genetic variants were

TABLE 21.1
Initial GWAS of T2D

Study	Cases	Controls	Population	Platform	Reference
WTCCC	1,924	2,938	United Kingdom	Affymetrix 500K	WTCCC (2007)
DGI	1,464	1,467	Finland, Sweden	Affymetrix 500K	Saxena et al. (2007)
deG	1,399	5,275	Iceland	Illumina 300K	Steinthorsdottir et al. (2007)
FUSION	1,161	1,174	Finland	Illumina 300K	Scott et al. (2007)
DGDG	694	645	France	Illumina 300K	Sladek et al. (2007)
DiaGen	500	497	East Finland, Germany, United Kingdom, Ashkenazi	Illumina 300K	Salonen et al. (2007)
Pima	300	334	Pima Indians	Affymetrix 100K	Hanson et al. (2007)
MAP	281	280	Mexican Americans	Affymetrix 100K	Hayes et al. (2007)
OOA	124	295	Amish	Affymetrix 100K	Rampersaud et al. (2007)
FHS	91	996	Massachusetts	Affymetrix 100K	Florez et al. (2007)
BBJ	194	1,558	Japan	Custom set of ~268K SNPs	Unoki et al. (2008)
MGP	187	1,504	Japan	JSNP Genome Scan 100K SNPs	Yasuda et al. (2008)
MA	679	697	France	Illumina 300K	Rung et al. (2009)
JM	519	503	Japan	Illumina 550K	Takeuchi et al. (2009)

WTCCC, Wellcome Trust Case Control Consortium; DGI, Diabetes Genetics Initiative; deG, deCODE Genetics; FUSION, Finland-United States Investigation of NIDDM Genetics; DGDG, Diabetes Gene Discovery Group; MAP, Mexican American Population (Starr County, TX); OOA, Old Order Amish; FHS, Framingham Heart Study; BBJ, BioBank Japan; MGP, Millennium Genome Project; MA, Multistage Association; JM: Japanese Multistage.

independently detected associated with T2D. Two loci in genes *TCF7L2* and *CDKAL1* were confirmed by another GWAS published at almost same time. The other genetic marker (rs13266634) in gene *SLC30A8* (solute carrier family 30 (zinc transporter), member 8) was confirmed by another GWAS.

In the Finland-United States Investigation of Non-insulin-dependent Diabetes Mellitus Genetics Study (Scott et al., 2007), 1,161 Finnish T2D patients and 1,174 Finnish normal, glucose-tolerant controls were genotyped for an association analysis, and 41 genetic variants were observed to be significantly associated with a risk of T2D. Genetic variants in or near genes *PPARG*, *KCNJ11*, *TCF7L2*, *IGF2BP2*, *CDKAL1*, *CDKN2A*, *CDKN2B*, *SLC30A8*, *HHEX* (hematopoietically expressed homeobox), and *FTO* (fat mass and obesity associated) were also identified as associated with T2D in other GWAS.

In the Diabetes Gene Discovery Group study conducted in France (Sladek et al., 2007), 694 T2D cases were selected to have at least one affected first-degree relative and an age at onset under 45 years, and the 645 controls had a fasting blood glucose <5.7 mmol/L. By analyzing the genotypes of 1,339 subjects, a total of 28 genetic variants were identified as significant in association with a risk of T2D. Some of the 28 loci such as *TCF2L2* and *HHEX* were also identified as genetic markers for T2D in other GWAS.

In the DiaGen Consortium study (Salonen et al., 2007), 500 familial cases and 497 controls from east Finland, Germany, the United Kingdom, and Ashkenazi were genotyped. Association analysis on the genotypes of those subjects identified nine genetic variants associated with T2D significantly. Three SNPs (rs12255372, rs7901695, and rs7903146) were in the gene *TCF7L2*, which was confirmed to be associated with T2D in other GWAS.

A study conducted in Pima Indians (Hanson et al., 2007) used a low-density GWAS platform in which 100,000 SNPs were interrogated on 300 T2D patients with an age of onset

TABLE 21.2
Significant SNP ($p < 10^{-5}$) Identified in the Initial GWAS

Chromosome	Gene	SNP	*p* Value	Reference
1		rs1775368	6.7×10^{-5}	Hanson et al. (2007)
		rs1932397	7.1×10^{-5}	Scott et al. (2007)
		rs1932465	5.61×10^{-6}	Hayes et al. (2007)
		rs9286938	9.1×10^{-5}	Scott et al. (2007)
	CAMTA1	rs1193179	1.2×10^{-6}	Sladek et al. (2007)
	DDOST	rs640742	2.9×10^{-5}	Scott et al. (2007)
	KCNA10	rs17025978	6.6×10^{-6}	Scott et al. (2007)
	NOTCH2	rs10923931	8.7×10^{-8}	Saxena et al. (2007)
	PDF4B	rs4655595	1.33×10^{-5}	WTCCC (2007)
	RYR2	rs6670163	2.7×10^{-5}	Sladek et al. (2007)
	SLAMF8	rs2501354	8.1×10^{-5}	Scott et al. (2007)
	TLR5	rs1341987	4.34×10^{-5}	WTCCC (2007)
	TTC13	rs6541240	6.2×10^{-5}	Sladek et al. (2007)
2		rs10497681	6.8×10^{-5}	Hanson et al. (2007)
		rs1395931	6.39×10^{-5}	Rampersaud et al. (2007)
		rs1446732	7.38×10^{-5}	Rampersaud et al. (2007)
		rs6712932	6.25×10^{-6}	Salonen et al. (2007)
		rs930621	8.74×10^{-5}	Rampersaud et al. (2007)
		rs9309324	1.93×10^{-5}	WTCCC (2007)
	CENTG2	rs4663576	1.95×10^{-5}	Rung et al. (2009)
	CXCR4	rs932206	4.6×10^{-6}	Sladek et al. (2007)
	GULP1	rs11688935	8.35×10^{-5}	WTCCC (2007)
	IRS1	rs1515114	2.05×10^{-5}	Rung et al. (2009)
	IRS1	rs2713539	2.32×10^{-5}	Rung et al. (2009)
	IRS1	rs2943641	1.84×10^{-5}	Rung et al. (2009)
	IRS1	rs2943645	4.52×10^{-5}	Rung et al. (2009)
	MSH6	rs3136279	6.81×10^{-5}	Rampersaud et al. (2007)
	PARD3B	rs17248501	7.62×10^{-5}	WTCCC (2007)
	PLB1	rs2338545	6.3×10^{-5}	Scott et al. (2007)
	RBMS1	rs6718526	1.15×10^{-5}	WTCCC (2007)
	TMEFF2	rs10497723	8.45×10^{-6}	Hayes et al. (2007)
3		rs10513440	9.01×10^{-5}	WTCCC (2007)
		rs11706900	7.1×10^{-6}	Saxena et al. (2007)
		rs11714343	9.6×10^{-5}	Scott et al. (2007)
		rs16846689	1.8×10^{-5}	Saxena et al. (2007)
		rs16851397	3.8×10^{-5}	Saxena et al. (2007)
		rs358806	3.05×10^{-6}	WTCCC (2007)
		rs823968	6.7×10^{-5}	Scott et al. (2007)
		rs9290075	7.0×10^{-5}	Hanson et al. (2007)
		rs9870410	3.8×10^{-5}	Scott et al. (2007)
	IGF2BP2	rs4376068	3.8×10^{-5}	Unoki et al. (2008)
		rs6769511	2.4×10^{-5}	Unoki et al. (2008)
	LOC131149	rs7651936	4.1×10^{-5}	Sladek et al. (2007)
	LOC646736	rs17533293	9.08×10^{-5}	Rung et al. (2009)
	MAP3K13	rs4687299	4.0×10^{-5}	Scott et al. (2007)
	PTPRG	rs739984	7.2×10^{-5}	Scott et al. (2007)

continued

TABLE 21.2 (CONTINUED)
Significant SNP ($p < 10^{-5}$) Identified in the Initial GWAS

Chromosome	Gene	SNP	*p* Value	Reference
	SLC6A20	rs13064991	5.5×10^{-5}	Sladek et al. (2007)
	TMEM108	rs13072106	8.7×10^{-5}	Scott et al. (2007)
4		rs1371251	7.59×10^{-5}	WTCCC (2007)
		rs6815973	1.61×10^{-5}	WTCCC (2007)
		rs6846031	9.84×10^{-5}	WTCCC (2007)
		rs7659604	9.42×10^{-6}	WTCCC (2007)
		SNP_A-4299379	7.2×10^{-5}	Saxena et al. (2007)
	ANKRD50	rs10518442	3.22×10^{-5}	Hayes et al. (2007)
	ATP8A1	rs13139219	7.8×10^{-5}	Scott et al. (2007)
	BMPR1B	rs17428564	1.4×10^{-5}	Saxena et al. (2007)
	LOC644419	rs282705	1.3×10^{-5}	Sladek et al. (2007)
	PPP2R2C	rs3796400	46.34×10^{-6}	Rung et al. (2009)
		rs3796403	3.50×10^{-6}	Rung et al. (2009)
	PRKG2	rs980720	1.25×10^{-5}	Rampersaud et al. (2007)
	SORCS2	rs886374	2.4×10^{-6}	Scott et al. (2007)
	SPRY1/ANKRD50	rs1498024	1.68×10^{-5}	Hayes et al. (2007)
	WFS1	rs10012946	1.07×10^{-6}	Rung et al. (2009)
		rs10804976	2.54×10^{-6}	Rung et al. (2009)
		rs1801212	1.10×10^{-6}	Rung et al. (2009)
		rs4689388	2.50×10^{-7}	Rung et al. (2009)
		rs734312	4.90×10^{-6}	Rung et al. (2009)
5		rs1030231	9.3×10^{-5}	Scott et al. (2007)
		rs10520926	2.6×10^{-5}	Hanson et al. (2007)
		rs501869	1.2×10^{-5}	Saxena et al. (2007)
	ARHGAP26	rs27779	2.5×10^{-5}	Scott et al. (2007)
6		rs10485249	7.96×10^{-5}	Rampersaud et al. (2007)
		rs910049	5.0×10^{-7}	Saxena et al. (2007)
	AHI1	rs1535435	1.86×10^{-5}	Salonen et al. (2007)
	CDKAL1	rs4712523	8.0×10^{-10}	Takeuchi et al. (2009)
			2.49×10^{-6}	Rung et al. (2009)
		rs4712524	5.7×10^{-6}	Unoki et al. (2008)
		rs6906327	1.61×10^{-6}	Rung et al. (2009)
		rs7754840	1.7×10^{-10}	Takeuchi et al. (2009)
		rs7756992	4.6×10^{-9}	Takeuchi et al. (2009)
			7.7×10^{-9}	Steinthorsdottir et al. (2007)
		rs9295475	1.2×10^{-5}	Unoki et al. (2008)
		rs9460546	3.0×10^{-5}	Unoki et al. (2008)
		rs9465871	3.34×10^{-7}	WTCCC (2007)
	GFRAL	rs7452656	3.15×10^{-5}	WTCCC (2007)
	LOC441171	rs9494266	2.67×10^{-5}	Salonen et al. (2007)
	QKI	rs1885732	7.64×10^{-5}	Rung et al. (2009)
	SYNE1	rs2673776	8.0×10^{-5}	Scott et al. (2007)
7		rs10954654	2.8×10^{-5}	Scott et al. (2007)
		rs557962	5.9×10^{-5}	Scott et al. (2007)
		rs612774	3.1×10^{-6}	Unoki et al. (2008)
		rs615545	5.9×10^{-6}	Scott et al. (2007)
	GPR154,AAA1	rs324978	9.7×10^{-5}	Saxena et al. (2007)
	Grb10	rs2237457	1.07×10^{-5}	Rampersaud et al. (2007)

TABLE 21.2 (CONTINUED)
Significant SNP ($p < 10^{-5}$) Identified in the Initial GWAS

Chromosome	Gene	SNP	*p* Value	Reference
	ICA1/NXPH1	rs757705	2.79×10^{-5}	Hayes et al. (2007)
	LOC642421	rs528957	7.8×10^{-5}	Scott et al. (2007)
	SLC13A1	rs2470984	9.0×10^{-5}	Scott et al. (2007)
8		rs2679765	3.37×10^{-5}	WTCCC (2007)
		rs2736010	5.83×10^{-5}	WTCCC (2007)
	CNBD1	rs1852027	7.6×10^{-5}	Scott et al. (2007)
	GPR20	rs7839244	6.8×10^{-5}	Scott et al. (2007)
	SLC30A8	rs13266634	3.3×10^{-6}	Steinthorsdottir et al. (2007)
			1.56×10^{-6}	Rung et al. (2009)
	TRPA1	rs13259803	3.0×10^{-7}	Unoki et al. (2008
9		rs10733336	8.3×10^{-5}	Saxena et al. (2007)
		rs10811661	3.6×10^{-5}	Saxena et al. (2007)
		rs2590504	8.84×10^{-5}	WTCCC (2007)
		rs7019589	7.49×10^{-5}	WTCCC (2007)
	CDKN2A/B	rs10811661	2.2×10^{-8}	Takeuchi et al. (2009)
		rs2383208	1.6×10^{-7}	Takeuchi et al. (2009)
	FAM69B	rs945384	3.6×10^{-5}	Sladek et al. (2007)
	OR13D1	rs10512332	4.60×10^{-5}	Hayes et al. (2007)
	VPS13A	rs2050831	8.4×10^{-5}	Sladek et al. (2007)
10		rs10509195	3.30×10^{-5}	Rampersaud et al. (2007)
		rs10509199	1.68×10^{-5}	Rampersaud et al. (2007)
		rs10509201	8.80×10^{-5}	Rampersaud et al. (2007)
		rs10829494	3.37×10^{-5}	WTCCC (2007)
		rs1916411	8.48×10^{-5}	Rampersaud et al. (2007)
		rs1916412	8.35×10^{-5}	Rampersaud et al. (2007)
		rs7474871	6.59×10^{-5}	WTCCC (2007)
	COL13A1	rs4082516	3.19×10^{-5}	Rampersaud et al. (2007)
	HHEX	rs1111875	1.2×10^{-5}	Sladek et al. (2007)
	HPSE2	rs1159006	4.01×10^{-5}	Hayes et al. (2007)
	PRKG1	rs11000542	9.72×10^{-5}	WTCCC (2007)
	SORBS1	rs1536558	4.77×10^{-5}	Hayes et al. (2007)
	STK32C	rs7910485	1.02×10^{-5}	Salonen et al. (2007)
		rs10885409	7.72×10^{-14}	Rung et al. (2009)
	TCF7L2	rs12255372	5.31×10^{-7}	Salonen et al. (2007)
			1.5×10^{-5}	Scott et al. (2007)
			2.18×10^{-22}	Rung et al. (2009)
		rs4506565	5.05×10^{-12}	WTCCC (2007)
		rs4918789	7.07×10^{-7}	Rung et al. (2009)
		rs7100927	5.2×10^{-8}	Sladek et al. (2007)
		rs7895307	7.71×10^{-6}	Rung et al. (2009)
		rs7900150	5.1×10^{-8}	Sladek et al. (2007)
		rs7901695	3.42×10^{-7}	Salonen et al. (2007)
		rs7903146	5.52×10^{-8}	Salonen et al. (2007)
			1.82×10^{-8}	Steinthorsdottir et al. (2007)
			1.2×10^{-5}	Scott et al. (2007)
		rs7904519	2.62×10^{-12}	Rung et al. (2009)
	ZNF239	rs9326506	2.99×10^{-5}	WTCCC (2007)

continued

TABLE 21.2 (CONTINUED)
Significant SNP ($p < 10^{-5}$) Identified in the Initial GWAS

Chromosome	Gene	SNP	*p* Value	Reference
11		rs10501281	5.3×10^{-5}	Scott et al. (2007)
		rs11021059	2.22×10^{-5}	WTCCC (2007)
		rs11226667	7.8×10^{-5}	Saxena et al. (2007)
		rs12273344	6.5×10^{-5}	Scott et al. (2007)
		rs516415	5.0×10^{-5}	Hanson et al. (2007)
		rs9300039	6.0×10^{-5}	Scott et al. (2007)
	C11ORF41	rs10836097	5.3×10^{-5}	Unoki et al. (2008)
	EXT2	rs11037909	1.8×10^{-5}	Sladek et al. (2007)
		rs1113132	3.7×10^{-5}	Sladek et al. (2007)
		rs3740878	1.8×10^{-5}	Sladek et al. (2007)
	KCNQ1	rs151290	7.4×10^{-5}	Yasuda et al. (2008)
		rs2237892	2.3×10^{-5}	Takeuchi et al. (2009)
		rs2237895	1.4×10^{-7}	Yasuda et al. (2008)
		rs2283228	5.7×10^{-5}	Unoki et al. (2008)
	MMP26	rs2499953	2.3×10^{-5}	Sladek et al. (2007)
	OR52B6	rs4910822	5.87×10^{-5}	Rung et al. (2009)
	ZBTB16	rs672849	1.5×10^{-5}	Hanson et al. (2007)
		rs686989	2.7×10^{-6}	Hanson et al. (2007)
12		rs11178531	3.75×10^{-5}	WTCCC (2007)
		rs1495377	6.52×10^{-6}	WTCCC (2007)
		rs1918416	4.9×10^{-5}	Scott et al. (2007)
	CORO1C	rs3825253	3.6×10^{-5}	Scott et al. (2007)
	DYRK2/IFNG	rs10492202	3.48×10^{-5}	Hayes et al. (2007)
	FLJ20674	rs4767658	4.1×10^{-5}	Scott et al. (2007)
	HIGD1C	rs12304921	7.07×10^{-6}	WTCCC (2007)
	METTL7A	rs17125088	4.90×10^{-5}	WTCCC (2007)
	PIK3C2G	rs12581163	6.07×10^{-5}	WTCCC (2007)
13		rs1287526	6.4×10^{-5}	Scott et al. (2007)
		rs17660169	2.3×10^{-5}	Saxena et al. (2007)
		rs9545903	7.2×10^{-5}	Scott et al. (2007)
14		rs1007383	7.59×10^{-5}	WTCCC (2007)
		rs11849174	5.4×10^{-5}	Scott et al. (2007)
		rs1211216	6.4×10^{-6}	Saxena et al. (2007)
		rs1449720	6.8×10^{-5}	Scott et al. (2007)
	LOC388015	rs3825569	3.7×10^{-5}	Scott et al. (2007)
	LOC646279	rs1256517	5.5×10^{-5}	Sladek et al. (2007)
15	*CCDC33*	rs2930291	4.40×10^{-5}	WTCCC (2007)
	ZFAND6	rs2903265	4.98×10^{-5}	WTCCC (2007)
16		rs10521095	2.3×10^{-5}	Scott et al. (2007)
		rs11647358	1.8×10^{-5}	Saxena et al. (2007)
		rs2099106	2.98×10^{-5}	WTCCC (2007)
	FTO	rs7193144	4.78×10^{-8}	WTCCC (2007)
		rs9939609	1.91×10^{-7}	WTCCC (2007)
	LOC729922	rs7499133	1.73×10^{-5}	Rung et al. (2009)
	MTHFSD	rs3751797	8.74×10^{-5}	Rampersaud et al. (2007)
17	*ACCN1*	rs2785061	5.6×10^{-5}	Saxena et al. (2007)
	KIAA1303	rs868432	1.73×10^{-5}	Rung et al. (2009)
	NLGN2	rs11078674	9.8×10^{-5}	Sladek et al. (2007)

TABLE 21.2 (CONTINUED)
Significant SNP ($p < 10^{-5}$) Identified in the Initial GWAS

Chromosome	Gene	SNP	*p* Value	Reference
	PRKCA	rs7207345	7.5×10^{-5}	Scott et al. (2007)
	SPECC1	rs1373147	2.79×10^{-5}	Rampersaud et al. (2007)
		rs2703813	8.12×10^{-5}	Rampersaud et al. (2007)
18	*CDH19*	rs508987	6.28×10^{-5}	WTCCC (2007)
	CDH2	rs9807662	7.5×10^{-5}	Saxena et al. (2007)
	CTDP1	rs70198	9.84×10^{-5}	WTCCC (2007)
	EPB41L3	rs1941011	4.96×10^{-5}	Hayes et al. (2007)
	SETBP1	rs616444	9.0×10^{-5}	Scott et al. (2007)
19	*LDLR*	rs6413504	8.2×10^{-5}	Sladek et al. (2007)
	ZNF350	rs2278419	4.1×10^{-5}	Sladek et al. (2007)
	ZNF615	rs11084127	1.1×10^{-5}	Sladek et al. (2007)
		rs11084128	1.3×10^{-5}	Sladek et al. (2007)
		rs1836002	1.5×10^{-5}	Sladek et al. (2007)
		rs1978717	7.5×10^{-6}	Sladek et al. (2007)
	ZNF649	rs8101509	2.2×10^{-5}	Sladek et al. (2007)
20	*SLC24A3*	rs6136651	1.81×10^{-5}	Hayes et al. (2007)
21		rs158081	2.36×10^{-5}	Salonen et al. (2007)
		rs200801	1.78×10^{-5}	Salonen et al. (2007)
		rs2254434	3.86×10^{-5}	Salonen et al. (2007)
		rs2831605	6.82×10^{-5}	Hayes et al. (2007)
	GRIK1	rs458685	6.54×10^{-5}	Hayes et al. (2007)
22		rs565979	7.0×10^{-5}	Scott et al. (2007)
	ELJ90680	rs10211998	4.1×10^{-5}	Sladek et al. (2007)
		rs5756371	5.1×10^{-5}	Sladek et al. (2007)
	FLJ27365	rs873492	9.6×10^{-5}	Sladek et al. (2007)
	SDF2L1	rs861844	3.21×10^{-5}	Hayes et al. (2007)

younger than 25 years and 334 control subjects aged greater than 45 years. Seven genetic variants were observed to be associated with T2D to a significant extent. Among those seven, two SNPs (rs672849 and rs686989) were in gene *ZBTB16* (zinc finger and BTB domain containing 16).

Hayes et al. (2007) conducted a study in which 281 Mexican Americans with T2B and 280 controls, randomly selected from Mexican Americans in the Starr county of Texas, were genotyped with a low density array (Affymetrix GeneChip Human Mapping 100K set). A total of 14 genetic variants were associated to a significant degree from allelic association exact tests. However, most of the associated SNPs, as well as the genes containing or near those SNPs, were not observed in the top significant SNPs from other GWAS of T2D. However, some of the less statistically significant SNPs ($p < 0.01$) were observed to be associated with T2D in other GWAS, such as a few SNPs near genes *UBQLNL* (ubiquilin-like protein), *RALGPS2* (Ral GEF with PH domain and SH3 binding motif 2), and *EGR2* (early growth response 2).

In an Old Order Amish study (Rampersaud et al., 2007), DNA samples from 124 T2D case subjects and 295 control subjects were collected from the Old Order Amish families and were genotyped. Likelihood ratio tests with age and sex as covariates identified 16 genetic variants significantly associated with T2D. The most significant association was for SNP rs2237457, which is located in gene *Grb10* (growth factor receptor-bound protein), an adaptor protein that regulates insulin receptor signaling. However, most of them were not detected from other GWAS of T2D.

They could be false associations because the sample size is very small, and the statistical power may not be high enough to provide reliable results.

Florez et al. (2007) studied a total of 1,087 Framingham Heart Study family members using the Affymetrix GeneChip Human Mapping 100K set. Cox proportional hazard survival analysis was used to detect association of genetic variants with a risk of T2D. No single genetic variant was identified to be associated significantly with T2D.

Unoki et al. (2008) used a customized set of 268,086 SNPs, which covered approximately 56% of common SNPs in the Japanese, for genotyping 194 patients with T2D and diabetic retinopathy and 1,558 controls collected in the BioBank Japan. A total of nine genetic variants were identified as significant loci for a predisposition of T2D. Some of them are in genes that were discovered to be associated with T2D from other GWAS, such as *IGF2BP2* (rs4376068 and rs6769511) and *CDKAL1* (rs4712524, rs9295475, and rs9460546). In addition, new genetic markers were identified, e.g., SNP rs2283228 in gene *KCNQ1* (potassium voltage-gated channel, KQT-like subfamily, member 1) on chromosome 11.

In the Millennium Genome Project (Yasuda et al., 2008), a set of 100,000 SNPs from a collection of standard Japanese SNPs were genotyped by multiplex polymerase chain reaction-based Invader analysis of 187 patients with T2D and two groups of controls: one group of 752 individuals representing the general Japanese population and another group of 752 patients of other diseases (Alzheimer's disease, gastric cancer, hypertension, and asthma). Two genetic markers (rs151290 and rs2237895) in gene *KCNQ1* were identified to be associated with T2D significantly by two association analyses using the two different groups of controls.

In a study called Multistage Association conducted by Rung et al. (2009), 1,376 French T2D cases and controls were genotyped using Illumina Hap300 and Human-1 bead arrays. Association analysis identified a total of 25 genetic variants significantly associated with T2D. Some of the associated SNPs confirmed genetic loci of T2D identified from other GWAS, such as genetic markers in genes *TCF7L2* and *CDKAL1*. New genetic loci were also identified to be associated with T2D, such as four SNPs (rs1515114, rs2713539, rs2943641, and rs2943645) in gene IRS1 (insulin receptor substrate 1).

In a Japanese multistage study (Takeuchi et al., 2009), 519 Japanese T2D cases and 503 controls were genotyped using the Infinium HumanHap550 BeadArray from Illumina. Association analysis was conducted using the Cochran-Armitage trend test. A total of six genetic variants in three genes (*CDKAL1*, *CDKN2A/CDKN2B*, and *KCNQ1*) were identified to be associated with T2D, confirming those genetic markers from other GWAS.

21.3.3 Replication of Genetic Markers of T2D

Given the complexity of a GWAS, multiple sources of Type I and Type II errors can arise in the associated SNPs (Hong et al., 2009). To ensure the validity of GWAS findings, the genetic markers identified from an initial GWAS are required to be replicated in at least one independent study. Replication can be conducted in the same or a different population with the same genetic background of the initial GWAS. Different GWAS used diverse strategies to select genetic markers for replication. In most GWAS, the genetic markers that were selected for replications were statistically significant (such as $p < 0.01$) rather than genome-wide significant (that would take into account the multiple testing). In addition to statistical consideration, biological knowledge and results of other GWAS are also used to guide the selection of SNPs for replication. Replication is different from technical validation of the genotype data; the latter requires genotyping a small set of SNPs, normally on a selected group of biological samples in the original GWAS, using a different technology than was used in the initial GWAS. In contrast, the sample size in replication is much larger than that of initial GWAS. A large sample size is necessary to achieve a sufficient statistic power because genome-wide significance levels should be used to count for multiple tests in an initial GWAS. A small number of genetic markers are usually used in a replication study. Thus, there are no multiple testing problems in replications, and a conventional statistical significance level could be acceptable. Actually, most of the replication studies used conventional significance levels, such as $p < 0.01$.

An emerging approach for replication of GWAS is the use of meta-analysis in which a set of initial GWAS data are combined for association analysis (Zeggini et al., 2008; Dupuis et al., 2010; Saxena et al., 2010; Voight et al., 2010). By combining the results of different GWAS of T2D, a meta-analysis not only can confirm significant genetic variants identified from initial GWAS but also can gain additional statistical power for the discovery of new genetic markers associated with T2D.

To date, there are a total of 78 SNPs in or near 58 genes confirmed in at least one replication study or meta-analysis (Table 21.3). Among the 58 genes, 17 genes had genetic markers replicated in at least two independent replication studies. One of the most encouraging and successful findings from T2D GWAS came from the successful replication of a genetic marker (SNP rs7903146) in gene *TCF7L2* in six different replication studies (Saxena et al., 2007; Scott et al., 2007; Sladek et al., 2007; Rung et al., 2009; Takeuchi et al., 2009). *TCF7L2* encodes a transcription factor that is involved in the Wnt-signalling pathway (Huelsken and Behrens, 2002). It has been observed that alterations of *TCF7L2* expression disrupt pancreatic islet function through dysfunction of proglucagon gene expression, leading to reduced insulin secretion and enhanced risk of T2D (Lyssenko et al., 2007). The previously established T2D susceptibility loci at two other genes, *PPARG* (SNP rs1801282) and *KCNJ11* (SNP rs5219), were also confirmed in multiple replication studies.

The first wave of T2D GWAS and subsequent replications not only confirmed some known genetic loci of T2D but also identified some entirely new T2D susceptibility loci, such as variants in or near genes *FTO*, *HHEX*, *CDKAL1*, *IGF2BP2*, *SLC30A8*, and *CDKN2B*. Except for variants in gene *FTO*, which exert their T2D effect through a primary impact on body mass index (Frayling et al., 2007), the genetic variants in or near the other five genes exert their primary effect on insulin secretion (Grarup et al., 2007). Genetic variants in genes *EXT2* (exostoses, multiple 2) and *LOC387761* were identified and replicated in a French population (Sladek et al., 2007). They have not been replicated in other GWAS and thus are less likely to represent genuine genetic signals. Although it is a convention to label the susceptibility loci according to the strongest signal (e.g., *FTO*), it is important to note that these genetic markers do not imply a confirmed mechanistic link. What GWAS and subsequent replication studies deliver are T2D-associated genetic variants, and extensive fine-mapping and functional studies are required to identify the variants that really cause T2D and to define the molecular and cellular mechanisms of T2D (McCarthy et al., 2008).

The effect size of genetic variants of T2D is not large, and only a very small portion of the predisposition to T2D could be explained by the genetic markers identified from individual GWAS of T2D. Efforts to find additional T2D-susceptibility loci with anticipated modest effect sizes are required. One approach is to increase the sample size in GWAS, an expensive undertaking. The most effective strategy for this has been to combine existing GWAS data through meta-analysis, and many new genetic markers of T2D were identified (Table 21.3). Taking the results from the Diabetes Genetics Replication and Meta-Analysis (DIAGRAM) consortium as an example (Zeggini et al., 2008), integration of data from three initial GWAS (Saxena et al., 2007; Scott et al., 2007; Zeggini et al., 2007) increased the sample size to 4,500 cases and 5,500 controls. Imputation algorithm (a statistical approach to predict genotypes based on reference samples such as HapMap samples) was also used to infer genotypes at additional SNPs that were not directly genotyped in the original GWAS, thereby extending the analysis to a total of 2.2 million SNPs across the genome. A total of 69 SNPs showed that associations in the meta-analysis were genotyped in an initial replication set of 22,426 subjects. The top 11 SNPs (representing ten independent loci) emerging from this second analysis were then further evaluated in 57,000 subjects, with most being replicated and six previously unknown loci detected.

TABLE 21.3
Genetic Variants Confirmed in Replication Studies or by Other Initial GWAS

Chromosome	Nearest Gene	SNP	Odds Ratio	*p* Value	Reference
1	*ADAM30*	rs2641348	1.1	1.2×10^{-3}	Zeggini et al. (2008)
	LOC646538	rs10493685	3.65	4.3×10^{-3}	Hanson et al. (2007)
	NOTCH2	rs10923931	1.09	2.9×10^{-3}	Zeggini et al. (2008)
	PKN2	rs6698181	1.09	9.7×10^{-4}	Saxena et al. (2007)
	PROX1	rs340874	1.07	7.2×10^{-10}	Dupuis et al. (2010)
2	*BCL11A*	rs10490072	1.08	1.4×10^{-5}	Zeggini et al. (2008)
		rs243021	1.08	6.2×10^{-11}	Voight et al. (2010)
	GCKR	rs1260326	0.79	1.0×10^{-2}	Vaxillaire et al. (2008)
		rs780094	1.06	1.3×10^{-9}	Dupuis et al. (2010)
	IRS1	rs2943641	1.16		Rung et al. (2009)
		rs7578326	1.10	2.2×10^{-15}	Voight et al. (2010)
	THADA	rs7578597	1.15	1.6×10^{-3}	Zeggini et al. (2008)
3	*ADAMTS9*	rs4607103	1.1	1.0×10^{-4}	Zeggini et al. (2008)
	ADCY5	rs11708067	1.12	9.9×10^{-21}	Dupuis et al. (2010)
	IGF2BP2	rs1470579	1.19	2.1×10^{-9}	Saxena et al. (2007)
		rs4402960	1.18	5.5×10^{-9}	Saxena et al. (2007)
			1.14	2.5×10^{-5}	Takeuchi et al. (2009)
			1.08	2.2×10^{-1}	Scott et al. (2007)
			1.09	1.8×10^{-2}	Zeggini et al. (2007)
	LOC100129403	rs9290075	1.39	1.34×10^{-2}	Hanson et al. (2007)
	PPARG	rs1801282	1.11	6.1×10^{-3}	Saxena et al. (2007)
			1.08	3.3×10^{-1}	Scott et al. (2007)
	SGOL1	rs1500415	3.69	4.89×10^{-2}	Hanson et al. (2007)
	SYN2, PPARG	rs17036101	1.13	4.5×10^{-3}	Zeggini et al. (2008)
	ZNF385D	rs424695	1.63	1.49×10^{-2}	Hanson et al. (2007)
4	*ELJ39370*	rs17044137	1.09	3.1×10^{-3}	Saxena et al. (2007)
	WFS1	rs4689388	1.07		Rung et al. (2009)
5	*MSNL1*	rs10520926	1.36	3.35×10^{-2}	Hanson et al. (2007)
	ZBED3	rs4457053	1.07	2.7×10^{-7}	Voight et al. (2010)
6	*AHI1*	rs1535435	1.287	2.0×10^{-4}	Salonen et al. (2007)
	CDKAL1	rs10946398	1.14	8.3×10^{-5}	Zeggini et al. (2007)
		rs4712523	1.17		Rung et al. (2009)
			1.23	4.0×10^{-12}	Takeuchi et al. (2009)
		rs7754840	1.06	2.4×10^{-2}	Saxena et al. (2007)
			1.08	2.0×10^{-1}	Scott et al. (2007)
		rs7756992	1.21	5.4×10^{-5}	Steinthorsdottir et al. (2007)
	LOC441171	rs9494266	1.308	5×10^{-5}	Salonen et al. (2007)
	VEGFA	rs9472138	1.07	1.5×10^{-3}	Zeggini et al. (2008)
7	*DGKB-TMEM195*	rs2191349	1.06	1.1×10^{-8}	Dupuis et al. (2010)
	GCK	rs4607517	1.07	5.0×10^{-8}	Dupuis et al. (2010)
	JAZF1	rs864745	1.08	8.1×10^{-5}	Zeggini et al. (2008)
	KLF14	rs972283	1.06	6.4×10^{-6}	Voight et al. (2010)
8	*MMP16*	rs6994019	3.49	1.89×10^{-2}	Hanson et al. (2007)
	TP53INP1	Rs896854	1.05	2.2×10^{-5}	Voight et al. (2010)
			1.12	1.4×10^{-1}	Saxena et al. (2007)
	SLC30A8	rs13266634	1.24	5.8×10^{-13}	Takeuchi et al. (2009)
			1.14	2.6×10^{-2}	Scott et al. (2007)

TABLE 21.3 (CONTINUED)
Genetic Variants Confirmed in Replication Studies or by Other Initial GWAS

Chromosome	Nearest Gene	SNP	Odds Ratio	*p* Value	Reference
			1.09	7.3×10^{-2}	Steinthorsdottir et al. (2007)
			1.18	6.1×10^{-8}	Sladek et al. (2007)
			1.12	1.2×10^{-3}	Zeggini et al. (2007)
9	*CDKN2B*	rs2383208	1.33	4.8×10^{-22}	Takeuchi et al. (2009)
		rs10811661	1.22	1.5×10^{-2}	Scott et al. (2009)
			1.16	2.2×10^{-5}	Saxena et al. (2007)
			1.18	1.7×10^{-4}	Zeggini et al. (2007)
		rs564398	1.12	8.6×10^{-4}	Zeggini et al. (2007)
	CHCHD9	rs13292136	1.08	2.4×10^{-4}	Voight et al. (2010)
10	*CDC123*	rs12779790	1.11	5.4×10^{-5}	Zeggini et al. (2008)
	HHEX	rs1111875	1.13	6.0×10^{-3}	Saxena et al. (2007)
			1.21	2.6×10^{-9}	Takeuchi et al. (2009)
			1.06	3.4×10^{-1}	Scott et al. (2007)
			1.19	3.0×10^{-6}	Sladek et al. (2007)
			1.08	2.0×10^{-2}	Zeggini et al. (2007)
		rs7923837	1.22	7.5×10^{-6}	Sladek et al. (2007)
	TCF7L2	rs7903146	1.36		Rung et al. (2009)
			1.4	3.9×10^{-28}	Saxena et al. (2007)
			1.59	5.3×10^{-11}	Takeuchi et al. (2009)
			1.3	3.5×10^{-4}	Scott et al. (2007)
			1.65	1.5×10^{-34}	Sladek et al. (2007)
11	*EXT2*	rs11037909	1.27	1.8×10^{-4}	Sladek et al. (2007)
		rs1113132	1.15	3.3×10^{-4}	Sladek et al. (2007)
		rs3740878	1.26	1.2×10^{-4}	Sladek et al. (2007)
	FANCF	rs10500938	1.65	1.1×10^{-3}	Hanson et al. (2007)
	KCNJ11	rs5219	1.14	2.6×10^{-6}	Saxena et al. (2007)
			1.02	3.01×10^{-1}	Takeuchi et al. (2009)
			1.04	5.5×10^{-1}	Scott et al. (2007)
	KCNQ1	rs2074196	1.32	2.1×10^{-7}	Yasuda et al. (2008)
		rs2237892	1.36	8.0×10^{-23}	Takeuchi et al. (2009)
			1.39	1.6×10^{-12}	Yasuda et al. (2008)
		rs2237895	1.25	4.7×10^{-5}	Yasuda et al. (2008)
			1.32	7.3×10^{-9}	Unoki et al. (2008)
		rs2237897	1.41	6.8×10^{-13}	Unoki et al. (2008)
		rs231362	1.07	3.2×10^{-9}	Voight et al. (2010)
		rs2283228	1.24	7×10^{-6}	Unoki et al. (2008)
	LOC387761	rs7480010	1.14	1.1×10^{-4}	Sladek et al. (2007)
	MTNR1B	rs10830963	1.12	2.0×10^{-2}	Lyssenko et al. (2009)
			1.23	3.6×10^{-4}	Sparso et al. (2009)
		rs1387153	1.08	4.4×10^{-10}	Voight et al. (2010)
			1.15	6.3×10^{-5}	Bounatia-Naji et al. (2009)
	RPL9P23	rs9300039	1.45	2.7×10^{-4}	Scott et al. (2007)
	CENTD2	rs1552224	1.14	3.2×10^{-18}	Voight et al. (2010)
	ZBTB16	rs672849	1.34	1.64×10^{-2}	Hanson et al. (2007)
		rs686989	1.27	3.33×10^{-2}	Hanson et al. (2007)

continued

TABLE 21.3 (CONTINUED)
Genetic Variants Confirmed in Replication Studies or by Other Initial GWAS

Chromosome	Nearest Gene	SNP	Odds Ratio	*p* Value	Reference
12	*DCD*	rs1153188	1.07	3.1×10^{-3}	Zeggini et al. (2008)
	LOC728114	rs1859441	1.86	2.7×10^{-3}	Hanson et al. (2007)
	PPFIA2	rs10506855	1.79	3.35×10^{-2}	Hanson et al. (2007)
	HMGA2	rs1531343	1.08	1.1×10^{-4}	Voight et al. (2010)
	HNF1A	rs7957197	1.05	4.6×10^{-4}	Voight et al. (2010)
	TSPAN8, LGR5	rs7961581	1.06	9.8×10^{-3}	Zeggini et al. (2008)
15	*ZFAND6*	rs11634397	1.05	1.2×10^{-5}	Voight et al. (2010)
	PRC1	rs8042680	1.06	1.6×10^{-6}	Voight et al. (2010)
16	*FTO*	rs8050136	1.18	6.3×10^{-3}	Scott et al. (2007)
			1.22	5.4×10^{-7}	Zeggini et al. (2007)
17	*HNF1B*	rs4430796	0.86	2.1×10^{-3}	Gudmundsson et al. (2007)
		rs7501939	0.88	4.5×10^{-3}	Gudmundsson et al. (2007)
19	*PEPD*	rs10425678	1.1	2.0×10^{-3}	Takeuchi et al. (2009)
X	*DUSP9*	rs5945326	1.32	2.3×10^{-5}	Voight et al. (2010)

21.4 CHALLENGES OF T2D GWAS

T2D GWAS are primarily motivated by the expectation of finding robust associations from which novel insights into T2D mechanisms and new avenues for clinical treatment could be generated. However, the task of moving from T2D-associated genetic variants to uncovering of T2D mechanisms will be a huge undertaking because the T2D GWAS is a complicated process (Figure 21.2), and there are a number of challenges.

21.4.1 Challenges in Study Design of T2D GWAS

Although there is increasing interest in the application of GWAS methodologies to population-based cohorts, the initial T2D GWAS used case-control study designs in which allele frequencies in patients with T2D were compared with those in a healthy group. The initial T2D GWAS and subsequent replication studies have demonstrated that the T2D risk attributions of individual genetic variants are usually small. Therefore, sample sizes in T2D GWAS should be large to achieve adequate statistical power. However, current T2D GWAS (Table 21.1) used less than 2,000 cases, and some of them are with only several hundred T2D patients. A small sample size reduces the statistical power, which may mask true positive associations. Although a consensus on the minimum sample size for GWAS has not been reached, it is clear that the larger the sample size, the better, all other things (including quality) being equal. Although there were efforts to increase sample size by combining initial T2D GWAS data and to conduct meta-analysis on the combined data (Zeggini et al., 2008; Dupuis et al., 2010; Saxena et al., 2010; Voight et al., 2010), a large sample size still remains an most important factor to obtain robust results for T2D GWAS.

Another challenge in the design of T2D GWAS is the proper selection of participants. Improper selection for a T2D GWAS is a potential source of obfuscation and variability in experimental design. Misclassifying participants into T2D and control groups can markedly reduce study power and result in spurious associations, particularly when a large number of non-T2D individuals are misclassified as T2D (Pearson and Manolio, 2008). When selecting T2D patients, careful participant ascertainment is necessary. However, a bias in selection of cases in T2D GWAS may arise if individuals are selected to overly minimize phenotypic heterogeneity by focusing on extreme and/or familial cases, resulting in decreasing statistical power (Howson et al., 2005). The results of such studies may not be applicable to other study populations. Controls should be selected from the same population as T2D patients. Optimal selection of controls remains controversial, although the accumulating empirical data indicate that many commonly expressed concerns have been overstated (McCarthy et al., 2008). Effectiveness of non-population-based controls was demonstrated by the Wellcome Trust Case Control Consortium study, in which the 2,938 UK population-based controls (as *common controls*) were compared with 1,924 T2D patients. This suggests that the initial identification of association with T2D may be resistant to these biases of using nonpopulation controls, especially given subsequent successful replication of some associations in studies using more traditional control groups (Scott et al., 2007). Of more concern in the improper selection of cases and controls in T2D GWAS may be nongenetic covariates such as obesity (Frayling et al., 2007), which may be confounded with the outcome and subsequently generate false positive associations. Improper population stratification of T2D patients and controls is an additional potential source of spurious associations related to T2D GWAS design. Population stratification (imbalances in populations in cases relative to controls) inflates the Type I error rate around variants that are informative about the population substructure (Price et al., 2006). Several statistical tools have been developed to correct for population stratification (Price et al., 2006; Cardon and Palmer, 2003) that are now incorporated into rigorous T2D GWAS analyses.

21.4.2 Challenges in Genotyping Experimental Settings

A fundamental component of a GWAS is the platform (Figure 21.2) that determines genotypes of SNPs in T2D patients and control subjects. Genotyping errors generated in this step, especially if distributed differentially between T2D patients and controls, are potential sources of spurious associations in T2D GWAS findings. Many challenges exist for correcting or reducing errors or biases introduced in genotyping experiments.

The debates about SNP selection (Barrett and Cardon, 2006; Pe'er et al., 2006) that dominated the early discussions on GWAS have been resolved mostly. The issue is now simplified to the choice of genotyping platforms, mainly between two leaders (Affymetrix and Illumina). Despite the fact that these two companies offer similar coverage of common variants in the HapMap (McCarthy et al., 2008), differences in SNP coverage between Affymetrix and Illumina exist and are a potential source of Type II errors (losing significant associations). Because the designs of probe sets on SNP arrays, as well as the experimental protocols, are quite different between these two platforms, the consistency of common SNPs interrogated in both platforms is vital for obtaining reliable T2D GWAS results. Inconsistency of this type of SNPs may cause both type I errors and type II errors in T2D GWAS results (Hong et al., 2009). Evaluation of variations between platforms has been reported (Hong et al., 2010b) and is expected to be thoroughly assessed in the future.

Another challenge related to coverage of genetic markers in the commercially available genotyping platforms is the variations in population used to derive the SNPs. For example, the Illumina 300K array captured 75% of HapMap common SNPs with an allelic correlation of 80% (a measure of the linkage disequilibrium between two SNPs) in subjects of European ancestry, but the coverage was only 28% in subjects of African ancestry. Equivalent Affymetrix arrays had lower coverage of common SNPs for Europeans but higher coverage for Africans (Barrett and Cardon, 2006).

It was also observed that the majority of SNPs assessed in commercially available genotyping platforms do not affect protein structure and appear unlikely to affect gene expression so that using commercially available chips limits the discovery power of T2D GWAS to locating chromosomal regions rather than the genetic variants that are responsible for causing the disease. The initial discovery from T2D GWAS may be sufficient for prognostic modeling, but understanding T2D mechanism will require more in-depth studies such as fine mapping or sequencing for the discovery of functional variants (Frazer et al., 2009).

Consistency between different generations of the same genotyping platform is another concern and remains a challenge for T2D GWAS. Different generations of genotyping platforms from both

Affymetrix and Illumina have been delivered to the market and used in T2D GWAS. For example, the Affymetrix GeneChip Human Mapping 500K array set (Affy500K) has been used in the published T2D GWAS (Takeuchi et al., 2009; Wellcome Trust Case Control Consortium, 2007; Saxena et al., 2007), but recently Affymetrix released the Genome-Wide Human SNP 6.0 array (Affy6) to the market. Because it is important to know whether genotypes determined with these two generations of SNP array are the same, we evaluated their consistency (Hong et al., 2010b). Genotypes of common SNPs were interrogated in both arrays using the 270 HapMap samples and determined using the same calling algorithm, Birdseed. In this case, 482,215 SNPs are common to both arrays. The scatter plot in Figure 21.3a compares the missing call rates per SNP (percentage of SNPs that are not assigned to one of the three genotypes: homozygote, heterozygote, and variant homozygote) in the Affy500K (x-axis) and in the Affy6 (y-axis) where each of the points represents one of the common SNPs. When a SNP is located on or near the diagonal line, it indicates that the missing call rate for the SNP is consistent between the Affy500k and the Affy6. From Figure 21.3a, it can be seen that a large number of SNPs are not consistent in missing call rates, some of which are very different between these two arrays. The scatter plot in Figure 21.3b depicts the missing call rates of the 270 HapMap samples based on the common SNPs in the Affy500K (x-axes) and Affy6 (y-axes) arrays. From Figure 21.3b, it can be seen that the inconsistencies in missing call rates per sample are much smaller than the missing call rates per SNP (Figure 21.3a). Moreover, it was observed that the missing call rates from the newer generation, Affy6 array, are slightly lower than the missing call rates from the older generation, Affy500K array. The p values of a paired two-sample t test for comparing the missing call rates per SNP (Figure 21.3a) and of missing call rates per sample (Figure 21.3b) were 0 and 5.057×10^{-60}, respectively, indicating that the difference of missing call rates per SNP and per sample between the Affy500K and the Affy6 are statistically significant.

The objective of a T2D GWAS is to identify the genetic markers associated with T2D. It is critical to assess how the inconsistency between different generations of SNP arrays can propagate errors to the list of associated SNPs identified in the downstream association analysis. To mimic case control in T2D GWAS, three association analyses were conducted for genotypes obtained from the Affy6 and Affy500k data of 270 HapMap samples. Each of the three population groups of the HapMap samples (European, Asian, and African) were set, in turn, as cases, whereas the other two groups were set as control. Association analyses were conducted to identify SNPs that can differentiate the case group from the control groups. The lists of significantly associated SNPs,

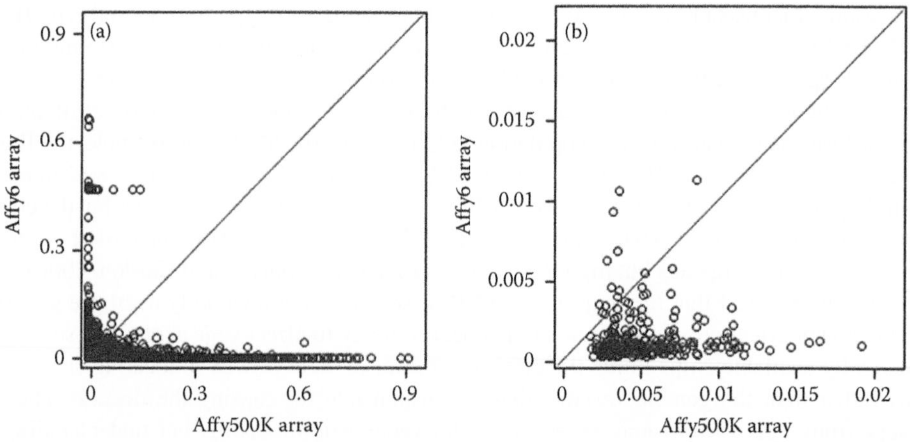

FIGURE 21.3 Comparison of genotype calls between SNP arrays. The missing call rates per SNP (a) and per sample (b) between arrays Affy500K and Affy6 were plotted for comparison. The diagonal lines indicate the locations of SNPs (a) and samples (b) when their missing call rates are exactly the same between these two arrays.

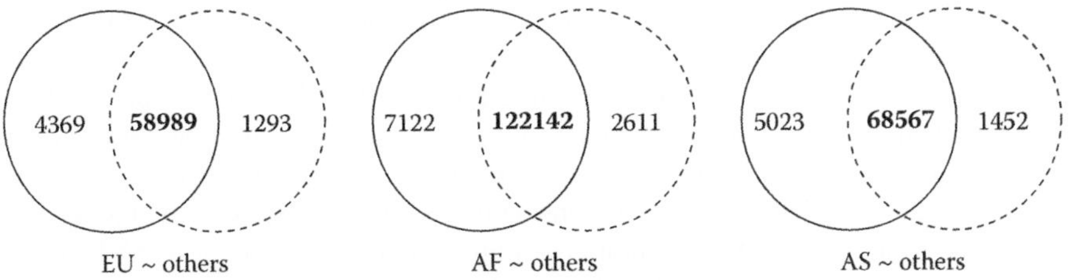

FIGURE 21.4 Venn diagrams for comparisons of lists of associated SNPs from the Affy500K and the Affy6 for assessing propagations of the inconsistency in genotypes between the two arrays to associated SNPs. The significantly associated SNPs identified using genotypic association test from the 482,251 common SNPs were compared between the Affy500K (numbers in the solid-line circles) and the Affy6 (numbers in the dotted-line circles). The numbers in bold font are the associated SNPs from both arrays, and the numbers in normal font are the SNPs only significant from one array (the Affy500K or the Affy6). EU ~ others, the association analyses results for European versus others; AF ~ others, the association analyses results for African versus others; AS ~ others, the association analyses results for Asian versus others.

identified using genotypic models from the Affy500K and Affy6 data, were compared using Venn diagrams (Figure 21.4). It is clear that for all case-control frameworks, the inconsistency in genotypes between the two SNP arrays influenced the downstream association analyses, resulting in different lists of associated SNPs and thus demonstrating that variation of SNP arrays is a potential challenge to reduce type I and II errors in T2D GWAS.

Reproducibility of genotyping experiments across laboratories and genotyping platforms is another concern in T2D GWAS, evidenced by many of the genetic variants associated with T2D being replicated in only one replication study (Table 21.3). Realizing the high cost for genotyping at this time, none of the current T2D GWAS results have been confirmed with the same samples and genotyping technologies by other laboratories. Although some of the associated SNPs from T2D GWAS were successfully replicated in multiple replications with different samples from the same population and different genotyping technologies, the reproducibility of genotyping across laboratories and across genotyping platforms remains a clear challenge at the moment.

Efforts to detect, prevent, and eradicate sources of technical errors and biases in genotyping in T2D GWAS are essential for improving the quality of data and gaining confidence in the results. We evaluated the robustness of current Affymetrix genotyping technologies that support GWAS. A reasonable reproducibility was observed at both the raw intensity levels and the genotype/copy-number variants (CNVs) derived from the raw intensity data (Hong et al., 2010b). It was observed that the discordance in called genotypes was generally small at the specimen level, an average of 1.14% between technical replicates. Because it was found that inconsistency in genotypes not only propagates to the downstream association analysis but also amplifies in the lists of significant associated SNPs (Hong et al., 2010b), the potential for errors caused by a small technical fluctuation of genotyping suggests that using technical replicates in T2D GWAS would improve quality of the data and increase the reliability of the findings.

21.4.3 Challenges in Genotype Calling

In T2D GWAS, a genotype-calling algorithm assigns genotypes for SNPs from the raw intensity data prior to downstream association analysis for identifying T2D-associated genetic variants. Many algorithms have been developed. One of the fundamental questions in T2D GWAS is how consistent the different genotype-calling algorithms are, although each of the algorithms reported a high successful call rate and accuracy. We evaluated the concordance of genotype calls from three Affymetrix algorithms (DM, BRLMM, and Birdseed) that were released along with their recent three generations of SNP arrays, and assessed potential spurious associations caused by genotyping-calling algorithms (Hong et al., 2010b). Discordance in genotypes was found in both missing and successful calls. Our observations suggest that there is room for improvements in both call rates and accuracy of genotype-calling algorithms.

Different versions of a same genotype-calling algorithm have been developed and are in use for GWAS. For example, Birdseed is a model-based clustering algorithm that converts continuous intensity data to discrete genotype data. It has two different versions (versions 1 and 2). Version 1 fits a Gaussian mixture model into a two-dimensional space using SNP-specific models as starting points to start the expectation maximization algorithm, whereas version 2 uses SNP-specific models in a pseudo-Bayesian fashion, limiting the possibility of arriving at a genotype clustering that is very different from the supplied models. Therefore, inconsistent genotypes may be called from a same set of raw intensity data using different versions of the same algorithm, especially when intensity data does not fit the SNP-specific model perfectly. The inconsistency in genotype results caused by different versions of genotype-calling algorithm has the potential to cause both type I and II errors. A thorough evaluation of inconsistency between different versions of genotype-calling algorithms and its effect on significantly T2D-associated SNPs has not yet been reported.

Genotype calling is a complicated process that usually requires that many user-specified parameters be adjusted for a particular algorithm. For example, it should be decided whether or not normalization is conducted before a genotype-calling algorithm is applied to determine genotypes based on the intensity data and which normalization method, if required, should be used. There are also many algorithm-specific parameters that need to be set. BLRMM, as an example, first derives an initial guess for each SNP's genotype using the DM algorithm (Di et al., 2005) and then analyzes across SNPs to identify cases of nonmonomorphisms. This subset of nonmonomorphism SNPs is then used to estimate a prior distribution on cluster centers and variance-covariance matrices. This subset of SNP genotypes is revisited, and the clusters and variances of the initial genotype guesses are combined with the prior distribution information of the SNPs in an *ad hoc* Bayesian procedure to derive a posterior estimate of cluster centers and variances. Genotypes of SNPs are called according to their Mahalanobis distances from the three cluster centers, and confidence scores are assigned to the calls. Therefore, the parameters with which to specify the p value cutoff for DM algorithm to seed clusters (default is set to 0.17) and the number of probe sets to be used for determining prior distribution (default is set to 10,000), influence the prior distribution on cluster centers and variance-covariance matrices. Different parameter settings may cause inconsistent genotype calls, and the inconsistency, in turn, may propagate to the downstream association analysis. For example, we investigated the effect of changing the confidence threshold in the genotype-calling algorithm BRLMM (Hong et al., 2010a). The comparisons of missing call rates per SNP and per sample when using BRLMM at confidence thresholds of 0.17, 0.30, 0.45, and 0.60 on the data set of 270 HapMap samples showed that inconsistent genotype calls were generated from the exact same raw data (Figure 21.5). The Pearson-correlation coefficients of corresponding comparisons were calculated and shown on the top of the scatter plots. t tests were performed to determine whether the two sets of missing call rates from a normal distribution could have the same mean when the standard deviations are unknown but assumed equal. The resulting p values for the comparisons were less than 0.0001, indicating that missing call rates per sample and per SNP are statistically different. Furthermore, it was observed that the inconsistency (defined as $1 - r$) of missing call rates were positively related to the corresponding differences of thresholds and negatively related to the sum of thresholds of the compared calling experiments (Figure 21.6).

A T2D GWAS usually involves the analyses of thousands of samples that generate thousands of raw data files (i.e., CEL files). The raw data file for one sample is tens of MB in size. It is often not practical to analyze all CEL files in one single batch on a single computer. The samples are therefore divided into many batches for genotype calling. The variation in ways to divide samples into different batches for genotype calling has potential to generate disparities in called genotypes that, in turn, cause spurious associations in T2D GWAS. The effects on genotype calls caused by changing the number and specific combinations of CEL files in batches and propagation of the effects to the downstream association analysis have been assessed using the 270 HapMap samples and the calling

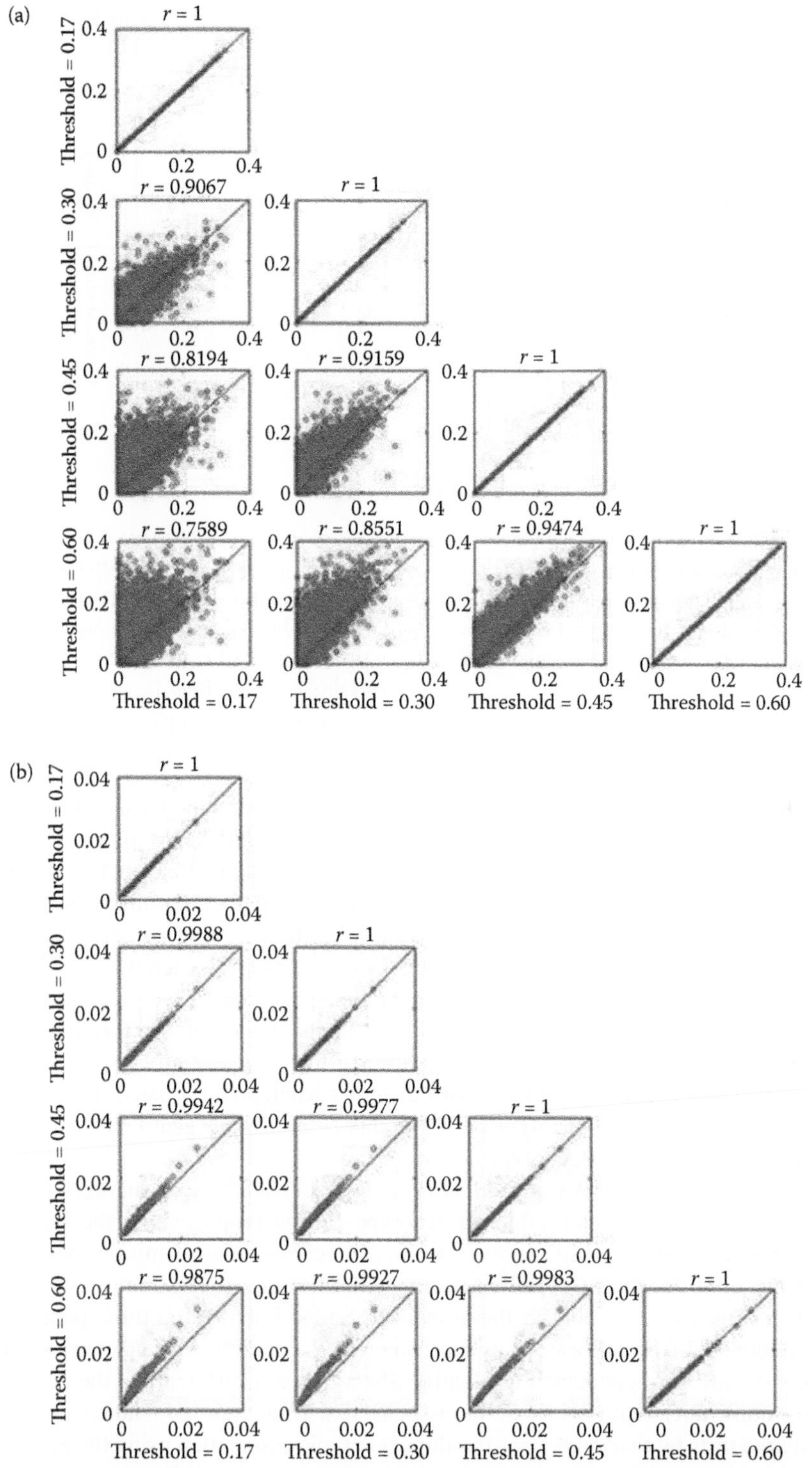

FIGURE 21.5 The missing call rates per SNP (a) and per sample (b) from genotype-calling results of BRLMM with different thresholds for the Affy500K raw data of the 270 HapMap samples were plotted for pair-wise comparison. The diagonal lines indicate that the missing call rates were the same in the two compared calling results. The Pearson correlation coefficients between the missing call rates of the two compared calling results were given on the top of corresponding scatter plots.

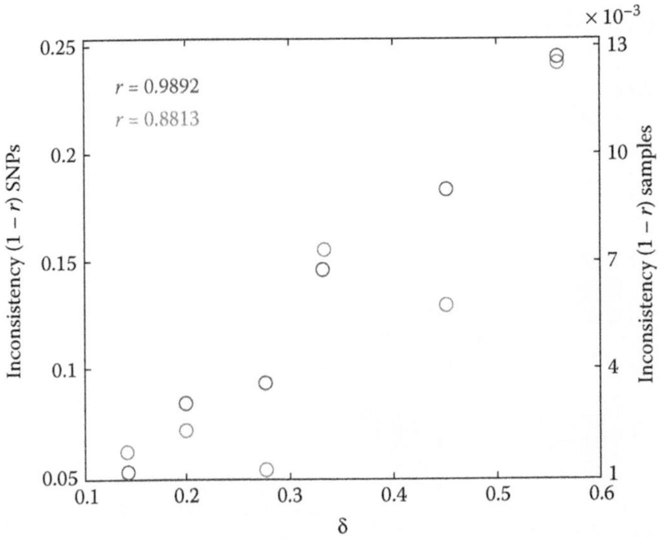

FIGURE 21.6 (See color insert.) Inconsistency $(1-r)$ of the missing call rates per SNP (left, *y*-axis and blue points) and per sample (right, *y*-axis and red points) from genotype-calling results of BRLMM with different confidence thresholds for the Affy500K raw data of the 270 HapMap samples were plotted against δ. The Pearson correlation coefficient, *r*, was calculated between the missing call rates of the two compared calling results *i* and *j*. The δ was defined and calculated as $\delta = \frac{|\text{Threshold}^{i} - \text{Threshold}^{j}|}{(\text{Threshold}^{i} + \text{Threshold}^{j})}$.

algorithm BRLMM (Hong et al., 2008). Three experiments were designed and conducted in order to assess the effect of batch size. In the first experiment (BS1), the 270 HapMap samples were divided into three batches based on their population groups: 90 Europeans, 90 Asians, and 90 Africans. The genotypes were called separately by BRLMM using the default parameter setting suggested by Affymetrix. The second experiment (BS2) used a batch size of 45 samples. Genotypes were called from the CEL files from 90 European samples in two batches, each with 45 CEL files using BRLMM with the same parameter settings as in the first experiment. The procedure was repeated for the Asian and African samples. In the third experiment (BS3), the batch size was 30 samples from each population group. To evaluate the batch-size effect on the genotypes called, concordance of genotypes called between experiments with different batch sizes is listed in Table 21.4. It can be seen that batch size affected genotype calls and that heterozygous genotype concordances were more affected than homozygous genotype concordances.

The objective of a T2D GWAS is to identify T2D-associated genetic markers. It is critical to assess whether and how the batch effect in genotype-calling propagates to the significant SNPs identified in the downstream association analysis. Three case-control mimic association analyses were conducted for each of the calling results with different batch sizes to assess propagation of batch effect in genotype calling to the associated SNPs. Each of the three population groups (European, Asian, and African) was set as case, whereas the other two groups were set as controls. Association analyses were conducted to identify SNPs that can differentiate the case group from the control groups. Different lists of SNPs associated with a same population group, identified from chi-squared tests by two degrees of freedom genotypic association using the genotype calling results with different batch sizes were compared using Venn diagrams. The comparisons of associated SNPs obtained from calling results with different batch sizes are given in Figure 21.7. It is clear that the batch-size effect on genotype calling propagated into the downstream association analyses. Moreover, it was observed that the larger the differences in two batch sizes, the fewer associated SNPs were shared by the two batch sizes.

TABLE 21.4
Concordance of Genotype Calls between Batch Sizes

Comparison		BS1 *versus* BS2	BS1 *versus* BS3	BS2 *versus* BS3
Successful calls for both	SNPs	134,258,764	134,187,584	134,265,847
	%	99.338	99.285	99.343
Concordant calls (all)	SNPs	134,248,899	134,187,584	134,253,973
	%	99.993	99.986	99.991
Concordant calls (hom)	SNPs	98,179,772	98,136,394	98,204,063
	%	99.997	99.993	99.995
Concordant calls (het)	SNPs	36,069,127	36,031,744	36,049,910
	%	99.981	99.964	99.980

Successful calls for both, SNP genotypes successfully called in both of the compared experiments; concordant calls (all), same genotype called in both of the compared experiments; concordant calls (hom), homozygous genotype called in both of the compared experiments; concordant calls (het), heterozygous genotype called in both of the compared experiments.

21.4.4 Challenges in Association Analysis

Following the generation of genotype data, there are two independent and related processes in downstream analyses: quality assessment (QA) and quality control (QC) of genotype data and identification of significantly associated genetic markers for a given phenotypic trait (Hong et al., 2009).

The purpose of QA/QC is to remove markers and samples with low quality, thus improving T2D GWAS results by decreasing both type I and II errors. Currently, there are challenges on how to apply QC/QA in GWAS as criteria for removing low-quality markers, and samples have to be made arbitrarily and variably to trade off sensitivity and specificity. Consequently, subsequent analyses results may also be affected by the quality-control process. For example, the influence of different p value cutoffs for determining the departure from Hardy-Weinberg equilibrium (HWE) on the statistical test for association analysis has not been assessed. HWE test is an imprecise tool for quality control purposes. Statistically testing the departures from HWE of SNPs is underpowered for detecting genotyping errors (Cox and Kraft, 2006). Determining whether redundant markers exist

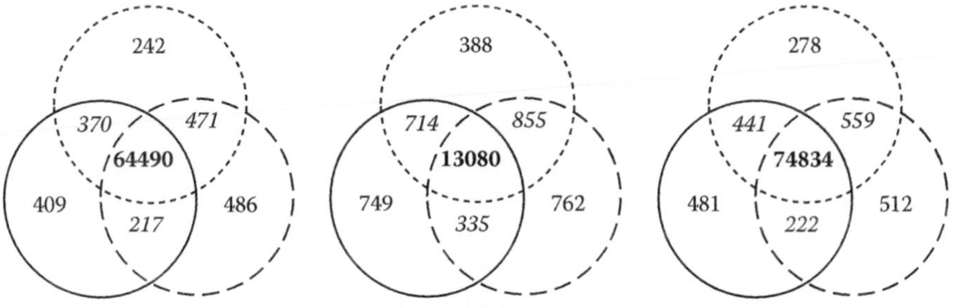

FIGURE 21.7 Venn diagrams for comparisons of the significantly associated SNPs identified using the genotype-calling results with different calling batch sizes. The numbers in circles are the significantly associated SNPs identified in association analyses using calling results from different batch sizes: solid-line circles for BS1, dotted-line circles for BS2, and dashed-line circles for BS3. The numbers in bold font represent the associated SNPs shared by all three batch sizes, the numbers in italic font represent the associated SNPs shared only by two batch sizes, and the numbers in normal font are the associated SNPs identified only by the corresponding batch sizes. EU ~ others, the association analyses results for European versus others; AF ~ others, the association analyses results for African versus others; AS ~ others, the association analyses results for Asian versus others. BS1, BS2, and BS3 are defined in the text.

and should be removed and the most appropriate method and criterion for doing so also should be investigated to see whether they affect T2D GWAS results. Samples with extremely high identity-by-state (IBS) may indicate duplicated samples. Whether the samples with high IBS affect the association analyses results has not been characterized and quantified and also should be investigated in the future.

The downstream association analysis is typically a statistical test conducted for each individual SNP with the objective of identifying those SNPs that exhibit a significant association with T2D. The procedure consists of testing, for each SNP, the null hypothesis that there is no association between the SNP and T2D against the alternative hypothesis that there is an association. The open challenge is how to select a statistic test that is both sensitive and specific with small type I and II errors. There are two possible types of associations:

1. Single-locus association between a single SNP and T2D
2. Multiloci association between a combination of SNPs and T2D

Current T2D GWAS are focused on single-locus association. There are two different types of statistical tests for determining significant single-locus associations: one degree of freedom allelic association test and two degrees of freedom genotypic association test.

From a statistical point of view, association analysis methods in T2D GWAS can be divided into two categories: frequency-based methods and Bayesian methods. A frequency-based method is to test hypotheses that a genetic variant is associated with risk of T2D and weighs the evidence against the null hypothesis by the p value that is defined as the probability of observing a stronger association than that estimated from the data, when there is indeed no association with T2D. In the frequency-based method, the decision to reject the null hypothesis of no association with T2D is typically based on controlling the probability of the type I error, and this is done by imposing a threshold on the p value. It does not assess per se whether the null hypothesis is true or false. The challenge for using these types of methods is to decide what p value should be used to estimate an association with T2D at a significance level.

A Bayesian method is grounded in a very different conceptual framework from the frequency-based methods and is becoming more popular in genetic epidemiology (Beaumont and Rannala, 2004). The principle of a Bayesian test of association with the risk of T2D is to first assume prior probabilities on the two hypotheses of no association and association, then use the genetic data to update the prior probabilities of the two hypotheses into their posterior probabilities. The decision to reject the null hypothesis is then based on an appropriate threshold on the odds of the posterior probabilities or, equivalently, on the posterior probability of the null hypothesis. In the Bayesian method, the decision to reject the null hypothesis is based directly on the probability that the null hypothesis is false, given the evidence provided by the data. It uses an explicit assessment of the likelihood of parameters and hypotheses. The choice of a best threshold can be based on trading off sensitivity and specificity of the Bayesian decision rule, posing a challenge for this type of method.

Many statistical methods can be used for a given genetic association analysis, such as chi-squared test, Fisher's exact test, and Cochran-Armitage test for case-control studies. The variation in selection of statistical models is another potential source of spurious associations generated in downstream association analysis of T2D GWAS. The level of type I errors and type II errors caused by different statistical methods remains a challenge at the moment and requires systematical evaluations in the future.

Another big challenge in an association analysis of T2D GWAS comes from the multiple tests. For example, when a threshold p value is adopted to accept a significant association with T2D in testing one genetic variant, the probability of one or more type I errors in testing multiple genetic variants increases with the number of genetic variants tested. The Bonferroni correction attempts to reduce the significance of each test by dividing the usual significance level by the number of genetic variants tested. It is too conservative and reduces the power dramatically and unnecessarily.

Therefore, controlling the false-discovery rate (FDR) rather than the overall false-positive rate was proposed as a less conservative method. FDR is the proportion of false-positive associations among the detected significant associations and can be controlled for using a simple algorithm (Benjamini et al., 2001). In addition, a variety of solutions for correcting multiple tests were proposed (Pawitan et al., 2006; Scheid and Spang, 2004). The correction does not change the rank of the p values of the genetic variants tested but simply provides an additional guidance as to which associations are most significant across the entire T2D GWAS.

21.4.5 Challenges in Validation and Interpretation of T2D GWAS Findings

Because of the inflated false-positive rate caused by multiple testing, the issue of population stratification that can confound associations and technical errors that can be committed during the collection of DNA samples, validating the findings from T2D GWAS is necessary to ensure their validity. Replication is necessary and highly recommended. It has been suggested that a larger sample size from an independent study population, with the same genetic background of samples of initial T2D GWAS, should be used for a replication study. The associations of the same genetic variants with the same genetic model should be reported. The benefit of inclusion of proxy SNPs that are in high linkage disequilibrium (LD) in replication studies is still not clear, and there are debates on it: some recommend this practice as further evidence of a real association (Chanock et al., 2007), whereas others recommend against it because it is unnecessary (McCarthy et al., 2008). The large sample size of the initial T2D GWAS is necessary to achieve appropriate statistical power at a genome-wide significance level. However, a replication study is limited to a small number of genetic markers to be tested. Therefore, it is generally considered that there is no multiple testing problem, and traditional significance levels should be acceptable, but the validity of this hypothesis has not been proved.

The T2D GWAS and the subsequent replication studies have identified tens of T2D-susceptibility variants so far (Table 21.3). Apart from genetic markers in gene *TCF7L2*, per-allele effect sizes are, at best, modest (mostly in the range with an odds ratio of less than 1.2), which explains the massive sample sizes required for robust associations. One open challenge, for most of these loci, is that the causal genetic variants have yet to be identified with any certainty. Although it is likely, for reasons of statistical power, that the causal variants in these regions are in high LD to those of the genetic variants showing significant association, this is not inevitable. For example, if an associated SNP detected in a T2D GWAS has an allele frequency of 30% in T2D patients and of 29% in control subjects, in principle, a causal genetic variant with a high LD to the SNP might have an allele frequency of 9% in T2D and only 6% in controls. The causal genetic variant would have an effect size much larger than that of the SNP detected in the original T2D GWAS.

It is expected that, on occasion, genes other than the most associated genomic region turn out to be responsible for T2D susceptibility. Nevertheless, compellable functions of some of the genomic loci identified by T2D GWAS, such as *SLC30A8* (Sladek et al., 2007), are clear. For example, based on the patterns of flanking recombination hot spots, the causal genetic variants lie within a 400-kb region on chromosome 10q that contains the coding regions of three genes, two of which, HHEX and IDE, have strong biological claims for a role in T2D pathogenesis. However, the functions of some genetic variants identified from T2D GWAS are less obvious. For example, several newly identified T2D-susceptibility loci locate in about a 15-kb interval on chromosome 9p21 that lies some 200 kb from the nearest protein coding genes, *CDKN2A* and *CDKN2B*, that encode cyclin-dependent kinase inhibitors primarily known for their role in the development of cancers (Finkel et al., 2007).

One of the challenges in interpreting T2D GWAS findings arises from the interesting observation that some genetic variants and the genes implicated in T2D susceptibility are also associated with other phenotypic traits (Prokopenko et al., 2008). For example, genetic variants in the genomic region 9p21, near genes *CDKN2A* and *CDKN2B*, identified from T2D GWAS and replication studies

(Takeuchi et al., 2009; Saxena et al., 2007; Scott et al., 2007; Zeggini et al., 2007), are associated with predisposition of coronary artery disease and of pathological dilatation of the major blood vessels (Helgadottir et al., 2007; McPherson et al., 2007). Genetic variants at HNF1B (rs4430796 and rs7501939) and JAZF1 (rs864745) have been proven to be T2D associated (Gudmundsson et al., 2007; Zeggini et al., 2008) and also have clear effects on susceptibility to prostate cancer (Thomas et al., 2008; Gudmundsson et al., 2007). Is pleiotropy the explanation? The question remains unanswered, right now, at least to the authors.

21.5 FUTURE PERSPECTIVES

T2D GWAS, as well as the subsequent replication studies, identified and validated a variety of genetic variants from multiple chromosomes (Table 21.3) associated with the risk of T2D, providing the first genome-wide perspective of the landscape of T2D susceptibility. However, these T2D-associated genetic variants have very small effects on the variability of T2D and explain a rather small portion of the heritability. Therefore, there is a long way for us to completely understand the mechanisms of T2D, to dissect the genetic architecture of T2D, and ultimately to prevent and provide efficient treatment of T2D. To help us to achieve the ultimate goal, advances in the following aspects are expected in the near future.

Some T2D-associated loci are also known to harbor rare mutations that are causal for infrequent, monogenic forms of diabetes. Therefore, rare mutations rather than common genetic variants, or a combination of both, might account for the unexplained variability of T2D, or at least part of it. Because most of the rare mutations are probably unknown and the SNP arrays that are commercially available are mainly designed to capture common variants catalogued through the HapMap project, deep sequencing might be the best tool to unveil the truth.

In addition to common genetic variants and rare mutations, structural variants, usually CNVs, are expected to be explored for dissecting genetic architecture of T2D. CNVs are defined as inherited duplications and deletions of kilobase to megabase lengths of DNA fragments, and they have been shown to be present in various numbers in all individuals. The importance of identifying areas of CNV for proper genotyping of a SNP in the context of a GWAS has been demonstrated (Ionita-Laza et al., 2009). Armed with more complete inventories of the sites of common CNVs and new tools that will enable genome-wide CNV genotyping, CNVs will take an increasingly prominent role alongside SNPs as targets of T2D GWAS.

Next-generation sequencing is fundamentally changing the way in which genomic information of individuals at the DNA level is being obtained for better understanding of the human genome (Fullwood et al., 2009). Resequencing of the newly identified T2D-susceptibility loci is expected to reveal additional instances of lower-frequency, higher-penetrance alleles and might contribute to more effective functional analyses of these regions.

Because the genetic variants associated with T2D can only explain a very small portion of the predisposition of T2D, it could be speculated that gene-gene and gene-environment interactions are responsible, for the most part, for the unexplained T2D susceptibility, even without clear evidence at the moment. Quantitatively assessing the interactions is notoriously difficult. Therefore, such new efforts will remain a formidable challenge in the new future. However, there is an increasing interest in assessing epigenetic modification. Authors hope that better understanding the epigenome, particularly in the context of genetic variation, will generate a direct and quantitative link between putative environmental influences and T2D.

REFERENCES

Barrett, J. C., Cardon, L. R. (2006). Evaluating coverage of genome-wide association studies. *Nat. Genet.* 38: 659–662.

Beaumont, M. A., Rannala, B. (2004). The Bayesian revolution in genetics. *Nat. Rev. Genet.* 5: 251–261.

Benjamini, Y., Drai, D., Elmer, G., et al. (2001). Controlling the false discovery rate in behavior genetics research. *Behav. Brain Res.* 125: 279–284.

Bloomgarden, Z. T. (2004). Type 2 diabetes in the young: The evolving epidemic. *Diabetes Care* 27: 998–1010.

Bouatia-Naji, N., Bonnefond, A., Cavalcanti-Proença, C., et al. (2009). A variant near MTNR1B is associated with increased fasting plasma glucose levels and type 2 diabetes risk. *Nat. Genet.* 41: 89–94.

Cardon, L. R., Palmer, L. J. (2003). Population stratification and spurious allelic association. *Lancet* 36: 598–604.

Chanock, S. J., Manolio, T., Boehnke, M., et al. (2007). Replicating genotype-phenotype associations. *Nature* 447: 655–660.

Cooke, D. W., Plotnick, L. (2008). Type 1 diabetes mellitus in pediatrics. *Pediatr. Rev.* 29: 374–384.

Cox, D. G., Kraft, P. (2006). Quantification of the power of Hardy-Weinberg equilibrium testing to detect genotyping error. *Hum. Hered.* 61: 10–14.

Di, X., Matsuzaki, H., Webster, T. A., et al. (2005). Dynamic model based algorithms for screening and genotyping over 100K SNPs on oligonucleotide microarrays. *Bioinformtics* 21: 1958–1963.

Dupuis, J., Langenberg, C., Prokopenko, I., et al. (2010). New genetic loci implicated in fasting glucose homeostasis and their impact on type 2 diabetes risk. *Nat. Genet.* 42: 105–116.

Finkel, T., Serrano, M., Blasco, M. A., et al. (2007). The common biology of cancer and ageing. *Nature* 448: 767–776.

Florez, J. C., Hirschhorn, J., Altshuler, D., et al. (2003). The inherited basis of diabetes mellitus: Implications for the genetic analysis of complex traits. *Annu. Rev. Genomics Hum. Genet.* 4: 257–291.

Florez, J. C., Manning, A. K., Dupuis, J., et al. (2007). A 100K genome-wide association scan for diabetes and related traits in the Framingham Heart Study: Replication and integration with other genome-wide datasets. *Diabetes* 56: 3063–3074.

Frayling, T. M., Timpson, N. J., Weedon, M. N., et al. (2007). A common variant in the FTO gene is associated with body mass index and predisposes to childhood and adult obesity. *Science* 316: 889–894.

Frazer, K. A., Murray, S. S., Schork, N. J., et al. (2009). Human genetic variation and its contribution to complex traits. *Nat. Rev. Genet.* 10: 241–251.

Fullwood, M. J., Wei, C. L., Liu, E. T., et al. (2009). Next-generation DNA sequencing of paired-end tags (PET) for transcriptome and genome analyses. *Genome Res.* 19: 521–532.

Gloyn, A. L. (2003). The search for type 2 diabetes genes. *Ageing Res. Rev.* 2: 111–127.

Grarup, N., Rose, C. S., Andersson, E. A., et al. (2007). Studies of association of variants near the HHEX, CDKN2A/B, and IGF2BP2 genes with type 2 diabetes and impaired insulin release in 10,705 Danish subjects: Validation and extension of genome-wide association studies. *Diabetes* 56: 3105–3111.

Gudmundsson, J., Sulem, P., Steinthorsdottir, V., et al. (2007). Two variants on chromosome 17 confer prostate cancer risk, and the one in TCF2 protects against type 2 diabetes. *Nat. Genet.* 39: 977–983.

Hansen, L. (2003). Candidate genes and late-onset type 2 diabetes mellitus: Susceptibility genes or common polymorphisms? *Dan. Med. Bull.* 50: 320–346.

Hanson, R. L., Bogardus, C., Duggan, D., et al. (2007). A search for variants associated with young-onset type 2 diabetes in American Indians in a 100K genotyping array. *Diabetes* 56: 3045–3052.

Hayes, M. G., Pluzhnikov, A., Miyake, K., et al. (2007). Identification of type 2 diabetes genes in Mexican Americans through genome-wide association studies. *Diabetes* 56: 3033–3044.

Helgadottir, A., Thorleifsson, G., Manolescu, A., et al. (2007). A common variant on chromosome 9p21 affects risk of myocardial infarction. *Science* 316: 1491–1493.

Hong, H., Su, Z., Ge, W., et al. (2008). Assessing batch effects of genotype calling algorithm BRLMM for the Affymetrix GeneChip Human Mapping 500K Array Set using 270 HapMap samples. *BMC Bioinformatics* 9: S17.

Hong, H., Shi, L., Fuscoe, J. C., et al. (2009). Potential sources of spurious associations and batch effects in genome-wide association studies. In Scherer, A. (Ed.), *Batch effects and noise in microarray experiments: Sources and solutions* (pp. 191–201). West Sussex, UK: John Wiley & Sons.

Hong, H., Su, Z., Ge, W., et al. (2010a). Evaluating variations of genotype calling: A potential source of spurious associations in genome-wide association studies. *J. Genet.* 89: 55–64.

Hong, H., Shi, L., Su, Z., et al. (2010b). Assessing sources of inconsistencies in genotypes and their effects on genome-wide association studies with HapMap samples. *Pharmacogenomics J.* 10: 364–374.

Howson, J. M., Barratt, B. J., Todd, J. A., et al. (2005). Comparison of population- and family-based methods for genetic association analysis in the presence of interacting loci. *Genet. Epidemiol.* 29: 51–67.

Hu, F. B. (2003). Sedentary lifestyle and risk of obesity and type 2 diabetes. *Lipids* 38: 103–108.

Huelsken, J., Behrens J. (2002). The Wnt signalling pathway. *J. Cell Sci.* 115: 3977–3978.

Ionita-Laza, I., Rogers, A. J., Lange, C., et al. (2009). Genetic association analysis of copy-number variation (CNV) in human disease pathogenesis. *Genomics* 93: 22–26.

Klein, R. J., Zeiss, C., Chew, E. Y., et al. (2005). Complement factor H polymorphism in age-related macular degeneration. *Science* 308: 385–389.

Lander, E. S. (1996). The new genomics: Global views of biology. *Science* 274: 536–539.

Lean, M. E. (2000). Obesity: Burdens of illness and strategies for prevention or management. *Drugs Today* 36: 773–784.

Lyssenko, V., Lupi, R., Marchetti, P., et al. (2007). Mechanism by which common variants in the TCF7L2 gene increase risk of type 2 diabetes. *J. Clin. Invest.* 117: 2155–2163.

Lyssenko, V., Nagorny, C. L., Erdos, M. R., et al. (2009). Common variant in MTNR1B associated with increased risk of type 2 diabetes and impaired early insulin secretion. *Nat. Genet.* 41: 82–88.

McCarthy, M. I. (2003). Growing evidence for diabetes susceptibility genes from genome scan data. *Curr. Diab. Rep.* 3: 159–167.

McCarthy, M. I., Abecasis, G. R., Cardon, L. R., et al. (2008). Genome-wide association studies for complex traits: Consensus, uncertainty and challenges. *Nat. Rev. Genet.* 9: 356–369.

McPherson, R., Pertsemlidis, A., Kavaslar, N., et al. (2007). A common allele on chromosome 9 associated with coronary heart disease. *Science* 316: 1488–1491.

Neel, J. (1962). Diabetes mellitus: a thrifty genotype rendered detrimental by "progress"? *Am. J. Hum. Genet.* 14: 353–362.

Pawitan, Y., Calza, S., Ploner. A., et al. (2006). Estimation of false discovery proportion under general dependence. *Bioinformatics* 22: 3025–3031.

Pe'er, I., Bakker, P. I., Maller, J., et al. (2006). Evaluating and improving power in whole-genome association studies using fixed marker sets. *Nat. Genet.* 38: 663–667.

Pearson, T. A., Manolio, T. A. (2008). How to interpret a genome-wide association study. *JAMA* 299: 1335–1344.

Price, A. L., Patterson, N. J., Plenge, R. M., et al. (2006). Principal component analysis corrects for stratification in genome-wide association studies. *Nat. Genet.* 38: 904–909.

Prokopenko, I., McCarthy, M. I., Lindgren, C. M. (2008). Diabetes: New genes, new understanding. *Trends Genet.* 24: 613–621.

Rampersaud, E., Damcott, C. M., Fu, M., et al. (2007). Identification of novel candidate genes for type 2 diabetes from a genome-wide association scan in the Old Order Amish: Evidence for replication from diabetes-related quantitative traits and from independent populations. *Diabetes* 56: 3053–3062.

Rung, J., Cauchi, S., Albrechtsen, A., et al. (2009). Genetic variant near IRS1 is associated with type 2 diabetes, insulin resistance and hyperinsulinemia. *Nat. Genet.* 41: 1110–1115.

Salonen, J. T., Uimari, P., Aalto, J., et al. (2007). Type 2 diabetes whole-genome association study in four populations: The DiaGen Consortium. *Am. J. Hum. Genet.* 81: 338–345.

Saxena, R., Voight, B. F., Lyssenko, V., et al. (2007). Genome-wide association analysis identifies loci for type 2 diabetes and triglyceride levels. *Science* 316: 1331–1336.

Saxena, R., Hivert, M. F. Langenberg, C., et al. (2010). Genetic variation in GIPR influences the glucose and insulin responses to an oral glucose challenge. *Nat. Genet.* 42: 142–148.

Scheid, S., Spang, R. (2004). A stochastic downhill search algorithm for estimating the local false discovery rate. *IEEE/ACM Trans. Comput. Biol. Bioinform.* 1: 98–108.

Scott, L. J., Mohlke, K. L., Bonnycastle, L. L., et al. (2007). A genome-wide association study of type 2 diabetes in Finns detects multiple susceptibility variants. *Science* 316: 1341–1345.

Shaw, J., Chisholm, D. (2003). Epidemiology and prevention of type 2 diabetes and the metabolic syndrome. *MJA* 179: 379–383.

Sladek, R., Rocheleau, G., Rung, J., et al. (2007). A genome-wide association study identifies novel risk loci for type 2 diabetes. *Nature* 445: 881–885.

Sparsø T, Bonnefond A, Andersson E., et al. (2009). G-allele of intronic rs10830963 in MTNR1B confers increased risk of impaired fasting glycemia and type 2 diabetes through an impaired glucose-stimulated insulin release: Studies involving 19,605 Europeans. *Diabetes* 58: 1450–1456.

Steinthorsdottir, V., Thorleifsson, G., Reynisdottir, I., et al. (2007). A variant in CDKAL1 influences insulin response and risk of type 2 diabetes. *Nat. Genet.* 39: 770–775.

Takeuchi, F., Serizawa, M., Yamamoto, K., et al. (2009). Confirmation of multiple risk loci and genetic impacts by a genome-wide association study of type 2 diabetes in the Japanese population. *Diabetes* 58: 1690–1699.

The International HapMap Consortium. (2005). A haplotype map of the human genome. *Nature* 437: 1299–1320.

The International HapMap Consortium. (2007). A second generation human haplotype map of over 3.1 million SNPs. *Nature* 449: 851–862.

Thomas, G., Jacobs, K. B., Yeager, M., et al. (2008). Multiple loci identified in a genome-wide association study of prostate cancer. *Nat. Genet.* 40: 310–315.

Timpson, N. J., Lindgren, C. M., Weedon, M. N., et al. (2009). Adiposity-related heterogeneity in patterns of type 2 diabetes susceptibility observed in genome-wide association data. *Diabetes* 58: 505–510.

Unoki, H., Takahashi, A., Kawaguchi, T., et al. (2008). SNPs in KCNQ1 are associated with susceptibility to type 2 diabetes in East Asian and European populations. *Nat. Genet.* 40: 1098–1102.

van Dam, R. M. (2003). The epidemiology of lifestyle and risk for type 2 diabetes. *Eur. J. Epidemiol.* 18: 1115–1125.

Vaxillaire, M., Cavalcanti-Proença, C., Dechaume, A., et al. (2008). The common P446L polymorphism in GCKR inversely modulates fasting glucose and triglyceride levels and reduces type 2 diabetes risk in the DESIR prospective general French population. *Diabetes* 57: 2253–2257.

Voight, B. F., Scott, L. J., Steinthorsdottir, M., et al. (2010). Twelve type 2 diabetes susceptibility loci identified through large-scale association analysis. *Nat. Genet.* 42: 579–589.

Wellcome Trust Case Control Consortium (2007). Genome-wide association study of 14,000 cases of seven common diseases and 3,000 shared controls. *Nature* 447: 661–678.

Wild, S., Roglic, G., Green, A., Sicree, R., et al. (2004). Global prevalence of diabetes: Estimates for 2000 and projections for 2030. *Diabetes Care* 27: 1047–1053.

World Health Organization (2006). Definition and Diagnosis of Diabetes Mellitus and intermediate hyperglycema. http://www.who.int/diabetes/publications/Definition%20and%20diagnosis%20of%20diabetes_new.pdf

Yasuda, K., Miyake, K., Horikawa, Y., et al. (2008). Variants in KCNQ1 are associated with susceptibility to type 2 diabetes mellitus. *Nat. Genet.* 40: 1092–1097.

Zeggini, E., Weedon, M. N., Lindgren, C. M., et al. (2007). Replication of genome-wide association signals in UK samples reveals risk loci for type 2 diabetes. *Science* 316: 1336–1341.

Zeggini, E., Scott, L. J., Saxena, R., et al. (2008). Meta-analysis of genome-wide association data and large-scale replication identifies additional susceptibility loci for type 2 diabetes. *Nat. Genet.* 40: 638–645.

Zondervan, K. T., Cardon, L. R. (2007). Designing candidate gene and genome-wide case-control association studies. *Nat. Protoc.* 2: 2492–2501.

22 Cardiac Channelopathies

Applications of Genomics and Proteomics in Diagnosis and Therapy

Cedric Viero
Cardiff University
Cardiff, United Kingdom

CONTENTS

22.1 GENERALITIES ON CARDIAC PHYSIOLOGY

The heart is a powerful pump that enables the circulation of oxygen and nutrients to every single cell of the organism via the blood vessels. It is composed of four chambers, and communication between them is rhythmically regulated to ensure an efficient and regular exchange of blood throughout the body. This rhythmic activity is controlled by electrical signals evoked by spontaneously electrically active cells, referred to as pacemaker cells, and transmitted to muscle cells that in turn contract in a synchronous manner. When the electrical signal reaches a muscle cell, it induces a depolarization at the cell membrane level, driven by the opening of sodium channels that are responsible for the rapid upstroke of the resultant action potential (AP), and this ultimately leads to the opening of L-type voltage-gated calcium channels (L-VGCC). This allows the influx of calcium into the cytosol of the cell, which in turn can activate calcium release channels (ryanodine receptors or RyR type 2) localized on the membrane of intracellular stores, i.e., the sarcoplasmic reticulum (SR). RyRs are huge macromolecular complexes that allow for the massive and rapid release of calcium ions from the SR into the cytosol (millions of ions per second). This important mechanism of calcium signaling amplification between L-VGCC and RyR has been termed calcium-induced calcium release. The resulting increase of intracellular calcium concentration leads eventually to the shortening of muscle cells (i.e., contraction) by a calcium-dependent conformational change of the proteins constituting the myofilaments. This phenomenon is known as excitation-contraction coupling (or ECC; Figure 22.1). The termination of this process occurs by reuptake into the SR and mitochondria (by the sarco/endoplasmic reticulum calcium ATPase [SERCA] pump and by the mitochondrial calcium uniporter, respectively) and extrusion (efflux through the plasma membrane calcium ATPase and by the sodium/calcium exchanger) of calcium. Hence muscle relaxation can take place.

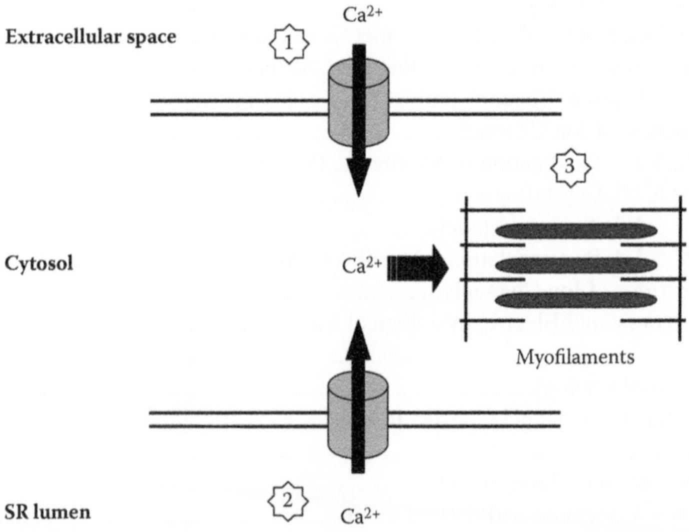

FIGURE 22.1 Schematic principle of the excitation-contraction coupling in the heart. When calcium ions enter a cardiac muscle cell via dihydropyridine receptors (1), the localized increase in calcium concentration activates calcium-release channels (ryanodine receptor or RyR) on the sarcoplasmic recticulum (SR) membrane (2). The resultant global and massive elevation of calcium concentration leads to the movement of myofilaments (3) and thus to the shortening of the cell.

Therefore, a series of three subsequent signals is needed during ECC to elicit a proper contraction/relaxation cycle of the myocardium: an AP induces calcium influx and release, which leads to shortening of the myofibrils of the contractile apparatus. The relaxation phase follows the same sequential activation: end of the electrical signal (repolarization of the cell membrane), leading to a decrease of the intracellular calcium concentration by calcium efflux and reuptake, which in turn terminates the shortening of the myocyte with contractile units recovering their initial length. The repolarization phase is mainly supported by the opening of potassium channels that extrude positive charges from the cell, thus allowing the membrane potential to return to its initial value. For an extensive review on ECC, readers should refer to Donald Bers's article (Bers, 2002).

22.2 GENERALITIES ON ION CHANNELS AND CARDIAC DISEASES

22.2.1 What Are the Ion Channels of the Heart?

An ion channel is a transmembrane protein forming a hydrophilic pore that enables the translocation of ions from one cellular compartment to another. Usually channels are integral membrane proteins or composed of several subunits that arrange in a circular form around a central hole providing the pathway for ion movement through the lipid membrane. In addition to the pore, the other important parts constituting a channel are the selectivity filter (a region dedicated to the recognition of specific ions), one or several gates that physically open and close the channel, and usually a voltage sensor that conveys the stimulation to the gate.

In the heart, sodium channels are extremely dynamic molecules made of integral membrane proteins exhibiting four homologous domains. Each domain contains six transmembrane segments. The S4 segment is composed of positive charges (lysine and arginine residues) and serves as voltage sensor. The linker between domains III and IV is critical for the fast inactivation property (Balser, 2001). Sodium channels localized at the plasma membrane level of the cardiomyocytes.

Potassium channels are tetrameric proteins formed by the assembly of four α subunits. Each subunit displays six transmembrane segments (Figure 22.2). In the myocardium, one can distinguish between five types of potassium currents: three inward rectifying currents: IK1, IKAch (modulated by the muscarinic acetylcholine receptor), and IK_{ATP} (ATP-sensitive potassium channels) and two voltage-activated currents: IK (slowly activating delayed rectifier) and Ito (transient outward current) (Deal, England, and Tamkun, 1996). Potassium channels also mainly localized at the plasma membrane, although IKATP and calcium-activated potassium channels are found in the mitochondria

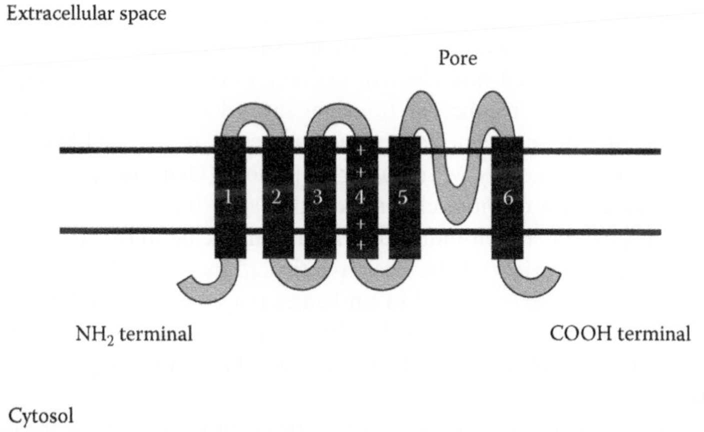

FIGURE 22.2 General structure of a typical voltage-gated potassium channel. The transmembrane segments of the α subunit of a potassium channel are depicted in this cartoon. The voltage sensor is represented by the positive charges. Four subunits have to assemble to form a functional channel.

(Garlid et al., 1997, 2003). Moreover, trimeric intracellular cation channels (or TRICs), permeable to potassium, are on the SR membrane and would provide the counter current during calcium release from the SR (Yazawa et al., 2007).

Calcium currents are the most predominant actors of cardiac excitation. There are two types of plasma membrane calcium channels expressed in heart cells: L-type and T-type voltage-gated calcium channels (L-VGCC and T-VGCC). T-VGCC can be activated by smaller depolarizations. L-VGCCs inactivate slowly and therefore provide a long-lasting calcium current (Reuter, Kokubun, and Prod'hom, 1986). T-VGCCs inactivate rapidly and are responsible for a transient current, which is particularly important in pacemaker cells. VGCC are composed of four domains, each one containing six transmembrane segments. While S4 segments represent the voltage sensors, VGCC can be modulated by additional subunits, namely α (from the extracellular side), β (from the intracellular side), γ (localized on the plasma membrane), and δ (also on the plasma membrane).

There is another class of calcium channels: the intracellular RyR and inositol-trisphosphate receptor ($InsP_3R$), which support the calcium release from the SR. RyR is an ubiquitous large protein of 565 kDa (Fleischer, 2008). The major isoform expressed in the heart is RyR type 2. In contrast to the latter one, $InsP_3Rs$ are more expressed at the vicinity of the nuclei of cardiac myocytes (Wu et al., 2006) and play a critical role in atrial myocytes, especially the isoform type 2 (Li et al., 2005; Lipp et al., 2000). The RyR general structure resembles a "mushroom" with a large cytosolic cap (where the regulations take place) and a relatively small transmembrane part constituting the pore of the channel (Zissimopoulos and Lai, 2007). In this respect, $InsP_3Rs$ share many features with their close relatives, the RyR channels.

Cardiac chloride channels are less known, but recent studies have identified several chloride channel genes in the heart, including cystic fibrosis transmembrane conductance regulator (CFTR), chloride channels 2 and 3 (ClC-2 and ClC-3), chloride channel accessory or calcium-activated chloride channel regulator (CLCA), Bestrophin, and transmembrane protein 16A (TMEM16A), which is also a calcium-activated choride channel (Duan, 2009). Bestrophin is a promising molecule encoding calcium-activated chloride channels with four transmembrane segments (TMSs). CLCA is described with five TMSs. CFTR encodes 12 TMSs with nucleotide-binding domains and a regulatory domain joining the two main motifs. The ClC channel is assumed to be 10–12 TMSs. However, crystallographic analysis suggests that the 18 α-helices of ClC are folded complexes (Suzuki, Morita, and Iwamoto, 2006). Cardiac chloride channels exhibit the particular feature to evoke both outward and inward currents; therefore, they can play a role during the depolarization and repolarization phases of the AP (Duan, 2009). It is noteworthy that they have the ability to shorten the AP on the one hand and can contribute to triggering the next AP on the other hand, hence taking part in arrhythmogenesis in diseased hearts, as we will describe in the next section.

In addition, calcium-activated nonselective cationic currents have recently been characterized and implicated in the occurrence of transient inward currents involved in certain forms of arrhythmias, and one of the identified molecular actors is the TRPM4 (transient receptor potential cation channel, subfamily M, member 4) protein (Demion et al., 2007). Moreover, a stretch-activated nonselective calcium-channel conductance has been identified in atrial myocytes as being responsible for stretch-induced atrial natriuretic peptide exocytosis (Zhang, Youm, and Earm, 2008). A splice variant of the stretch-activated two-pore-domain potassium-channel TREK 1 has been found to be expressed in rat hearts (Li et al., 2006). Furthermore, the family of transient receptor potential canonical (TRPC) channels is present in the heart, and the coordination between both subfamilies TRPC1/4/5 and TRPC3/6/7 regulates calcium-induced hypertrophy (Wu et al., 2010).

Hence a wide range of ion channels takes part in the electrical activity of the cardiac muscle in order to preserve the regular pace of contractions. However, deregulations of channel gating and ion translocation can lead to dramatic impairments of cardiac rhythmicity called arrhythmias that may eventually induce heart failure and sudden death.

22.2.2 Cardiac Diseases Involving Ionic Deregulations

Pathologies of the heart include a variety of diverse dysfunctions affecting the vascular system, the myocardium (or cardiac muscle), the pericardium (sac containing the heart and the great vessels), the electrical conduction pathway, nonmuscle and nonneuronal cells, or the valves. One can distinguish between different types of cardiomyopathies: inflammation (reaction of the heart to an infection by activation of B- and T-cells; it often accompanies other cardiac dysfunctions), ischemia (decrease of blood supply), atherosclerosis (thickening of the wall of arteries by accumulation of macrophages, cholesterol, and triglycerides), hypertrophy (increase of the size of muscle cells), fibrosis (proliferation of cardiac fibroblasts), arrhythmias (impaired electrical activity), heart failure (insufficiency of the heart to pump blood properly), sudden cardiac death (consequence of irreversible ischemia, hypertrophy, or arrhythmia), and congenital heart malformation (alteration of the structure of the heart and vessels appearing at birth).

A specific category of heart disease has for origin the dysfunction of ion channels and is referred to as channelopathies (Kass, 2005). This is a term used to describe the acquired and genetic diseases affecting the function of ion channel proteins. The alteration of various types of ion channels is involved in the genesis of these cardiac pathologies, and the main outcome is the generation of arrhythmias. Depending on the type of channel affected, leading to either an outward or an inward current, these dysfunctions have the ability to cause early afterdepolarizations or delayed afterdepolarizations.

Early afterdepolarizations (EADs) interrupt the development of the AP and occur before normal repolarization is completed. Typically they are due to augmented opening of calcium channels at the plasma membrane level or are due to the opening of sodium channels. Early afterdepolarizations can result in torsades de pointes (irregular ventricular heartbeats), tachycardia, and other arrhythmias.

Afterhyperpolarizations follow an AP and are mediated by voltage-gated sodium or chloride channels. This phenomenon requires the fast closing of potassium channels to limit repolarization.

Delayed afterdepolarizations (DADs), on the other hand, begin after repolarization is completed, but before another AP would normally occur. They are due to elevated cytosolic calcium concentrations. The overload of the SR may cause spontaneous calcium release during repolarization, causing the released calcium to exit the cell through the sodium/calcium exchanger, which results in a net depolarizing current.

22.3 EXAMPLES OF CARDIAC CHANNELOPATHIES

22.3.1 Brugada Syndrome

The first clinical description of this pathology was made by Brugada and Brugada in 1992. Although the heart structure of the patients remains intact, this syndrome can lead to sudden death. Indeed it is an inherited autosomal dominant disease affecting the *SCN5A* gene that encodes the cardiac sodium channel α subunit (Liang et al., 2010). The *SCN5A* mutation explains at least 18–30% of the cases. The causes responsible for the other cases are still unknown.

22.3.2 Long QT Syndrome

This particular pathology is a congenital disorder characterized by a prolongation of the QT interval observable on an electrocardiogram (ECG; Figure 22.3). Indeed all forms of long QT (LQT) syndrome involve an abnormal repolarization of the heart that is delayed, and this increases the risk of occurrence of episodes of torsades de pointes (Ackerman, 2004).

There are several forms of LQT syndromes. So far 12 different syndromes have been characterized, according to the particular gene mutated. Mostly the genes affected code for potassium channels such as KCNQ1 (LQT1), KCNH2 (LQT2), or sodium channels such as SCN5A (LQT3).

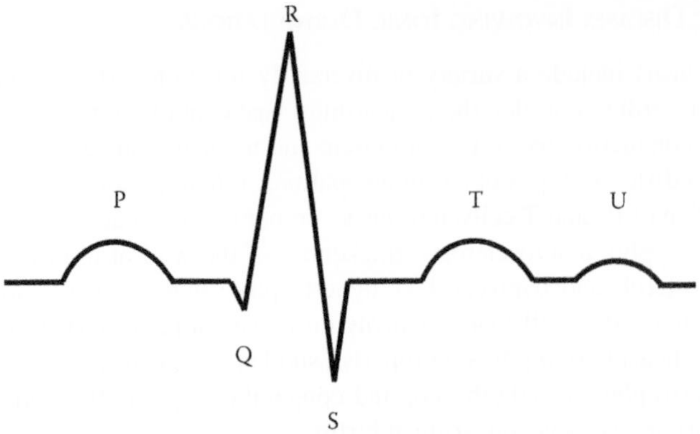

FIGURE 22.3 Electrocardiogram trace. Electrocardiography is an interpreted recording of the electrical activity of the heart over time. The study of the tracing characteristics gives insights into the activity of the heart during a cardiac cycle. One can usually distinguish a P wave (atrial depolarization), a QRS complex (ventricular depolarization), a T wave (ventricular repolarization), and a U wave (repolarization of the Purkinje fibers).

In addition, LQT8 was shown to be due to the α subunit of the calcium channel Cav1.2 encoded by the gene *CACNA1c* and can lead to Timothy's syndrome (characterized by malformations of the heart, fingers, and toes and by autism). Other peculiar genes involved can be ankyrin B, caveolin 3, and *AKAP 9*.

It is believed that the so-called EADs seen in LQT syndromes are due to reopening of L-type calcium channels during the plateau phase of the cardiac AP. Since adrenergic stimulation can increase the activity of these channels, the risk of sudden death in individuals with LQT is high during elevated adrenergic states (i.e., exercise, excitement), especially because repolarization is impaired. Normally during adrenergic states, repolarizing currents will also be enhanced to shorten the AP. In the absence of this shortening and the presence of increased L-type calcium current, EADs may arise. Adrenergic stimulation activates sodium/potassium pumps at the plasma membrane level (exchanging three for two ions, respectively), therefore creating a net outward repolarizing current that decreases heart rate (Tan et al., 1995). The role of intracellular calcium overload and of the sodium/calcium exchanger in the generation of EADs is still under great debate.

22.3.3 Short QT Syndrome

Again, although this genetic disease does not lead to morphological abnormalities of the heart, this autosomal dominant disease can induce atrial fibrillations, palpitations, syncope, and ventricular fibrillations, causing sudden death. So far, mutations in three potassium channels related to short QT syndrome have been indentified: KCNH2, KCNJ2, and KCNQ1. The activity of outward channels being increased, the plateau phase of the AP will be shortened, which results in an overall shortening of refractory periods and of the QT interval on the ECG (Gaita et al., 2004). Short QT syndrome is still poorly characterized, and huge efforts need to be made in order to be able to understand its etiology and establish a proper diagnosis.

22.3.4 Catecholaminergic Polymorphic Ventricular Tachycardia

The particular feature of this pathology is the occurrence of syncope during exercise or emotional stress. Another important characteristic is that mutations in that case affect intracellular receptor complexes known as RyRs as described above. The two currently identified genes associated with catecholaminergic polymorphic ventricular tachycardia (CPVT) are *RyR2*

TABLE 22.1
Examples of Cardiac Channelopathies

Ion Channels	Diseases	Drugs
Sodium channels	Brugada syndrome	Class I – Local anesthetics
Potassium channels	Long QT, Short QT	Class III – HERG blockers
L-type calcium channels	Long QT	Class IV – Dihydropyridines
Ryanodine receptors	CPVT	JTV519 – Flecainide

This simplified table summarizes specific features for channels affected by ionic deregulations in the heart. Specific references to related works can be found in the dedicated sections of this chapter.

(autosomal dominant; responsible for approximately 50–55% of cases of CPVT (Priori et al., 2002) and *CASQ2* (autosomal recessive). The latter encodes the calsequestrin protein. It serves as an auxiliary molecule to *RyR2* by storing calcium ions with its calcium-binding sites (40–50 calcium ions per molecule would then be sequestered within the lumen of the SR [Mitchell, Simmerman, and Jones, 1988]).

The cellular mechanisms accompanying this genetic defect (mutations of *RyR2*) that induces ventricular tachycardia are mainly based on the hypothesis of "leaky" *RyR2* channels (George, Higgs, and Lai, 2003; Jiang et al., 2002; Wehrens et al., 2003). The resulting progressive increase of the cytosolic free calcium concentration creates an electrical instability, which triggers DAD and arrhythmias.

In the case of mutations of *CASQ2*, *RyR2* affinity to luminal calcium would be increased, hence enhancing the activity of intracellular calcium release. This might be due to the decrease of calcium-buffering action by calsequestrin or the impairment of a proper interaction between RyR and calsequestrin (di Barletta et al., 2006; Viatchenko-Karpinski et al., 2004).

Although there seems to be a consensus view regarding the calcium leakage from RyR2 as an underlying mechanism for CPVT, the molecular elements controling the leak and the stability of RyR are still under debate. Currently three hypotheses have been proposed:

1. A defect in the binding with its accessory protein FKBP12.6 would lead to instability of the channel (Wehrens et al., 2003).
2. A change in the threshold for store overload-induced calcium release would be responsible for the different sensitivity to luminal calcium and thus produce spontaneous spillover (Jiang et al., 2004, 2005).
3. Alterations in domain-domain interactions would unzip the channel stable conformation, which would increase RyR activity (George et al., 2006).

These proposed mechanisms accounting for CPVT could also be common to the ones implicated in sudden cardiac death, other arrhythmias, and heart failure (Yano, 2008). Table 21.1 summarizes the four examples presented above.

22.4 GENOMICS OF CARDIAC CHANNELOPATHIES

22.4.1 GENETICS

Clearly the sequencing of the human genome brought a new perspective in the understanding of the occurrence of particular inherited diseases and congenital syndromes, especially with regards to genetic pathologies affecting ion channels. Functional studies represent a limited approach, and therefore the genetic screening of large and adequately matched control populations for absence of the putative mutations is important to prove disease causality. Recent works have reported the typing of 150–200 controls (300–400 chromosomes) for putative mutations with a prevalence of 1% by power analysis (Jurkat-Rott and Lehmann-Horn, 2005). Another algorithm recommends the exclusion of the putative mutation in ethnically matched control chromosomes (Jurkat-Rott and Lehmann-Horn, 2004). The equation proposes to test 460 control chromosomes (230 control individuals) to ensure an error of 1% and a mutation present on 1% of tested patient chromosomes (Jurkat-Rott and Lehmann-Horn, 2005).

The first accurate description of the genomic structure of an ion channel gene responsible for cardiac arrhythmias was carried out by Wang et al. in 1996 using *SCN5A* encoding a sodium channel and whose mutations can cause long QT3 as reported above. The exact characterization of the gene locus on the chromosome, its size, its exon/intron composition, and polymorphisms provide a valuable and huge amount of information to identify mutations related to cardiac channelopathies or to other phenotypes and identify similar mutations in genes from the same family. The analysis of polymorphisms (such as uncommon nucleotide repetitions) results in useful tools for linkage studies in order to discover LQT families connected to a particular chromosome or discover any other possible links. Moreover, the investigation of common DNA variations is essential to avoid false-positive diagnosis, and this should be carried out on large populations from diverse ethnic origins (Priori and Napolitano, 2006). Genetic analysis is the basis for developing genetic testing, which is particularly helpful to screen families and provide counseling, in situations where there are polymorphisms that correlate to genotype-phenotype deficits or aberrant cardiac function (Priori and Napolitano, 2006).

22.4.2 Molecular Biology

Knowing the exact nucleotide sequence of a gene encoding a particular ion channel is obviously extremely useful to spot single mutations leading to direct change in the structure (folding or conformation) of the main subunit forming the pore of the channel. It also enables the recognition of regulatory sequences (phosphorylation, binding, and interaction sites) that might be involved directly or indirectly in the onset of channelopathies.

Basic science and pathophysiological studies benefitted greatly from the finding of full nucleotide sequences for cardiac ion channels and from the cloning of corresponding cDNA. Huge progress in purification methods of membrane proteins has supported this major advance.

The determination of DNA sequences has been crucial in establishing the techniques of mutagenesis, transfection, and expression in heterologous systems and in their reconstitution in artificial membranes for electrophysiological studies. Furthermore, from an ethical point of view, access to human biopsies to extract channel proteins is extremely limited, and therefore the deciphering and cloning of human DNA sequences were undoubtedly helpful in providing human material for basic science research. An example of this is the case of *RyR2* channels. The molecular cloning of human *RyR2* led to the identification of two alternative splice variants, which exhibit different effects on the apoptosis level in a cardiac myocyte cell line (George et al., 2007). Indeed the discovery of splice variants adds more complexity to the function of particular channels, because one gene can encode for many proteins with diverse functions.

22.4.3 Regulation of Cardiac Ion Channel Gene Expression

The control of gene expression in the heart and the related deregulation affecting ion channels is a vast topic requiring a separate chapter. Here we decided to concentrate on novel aspects of this regulation, limitations, and perspectives for genomic studies and diagnosis.

Considering the analysis of changes in gene expression, one has to bear in mind that there is a differential distribution of ion channel expression within the heart (Schram et al., 2002). This feature is the result of a regionalization of cardiac electrical properties to ensure a rapid and controlled propagation of electrical signals leading to rhythmic and powerful contractions. Hence every area of the heart is specifically designed to fulfill a particular role in the conduction system, and therefore the corresponding tool kit of ion channels will differ from one region to another. For instance, T-type VGCC $Ca_v3.1$ and $Ca_v3.2$ encode I_{CaT} α subunits, and $Ca_v3.1$ mRNA expression is 30-fold greater in mouse sinoatrial node (SAN) than in mouse atrium (Bohn et al., 2000). $Ca_v3.2$ expression is lower than $Ca_v3.1$ expression, but it is also greater in the SAN (Bohn et al., 2000). By contrast the expression of T-type VGCCs is rather low in neonatal ventricles but may increase during pathological conditions such as aldosterone stimulation (Lalevée et al., 2005) or mineralocorticoid receptor activation (Maturana et al., 2009). This phenomenon is referred to as reexpression of fetal genes during cardiac pathologies that could confer automaticity to ventricular cells and would represent a proarrhythmogenic condition if occurring *in vivo*.

MicroRNAs are short nucleotide sequences (~22 nucleotides) regulating complementary sequences of messenger RNAs at the posttranscriptional level (Barh, Bhat, and Viero, 2010). Mostly they play a role in gene silencing (transcript degradation and sequestration, translational suppression) but not exclusively. Specific microRNAs are expressed in cardiac tissues, and some of them have been shown to be involved in the generation of arrhythmias such as miR-1, which targets the expression of the potassium channel subunit Kir2.1 and connexin 43 (Yang et al., 2007). Moreover miR-133 would decrease the expression of HERG potassium channel, leading to long QT syndrome in diabetic conditions (Xiao et al., 2007). Although it is now evident that microRNAs modulate ion channels and participate in the induction of cardiac pathologies, it is still difficult to envisage using microRNAs as therapeutic targets because of the complexity of the related pathways. Indeed, some microRNAs target various channel genes and display dual effects, whereas other microRNAs target same genes but with opposite effects (Luo et al., 2010).

22.4.4 Functional Genomics of Cardiac Ion Channel Genes

Following completion of the sequencing of various genomes, including that of the human, the complete repertoire of ion channel genes has been elucidated for different species. How transcripts issued from this gene collection are expressed and modulated in relation to variable physiological and pathological situations is the subject of functional or physiological genomics. Specialized microarrays (IonChips) comprising probes for the ensemble of ion channel and regulatory genes were developed as an alternative to whole-genome DNA chips. Although whole-genome microarrays provide a huge amount of data and important information on the onset of a particular pathology (Yang et al., 2000), the analysis and interpretation of the results have been found to be extremely complex in some cases in the light of interfering parameters and the lack of detailed screenings concerning specific types of molecules. Therefore the alternative of specialized microarrays was developed where the focus can be placed either on cellular functions, organs, or molecular types for which only a small subset of genes is involved. Obviously, the changes observed in transcript expression have to be correlated with assessments of protein expression by Western blots for instance. In the case of cardiac ion channels, functional tests as patch clamp and electrocardiography can also be performed to validate mRNA expression modifications seen at the microarray level (Le Bouter et al., 2003). Physiological genomics of cardiac ion channel genes is a growing field that, in combination with genetics, should markedly increase our comprehension of the molecular mechanisms leading to arrhythmias (Demolombe et al., 2005).

22.5 PROTEOMICS OF CARDIAC CHANNELOPATHIES

22.5.1 Structure of Ion Channels

22.5.1.1 Purification of Membrane Proteins

The first step in the determination of the structure of ion channels is to isolate and purify these proteins from biological membranes. This is realized by the use of specific detergents at very definite concentrations (depending on the type of channel) and on the tissue from which the channel is isolated (Linke, 2009). For a thorough methodology, readers are advised to find some relevant information in the review by Lin and Guidotti (2009).

22.5.1.2 Crystallography

When Roderick MacKinnon unveiled the three-dimensional (3D) crystal structure of a bacterial potassium channel in 1998 (Doyle et al., 1998), a new world of understanding and perspectives had been entered in terms of ion channel biochemistry and biophysics. For the first time, one was able to relate a reliable channel structure to its function and to explain the mechanisms of ion conduction and selectivity. This ability obviously greatly assisted the development of research in channelopathies. Recently, this knowledge on ion-channel structural biology has served as a strong basis for the establishment of large-scale studies and genomics approaches for investigations into the structure of membrane proteins (Love et al., 2010) and should definitely lead to the collection of a huge amount of data on cardiac ion channel structural determinants.

22.5.1.3 Analogy Models

In some cases, high resolution 3D crystallographic data cannot be obtained because of methodological limitations, such as when the protein size does not allow the process leading to the creation of crystals. This is sadly the situation concerning RyR2 with its enormous size of 565 kDa, which impairs any current crystallization attempts. Alternatively analogy models can be developed to overcome such issues. This strategy was used to predict the structure of the pore-forming region of RyR2 by comparison with the pore of a potassium channel, KcsA (Welch et al., 2004). The analogy procedure was also successfully employed in the case of cardiac sodium channels. Both comparisons with the crystal structure of potassium channels and with particular flexible enzymatic motifs shed light on the relation structure-function of cardiac sodium channels before any crystallization attempt (Balser, 1999). Moreover, while the structural components of the selectivity filter were not completely characterized, sodium and plasma membrane calcium channels appeared to present structural analogies of the pore regions according to early data (Marban, Yamagishi, and Tomaselli, 1998).

22.5.1.4 Posttranslational Modifications

The concept of channelopathies is not restricted to genetic disorders; notably, changes in the expression or posttranslational modification of ion channels underlie the fatal arrhythmias associated with heart failure (Marbán, 2002). One of the most influential posttranslational changes is the modification of cysteines by nitric oxide (NO), referred to as *S*-nitrosylation. NO signaling regulates the redox state of cardiac cells, particularly of proteins involved in ECC, such as RyR2. When this signaling is impaired, ischemia, heart failure, or atrial fibrillation can occur (Gonzalez et al., 2009). Two other mechanisms of posttranslational modifications are the ubiquitylation and SUMOylation of channels leading to the modulation of their degradation and regulation (Rougier, Albesa, and Abriel, 2010). Sodium channels can also undergo deglycosylation, which is decisive for the onset of cardiac arrhythmias in heart failure (Ufret-Vincenty et al., 2001). In this regard, Montpetit et al. (2009) discussed the recent findings in the modulation of cardiac electrical signaling by controlled and unusual glycosylation. Furthermore posttranslational modifications of β subunits of VGCC have been reported (Chien and Hosey, 1998).

22.5.2 Functions of Ion Channels

The general and common role of all ion channels is the regulation of the electrolyte flows between membranes to maintain ionic homeostasis and electrical potentials or to trigger electrical signals that will be transduced into changes in phosphorylation level and gene and protein expression.

Beside this, they fulfill various biological functions within a cell, ranging from contraction to energetic metabolism. However, it is difficult to relate one single molecule such as an ion channel to one particular biological process, because the latter often involves a network of coordinated molecules. In this regard, the proteomic approach is a very useful tool to unravel the molecular partners of ion channels, their interactions, and their relative influences within a specific biological functional system. Usually, proteomic analysis requires the combination of high-throughput techniques to identify proteins at a large scale, these include: two-dimensional gel electrophoresis, mass spectroscopy, and bioinformatics analysis. Zlatkovic et al. (2009) identified the expression profile related to ATP-sensitive potassium channels (K_{ATP}) during hypertension-induced heart failure using a proteomic approach. Indeed, the absence of K_{ATP} was accompanied with a remodeling of gene expression leading to contractile dysfunctions, fibrosis, and hypertrophy.

To fulfill their roles of generating ionic flows, functional ion channels have to be inserted into cell membranes (plasma or organelle membranes). This implies that a sufficient number of channel proteins or more precisely wild-type and functional subunits should be present at the membrane level. The example of LQT2 shows that impairment of protein trafficking results in a reduction in the expression of HERG channels to the cell surface membrane. Mutant channels are sequestered in the endoplasmic reticulum as glycosylated proteins (Delisle et al., 2004). It is noteworthy that some of the mutant subunits have dominant-negative effects. They are still able to coassemble with wild-type subunits, but the resultant multimers remained trapped in the endoplasmic reticulum (Ficker et al., 2000). Other defects in channel trafficking were found in different channelopathies such as Brugada (cardiac sodium channel [Baroudi et al., 2002]) and Andersen (*KCNJ2* gene [Bendahhou et al., 2003]) syndromes. Interestingly, successful strategies for rescuing these dysfunctions in channel trafficking were developed *in vitro*:

1. Reduction of cell culture temperature (Zhou, Gong, and January, 1999)
2. Incubation in glycerol at the molar range (Delisle et al., 2004)
3. Incubation with drugs that bind to and cause HERG channel block or their analogues (Ficker et al., 2002)
4. Incubation with thapsigargin, the SERCA inhibitor (Delisle et al., 2003)

These different methods, although not specific, appear to be promising and certainly deserve further investigations to determine whether they be employed *in vivo*. Although the trafficking would be rescued in that case, the expression of normal channels at the cell surface would not be improved.

22.5.3 Proteomic and Electrophysiological Biomarkers

In pathologies such as cancer, the discovery and development of reliable biomarkers is essential for drug screening and diagnosis (Barh, 2009). Likewise, cardiac ion channel diseases have drawn much attention over the last decade at both clinical and pharmaceutical levels, and the related search for markers is an increasing field of interest.

Ion channels themselves can serve as markers to identify arrhythmic patterns. This is the case for L-VGCC in atrial myocytes. The decrease of the related current is observed in patients in sinus rhythm and is associated with a high risk of atrial fibrillation (Dinanian et al., 2008).

Patients with heart failure display a less powerful heart-rate response during exercise, which constitutes a risk factor for mortality. Researchers established the clinical relevance of Kir6.2 E23K as a biomarker for impaired stress performance and thereby underscored the essential role of K_{ATP} channels in human cardiac physiology (Olson and Terzic, 2010).

To date, the best diagnostic tool continues to be electrocardiography, which represents a therapeutic biomarker that is extremely useful for pediatric and adult cardiac diseases (Heier et al., 2010). Indeed, it is vital for pharmaceutical industries to be able to assess the toxicity of drugs in order to detect any potential arrhythmic risk. Three classes of antiarrhythmic drugs were developed and are currently employed to target cardiac ion channels. These include class I (Ia, Ib, and Ic) drugs that affect sodium channels; class III compounds that interact with potassium channels; and class IV drugs that have calcium channel-blocking actions (Table 22.1).

The main clinical ECG biomarker is the measure of the interval between the Q wave and the T wave (QT interval). Moreover, the electrophysiological biomarkers routinely used are the AP duration, the effective refractory period, and amplitude of the ionic conductance and gating properties of ion channels, T wave amplitude, and ST-segment elevation. Recently, a lot of emphasis has been put on the establishment of computer modeling and simulations to enable the investigator to mimic what can happen at the cellular, tissue, and whole ventricular level, and predict the effects of drug interactions on heart mechanisms (Corrias et al., 2010). Another newly introduced method of exploring potential therapeutic targets is the atomic force microscopy as described in the review by Lal and Arnsdorf (2010).

Improving early diagnosis would improve the patients' quality of life by reducing the use of invasive treatments and also would reduce costs for national health systems. Patient blood samples are screened to detect valid and reliable genetic biomarkers (Marban, Yamagishi, and Tomaselli, 1998). They are particularly interested in messenger RNA and microRNA. This approach is still under investigation and should bring a huge amount of information in the field of genomic diagnosis.

22.6 DIAGNOSIS

22.6.1 Electrophysiology

Conventional methods of detecting anomalies in cardiac ion channel function in patients include clinical individual and family history, physical examination, and electrocardiography (Marbán, 2002). As stated above, these channel dysfunctions appear without structural cardiac abnormalities. ECG recordings provide a huge amount of information, and the parameters traditionally investigated are: P wave, T wave, U wave, QRS vector alternans, and the occurrence of ventricular tachycardia. It is essential to record resting ECGs, as well as ECGs during physical activity and acute emotional stress because the incidence of some of the channelopathies is only observed during activation of the catecholaminergic system as is the case with CPVT (Priori et al., 2002).

22.6.2 Cardiac Ion Channel Genetic Testing

The development of genetic testing in clinical practice is increasing, especially in the field of inherited arrhythmogenic diseases. Correlating genotype and phenotype for a specific pathology is of peculiar importance. In this respect, Priori and Napolitano (2006) established a score system that takes into account the differences in various heart dysfunctions. This score reflects an estimation of the relationship between cost and benefit regarding the use of genetic analysis for every pathology. Three groups have been identified:

1. Disorders where genetic testing is necessary and advised (value ≥ 3).
2. Disorders where genetic testing could be useful and might be planned (value between 1 and 3).
3. Genetic analysis is carried out for research purposes (value ≤ 1).

The rationale behind such a study is to support a more effective introduction of genetic analysis in clinical practice in prioritizing inherited cardiac pathologies. This approach of a systematic scoring would also appear to be more convincing for the stakeholders, so that genetic analysis could become reimbursed in a more consistent way.

Genetic testing of cardiac ion channel mutations in persons with suspected channelopathies is essential in order to determine the risk for sudden cardiac death. The FAMILION® test (PGx Health™) is a genetic test designed to detect mutations in inherited cardiac channelopathies, such as LQT syndrome. It is proposed for use in determining the diagnosis of LQT syndrome and for counseling of individuals affected by it. The test relies on the analysis of the DNA taken from a blood sample for identification of mutations in five cardiac ion channel genes that have been associated with disorders (Kapplinger et al., 2009).

A recent clinical study carried out in four European countries showed that mutation of *SCN5A* in Brugada syndrome patients was not a reliable predictor of arrhythmic events (Probst et al., 2010). Still, for Brugada syndrome patients, genetic analysis provides negative results for 70–80% of them (Kapplinger et al., 2010). This is partly attributable to the limitation in clinical and genetic data for this particular syndrome (Zhou and Wang, 2010).

The development of clinical genomics represents an exciting and challenging perspective, and several studies emerged over the recent years to provide evidence of the involvement of channel transcript remodeling in heart diseases (Borlak and Thum, 2003; Lamirault et al., 2006). The limitations of this approach include the restriction in accessing human biopsies due to obvious ethical concerns and the issue of interpreting transcript expression changes into protein functions (Demolombe et al., 2005). Therefore, a strong emphasis should be put on improving large-scale investigations of ion-channel proteins in the heart. However, again limitation in the use of human tissues and the difficulty of studying membrane proteins prevent any rapid progress in this field of research.

22.7 THERAPY

22.7.1 Conventional Management

Whether the treatment of manifestations of channel dysfunction is based on the implantation of a cardioverter defibrillator in case of a history of cardiac arrest or the administration of specific drugs very much depends on the type of channel disorder. For instance, infusion of isoproterenol has been quite successful in reversing electrical storms in Brugada syndrome (Shimizu, Aiba, and Antzelevitch, 2005). In contrast, beta-blockers would be indicated in the case of CPVT mutations (Rosso et al., 2007). Oral quinidine was also effective in the treatment of electrical storms (Jongman et al., 2007). Alternatively, a more invasive technique such as left cardiac sympathetic denervation can be employed to decrease the quantity of catecholamines activating the heart when beta-blocker treatment fails to control symptoms in CPVT patients (Wilde et al., 2008).

22.7.2 Issues Associated with Drug Design

As stated by Ouzounian et al. (2007), the progress in pharmacogenomics opens a new era in which the concept of the *miracle drug* fixing everything all the time has moved to the concept of *personalized drugs* able to target a specific disorder in a particular patient at a certain time. Within this novel paradigm, all related disciplines (genomics, proteomics, pharmacology, electrophysiology, and clinical trials) join their efforts and work together to identify and validate potent compounds to tackle channelopathies, especially in the heart.

The most well-known class of antiarrhythmic drugs is the local anesthetic group or class I. These compounds are blockers of the sodium current and have thus been successfully used to treat cardiac arrhythmias. However, some of them were found to have proarrhythmic effects (Echt et al., 1991). Structurally they share a common organization with hydrophilic and hydrophobic domains linked by an amide or ester (Balser, 1999). Their mechanism of action is based on different low- and high-affinity binding sites to three distinct conformational gating states of the cardiac sodium channel:

closed, open, and inactivated. This explains the peculiar property of time and voltage dependence of the effect of local anesthetics.

Specific residues in the S6 segment of domain IV appear to be pivotal in the use-dependent block by local anesthetics (lidocaine, etidocaine, and phenytoin), especially two particular amino acids as identified by the mutations F1764A and Y1771A in rat brain IIA sodium channels in which drug affinity for the inactivated channel conformation was decreased (Ragsdale et al., 1994, 1996). Interestingly, residues from similar regions of S6 are involved in the block of potassium channels by quinidine (Yeola et al., 1996). Moreover, the block of calcium channels by dihydropyridine and phenylalkylamine seems to require amino acids in the segment S6 of domain IV (Hockerman et al., 1995; Schuster et al., 1996). In addition, residues in the S5 segments of domains III and IV participate in the block of calcium channels by dihydropyridine (Ito, Klugbauer, and Hofmann, 1997; Peterson, Tanada, and Catterall, 1996). For a detailed study of class I drug action in sodium channels with all their molecular and biophysical implications, readers are kindly invited to refer to the excellent review of Balser (1999). Hence, although specific binding sites can be identified for the drug interaction with channels, common mechanisms most likely account for the block of most of the main ion channels. This renders the design of specific drugs particularly difficult.

Intriguingly, the local anesthetic flecainide showed very effective blocking effects on RyR2 (Hilliard et al., 2010) and can therefore be considered as a promising therapeutic agent to counter CPVT in animal models and human patients (Watanabe et al., 2009). Nonetheless, the molecular mechanism underlying the inhibition of calcium release from RyR2 is not fully understood.

Another noteworthy example in RyR2 is the compound JTV519 (also known as K201). It is said to stabilize the calcium leak in RyR, and this hypothesis was confirmed in some models of CPVT-causing mutations *in vitro* (Lehnart et al., 2004). Nevertheless, the use of a CPVT R4496C knock-in mouse model revealed the lack of a significant antiarrhythmic effect of this drug (Liu et al., 2006).

The pharmacological approach of targeting ion channels with specific compounds depends on the ability to determine the right dose to administer to the patient in order to reach the active concentration allowing the optimal effect to occur. Obviously this is not an easy task to perform because drugs can be degraded or not even reach their final goal at the expected concentration. According to their specific pK_a and to the pH value at the target locus, the degree of ionization will differ from one compound to another. In this respect, a drug having to cross the phospholipid plasma membrane to be active might not be in sufficient quantity where the ion channel sits. With all these problems and challenges (specificity, toxicity, active forms, and complex disorders in which the entire cell has undergone a drastic remodeling), it might be sensible to think about complementary and more global techniques enabling the cell itself to provide its own tools to counteract anomalies in ionic currents and cardiac rhythm.

22.7.3 Gene Therapy

One of the new strategies of treatment that can be established out of our understanding of the molecular basis of channel dysfunction is gene therapy (Marbán, 2002). With technologies increasingly improving, methods of delivering genes to cardiac muscle are nowadays available and successfully employed (Lyon et al., 2008; Neyroud et al., 2002). The most efficient way of transferring genes into cardiac tissues is to inject adenoviral vectors expressing the genes of interest (Hoppe, Marbán, and Johns, 2001; Viero et al., 2008).

Mazhari et al. (2002) expressed the regulatory protein KCNE3 interacting with the potassium channel KCNQ1 in the ventricles of guinea pigs. This resulted in an increased outward current, a reduction of the AP duration, and a shortening of the QT interval. However, the authors clearly claim that a beneficial effect on cardiac function can only be achieved if the ectopic expression is homogenous throughout the ventricles.

Alternatively, gene therapy does not have to directly target ion channels but can also affect signaling pathways. Donahue et al. (2000) concentrated on the atrioventricular node in which they overexpressed the protein Gα_i subunit, thus inducing a local beta-blockade that decreased the cardiac rate during atrial fibrillation.

Other compelling targets of gene therapy in the heart are proteins involved in calcium handling such as SERCA2 and parvalbumin (Sakata et al., 2007). The delivery of *SERCA1a* gene also strongly improves myocardial function in rabbits (Logeart et al., 2006). Ziolo et al. (2005) addressed the issue of increasing *SERCA* function as well by injecting a dominant-negative mutant phospholamban (SERCA accessory protein) adenovirus. Gene therapy certainly becomes one of the most elegant ways to work toward personalized medicine.

22.7.4 Cell Therapy

An exciting area of research is stem cell technology and the possibility of achieving transplantations as a regenerative approach for cardiovascular diseases. Recently mice lacking the K_{ATP} gene, which led to the onset of cardiomyopathy and ventricular arrhythmias, were generated. Epicardial injection of embryonic stem cells resulted in an increase of contractile and electrical function (Yamada et al., 2008).

This advanced technique is not restricted to the experimental level, but clinical trials involving stem cell therapy have been carried out all around the world, and the results appear to be promising. Studies show that patients maintain a safe health profile, whereas small, although significant improvements in cardiac function and structural remodeling were observed (Gersh et al., 2009). Indeed, important features for the cardiogenic development of pluripotent stem cells have been revealed by the combination of transcriptomic and computational biology (Chiriac et al., 2010). The limitations of stem cell therapy, however, comprise the uncertainty about the long-term fate of a transplanted cell and the poor understanding we have regarding the orientation that the cellular layers should adopt within the heart to optimize their beneficial effect.

22.8 CONCLUDING REMARKS

Because the scope of this chapter is broad, the citations and discussion were selective, and the reader is referred to more thorough analysis of particular subtopics. This chapter has been designed to target both experts in cardiac electrophysiology and readers interested in general genomic and proteomic regulations and technologies. Therefore, very basic aspects of cardiac physiology and pathology are presented here, whereas an overview of the current knowledge on cardiac channelopathies and perspectives regarding the use of genomics and proteomics data for diagnosis and therapy can also be found.

Channelopathies consist of various disorders, inherited or acquired. Indeed, it is said that heart failure represents a form of LQT syndrome (Marbán, 1999). Gaining some understanding of the mechanisms underlying these ion channel pathologies will surely give more insight into the onset of heart failure and sudden death, but for this, both researchers and clinicians have to rely on valid large-scale studies enabled by the methods offered by genomics and proteomics (Figure 22.4). High-throughput investigations might also provide evidence for genetic or epigenetic alterations of channels thus far less explored or less implicated in diseases such as $InsP_3R$, TRIC, TRP, chloride channels, and T-type VGCC.

The future studies in channelopathies of the heart will be based on global approaches leading to personalized medicine. In order to be able to provide treatment better adjusted to the individual, integrative technologies such as computational medicine and cell therapy will be employed (Figure 22.4).

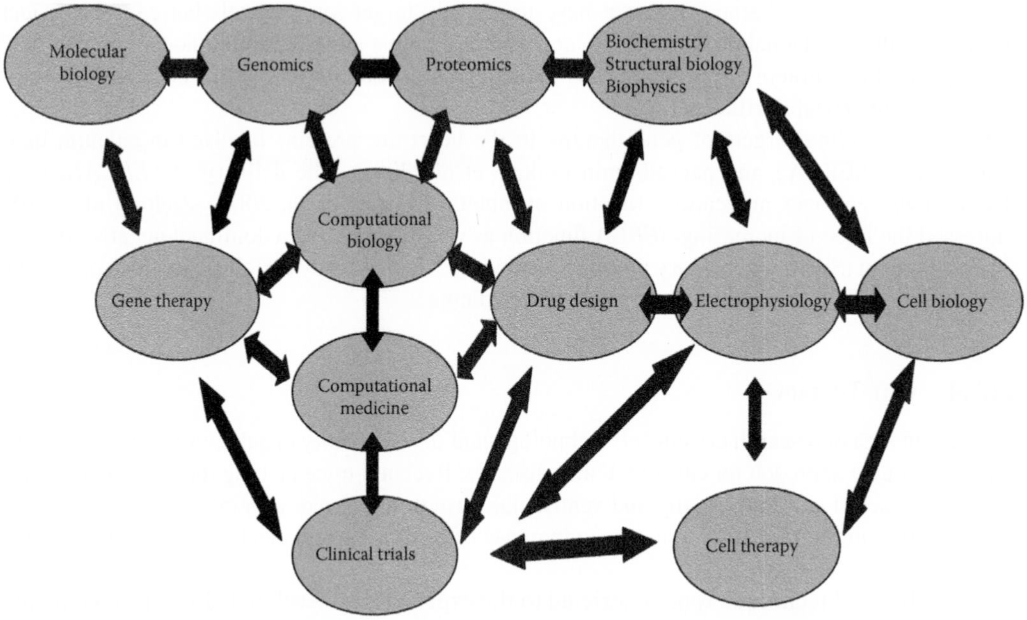

FIGURE 22.4 Strategies employed to tackle cardiac channelopathies. This diagram displays the links between the various disciplines involved in the investigation and treatment of ion channel disorders in the heart. For clarity, not all the direct connections are shown here. For instance, *cell biology* is also directly related to *molecular biology* and *drug design*.

REFERENCES

Ackerman, M. (2004). Cardiac channelopathies: It's in the genes. *Nat. Med.* 10: 463–464.

Balser, J. (1999). Structure and function of the cardiac sodium channels. *Cardiovasc. Res.* 42: 327–338.

Balser, J. (2001). The cardiac sodium channel: Gating function and molecular pharmacology. *J. Mol. Cell. Cardiol.* 33: 599–613.

Barh, D. (2009). Biomarkers, critical disease pathways, drug targets, and alternative medicine in male breast cancer. *Curr. Drug Targets* 10: 1–8.

Barh, D., Bhat, D., Viero, C. (2010). miReg: A resource for microRNA regulation. *J. Integr. Bioinform.* 7: 144.

Baroudi, G., et al. (2002). Expression and intracellular localization of an SCN5A double mutant R1232W/T1620M implicated in Brugada syndrome. *Circ. Res.* 90: E11–E16.

Bendahhou, S., et al. (2003). Defective potassium channel Kir2.1 trafficking underlies Andersen-Tawil syndrome. *J. Biol. Chem.* 278: 51779–51785.

Bers, D. (2002). Cardiac excitation-contraction coupling. *Nature* 415: 198–205.

Bohn, G., et al. (2000). Expression of T- and L-type calcium channel mRNA in murine sinoatrial node. *FEBS Lett.* 481: 73–76.

Borlak, J., Thum, T. (2003). Hallmarks of ion channel gene expression in end-stage heart failure. *FASEB J.* 17: 1592–1608.

Brugada, P., Brugada, J. (1992). Right bundle branch block, persistent ST segment elevation and sudden cardiac death: A distinct clinical and electrocardiographic syndrome: A multicenter report. *J. Am. Coll. Cardiol.* 20: 1391–1396.

Chien, A., Hosey, M. (1998). Post-translational modifications of beta subunits of voltage-dependent calcium channels. *J. Bioenerg. Biomembr.* 30: 377–386.

Chiriac, A., et al. (2010). Cardiogenic induction of pluripotent stem cells streamlined through a conserved SDF-1/VEGF/BMP2 integrated network. *PLoS One* 5: e9943.

Corrias, A., et al. (2010). Arrhythmic risk biomarkers for the assessment of drug cardiotoxicity: From experiments to computer simulations. *Philos. Transact. A Math. Phys. Eng. Sci.* 368: 3001–3025.

Deal, K., England, S., Tamkun, M. (1996). Molecular physiology of cardiac potassium channels. *Physiol. Rev.* 76: 49–67.

Delisle, B., et al. (2003). Thapsigargin selectively rescues the trafficking defective LQT2 channels G601S and F805C. *J. Biol. Chem.* 278: 35749–35754.
Delisle, B., et al. (2004). Biology of cardiac arrhythmias: Ion channel protein trafficking. *Circ. Res.* 94: 1418–1428.
Demion, M., et al. (2007). TRPM4, a Ca^{2+}-activated nonselective cation channel in mouse sino-atrial node cells. *Cardiovasc. Res.* 73: 531–538.
Demolombe, S., et al. (2005). Functional genomics of cardiac ion channel genes. *Cardiovasc. Res.* 67: 438–447.
di Barletta, M., et al. (2006). Clinical phenotype and functional characterization of CASQ2 mutations associated with catecholaminergic polymorphic ventricular tachycardia. *Circulation* 114: 1012–1019.
Dinanian, S., et al. (2008). Downregulation of the calcium current in human right atrial myocytes from patients in sinus rhythm but with a high risk of atrial fibrillation. *Eur. Heart J.* 29: 1190–1197.
Donahue, J., et al. (2000). Focal modification of electrical conduction in the heart by viral gene transfer. *Nat. Med.* 6: 1395–1398.
Doyle, D., et al. (1998). The structure of the potassium channel: Molecular basis of K^+ conduction and selectivity. *Science* 280: 69–77.
Duan, D. (2009). Phenomics of cardiac chloride channels: The systematic study of chloride channel function in the heart. *J. Physiol.* 587: 2163–2177.
Echt, D., et al. (1991). Mortality and morbidity in patients receiving encainide, flecainide, or placebo: The Cardiac Arrhythmia Suppression Trial. *N. Engl. J. Med.* 324: 781–788.
Ficker, E., et al. (2000). Retention in the endoplasmic reticulum as a mechanism of dominant-negative current suppression in human long QT syndrome. *J. Mol. Cell. Cardiol.* 32: 2327–2337.
Ficker, E., et al. (2002). The binding site for channel blockers that rescue misprocessed human long QT syndrome type 2 ether-a-gogo-related gene (HERG) mutations. *J. Biol. Chem.* 277: 4989–4998.
Fleischer, S. (2008). Personal recollections on the discovery of the ryanodine receptors of muscle. *Biochem. Biophys. Res. Commun.* 369 : 195–207.
Gaita, F., et al. (2004). Short QT syndrome: Pharmacological treatment. *J. Am. Coll. Cardiol.* 43: 1494–1499.
Garlid, K., et al. (1997). Cardioprotective effect of diazoxide and its interaction with mitochondrial ATP-sensitive K^+ channels: Possible mechanism of cardioprotection. *Circ. Res.* 81: 1072–1082.
Garlid, K., et al. (2003). Mitochondrial potassium transport: The role of the mitochondrial ATP-sensitive K(+) channel in cardiac function and cardioprotection. *Biochim. Biophys. Acta* 1606: 1–21.
George, C., Higgs, G., Lai, F. (2003). Ryanodine receptor mutations associated with stress-induced ventricular tachycardia mediate increased calcium release in stimulated cardiomyocytes. *Circ. Res.* 93: 531–540.
George, C., et al. (2006). Arrhythmogenic mutation-linked defects in ryanodine receptor autoregulation reveal a novel mechanism of Ca^{2+} release channel dysfunction. *Circ. Res.* 98: 88–97.
George, C., et al. (2007). Alternative splicing of ryanodine receptors modulates cardiomyocyte Ca^{2+} signaling and susceptibility to apoptosis. *Circ. Res.* 100: 874–883.
Gersh, B., et al. (2009). Cardiac cell repair therapy: A clinical perspective. *Mayo Clin. Proc.* 84: 876–892.
Gonzalez, D., et al. (2009). *S*-Nitrosylation of cardiac ion channels. *J. Cardiovasc. Pharmacol.* 54: 188–195.
Heier, C., et al. (2010). Development of electrocardiogram intervals during growth of FVB/N neonate mice. *BMC Physiol.* 10: 16.
Hilliard, F., et al. (2010). Flecainide inhibits arrhythmogenic Ca^{2+} waves by open state block of ryanodine receptor Ca^{2+} release channels and reduction of Ca^{2+} spark mass. *J. Mol. Cell. Cardiol.* 48: 293–301.
Hockerman, G., et al. (1995). Molecular determinants of high affinity phenylalkylamine block of L-type calcium channels. *J. Biol. Chem.* 270: 22119–22122.
Hoppe, U., Marbán, E., Johns, D. (2001). Distinct gene-specific mechanisms of arrhythmia revealed by cardiac gene transfer of two long QT disease genes, HERG and KCNE1. *Proc. Natl. Acad. Sci. U.S.A.* 98: 5335–5340.
Ito, H., Klugbauer, N., Hofmann, F. (1997). Transfer of the high affinity dihydropyridine sensitivity from L-type to non-L-type calcium channel. *Mol. Pharmacol.* 52: 735–740.
Jiang, D., et al. (2002). Enhanced basal activity of a cardiac Ca^{2+} release channel (ryanodine receptor) mutant associated with ventricular tachycardia and sudden death. *Circ. Res.* 91: 218–225.
Jiang, D., et al. (2004). RyR2 mutations linked to ventricular tachycardia and sudden death reduce the threshold for store-overload-induced Ca^{2+} release (SOICR). *Proc. Natl. Acad. Sci. U.S.A.* 101: 13062–13067.
Jiang, D., et al. (2005). Enhanced store overload-induced Ca^{2+} release and channel sensitivity to luminal Ca^{2+} activation are common defects of RyR2 mutations linked to ventricular tachycardia and sudden death. *Circ. Res.* 97: 1173–1181.
Jongman, J., et al. (2007). Electrical storms in Brugada syndrome successfully treated with isoproterenol infusion and quinidine orally. *Neth. Heart. J.* 15: 151–155.

Jurkat-Rott, K., Lehmann-Horn, F. (2004). Periodic paralysis mutation MiRP2-R83H in controls: Interpretations and general recommendation. *Neurology* 62: 1012–1015.

Jurkat-Rott, K., Lehmann-Horn, F. (2005). Muscle channelopathies and critical points in functional and genetic studies. *J. Clin. Invest.* 115: 2000–2009.

Kapplinger, J., et al. (2009). Spectrum and prevalence of mutations from the first 2,500 consecutive unrelated patients referred for the FAMILION long QT syndrome genetic test. *Heart Rhythm* 6: 1297–1303.

Kapplinger, J., et al. (2010). An international compendium of mutations in the SCN5A-encoded cardiac sodium channel in patients referred for Brugada syndrome genetic testing. *Heart Rhythm* 7: 33–46.

Kass, R. (2005). The channelopathies: Novel insights into molecular and genetic mechanisms of human disease. *J. Clin. Invest.* 115: 1986–1989.

Lal, R., Arnsdorf, M. (2010). Multidimensional atomic force microscopy for drug discovery: A versatile tool for defining targets, designing therapeutics and monitoring their efficacy. *Life Sci.* 86: 545–562.

Lalevée, N., et al. (2005). Aldosterone increases T-type calcium channel expression and in vitro beating frequency in neonatal rat cardiomyocytes. *Cardiovasc. Res.* 67: 216–224.

Lamirault, G., et al. (2006). Gene expression profile associated with chronic atrial fibrillation and underlying valvular heart disease in man. *J. Mol. Cell. Cardiol.* 40: 173–184.

Le Bouter, S., et al. (2003). Microarray analysis reveals complex remodeling of cardiac ion channel expression with altered thyroid status: Relation to cellular and integrated electrophysiology. *Circ. Res.* 92: 234–242.

Lehnart, S., et al. (2004). Sudden death in familial polymorphic ventricular tachycardia associated with calcium release channel (ryanodine receptor) leak. *Circulation* 109: 3208–3214.

Li, X., et al. (2005). Endothelin-1-induced arrhythmogenic Ca^{2+} signaling is abolished in atrial myocytes of inositol-1,4,5-trisphosphate (IP3)-receptor type 2-deficient mice. *Circ. Res.* 96: 1274–1281.

Li, X. T., et al. (2006). The stretch-activated potassium channel TREK-1 in rat cardiac ventricular muscle. *Cardiovasc. Res.* 69: 86–97.

Liang, P., et al. (2010). Genetic analysis of Brugada syndrome and congenital long-QT syndrome type 3 in the Chinese. *J. Cardiovasc. Dis. Res.* 1: 69–74.

Lin, S., Guidotti, G. (2009). Purification of membrane proteins. *Methods Enzymol.* 463: 619–629.

Linke, D. (2009). Detergents: An overview. *Methods Enzymol.* 463: 603–617.

Lipp, P., et al. (2000). Functional InsP3 receptors that may modulate excitation-contraction coupling in the heart. *Curr. Biol.* 10: 939–942.

Liu, N., et al. (2006). Arrhythmogenesis in catecholaminergic polymorphic ventricular tachycardia: Insights from a RyR2 R4496C knock-in mouse model. *Circ. Res.* 99: 292–298.

Logeart, D., et al. (2006). Percutaneous intracoronary delivery of SERCA gene increases myocardial function: A tissue Doppler imaging echocardiographic study. *Am. J. Physiol. Heart Circ. Physiol.* 291: H1773–H1779.

Love, J., et al. (2010). The New York Consortium on Membrane Protein Structure (NYCOMPS): A high-throughput platform for structural genomics of integral membrane proteins. *J. Struct. Funct. Genomics* 11: 191–199.

Luo, X., et al. (2010). Regulation of human cardiac ion channel genes by microRNAs: Theoretical perspective and pathophysiological implications. *Cell. Physiol. Biochem.* 25: 571–586.

Lyon, A., et al. (2008). Gene therapy: Targeting the myocardium. *Heart* 94: 89–99.

Marban, E., Yamagishi, T., Tomaselli, G. (1998). Structure and function of voltage-gated sodium channels. *J. Physiol.* 508: 647–657.

Marbán, E. (1999). Heart failure: The electrophysiologic connection. *J. Cardiovasc. Electrophysiol.* 10: 1425–1428.

Marbán, E. (2002). Cardiac channelopathies. *Nature* 415: 213–218.

Maturana, A., et al. (2009). Role of the T-type calcium channel CaV3.2 in the chronotropic action of corticosteroids in isolated rat ventricular myocytes. *Endocrinology* 150: 3726–3734.

Mazhari, R., et al. (2002). Ectopic expression of KCNE3 accelerates cardiac repolarization and abbreviates the QT interval. *J. Clin. Invest.* 109: 1083–1090.

Mitchell, R., Simmerman, H., Jones, L. (1988). Ca^{2+} binding effects on protein conformation and protein interactions of canine cardiac calsequestrin. *J. Biol. Chem.* 263: 1376–1381.

Montpetit, M., et al. (2009). Regulated and aberrant glycosylation modulate cardiac electrical signaling. *Proc. Natl. Acad. Sci. U.S.A.* 106: 16517–16522.

Neyroud, N., et al. (2002). Gene delivery to cardiac muscle. *Methods Enzymol.* 346: 323–334.

Olson, T., Terzic, A. (2010). Human K(ATP) channelopathies: Diseases of metabolic homeostasis. *Pflugers Arch.* 460: 295–306.

Ouzounian, M., et al. (2007). Predict, prevent and personalize: Genomic and proteomic approaches to cardiovascular medicine. *Can. J. Cardiol.* 23 (suppl.): 28A–33A.

Peterson, B., Tanada, T., Catterall, W. (1996). Molecular determinants of high affinity dihydropyridine binding in L-type calcium channels. *J. Biol. Chem.* 271: 5293–5296.

Priori, S., Napolitano, C. (2006). Role of genetic analyses in cardiology: Part I: Mendelian diseases: Cardiac channelopathies. *Circulation* 113: 1130–1135.

Priori, S., et al. (2002). Clinical and molecular characterization of patients with catecholaminergic polymorphic ventricular tachycardia. *Circulation* 106: 69–74.

Probst, V., et al. (2010). Long-term prognosis of patients diagnosed with Brugada syndrome: Results from the FINGER Brugada Syndrome Registry. *Circulation* 121: 635–643.

Ragsdale, D., et al. (1994). Molecular determinants of state-dependent block of Na^+ channels by local anesthetics. *Science* 265: 1724–1728.

Ragsdale, D., et al. (1996). Common molecular determinants of local anesthetic, antiarrhythmic, and anticonvulsant block of voltage-gated Na+ channels. *Proc. Natl. Acad. Sci. U.S.A.* 93: 9270–9275.

Reuter, H., Kokubun, S., Prod'hom, B. (1986). Properties and modulation of cardiac calcium channels. *J. Exp. Biol.* 124: 191–201.

Rosso, R., et al. (2007). Calcium channel blockers and beta-blockers versus beta-blockers alone for preventing exercise-induced arrhythmias in catecholaminergic polymorphic ventricular tachycardia. *Heart Rhythm* 4: 1149–1154.

Rougier, J., Albesa, M., Abriel, H. (2010). Ubiquitylation and SUMOylation of cardiac ion channels. *J. Cardiovasc. Pharmacol.* 56: 22–28.

Sakata, S., et al. (2007). Restoration of mechanical and energetic function in failing aortic-banded rat hearts by gene transfer of calcium cycling proteins. *J. Mol. Cell. Cardiol.* 42: 852–861.

Schram, G., et al. (2002). Differential distribution of cardiac ion channel expression as a basis for regional specialization in electrical function. *Circ. Res.* 90: 939–950.

Schuster, A., et al. (1996). The IVS6 segment of the L-type calcium channel is critical for the action of dihydropyridines and phenylalkylamines. *EMBO J.* 15: 2365–2370.

Shimizu, W., Aiba, T., Antzelevitch, C. (2005). Specific therapy based on the genotype and cellular mechanism in inherited cardiac arrhythmias: Long QT syndrome and Brugada syndrome. *Curr. Pharm. Des.* 11: 1561–1572.

Suzuki, M., Morita, T., Iwamoto, T. (2006). Diversity of Cl(–) channels. *Cell. Mol. Life Sci.* 63: 12–24.

Tan, H., et al. (1995). Electrophysiologic mechanisms of the long QT interval syndromes and torsade de pointes. *Ann. Intern. Med.* 122: 701–714.

Ufret-Vincenty, C., et al. (2001). Role of sodium channel deglycosylation in the genesis of cardiac arrhythmias in heart failure. *J. Biol. Chem.* 276: 28197–28203.

Viatchenko-Karpinski, S., et al. (2004). Abnormal calcium signaling and sudden cardiac death associated with mutation of calsequestrin. *Circ. Res.* 94: 471–477.

Viero, C., et al. (2008). A primary culture system for sustained expression of a calcium sensor in preserved adult rat ventricular myocytes. *Cell Calcium* 43: 59–71.

Wang, Q., et al. (1996). Genomic organization of the human SCN5A gene encoding the cardiac sodium channel. *Genomics* 34: 9–16.

Watanabe, H., et al. (2009). Flecainide prevents catecholaminergic polymorphic ventricular tachycardia in mice and humans. *Nat. Med.* 15: 380–383.

Wehrens, X., et al. (2003). FKBP12.6 deficiency and defective calcium release channel (ryanodine receptor) function linked to exercise-induced sudden cardiac death. *Cell* 113: 829–840.

Welch, W., et al. (2004). A model of the putative pore region of the cardiac ryanodine receptor channel. *Biophys. J.* 87: 2335–2351.

Wilde, A., et al. (2008). Left cardiac sympathetic denervation for catecholaminergic polymorphic ventricular tachycardia. *N. Engl. J. Med.* 358: 2024–2029.

Wu, X., et al. (2006). Local InsP3-dependent perinuclear Ca^{2+} signaling in cardiac myocyte excitation-transcription coupling. *J. Clin. Invest.* 116: 675–682.

Wu, X., et al. (2010). TRPC channels are necessary mediators of pathologic cardiac hypertrophy. *Proc. Natl. Acad. Sci. U.S.A.* 107: 7000–7005.

Xiao, J., et al. (2007). MicroRNA miR-133 represses HERG K^+ channel expression contributing to QT prolongation in diabetic hearts. *J. Biol. Chem.* 282: 12363–12367.

Yamada, S., et al. (2008). Embryonic stem cell therapy of heart failure in genetic cardiomyopathy. *Stem Cells* 26: 2644–2653.

Yang, B., et al. (2007). The muscle-specific microRNA miR-1 regulates cardiac arrhythmogenic potential by targeting GJA1 and KCNJ2. *Nat. Med.* 13: 486–491.

Yang, J., et al. (2000). Decreased SLIM1 expression and increased gelsolin expression in failing human hearts measured by high-density oligonucleotide arrays. *Circulation* 102: 3046–3052.

Yano, M. (2008). Ryanodine receptor as a new therapeutic target of heart failure and lethal arrhythmia. *Circ. J.* 72: 509–514.

Yazawa, M., et al. (2007). TRIC channels are essential for Ca^{2+} handling in intracellular stores. *Nature* 448: 78–82.

Yeola, S., et al. (1996). Molecular analysis of a binding site for quinidine in a human cardiac delayed rectifier K^+ channel: Role of S6 in antiarrhythmic drug binding. *Circ. Res.* 78: 1105–1114.

Zhang, Y., Youm, J., Earm, Y. (2008). Stretch-activated non-selective cation channel: A causal link between mechanical stretch and atrial natriuretic peptide secretion. *Prog. Biophys. Mol. Biol.* 98: 1–9.

Zhou, P., Wang, J. (2010). Genetic testing for channelopathies: More than ten years progress and remaining challenges. *J. Cardiovasc. Dis. Res.* 1: 47–49.

Zhou, Z., Gong, Q., January, C. (1999). Correction of defective protein trafficking of a mutant HERG potassium channel in human long QT syndrome: Pharmacological and temperature effects. *J. Biol. Chem.* 274: 31123–31126.

Ziolo, M., et al. (2005). Adenoviral gene transfer of mutant phospholamban rescues contractile dysfunction in failing rabbit myocytes with relatively preserved SERCA function. *Circ. Res.* 96: 815–817.

Zissimopoulos, S., Lai, F. A. (2007). Ryanodine receptor structure, function and pathophysiology. *New Comp. Biochem.* 41: 287–342.

Zlatkovic, J., et al. (2009). Proteomic profiling of KATP channel-deficient hypertensive heart maps risk for maladaptive cardiomyopathic outcome. *Proteomics* 9: 1314–1325.

23 Towards an Omic Perspective on Infectious Disease and Its Therapy

Integrating Immunoinformatics, Immunomics, and Vaccinomics

Matthew N. Davies
Institute of Psychiatry
London, United Kingdom

Darren R. Flower
Aston University
Birmingham, United Kingdom

CONTENTS

23.1 INTRODUCTION AND BACKGROUND

Genomics long ago changed the conceptual landscape that characterized bioscience; the clear and present implication being that we can as scientists understand biological function both faster and at a deeper level than ever before. Experimental science is continually grasping at the runaway stallion, devising and exploiting postgenomic strategies to exploit the data generated by the pullulating deluge of information immanent within genomics.

Postgenomic approaches aimed at elucidating function are many and include next-generation genomic sequencing, transcriptomics, proteomics, and the analysis of protein-protein interactions. This initiative relies primarily on high-throughput techniques to measure mRNA (the transcriptome), protein (the proteome), and metabolite (the metabolome) components of cells, tissues, organs, and whole organisms. The principal underlying idea is that of undertaking work in parallel, addressing questions not individually, through single, yet complex and convoluted experiments but rather *en masse* through sophisticated procedures examining not isolated biological phenomena but rather thousands, or even tens of thousands, of such phenomena. Underpinning all such attempts at understanding data on such a vast scale are informatic techniques, primarily bioinformatics; both analytical bioinformatics and predictive bioinformatics.

In the past decade or two, riding on the back of genomics' success, there has been an explosion in the elaboration of -omes, and the development of so-called omic technologies targeting at exploring such -omes, each with its own particular focus. Within this mass of harmonious and conflicting definitions, the website http://www.genomicglossaries.com/content/omes.asp lists innumerable distinct -omes and omics. For the sake of context, let us quickly list a few of the most useful, pertinent, and germane to our present discussion.

23.1.1 The Genome

Originally, *genome* meant the complete haploid set of chromosomes of an organism, but later the word took on other meanings, such as the complete set of genes of an organism or organelle. It has come to mean something that symbolizes the DNA sequence, or rather sequences, encoding an organism: the essence of self, much in the same way that DNA is used within the vernacular. Dictionaries trace the use of the word genome to at least 1926, but it may have been in use for some time prior to that.

The number of sequenced genomes is now large and ever increasing. Indeed, just trying to track such changes is futile, so quickly do new genomes appear. In just a few years, a genome sequence has gone from a true achievement able to stop the scientific world in its tracks to something mundane and workmanlike. Eventually, genomic sequencing may take its place as a workaday laboratory exercise akin to running a Western blot. Soon it may become the subject of a single postgraduate student's thesis. A little later, and an undergraduate might need to sequence ten or twenty genomes just to finish her final year project.

Alongside the genome sits the epigenome and thus epigenetics. Epigenetics is, to some extent at least, the nascent science of how genomes, through the medium of the organisms they code for, interact with their environment. It is a phenomenon that has an immense and previously unrecognized impact upon the nature of self at the molecular level. How a phenotype is contingent upon the interplay between environment, organism, and gene was an idea initially propounded by Conrad Waddington (1905–1975). Several molecular mechanisms combine to effect epigenetics (Doerfler, 2008). These include DNA methylation, so-called histone remodeling, and genomic imprinting. This occurs in mammals and higher plants, where there is a significant maternal investment in each offspring. In imprinting, an allele from one parent is silenced. It happens for only a few genes. This mechanism probably evolved through competition over the allocation of resources to descendents, yet because only a single copy of each imprinted gene is inherited, it is sensitive to any epigenetic modification induced by environmental change. Imprinted genes are typically involved in mediating metabolism and nutrient processing.

Epigenetic changes to gene expression are transmitted by so-called non-Mendelian mechanisms of inheritance. The evolutionary rationale for epigenetic inheritance suggests that it affords a rapid way of adapting to transient environmental changes without needing underlying genes to undergo Darwinian selection. Epigenetic mechanisms are themselves the product of genome evolution, creating alternative, additional mechanisms able to accelerate an organism's ability to adapt, survive, and, most importantly, to propagate itself by reproducing. Epigenetic alterations of gene expression may create novel phenotypes that, in turn, act through many generations, exerting selective pressure on some genes, stimulating long-term changes in the genome. Inheritance via the epigenetic mechanism may be viewed not as an evolutionary alternative complementing natural selection by mutation, but rather as a driving force behind durable and long-lasting genetic change.

Beyond the world of the genome and the epigenome lies the world of the transcriptome (the dynamic complement of messenger RNA), proteome (the dynamic complement of proteins), peptidome (the dynamic complement of peptides present in the cell generated by germ-line encoding and proteolysis), glycome (the dynamic complement of carbohydrate and glyocproteins), and metabolome (the dynamic complement of metabolites and other cellular small molecules). Distinct proteins have different properties and thus different functions in different contexts. Identifying and elaborating the functions of innumerable gene products using either high-throughput approaches or through traditional biochemistry is and will prove a much more difficult, even daunting affair.

23.1.2 Transcriptome

The transcriptome is the complement of messenger RNAs (mRNAs) transcribed from a genome. This is a dynamic set of molecules that are constantly changing with time in response to the conditions experienced by the cell. The development of transcriptomics has occurred through the use of microarray technology, which is capable of determining mRNA expression profiles. It was once thought that life could be understood by identifying each and every protein and then determining its function; now it has emerged that this was an oversimplified view of how cells behave. Genes have many promoters, and their expression is tightly regulated by a complex and integrated system.

23.1.3 Proteome

The dictionary defines *proteome* as meaning the entire complement of proteins that is (or can be) expressed by a cell, tissue, or organism and traces use of this word to 1995, which is the point it appears in the primary literature, but was extant long before that. Proteomics is the science of the proteome. The proteome constantly changes, ever in flux. The proteome is, conceptually, biology; proteins make nature happen. Genes are the quintessence of inheritance, yet it is only through the medium of the proteome that they are able to propagate themselves.

To summarize: the genome of an organism is the sequence of DNA that encodes it. Part of a genome will code for genes; these genes will in turn make mRNA, which will make proteins. These proteins will undertake most, but not all, duties within the cell. Part of the genome regulates the expression of DNA as mRNA and proteins, rather than coding for them directly.

However, despite the enormous temporal and pecuniary investment in the science of genomics, even simple questions go unanswered. How many genes are there in human beings? This should be a simple and straightforward question but appears to be neither (Southan, 2004). The putative, predicted, and promulgated size of the human genome has been deflating for years: decreasing from early and inaccurate figures in excess of 100,000 first to about 40,000 genes and then to figures near 20,000. A trustworthy and still-current estimate from 2006 puts human protein-coding genes in the region of 25,043 (Nordstrom et al., 2006), whereas a 2007 estimate places the value at about 20,488, with around 100 genes yet to be filed, stamped, indexed, debriefed, and numbered (Clamp et al., 2007).

The proteome is much larger, however. This fact is what makes the human genome as small as it presently seems to be. The proteome is generated and regulated by the genome both directly and indirectly, by epigenetic mechanisms, and—crucially—by the proteome acting upon itself, as well as simultaneously acting upon the genome and transcriptome. Because the proteome is larger, it comes closer to and then exceeds the original estimates for the size of genome. Coupled to that is the ability of one protein to undertake several functions (Jeffery, 1999, 2003, 2004, 2009). In addition, other proteins have different functions in different cellular compartments; this means that the functional diversity implicit within the proteome is large enough for us to hold faith with the notion that the function, action, and behavior of organisms, notably ourselves, arises solely from their physical structure, as encoded by the genome, without the need to invoke vitalism or other ideas stemming from intelligent design.

The size of the proteome versus the direct readout of open reading frames (ORFs) is partly mediated by the phenomenon of splice variants (Keren et al., 2010), as well as the existence of inteins (protein splicing elements) catalyzing their own excision from flanking sequences (called exteins), making new proteins wherein exteins are linked by peptide bond (Perler et al., 1994). Most alternative-splicing occurs within coding regions. The most commonly seen such event is exon skipping; here an exon or set of continuous exons are permuted into distinct mRNAs. Less frequent are the use of donor and acceptor sites and intron retention. Thus, alternative splicing and RNA editing have come to replace the presupposed one-to-one correspondence of gene to protein, and in its stead we see that one gene can be manifest as a plethora of alternate proteins. Alternative splicing is major factor driving proteomic diversity and provides a partial rationale for the gulf between the small size of the human genome and its implicit and increasingly explicit complexity.

Other diversification mechanisms are numerous and subsume posttranslational modifications, cleavage of precursors, and other types of proteolytic activation. Phosphorylation, for example, was discovered more than 40 years ago, and kinase and phosphatase cascades have subsequently been central to deciphering cellular mechanisms. Moreover, glycosylation, which labels proteins with both linear and branched carbohydrate chains, mediates a dauntingly complex modification system that regulates the ultimate function and cellular location of proteins within or beyond the cell, including proteolytic half-life of circulating protein. Ubiquitination targets proteins for degradation. Reversible limited ubiquitination can modify the activity of certain proteins. Ubiquitination can also act in an apoptotic manner: the initial protein activation through adding the first ubiquitins is followed by a shut down process whereby the proteasome recognizes this as a target for elimination. Other examples of relevant posttranslational modifications include *O*-GlcNAcylation and poly(ADP-ribosyl)ation. Many of these mechanisms seem to modulate and modify each other, compounding their ability to increase the number of potential proteins. The estimated size of the proteome is perhaps 100 to 1,000 times the number of genes. The proteome varies according to cell type and the cell's functional state.

The amount of information residing within the postgenomic landscape is both a daunting challenge and a wonderful opportunity. This idea is typified most completely by the current status of immunology. A deep understanding of the immune system is vital if we are to understand infectious disease and how to combat it. Disease remains a significant cause of human mortality globally. It is likewise the most significant source of preventable death: in contrast to other causes, disease can be attacked systematically using vaccines and drugs and, in addition to the work of physicians and surgeons, by improving drinking water and sanitation. Although it may be that the greatest benefit to man has come through improved sanitation and drinking water, it is almost self-evident that vaccines have made almost as significant and enduring a contribution.

Immunological mechanisms are also vital in areas other than infectious disease. Cancer and the augmentation of the immune response to carcinomas and cancer antigens are pivotal areas for future exploration. Likewise, the inappropriate response of the immune system to self-proteins, as in allergy and autoimmune diseases, is another area where immunomodulators mediating immunotherapy can be effective. Vaccine discovery and development are important parts of global and local publically funded efforts in health-improvement programs.

Postgenomic science offers many potent tools to the study of immunology: not least immunoinformatics and immunomics, the theoretical and experimental components of the postgenomic assault on immunology. Immunoinformatics is the application of advanced informatics techniques to molecules of the immune systems, much as bioinformatics is the application of advanced informatics techniques to biological molecules generally. Immunomics is a loosely defined discipline residing at the interface between the host immune proteome and a potentially pathological proteome derived from pathogenic microorganisms or from the self. It is a modern high throughput approach, and immunomics habitually involves the identification of antigens, epitopes, and adjuvants (Figure 23.1).

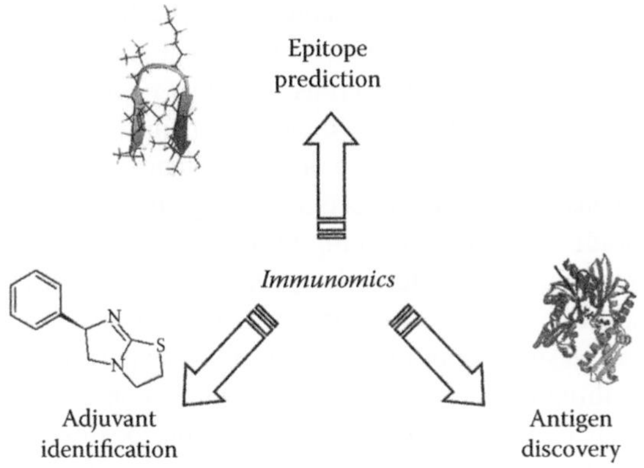

FIGURE 23.1 The scope of immunomics.

The immunome depends upon the host as well as the pathogen. A peptide is not antigenic if the immune system does not recognize or respond to it. A pertinent example of this is the restriction of T-cell responses by the major histocompatibility complex (MHC). The peptide specificity of an MHC allele may overlap with other expressed alleles, yet the totality of the specificity of all alleles will not completely envelop the entire sequence space of all potential peptides. Thus, peptides that bind to any allelic MHC variants may be antigenic. The ability to define the specificity of different MHCs computationally, which we may call *in silico* immunomics or *in silico* immunological proteomics for want of a more succinct term, is an important, yet eminently realizable, goal of immuninformatics.

At the heart of attempts to understand the immunome are attempts to understand the notion of the immunological self and thus define the terms self and nonself and give them more than a notional reality. Before beginning such an exploration, we should perhaps ask one important question: Why is this important? Each year, two million die from infections that vaccines could and should prevent; 27 million children annually fail to receive basic vaccines. Understanding self, and particularly the discrimination between self and nonself, lies at the heart of our attempts to understand the way vaccines work and how we may be able to deceive the immune system into misinterpreting the signals it receives. Thus, we turn to this before examining the contributions made by these other postgenomic techniques.

23.2 THE IMMUNOLOGICAL SELF

The term *self* can have several meanings; some are harmonious, and some are in conflict. Properly differentiating between the self, as an unrealized, possibly unrealizable, concept able to encompass the physical and psychological manifestations of identity and the perhaps more mundane recognition of internal self (*me*) versus external nonself (*you*) is a formidable problem indeed. Fascinating parallels exist between the immune self and the many definitions arising from the humanities, psychology, and molecular science. Although we should not let these parallels distract us, we should nonetheless acknowledge them.

Self—the word and concept embodied by the word—has many meanings, and each of those meanings can host many related exegeses. Definition of a self pertinent to immunology relates to the ability of the immune system to identify molecules, cells, and organs as belonging to the host and to differentiate itself from nonself: molecules, cells, and organs with an exogenous and potentially pathogenic origin. However, rather than say definition in the singular we should rather say definitions in the plural, for within immunology there are many definitions of self.

Arguably, the most famous current definition was provided by Polly Matzinger's danger model (Matzinger, 2001). This proposes that the immune system reacts to danger signals, be they of external origin or from injured cells. Thus, the danger model effaces the immune self, replacing it with the idea of danger signals: any molecular signal of whatever origin that is itself dangerous or can act as a flag for the presence of other dangerous substances could act in this way. Such a model is simple and seemingly compelling. There are counterarguments, such as the role of inflammation, which is seen both as a cause and as an outcome of the mechanisms labeled as danger theory.

Clearly, both self and nonself can encode recognition signals. The self is thus encoded as being part of the host, and nonself is encoded as being part of some identified nonhost, for example a particular bacterium or set of bacteria. Alternatively, theories can be formulated wherein either but not both the self or the nonself can be seen as an empty placeholder. For example, a self could be identified as something possessing one or more signals of being part of the host, and nonself would be identified as being anything else. An entity is thus seen as nonself if it lacks a self signal. The reverse could hold. Nonself could be identified as something possessing one or more signals of being part of a specific nonhost organism, and self would be identified as being anything else. An entity is thus seen as self if it lacks a nonself signal.

A so-called self-positive model is presented by Burnet's clonal selection theory (Ada, 2008). This posits that T-cells that react with host epitopes are eliminated during development, allowing only T-cells without reactivity to host epitopes to persist, and are thus able to engage the pathogenic molecular products. Thus, the alien is removed by those T-cells that have failed to be eliminated. Thus, self is an empty set, and only cells reactive to nonself are retained. As with all theories, there are problems with Burnet's ideas. In a sense, the self manifests itself as an implicit background against which alien substances—effectively nonself—can be identified: nonself stands out as distinct from this background and then thus be seen as different and needing to be eliminated.

Self has within immunology no context-independent meaning. Self exists only as the complement of the nonself. Thus, one reading of the clonal selection theory holds it to be purely differential and negative. However, difference and similarity are poorly formulated concepts in philosophy, and this undermines many a theory; for by *similarity*, philosophers will often mean a lack of identity. A more subtle and nuanced version of Burnet's theory will present it as deleted self (a bounded set of molecular patterns) verses nonself (an open but incomplete repertoire of other—and distinct—molecular patterns).

The largely discredited network theory of the immune system by Jerne (Jerne, 1971, 1980) introduced, or at least systematized, many now familiar terms: epitope, paratope, allotype, and idiotype. Jerne's contention was that the immune system was composed of a network or networks that were themselves composed of a large and complicated networks of paratopes recognizing groups of idiotopes and networks of idiotopes recognized by groups of paratopes.

Irun Cohen argues that immune system can recognize the self as well as the nonself (Cohen, 1993, 1996, 2007). There is no pivotal difference between self and nonself, and these qualities are contingent upon the interpretation made of them by the immune system. This interpretation is guided by genetic and somatic factors. Evolution has generated inflammatory mechanisms for coping with infection. This is innate immunity. Janeway clarifies and exemplifies this when he argues that immunity discriminates between noninfectious self and infectious nonself (Medzhitov and Janeway, 1996, 1997, 1998, 2000a, 2000b, 2002). The somatic component is provided by the life history of the host organism. Mammals evolve more slowly than microorganisms, and so pathogens would always win an immunological arms race. Cohen's arguments suggest that immunity is a highly orchestrated and contextual system that transcends the simple dichotomy between self and nonself.

Most theories, and the abstractions from which they grow, are built around certain key ideas that do indeed illuminate the issues at hand, yet it is only by addressing the molecular mechanisms at the heart of immunology that we make this discourse of practical relevance and utility. Whether certain entities are viewed as self or as nonself emerges in context. The same agent is ignored in one context, yet may be attacked in another. The world is interested in vaccines and drugs that save lives, reduce morbidity, and facilitate economic stability and growth not in esoteric and recondite

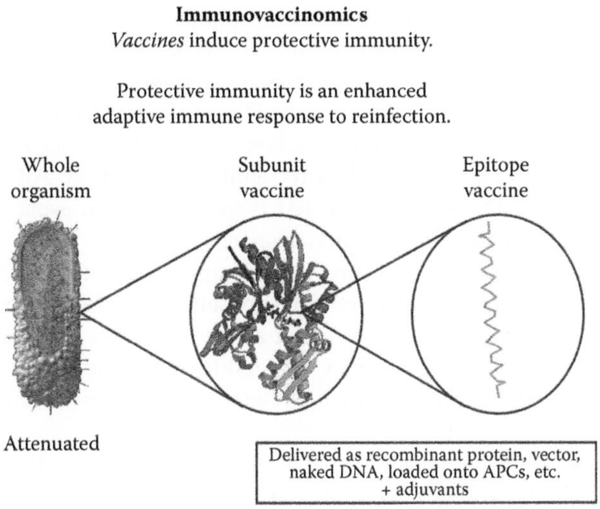

FIGURE 23.2 Types of vaccine.

discussions such as these. Vaccines come in many guises, but principal among these are the attenuated or killed whole pathogen vaccines, vaccines based on whole protein antigens, and epitope-based vaccines (Figure 23.2). To reach a useful and a useable definition of the immune self, we need to look at the immune system as an emergent process arising ultimately from the interactions of molecules.

The self can, in the language of the late twentieth and early twenty-first centuries, be succinctly defined as a product or, more properly, a combination of genetic, epigenetic, and environmental factors working synergistically to create and define a whole. This whole begins but does not end with the genome. We cannot simply reduce everything to genes received from our ancestors and antecedents via germ-line cells. Beyond genetic variation, there is epigenetic variation. Both genetics and epigentics are combined with effects from the environment—such as environmental chemicals and diet, including sex disruptors and toxic pollutants—to elaborate and make manifest the fully autonomous systems from which biology is composed: cells, organs, and whole organisms (Figure 23.3).

Depending on your particular perspective, self is, ultimately, either a signal or something that is recognized. Likewise, ultimately, nonself is also simply just a signal or something recognized.

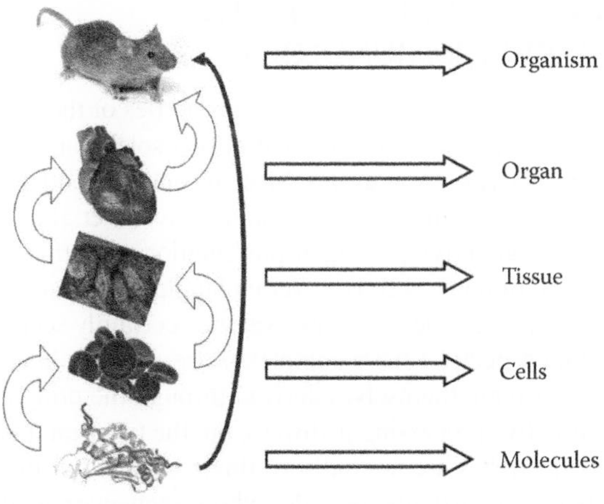

FIGURE 23.3 The length-scale hierarchy of systems immunomics.

That these signals are recognized and responded to within the context of a complex and sometimes confusing system, exhibiting confounding behavior on several levels, is irrelevant.

In sense then, although how real this sense may be is open to question, the self is both molecules—peptides and proteins—and signals. The self and the nonself are merely, yet not solely, molecules that are recognized—or more properly bound—by other molecules. That is all that self and nonself ultimately are: molecules and their recognition by the host. All the rest is just waffle, confusion, and obfuscation. The signal in the self is the recognition event—MHC- or antibody-mediated—which triggers the immune system to respond.

It was once felt that the MHC would provide a simple, straightforward, unambiguous, and unequivocal criterion able to discriminate self from nonself. Class I MHC molecules are expressed by almost every nucleated cell and are able to act as the mediators of signals acting to identify self. This is realized in the ternary complex of peptide-MHC and T-cell receptor (TCR) that is the necessary preliminary to the activation of the T-cell and thus the initiation of concomitant immune responses.

Scientists with a philosophical bent often lambaste materialist, reductionist approaches to the issue of self and nonself. They argue that only under certain conditions is there an unambiguous correspondence between a sign or signal and the thing being signified. They argue, and argue with some force and justification, that the meaning or meanings of self are seldom exhausted by an inductively derived catalogue of instances of direct correspondence between signified and sign.

The self, and thus nonself, is dynamic. Under certain constraints and in response to certain stimuli, the immune system is capable of attacking host constituents. Autoimmunity is immunity turning against the self it is meant to defend. Autoimmunity is usually associated with disease but can also be a normative function of homeostatic maintenance and control. Thus, self is not a stable and unchanging entity but clearly depends strongly upon context and is prone to influence by its environment.

The fundamental molecular mechanisms underlying cellular and humoral immunity are quite different. T-cell immunity is mediate by the molecular recognition of peptides bound to MHC molecules, essentially short denatured fragments excised from proteins via proteolytic degradation. B-cell-mediated immunity is made manifest by antibody recognition of a protein antigen's three-dimensional structure. Thus, the molecular recognition events at the heart of cellular immunity are essentially conformation independent and are instead mediated by recognition of amino acid sides within the context of a peptide-MHC complex. Humoral immunity is, by contrast, highly dependent on the conformation of a folded protein.

23.3 THE MOLECULAR DEFINITION OF SELF: CELLULAR ADAPTIVE IMMUNOLOGY

Traditionally, when attempting to analyze and predict properties of the cellular response, immunoinformaticians have centered their attention and their efforts solely on the specificity of MHC molecules (Davies and Flower, 2007; Flower, 2003). More recently, attention has turned to the richer, deeper, more challenging world of antigen presentation (Vivona et al., 2008). Our understanding of the manifold mechanisms underlying antigen presentation, and thus the manifestation of the molecular components of the immune self, is as yet incomplete and partial. These mechanisms, as we currently picture them, are by no means simple, yet they certainly seem clever. As with all exciting science, many important aspects of these complex processes remain controversial.

MHCs bind peptides, which are themselves derived through the proteolytic degradation of proteins. There are many alternative processing pathways, but the two best understood are the classical class I and classical class II. MHCs are not indiscriminate binders but importantly exhibit a finely tuned yet complex specificity for particular peptides whose sequences are composed of the 20 commonly occurring amino acids. MHCs also display a wider specificity, which is in itself quite catholic in terms of the molecules they can bind; MHCs are not restricted solely to peptides, they also bind a variety of other molecules.

The immunological peptide repertoire or immune peptidome is compounded in various ways, perhaps most significantly through the effect of posttranslational modifications (PTMs). Such PTMs include phosphorylation, lipidation, and, most importantly, glycosylation. Glycosylated proteins can be targets for binding by cell surface receptors based on sugar-binding leptin domains. Glycosylated epitopes can also be bound by TCRs and antibodies. Lipids can act as epitopes directly through their presentation by CD1. PTMs can also be transitory, such as phosphorylation, or more permanent, such as modified amino acids. Many of these can be part of functional epitopes recognized by the immune system.

To these can be added a wide range of synthetically modified peptides, which are bound by MHCs and recognized by T-cells, as well as natural and synthetic small molecules, which can also bind MHCs. Small molecule drug-like compounds bind MHCs; this can mediate pathological effects and has important implications in behavior-modifying odor recognition.

Class I MHC molecules make available to immune surveillance markers that sample important intracellular changes such as viral infection, the presence of intracellular bacteria, or malignant T-cellular transformation as seen in tumor cells. The flagging or signaling of such profound cellular events ensures the induction of an appropriate immune response by circulating CD8+ T-cells. By contrast, class II MHC molecules reveal to circulating CD4+ T-cell-mediated immune surveillance markers that sample extracellular events.

The natural repertoire of class I MHC-presented peptides is rather broader than is widely supposed (Lin et al., 2008; Heath et al., 2004; Vyas et al., 2008). MHC class I ligands are derived primarily from degraded endogenously expressed intracellular proteins. Intracellular peptide fragments arise from two sources: self-peptides derived from the host genome and proteins from external sources such as pathogenic microbes, principally those originating from viral infection. This seeming simplicity masks several layers of complexity.

There are several distinct steps in class I antigen processing and presentation. Antigens are initially acquired from proteins with errors, which may result from misincorporation or premature termination. Misfolded proteins are then targeted for degradation by being tagged with ubiquitin. The proteasome then proteolytically digests ubiquitylated proteins in a stochastic manner into a population of relatively short peptides. Subsequently, digested peptides are translocated from the cytoplasm into the endoplasmic reticulum by the transmembrane transporter associated with antigen processing (TAP). Once in the ER, peptides are bound by newly formed class I MHC molecules. Class I MHC heavy chains and β_2-microglobulin are both synthesized in the ER. Fully loaded trimeric complexes of class I heavy chains, β_2-microglobulin, and peptide permit optimal folding and glycosylation and are transported via the Golgi complex to the cell surface.

MHC class I ligands are derived primarily from endogenous proteins and were previously thought to be 8–11 amino acids long. However, there is now much evidence that long peptides (13–15 amino acids and above) are also presented by class I MHCs in many—possibly all—vertebrates including human, mouse, cattle, and horse. The repertoire of presented class I peptides is expanded through aberrant transcription and translation of viral and self-proteins. Various mechanisms pertain such as readthrough, which leads to alternative open reading frames or alternative splicing, which generates protein isotypes with different sequences at exon-exon boundaries. Both of these have been shown to generate immunogenic epitopes. Autoantigen and self-tumor antigen transcripts for example experience elevated rates of alternative splicing.

Intracellular proteins, including newly synthesized proteins, are degraded quickly, producing large amounts of short peptides. Nonfunctional proteins, or defective ribosomal products, result from errors in translation and processing. They form a significant proportion of newly synthesized proteins, which is rapidly digested by the proteasome. Viruses can invade host cells and generate viral proteins and bacteria can inject protein into the host cell via the type III secretion system; both are also degraded by the host.

Intracellular protein degradation is mediated by a multiprotein proteolytic complex called the proteasome. A whole variety of protein such as heat-denatured proteins; incorrectly assembled, mistranslated, or misfolded proteins; and regulatory proteins with limited half-lives are targeted by the proteasome. For antigenic proteins, it favors oxidized protein substrates, because about 75% of oxidized intracellular proteins are degraded by proteasomes, and the 20 S proteasome prefers partially denatured oxidized proteins.

After peptides are degraded by the proteasome, they are transferred into the lumen of the endoplasmic reticulum (ER). The translocation process from cytosol to ER consumes ATP. The TAP is required for peptide transit. TAP also has the ability to interact with peptide-free class I HLA molecules in the ER. After peptides associate with class I HLA molecules, the resulting complexes are released from TAP and are then delivered to the cell surface.

Immunology dogma suggests that newly synthesized class I MHC molecules are not stable in a peptide-free state and are retained in the ER in a partially folded form. Several chaperone proteins are needed to complete MHC folding. Formation of the MHC-peptide complex is quite intricate and complicated, and it is facilitated by a variety of proteins including tapasin, calreticulin, and ERp57. Once complexed to peptide and β_2-microglobulin, the MHC protein leaves the ER and is transported to the cell surface. The peptide-binding process is considered as the rate-limiting step of MHC protein assembly, because only a fraction of the peptides are able to bind to MHC.

However, there are a number of other processing routes that complicate the simple picture outlined above. Peptides cleaved by the proteasome are 3–25 amino acids long, whereas most class I MHC epitopes are less than 15 amino acids long. Only about 15% of those peptides that are degraded by the proteasome are of the appropriate length for class I MHC binding. 70% of peptides are too short, and 15% are too long. Long peptides may be trimmed to the correct size by various cellular peptidases.

Analysis of peptide extracted from proteasome-inhibited cells suggests that nonproteasome cytoplasmic proteases are involved in antigen processing pathway. Peptides are digested in the cytosol by several peptidases such as leucine aminopeptidase, tripeptidyl peptidase II (TPPII), thimet oligopeptidease, bleomycin hydrolase, and puromycin-sensitive aminopeptidease.

TPPII was suggested to be a peptide supplier because of its ability to cleave peptides *in vitro* and its upregulation in cells surviving partial proteasome inhibition. Recently, an enzyme located in the lumen in ER and named ERAAP (ER aminopeptidase associated with antigen processing) or ERAP1 was proven to be responsible for the final trim of the N termini of peptides presented by MHC class I molecules. However, currently there is insufficient quantitative data about the role of these proteases to allow a precise bioinformatic evaluation of their impact on the antigen processing pathway. Alternatives to TAP are also now emerging, such as Sec61, which also effects retrograde transport back into the cytoplasm from the ER.

Class II MHC expression is believed to be restricted primarily to professional antigen-presenting cells (APCs), including macrophages and dendritic cells (DCs). In the MHC class II processing pathway, following the receptor-mediated endocytosis of exogenous antigens by APCs, presented proteins are targeted to the multicompartment lysosomal-endosomal apparatus, passing first into endosomes and then into late endosomes, ending up in lysosomes. While in transit, antigens are proteolytically fragmented into peptides by cathepsins. Before final cell-surface presentation, peptides are bound by class II MHCs. MHC class II ligands have a more variable length of 9–25 amino acids and are derived mainly from exogenous proteins. Peptide-bound class II MHC molecules are ultimately translocated to the cell surface where they are available for immune surveillance by $CD4^+$ T-cells.

External, extracellular antigen is endocytosed or phagocytosed by APCs and is directed into the phagosome. These membrane-bound organelles maturate, undergoing sequential modification and ultimately fusing with lysosomes to create phagolysosomes, where ingested and degraded antigen encounter class II MHC molecules. MHC class II molecules that are contained in the lysosomes are loaded with peptide fragments of the bacteria or viruses that are formed by lysosomal proteases.

Both MHC class II molecules and the tetraspanin member CD63 are specifically recruited to pathogen-containing phagosomes. Subsequently, MHC class II molecules, as well many other lysosomal proteins including tetraspanins, are transported in endolysosomal tubules to the cell surface. Surface MHC class II molecules can be found in membrane microdomains with other costimulatory proteins.

Peptide display is also dependent on the processing of proteins into peptides within the endosomal/lysosomal compartments. Here proteins, including those derived from pathogens, are degraded by cathepsins, a particular type of protease. Class II MHCs then bind these peptides and are subsequently transported to the cell surface. The peptide specificity evinced by cathepsins has also been studied, resulting in the publication of simple cleavage motifs. However, more work is needed before accurate predictive methods are realized. Much still remains to be discovered in terms of the mechanism of class II presentation.

Whole pathogenic microbes exist transiently in the interstitial space. Such pathogens are phagocytosed by DCs. In DCs, antigen presentation is tightly controlled by Toll-like-receptor (TLR) signaling pathways. Cell-surface TLRs are activated by phagocytosis of bacteria. DC activation by TLR induces endolysosomal tubule formation and phagosome maturation. These subcellular structures contain a plethora of proteins, including class II MHC molecules, and effect protein delivery to the cell surface. Phagocytosis is restricted to professional APCs. It takes up microorganisms and apoptotic bodies and much else besides and shuttles them into phagosomes.

Another complicating feature germane to this discussion is the issue of cross-presentation. Class I and class II pathways are not distinct but are actually interconnected or, at least, leaky, allowing peptides from one pathway access to the other, whereas in endosomal processing, endoproteolysis is followed by N- and C-terminal trimming. Many peptides escape from the endosome and pass into the cytoplasm where they enter the familiar class I pathway comprising the proteasome, TAP, and MHC. Likewise, endogenously expressed cytosolic and nuclear antigens access MHC class II via a number of intracellular autophagic pathways, including macroautophagy, microautophagy, and chaperone-mediated autophagy (Gannage and Munz, 2009; Crotzer and Blum, 2009). Macroautophagy is characterized by the formation of autophagosomes, double-membrane structures, which capture proteins from the cytoplasm, processing them via acidic proteases, and eventually fuse with lysosomes. Microautophagy involves lysosomal invagination, which directly sequesters cytoplasmic proteins.

Taken together, all of these mechanisms greatly increase the potential size and diversity of the immunogenic repertoire—or immunome—of reactive peptides. Thus, one may argue, and argue cogently, that, in the face of such complexity, the only realistic way to address this potential enormity of the peptide repertoire is via computational analysis and prediction.

23.4 THE MOLECULAR DEFINITION OF SELF: HUMORAL ADAPTIVE IMMUNITY

Self is also mediated by recognition of so-called B-cell epitopes: regions of a protein surface bound specifically by soluble or membrane-bound antibody molecules. The immunoprotection offered by almost all vaccines is mediated completely or predominantly through the induction of antibodies, which act mostly in infection at the bacteremic or viremic stage. Humoral immunogenicity, as mediated by soluble or membrane-bound cell surface antibodies through B-cells generate antibodies when stimulated by helper T-cells as part of the adaptive immune response. The antibodies act to bind and neutralize pathogenic material from a virus or bacterium. Individual antibodies are composed of two sets of heavy and light chains (Figure 23.1). Each B-cell produces a unique antibody because of the effects of somatic hypermutation and gene segment rearrangement. Cells of the primary repertoire bearing antibodies able to recognize antigen expand clonally: an iterative process of directed hypermutation and antigen-mediated selection. This facilitates the rapid maturation of antigen-specific antibodies with a high affinity for a specific epitope. An appropriate B-cell that deals with a specific infection is selected and cloned to deal with the primary infection, and a population of the B-cell is then maintained in the body to combat secondary infection. It is the capacity to produce a huge variety of different antibodies that allows the immune system to deal with a broad range of infections.

B-cell epitopes can be linear (also called continuous) or discontinuous. Linear epitopes are single, short, continuous subsequences within an antigen. Discontinuous epitopes are groups of individual, isolated residues forming patches on the surface of the antigen. The verity and exegesis of an epitope depends on the nature of their experimental determination. Linear epitopes are identified using some kind of experimental screening procedure, usually PEPSCAN, whereby overlapping sequences are assayed against preexisting ex vivo antibodies. Discontinuous epitopes are usually identified from the structure of an antigen, typically one derived experimentally by x-ray crystallography or multidimensional nuclear magnetic resonance. Discontinuous epitopes are also identified by making site-directed mutants of the antigen and testing them for their effect on antibody binding.

Sequence-based B-cell epitope prediction methods are limited to the identification of linear epitopes. If we look back a decade or two, then most predictors of either T-cells or B-cell epitopes were based on identifying maximally valued regions of sequences—essentially looking for peaks, or troughs, in some form of a propensity plot. This was long ago shown to be inappropriate for T-cell epitopes, and consequently many advanced methods for T-cell epitopes prediction have arisen. However, many—or perhaps that should be most, if not actually all—B-cell epitope prediction methods continue to rely, wholly or in part, on finding such peaks. However, no single property is known that is able to predict linear or discontinuous epitope location with any reliability or accuracy. Most prediction methods use properties related to surface exposure, such as accessibility, hydrophilicity, flexibility/mobility, and loop and turn structures, because it is believed that epitopes, at least for nondenatured proteins, must be solvent-accessible if antibody binding is to occur.

Early methods adopted a sliding-window method, adapting well-used hydropathy scaled to identify maximal property peaks. A correctly predicted epitope will correspond to a peak within a few residues of an antigenic residue. Short window sizes were required to reduce erratic peak values; the optimal being six residues. Longer window sizes performed less well, perhaps because of an increasing probability of including hydrophobic residues. Using stringent data from peer-reviewed papers only, Blythe and Flower (2005) have explored the veracity and accuracy of sequence-based B-cell epitope prediction. Using 484 amino acid scales and 50 epitope-mapped protein sequences, as defined using polyclonal antibodies, the analysis indicated that the underlining method had little validity. Framed in terms of receptor operator curve plots, the best method produced predictions little different from random.

The poor performance demonstrated by B-cell epitope prediction algorithms is troubling. No explanation seems overly convincing. It is unlikely that the available methodology is to blame, because data-mining techniques have proved much more successful in other areas. The explanation favored here again targets the experimental data as the source of the problem. The most widely available data derives from PEPSCAN, and there are reasons to suspect that this is not what it seems or what people believe it to be. Experimentally derived epitopes are identified by assayed against preexisting antibodies with an affinity for whole antigens.

However, when such epitopes are mapped back onto antigen structures, their locations are scattered randomly through the protein. They do not form discrete patches as one would expect if they are simple mimics of crystallographically identified discontinuous epitopes. These *in situ* epitopes can be exposed or completely buried and thus inaccessible to antibody binding and also in every state in between. If we compare the conformation adopted by antibody-bound peptides with those *in situ* in the intact antigen, we see that they are typically very different. However, if we compare antibodies in intact antigen and in whole antigen-antibody complexes, they are very similar. Thus, the recognition of epitopes in a PEPSCAN analysis requires explanations other than the simplest one of a one-to-one correspondence. One explanation could be that the preformed antibody recognizes denatured antigen *in vivo*. Another explanation is that the isolated peptide adopts a conformation that is able to mimic the surface features of a discontinuous epitope

23.5 IMMUNOMICS

Immunomics is an evolving word, with several extant definitions, but a key meaning, and the meaning we chose here, is this: immunomics is the combined application of methods at the level of proteomics, transcriptomics, and bioinformatics to several key problems arising in immunology and vaccinology, principally identifying the components of the immunogenic set of antigens existing at the interface of host and pathogen. In this regard, if in no other, immunomics should help us to address vaccine discovery on a genomic scale. Genome sizes for pathogens range between a few hundred for viruses, to a few thousand for bacteria, and up to a few tens of thousands for parasitic microorganisms. Interestingly, amongst the largest number of genes currently known for any organism, is that of the protozoan *Trichomonas vaginalis,* which is responsible for trichomoniasis (Carlton et al., 2007); this very large, if repetitive, genome is at least comparable in size to the human genome and is possibly over three times as large. Leveraging these technologies should allow this rare superfluity to be reduced to a manageably short list of candidates, perhaps no more than ten or twenty. Such candidates would then require channeling through a set of subsequent processes including recombinant expression, purification, and testing for immunogenicity and protective efficacy (Vivona et al., 2008). Thus, immunomics and vaccinomics overlap significantly with reverse vaccinology.

Reverse vaccinology is a principal means of identifying subunit vaccines and involves a considerable computational contribution (Tettelin, 2009; Vivona et al., 2008). Conventional experimental approaches cultivate pathogens under laboratory conditions, dissecting them into their components, with proteins displaying protective immunity identified as antigens. However, it is not always possible to cultivate a particular pathogen in the lab, nor are all proteins expressed during infection easily expressed *in vitro*, meaning that candidate vaccines can be missed. Reverse vaccinology, by contrast, analyzes pathogen genomes to identify potential antigens and is typically more effective for prokaryotic than eukaryotic organisms.

Genome analysis using gene identification methods has identified numerous ORFs. Often the function of such ORFs remains cryptic and unknowable. Once all of the ORFs have been identified, proteins with the characteristics of secreted or surface molecules must be identified. Unlike the relatively straightforward task of identifying ORFs, selecting proteins prone to immune system surveillance is challenging.

Immunomics combines proteomic techniques (multidimensional polyacrylamide gel electrophoretic techniques and sensitive biological mass spectrometry), transcriptomics (microarrays and next-generation sequencing), with support from immunoinformatics (burgeoning protein and DNA databases), to effect the high-throughput identification of protein populations from different sources.

Clearly immunomics is more than a few, albeit sophisticated, techniques, such as the yeast two-hybrid system (Serebriiskii et al., 2001; Toby and Golemis, 2001) and so-called protein arrays. The cutting edge of immunomics offers to immunology and immunologists much that is of value. Because the identification of putative whole protein subunit vaccines is a pivotal objective of immunovaccinology, immunomics should help deliver vaccines capable of eliciting significant responses from both humoral and cellular immunity, while circumventing the toxicity problems of vaccines created from whole microbes (Vivona et al., 2008).

In what follows, we will briefly survey examples of viral, bacterial, fungal, and parasite immunomics. Let us start with host-virus interactions. Toda et al. (2000) profiled proliferative B-lymphoblastoid cell lines infected with Epstein-Barr virus using proteomic techniques, identifying 16-kDa protein phosphoprotein stathmin. Diaz and co-workers (Diaz et al., 2002; Greco et al., 2001) examined ribosomal modifications induced by herpes simplex virus type 1 and showed that virus infection induces unusual phosphorylation of proteins of the small ribosomal subunit.

Rodriguez et al. (2001) used electrophoresis to examine protein expression patterns in Vero cells infected with African swine fever virus (ASFV) attenuated strain BA71V and porcine alveolar macrophages cells treated with the ASFV virulent strain E70. Comparison of infected and noninfected proteomes indicates that ASFV inhibits protein synthesis for 65% of cellular proteins, whereas expression of a few proteins increases two-fold.

Bacteria have special features, which are generally lacking in other organisms, that facilitate immunomic analysis. This results from the abundance of extant information on bacterial genomes, coupled to their relatively simple gene regulation, and experimental tractability. Within the context of bacterial immunovaccinology, one of the main objectives is to determine the differences between two bacterial species or the same species in different environments or growth phases.

Immunomic approaches have been applied successfully to bacteria, including the determination of the proteomes for several bacterial species: *Salmonella typhimurium* (O'Connor et al., 1997), *Bacillus subtilis* (Hecker and Engelmann, 2000), and *Mycoplasma pneumoniae* (Ueberle et al., 2002). In another study, *Chlamydia pneumonia* was cultured in Hep-2 cells, and approximately 167 proteins from it were separated by electrophoresis (Vandahl et al., 2001, 2002, 2004). The proteins identified included 31 hypothetical proteins including several involved in the type III secretion apparatus.

Directly addressing the identification of vaccine targets, Chakravarti et al. (2000) analyzed the *Helicobacter pylori* genome. Proteins were identified from outer membrane preparations using proteomic technologies. An outer membrane fraction, purified from disrupted cells, was treated with Triton X-100, centrifuged, treated with detergent, and separated eletrophoretically. With those proteins, which reacted against monoclonal antibodies and were identified by mass fingerprinting, Haas et al. (2002) compared the reactivity of sera from *H. pylori*-infected patients and several control groups to identify proteins from *H. pylori* strain HP 26695, isolating 310 proteins.

Parasitic infections are also responsible for causing serious and endemic disease, particularly in the developing world. Vaccines against malaria and other parasites have not been successful, yet vaccines that control livestock parasites, such as nematode and trematode infections, have been more useful (Dalton et al., 2003; Dalton and Mulcahy, 2001).

Jeffries et al. (2001) analyzed *Fasciola hepatica* using proteomics and found several proteins including cathepsin L proteases and other enzymes involved in protecting the parasite from the host immune response: thioredoxin peroxidase, superoxide dismutase, and glutathione *S*-transferases, yet vaccines against true protozoans are proving more recalitrant, particularly so for malaria. This disease is caused by *Plasmodium falciparum*, and it kills over two million people annually. Other parasitic diseases include leishmaniasis and schistosomiasis. Leishmaniasis affects around 2 million annually, with approximately 350 million at risk globally. Vaccines for malaria and leishmaniasis have been taken to clinical trials, whereas vaccines for schistosomiasis are in late clinical trials. The control of leishmaniasis remains a problem, and no viable vaccines exist for the disease.

23.6 IMMUNOINFORMATICS

For immunoinformatics, the focus remains the epitope, the immunological quantum that lies at the figurative heart of immunology, both as a science of description and as a science of action. Immunology studies immune phenomena and their underlying molecular mechanisms in order to reach a full and proper understanding of these processes in both practical and philosophical terms. Because immune systems are not uncomplicated things, much laborious effort is still expended in enumerating and elaborating the copious detail implicit in such phenomena, yet immunology, in the guise of vaccinology, is also a science of action, and much effort continues to be expended in trying to understand and manipulate nonself. Nonself, when manipulated properly, can yield effective and practically useful vaccines. The word vaccine can refer to all molecular or supramolecular agents able to stimulate specific, protective immunity against pathogenic microbes and the disease they cause. Vaccines act to mitigate the effects of subsequent infection as well as blocking the capability of pathogens to injure and kill their host organisms.

Complex microbial pathogens, such as *Mycobacterium tuberculosis*, can interact within the immune system in a multitude of ways (Moise et al., 2007; McMurry et al., 2005). For a vaccine to be effective, it must invoke a strong response from both T-cells and B-cells; therefore epitope mapping is a central issue in their design. *In silico* prediction methods can accelerate epitope discovery greatly. B-cell and T-cell epitope mapping has led to the predictive scanning of pathogen genomes for potential epitopes. There are over 4,000 proteins in the tuberculosis genome; this means that experimental analysis of host-pathogen interactions would be prohibitive in terms of time, labor, and expense (Figure 23.4).

T-cell epitopes are antigenic peptide fragments derived from a pathogen that, when bound to a MHC molecule, interact with T-cell receptors after transport to the surface of an antigen-presenting cell. If sufficient quantities of the epitope are presented, the T-cell may trigger an adaptive immune response specific for the pathogen. MHC class I and class II molecules form complexes with different types of peptide. The class I molecule binds a peptides of 8–15 amino acids in length within a single closed groove. The peptide is secured largely through interactions with anchoring residues at the N and C termini of the peptide, whereas the central region is more flexible. Class II peptides vary in length from 12 to 25 amino acids and are bound by the protrusion of peptide side chains into cavities within the groove and through a series of hydrogen bonds formed between the main chain peptide atoms and the side chains atoms of the MHC molecule. Unlike the Class I molecule, where the binding site is closed at either end, the peptide is able to extend from either of the two open ends of the binding groove.

Using peptide sequence data, experimentally determined IC_{50} and BL_{50} affinity data has been used to devise many MHC-peptide binding prediction algorithms, able to separate binders from nonbinders. Such methods include motif-based systems, support vector machines (SVMs) (Liu et al., 2006; Wan et al., 2006), hidden Markov models (HMMs) (Zhang et al., 2006; Noguchi et al., 2002; Mamitsuka, 1998), quantitative structure-activity relationship analysis (Doytchinova and Flower, 2003; Doytychinova et al., 2005), and structure-based approaches (Wan et al., 2004, 2005a, 2005b,

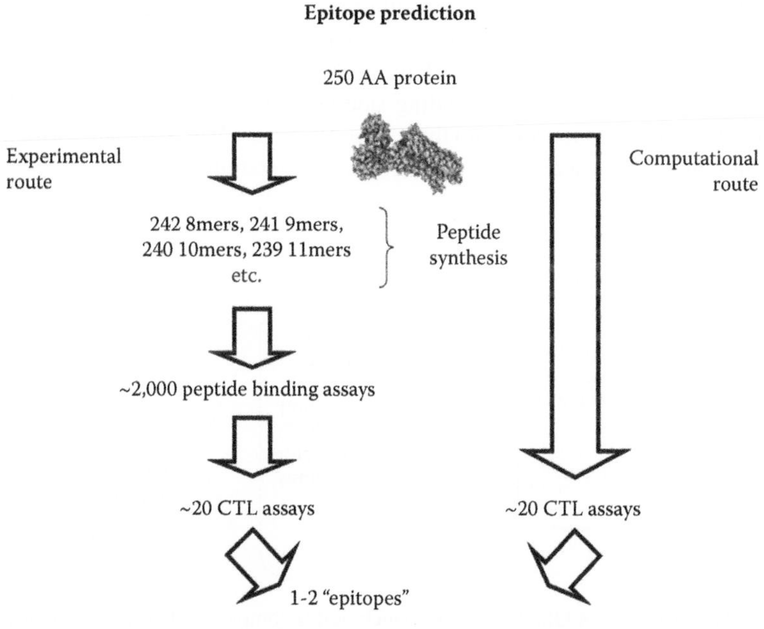

FIGURE 23.4 Experimental versus *in silico* epitope mapping.

2008). MHC-binding motifs are an easily understood epitope identification method, although such motifs invariably generate numerous false positives and numerous false negatives.

SVMs are machine learning algorithms based on statistical theory that seeks to separate data into two distinct classes (in this case binders and nonbinders). HMMs are statistical models where the system being modeled is assumed to be a Markov process with unknown parameters. In a HMM, the internal state is not directly observed, but those variables influencing its state are and can determine those sequences with binder-like characteristics.

QSAR-based techniques have also been utilized in the identification of peptide interactions with class I MHC, incrementally optimizing and improving the individual residue-to-residue interactions within the binding groove. This has led to the design of so-called superbinders that minimize the entropic disruption in the groove and are therefore able to stabilize even disfavored residues within so-called anchor positions. QSAR techniques that account for complex biological properties, such as peptide binding to MHCs, are typically based on combining physicochemical properties, either themselves predicted or directly measured (Figure 23.5).

Finally, molecular dynamics has been used to quantify the energetic interactions between the MHC molecule and peptide for both class I and class II by analysis of the three-dimensional structure of the MHC-peptide complex. Several programs are available that can help design and optimize vaccines, and many more programs are available for T-cell epitope prediction.

The word antigen has a wide meaning in immunology. We use it here to mean a protein, specifically one from a pathogenic microorganism, that evokes a measurable immune response. The automated identification of potentially immunogenic antigens from genome sequences is another facet of immunoinformatic support for immunomic discovery of vaccines (Figure 23.6).

Proteins antigens in bacteria are often acquired in groups, through a process summarized in the phrase *horizontal transfer*. Such groups are known as pathogenicity islands. The unusual G+C content of genes and particularly large gene clusters is tantamount to a signature characteristic of genes acquired via the recondite mechanisms typical of so-called horizontal transfer. Computational genome analyses at the nucleic acid level can thus allow the discovery of such regions, helping to identify potentially large sections of sequence as pathogenicity islands and thus implicating certain protein ORFs as encoded virulence genes.

Perhaps the most obvious antigens are virulence factors (VF): proteins that enable a pathogen to colonize a host or induce disease. Analysis of pathogens—such as *Vibrio cholerae* or *Streptococcus pyogenes*—has identified coordinated systems of toxins and virulence factors that may comprise over 40 distinct proteins. Traditionally, VFs have been classified as adherence/colonization factors, invasions, exotoxins, transporters, iron-binding siderophores, and miscellaneous cell surface factors. A broader definition, groups VFs into three: true VF genes; VFs associated with the expression of true VF genes; and VF lifestyle genes required for colonization of the host.

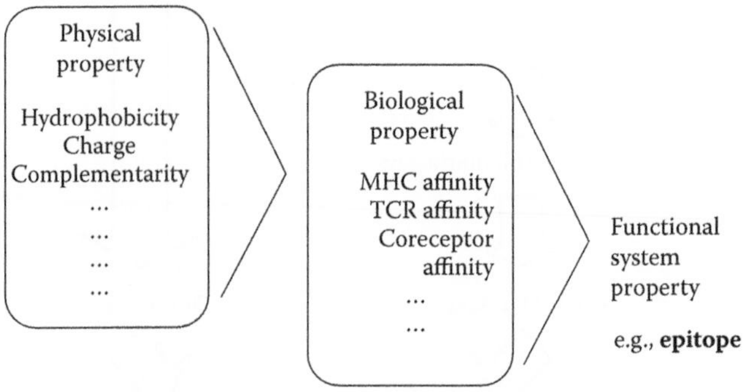

FIGURE 23.5 The hierarchy of QSAR analysis, which builds complex and cryptic biological property prediction on the firmer basis of physicochemical measurement and prediction.

FIGURE 23.6 Identifying antigens from genome sequences.

Several databases exist that archive VFs. The Virulence Factors Database puts an emphasis on functional and structural biology and can be searched using text, sequence-searching, or functional queries (Yang et al., 2008; Chen et al., 2005). TVFac (Los Alamos National Laboratory Toxin & Virulence Factor Database) contains information on over 250 organisms and separate records for virulence factors. *Candida albicans* Virulence Factor (CandiVF) contains VFs that may be searched using BLAST (Tongchusak et al., 2008). PHI-BASE seeks to integrate VFs from a variety of pathogens (Winnenburg et al., 2006, 2008; Baldwin et al., 2006).

Obviously, antigens need not be virulence factors, and another nascent database is intending to capture a wider tranche of data. We are helping to develop the AntigenDB database (Ansari et al., 2010) (http://www.imtech.res.in/raghava/antigendb), which will aid considerably this endeavor. Historically, antigens have been supposed to be secreted or exposed membrane proteins accessible to surveillance of the immune system. Subcellular location prediction is thus a key approach to predicting antigens. There are two basic kinds of prediction method: manual construction of rules of what determines subcellular location and the application of data-driven machine learning methods, which determine factors that discriminate between proteins from different known locations. Accuracy differs markedly between different methods and different compartments, mostly because of a paucity of data. The data used to discriminate between compartments include the amino acid composition of the whole protein; sequence-derived features of the protein, such as hydrophobic regions; the presence of certain specific motifs; or a combination thereof.

Different organisms evince different locations. PSORT is a knowledge-based, multicategory prediction method, composed of several programs, for subcellular location; it is often regarded as a gold standard (Gardy et al., 2003; Nakai and Horton, 1999). PSORT I predicts 17 different subcellular compartments and was trained on 295 different proteins, whereas PSORT II predicts 10 locations and was trained on 1,080 yeast proteins. Using a test set of 940 plant proteins and 2,738 nonplant proteins, the accuracy of PSORT I and II was 69.8 and 83.2%, respectively. There are several specialized version of PSORT. iPSORT deals specifically with secreted, mitochondrial, and chloroplast locations; its accuracy is 83.4% for plants and 88.5% for nonplants. PSORT-B only predicts bacterial subcellular locations. It reports precision values of 96.5% and recall values of 74.8%. PSORT-B is a multicategory method that combines six algorithms using a Bayesian network.

Among binary approaches, arguably the best method is SignalP, which employs neural networks and predicts N-terminal Spase-I-cleaved secretion signal sequences and their cleavage site (Emanuelsson et al., 2007; Bendtsen et al., 2004). The signal predicted is the type II signal peptide common to both eukaryotic and prokaryotic organisms, for which there is wealth of data, in terms of both quality and quantity. A recent enhancement of SignalP is a hidden Markov model version able to discriminate uncleaved signal anchors from cleaved signal peptides.

One of the limitations of SignalP is overprediction, because it is unable to discriminate between several very similar signal sequences, regularly predicting membrane proteins and lipopteins as type II signals. Many other kinds of signal sequence exist. For example, a number of methods have been developed to predict lipoproteins. The prediction of proteins that are translocated via the TAT-dependent pathway is also important but is not addressed yet in any depth.

We have developed VaxiJen (Doytchinova and Flower, 2007a, 2007b) (http://www.jenner.ac.uk/VaxiJen), which implements a statistical model able to discriminate between candidate vaccines and nonantigens, using an alignment-free representation of the protein sequence. Rather than concentrate on epitope and nonepitope regions, the method used bacterial, viral, and tumor protein datasets to derive statistical models for predicting whole protein antigenicity. The models showed prediction accuracy up to 89%, indicating a far higher degree of accuracy than has, for example, been obtained previously for B-cell epitope prediction. Such a method is an imperfect beginning; future research will yield significantly more insight as the number of known protective antigens increases.

However, manifestation of the immunogenicity of a pernicious, pathogenic antigen, as opposed to that of an isolated epitope, arises from the complicated mixing of numerous intertwined extrinsic and intrinsic factors, operating at various temporal and length scales. Such factors include protein-side properties, arising exclusively from the antigen itself, coupled to pathogen-side and host-side properties. Protein-side properties include the state of aggregation exhibited by a candidate vaccine and any posttranslational danger signals that the protein might possess. Host-side properties are intrinsic to the host immune system and include the possession of B-cell epitopes or T-cell epitopes, and possession of pathogen-associated molecular patterns (PAMPs), which are recognized by the innate immune system. Pathogen-side properties are intrinsic to the whole pathogen and include an antigen's expression level, the rate spectrum of its expression and its secretion, as well as its location within the cell (Figure 23.7). A list of immunoinformatics servers and databases for immunomics and vaccinomics is included in Table 23.1.

23.7 DISCUSSION

In this chapter, we have concerned ourselves with postgenomic approaches to the study of immunology, specifically the exploration of host-pathogen interaction and, in particular, what impact this has on the design and discovery of vaccines. This is the realm of immunomics and immunoinformatics. Postgenomics builds on our success in elaborating the structure of genomes and enumerating them across the phyletic spectrum. However, the current so-called human genome sequence is a composite derived from five individuals, yet the genome of every individual organism is different, and each genome can result in a diversity of phenotypes, each dependent, in part, on the environment. Each unique human individual is subtly different, and such differences arise from a combination

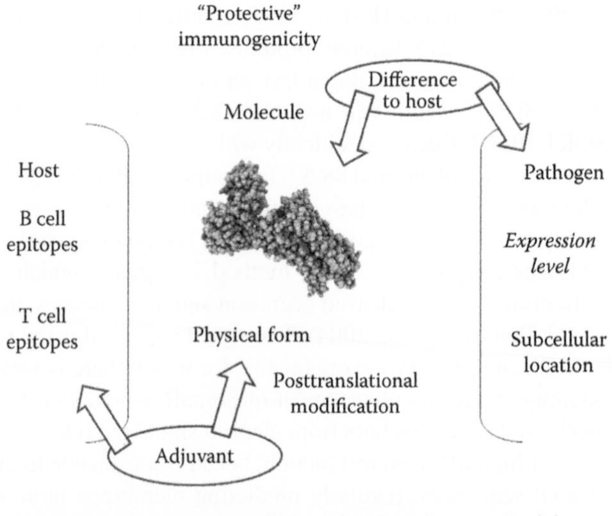

FIGURE 23.7 The systems biology and the prediction of antigen immunogenicity.

TABLE 23.1
Immunoinformatic Databases and Servers for Immunomics and Vaccinomics

	Host databases
IMGT/HLA	http://www.ebi.ac.uk/imgt/hla/allele.html
IMGT/TR	http://imgt.cines.fr/textes/IMGTrepertoire
IPD Database	http://www.ebi.ac.uk/ipd/index.html
Kabat	http://www.kabatdatabase.com
VBASE	http://www.vbase2.org
ABG	http://www.ibt.unam.mx/vir
V BASE	http://vbase.mrc-cpe.cam.ac.uk
	Pathogen databases
APB	http://www.engr.psu.edu/ae/iec/abe/database.asp
APDD	http://psychro.bioinformatics.unsw.edu.au/pathogen/index.php
ARS	http://www.ars.usda.gov/research/projects/projects.htm?accn_no=406518
BROP	http://www.brop.org
EDWIP	http://cricket.inhs.uiuc.edu/edwipweb/edwipabout.htm
FPPD	http://fppd.cbio.psu.edu
LEGER	http://leger2.gbf.de/cgi-bin/expLeger.pl
ORALGEN	http://www.oralgen.lanl.gov
Pathema	http://www.tigr.org/pathema/index.shtml
ShiBASE	http://www.mgc.ac.cn/ShiBASE
STDGen	http://www.stdgen.lanl.gov
VBI	http://phytophthora.vbi.vt.edu
VIDIL	http://insectweb.inhs.uiuc.edu/Pathogens/VIDIL/index.html
VFDB	http://zdsys.chgb.org.cn/VFs/main.htm
CandiVF	http://research.i2r.a-star.edu.sg/Templar/DB/CandiVF
TVfac	http://www.tvfac.lanl.gov
PRINTS	http://www.jenner.ac.uk/BacBix3/PPprints.htm
ClinMalDB-USP	http://malariadb.ime.usp.br/malaria/us/bioinformaticResearch.jsp
Fish Pathogen Database	http://dbsdb.nus.edu.sg/fpdb/about.html
PHI-BASE	http://www.phi-base.org
	T-cell databases
AntiJen	http://www.jenner.ac.uk/antijen/aj_tcell.htm
EPIMHC	http://bio.dfci.harvard.edu/epimhc
FIMM	http://research.i2r.a-star.edu.sg/fimm
HLA ligand Database	http://hlaligand.ouhsc.edu/index_2.html
HIV immunology	http://www.hiv.lanl.gov/immunology
HCV immunology	http://hcv.lanl.gov/content/immuno/immuno-main.html
IEDB	http://epitope2.immuneepitope.org/home.do
JenPep	http://www.jenner.ac.uk/jenpep2
MHCBN	http://www.imtech.res.in/raghava/mhcbn
MHCPEP	http://wehih.wehi.edu.au/mhcpep
MPID-T	http://surya.bic.nus.edu.sg/mpidt
SYFPEITHI	http://www.syfpeithi.de
	B-cell databases
AntiJen	http://www.jenner.ac.uk/antijen/aj_bcell.htm
BCIPEP	http://www.imtech.res.in/raghava/bcipep
CED	http://web.kuicr.kyoto-u.ac.jp/~ced

continued

TABLE 23.1 (CONTINUED)
Immunoinformatic Databases and Servers for Immunomics and Vaccinomics

EPITOME	http://www.rostlab.org/services/epitome
IEDB	http://epitope2.immuneepitope.org/home.do
HaptenDB	http://www.imtech.res.in/raghava/haptendb
HIV immunology	http://www.hiv.lanl.gov/immunology
HCV immunology	http://hcv.lanl.gov/immuno
	Prediction servers
ABCpred	http://www.imtech.res.in/raghava/abcpred
AllerPredict	http://sdmc.i2r.a-star.edu.sg/Templar/DB/Allergen
BcePred	http://www.imtech.res.in/raghava/bcepred/bcepred_submission.html
BIMAS	http://thr.cit.nih.gov/molbio/hla_bind
ELF	http://www.hiv.lanl.gov/content/hiv-db/ELF/epitope_analyzer.html
EpiPredict	http://www.epipredict.de/Prediction/prediction.html
EpiVax	http://www.epivax.com
CTLPred	http://www.imtech.res.in/raghava/ctlpred
IEDB binding, MHC class I	http://www.immuneepitope.org/analyze/html/mhc_binding.html
IEDB binding, MHC class II	http://www.immuneepitope.org/tools/matrix/iedb_input
MHC-Thread	http://www.csd.abdn.ac.uk/~gjlk/MHC-Thread
MHCPred	http://www.jenner.ac.uk/MHCPred
MHC2Pred	http://www.imtech.res.in/raghava/mhc2pred
MMBPred	http://www.imtech.res.in/raghava/mmbpred
MotifScan	http://www.hiv.lanl.gov/content/immunology/motif_scan
NetCTL	http://www.cbs.dtu.dk/services/NetCTL
NetMHC	http://www.cbs.dtu.dk/services/NetMHC
PAProC II	http://www.paproc.de
PREDEPP	http://margalit.huji.ac.il
ProPred	http://www.imtech.res.in/raghava/propred
ProPred-I	http://www.imtech.res.in/raghava/propred1
RankPep	http://bio.dfci.harvard.edu/Tools/rankpep.html
SVMHC	http://www-bs.informatik.uni-tuebingen.de/SVMHC
SYFPEITHI	http://www.syfpeithi.de/Scripts/MHCServer.dll/EpitopePrediction.htm

of genetic, environmental, and epigenetic factors. At the genetic level, most differences result from variant genes and from heritable diversity in gene regulatory mechanisms. Variant genes can include faulty genes that lead to genetic diseases and dominant and recessive alleles.

In 2007, the first individual human genomes were sequenced and published (Levy et al., 2007). Thus, James Watson and J. Craig Venter became the first of thousands—perhaps in time millions—to know the whole sequence of their own DNA. Such self-knowledge will, many hope, be a significant component driving the development of personalized medicine. One of the most interesting aspects of personalized medicine is the emergent field of vaccinomics. Vaccinomics conflates the disciplines of immunogenetics and immunogenomics, leveraging our burgeoning knowledge of immune genetics to enhance our understanding of immune responses to vaccines or vaccination. Vaccinomics is thus poised to foment the development of actualized personalized vaccines, through an exacerbated understanding of immune genotype and how this underlies the phenotype of response, providing a means to manipulate and even design vaccines to moderate and direct immune responses and obviate pathological reactions to vaccines and their components.

Vaccinomics has been defined as "predictive or individualized vaccinology" and involves the identification from a short list of possibilities for the most suitable prophylactic or therapeutic vaccines for a specific individual. Such a vaccine might be targeted at tumors and include DNA coding for a protein adjuvant, together with DNA coding for proteins overexpressed on cancer cells. Vaccinomics also encompasses the fields of immunogenetics and immunogenomics as applied to understanding the mechanisms of heterogeneity in immune responses to vaccines. Working at both the individual level and at the level of whole populations, vaccinomics examines the influence of genetic differences on the diversity of humoral, cell-mediated, and innate immune responses to vaccine moieties, vaccination protocols, and adjuvants.

In this chapter, we have striven to explore and to examine aspects of the immune self and its discrimination from nonself. The immune self is often said to be a conceptualization of immune processes but can also be thought of in terms of molecular patterns recognized by immune agents: epitopes primarily, be they generic epitopes recognized by innate immunity or lipids, carbohydrates, and peptides recognized by T-cells and antibodies.

Immunology, as the study of the immune system, is engaged in the exquisite dissection of these components and, through inductive or synthetic reductionism, a proper exploration of its behavior, yet the immune system as a definable entity is primarily a human construct, which is not to say that host immune responses are unreal; they are no anthropomorphized phantom: they exist. The immune system is, after all, what stands between us and death from infectious diseases. However, the immune system is a concept rather than a discrete and compartmentalized organ. The components of the immune system exist at many length scales: molecules, cells, tissues, and organs. It arises spontaneously at many, many levels from out of the whole organism of which the host is comprised. In time, synthetic reductionism will allow a still elusive predictive understanding of immune processes to emerge from this century long endeavor.

In the MHC class I pathway, for example, peptides are created not just by the proteasome but also by a tranche of other proteases, like tripeptidyl peptidase II, in the cytosol; nor is TAP the only transporter into the ER, and indeed some work in a retrograde fashion. This and many other agents within the cell are involved in degrading and presenting peptides to inspection by T-cells. For example, N-terminal peptidases in the ER trim these peptides. Thus, it may soon be possible to model the whole of the presentation pathway and use it as a transparent gray box to allow us to generate an *in silico* picture of the cell surface complement of MHC-bound peptides as a surrogate of cellular self (Figure 23.8).

The identification of the agents and objects involved in mediating the immune response is a tractable if laborious task and a task with many spectacular successes to its credit. That this task is a long way from being completed is a well understood if poorly articulated assertion. However,

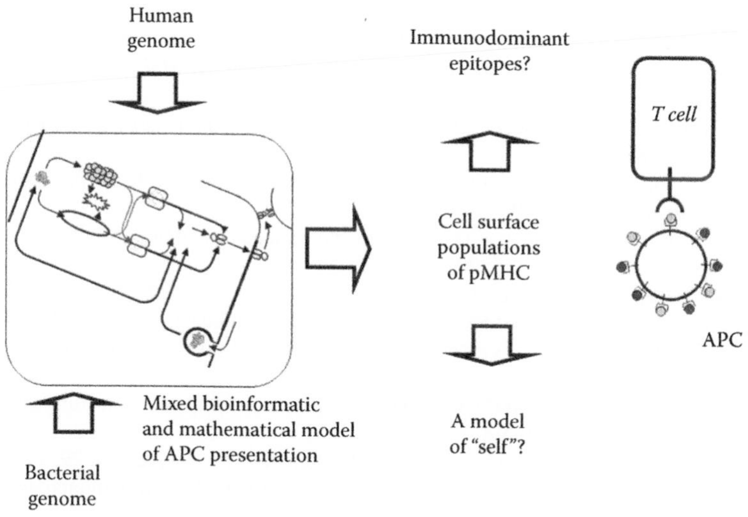

FIGURE 23.8 Modeling cellular self.

assigning human descriptions to molecules and cells, such as antibodies and lymphocytes, is at best a risky business. We are bound by our partial ignorance, not our partial knowledge. An equally difficult yet equally tractable task is charting how these agents work together to make manifest immunity at the whole organism level. However, defining the underlying basis of these behavioral patterns is a task whose difficulty is not to be underestimated. However, the immune self is not a real entity as lymphocytes, cytokines, or the thymus can be thought of as real entities; rather, the immune self is a context-dependent construction. Self or nonself is not defined by reference to a single specific entity; rather it can be defined as the mutable response by a complex system to a diverse collection of molecular entities.

The boundaries of self and nonself are defined by the recognition properties of the immune system, which can be defined and understood in terms of the physical properties of the molecules composing the immune system. Thus, the properties embodied in the concepts of self and nonself are properly emergent. There is no simple positive definition of self and no simple negative definition either. Neither genetic reductionism nor clonal selection theory has all the answers. In a not wholly satisfying way, a practically useful definition self emerges from the sum of its parts.

It may be that we ultimately find it helpful to consider the terms self and nonself as physically embodied entities rather than as phantom signals and to think that these entities are determined by the mechanisms of the immune system. For nonself, some of these entities are created and then recognized by the immune system (processed and presented T-cell and B-cell epitopes), and others, such as PAMPs and B-cell epitopes, are simply recognized.

Ultimately then, this is what immunomics is about and what immunomics will help deliver: a comprehensive and cohesive definition of the elusive immune self. Not a definition framed in pretentious and grandiose pseudophilosophical terms but a definition that can be thought of as a real, grubby, down-and-dirty set of molecules recognized by other molecules. Thus, the immune self and thus the immune nonself will be reduced to practice: identified, tagged, and targeted for destruction. Leveraging the power of immunoinformatics to draw meaning from seeming nonsense, immunomics is set to usher in a new world of rational vaccine design that will in its turn actualize the an era of personalized, tailor-made vaccines targeting individuals, not populations. This is the exciting world of vaccinomics.

REFERENCES

Ada, G. (2008). The enunciation and impact of Macfarlane Burnet's clonal selection theory of acquired immunity. *Immunol. Cell Biol.* 86: 116–118.

Ansari, H. R., Flower, D. R., Raghava, G. P. (2010). AntigenDB: An immunoinformatics database of pathogen antigens. *Nucleic Acids Res.* 38: D847–D853.

Baldwin, T. K., Winnenburg, R., Urban, M., et al. (2006). The pathogen-host interactions database (PHI-base) provides insights into generic and novel themes of pathogenicity. *Mol. Plant Microbe Interact.* 19: 1451–1462.

Bendtsen, J. D., Nielsen, H., Von Heijne, G., et al. (2004). Improved prediction of signal peptides: SignalP 3.0. *J. Mol. Biol.* 340: 783–795.

Blythe, M. J., Flower, D. R. (2005). Benchmarking B cell epitope prediction: Underperformance of existing methods. *Protein Sci.* 14: 246–248.

Carlton, J. M., Hirt, R. P., Silva, J. C., et al. (2007). Draft genome sequence of the sexually transmitted pathogen *Trichomonas vaginalis*. *Science* 315: 207–212.

Chakravarti, D. N., Fiske, M. J., Fletcher, L. D., et al. (2000). Application of genomics and proteomics for identification of bacterial gene products as potential vaccine candidates. *Vaccine* 19: 601–612.

Chen, L., Yang, J., Yu, J., et al. (2005). VFDB: A reference database for bacterial virulence factors. *Nucleic Acids Res.* 33: D325–D328.

Clamp, M., Fry, B., Kamal, M., et al. (2007). Distinguishing protein-coding and noncoding genes in the human genome. *Proc. Natl. Acad. Sci. U.S.A.* 104: 19428–19433.

Cohen, I. R. (1993). The meaning of the immunological homunculus. *Isr. J. Med. Sci.* 29: 173–174.

Cohen, I. R. (1996). Kadishman's tree, Escher's angels, and the immunological homunculus. *Isr. J. Med. Sci.* 32: 44–50.

Cohen, I. R. (2007). Biomarkers, self-antigens and the immunological homunculus. *J. Autoimmun.* 29: 246–249.

Crotzer, V. L., Blum, J. S. (2009). Autophagy and its role in MHC-mediated antigen presentation. *J. Immunol.* 182: 3335–3341.

Dalton, J. P., Mulcahy, G. (2001). Parasite vaccines: A reality? *Vet. Parasitol.* 98: 149–167.

Dalton, J. P., Brindley, P. J., Knox, D. P., et al. (2003). Helminth vaccines: From mining genomic information for vaccine targets to systems used for protein expression. *Int. J. Parasitol.* 33: 621–640.

Davies, M. N., Flower, D. R. (2007). Harnessing bioinformatics to discover new vaccines. *Drug Discov. Today* 12: 389–395.

Diaz, J. J., Giraud, S., Greco, A. (2002). Alteration of ribosomal protein maps in herpes simplex virus type 1 infection. *J. Chromatogr. B Analyt. Technol. Biomed. Life Sci.* 771: 237–249.

Doerfler, W. (2008). In pursuit of the first recognized epigenetic signal: DNA methylation: A 1976 to 2008 synopsis. *Epigenetics* 3: 125–133.

Doytchinova, I., Flower, D. (2003). The HLA-A2-supermotif: A QSAR definition. *Org. Biomol. Chem.* 1: 2648–2654.

Doytchinova, I. A., Flower, D. R. (2007a). Identifying candidate subunit vaccines using an alignment-independent method based on principal amino acid properties. *Vaccine* 25: 856–866.

Doytchinova, I. A., Flower, D. R. (2007b). VaxiJen: A server for prediction of protective antigens, tumour antigens and subunit vaccines. *BMC Bioinformatics* 8: 4.

Doytchinova, I. A., Walshe, V., Borrow, P., et al. (2005). Towards the chemometric dissection of peptide-HLA-A*0201 binding affinity: Comparison of local and global QSAR models. *J. Comput. Aided Mol. Des.* 19: 203–212.

Emanuelsson, O., Brunak, S., Von Heijne, G., et al. (2007). Locating proteins in the cell using TargetP, SignalP and related tools. *Nat. Protoc.* 2: 953–971.

Flower, D. R. (2003). Towards in silico prediction of immunogenic epitopes. *Trends Immunol.* 24: 667–674.

Gannage, M., Munz, C. (2009). Autophagy in MHC class II presentation of endogenous antigens. *Curr. Top. Microbiol. Immunol.* 335: 123–140.

Gardy, J. L., Spencer, C., Wang, K., et al. (2003). PSORT-B: Improving protein subcellular localization prediction for Gram-negative bacteria. *Nucleic Acids Res.* 31: 3613–3617.

Greco, A., Bienvenut, W., Sanchez, J. C., et al. (2001). Identification of ribosome-associated viral and cellular basic proteins during the course of infection with herpes simplex virus type 1. *Proteomics* 1: 545–549.

Haas, G., Karaali, G., Ebermayer, K., et al. (2002). Immunoproteomics of *Helicobacter pylori* infection and relation to gastric disease. *Proteomics* 2: 313–324.

Heath, W. R., Belz, G. T., Behrens, G. M., et al. (2004). Cross-presentation, dendritic cell subsets, and the generation of immunity to cellular antigens. *Immunol. Rev.* 199: 9–26.

Hecker, M., Engelmann, S. (2000). Proteomics, DNA arrays and the analysis of still unknown regulons and unknown proteins of *Bacillus subtilis* and pathogenic gram-positive bacteria. *Int. J. Med. Microbiol.* 290: 123–134.

Jefferies, J. R., Campbell, A. M., Van Rossum, A. J., et al. (2001). Proteomic analysis of *Fasciola hepatica* excretory-secretory products. *Proteomics* 1: 1128–1132.

Jeffery, C. J. (1999). Moonlighting proteins. *Trends Biochem. Sci.* 24: 8–11.

Jeffery, C. J. (2003). Moonlighting proteins: Old proteins learning new tricks. *Trends Genet.* 19: 415–417.

Jeffery, C. J. (2004). Molecular mechanisms for multitasking: Recent crystal structures of moonlighting proteins. *Curr. Opin. Struct. Biol.* 14: 663–668.

Jeffery, C. J. (2009). Moonlighting proteins: An update. *Mol. Biosyst.* 5: 345–350.

Jerne, N. K. (1971). The somatic generation of immune recognition. *Eur. J. Immunol.* 1: 1–9.

Jerne, N. K. (1980). Generating diversity of T-cells by somatic mutation of germ-line genes. *Dev. Comp. Immunol.* 4: 383–384.

Keren, H., Lev-Maor, G., Ast, G. (2010). Alternative splicing and evolution: Diversification, exon definition and function. *Nat. Rev. Genet.* 11: 345–355.

Levy, S., Sutton, G., Ng, P. C., et al. (2007). The diploid genome sequence of an individual human. *PLoS Biol.* 5: e254.

Lin, M. L., Zhan, Y., Villadangos, J. A., et al. (2008). The cell biology of cross-presentation and the role of dendritic cell subsets. *Immunol. Cell Biol.* 86: 353–362.

Liu, W., Meng, X., Xu, Q., et al. (2006). Quantitative prediction of mouse class I MHC peptide binding affinity using support vector machine regression (SVR) models. *BMC Bioinformatics* 7: 182.

Mamitsuka, H. (1998). Predicting peptides that bind to MHC molecules using supervised learning of hidden Markov models. *Proteins* 33: 460–474.

Matzinger, P. (2001). Essay 1: The danger model in its historical context. *Scand. J. Immunol.* 54: 4–9.

McMurry, J., Sbai, H., Gennaro, M. L., et al. (2005). Analyzing *Mycobacterium tuberculosis* proteomes for candidate vaccine epitopes. *Tuberculosis* 85: 95–105.

Medzhitov, R., Janeway, C. A., Jr. (1996). On the semantics of immune recognition. *Res. Immunol.* 147: 208–214.

Medzhitov, R., Janeway, C. A., Jr. (1997). Innate immunity: The virtues of a nonclonal system of recognition. *Cell* 91: 295–298.

Medzhitov, R., Janeway, C. A., Jr. (1998). Innate immune recognition and control of adaptive immune responses. *Semin. Immunol.* 10: 351–353.

Medzhitov, R., Janeway, C., Jr. (2000a). Innate immune recognition: Mechanisms and pathways. *Immunol. Rev.* 173: 89–97.

Medzhitov, R., Janeway, C., Jr. (2000b). The Toll receptor family and microbial recognition. *Trends Microbiol.* 8: 452–456.

Medzhitov, R., Janeway, C. A., Jr. (2002). Decoding the patterns of self and nonself by the innate immune system. *Science* 296: 298–300.

Moise, L., Mcmurry, J., Rivera, D. S., et al. (2007). Progress towards a genome-derived, epitope-driven vaccine for latent TB infection. *Med. Health R.I.* 90: 301–303.

Nakai, K., Horton, P. (1999). PSORT: A program for detecting sorting signals in proteins and predicting their subcellular localization. *Trends Biochem. Sci.* 24: 34–36.

Noguchi, H., Kato, R., Hanai, T., et al. (2002). Hidden Markov model-based prediction of antigenic peptides that interact with MHC class II molecules. *J. Biosci. Bioeng.* 94: 264–270.

Nordstrom, K. J., Mirza, M. A., Larsson, T. P., et al. (2006). Comprehensive comparisons of the current human, mouse, and rat RefSeq, Ensembl, EST, and FANTOM3 datasets: Identification of new human genes with specific tissue expression profile. *Biochem. Biophys. Res. Commun.* 348: 1063–1074.

O'Connor, C. D., Farris, M., Fowler, R., et al. (1997). The proteome of *Salmonella enterica* serovar typhimurium: Current progress on its determination and some applications. *Electrophoresis* 18: 1483–1490.

Perler, F. B., Davis, E. O., Dean, G. E., et al. (1994). Protein splicing elements: Inteins and exteins: A definition of terms and recommended nomenclature. *Nucleic Acids Res.* 22: 1125–1127.

Rodriguez, J. M., Salas, M. L., Santaren, J. F. (2001). African swine fever virus-induced polypeptides in porcine alveolar macrophages and in Vero cells: Two-dimensional gel analysis. *Proteomics* 1: 1447–1456.

Serebriiskii, I. G., Toby, G. G., Finley, R. L., Jr., et al. (2001). Genomic analysis utilizing the yeast two-hybrid system. *Methods Mol. Biol.* 175: 415–454.

Southan, C. (2004). Has the yo-yo stopped? An assessment of human protein-coding gene number. *Proteomics* 4: 1712–1726.

Tettelin, H. (2009). The bacterial pan-genome and reverse vaccinology. *Genome Dyn.* 6: 35–47.

Toby, G. G., Golemis, E. A. (2001). Using the yeast interaction trap and other two-hybrid-based approaches to study protein-protein interactions. *Methods* 24: 201–217.

Toda, T., Sugimoto, M., Omori, A., et al. (2000). Proteomic analysis of Epstein-Barr virus-transformed human B-lymphoblastoid cell lines before and after immortalization. *Electrophoresis* 21: 1814–1822.

Tongchusak, S., Brusic, V., Chaiyaroj, S. C. (2008). Promiscuous T cell epitope prediction of *Candida albicans* secretory aspartyl protienase family of proteins. *Infect. Genet. Evol.* 8: 467–473.

Ueberle, B., Frank, R., Herrmann, R. (2002). The proteome of the bacterium *Mycoplasma pneumoniae*: Comparing predicted open reading frames to identified gene products. *Proteomics* 2: 754–764.

Vandahl, B. B., Birkelund, S., Christiansen, G. (2002). Proteome analysis of *Chlamydia pneumoniae*. *Methods Enzymol.* 358: 277–288.

Vandahl, B. B., Birkelund, S., Christiansen, G. (2004). Genome and proteome analysis of *Chlamydia*. *Proteomics* 4: 2831–2842.

Vandahl, B. B., Birkelund, S., Demol, H., et al. (2001). Proteome analysis of the *Chlamydia pneumoniae* elementary body. *Electrophoresis* 22: 1204–1223.

Vivona, S., Gardy, J. L., Ramachandran, S., et al. (2008). Computer-aided biotechnology: From immuno-informatics to reverse vaccinology. *Trends Biotechnol.* 26: 190–200.

Vyas, J. M., Van Der Veen, A. G., Ploegh, H. L. (2008). The known unknowns of antigen processing and presentation. *Nat. Rev. Immunol.* 8: 607–618.

Wan, J., Liu, W., Xu, Q., et al. (2006). SVRMHC prediction server for MHC-binding peptides. *BMC Bioinformatics* 7: 463.

Wan, S., Coveney, P., Flower, D. R. (2004). Large-scale molecular dynamics simulations of HLA-A*0201 complexed with a tumor-specific antigenic peptide: Can the alpha3 and beta2m domains be neglected? *J. Comput. Chem.* 25: 1803–1813.

Wan, S., Coveney, P. V., Flower, D. R. (2005a). Molecular basis of peptide recognition by the TCR: Affinity differences calculated using large scale computing. *J. Immunol.* 175: 1715–1723.

Wan, S., Coveney, P. V., Flower, D. R. (2005b). Peptide recognition by the T cell receptor: Comparison of binding free energies from thermodynamic integration, Poisson-Boltzmann and linear interaction energy approximations. *Philos. Trans. A Math. Phys. Eng. Sci.* 363: 2037–2053.

Wan, S., Flower, D. R., Coveney, P. V. (2008). Toward an atomistic understanding of the immune synapse: Large-scale molecular dynamics simulation of a membrane-embedded TCR-pMHC-CD4 complex. *Mol. Immunol.* 45: 1221–1230.

Winnenburg, R., Baldwin, T. K., Urban, M., et al. (2006). PHI-base: A new database for pathogen host interactions. *Nucleic Acids Res.* 34: D459–D464.

Winnenburg, R., Urban, M., Beacham, A., et al. (2008). PHI-base update: Additions to the pathogen host interaction database. *Nucleic Acids Res.* 36: D572–D576.

Yang, J., Chen, L., Sun, L., et al. (2008). VFDB 2008 release: An enhanced web-based resource for comparative pathogenomics. *Nucleic Acids Res.* 36: D539–D542.

Zhang, C., Bickis, M. G., Wu, F. X., et al. (2006). Optimally-connected hidden Markov models for predicting MHC-binding peptides. *J. Bioinform. Comput. Biol.* 4: 959–980.

24 NeuroAIDS and Omics of HIV Vpr

Ashish Swarup Verma, Udai Pratap Singh, and Priyadarshini Mallick
Amity University
Noida, India

Premendra Dhar Dwivedi
Indian Institute of Toxicology Research
Lucknow, India

Anchal Singh
Kirksville College of Osteopathic Medicine
Kirksville, Missouri

CONTENTS

24.1 OMICS IN CURRENT PERSPECTIVE

The advent of molecular biology helped us to understand the role of gene(s) in any cell or in any organism. This was the start of the *one-gene-at-a-time* approach to understand the significance of specific gene(s) in biological response. With time, biological sciences have seen unprecedented development. In the past 20 years, high-throughput technologies have gained enormous importance, producing enormous amounts of data in minimal time with a minimal quantity of biological material. The data generated through high-throughput technology is suggestive of a holistic approach, where one can monitor alterations or effects of one indicator on various gene(s) and its product(s) simultaneously. In simple terms, omics is an approach that has reversed the way we study any biological system. Omics is a way to study the multifaceted aspects of a biological system, which is contrary to the established *one-gene-at-a-time* approach.

Etymological analysis suggests that the *-ome* suffix is derived from Sanskrit word *OM*. OM is described as "completeness and fullness" in Sanskrit (Lederberg and McCray, 2001). In a similar fashion, the study of biological systems in totality has evolved as omics. Later, omics branched out with specific systems, e.g., genomics, proteomics, transcriptomics, etc.

In 1920, Hans Winkler, a botanist from Germany was the first one to combine the word *gene* and the suffix *-ome* and create the term *genom(e)* (Winkler, 1920). A catchy word, genomics, was later coined by Victor McKusick and Frank Ruddle (1987). McKusick and Ruddle cofounded a new journal entitled *Genomics*. At that time, genomics was used to study linear gene mapping, DNA sequencing, and comparative study of genomes among different species.

Omics as a discipline of biological science started a new era, which defines how we can study life sciences with its new meaning. Omics has helped researchers to study organisms at levels of genomes (complete genetic sequences) (Parfrey et al., 2008), transcriptomes (total mRNA) (Velculesen et al., 1997), proteomes (protein expressions) (Anderson and Anderson, 1998), metabolomes (metabolic network) (Oliver et al., 1998), and interactomes (protein-protein interaction) (Sanchez et al., 1999). New subdisciplines of omics are continuously emerging as their new applications are being recognized. In a real sense, genome sequencing only tells us what a cell can do, but in order to understand what a cell is doing, we must use microarray technologies (Kulesh et al., 1987).

Microarray tests can examine mRNA expression of thousands of genes, which gives a global purview of cell functions under single or multiple conditions. Usually, profiling of expressed transcripts provides sufficient information as rough estimates for protein expression in predefined condition(s). The easiest solution for these problems is high-throughput profiling, even at a protein level. Protein profiling is helpful to identify, characterize, and quantify proteins expressed or produced by cells under experimental or natural conditions like diseases (Aebersold and Mann, 2003; Ong and Mann, 2005).

Omics always is a means to produce data in bulk; therefore data generated through a high-throughput method is certainly a rich source of information regarding any biological system. Various immunodeficiency viruses like feline immunodeficiency virus (FIV), human immunodeficiency virus, types 1 and 2 (HIV-1 and -2), simian human immunodeficiency virus (SHIV), and simian immunodeficiency virus (SIV) are known to have detrimental effects on their host; therefore implications of omics approaches can be highly useful to understand the disease processes, which ultimately will help to develop better therapeutic solutions.

24.2 DISCOVERY AND ORIGIN OF HUMAN IMMUNODEFICIENCY VIRUS

HIV has emerged as unique infection ~30 years ago in 1981. Devastation of the immune system is a hallmark of HIV infection, which leads to the fatal clinical condition known as acquired immunodeficiency syndrome (AIDS). AIDS is the ultimate chapter in the life of an HIV-seropositive patient. AIDS can be clinically defined as a condition in which $CD4^+$ T-lymphocytes counts are <200 cells/μl of blood along with presence of AIDS-defining illnesses like HIV-associated dementia (HAD), HIV wasting syndrome, Kaposi's sarcoma, and others (Power et al., 2009; Centers for Disease Control and Prevention, 1992). A common perception is that HIV is of only one type, but HIV is of two types HIV-1 and HIV-2. Between these two, HIV-1 is globally prevalent with fatal consequences. However, HIV-2 is a less-pathogenic, slowly progressing virus that is mostly confined to African countries, although a few cases of HIV-2 have been reported in other parts of the world too.

24.2.1 Discovery of HIV

The first report of HIV infection was in gay men from Los Angeles (Gottlieb et al., 1981). These patients died from a rare disease known as *Pneumocystis carinii* pneumonia (PCP). PCP usually occurs only in immunodeficient patients. Initially, because HIV was not discovered, these patients could not be confirmed as being infected with HIV. Later, upon the discovery of HIV, these patients were confirmed to be infected with HIV. This infection was noticed for the first time in gay men, but the majority of early cases were also reported in gay men. The unique association of HIV with gay men made HIV infections quite controversial because of religious opinions. Sooner than expected, HIV infections were reported from all walks of life, as well as from all corners of the globe. Ever-increasing numbers of HIV-infected patients raised a feeling that HIV infections were spreading like an epidemic. This was the main reason serious and sincere efforts were made to control HIV infection (Verma et al., 2009).

Luc Montagnier from Pasteur Institute in France was the first scientist to identify the causative organism for AIDS in 1983, and his group named this virus lymphadenopathy-associated virus. In 1984, Dr. Robert Gallo's group at the National Cancer Institute in the United States reconfirmed the findings of Montagnier, but Gallo's group called this virus human T-cell lymphotropic virus ($HTLV_{III}$). Both discoveries were reported almost at the same time, therefore a dispute arouse regarding who discovered HIV. The bitter dispute became impossible to resolve; therefore a meeting between the heads of state was warranted. Ronald Regan, then-president of the United States and Jacque Chiraq, then-president of France, met to resolve this dispute. They concluded that Robert Gallo and Luc Montagnier both were equal contributors in the discovery of HIV. Finally, the importance of Luc Montagnier's contribution toward the discovery of HIV was recognized in 2008, when he was awarded the Nobel Prize for Medicine and Physiology. The Centers for Disease Control and Prevention (CDC) for the first time defined AIDS as a disease, whereas HIV was the name adopted by the International Committee on Taxonomy of Viruses in 1986. Before 1986, one can easily find various names for the same virus in literature.

24.2.2 Origin of HIV

The origin of HIV is still uncertain and debatable. There is a common belief that it mutated from SIV. SIV commonly infects monkeys and certain subgroups of chimpanzees in West Central Africa. As per one hypothesis, SIV adapted for humans and was transmitted from monkeys to humans. HIV is a member of the *Retroviridae* family. Other viruses from the *Retroviridae* family are known to infect other animals like cats, sheep, horse, and cattle.

24.3 NEUROAIDS

HIV may directly infect and injure both the central nervous system (CNS) and the peripheral nervous system (PNS). Neurosusceptibility of HIV could cause severe neurotoxicity, which may lead to the onset of various neuropsychiatric complications. Neuropsychiatric complications associated with neuroAIDS are due to the neurotoxicity and neuroinvasive nature of HIV. HAD, HIV-associated encephalopathy, and HIV wasting syndrome are some of the common names used in literature for neuroAIDS. Some of the common disorders related to neuroAIDS are listed in Table 24.1.

TABLE 24.1
Common Neuropsychiatric Disorders among NeuroAIDS Patients

Addiction
Anxiety
Depression
Epilepsy
Mania
Mood disorders
Neurocognitive impairment
Neuropathic pain
Physical disability
Seizures

A reduction in the incidence of opportunistic infections has been observed in HIV patients who are under effective antiretroviral treatment, which is a major contributing factor for the improvement in their immune status (McCombe et al., 2009). There is a reduction in opportunistic infection in long-term HIV seropositives even though neuropsychiatric complications are reported to be as high as 50% or more among long-term HIV seropositives (McArthur et al., 2005).

24.4 EPIDEMIOLOGY

The ever-increasing number of HIV incidences was the reason for the United Nations to recognize the urgency for intervention and for the creation of a new institution, UNAIDS, under the umbrella of United Nations. The name combines *UN* for United Nations and *AIDS* for acquired immunodeficiency syndrome. The major responsibility of UNAIDS was to monitor, support, and develop strategies for fighting against HIV and AIDS at a global level. Since then, UNAIDS has published an annual report about the status of HIV infections every year, which includes updates at both global and national levels. In addition, these reports provide new insights about HIV and AIDS. The UNAIDS Annual Report is considered to be the most authentic documentation about HIV infection; the most recent one was published in 2009 (UNAIDS, 2009).

As per the UNAIDS report (2009), there are ~33.4 million people living with HIV with an expected range of ~31.1–35.8 million. Men and women are equally vulnerable to HIV infections, with more than 2.1 million children (up to 15 years of age) expected to be living with HIV until 2008. The African continent is still the worst affected region because more than two-thirds (~67%) of HIV infections are reported in African nations. African countries also have the highest incidences of new infections, along with highest mortality caused by HIV and AIDS. Efficient measures to control HIV infections are still not in place in African countries. A closer analysis of data reveals that on a daily basis ~7,400 individuals get infected with HIV, whereas ~5,500 individuals die because of HIV infections and AIDS (Verma et al., 2010a). These numbers are indicative of poor health care, poor living conditions, and lack of resources for efficient monitoring of HIV infections. On an annual basis, these numbers suggest that the HIV epidemic is still a serious challenge for public health. Sincere efforts are urgently needed to address this issue (Table 24.2). From an epidemiological point of view, without any doubt, every new UNAIDS report does suggest an improvement in status of HIV infections. Recent epidemiological assessments have some encouraging facts, which are as follows:

1. Since 1996, the prevalence of HIV infection remains same.
2. Long-term HIV seropositives and new infections are the main contributors for ongoing accumulation of HIV-infected patients.
3. There is a decline in HIV prevalence in specific nations.
4. Decreased HIV-related deaths are due to improvement in health-care facilities and improved antiretroviral medications.
5. New HIV infections are declining at global level.

As per one estimate, so far, >65 million people have been infected with HIV, out of which >35 million individuals are still living with HIV infections at the present time. Higher incidences of neuropsychiatric signs and symptoms have been observed among long-term HIV-seropositive patients. As per some estimates, ≥50% of HIV seropositives (long-term survivors) are presented with neurological complications (McArthur et al., 2005). These neurological complications are associated either with the CNS or the PNS or both. Neuropsychiatric signs and symptoms get more pronounced clinically with the progression of HIV infections towards AIDS. Certainly, HIV is a contributing factor for neuroAIDS, but the contribution of individual genes towards neuroAIDS can be neither denied nor ignored. Various omics approaches may offer more insights about the role of individual genes in neuroAIDS.

TABLE 24.2
Estimate for HIV/AIDS Prevalence at Global Level

	Living with HIV	New Infections*	Death*
	←	(In Millions)	→
Adults	31.3 (29.2–33.7)	2.3 (2.0–2.5)	2.0 (1.4–2.1)
Children	2.1 (1.2–2.9)	0.43 (0.24–0.61)	0.28 (0.15–0.41)
Total	**33.40 (31.1–35.8)**	**2.7 (2.4–3.0)**	**2.0 (1.7–4.0)**
	Sub-Saharan Africa (Worst Affected Area)		
Total	22.4 (20.8–24.1) [~67%]	1.9 (1.6–2.2) [~70%]	1.4 (1.1–1.7) [~70%]
	Oceania (Least Affected Area)		
Total	0.06 (0.05–0.07) [~0.001%]	0.004 (0.003–0.005) [~0.001%]	0.002 (0.001–0.003) [~0.001%]

Source: Adapted from *UNAIDS Report*, 1–100, 2009.
The values given in parentheses are the ranges. The values given in square brackets are the percentages of the global estimate.
* Annual estimate.

24.5 HIV

24.5.1 Structure of HIV

HIV is roughly spherical in structure, measuring about 120 nm in diameter. Taxonomically, HIV belongs to the *Lentivirus* genus and the *Retroviridae* family. Being a member of the *Retroviridae* family, the genetic material in HIV is a single-stranded, positive-sense RNA. HIV does not have any DNA in its structure. To utilize RNA as a genetic material, HIV has reverse-transcriptase (RT) enzyme to transfer genetic information to its new progeny. The genome size of HIV is ~9.8 kb with nine genes excluding long terminal repeat (*LTR*). These nine genes produce 15 viral proteins. These proteins/genes can be divided into three categories based on their functions. The structural proteins are Tat and Rev; the regulatory proteins are Gag, Env, and Pol; whereas Nef, Vpr, Vif, and Vpu serve as accessory proteins. A precursor protein is encoded by *pol*, which gives rise to four different smaller cleaved products, viz., p10, p32, p51, and p64. Of these, p10 acts as protease that cleaves gag precursor, p32 acts as integrase enzyme and is helpful for integration of proviral DNA into host genome, and p51 and p64 both have reverse transcriptase activity. p64 has an additional function to act as RNAse (p15) (Figure 24.1) (Abbas et al., 2008).

24.5.2 Comparative Omics of HIV

It is beyond the scope of this chapter to compare the genomes of all possible immunodeficiency viruses. Here we have compared the genomic organization of HIV-1, HIV-2, SHIV, SIV, and FIV. A common feature among these immunodeficiency viruses is that their genomic sizes are similar; in general, it can be said that their genome sizes are ~10 kb. Among these viruses, the genome of HIV-1 is the smallest, *i.e.*, 9,181 bp, whereas HIV-2 is the largest with 10,359 bp; the genomic sizes

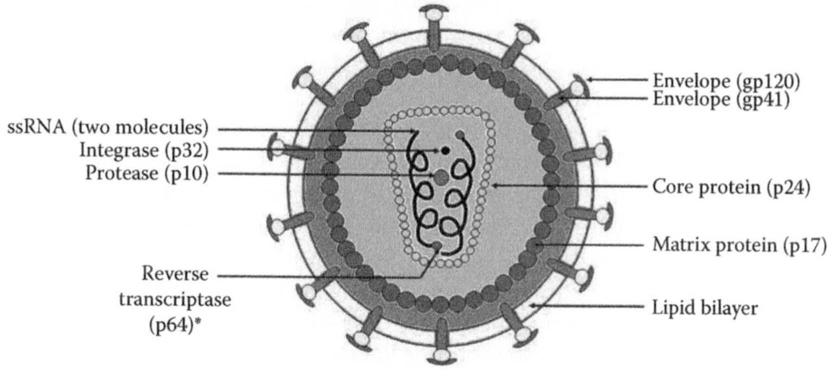

FIGURE 24.1 Structure of HIV-1. This is a graphical representation of a cross-section of HIV-1. The outer layer is known as the envelope. It consists of the lipid bilayer and comprises gp120 and gp41. The layer next to the envelope consists of matrix protein (p17), which is followed by a layer that consists of core protein (p24). At the center of virion, it has two molecules of single-stranded RNA (ssRNA) and other enzymes. These enzymes are protease (p10), integrase (p32), and reverse transcriptase (p64). The reverse-transcriptase enzyme also contains RNase H (p15). The specific name of each individual protein along with its location is shown, whereas the molecular mass of each specific protein is presented in parentheses. *, polymerase (which contains reverse transcriptase, RNase H, integrase, and protease) is not shown in this figure.

of the other viruses lie between these two limits. A closer analysis suggests that genomic size among these viruses varies by ~1 kb, even though their biology, pathogenesis, and host specificity vary widely. For example, HIV-1 and HIV-2 are very closely associated viruses, and their genome sizes have differences of about 1,178 bases, but their pathogenicity varies tremendously. At the genomic level, HIV-2 contains *vpx* apart from *vpr* (Figure 24.2).

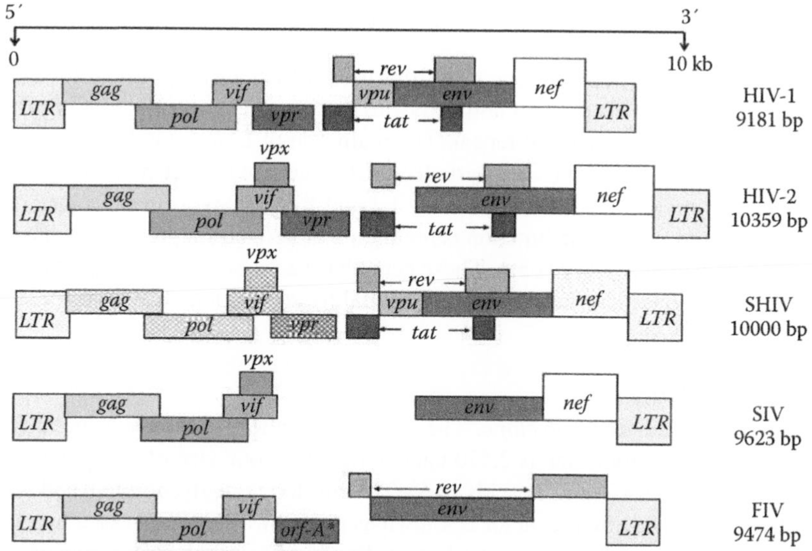

FIGURE 24.2 **(See color insert.)** Genomic organization of five different immunodeficiency viruses. HIV-1, human immunodeficiency virus, type 1; HIV-2, human immunodeficiency virus, type 2; SHIV, simian human immunodeficiency virus; SIV, simian immunodeficiency virus; FIV, feline immunodeficiency virus; *LTR*, long terminal repeat; *gag*, group-specific antigen; *pol*, polymerase; *vpr*, viral protein R; *vpu*, viral protein U; *tat*, transactivator of transcription; *env*, envelope; *nef*, negative factor; *rev*, regulator of expression of viral proteins; *vpx*, viral protein X; *orf-A*, open reading frame A. The figure is not to scale. In the case of SHIV, the checkered box represents genes of simian origin. In the case of FIV, the asterisk represents the Vpr equivalent.

The question is whether the addition of *vpx* in the genome is counteractive toward pathogenicity of HIV-2; we do not know the answer. This question, however, can be answered using the strategy of omics. SHIV is a chimeric virus, and its genome is larger than HIV-1 but smaller than HIV-2. The larger size of this virus is probably attributable to the addition of new genes into its genome from HIV. Even though SHIV is not showing all possible effects of HIV-1, either it is due to variation in its genome or perhaps due to variation in host specificity. If these problems are host-specific, probably they can be answered with the application of genomics and proteomics. If we look at the genome organization of SIV, it lacks *vpr* and *tat*, genes that are accessory and regulatory in nature, respectively. Are they responsible for their unique behavior with host specificity? Further studies are warranted, because at this moment we do not have any conclusive answer. At genomic levels, FIV does not have *tat*, but it has *orfA* in place of *vpr*. FIV infection cannot be induced in laboratory cats, whereas in natural conditions, cats are known to be infected with FIV. This suggests that concurrent infections are necessary for FIV infections. In the case of FIV pathogenesis, the question is do the FIVs use machinery of other pathogens to show their pathogenic effect in cats? Again we do not have a conclusive answer. In a similar fashion, there are always some variations in proteins that are produced by these different immunodeficiency viruses. Do these variations have any significant role in their pathogenesis? In the present circumstances, it can be said that omics is a means to explore the possible answers to these questions about the variations in pathogenicity of different viruses.

24.5.3 Molecular Biology of HIV

HIV-1 has a small genome of size ~9.8 kb with nine genes, which in turn produce 15 different proteins. These 15 proteins can be divided in three subcategories:

1. Structural proteins Gag, Env, and Pol
2. Regulatory proteins Tat and Rev
3. Accessory proteins Nef, Vpr, Vif, and Vpu (Figure 24.2)

Reverse transcriptase is characteristic of retroviruses, which are responsible for the conversion of viral RNA into cDNA. This process faithfully carries the genetic information of viruses. *LTR* is another gene known as long terminal repeat. The main role of *LTR* is integration of provirus into host cell DNA. Synthesis of new virions starts from *LTR* because transcriptional factors of the host cell bind to the *LTR* region of the viral genome. Tat and Rev belong to the category of regulatory proteins in HIV genome, and their function is regulation of new virion production. The third group of genes are specified as accessory genes. These genes have accessory roles to assist in efficient production of new virion, but these genes are not essential for replication. The proteins of third group are Nef, Vif, Vpr, and Vpu.

24.5.3.1 Envelope (Env)

The name Env is derived from *envelope*. The genomic location of the envelope is from nucleotides 5771 to 8341, and its total size is 2,570 nucleotides. The total size of Env precursor protein is 160 kDa, which comprises 856 amino acids (AA). The envelope is a glycoprotein and is synthesized as a polyprotein precursor. Env precursor is known as gp160, which is processed by a cellular protease rather than viral protease during trafficking of Env to cell surface. Processing of gp160 results in production of two proteins: surface Env, known as glycoprotein gp120, and a transmembrane glycoprotein known as gp41. gp120 contains determinants that interact with T-cell receptors and coreceptors. gp120 also has conserved and variable regions, and each of these regions has five domains, C1–C5 and V1–V5. gp41 consists of three domains: a domain essential for membrane fusion known as ectodomain, an anchor sequence that is a transmembrane protein, and a cytoplasmic tail. gp120 primarily binds with target cell because of coreceptor interaction, which induces conformational changes in gp41. Conformational changes in gp41 promote formation of gp120/gp41 glycoprotein complex and fusion of virus to host cell membrane. At the heart of fusion reaction is the ectodomain of gp41; this region contains a highly hydrophobic N terminus (fusion peptide) and two heptad repeat motifs, referred as N-helix and C-helix (Freed, 2001). Demented and nondemented patients with AIDS differ in brain-derived HIV Env sequences. To replicate in cells of CNS, HIV selects envelopes with reduced dependence on CD4 and increased fusion activity (Thomas et al., 2007).

24.5.3.2 Group-Specific Antigen (*Gag*)

The group-specific antigen gene (*Gag*) is localized from nucleotides 336 to 1838 and comprises 1,502 nucleotides. It produces a protein of 55 kDa containing 500 AA. *Gag* gene encodes for Pr55Gag, which is a polyprotein precursor. The major role of this polyprotein precursor is completion of viral assembly at the end of replication. Viral proteases cleave Pr55Gag into smaller proteins. Proteolytic cleavage of Pr55Gag results in formation of four proteins, viz., p17, p24, p7, and p6, as well as two spacer peptides, p2 and p1. p17 is also known as matrix protein (MA), p24 is known as capsid protein (CA), p7 is known as nucleocapsid (NA), and p6 is known as Vpr-binding protein. The N terminus of MA contains 131 AA residues, and it is mainly responsible for targeting and binding to the plasma membrane. Viral particle assembly takes place at the plasma membrane of infected cells (Freed, 1998).

24.5.3.3 Long Terminal Repeat (*LTR*)

LTR means *long terminal repeat. LTR* is present on either side of viral genome. It serves as the initiation site for transcription of viral genome, and it harbors *cis*-acting elements, which are required for RNA synthesis. *LTR* consists of three regions: U3 (for unique, 3' end), R (for repeated), and U5 (for unique, 5' end). Transcription initiates at the U3/R junction. Various elements present in U3 help in direct binding of RNA polymerase II (Pol II) to DNA templates. Transcription initiation from *LTR* leads to generation of >30 copies of viral RNAs (Purcell and Martin, 1993). Newly synthesized viral RNA falls into three major classes:

1. Unspliced RNAs, which function as precursors for Gag and Gag-Pol polyprotein
2. Partially spliced mRNAs (~5 kb), which encode Env, Vif, Vpu, and Vpr proteins
3. Small but multiple spliced mRNAs (1.7–2.0 kb), which encode for Rev, Tat, and Nef

The basal transcriptional activity from *LTR* is very low, but the transcription rate is greatly increased in the presence of Tat, which acts as transactivator of transcription (Fisher et al., 1986).

24.5.3.4 Negative Factor (Nef)

The word Nef has evolved from *negative factor.* Its genomic location is from nucleotides 8343 to 8963, and the total length of the *nef* gene is 620 nucleotides. Nef is an accessory protein of HIV and contains about 206 AA with a molecular mass of 27 kDa. Viral proteases cleave Nef into a 20-kDa form by removing a 7-kDa polypeptide chain. Nef is expressed during early stages of replication in host cells. Nef downregulates expression of major histocompatibility complex class II and cell receptors like CD4, CD8, CCR5, CXCR4, etc. Due to these reasons, Nef is considered as an important protein for *in vivo* pathogenesis (Kirchhoff et al., 1995; Forshey and Aiken, 2003). Originally, Nef was described as a negative factor for viral replication (Ahmad and Venkatesan, 1988), but later it turned out to be a positive factor for *in vivo* pathogenesis (Cullen, 1998). So far, three important functions of Nef protein have been observed in *in vitro* studies:

1. It perturbs endocytosis.
2. It modulates signal transduction pathways in infected cells (Greenway et al., 2000; Arora et al., 2002).
3. It enhances viral infectivity.

Nef gets incorporated into HIV virions during formation of new virus particles (Kotov et al., 1999). Nef is localized in the viral core, where it is present in very low quantities (Kramer-Hammerle et al., 2005a). It has been suggested that Nef plays a major role in HIV-1 disease progression toward AIDS (Gorry et al., 2007) by supporting fusion of HIV-1 to target cells (Schaeffer et al., 2001).

Nef seems to play a significant role in alterations of CNS functioning, because Nef is found to be present in higher levels in astrocytes, causing alterations in their growth. Nef can alter the electrophysiology of neurons and induce inflammatory mediators from monocytes.

24.5.3.5 Polymerase (Pol)

The name Pol is derived from *polymerase.* Its genomic location is from nucleotides 1839 to 4642 and consists of 2,803 nucleotides. Pol protein is of 112 kDa and consists of 935 AA. The *pol*-encoded enzymes are initially synthesized as part of a large polyprotein precursor, Pr160Gag-Pol, whose synthesis results from a rare frameshifting event during Pr55Gag translation. The individual *pol*-encoded enzymes, viral protease (PR or p10), RT (p64), and integrase (IN or p32) are cleaved from Pr160Gag-Pol by viral proteases. In HIV, RT is a heterodimer of two subunits, i.e., of 64 kDa (p64) and 51 kDa (p51). These two subunits are derived from the same region of Pr160Gag-Pol precursor protein. Integrase contains a single polypeptide with a molecular mass of 32 kDa. Following nuclear import of viral preintegration complex, 32-kDa integrase protein promotes insertion of linear but double-stranded proviral DNA into the chromosome of the host cell (Brown, 1997). The integrated DNA, referred to as a *provirus*, behaves like any other cellular gene. Integration is an absolutely necessary step for viral replication, because integrase mutant viruses fail to spread infections.

24.5.3.6 Regulator of Expression of Viral Proteins (Rev)

The name Rev is derived from *regulator of expression of viral proteins.* Rev is encoded by two exons, both of which are essential to produce functional protein (Pollard and Malim, 1998). The genomic location of *rev* is from nucleotides 5516 to 5591 and nucleotides 7925 to 8199; therefore *rev* comprises 75- and 274-nucleotide sequences, respectively. The collective length of *rev* is 349 nucleotides. Rev is a regulatory protein and is essential for the regulation of viral replication. The molecular mass of Rev protein is 19 kDa, which contains 116 AA. In normal conditions, mRNAs are fully spliced before their transport out of nucleus. Rev downregulates posttranscriptional splicing of viral mRNAs. Rev contains two functional domains: the arginine-rich domain, which binds with viral RNA and support nuclear localization, and the leucine-rich domain, which is hydrophobic in nature and mediates nuclear export of viral RNA.

24.5.3.7 Transactivator of Transcription (Tat)

The name Tat is derived from *transactivator of transcription.* Tat is a regulatory protein that is essential for regulation of viral replication. Genomic location of *tat* is from nucleotides 5377 to 5591 and nucleotides 7925 to 7970. *tat* comprises 214- and 45-nucleotide sequences, respectively. Tat is an 86–110 AA-long protein with molecular mass of 16 kDa. Tat is a nonstructural protein and is the product of two exons. Tat protein consists of several functional domains:

1. An activation domain, which consists of 48 AA at the N terminus along with an acidic domain in it
2. A highly basic RNA-binding domain
3. An overlapping nuclear localization signal

The acidic domain is a cystine-rich region with a hydrophobic core element. Secreted Tat may be taken up by neighboring cells; therefore Tat affects both infected and noninfected cells (Kumar et al., 1999; McManus et al., 2000). Apart from peripheral circulation, Tat has been reported in the brains of HIV-infected individuals with known CNS pathology (Del Valle et al., 2000). Tat induces apoptosis of neurons both *in vivo* and *in vitro* via oxidative stress pathway. Neuronal excitotoxicity is also commonly noticed in the presence of Tat.

24.5.3.8 Viral Infectivity Factor (Vif)

Vif is considered an accessory protein, and the name is derived from *viral infectivity factor. vif* is a conserved sequence among lentiviruses. Vif has a major role in the production of infection competent new virions from infected cells. Genomic location of *vif* is from nucleotides 4587 to 5165, which comprises 578 nucleotides. The molecular mass of Vif protein is 23 kDa and is made up of 192 AA. Vif is expressed in a late stage of HIV replication and localizes in the cytoplasm of infected cells (Lake et al., 2003). *vif* was first identified in 1987 and considered to be a gene responsible for viral infectivity, but its mechanism of action is not clear. Vif inhibits premature processing of Pr55Gag in cell cytoplasm. The proteolytic activity of viral protease exposes or conceals the functional domains of Vif, which facilitates virion maturation or infectivity of target cells (Khan et al., 2002). Thus, Vif appears to act during viral assembly in virus-producing cells or subsequently in virion maturation to produce virions competent for reverse transcription in the target cell.

24.5.3.9 Viral Protein R (Vpr)

The function of Vpr is discussed in Section 24.6.

24.5.3.10 Viral Protein U (Vpu)

The genomic location of *vpu* is from nucleotides 5608 to 5856; *vpu* consists of 248 nucleotides and produces a protein of 16 kDa that consists of 82 AA. Viral protein U is type 1 integral membrane phosphoprotein unique to HIV-1 and SIVcpz lentiviruses (Strebel et al., 1988). Vpu enhances pathogenesis *in vivo*, even though it is not an essential protein (Deora and Ratner, 2001). Vpu protein usually gets expressed during replication in host cell. Vpu cannot be detected in virions because Vpu protein does not get packaged into viral particle (Bour and Strebel, 2000). Vpu consists of two domains: a hydrophobic N-terminal transmembrane domain that anchors Vpu to cell membrane and a hydrophilic C-terminal domain that lines up with the membrane surface of infected cells (Marassi et al., 1999). Vpu performs two major functions during HIV-1 replication: it enhances the release of viral particles and promotes CD4 degradation. CD4 degradation by Vpu is mediated through host ubiquitin/proteasome pathway. The outcome of Vpu induced CD4 degradation is to liberate gp160 from Env/CD4 complexes in endoplasmic reticulum.

24.5.3.11 Viral Protein X (Vpx)

The name Vpx is derived from *viral protein X*. Its genomic location is from nucleotides 5898 to 6239 and comprises of 341 nucleotides. Vpx protein is 16 kDa in size and consists of 113 AA. *vpx* is present in HIV-2 and SIV but not in HIV-1. *vpx* should be considered as the equivalent of *vpr* in HIV-2 and SIV, because *vpx* has considerable sequence homology with *vpr* of HIV. Like Vpr, Vpx also gets incorporated at relatively high levels into virions because of interaction of Vpx with the C terminus of Gag. Vpx plays an important role in infection of nondividing cells like Vpr with the exception that Vpx does not induce cell-cycle arrest like Vpr (Fletcher et al., 1996).

24.6 VIRAL PROTEIN R (VPR)

24.6.1 Vpr and Overview

Vpr consists of 96 AA; therefore the predicted molecular mass of Vpr is 12.7 kDa. However, Vpr migrates as a 14/15-kDa band during electrophoresis; this increase in real molecular mass of Vpr is due to posttranslational modifications. Vpr is an accessory protein of HIV, which is not essential for viral replication. Truncation of the open reading frame in *vpr* has resulted in production of slowly replicating viral progeny (Ogawa et al., 1989). This is the rationale for the name Vpr, which is from *viral protein, regulatory*. Vpr undergoes oligomerization to form a hexamer of 100 kDa. 1–42 AA at N-terminal of Vpr are responsible for oligomerization, of which 36–42 AA are crucial. On the basis of oligomerization, Vpr protein can be divided into two parts: the trypsin-resistant part, which lies between 1 and 73 AA and is highly resistant to trypsin digestion, and the trypsin-sensitive part, which lies between 74 and 96 AA and is easily digestible by trypsin.

Vpr consists of three domains that expand from 1 to 96 AA. The first domain extends from AA 1 to 42, the second domain extends from AA 43 to 82, whereas the third domain extends from AA 77 to 96. AA 17–34 of the first domain forms an amphipathic α-helix (DiMarzio et al., 1995; Yao et al., 1995), which forms an ion channel. The cytopathic effects of Vpr are attributable to its ability to form ion channels, which disturb the osmolarity of cells. The second domain of Vpr contains the HS/FRIG motif (71–82 AA), which is a leucine/isoleucine-rich region (60–81 AA). The second domain of Vpr forms a leucine zipper-like structure that interacts with host cell proteins (Zhao et al., 1994). The third domain of Vpr protein lies in the C terminus, which contains six arginine residues and is known as a protein transduction domain because it is responsible for transduction (Kichler et al., 2000; Coeytaux et al., 2003). The serine residues of the third domain are responsible for nuclear localization signals to Vpr (Lu et al., 1993). The C terminus of Vpr is also responsible for some of the biological functions such as apoptosis, cell-cycle arrest, and defects in mitosis (Romani and Engelbrecht, 2009).

Vpr is usually detected at an early stage of infection because Vpr is present in virions and is released into host cells during infection. However, at late stages of replication, Vpr can be detected in higher quantity in infected cells. Vpr incorporation into new virions is possible because of direct interaction of Vpr at the C terminus of Gag precursor (Bachand et al., 1999; Selig et al., 1999). Efficient incorporation of Vpr into virion requires a ratio of 1:7 for Vpr/Gag that means each virion may have ~275 molecules of Vpr in it. Vpr is an accessory protein and is therefore not crucial for replication, but it has been implied in various biological functions like transcription of new viral genome (Sawaya et al., 2000), apoptosis induction, disruption of cell-cycle control, induction of defects in mitosis (Chang et al., 2004), nuclear transport of preintegration complex (PIC) (Vodicka et al., 1998), facilitation of reverse transcription (Rogel et al., 1995), suppression of immune activation (Ramanathan et al., 2002), and reduction of the HIV mutation rate (Jowett et al., 1999). Vpr is capable of breaching cell membranes, and this property supports entry of extracellular Vpr into uninfected cells (Sherman et al., 2002a).

24.6.2 Comparative Omics of Vpr

We have compared Vpr from HIV-1, HIV-2, SHIV, SIV, and FIV, and the comparison is detailed in Table 24.3. In HIV-1, HIV-2, and SHIV, this gene is known as *vpr*; however, *vpr* in SHIV is of simian origin. In the case of SIV, the gene equivalent to *vpr* is known as *vpx*, whereas the *vpr* equivalent in FIV is cited in the literature as *orf-A/orf-2* (Figure 24.2). In general, the size of the *vpr* gene varies from 237 to 357 nucleotides. The smallest size of *vpr* is 237 nucleotides in FIV, whereas the largest one is 357 nucleotides in the case of SIV. In general, the genomic location of *vpr/vpx/orf-A* or *orf-2* is after nucleotide sequence number 5000, and it ends before the 6600 nucleotide sequence, implying that the start codon and the stop codon both lie in between these nucleotide numbers. Detailed information about these genes can be easily extracted from different databases by using a unique gene identifier given to the *vpr* gene from different immunodeficiency viruses. The molecular mass of Vpr protein varies from 9.6 to 14 kDa. The Vpr protein of HIV-1 has the highest molecular mass of 14 kDa, whereas the molecular mass of Vpr-like proteins is smallest in SIV (9.6 kDa) and FIV (9.65 kDa) (Table 24.3). What we still do not know is how crucial are all of these differences in *vpr* gene and proteins for their biological properties. Detailed omic studies may offer some insight on this issue. Nucleotide sequences for *Vpr* from five different immunodeficiency viruses were retrieved from the NCBI along with the location in their respective genomes, and their identifiers are shown in parentheses (Figure 24.3a).

TABLE 24.3
Comparison of Vpr (Gene and Protein) of Five Different Immunodeficiency Viruses

Features	HIV-1	HIV-2	SHIV	SIV	FIV
		Gene			
Gene/variant	*vpr*	*vpr*	*vpr*[#]	*vpx*	*orf-A/orf-2*
Locus	HIV1gp4	HIV2gp4	SIVgp05	SIVgp3	FIVgp4
Gene size[†]	292	264	306	357	237
Start codon	5105	6239	5634	5683	5992
Stop codon	5396	6502	5939	6039	6228
Accession number	NC_001802.1	NC_001722.1	NC_001870.1	NC_001549.1	NC_001482.1
GeneID	155807	1724718	1446396	1490006	1724707
		Protein			
Protein	vpr	vpr	vpr[#]	vpx	FIVgp4[††]
Total amino acids	96	87	101	118	78
Protein molecular mass (kDa)	14	10.1	11.5	9.6	9.7
Accession number	NP_057852.2	NP_056841.1	NP_046126.1	NP_054371.1	NP_040975.1

HIV-1, human immunodeficiency virus-1; HIV-2, human immunodeficiency virus-2; SHIV, simian human immunodeficiency virus; SIV, simian immunodeficiency virus; FIV, feline immunodeficiency virus; [†], gene size is calculated from NCBI databank; [#], in SHIV, vpr is of simian origin; [††], in case of FIV, vpr protein is known as FIVgp4.

Later, these sequences were aligned with Clustal W software, and it was found that the frequency for homology of different nucleotides varied among genomes of five different immunodeficiency viruses. We have noticed that in total, 42 nucleotides were conserved in *vpr* among all immunodeficiency viruses. Out of four nucleotides, adenine is the most conserved nucleotide and is present in 48% of the conserved positions of *vpr* gene. The least conserved nucleotide is cytosine, which occupies 5% of the conserved positions, respectively. Guanine and thymine are among moderately conserved nucleotides, occurring at 17 and 31% of the conserved positions. With closer analysis, we found that there are around 34 clusters of conserved nucleotide sequences among HIV-1 and HIV-2. Around 16 of these clusters have less than two nucleotides, whereas the other 14 clusters contain more than two nucleotides. These conserved clusters are highlighted in Figure 24.3b. At this moment, it is not clear how these variations and similarities in the nucleotide sequences of *vpr* among different immunodeficiency viruses may affect their biology. Amino acid sequences for Vpr from five different immunodeficiency viruses were retrieved from NCBI, and their identifiers are shown in parentheses (Figure 24.4a).

When these amino acid sequences were aligned with Clustal W software, we noticed that three amino acids are identical, and ten amino acid residues are conserved in nature, whereas nine amino acid residues are semiconservative in nature. After closer analysis for amino acid sequences among HIV-1 and HIV-2, we have found that 37 amino acid residues are similar in these two immunodeficiency viruses. Out of which there are four clusters of two and three amino acid residues and one cluster of four amino acid residues (Fig 24.4b).

24.6.3 Biology of Vpr

24.6.3.1 Vpr and HIV Infection

Vpr supports an increase of viral infection in all cell types studied, so far. Vpr gets localized into the nuclei of infected cells (Lu et al., 1993). During the late stages of virus replication, Vpr incorporates into virions. Vpr localization into the nucleus of infected cells is because of two nuclear

(a)

Human immunodeficiency virus-1 (NC_001802.1)

^{5105}ATGGAACAAGCCCCAGAAGACCAAGGGCCACAGAGGGAGCCACACAATGAATGGACACTAGAGCTT
TTAGAGGAGCTTAAGAATGAAGCTGTTAGACATTTTCCTAGGATTTGGCTCCATGGCTTAGGGCAACATA
TCTATGAAACTTATGGGGATACTTGGGCAGGAGTGGAAGCCATAATAAGAATTCTGCAACAACTGCTGTT
TATCCATTTTCAGAATTGGGTGTCGACATAGCAGAATAGGCGTTACTCGACAGAGGAGAGCAAGAAATG
GAGCCAGTAGATCCTAG$_{5396}$

Human immunodeficiency virus-2 (NC_001722.1)

^{6239}ATGACTGAAGCACCAACAGAGTTTCCCCCAGAAGATGGGACCCCACGGAGGGACTTAGGGAGTGACT
GGGTAATAGAAACTCTGAGGGAAATAAAGGAAGAAGCCTTAAGACATTTTGATCCCCGCTTGCTAATTGC
TCTTGGCTACTATATCCATAATAGACATGGAGACACCCTTGAAGGCGCCAGAGAGCTCATTAAAACCCTA
CAACGAGCCCTCTTCGTGCACTTCAGAGCGGGATGTAACCGCTCAAGAATTGGCTAA$_{6502}$

Simian human immunodeficiency virus (NC_001870.1)

^{5634}ATGGAAGAAAGACCTCCAGAAAATGAAGGACCACAAAGGGAACCATGGGATGAATGGGTAGTGGAG
GTTCTGGAAGAACTGAAAGAAGAAGCTTTAAAACATTTTGATCCTCGCTTGCTAACTGCACTTGGTAATC
ATATCTATAATCGTCACGGAGACACTCTAGAGGGAGCAGGAGAACTCATTAGAATCCTCCAACGAGCGC
TCTTCATGCATTTCAGAGGCGGATGCATCCACTCCAGAATCGGCCAACCTGGGGGAGGAAATCCTCTCTC
AGCTATACCGCCCTCTAGAAGCATGCTGTAG$_{5939}$

Simian immunodeficiency virus (NC_001549.1)†

^{5683}ATGGCATCAGGAAGAGATCCAAGAGAACCTTTACCAGGATGGCTGGAAATCTGGGATCTAGACAGGG
AGCCATGGGACGAATGGCTACAAGACATGCTCAGGGATCTAAACGAAGAAGCCAGAAGGCACTTTGGAA
TGAACATGCTAATCCGAGTATGGAATTACTGTGTAGAGGAGGGAAGGAGACATAATACCCCATGGAATG
AGATAGGCTACAAGTACTATAGAATTGTTCAAAAGTCTATGTTTGTACATTTCAGATGTGGTTGTAGAAG
GAGAGGACCTTTTTCCCCTTACGAAGAGAGGAGAAATGGACAAGGAGGAGGAGCCCCACCCCCTCCTCC
AGGACTTGCATAG$_{6039}$

Feline immunodeficiency virus (NC_001482.1)@

^{5992}ATGGAAGACATAATAGTATTATTCAATAGGGTCACTGAGAAACTAGAAAAAGAATTAGCTATCAGAA
TATTTGTATTAGCACATCAATTAGAAAGGGACAAAGCTATTAGATTACTACAAGGATTATTTTGGAGATA
TAGATTTAAGAAACCCCGAGTAGATTATTGTTTATGTTGGTGGTGTTGCAAATTCTATTATTGGCAGTTGC
AATCTACATTATCAATAACTACTGCTTAG$_{6228}$

(b)

```
HIV-1   ------------------------------ATGGAACAAGCCCCAGAAGACCAAGGGCCA (30)
HIV-2   ---------------ATGACTGAAGCACCAACAGAGTTTCCCCCAGAAGATGGGACCCCA (45)
SHIV    ---------------------------ATGGAAGAAAGACCTCCAGAAAATGAAGGACCA (33)
SIV     ATGGCATCAGGAAGAGATCCAAGAGAACCTTTACCAGGATGGCTGGAAATCTGGGATCTA (60)
FIV     ------------------------------------------------------------

HIV-1   CAGAGGGAGCCACACAATGAATGGACACTAGAGCTTTTAGAGGAGCTTAAGAATGA--AG (88)
HIV-2   CGGAGGGACTTAGGGAGTGACTGGGTAATAGAAACTCTGAGGGAAATAAAGGAAGA--AG (103)
SHIV    CAAAGGGAACCATGGGATGAATGGGTAGTGGAGGTTCTGGAAGAACTGAAAGAAGA--AG (91)
SIV     GACAGGGAGCCATGGGACGAATGGCTACAAGACATGCTCAGGGATCTAAACGAAGA--AG (118)
FIV     --ATGGAAGACATAATAGTATTATTCAATAGGGTCACTGAGAAACTAGAAAAAGAATTAG (58)
            ** *   *       * *    *   *      *     *    **  *  *  **

HIV-1   CTGTTAGACATTTTCCTAGGATTTGGCTCCATGGCTTAGGGCAACA-TATCTATGAAACT (147)
HIV-2   CCTTAAGACATTTTGATCCCCGCTTGCTAATTGCTCTTGGCTACTA-TATCCATAATAGA (162)
SHIV    CTTTAAAACATTTTGATCCTCGCTTGCTAACTGCACTTGGTAATCA-TATCTATAATCGT (150)
SIV     CCAGAAGGCACTTTGGAATGAACATGCTAATCCGAGTATGGAATTACTGTGTAGAGGAGG (178)
FIV     CTATCAGA-ATATTTGTATTAGCACATCAATTAGAAAGGGACAAAGCTAT-TAGATTACT (116)
        *    *   *  **                         *  *    * *  *

HIV-1   TATGGGGATA-------CTTGGGCAGGAGTGGAAGCCATAAT----AAGAATTCTGCAAC (196)
HIV-2   CATGGAGACA-------CCCTTGAAGGCGCCAGAGAGCTCAT----TAAAACCCTACAAC (211)
SHIV    CACGGAGACA-------CTCTAGAGGGAGCAGGAGAACTCAT----TAGAATCCTCCAAC (199)
SIV     GAAGGAGACATAATACCCCATGGAATGAGATAGGCTACAAGTACTATAGAATTGTTCAAA (238)
FIV     ACAAGGATTA-------TTTTGGAGATATAGATTTAAGAAAC----CCCGAGTAGATTAT (165)
            *    *            *                           *       *

HIV-1   AACTGCTGTTTATCCATTTTCAGAATTGGGTGTCGACATAGCAGAATAGGCGTTACTCGA (256)
HIV-2   GAGCCCTCTTCGTGCACTT-CAGAGCGGGATGTAACCGCTCAAGAATTGGC-TAA----- (264)
SHIV    GAGCGCTCTTCATGCATTT-CAGAGGCGGATGCATCCACTCCAGAATCGGC-CAACCTGG (257)
SIV     AGTCTATGTTTGTACATTT-CAGATGTGGTTGTAGAAGGAGAGGACCTTTTTCCCCTTAC (297)
FIV     TGTTTATGTTGGTGGTGTTGCAAATTCTATTATTGGCAGTT-GCAATCTACATTATCAAT (224)
              * **   *    ** ** *      *             *

HIV-1   CAGAGGAGAGCAAGAAATGGAGCCAGTAGATCCTAG------------------------ (292)
HIV-2   ------------------------------------------------------------
SHIV    GGGAGGAAATCCTCTCTCAGCTATACCGCCCTCTAGAAGCATGCTGTAG----------- (306)
SIV     GAAGAGAGGAGAAATGGACAAGGAGGAGGAGCCCCACCCCCTCCTCCAGGACTTGCATAG (357)
FIV     AACTACTGCTTAG----------------------------------------------- (237)
```

FIGURE 24.3 Nucleotide sequence of vpr. (a) Nucleotide sequence of *vpr* from five different immunodeficiency viruses. The accession number of each sequence is given in parentheses. The position of the start codon is given as a superscript number, while the location of the stop codon is given as a subscript number. Sequences were retrieved from NCBI database (http://ncbi.nlm.nih.gov). †, in SIV it is known as *vpx*; @, in FIV it is known as *orf*-A/*orf*-2. (b) *vpr* nucleotide sequences from five different immunodeficiency viruses were aligned using Clustal W software. *, sequence homology. Nucleotide sequences similar between HIV-1 and HIV-2 are highlighted.

(a)

Human immunodeficiency virus-1 (NP_057852.2)

MEQAPEDQGPQREPHNEWTLELLEELKNEAVRHFPRIWLHGLGQHIYETYGDTWAGVEAIIRILQQLLFIHF
RIGCRHSRIGVTRQRRARNGASRS

Human immunodeficiency virus-2 (NP_056841.1)

MTEAPTEFPPEDGTPRRDLGSDWVIETLREIKEEALRHFDPRLLIALGYYIHNRHGDTLEGARELIKTLQRAL
FVHFRAGCNRSRIG

Simian human immunodeficiency virus (NP_046126.1)

MEERPPENEGPQREPWDEWVVEVLEELKEEALKHFDPRLLTALGNHIYNRHGDTLEGAGELIRILQRALFM
HFRGGCIHSRIGQPGGGNPLSAIPPSRSML

Simian immunodeficiency virus (NP_054371.1)†

MASGRDPREPLPGWLEIWDLDREPWDEWLQDMLRDLNEEARRHFGMNMLIRVWNYCVEEGRRHNTPWN
EIGYKYYRIVQKSMFVHFRCGCRRRGPFSPYEERRNGQGGGAPPPPPGLA

Feline immunodeficiency virus (NP_040975.1)@

MEDIIVLFNRVTEKLEKELAIRIFVLAHQLERDKAIRLLQGLFWRYRFKKPRVDYCLCWWCCKFYYWQLQS
TLSITTA

(b)

```
HIV-1           MEQA-----PEDQGP---------QREPHNEWTLELLEELKNEAVRHFPRIWLHGLGQHIY (47)
HIV-2           MTEAPTEFPPEDGTP---------RRDLGSDWVIETLREIKEEALRHFDPRLLIALGYYIH (52)
SHIV            MEER----PPENEGP---------QREPWDEWVVEVLEELKEEALKHFDPRLLTALGNHIY (48)
SIV             MASGR---DPREPLPGWLEIWDLDREPWDEWLQDMLRDLNEEARRHFGMNMLIRVWNYCV (57)
FIV             MEDII---VLFNRVT---------EKLEKELAIRIF-VLAHQLERDKAIRLLQGL-FWRY (46)
                * .        :  .        ..  .:     : : .:  :.     *  :

HIV-1           E---TYGDTWAGVE-AIIRILQQLLFIHFRIGCRH----SRIGVTRQRRARNGASRS--- (96)
HIV-2           N---RHGDTLEGAR-ELIKTLQRALFVHFRAGCNR----SRIG----------------- (87)
SHIV            N---RHGDTLEGAG-ELIRILQRALFMHFRGGCIH----SRIGQPGGGNPLSAIPPSRSM (100)
SIV             EEGRRHNTPWNEIGYKYYRIVQKSMFVHFRCGCRRRGPFSPYEERRNGQGGGAPPPPPGL (117)
FIV             R--------FKKPRVDYCLCWWCCKFYYWQLQSTL-----SITTA--------------- (78)
                .                       * :::  .

HIV-1           -
HIV-2           -
SHIV            L (101)
SIV             A (118)
FIV             -
```

FIGURE 24.4 Amino-acid sequence for Vpr. (a) Amino-acid sequences of Vpr from five different immunodeficiency viruses. The accession number of each amino-acid sequence is given in parentheses. The sequences were retrieved from the NCBI database (http://ncbi.nlm.nih.gov). †, in SIV it is known as *vpx*; @, in FIV it is known as *orf*-A/*orf*-2. (b) Amino-acid sequences for Vpr have been aligned using Clustal W software. *, identical sequences; :, conserved substitutions; and ., semiconserved substitutions. Amino acid residues similar between HIV-1 and HIV-2 are highlighted.

localization signals (Sherman et al., 2001). Thus, active nuclear transport of Vpr may be a necessity for Vpr pathogenicity. Extracellular Vpr has been shown to increase viral replication in latently infected T lymphocytes (Levy et al., 1994). Vpr enhances viral transcription and production of new virions in monocytic and macrophage cell types. Infection of nondividing cells like monocytes and macrophages is critically dependent on Vpr. A productive HIV infection in these cells is due to translocation of viral PIC from cytoplasm to nucleus (Emerman et al., 1996). PIC is considered to have nucleophilic characteristics, because of nuclear localization signals present in Gag and Vpr, which support migration of PIC (Rouzic and Benichou, 2005).

24.6.3.2 Vpr and Immune Suppression

A competent immune system helps to eliminate any threat to the body, no matter in which form a threat is presented to the host. As we know, immune suppression is a hallmark of HIV infection leading to AIDS. At early stages of HIV infection, immune activation has been observed. This activation is the rationale used for diagnosis of HIV infection, which is based on antibodies against HIV being present in host circulation. As HIV infection progresses, there is a continuous decline in immune response. This suppression of immune response could be either due to HIV itself or due to various HIV proteins produced by HIV during the course of infection.

Vpr has also been found to play an active role in immune suppression. Vpr suppresses T-helper type 1 (Th_1) and antigen-specific cytotoxic T-lymphocyte responses in host. Vpr is known to induce a powerful immunosuppression, because it suppresses both humoral-mediated immune response as well as cell-mediated immune response by suppressing various types of immune cells like T-cells, monocytes, macrophages, and natural killer (NK) cells. Vpr suppresses specific antibody production against HIV by inhibiting T-cell clonal expansion via induction of G2 arrest (Huard et al., 2008) and suppressing T-cell proliferation. Vpr downregulates expression of CD40, CD80, CD83, and CD86 on macrophages and dendritic cells under *in vitro* conditions. Vpr also reduces the capability of monocytes to mature into dendritic cells (Muthumani et al., 2004). These effects of Vpr ultimately suppress antigen processing and presentation. Vpr impairs NK cell activity *in vitro* by reducing γ-interferon production (Majumder et al., 2008). Vpr suppresses T-cell activation and downregulates production of proinflammatory cytokines (tumor necrosis factor [TNF]-α and interleukin [IL]-12) and chemokines (RANTES, MIP-1α, and MIP-1β).

24.6.3.3 Vpr and Host Cell Responses

During HIV infection, the role of host innate immunity is exemplified by induction of cellular heat-shock proteins (HSPs) (Brenner et al., 1995). For example, HSP27 and HSP70 mRNA transcripts can be detected within 3–8 h following HIV infection, but this expression is transient in nature, and it gets downregulated within 24 h postinfection. The expression of HSPs is concomitant with appearance of full-length viral mRNA (Iordanskiy et al., 2004). At a functional level, HSP27 acts as an antagonist for Vpr-specific cell-cycle arrest and apoptosis (Liang et al., 2007). HSP70 binding to Vpr may have two implications: HSP70 has neutralizing effect on Vpr activity, and HSP70 may be responsible for a reduction in the nuclear import activity of Vpr, which ultimately reduces HIV-1 replication. HSP70 regulates Vpr-dependent G_2 arrest and apoptosis; therefore, it can be concluded that HSP70 plays a role as an anti-HIV factor. Apart from HSPs, other cellular proteins like heat shock factor 1 (Benko et al., 2007), elongation factor 2 (Zelivianski et al., 2006), and glycogen-synthase kinase-3 also respond to HIV infection as well as Vpr expression.

24.6.3.4 Vpr and Viral PIC

After the release of viral protein and genome into the cytoplasm of an infected cell, synthesis of proviral DNA starts. For productive infection of HIV, it is essential for proviral DNA to integrate into cellular DNA. Translocation of HIV-PIC to nucleus was one of the first recognized activities mediated by Vpr. To increase the efficiency for the transport of cDNA, the preintegration complex is formed. HIV replication in nondividing cells (or terminally differentiated cells like macrophages, monocytes, and incompletely activated $CD4^+$ T-lymphocytes) requires transportation of proviral DNA (in context to viral PIC) from cytoplasm to nucleus, where Vpr plays a key role in transportation of PIC (Emerman et al., 1994; Heinzinger et al., 1994). The PIC is composed of various viral proteins, such as Vpr, RT, IN, NA, and MA, in addition to viral nucleic acids. In association with cytoplasmic microtubules, Vpr directs PIC towards nucleus, where PIC localizes in perinuclear region close to centrosomes. Cytoplasmic dynein also supports PIC movement towards nucleus.

Classical nuclear localization sequence (cNLS) is a short peptide region rich in lysines and arginines, responsible for PIC transportation. cNLS binds with adaptor proteins like importin-α (*i.e.*, karyopherin-α or NLS receptor-α). cNLS-importin-α complex binds with importin-β receptor (*i.e.*, karyopherin-β or NLS receptor-β) and interacts with nuclear pore complex, which leads to nuclear translocation of PIC. Most studies have reported that Vpr expressed without other viral proteins localizes predominantly to nuclear envelope (Lu et al., 1993; Chen et al., 1999). Two hypotheses for the mode of action of Vpr in nuclear import have been proposed: Vpr targets PIC to the nucleus via a distinct, importin-independent pathway (Gallay et al., 1996; Jenkins et al., 1998), and Vpr modifies cellular importin-dependent import machinery.

24.6.3.5 Vpr and G_2/M Arrest

G_2/M works as a checkpoint for mitosis, and the main purpose of this checkpoint is to prevent the transfer of mistakes that occurred during DNA replication to the progeny of mitosis. This step indirectly supports viral replication and prolongs cell survival, for which Vpr plays a significant role by making sure that virus gets replicated without inducing the apoptosis of infected cells. During HIV infection, the viral genome has to be integrated with host genomic DNA. Integration of viral genome to host genomic DNA is recognized as an error by the molecular check system of host cell. This could be a reason to induce the apoptosis of infected cells so that cells with errors should be eliminated, which will not support HIV infection, but HIV Vpr has a unique ability to initiate G_2/M arrest, which supports survival of infected cells until the completion of HIV replication. Therefore, for survival, HIV has evolved a mechanism either to control or block the process of molecular corrections. This is achieved by G_2/M arrest of infected cells caused by the action of Vpr. Vpr exposure to host cells can take place by the following three means: Vpr coming from virion, the presence of extracellular Vpr, and de novo generation of Vpr in infected cells. Exposure of Vpr to infected cells by any or all of these methods can cause G_2/M arrest (Sherman et al., 2002b). Vpr induces replication stress, which can trigger cell-cycle arrest, an important cellular event to support HIV infectivity.

Provirus transcription is high during G_2. Therefore G_2 arrest provides replication advantages to HIV (Goh et al., 1998; Belzile et al., 2007), and it is supported by the activity of Vpr. Vpr stops cell-cycle progression from G_2 to M by degradation of some cellular proteins, which have not been characterized yet (Wen et al., 2007). Additionally, G_2 arrest induces apoptosis rather than necrosis, as a result of which the neighboring infected cells do not experience a hostile environment.

The interaction of Vpr with cellular DNA repair proteins like HHR23A and uracil DNA glycosylase interfere with DNA repair process, which helps in cell-cycle arrest at the G_2/M stage (Elder et al., 2002). G_2 arrest helps to improve HIV protein expression during polyadenylation and translation (Brasey et al., 2003) in addition to transcription.

Vpr up regulates expression of gene known as human survivin, which produces survivin protein (Zhu et al., 2003). Survivin protein extends the survival time of cells undergoing apoptosis. This action provides enough time for the virus to complete replication in cells undergoing apoptosis. This mechanism is helpful in increasing infectivity. Finally, these cells undergo apoptosis as induced by Vpr.

24.6.3.6 Vpr and Apoptosis

Physiologically apoptosis is a homeostatis mechanism used by multicellular organisms for the removal of defective cells from tissue without having any adverse effect. Although apoptosis of HIV-infected cells promotes HIV infection, in addition the loss of T-cells caused by apoptosis reduces the immune competence of the host. Therefore, apoptosis of T-cells leads to immune dysfunction, which is a hallmark feature of AIDS. Various stimuli induce apoptosis, but mostly apoptosis is regulated by two death-signaling pathways: extrinsic pathways and intrinsic pathways. Both pathways of apoptosis are mediated by caspases.

The three forms of Vpr (virion-associated, de novo synthesized, and extracellular) have been found to induce apoptosis to target cells (Stewart et al., 1997). A conserved motif at the C terminus of Vpr, which spans from AA 71 to 82 is responsible for the induction of apoptosis. Vpr induces the intrinsic pathway of apoptosis, with the release of different mitochondrial intermembrane proteins, viz., adenine nucleotide translocator (ANT), apoptosis-inducing factor, cytochrome *c*, procaspases, and HSPs. Releases of these proteins are essential to activate caspases and DNAses. Evidence suggests that Vpr permeabilizes into mitochondrial membrane by binding with permeability transition pore complex (PTPC) through a process known as mitochondrial outer membrane permeabilization (MOMP) (Green and Kroemer, 2004). Vpr-induced apoptosis cannot occur in the absence of PTPC. Vpr crosses outer mitochondrial membrane through the voltage-dependent anion channel and binds with ANT. Vpr uses MOMP by binding with ANT of inner mitochondrial membrane (Jacotot et al., 2001; Brenner and Kroemer, 2003). These steps cause depolarization and swelling of inner mitochondrial membrane. Changes in mitochondrial membrane induce release of apoptosis factors. Only a small fraction of Vpr molecules is sufficient to induce apoptosis.

24.5.3.7 Vpr and Modulation of Gene Expression

Vpr protein is produced along with the expression of host genomic DNA; therefore, it affects functioning of infected cells (Balasubramanyam et al., 2007). Alteration in host cell functioning is attributable to changes in gene expression induced by either direct or indirect effects of Vpr. Vpr alters cell proliferation and differentiation. Vpr downregulates activity of p53 and Spl transcription factors, which enhances promoter activity of *LTR* (Sawaya et al., 1998). Under normal conditions, glucocorticoids induce antiinflammatory and immunosuppressive effects. Vpr mimics these effects of glucocorticoids through its interaction with glucocorticoid receptor (GR) (Ramanathan et al., 2002). Interaction of Vpr with GR has direct effect on virus replication and gene expression.

24.6.3.8 Vpr and RT Fidelity

After fusion of virion with host cell, viral genome and proteins are released into host-cell cytoplasm. Reverse transcription of viral RNA takes place in cytoplasm within a large nucleoprotein complex known as reverse transcription complex (RTC). RTC consists of two copies of viral RNA along with various viral proteins like reverse transcriptase, integrase, Vpr, and a few molecules of matrix protein (Fassati and Goff, 2001; Nermut and Fassati, 2003). Initial reverse transcription of viral RNA starts within the viral particle itself; the rest of the process gets completed in the cytosol of target cells after viral entry. The presence of Vpr in RTC suggests that Vpr may have an important role for the completion of virus life cycle in infected cells. Mutations are very common during the RT process, because RT lacks proofreading ability. Excessive mutation can be detrimental to HIV; however, Vpr compensates, maintaining a significant level of HIV replication by reducing or controlling mutations during replication.

24.7 HIV REPLICATION

The replication of HIV is quite different from replication of other viruses, because here the genetic material is RNA instead of DNA. Any virus, including HIV too, in principle is simply a bag of nucleoprotein complex without any sign of life. Viruses including HIV can replicate only in a living host because they use the host's synthetic machinery to their advantage. The process of HIV replication can be divided into five different stages (Figure 24.5) (Walker, 2007; Abbas et al., 2008; Verma et al., 2009):

1. Viral entry
2. Reverse transcription
3. Integration of proviral DNA
4. Transcription and translation
5. Completion and release

24.7.1 Viral Entry

The initial step for viral infection is attachment of HIV to specific receptors on host cells followed by viral entry. The process of viral attachment starts with the interaction of two noncovalently associated viral glyoproteins, gp120 and gp41. The gp120 of HIV binds with high affinity to CD4

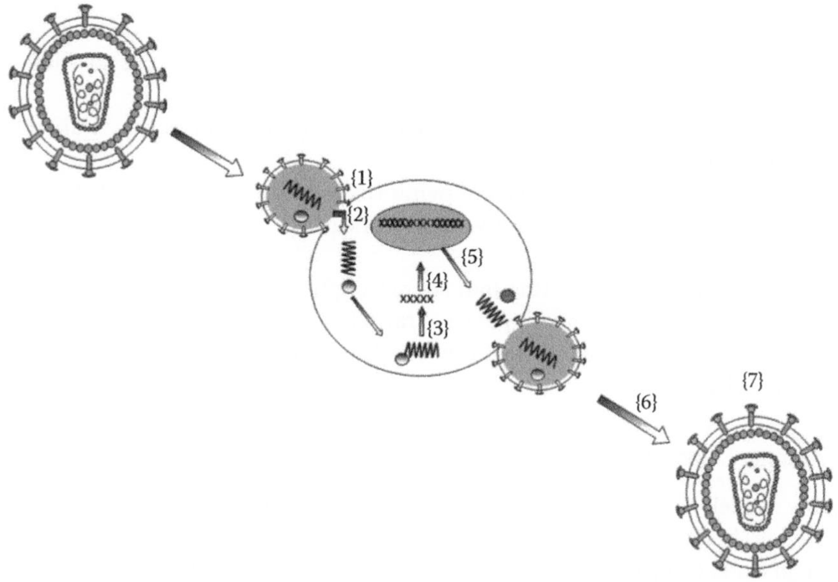

FIGURE 24.5 Schematic representation of HIV replication. {1} Attachment of HIV to host cell. {2} Fusion and release of HIV components into host cell. {3} Conversion of viral RNA into cDNA with the help of reverse-transcriptase enzyme. {4} Integration of cDNA into host genomic DNA with the help of integrase enzyme. {5} Completion of synthesis of viral RNA and other polypeptide. {6} Release of new virus. {7} New virion. , viral RNA; XXX, cDNA; , protease; , reverse transcriptase.

receptor present on T-cells. CD4 is a 58-kDa glycoprotein present on the cell surface of >60% of T-lymphocytes. CD4 receptors act as primary receptors for HIV entry. CD4 receptors are also present on other cells like monocytes, macrophages, dendritic cells, and microglial cells. Studies with soluble CD4 receptors have confirmed that CD4 is the primary receptor for HIV-1, HIV-2, and SIV. Viral attachment leads to fusion of viral envelope with plasma membrane of host cells. This fusion takes place in association with gp41. The fusion of virion with the plasma membrane of the host cell supports the release of viral genome and proteins into host cells. CXCR4 and CCR5 are other coreceptors that are important for viral entry, although their expression varies with cell type.

24.7.2 Reverse Transcription

After release of viral genome into cytoplasm, reverse-transcriptase enzyme starts synthesizing cDNA from viral RNA. The newly synthesized cDNA is known as *proviral DNA*. Because the RT enzyme lacks proofreading ability, the process of reverse transcription is highly error prone. High vulnerability for error during DNA synthesis is the major cause for the enormously high mutation rate in HIV. The majority of mutations help HIV survival, whereas some mutations can be detrimental to HIV itself. Drug resistance in HIV is also attributable to the ability of HIV to mutate frequently.

24.7.3 Integration of Proviral DNA

After synthesis of proviral DNA, it has to be integrated with host cell DNA. Integration of proviral DNA into the host's genomic DNA is mediated through a viral enzyme known as integrase. *LTR* is also considered essential for HIV integration.

24.7.4 Transcription and Translation

After integration of proviral DNA into the host-cell genome, there are two possibilities:

1. Viral DNA remains dormant in host cell and does not initiate an active viral replication, which is known as *latent HIV infection*.
2. In other circumstances, integration of proviral DNA causes activation of different cellular transcription factors required for active viral replication.

Viral *LTR* binds with host-cell NF-κB to initiate transcription of viral RNA. At this stage, Tat and Rev viral proteins are produced and regulate viral replication in activated but infected T-cells. During the late stages of viral replication, Env and Gag proteins are produced and complete the formation of new virions.

24.7.5 Assembly and Release

The assembly and release of virions is the final stage of viral replication, which supports the spread of viral infection. In this phase, Env glycoprotein passes from endoplasmic reticulum and is transported through Golgi complexes. Env protein (gp160) is cleaved by proteases leading to production of gp120 and gp41. After cleavage, gp120 and gp41 both migrate towards the plasma membrane, where gp41 anchors gp120 to plasma membrane of infected cells. Gag and Gag-Pol proteins help in the budding of new virions. Viral proteases cleave polyproteins into different functional HIV proteins and enzymes. A better understanding of various aspects of HIV replication with the application of omics tools may be useful in identifying novel drug targets and designing new drugs.

24.8 CLINICAL STAGES OF HIV INFECTION

HIV infection starts with seroconversion and terminates into full-blown AIDS. Enormous variations have been observed among different clinical end points during progression of the disease, which causes redundancy of various diagnostic tests. There was a need to establish a standard classification for different clinical stages of HIV infections. CDC has therefore developed a system of classification to designate different clinical stages among HIV seropositives (CDC, 1992).

Every clinical stage has different end point determinations, which are helpful in deciding on a course of treatment. Presently, there are hardly any systematic studies that demonstrate a correlation between different clinical stages of HIV-infected patients and omics. More studies with larger sample sizes may be very helpful in gaining insight into understanding the triggers for transition of one clinical stage to another. Omics data will be able to indicate unique salient features for different clinical stages, although it remains questionable as to whether or not this information can be used for further improvement in the health status of HIV-seropositive patients.

24.8.1 Stage I: Primary HIV Infection or Seroconversion

The primary HIV infection stage is also known as seroconversion. At this stage, antibodies against HIV are present. Mild symptoms like fever, malaise, diarrhea, lymphoadenopathy, sore throat, headaches, etc., are reported in >65% cases. Diagnosis is done by detecting HIV antigens and anti-HIV antibodies in a blood sample. Confirmatory tests are recommended for patients with negative results with the history of high-risk behavior. The primary HIV infection stage lasts about 10–12 weeks.

24.8.2 Stage II: Asymptomatic Stage

During the asymptomatic phase, antibodies against HIV can be detected regularly in blood. Viral replication does not stop but slows down tremendously. $CD4^+$ counts can remain normal or above normal, i.e., 350×10^6 cells/L during the asymptomatic stage. HIV-seropositive patients can remain in the asymptomatic phase for ≥10 years, this stage can be prolonged, if the correct choices of antiretroviral treatments are made. The main objective of treatment at this stage is to maintain normal immune functions in patients by maximally suppressing viral replication. Minimization of viral replication through therapeutic intervention prevents/or slows down erosion of the immune system.

24.8.3 Stage III: Persistent Generalized Lymphoadenopathy

In this stage, HIV patients look otherwise healthy; nonspecific adenopathy may persist, but lymph node biopsy is not recommended as routine. Up to 3 months, indications from biopsies in HIV patients are similar to those of any non-HIV patients.

24.8.4 Stage IV: Symptomatic HIV Infection

A rapid decline of immune competence is observed due to enhanced HIV replication, which leads to rapid progression of the disease. The crucial triggers to increase viral replication at this stage are still not clear. Some of the common constitutional symptoms among HIV patients during the symptomatic phase include fever, malaise, etc. These constitutional symptoms can be treated easily, although HIV replication at this stage is difficult to control. Choices of therapeutic intervention for further improvement in the health status of patients are either very limited or not very useful for patients. Ultimately these patients start showing neurological symptoms (neuroAIDS), AIDS-defining cancer, and various opportunistic infections. These are the signs and symptoms of terminal illness, which require counseling of patient about the final outcome of disease.

24.9 CLADE DIVERSITY

There is a general perception that HIV is a single virus, but in reality this assumption is absolutely wrong. Genomic variation in HIV is known as *genetic diversity* or *genetic variation.* Surprisingly, genetic variations frequently occur even in different compartments of the body of an individual patient infected with HIV. Among the scientific community, an opinion is evolving that HIV is becoming a supervariable virus because the variations observed in HIV are comparatively higher than those in other viruses famous for their variability. Genetic diversity is higher in peripheral circulation compared with the CNS. The precise reason for this genetic diversity in HIV is not well understood except that the reverse-transcriptase enzyme lacks proofreading ability. The absence of proofreading ability is the main reason for the evolution of new HIV subtypes. In general, these variations are very helpful for viral survival, but excessive mutations in HIV genome can be detrimental to virus, itself. A high replication rate of HIV also contributes for high genetic variability. As per one estimate, HIV can replicate a billion times in 24 h in an infected host. If we consider that only 1% is the error rate, more than one million new viral progeny (which are different from the original) can be produced in a day (Boehringer Ingelheim, 2005). This scenario is just for 24 h of replication in a single patient, who will be surviving for years after the initial infection, producing an insurmountable number of new progeny. Many questions still remain unanswered. What are the implications of these variations for the host? How is this process beneficial to the virus itself? How and up to what extent do different HIV genes get affected by these variations? Do they take advantage of their variability for alteration in susceptibility? What effects do these variations have on therapeutic strategies? What are the implications of these variations for prophylactic measures like vaccine development? Which gene is more susceptible to variation and why?

To understand the variability, HIV strains have been divided into three main categories: M (major), N (new), and O (outlying). These three strains account for all HIV infections across the

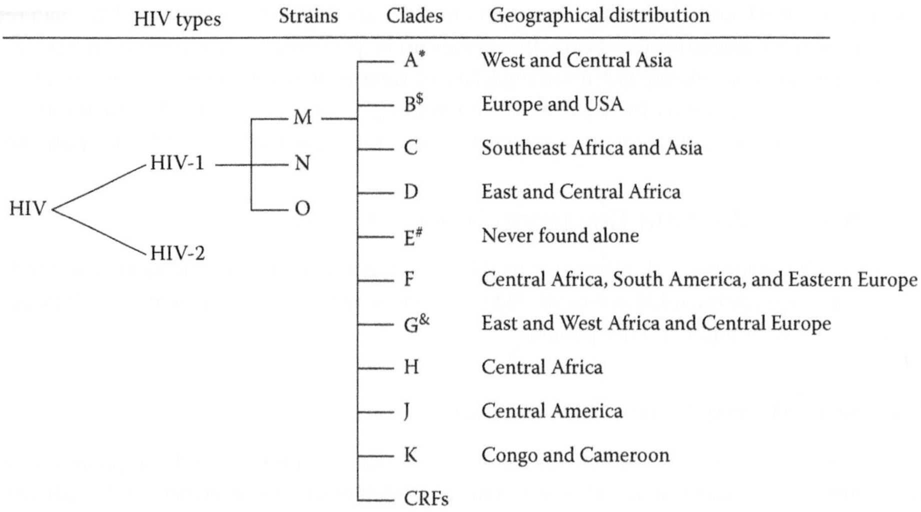

FIGURE 24.6 Different strains and clades of HIV. M, N, O, and P are viral strains whereas, A–H, J, and K are clades of HIV. CRFs, circulating recombinant forms; *, recombinant variant; $, >40% new infections in Europe are non-B because of increased migration; #, E is always present as A/E recombinant; &, A/G variant also exists.

globe. Distribution of these strains vary from one geographical location to another, but the M type is the main strain, which accounts for >90% HIV infections. Classification of M, N, and O is based on variations in the *gag* gene. The M strain of HIV is further divided into 11 major groups called *clades*. Clade classification is based on similarity as well as variability among these groups. These clades belonging to the M strain are designated as A–H, J, and K. Somehow clades have shown geographical preferences (Figure 24.6) (Spira et al., 2003).

In general, clade B is prevalent in developed parts of the world like North America and Europe, whereas Clade C is more prevalent in poor communities in Africa and Asia. Some clades are also know as recombinant (A and A/G), whereas some are mosaic form (A/E). This is not an exhaustive list of clades. A high possibility of evolution of new clades cannot be ruled out because of frequent international travel, which could be a reason for clade mixing and evolution of new clades.

HIV-1 has nine genes, and the question is, which gene(s) contribute for such a huge variability? There is no precise answer for this question, because RT enzyme itself could be responsible for variation in all nine genes, although the vulnerability of different genes to mutation is dependent upon various factors like nucleotide sequences, gene location, and the efficacy of individual gene transcription and translation. Variability is an advantage for the evolution and survival of virus and conversely a disadvantage too, which can make them vulnerable to extinction. Usually, clade diversity classification is based on variation in the *env* gene. So far, *env* is found to be highly variable gene in HIV, and ~20–50% variation is the criteria for their classification. Env protein among M and O type may vary up to 20–50%, whereas the N subtype turns out to be equidistant from M and O subtypes. Interclade variation of *env* has been reported to be ~20–30%. However, intraclade variation of *env* is ~10–15% in the M subtype.

Another important gene in HIV is *pol*, which encodes for RT and protease enzyme. It has been reported that *pol* is 2–3 times less divergent compared with *env*. Possibly the decreased vulnerability of the *pol* gene to variation is to avoid extinction of virus, because *pol* is the most crucial gene for viral replication. Excessive mutation of the *pol* gene would become a threat to HIV survival. Knowing the importance of the *pol* gene, we would like to emphasize that the intraclade difference for *pol* among clade B is around 3.5–5.6%, which is comparatively a lot less than *env*. The knowledge about variation in *pol* is important for the development of treatment protocols for HIV, because the majority of antiretroviral drugs target RT activity. An in-depth investigation to understand nucleotide variation and how nucleotide changes are getting translated ultimately into variations among amino acid is important, because these proteins are finally responsible for biological activity.

Research on clade diversity of structural genes has received the most attention so far. HIV also has different accessory genes in its genome. As we know, these genes are not essential, but accessory genes play an important role in the maintenance of HIV infections by assuring efficient and sufficient viral replication. In this respect, studies of variability among various accessory genes are also important. *nef* has been reported to vary ~14–24% among different clades, whereas the closest similarity has been shown between clades B and D. At present, it is difficult to predict the importance of variations in *nef* with reference to its role in disease progression. Similarly, a possibility of variation in *vpr* is also expected and predicted, but we do not have sufficient information yet.

Further studies in omics can offer us a better understanding about the significance of different genes for the evolution of new clades; probably these studies can shed some light on the compelling reasons for the evolution of new clades. International repositories for viral clades could prove to be a useful resource for studying their impact on pathogenesis, the competition among different clades, and the reasons for the cellular susceptibility of specific clades. It is very intriguing that clades have geographical preferences. What is the role of the biology and physiology of natives of those localities in the geographical preference of virus? This approach may also be useful in opening up new avenues for neuroAIDS research, both to understand pathogenesis and to design new therapeutic strategies.

Recently, new schools of thought and evidence suggest that clades themselves might have important roles in HIV transmission, even though the biology behind this kind of conclusion is still a matter of active debate and warrants in-depth study. It has been claimed by a group of scientists that subtype C is more transmittable via a heterosexual route compared with the B subtype, although some groups contradict this opinion. The role of clade diversity has been noticed in vertical transmission, too. Subtype D is more transmittable via a vertical route compared with the A subtype. Various conflicting claims have been cited in literature about HIV transmission via a vertical route. Further studies are warranted to validate these claims.

24.10 NEUROINFLAMMATION, NEUROTOXICITY, AND NEUROAIDS

Enormous improvement in antiretrovirals has been seen. Still, the fact remains that HIV infection can only be controlled and cannot be cured. Nowadays, with the proper use of medication, there is an average increase of >20 years in the life span of HIV patients. The same patients two decades ago would not have survived for more than 2–4 years after their initial diagnosis. Improved antiretroviral treatments have slowed HIV replication tremendously, but a basal level of HIV replication continues even in the presence of antiretroviral treatment. This minimal or barely detectable level of HIV replication leads to subchronic HIV infection and delays the onset of AIDS for >10–20 years depending upon host susceptibility. Low levels of HIV replication can cause incomplete viral replication and apoptosis of infected cells; as a result, incomplete viral components, viz., *tat*, *nef*, *vpr*, etc., or their proteins (complete or incomplete), are released into peripheral circulation. These viral components (both proteins and nucleotides) find their entry into brain, resulting in chronic accumulation of various virotoxins. This is one of the most plausible reasons for the routinely observed dementia with early onset (~30–45 years of age) among HIV patients, whereas the symptoms of dementia in the general population (non-HIV-infected) are observed at a later age (~60 years and above). Therefore, it can be easily assumed that virotoxins may have a crucial role in neuroinflammation and neurotoxicity (Adie-Biassette et al., 1995; An et al., 1996; Gray et al., 2000, Verma et al., 2010b, 2010c).

HIV finds entry into the CNS and establishes a productive infection in the brain during the early stages of infection. HIV is detectable in perivascular macrophages and microglial cells

(Resnick et al., 1988). Neurons are nonpermissive for HIV infection, but infection of accessory cells of CNS is considered as one of the leading causes of neuronal loss. Necrosis and apoptosis both have been associated with neuronal loss in the brains of HIV-infected patients. Still, the causes of neuropathogenesis caused by HIV infections are not certain. A common consensus is that HIV-encoded proteins (virotoxins) and neurotoxic biomolecules secreted by hosts' activated astrocytes and glial cells in response to HIV infection may have crucial role in neuroinflammation. As mentioned, abnormalities of CNS and the release of viral components in circulation contribute towards neuronal death without any possibility of recovery (Jones and Power, 2006).

HIV patients undergoing antiretroviral therapy have an almost continuous presence of HIV virotoxins in their blood circulation. There is a high probability for the entry of these virotoxins into the brain by breaching/crossing the blood-brain barrier (BBB). The BBB is a continuous layer of tightly linked microvascular endothelial cells surrounding the brain that separates the CNS from the rest of the body. This makes the CNS the most protected organ/system in the body. The cerebrospinal fluid (CSF) is separated from peripheral circulation by the blood-CSF barrier known as the choroid plexus (Rapoport, 1976). Pathological conditions and injuries leave opportunities for pathogens, immune cells, and various biomolecules to enter into CNS. The five major components of CNS are different types of cells, viz., astrocytes, oligodendrocytes, neurons, microglia, and perivascular macrophages (Gonzalez-Scarano and Martin-Garcia, 2005; Ghafouri et al., 2006). Some of these cells have specific receptors for HIV, whereas most do not. However, the virus somehow manages to affect almost all of the CNS cells, as evidenced by numerous research reports. In the brain, astrocytes and microglial cells produce various proinflammatory cytokines in response to any foreign molecule like HIV virotoxins, so that the adverse effect of virotoxins can be minimized. Unfortunately, these proinflammatory cytokines, rather than playing a protective role in the brain, end up causing neuronal loss via apoptosis. Particularly for brain, necrosis, or apoptosis, both are harmful because neurons are unable to regenerate, making neuronal loss permanent. Vpr is an accessory protein of HIV, and it is synthesized at the late stages of HIV replication. Much evidence suggests that HIV Vpr might be a potentially toxic molecule that can mediate neuronal apoptosis in HIV-infected patients (Piller et al., 1998; Sabbah and Roques, 2005). Vpr is also essential for active HIV replication in macrophages (Emerman, 1996; Subbramanian et al., 1998). Vpr protein has been detected in the CSF and the serum of HIV-infected patients presented with neurological complications (Levy et al., 1994). Apoptosis has been reported in *in vitro* studies of neurons of different regions of rat brain, viz., hippocampus (Piller et al., 1998), cortical, and striatum (Sabbah and Roques, 2005). Neuronal apoptosis has also been demonstrated to be induced by Vpr in different human neuronal cell lines (Patel et al., 2000, 2002). Vpr can also induce cytokines, and it is also possible that these cytokines do have a neuroinflammatory effect on the brain.

24.10.1 Astrocytes

Astrocytes act as brain sentinels by regulating levels of neurotransmitters like glutamate. Astrocytes are responsible for maintaining brain homeostasis and possess receptors for various neurotransmitters. Proliferation of astrocytes has also been reported in HIV-infected brains (Brack-Werner, 1999; Dong and Benveniste, 2001). Astrocytes have failed to show any robust viral replication, and only a few astrocytes were found to be positive for HIV antigen.

24.10.2 Microglia and Perivascular Macrophages

Astrocytes, microglia, and perivascular macrophages (PM) are the main resident cells in the perivascular region of brain. Microglia and perivascular macrophages behave as resident immunocompetent cells, and these cells respond to unfavorable conditions. Microglia and

PM also respond to infected cells entering into the brain, including HIV-infected monocytes and T-cells. Infected migratory peripheral monocytes and T-cells enter into brain because of a breach in the BBB, and they may become the source of HIV entry in brain (Kaul et al., 2001; Anderson et al., 2002). HIV infections have been reported in microglial cells from adult, infant, and fetal brains *in vitro*. Microglial cells express receptors required for productive HIV infection, viz., CD4, CCR5, and other coreceptors like CCR3, CCR2b, CCR8, and CXCR6. The presence of these receptors and coreceptors make these cells susceptible to HIV infection. *In vitro* studies have shown that mixed microglial culture from human brain can retain replication-competent HIV up to a few months; however, HIV replication is at low levels (Albright et al., 2004).

Perivascular macrophages ($CD4^{+}45^{+}$; PMs) are flat and elongated cells that are mostly located adjacent to brain microvascular endothelial cells. PMs in the brain are continuously recycled by migratory monocytes entering into the brain via peripheral circulation. This recycling/replenishment of PMs can also be considered as a cost paid for the *opening-the-door* phenomenon. PMs are among the most infected cells of CNS, as confirmed by *in situ* immunohistochemistry, in HIV-1 and HIV-2 infections. PMs have been reported to have an ability for active viral replication (Wiley et al., 1986; Takahashi et al., 1996; Fischer-Smith et al., 2004).

24.10.3 Neurons

Neurons are the main effector cells for any cognitive and motor function; therefore, the importance of neurons in HIV-related neuropathogenesis cannot be denied. Neurons are nonpermissive for HIV, because neurons do not express receptors for HIV infection. The presence of HIV viral DNA and proteins in neurons have been reported in a few studies (Nuovo et al., 1994; Bagasra et al., 1996). Generally, when neurons are subjected to any adverse stimuli, they die either by apoptosis or by necrosis. Even *in vitro* studies failed to demonstrate a productive HIV infection in neurons. Neurons cannot regenerate, and that is a major contributing factor toward the failure to detect any HIV infection (Gonzalez-Scarano and Martin-Garcia, 2005).

24.10.4 Oligodendrocytes

Oligodendrocytes are a type of cell present in CNS and responsible for producing the myelin sheath surrounding neurons. Oligodendrocytes are CD4-negative cells; therefore these cells do not get infected with HIV, but some reports have suggested the presence of HIV nucleic acid in oligodendrocytes, which has been detected by *in situ* PCR (Nuovo et al., 1994; Bagasra et al., 1996; Neumann et al., 2001). A limited HIV infectivity has been reported *in vitro* with a specific HIV strain (Albright et al., 1996). It can be postulated that binding of HIV gp120 to galactosylceramide or other proteoglycans of oligodendrocytes may reduce myelin synthesis and can cause increase intracellular Ca^{2+} levels leading to apoptosis. These alterations in oligodendrocytes may modulate axonal conduction, which could lead toward HIV-induced neuropathogenesis.

24.10.5 Inflammatory Cascade

Neurodegeneration and neuronal loss has been commonly observed in HIV-seropositive patients, even though neurons do not get infected with HIV. Neuronal loss may be attributed to neurotoxicity. An exaggeration of neurotoxicity is possible because of excessive production of various proinflammatory and inflammatory biomolecules by different cells of CNS.

Various inflammatory biomolecules include different proinflammatory cytokines, inflammatory cytokines, reactive oxygen species, and certain other biomolecules with neurotoxic potential. These mediators cause neurotoxicity by altering neurotransmission leading to neuronal apoptosis (Panek and Benveniste, 1995; Boven et al., 1999). TNF-α, platelet-activating factor, nitric oxide (NO), and quinolinic acid also behave as neurotoxicants to induce neurotoxicity. *N*-Methyl-D-aspartate-type glutamate-associated neurotoxicity is induced by NO, which is produced by microvascular endothelial cells. An abnormally high level of NO synthase has also been reported in AIDS patients' brains. Similarly, ~40 times higher levels of NO synthase in the neurons of HIV-infected drug addicts have also been observed (Adamson et al., 1996; Minagar et al., 2002).

Proinflammatory cytokines and inflammatory cytokines have been found to play a crucial role in neurotoxicity. Proinflammatory cytokines like TNF-α, IL-1, and IFN-γ are found to be present in elevated level in AIDS patients (Yoshioka et al., 1995; Griffin, 1997). TNF-α is produced by macrophages and microglia and mainly affects oligodendrocytes (Wilt et al., 1995). TNF-α causes damage to the BBB, facilitating the entry of peripheral blood cells (Fiala et al., 1996). Elevated TNF-α levels have been reported in HIV patients. Astrocytes and perivascular macrophages are directly involved with HIV replication in CNS. The majority of CNS cells are incapable of productive HIV infection with the exception of microglia and perivascular macrophages. However, even those cell types that do not participate directly in HIV replication afflict neurotoxicity because of alteration in their normal cellular functioning.

24.10.6 Neuroinvasion

Neuroinvasion is known as the entry of HIV and HIV-related products into CNS. It is difficult to explain HIV neuroinvasion. Experimental evidence suggests HIV infectivity to the brain, which raises real concerns about HIV infectivity because of (Verma et al., 2010b; 2010c):

1. The absence of HIV receptors and coreceptors on most of the cells of the CNS
2. The presence of the blood-brain barrier
3. Adverse responses leading to neuronal death

Research has been conducted to understand the mechanism of HIV entry into the CNS via the BBB (Gonzalez-Scarano and Martin-Garcia, 2005; Ghafouri et al., 2006). On the basis of experimental evidence, three different modes of HIV infection to the CNS have been proposed: the Trojan horse hypothesis, direct HIV entry into brain, and transcytosis.

The most acceptable of these three hypotheses is the Trojan horse hypothesis, which suggests that HIV enters into the CNS as a copassenger during the entry of infected cells (T-cells and monocytes). Abundant numbers of T-cells and monocytes infected with HIV circulate in peripheral blood. Infected cells may cross the BBB to reach the CNS, where they may become a source of viral propagation (Haase, 1986) (Figure 24.7). This model is strongly supported by *in situ* hybridization and immunohistochemical staining (Wiley et al., 1986; Takahashi et al., 1996; Fischer-Smith et al., 2004). HIV-induced abnormalities in the BBB have been observed, but the mechanism of microvascular endothelial cell infection is still not clear because these cells lack receptors and coreceptors for HIV infections. These cells express CCR5 and CXCR4, whereas expression of CD4 on these cells is contradictory (Petito and Cash 1992; Mukhtar et al., 2002; Stins et al., 2003). HIV may gain entry into the CNS either by transcytosis through microvascular endothelial cells or by direct entry (Bomsel, 1997; Banks et al., 2001; Gonzalez-Scarano and Martin-Garcia, 2005; Kramer-Hammerle et al., 2005b). Still more studies are required to pinpoint the reasons for HIV infectivity to the brain. A genuine answer regarding the role of intercellular and intracellular interaction with reference to HIV infection can be found using omics approaches. More information and data about intercellular and intracellular interaction will be helpful in understanding the pathobiological reasons for neuroAIDS and may play a significant role in the development of new therapeutic strategies.

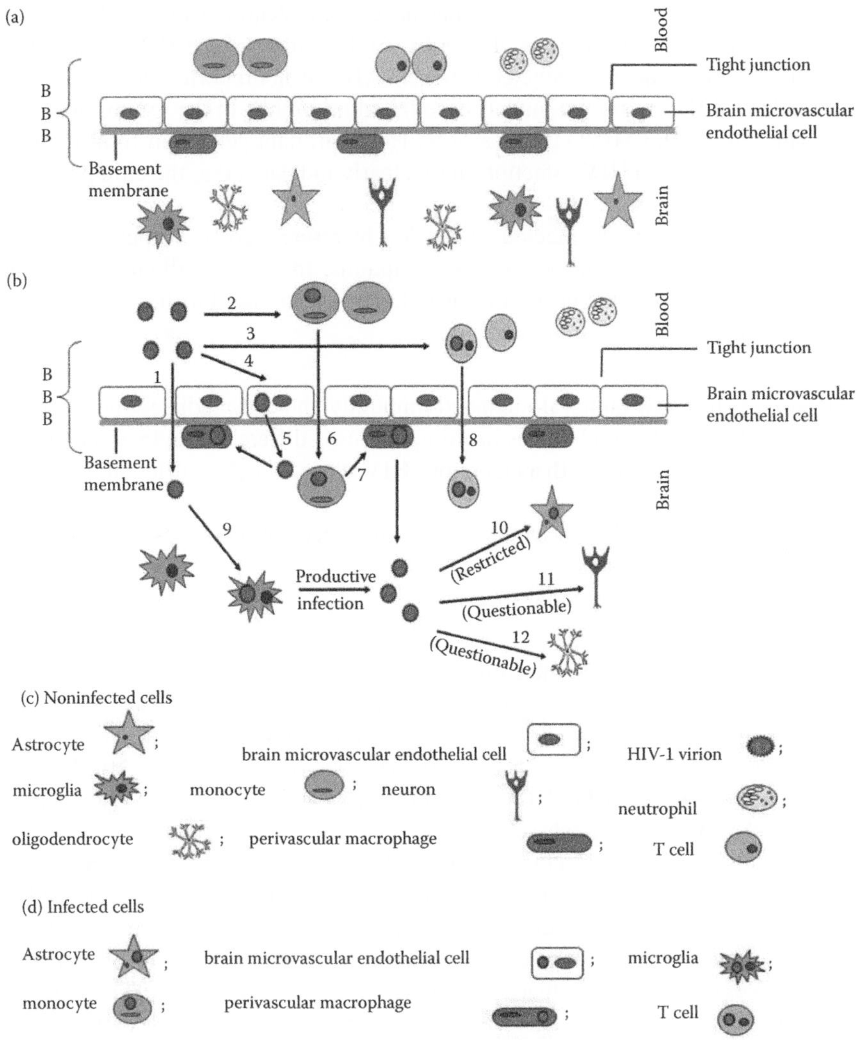

FIGURE 24.7 **(See color insert.)** Neuroinvasion of HIV. Schematic representation of various cellular components of blood-brain barrier (BBB). (a) Intact BBB along with its cellular components. Astrocytes, microglia, brain microvascular endothelial cells, perivascular macrophages, oligodendrocytes, etc., are commonly found cells in BBB. (b) Schematic representation of BBB, when BBB is breached due to HIV infection. Symbols used in (a) and (b) are defined in (c) and (d), respectively. This figure is also the depiction of the Trojan horse hypothesis, where (1) direct entry of HIV to the brain is caused by opening of the tight junction; (2) HIV infects the peripheral monocytes; (3) HIV infects the peripheral T-lymphocytes; (4) infection of HIV to the brain is caused by transcytosis via brain microvascular endothelial cells; (5) HIV is released from infected brain microvascular endothelial cells into the brain compartment; (6) direct entry of infected monocytes into the brain is caused by a breach of BBB; (7) infected monocytes differentiates into perivascular macrophages; (8) direct entry of infected T-lymphocytes into the brain is caused by a breach of BBB; (9) HIV directly infects microglia; (10) HIV infection to astrocytes is considered as a restricted infection; (11) HIV infection to neurons is still questionable; and (12) HIV infection to oligodendrocytes is still questionable.

24.11 ANIMAL MODELS FOR HIV RESEARCH

Knowledge of natural history of any disease is essential to evaluate the efficacy of any therapeutic intervention, for which a suitable animal model is essential. Unfortunately, the major hurdle for HIV research is that a suitable animal model is unavailable. There are numerous scientific problems with AIDS-related animal experimentation that have been identified by the Medical Research Modernization Committee (Anderegg et al., 2006). To date, most of the HIV-related studies have been done either *in vitro* or *in vivo* (infected patients). AIDS is not reported in nonhumans. Various attempts to replicate AIDS in animal models have failed miserably. In theory, chimpanzees are the only nonhumans suitable for HIV/AIDS research. However, data generated from the research on chimpanzee with reference to HIV infections have clearly indicated that the chimpanzee is not the right answer for HIV research.

So far, >100 chimpanzees were infected with HIV-1 by researchers over a period of more than 10 years; however, just two of them showed disease symptoms, and such small numbers are not useful to draw any conclusions about the utility of this data for human use. Further studies revealed that chimpanzees have the following limitations:

1. Chimpanzees show low humoral-mediated immunity and cell-mediated immunity.
2. The ratio of T4 and T8 lymphocytes in chimpanzees is different than that in humans.
3. These animals were treated with a laboratory HIV strain, which is different from naturally occurring HIV.
4. HIV replication is limited in chimpanzees, making vaccination successful, which may not be true in the case of humans.
5. The chimpanzee is an endangered species.

After realizing the flaws of using the chimpanzee as an animal model for HIV research, further studies were focused on monkeys. We reiterate that monkey has also turn out to be an unsuitable model for HIV research, because monkeys could not be infected with HIV. SIV can infect monkeys. The rhesus monkey can be artificially infected with the sooty mangabey SIV strain. The major issue with SIV is that it is closely related to HIV-2 rather than HIV-1. Antibodies against V-3 loop of HIV is characteristics of HIV infection, whereas SIV does not show this response.

A new virus, SHIV, has been developed by recombining the genes encoding the outer covering from HIV and the rest of the genome from SIV. SHIV has shown a lot of similarity with HIV and SIV and has been extensively used to understand pathogenesis in monkey models. Several other animal models for immunodeficiency viruses are available, for example, in cats. Cats can get infected with FIV only. FIV is closer to nonprimate lentiviruses, and it infects through CD9 rather than CD4. The major obstacle with the cat model is that cats cannot be infected with FIV in the laboratory, but in natural circumstances, cats are known to be infected with FIV, and general opinion is that for FIV infection in cats, there is an essential need for concurrent infections.

Mice are generally considered to be one of the most useful and suitable models for animal research. Unfortunately, mice are not good hosts for HIV at all because they do not express the coreceptors essential for HIV infection. Mice with severe combine immunodeficiencies are available, but we are unable to use this model to its full potential for HIV research.

In conclusion, the critical insights about AIDS, its mechanism, and its therapy have been derived exclusively from *in vitro* and *in vivo* studies on humans, whereas animal studies consistently fail to reproduce human AIDS. However, in the last decade, scientists have developed different transgenic animal models using mice and rats to investigate the importance of individual HIV genes, and studies are ongoing.

24.12 FUTURE DIRECTION

One of the most burning issues about HIV is that HIV infections can be controlled but cannot be cured. Even though lot of efforts are made in this direction, still we are far from curing HIV. One of the most compelling reasons is that we do not have any animal model to study the biology of the disease. Various animal models have been put to trial; all of them suffer from one limitation or another. Therefore, data generated from these animal models can only be used as an indicator rather than confirmatory data that can be applied to human patients. This is one of the major reasons that no matter how many times we work on a vaccine against HIV, we fail miserably. Sometimes, we have to stop some of the clinical trials for vaccines without even understanding the biology behind it. Numerous adverse effects are associated with the vaccine against HIV.

The one-gene-at-a-time approach to answer various questions about life processes has generated valuable information, this approach has a major role in studying the impact of an inducer (toxicant, pathogen, or xenobiotics) in a predecided format. This approach has helped us to understand the mechanism of action of a specific gene or protein, as well as its role in pathogenesis. A major drawback of this approach is that it fails to provide any useful information about how it affects other aspects, including changes both at cellular and molecular levels. As we know, the smallest functional unit of body, *i.e.*, the cell, does not work alone; it always works in coordination with complex networks and pathways, even at molecular levels. Therefore, the advent of high-throughput screening is a boon to provide answers to some of the most complicated aspects of the biology of life. High-throughput screening is a holistic approach used to observes and analyze the impact of an inducer. This can help us to develop better strategies to resolve some of these problems in HIV research.

The HIV genome is very small, consisting of 9 genes and producing only 15 proteins during its replication cycle, which can cause massive deleterious effects. Since the advent of HIV, all of these genes have been extensively studied using the one-gene-at-a-time approach, which provided valuable data at least to develop and design various medicines and treatment regimens. Still, we lack the information that can give us a feeling about what HIV is doing to the entire system either at whole-body levels or at cellular levels.

Vpr is considered as an accessory gene; its major role is to maintain a favorable environment for HIV replication. Various activities of this gene like antitumor activity, regulation of viral infectivity, apoptosis induction, and G_2/M arrest, have been confirmed so far; by adopting the one-gene-at-a-time approach, however, we still lack information about it in holistic terms. Certainly data generated with reference to approaches like omics, genomics, proteomics, metabolomics, transcriptomics, and others may be useful to provide some of the missing links. Having this information will help us to understand the role of Vpr in pathogenesis as well as be useful to evolve new strategies for therapeutic interventions.

The approach of omics has unique significance for HIV infection, because we can study HIV infection and the role of individual genes in animal models. The omics approach can shed some more light on infection, infectivity, and the role of individual genes, as well as interaction of different genes, which will be helpful to delineate their impact on the host. OMICS studies in relation to neuroAIDS will be important to understand its pathobiology. This new information will be helpful in developing new and efficient strategies either to control or cure it.

24.13 INTERNET RESOURCES AND OMICS

Recent trends related to the studies of omics cannot be imagined without the application of various tools available to us in the arena of bioinformatics. In the past couple of decades, bioinformatics has been of unprecedented significance for its utility to delineate the intricacies of biological systems. At present various resources are available on the Internet that can be useful for studies on the omics aspect of any subject, including HIV. One of the most useful and important resources for data mining of amino acid sequences as well as nucleotide sequences is http://www.ncbi.nlm.nih.gov. NCBI is considered to be one of the most validated and resourceful websites. NCBI also offers various options like primer designing, primer databases, and web tools based for the prediction of signal transduction pathways, which can be used to delineate the theoretical possibilities for different pathways that an individual viral protein can follow.

ACKNOWLEDGMENTS

The authors are thankful to Prof. S. M. P. Khurana, Director, Amity Institute of Biotechnology, Amity University, Noida, India, for providing necessary resources and facilities for completion of this manuscript. A. S. V., U. P.S., and A. S. are thankful to their students Mr. Shishir Agrahari, Ms. Shruti Rastogi, Mr. Pranshu Abhishek, and Mr. Chandi Gupta for their untiring efforts to perform literature searches. The authors are also grateful to Mr. Dinesh Kumar for his secretarial assistance and graphic designing.

REFERENCES

Abbas, K. A, Lichtman, A. H., Pillai, S. (2008). *Cellular and Molecular Immunology*, 6th edition, New Delhi: Elsevier.

Adamson, D. C., Wildemann, B., Sasaki, M., et al. (1996). Immunologic NO synthase: Elevation in severe AIDS dementia and induction by HIV-1 gp41. *Science* 274: 1917–1921.

Adie-Biassette, H., Levy, Y., Colombel, M., et al. (1995). Neuronal apoptosis in HIV infection in adults. *Neuropathol. Appl. Neurobiol.* 21: 218–227.

Aebersold, R., Mann, M. (2003). Mass spectrometry-based proteomics. *Nature* 422: 198–207.

Ahmad, N., Venkatesan, S. (1988). Nef protein of HIV-1 is a transcriptional repressor of HIV-1 LTR. *Science* 241: 1481–1485.

Albright, A. V., Strizki, J., Harouse, J. M., et al. (1996). HIV-1 infection of cultured human adult oligodendrocytes. *Virology* 217: 211–219.

Albright, A. V., Vos, R. M., Gonzalez-Scarano, F. (2004). Low-level HIV replication in mixed glial cultures is associated with alterations in processing of p55(Gag). *Virology* 325: 328–339.

An, S. F., Giometto, B., Scaravilli, T., et al. (1996). Programmed cell death in brains of HIV-1-positive AIDS and pre-AIDS patients. *Acta Neuropathol.* 91: 169–173.

Anderegg, C., Archibald,K., Bailey, J., et al. (2006). *A critical look at animal experimentation*. Cleveland, OH: Medical Research Modernization Committee.

Anderson, E., Zink, W., Xiong, H., et al. (2002). HIV-1-associated dementia: A metabolic encephalopathy perpetrated by virus infected and immune-competent mononuclear phagocytes. *J. Acquir. Immune Defic. Syndr.* 31 (Suppl. 2): S43–S54.

Anderson, N. L., Anderson, N. G. (1998). Proteome and proteomics: New technologies, new concepts, and new words. *Electrophoresis* 19: 1853–1861.

Arora, V. K., Fredericksen, B. L., Garcia, J. V. (2002). Nef: Agent of cell subversion. *Microbes Infect.* 4: 189–199.

Bachand, F., Yao, X. J., Hrimech, M., et al. (1999). Incorporation of Vpr into human immunodeficiency virus type 1 requires a direct interaction with the p6 domain of the p55 gag precursor. *J. Biol. Chem.* 274: 9083–9091.

Bagasra, O., Lavi, E., Bobroski, L., et al. (1996). Cellular reservoirs of HIV-1 in the central nervous system of infected individuals: Identification by the combination of in situ polymerase chain reaction and immunohistochemistry. *AIDS* 10: 573–585.

Balasubramanyam, A., Mersmann, H., Jahoor, F., et al. (2007). Effects of transgenic expression of HIV-1 Vpr on lipid and energy metabolism in mice. *Am. J. Physiol. Endocrinol. Metab.* 292: E40–E48.

Banks, W. A., Freed, E. O., Wolf, K. M., et al. (2001). Transport of human immunodeficiency virus type 1 pseudoviruses across the blood-brain barrier: Role of envelope proteins and adsorptive endocytosis. *J. Virol.* 75: 4681–4691.

Belzile, J. P., Duisit, G., Rougeau, N., et al. (2007). HIV-1 Vpr-mediated G2 arrest involves the DDB1-CUL4AVPRBP E3 ubiquitin ligase. *PLoS Pathol* 3: E85.

Benko, Z., Liang, D., Agbottah, E., et al. (2007). Antagonistic interaction of HIV-1 Vpr with Hsf-mediated cellular heat shock response and Hspl6 in fission yeast (*Schizosaccharomyces pombe*). *Retrovirology* 4: 16–23.

Boehringer Ingelheim Pharmaceuticals Incorporated (2005). NNRTI mode of action (animation).

Bomsel, M. (1997). Transcytosis of infectious human immunodeficiency virus across a tight human epithelial cell line barrier. *Nat. Med.* 3: 42–47.

Bour, S., Strebel, K. (2000). HIV accessory proteins: Multifunctional components of a complex system. *Adv. Pharmacol.* 48: 75–119.

Boven, L. A., van der Bruggen, T., Sweder van Asbeck, B., et al. (1999). Potential role of CCR5 polymorphism in the development of AIDS dementia complex. *FEMS Immunol. Med. Microbiol.* 26: 243–247.

Brack-Werner, R. (1999). Astrocytes: HIV cellular reservoirs and important participants in neuropathogenesis. *AIDS* 13: 1–22.

Brasey, A., Lopez-Lastra, M., Ohlmann, T., et al. (2003). The leader of human immunodeficiency virus type 1 genomic RNA harbors an internal ribosome entry segment that is active during the G_2/M phase of the cell cycle. *J. Virol.* 77: 3939–3949.

Brenner, B. G., Tao, Y., Pearson, E., et al. (1995). Altered constitutive and stress-regulated heat shock protein 27 expression in HIV type 1-infected cell lines. *AIDS Res. Hum. Retroviruses* 11: 713–717.

Brenner, C., Kroemer, G. (2003). The mitochondriotoxic domain of Vpr determines HIV-1 virulence. *J. Clin. Invest.* 111: 1455–1457.

Brown, P. O. (1997). Integration. In Coffin, J. M., Varmus, H. E. (Eds.), *Retroviruses* (pp. 161–203). Cold Spring Harbor, NY: Cold Spring Harbor Laboratory.

Centers for Disease Control and Prevention (1992). 1993 revised classification system for HIV infection and expanded surveillance definition for AIDS among adolescents and adults. *MMWR* 41 (RR-17): 1–19.

Chang, F., Re, F., Sebastian, S., et al. (2004). HIV-1 Vpr induces defects in mitosis, cytokinesis, nuclear structure, and centrosomes. *Mol. Biol. Cell* 15: 1793–1801.

Chen, M., Elder, R. T., Yu, M., et al. (1999). Mutational analysis of Vpr-induced G2 arrest, nuclear localization, and cell death in fission yeast. *J. Virol.* 73: 3236–3245.

Coeytaux, E., Coulaud, D., Le Cam, E., et al. (2003). The cationic amphipathic alpha-helix of HIV-1 viral protein R (Vpr) binds to nucleic acids, permeabilizes membranes, and efficiently transfects cells. *J. Biol. Chem.* 278: 18110–18116.

Cullen, B. R. (1998). HIV-1 auxiliary proteins: Making connections in a dying cell. *Cell* 93: 685–692.

Del Valle, L., Croul, S., Morgello, S., et al. (2000). Detection of HIV-1 Tat and JCV capsid protein, VP1, in AIDS brain with progressive multifocal leukoencephalopathy. *J. Neurovirol.* 6: 221–228.

Deora, A., Ratner, L. (2001). Viral protein U (Vpu)-mediated enhancement of human immunodeficiency virus type 1 particle release depends on the rate of cellular proliferation. *J. Virol.* 75: 6714–6718.

DiMarzio, P., Choe, S., Ebright, M., et al. (1995). Structure-function studies on HIV-1 Vpr suggest that the predicted alpha-helical amino-terminal domain controls virion incorporation and nuclear localization while cell cycle blocking is controlled by the carboxyterminus. *1995 Retroviruses Meeting* (p. 376). Cold Spring Harbor, New York.

Dong, Y., Benveniste, E. N. (2001). Immune function of astrocytes. *Glia* 36: 180–190.

Elder, R. T., Benko, Z., Zhao, Y. (2002). HIV-1 VPR modulates cell cycle G_2/M transition through an alternative cellular mechanism other than the classic mitotic checkpoints. *Front. Biosci.* 7: d349–d357.

Emerman, M. (1996). HIV-1, Vpr and the cell cycle. *Curr. Biol.* 6: 1096–1103.

Emerman, M., Bukrinsky, M., Stevenson, M. (1994). HIV-1 infection of non-dividing cells. *Nature* 369: 107–108.

Fassati, A., Goff, S. P. (2001). Characterization of intracellular reverse transcription complexes of human immunodeficiency virus type 1. *J. Virol.* 75: 3626–3635.

Fiala, M., Rhodes, R. H., Shapshak, P., et al. (1996). Regulation of HIV-1 infection in astrocytes: Expression of Nef, TNF-alpha and IL-6 is enhanced in coculture of astrocytes with macrophages. *J. Neurovirol.* 2: 158–166.

Fischer-Smith, T., Croul, S., Adeniyi, A., et al. (2004). Macrophage/microglial accumulation and proliferating cell nuclear antigen expression in the central nervous system in human immunodeficiency virus encephalopathy. *Am. J. Pathol.* 164: 2089–2099.

Fisher, A. G., Feinberg, M. B., Josephs, S. F., et al. (1986). The trans-activator gene of HTLV-III is essential for virus replication. *Nature* 320: 367–371.

Fletcher, T. M., 3rd, Brichacek, B., Sharova, N., et al. (1996). Nuclear import and cell cycle arrest functions of the HIV-1 Vpr protein are encoded by two separate genes in HIV-2/SIV(SM). *EMBO J.* 15: 6155–6165.

Forshey, B. M., Aiken, C. (2003). Disassembly of human immunodeficiency virus type 1 cores in vitro reveals association of nef with the subviral ribonucleoprotein complex. *J. Virol.* 77: 4409–4414.

Freed, E. O. (1998). HIV-1 gag proteins: Diverse functions in the virus life cycle. *Virology* 251: 1–15.

Freed, E. O. (2001). HIV-1 replication. *Somatic Cell. Mol. Genet.* 26: 13–33.

Gallay, P., Stitt, V., Mundy, C., et al. (1996). Role of the karyopherin pathway in human immunodeficiency virus type 1 nuclear import. *J. Virol.* 70: 1027–1032.

Ghafouri, M., Amini, S., Khalili, K., et al. (2006). HIV-1 associated dementia: Symptoms and causes. *Retrovirology* 3: 28.

Goh, W. C., Rogel, M. E., Kinsey, C. M., et al. (1998). HIV-1 Vpr increases viral expression by manipulation of the cell cycle: A mechanism for selection of Vpr in vivo. *Nat. Med.* 4: 65–71.

Gonzalez-Scarano, F., Martin-Garcia, J. (2005). The neuropathogenesis of AIDS. *Nat. Rev. Immunol.* 5: 69–81.

Gorry, P. R., McPhee, D. A., Verity, E., et al. (2007). Pathogenicity and immunogenicity of attenuated, nef-deleted HIV-1 strains in vivo. *Retrovirology* 4: 66.

Gottlieb, M. S., Schroff, R., Schanker, H. M. (1981). *Pneumocystis carinii* pneumonia and mucosal candidiasis in previously healthy homosexual men: Evidence of a new acquired cellular immunodeficiency. *N. Engl. J. Med.* 305: 1425–1431.

Gray, F., Adle-Biassette, H., Brion, F., et al. (2000). Neuronal apoptosis in human immunodeficiency virus infection. *J. Neurovirol.* 6: S38–S43.

Green, D. R., Kroemer, G. (2004). The pathophysiology of mitochondrial cell death. *Science* 305: 626–629.

Greenway, A. L., Holloway, G., McPhee, D. A. (2000). HIV-1 Nef: A critical factor in viral-induced pathogenesis. *Adv. Pharmacol.* 48: 299–343.

Griffin, D. E. (1997). Cytokines in the brain during viral infection: Clues to HIV-associated dementia. *J. Clin. Invest.* 100: 2948–2951.

Haase, A. T. (1986). Pathogenesis of lentivirus infections. *Nature* 322: 130–136.

Heinzinger, N., Bukinsky, M., Haggerty, S., et al. (1994). The Vpr protein of human immunodeficiency virus type 1 influences nuclear localization of viral nucleic acids in nondividing host cells. *Proc. Natl. Acad. Sci. U.S.A.* 91: 7311–7315.

Huard, S., Elder, R. T., Liang, D., et al. (2008). Human immunodeficiency virus type 1 Vpr induces cell cycle G_2 arrest through Srkl/MK2-mediated phosphorylation of Cdc25. *J. Virol.* 82: 2904–2917.

Iordanskiy, S., Zhao, Y., Dubrovsky, L., et al. (2004). Heat shock protein 70 protects cells from cell cycle arrest and apoptosis induced by human immunodeficiency virus type 1 viral protein R. *J. Virol.* 78: 9697–9704.

Jacotot, E., Ferri, K. F., El Hamel, C., et al. (2001). Control of mitochondrial membrane permeabilization by adenine nucleotide translocator interacting with HIV-1 viral protein R and Bcl-2. *J. Exp. Med.* 193: 509–519.

Jenkins, Y., McEntee, M., Weis, K., et al. (1998). Characterization of HIV-1 vpr nuclear import: Analysis of signals and pathways. *J. Cell Biol.* 143: 875–885.

Jones, G., Power, C. (2006). Regulation of neural cell survival by HIV-1 infection. *Neurobiol. Dis.* 21: 1–17.

Jowett, J. B., Xie, Y. M., Chen, I. S. Y. (1999). The presence of human immunodeficiency virus type 1 Vpr correlates with a decrease in the frequency of mutations in a plasmid shuttle vector. *J. Virol.* 73: 7132–7137.

Kaul, M., Garden, G. A., Lipton, S. A. (2001). Pathways to neuronal injury and apoptosis in HIV-associated dementia. *Nature* 410: 988–994.

Khan, M. A., Akari, H., Kao, S., et al. (2002). Intravirion processing of the human immunodeficiency virus type 1 Vif protein by the viral protease may be correlated with Vif function. *J. Virol.* 76: 9112–9123.

Kichler, A., Pages, J. C., Leborgne, C., et al. (2000). Efficient DNA transfection mediated by the C-terminal domain of human immunodeficiency virus type 1 viral protein R. *J. Virol.* 74: 5424–5431.

Kirchhoff, F., Greenough, T. C., Brettler, D. B., et al. (1995). Absence of intact nef sequences in a long-term survivor with nonprogressive HIV-1 infection. *N. Engl. J. Med.* 332: 228–232.

Kotov, A., Zhou, J., Flicker, P., et al. (1999). Association of Nef with the human immunodeficiency virus type 1 core. *J. Virol.* 73: 8824–8830.

Kramer-Hammerle, S., Hahn, A., Brack-Werner, R., et al. (2005a). Elucidating effects of longterm expression of HIV-1 Nef on astrocytes by microarray, promoter and literature analyses. *Gene* 358: 31–38.

Kramer-Hammerle, S., Rothenaigner, I., Wolff, H., et al. (2005b). Cells of the central nervous system as targets and reservoirs of the human immunodeficiency virus. *Virus Res.* 11: 194–213.

Kulesh, D. A., Clive, D. R., Zarlenga, D. S., et al. (1987). Identification of interferon-modulated proliferation-related cDNA sequences. *Proc. Natl. Acad. Sci. U.S.A.* 84: 8453–8457.

Kumar, A., Dhawan, S., Mukhopadhyay, A., et al. (1999). Human immunodeficiency virus-1-tat induces matrix metalloproteinase-9 in monocytes through protein tyrosine phosphatase-mediated activation of nuclear transcription factor NF-kappaB. *FEBS Lett.* 462: 140–144.

Lake, J., Carr, J., Feng, F., et al. (2003). The role of Vif during HIV-1 infection: Interaction with novel host cellular factors. *J. Clin. Virol.* 26: 143–152.

Lederberg, J., McCray, A. T. (2001). Ome sweet omics: A genealogical treasury of words. *The Scientist* 15: 8.

Levy, D. N., Refaeli, Y., MacGregor, R. R., et al. (1994). Serum Vpr regulates productive infection and latency of human immunodeficiency virus type 1. *Proc. Natl. Acad. Sci. U.S.A.* 91: 10873–10877.

Liang, D., Benko, Z., Agbottah, E., et al. (2007). Anti-vpr activities of heat shock protein 27. *Mol. Med.* 13: 229–239.

Lu, Y. L., Spearman, P., Ratner, L. (1993). Human immunodeficiency virus type 1 viral protein R localization in infected cells and virions. *J. Virol.* 67: 6542–6550.

Majumder, B., Venkatachari, N. J., O'Leary, S., et al. (2008). Infection with Vpr-positive human immunodeficiency virus type 1 impairs NK cell function indirectly through cytokine dysregulation of infected target cells. *J. Virol.* 82: 7189–7200.

Marassi, F. M., Ma, C., Gratkowski, H. et al. (1999). Correlation of the structural and functional domains in the membrane protein Vpu from HIV-1. *Proc. Natl. Acad. Sci. U.S.A.* 96: 14336–14341.

McArthur, J. C., Brew, B. J., Nath, A. (2005). Neurological complications of HIV infection. *Lancet Neurol.* 4: 543–555.

McCombe, J. A., Noorbakhsh, F., Buchholz, C., et al. (2009). NeuroAIDS: A watershed for mental health and nervous system disorders. *J. Psychiatry Neurosci.* 34: 83–85.

McKusick, V. A., Ruddle, F. H. (1987). Toward a complete map of the human genome. *Genomics* 1: 103–106.

McManus, C. M., Weidenheim, K., Woodman, S. E. (2000). Chemokine and chemokinereceptor expression in human glial elements: Induction by the HIV protein, Tat and chemokine autoregulation. *Am. J. Pathol.* 156: 1441–1453.

Minagar, A., Shapshak, P., Fujimura, R., et al. (2002). The role of macrophage/microglia and astrocytes in the pathogenesis of three neurologic disorders: HIV-associated dementia, Alzheimer disease, and multiple sclerosis. *J. Neurol. Sci.* 202: 13–23.

Mukhtar, M., Harley, S., Chen, P., et al. (2002). Primary isolated human brain microvascular endothelial cells express diverse HIV/SIV-associated chemokine coreceptors and DC-SIGN and L-SIGN. *Virology* 297: 78–88.

Muthumani, K., Hwang, D. S., Choo, A. Y., et al. (2004). HIV-1 Vpr inhibits the maturation and activation of macrophages and dendritic cells in vitro. *Int. Immunol.* 17: 103–116.

Nermut, M. V., Fassati, A. (2003). Structural analyses of purified human immunodeficiency virus type 1 intracellular reverse transcription complexes. *J. Virol.* 77: 8196–8206.

Neumann, M., Afonina, E., Ceccherini-Silberstein, F., et al. (2001). Nucleocytoplasmic transport in human astrocytes: Decreased nuclear uptake of the HIV Rev shuttle protein. *J. Cell Sci.* 114: 1717–1729.

Nuovo, G. J., Becker, J., Burk, M. W., et al. (1994). In situ detection of PCR-amplified HIV-1 nucleic acids in lymph nodes and peripheral blood in patients with asymptomatic HIV-1 infection and advanced-stage AIDS. *J. Acquir. Immune Defic. Syndr.* 7: 916–923.

Ogawa, K. (1989). Mutational analysis of the human immunodeficiency virus vpr open reading frame. *J. Virol.* 63: 4110–4114.

Oliver, S. G., Winson, M. K., Kell, D. B., et al. (1998). Systematic functional analysis of the yeast genome. *Trends Biotechnol.* 16: 373–378.

Ong, S. E., Mann, M. (2005). Mass spectrometry-based proteomics turns quantitative. *Nat. Chem. Biol.* 1: 252–262.

Panek, R. B., Benveniste, E. N. (1995). Class II MHC gene expression in microglia: Regulation by the cytokines IFN-gamma, TNF-alpha, and TGF-beta. *J. Immunol.* 154: 2846–2854.

Parfrey, L. W., Lahr, D. J., Katz, L. A. (2008). The dynamic nature of eukaryotic genomes. *Mol. Biol. Evol.* 25: 787–794.

Patel, C. A., Mukhtar, M., Pomerantz, R. J. (2000). Human immunodeficiency virus type 1 Vpr induces apoptosis in human neuronal cells. *J. Virol.* 74: 9717–9726.

Patel, C. A., Mukhtar, M., Harley, S., et al. (2002). Lentiviral expression of HIV-1 Vpr induces apoptosis in human neurons. *J. Neurovirol.* 8: 86–99.

Petito, C. K., Cash, K. S. (1992). Blood-brain barrier abnormalities in the acquired immunodeficiency syndrome: Immunohistochemical localization of serum proteins in postmortem brain. *Ann. Neurol.* 32: 658–666.

Piller, S. C., Jans, P., Gage, P. W., et al. (1998). Extracellular HIV-1 virus protein R causes a large inward current and cell death in cultured hippocampal neurons: Implications for AIDS pathology. *Proc. Natl. Acad. Sci. U.S.A.* 95: 4595–4600.

Pollard, V. W., Malim, M. H. (1998). The HIV-1 Rev protein. *Annu. Rev. Microbiol.* 52: 491–532.

Power, C., Boisse, L., Rornke, S., et al. (2009). NeuroAIDS: An evolving epidemic. *Can. J. Neurosci.* 36: 285–295.

Purcell, D. F., Martin, M. A. (1993). Alternative splicing of human immunodeficiency virus type 1 mRNA modulates viral protein expression, replication, and infectivity. *J. Virol.* 67: 6365–6378.

Ramanathan, M. P., Curley, E., Su, M., et al. (2002). Carboxyl terminus of hVIP/mov34 is critical for HIV-1-Vpr interaction and glucocorticoid-mediated signaling. *J. Biol. Chem.* 277: 47854–47860.

Rapoport, S. I. (1976). *Blood-brain barrier in physiology and medicine*. New York: Raven Press.

Resnick, L., Berger, J. R., Shapshak, P., et al. (1988). Early penetration of the blood-brain-barrier by HIV. *Neurology* 38: 9–14.

Rogel, M. E., Wu, L. I., Emerman, M. (1995). The human immunodeficiency virus type 1 vpr gene prevents cell proliferation during chronic infection. *J. Virol.* 69: 882–888.

Romani, B., Engelbrecht, S. (2009). Human immunodeficiency virus type 1 Vpr: Functions and molecular interactions. *J. Gen. Virol.* 90: 1795–1805.

Rouzic, E. L., Benichou, S. (2005). The Vpr protein from HIV-1: Distinct roles along the viral life cycle. 2: 11.

Sabbah, E. N., Roques, B. P. (2005). Critical implication of the (70–96) domain of human immunodeficiency virus type 1 Vpr protein in apoptosis of primary rat cortical and striatal neurons. *J. Neurovirol.* 11: 489–502.

Sanchez, C., Lachaize, C., Janody, F. et al. (1999). Grasping at molecular interactions and genetic networks in *Drosophila melanogaster* using FlyNets, an internet database. *Nucleic Acids Res.* 27: 89–94.

Sawaya, B. E., Khalili, K., Gordon, J., et al. (2000). Cooperative interaction between HIV-1 regulatory proteins Tat and Vpr modulates transcription of the viral genome. *J. Biol. Chem.* 275: 35209–35214.

Sawaya, B. E., Khalili, K., Mercer, W. E., et al. (1998). Cooperative actions of HIV-1 Vpr and p53 modulate viral gene transcription. *J. Biol. Chem.* 273: 20052–20057.

Schaeffer, E., Geleziunas, R., Greene, W. C. (2001). Human immunodeficiency virus type 1 Nef functions at the level of virus entry by enhancing cytoplasmic delivery of virions. *J. Virol.* 75: 2993–3000.

Selig, L., Pages, J. C., Tanchou, V., et al. (1999). Interaction with the p6 domain of the gag precursor mediates incorporation into virions of Vpr and Vpx proteins from primate lentiviruses. *J. Virol.* 73: 592–600.

Sherman, M. P., De Noronha, C. M., Heusch, M. I., et al. (2001). Nucleocytoplasmic shuttling by human immunodeficiency virus type 1 Vpr. *J. Virol.* 75: 1522–1532.

Sherman, M. P., De Noronha, C. M., Williams, S. A., et al. (2002a). Insights into the biology of HIV-1 viral protein R. *DNA Cell Biol.* 21: 679–688.

Sherman, M. P., Schubert, U., Williams, S. A., et al. (2002b). HIV-1 Vpr displays natural protein-transducing properties: implications for viral pathogenesis. *Virology* 302: 95–105.

Spira, S., Waniberg, M. A., Loemba, H., et al. (2003). Impact of clade diversity on HIV-1 virulence, antiretroviral drug sensitivity and drug resistance. *J. Antimicrob. Chemother.* 51: 229–240.

Stewart, S. A., Poon, B., Jowett, J. B., et al. (1997). Human immunodeficiency virus type 1 Vpr induces apoptosis following cell cycle arrest. *J. Virol.* 71: 5579–5592.

Stins, M. F., Pearce, D., Di Cello, F., et al. (2003). Induction of intercellular adhesion molecule-1 on human brain endothelial cells by HIV-1 gp120: Role of CD4 and chemokine coreceptors. *Lab. Invest.* 83: 1787–1798.

Strebel, K., Klimkait, T., Martin, M. (1988). A novel gene of HIV-1, vpu, and its 16-kDa product. *Science* 241: 1221–1223.

Subbramanian, R. A., Kessous-Elbaz, A., Lodge, R., et al. (1998). Human immunodeficiency virus type 1 Vpr is a positive regulator of viral transcription and infectivity in primary human macrophages. *J. Exp. Med.* 187: 1103–1111.

Takahashi, K., Wesselingh, S. L., Griffin, D. E., et al. (1996). Localization of HIV-1 in human brain using polymerase chain reaction in situ hybridization and immunocytochemistry. *Ann. Neurol.* 39: 705–711.

Thomas, E., Dunfee, R., Stanton, J., et al. (2007). Macrophage entry mediated by HIV Envs from brain and lymphoid tissues is determined by the capacity to use low CD4 levels and overall efficiency of fusion. *Virology* 360: 105–119.

UNAIDS (2009). AIDS epidemic update. *UNAIDS Report 2009* (pp. 1–100).

Velculesen, V. E., Zhang, L., Zhou, W., et al. (1997). Characterization of the yeast transcriptome. *Cell* 88: 243–251.

Verma, A. S., Bhatt, S. M., Singh, A., et al. (2009). HIV: An Introduction. In Chauhan, A. K., Varma, A. (Eds.), *Textbook on molecular biotechnology* (pp. 853–878). Delhi: I. K. International Publishing House.

Verma, A. S., Singh, U. P., Singh, A. (2010a). NeuroAIDS: A real concern. *IIOAB J.* 1: 28–31

Verma, A. S., Singh, A., Singh, U. P., et al. (2010b). NeuroAIDS in Indian scenario. In Gaur, R. K., et al. (Eds.), *Recent trends in biotechnology and microbiology* (pp. 155–167). New York: Nova Science Publishers Inc.

Verma, A. S., Singh, U. P., Dwivedi, P. D., et al. (2010c). NeuroAIDS: Role of cells of Central Nervous System (CNS). *J. Pharm. Bioall. Sci.* 4: 300–306.

Vodicka, M. A., Koepp, D. M., Silver, P. A., et al. (1998). HIV-1 Vpr interacts with the nuclear transport pathway to promote macrophage infection. *Genes Dev* 12: 175–185.

Walker B. D. (2007). AIDS and secondary immunodeficiency. In Male, D., et al. (Eds.), *Immunology* (pp. 311–324). Toronto, Canada: Mosby Elsevier.

Wen, X., Duus, K. M., Friedrich, T. D., et al. (2007). The HIV1 protein Vpr acts to promote G2 cell cycle arrest by engaging a DDB1 and cullin4A-containing ubiquitin ligase complex using VprBP/DCAF1 as an adaptor. *J. Biol. Chem.* 282: 27046–27057.

Wiley, C. A., Schrier, R. D., Nelson, J. A., et al. (1986). Cellular localization of human immunodeficiency virus infection within the brains of acquired immune deficiency syndrome patients. *Proc. Natl. Acad. Sci. U.S.A.* 83: 7089–7093.

Wilt, S. G., Milward, E., Zhou, J. M., et al. (1995). In vitro evidence for a dual role of tumor necrosis factor-alpha in human immunodeficiency virus type 1 encephalopathy. *Ann. Neurol.* 37: 381–394.

Winkler, H. (1920). *Verbreitung und ursache der parthenogenesis im pflanzen und tierreiche* [Spread and cause of parthenogenesis in the plant and animal kingdoms]. Jena, Germany: Verlag von Gustav Fischer.

Yao, X. J., Subbramanian, R., Rougeau, N., et al. (1995). Mutagenic analysis of HIV-1 Vpr: Role of a predicted N-terminal alpha helical structure on Vpr nuclear localization and virion-incorporation. *1995 Retroviruses Meeting* (p. 380). Cold Spring Harbor, NY.

Yoshioka, M., Bradley, W. G., Shapshak, P., et al. (1995). Role of immune activation and cytokine expression in HIV-1-associated neurologic diseases. *Adv. Neuroimmunol.* 5: 335–358.

Zelivianski, S., Liang, D., Chen, M., et al. (2006). Suppressive effect of elongation factor 2 on apoptosis induced by HIV-1 viral protein R. *Apoptosis* 11: 377–388.

Zhao, L. J., Mukherjee, S., Narayan, O. (1994). Biochemical mechanism of HIV-1 Vpr function: Specific interaction with a cellular protein. *J. Biol. Chem.* 269: 15577–15582.

Zhu, Y., Roshal, M., Li, F., et al. (2003). Upregulation of survivin by HIV-1 Vpr. *Apoptosis* 8: 71–79.

25 Epigenetics in Neuropsychiatry

Trevor Archer
University of Gothenburg
Gothenburg, Sweden

Kenneth Blum
University of Florida College of Medicine and McKnight Brain Institute
Gainesville, Florida

CONTENTS

25.1 INTRODUCTION

The notion of epigenetics offers a putative interface between genetic and environmental factors that interact to provide the phenotypic expression. The impact of the environment on gene expression (epigenetics) and the convergence of genes and environment along common biological pathways induce greater effects than either those of genes or environment in isolation (van Winkel et al., 2010). Transgenerational epigenetic inheritance, i.e., the survival of epigenetic modifications over generations, provides a process through which maternal nurturing behavior may influence the development and health of the offspring (Franklin and Mansuy, 2010a, 2010b; Franklin et al., 2010; Jablonka and Raz, 2009; Nadeau, 2009). In order to clarify the phenomena; modulating physiological phenotypes over generations, Ho and Burggren (2010) discuss:

1. How the concepts of epigenetics and maternal influence overlap with and are distinct from each other
2. Based on these notions, analyses that provide illustrations of existing animal physiological studies
3. A construct that provides integration of these concepts into the design underlying animal physiology studies

As described by Kappeler and Meaney (2010), the parental regulation of DNA methylation provides a plausible and prototypic candidate mechanism for parental effects on phenotypic variation. Mother-offspring interactions in rodents have demonstrated that parental signals influence the DNA methylation and set the cascade, leading to stable changes in gene expression in motion. In the mood disorders (bipolar and major depressive disorder), a broad range of studies have documented alterations of mitochondrial, oligodendrocytes, and myelin-related genes. These include signaling and olidendroglial-related genes in depression and γ-aminobutyric acid (GABA)-glutamate related genes in depression and suicide (Sequeira and Turecki, 2006). The epigenetic modifications of DNA methylation and chromatin remodeling are necessary, for example, for maintaining and regulating adult hippocampal neurogenesis but are associated, nevertheless, in the neuropsychiatric disorders. Epigenetic mechanisms contribute critical roles in encoding experience and environmental stimuli into stable, behaviorally meaningful alterations in gene expression (Hsieh and Eisch, 2010), but the adaptive aspect may take on a dysregulatory role. Early-life stressful events can provoke long-lasting *memory traces* on individuals' genes, thereby programming a persisting risk for depression; for example, postnatal stress in mice marks the arginine vasopressin gene, resulting in hyperactivation of the hypothalamic-pituitary-adrenal (HPA) axis together with behavioral and endocrine mobilizations. Murgatroyd et al. (2010a) describe how epigenetic memory evolves by a dual-step process that the epigenetic reader and writer, MeCP2, coordinates. The initial derepression of arginine vasopressin is driven by neuronal activity, causing Ca^{2+}/calmodulin kinase-dependent phosphorylation and dissociation of MeCP2; then subsequent hypomethylation at the AVP enhancer develops gradually to sustain derepression. The MeCP2 occupancy uncouples from the initial stimulus proceeding to the hard coding of early-life experience at the level of DNA methylation by a transition to a persistent irreversible epigenetic memory (Murgatroyd et al., 2009, 2010b).

The epigenetic regulation of gene expression has provided insights purporting to an eventual understanding of disorder causality (Foley et al., 2009). Epigenetics comprises heritable yet concomitantly variable modifications of genomic DNA that define gene expression. Epigenetic mechanisms may exert lasting control over gene expression without altering the genetic code; the focus is on how cellular traits may be inherited without alteration of the DNA sequence. Epigenetic regulation is governed by mechanisms that involve DNA methylation and covalent histone modification within gene promoters (Tsankova et al., 2007). For example, in individuals with chronic-fatigue syndrome, Falkenberg et al. (2010) found that the promoter polymorphism (rs6311) can affect both transcription-factor binding and promoter methylation. This, along with an individual's stress response, can affect the rate of HTR2A transcription in a genotype in a methylation-dependent manner (Malik and Roeder, 2010). Both DNA methylation and covalent histone modification are responsive to environmental influences and are putatively modifiable (Craig, 2005; Weaver et al., 2004; Zhang and Meaney, 2010). An individual's behavioral, physiological, and social milieu influence and are influenced by the epigenome, which is composed predominantly of chromatin and the covalent modification of DNA by methylation. Epigenetic patterns are sculpted during development to shape the diversity of gene expression programs in the organism, patterns that vary from cell type to cell type and remain potentially dynamic throughout the life span. Bagot and Meaney (2010) indicate that epigenetic remodeling occurs in response to the environmental activation of cellular signaling pathways associated with synaptic plasticity; they suggest that epigenetic marks are actively remodeled during early development in response to environmental events that regulate neural development and function. These epigenetic marks are subject to remodeling by environmental influences even at later stages in development. Variations in environmental exposures, e.g., early parental care, may impact epigenetic patterns, with implications for psychiatric health; because epigenetic programming defines the state of expression of genes, epigenetic differences could have the same consequences as genetic polymorphisms (Archer et al., 2010a). In contrast to genetic sequence differences, epigenetic alterations are potentially reversible. Nevertheless, epigenetic mechanisms mediating early-life adversity contribute to later-onset neurological dysfunction and disease (McGowan and Szyf, 2010).

The role of epigenetic processes in depressive disorders provides critical insights: McGowan et al. (2008) have shown an aberrant regulation of the protein synthesis machinery in postmortem suicide brains, thereby implicating the epigenetic modulation of rRNA in the pathophysiology of suicide. Traumatic adversity in early childhood sets the stage for later neuropsychiatric disorder, for example, by altering the HPA-stress response profiles, implicating epigenetic programming of glucocorticoid receptor expression and increasing the risk for mood disorder and suicide. McGowan et al. (2009) studied epigenetic differences in a neuron-specific glucocorticoid receptor (NR3C1) promoter between postmortem hippocampus obtained from suicide victims with a history of childhood abuse and those from either suicide victims with no childhood abuse or controls. They found decreased levels of glucocorticoid receptor mRNA, as well as mRNA transcripts bearing the glucocorticoid receptor 1F splice variant and increased cytosine methylation of an NR3C1 promoter. Patch-methylated NR3C1 promoter constructs that mimicked the methylation state in samples from abused suicide victims showed decreased NGFI-A transcription factor binding and NGFI-A-inducible gene transcription. Much attention has focused on common enzymatic modifications to chromatin structure that up- or downregulate gene expression in a manner transmissible to daughter cells. These mechanisms also regulate gene expression in humans and are sustained within individual cells (Brami-Cherrier et al., 2005, 2007, 2009; Tsankova et al., 2007). Chromatin is a complex of DNA, histones, and associated nonhistone proteins in the cell nucleus. DNA wraps around histone octamers, which are comprised of two histone (H2A and H2B) and H3 and H4 copies that supercoil to form a highly condensed chromatin structure that participates directly in gene expression (Kouzaridesy, 2007; Li et al., 2007; Luger and Richmond, 1998).

The deregulation of epigenetic mechanisms combines with genetic alterations in the development and progression of several Mendelian disorders, as exemplified by certain inherited diseases; for example, Rett syndrome, immunodeficiency-centromeric instability-facial anomalies syndrome, and facioscapulohumeral muscular dystrophy all result from altered gene silencing (Perini and Tupler, 2010). Neurodevelopmental conditions caused by particular genotypic conditions like heterochromatin dysregulation may provide certain trends in the eventual neuropsychiatric outcomes of epigenetics (Hahn et al., 2010). Heterochromatin, a repressive chromatin state characterized by densely packed DNA and low transcriptional activity, induces gene silencing essential for mediating developmental transitions. Heterochromatin presents further global functions in ensuring chromosome segregation and genomic integrity. Imprinted genes are expressed from only one of the two parental alleles. Hahn et al. (2010) have described how altered heterochromatic states can impair normal gene expression patterns, leading to the development of different diseases.

Genomic imprinting is a genetic phenomenon by which certain genes are expressed in a parent-of-origin-specific fashion, an inheritance process independent of the classical Mendelian inheritance. It refers to genes that are expressed one of the two parental alleles in a parent-of-origin-specific fashion. Imprinted genes are either expressed only from the allele inherited from the maternal parent (e.g. *H19* or *CDKN1C*) or from the allele inherited from the paternal parent (e.g., *IGF-2*). It is an epigenetic process that involves methylation and histone modifications in order to achieve monoallelic gene expression without altering the genetic sequence (Isles et al., 2006). Brideau et al. (2010) have shown that specific epigenetic features in mouse cells correlate with imprinting status in mice and have identified hundreds of additional genes that are predicted to be imprinted in the mouse. These epigenetic marks are established in the germ line and are maintained throughout all of the somatic cells of an organism. A consequence of genomic imprinting is that viable embryos must receive two haploid genome complements from parents of opposite sexes. Epigenetic modifications (DNA methylation and histone tail modifications) alter the conformation of chromatin fiber and therefore regulate the expression of the underlying genes that may lead to parental-specific expression (Davies et al., 2007, 2008; Wilkinson et al., 2007). Deletions, duplication, mutations, or alterations of imprinting of the only active allele, as well as uniparental disomy or loss of imprinting of the inactive allele, led to an unbalance (loss of function or gain of function) in the dosage of the gene product and may have phenotypic consequences (Gurrieri and Accadia, 2009). Grafodatskaya et al. (2010) have reported genes or genomic regions exhibiting abnormal epigenetic regulation in association with either the syndromic (15q11-13 maternal duplication) or the nonsyndromic forms of autism-spectrum disorders. Behavioral phenotypes linked with genetic syndromes have been studied extensively in order to generate rich descriptions of phenomenology, determine the degree of specificity of behaviors for a particular syndrome, and examine potential interactions between genetic predispositions for behavior and environmental influences.

The appropriate expression of imprinted genes is important for normal development, with numerous genetic diseases associated with imprinting defects including Angelman syndrome and Prader-Willi syndrome (Doe et al., 2009). Genomic imprinting refers to the parent-of-origin-specific epigenetic marking of a number of genes that leads to a bias in expression between maternally and paternally inherited imprinted genes (Isles et al., 2006), which, in some cases, results in monoallelic expression from one parental allele. Prader-Willi syndrome, a neurodevelopmental disorder caused by deletion/inactivation of a paternally expressed imprinted gene on chromosome 15q11-q13 (Nicholls and Knepper, 2001), is characterized by failure to thrive in infancy, pre- and postnatal hypotonia, hypophagia in early life followed by a voracious appetite for food (hyperphagia), obesity, short stature, hypogonadism, acromicria, psychomotor retardation, and mild learning impairments (Holm et al., 1993; Whittington et al., 2004, 2009). The syndrome offers a case for illustrative purposes: C15orf2 (Chromosome 15 open reading frame 2) is an intronless gene located in the Prader-Willi syndrome (PWS) chromosomal region on human chromosome 15 and present in several regions of the brain (Wawrzik et al., 2010). In PWS, there are specific relationships between apparently distinct aspects of the PWS behavioral phenotype and specific endophenotypic characteristics (Woodcock et al., 2009). It is known also that the genes in the cluster on chromosome 15q11-q13 contribute to autism-spectrum disorders and schizophrenia (Boer et al., 2002). Hogart et al. (2009) have found that particular genetic copy number changes combined with additional genetic or environmental influences on epigenetic mechanisms impact outcome and clinical heterogeneity of 15q11-13 duplication syndromes, as in the case of PWS. Relkovic et al. (2010) employed the deletion mouse model, PWS-IC$^{+/-}$, to study visuospatial attention, response control, locomotor activity, open-field behavior, and sensory motor gating in this condition. PWS-IC$^{+/-}$ mice showed increased acoustic startle and reduced prepulse inhibition, less locomotor activity, and attentional deficits. They discussed these observations from the perspective of frontal-lobe abnormalities contributing to the PWS clinical condition that appear to parallel those of schizophrenia-spectrum disorders.

25.2 EPIGENETICS OF DEPRESSIVE DISORDERS

As described by Renthal and Nestler (2009), animal models of depression and drug addiction have shown that alterations in gene expression are mediated, at least partially, by epigenetic mechanisms changing chromatin structure on specific gene promoters; these alterations are implicated in the etiopathogenesis of the respective disorders. The marked control exerted by chromatin remodeling on gene expression together with the potential stability of chromatin mechanisms imply that chromatin regulation offers a prime candidate for mediating those aspects of long-term neural plasticity that result ultimately in neuropsychiatric conditions (Renthal and Nestler, 2008). These histone modification epigenetic mechanisms and DNA methylation have been found to affect a diversity of pathways culminating in animal models of depression (cf. Schroeder et al., 2010). Early-life adversity and stress induces the adverse alterations of gene expression profiles, for instance involving the glucocorticoid receptor or brain-derived neurotrophic factor (BDNF), and may be reversible by epigenetic drugs. Serotonin transporter (5-HTT), which terminates the action of serotonin (5-HT) reuptake into presynaptic terminals, plays a key role in regulation of the neurotransmitter (Blakely et al., 1994; Flattem and Blakely, 2000). For instance, child sex abuse, linked to depressive disorders and other health problems (Jewkes et al., 2010; Kelly, 2010), may create long-lasting changes in methylation of the promoter region of 5-HTT in women (Beach et al., 2010). 5-HTT gene depletion may allow a window for observing the link between depression and life-long 5-HTT dysfunction (Canli and Lesch, 2007; Murphy and Lesch, 2008). 5-HTT-knockout mice exhibit behavioral profiles and biomarkers related to anxiety and depression (Holmes et al., 2003a, 2003b; Iritani et al., 2006; Lira et al., 2003). 5-HTT-knockout rats (Homberg et al., 2007; Olivier et al., 2008) show a behavioral phenotype with a strong similarity to 5-HTT-knockout mice (Kalueff et al., 2010).

Tryptophan hydroxylase is the rate-limiting enzyme of 5-HT biosynthesis, and the serotonin transporter (SLC6A4), involved in the reuptake of serotonin from the synaptic gap, is essentially involved in serotonergic signaling. Perroud et al. (2010) compared the levels of expression of serotonin-related genes between suicide completers and controls to identify genetic loci involved in their regulation. SLC6A4, TPH1, and TPH2 mRNA levels were measured in the ventral prefrontal cortex of 39 suicide completers and 40 matched controls. They found that TPH2 levels were found significantly increased in suicide completers and that the single nucleotide polymorphism, rs10748185, located in the promoter region of TPH2 significantly affected the levels of TPH2 mRNA expression. The influence of epigenetic processes upon serotonergic systems in affective disorders remains an important avenue of investigation.

A plethora of evidence derived from postmortem studies, animal studies, blood levels, and genetic studies have indicated that BDNF is involved in the pathogenesis of depression and in the mechanism of action of biological treatments for depression (Berton and Nestler, 2006; Groves, 2007; Molteni et al., 2009, 2010; Post, 2010). Early-life adversity and maltreatment induced long-lasting changes in methylation of BDNF DNA that resulted in altered BDNF gene expression in the adult prefrontal cortex; moreover, altered BDNF DNA methylation was observed in the offspring of females that had previously experienced the maltreatment regimen (Roth et al., 2010). Bocchio-Chiavetto et al. (2010) have obtained decreased serum BDNF levels in patients with major depression and also demonstrated that it downregulated the mature form of the neurotrophin BDNF, whereas meta-analyses showed a reduction of both BDNF serum and plasma levels in patients presenting with the disorder. In this respect, Calabrese et al. (2010) observed that that chronic treatment with the antidepressant duloxetine normalized the expression of BDNF mRNA-coding exon (IX) in the hippocampus and the prefrontal cortex of 5-HTT-knockout rats through the modulation of selected neurotrophin transcripts, the expression of which was upregulated by the antidepressant only in 5-HTT-knockout rats. On the other hand, the modulation of BDNF protein by duloxetine in the frontal cortex was abolished in mutant rats. In consensus with other findings, these results suggest that animals with a genetic defect of the serotonin transporter maintain the ability to show neuroplastic changes in response to antidepressant drugs. Because these animals show depression-like behavior, the region and isoform-specific increase of BDNF levels offer a mechanism activated by long-term antidepressant treatment to restore normal plasticity that is defective under genetic dysfunction of the serotonin transporter. Molteni et al. (2009) analyzed the expressed of BDNF and found that levels were reduced significantly in the hippocampus and prefrontal cortex of 5-HTT-knockout mice through transcriptional changes affecting different neurotrophin isoforms. The reduction of BDNF gene expression was partially due to epigenetic changes affecting the promoter regions of exons IV and VI. Furthermore, BDNF gene expression was decreased significantly in leukocytes of healthy volunteers carrying the S allele of the 5-HTTLPR (Molteni et al., 2009), implying that alterations both in the knockout mice and healthy humans may be linked to elevated vulnerability for mood disorders. Taken together, the emerging notion of altered (probably decreased) levels of BDNF protein and BDNF mRNA expression in affective disorders in the hippocampus of rodents (Deltheil et al., 2008; Guiard et al., 2008) seems intimately related to epigenetic considerations of disease pathogenesis (Lewis et al., 2010; McGuffin et al., 2010; Tsankova et al., 2006; Uher and McGuffin, 2008).

Because cohort studies provide information regarding the cause and consequence of epigenetic manipulation, Olsson et al. (2010) examined the association between risk for depression and buccal-cell (the inner lining of mouth or cheek, routinely shed and replaced by new cells) methylation. Epigenotyping was limited to promoter methylation of the serotonin transporter gene (*5-HTT*) with transcription-limiting variable number tandem repeat (location in a genome where a short nucleotide sequence is organized as a tandem repeat) and the *5-HTT* promoter (*5-HTTLPR*) genotyped. They drew a nested sample of 25 depressed and 125 nondepressed adolescents from an established longitudinal study of adolescent health but found no association between either depressive symptoms and either buccal cell *5-HTT* methylation or *5-HTTLPR*. Nevertheless, depressive symptoms were more common among individuals with elevated buccal cell *5-HTT* methylation carrying the *5-HTTLPR* short allele. Both complete and partial methylation of a *5-HTT* reporter gene in an expressing cell line reduced *5-HTT* activity (Philibert et al., 2008; Risch et al., 2009). Applying the data from a previously reported sample of 89 major depressive disorder patients, Brockmann et al. (2010) examined post hoc the effect of 5-HTTLPR status on resting state perfusion, using (99m) Tc-HMPAO-SPECT. Major-depression–disorder patients were stratified according to receptor polymorphism, using both a ciallelic (group A: L/L versus group B: S/S and S/L genotype) and a triallelic approach (group A': LA/LA versus group B': non-LA/LA genotype). The authors found no significant differences between both subgroups regarding age, gender, severity of depression, medication, or treatment response. Using the ciallelic approach, group B, compared with group A, revealed a significantly higher resting-state perfusion in medial prefrontal cortex. Additional region-of-interest analyses showed the relative overactivity of the amygdalae in group B. Similar effects were observed in the triallelic approach. The opposite contrasts (group A > group B) revealed no significant effects. In major-depression patients, it appears that the 5-HTTLPR gene polymorphism modulates resting state perfusion at major regions involved in the processing of mood (Butler et al., 2010; Wichers et al., 2008, 2009). Human observational studies (Wankerl et al., 2010) have found that in interaction with life stress, the short or S allele of the serotonin transporter gene-linked polymorphic region, 5-HTTLPR, on the basis of convergent evidence derived from different research fields, is associated with an enhanced risk for depression; furthermore, 5-HTTLPR is linked with different biological pathways regulating expressions of stress and depression.

The role of glucocorticoid receptors within a dysfunctional HPA axis feedback-regulation system has been implicated in the pathophysiology of major depressive disorder (Markopoulou et al., 2009; Juruena et al., 2006, 2009a, 2009b, 2010). Early childhood events, presumably adverse, were found to exert changes upon glucocorticoid receptors in both laboratory and clinical studies focused upon depressive disorders (Boyle et al., 2005; Chourbaji et al., 2008; Heim et al., 2008; Liu et al., 1997; McGowan et al., 2009; Weaver et al., 2004). It has been indicated that overactivity in the HPA axis may not necessarily be the result of depressiveness, but rather, part of the pathophysiology associated with the disorder (Pariante and Lightman, 2008; Tichomirowa et al., 2005). Hormonal challenge tests, such as the dexamethasone/corticotropin-releasing hormone test, have indicated elevated HPA activity (hypercortisolism) in at least a portion of patients with depression, although growing evidence has suggested that abnormally low HPA activity (hypocortisolism) has also been implicated in a variety of stress-related conditions (Rossi-George et al., 2009). For example, patients with Addison's disease, characterized by chronic adrenal insufficiency and hypocortisolism, have not only increased levels of anxiety and fear, and over-reaction to stimuli but decreased performance efficiency and need for social contact (Warmuz-Stangierska et al., 2010). It has been demonstrated repeatedly that stress reduces the expression of BDNF and that antidepressant treatments increase it (Kluge et al., 2010; Kubera et al., 2010; Zhang et al., 2010); furthermore, the glucocorticoid receptor interacts with the specific receptor of BDNF, TrkB, and excessive glucocorticoid interferes with BDNF signaling. Thus, alterations in BDNF functioning are involved in the structural changes and possibly impaired neurogenesis in the brain of depressed patients (Kunugi et al., 2010). The interaction of BDNF with the HPA axis is revealing, because the system is related intimately to stress and homeostasis restoration (Conrad et al., 2009). BDNF acts as an autocrine survival factor; in neurons supporting its own release at low BDNF concentrations and inhibiting it at high concentrations, it binds to high- and low-affinity receptors and exerts counteracting effects on the survival and apoptosis of cells (Peters et al., 2007).

Several studies have demonstrated that the epigenetic state of a gene can be established through early in life experience and is potentially reversible in adult life: maternal licking and grooming increases glucocorticoid receptor expression in the offspring via increased hippocampal serotonergic tone accompanied by increased histone-acetylase–transferase activity, histone acetylation, and DNA demethylation mediated by the transcription factor nerve growth factor-inducible protein-A (Weaver, 2009). It has been observed that the family inheritance of major depressive disorder in cord blood of newborn infants of depressed mothers presents an association between prenatal exposure to the mothers' depressed mood and the methylation status of the glucocorticoid receptor gene promoter exon 1F (Oberlander et al., 2008). Alt et al. (2010) have found, in postmortem brains of a group of major-depressive–disorder patients, that although glucocorticoid receptor was unaltered, glucocorticoid receptor α was decreased in the amygdale and cingulated cortex, and the transcription factor, NGFI-A, for exon 1F was downregulated in the patients' hippocampus (see the work of Turner et al. [2010] for further information on the transcriptional control of the glucocorticoid receptor). In consensus, it appears that the epigenetic state of a gene may be established through early-life experiences, with potential reversal in adult life. In this regard, epigenetic modifications in target gene promoters in response to environmental demand may ensure the stable, yet dynamic, regulation that mediates persistent changes in biological and behavioral phenotype over the lifespan (Weaver, 2009).

Deficits in myelination and glial-cell gene expression imply aberrant oligodendrite developmental and functional involvement in neuropsychiatric conditions (Åberg et al., 2006; Haroutunian et al., 2006). There appears to be a reduction in the expression of glial-specific genes and those involved in myelination among suicide victims presenting depression (Klempan et al., 2009c). There is a burgeoning body of information regarding the putative influence of myelination-related deficits in the etiopathogenesis of depressive disorders (Behan et al., 2010; Cotter et al., 2001, 2002; Hamidi et al., 2004; Uranova et al., 2004). Klempan et al. (2009a) have described a specific role of QKI, an oligodendrocyte-specific RNA-binding protein essential for cell development and myelination, as implied by its reduced expression and known interactions with genes involved in oligodendrocyte determination. The polyamines are involved in both brain morphological and glial processes (Krauss et al., 2007). The downregulation of SAT1 (spermidine/spermine N^1-acetyltransferase) in brain Brodmann areas 4, 8/9, and 11 and additional brain regions has been observed in suicide victims (Guipponi et al., 2009; Klempan et al., 2009b; Sequeira et al., 2006, 2007). Nevertheless, Fiori and Turecki (2010) found that the genetic and epigenetic factors studied in an analysis of suicide completers demonstrated little influence on the expression levels of spermine synthase and spermine oxidase, the two enzymes involved in polyamine metabolism; thus, those factors would seem not to be involved in the dysregulated expression of these genes in the sample of suicide victims. Despite the above uncertainty, the influences of epigenetic processes in major depressive disorder and suicide have gathered momentum: Poulter et al. (2008) showed that that DNA methyltransferase mRNA expression was altered in suicide brain; this alteration of expression in the frontopolar cortex was associated with increased methylation of a gene, the $GABA_A$ receptor $\alpha1$ subunit promoter region, whose mRNA expression was previously shown to be reduced.

Despite the burgeoning emergence of evidence for epigenetic regulation (dysregulation) underlying depressive and mood disorders, the methodological issues persist: genome-wide association studies (GWAS) have implicated several genes (e.g., *5-HTT*, tryptophan hydroxylase 2, BDNF) related to suicide behavior, but not all of the studies support these findings (Tsai et al., 2010). The explanatory power and pathway leading to the clinical translation of risk estimates for common variants reported in GWAS remain unclear because of the presence of rare and structural genetic variation. There remains too the neglect of a consistent definition for the suicide-behavior phenotype among existing studies; this inconsistency complicates the comparability of the available studies and the data pooling (Tsai et al., 2010).

25.3 EPIGENETICS OF SCHIZOPHRENIA

The epigenetic regulation of gene expression may be influenced by several factors originating from the environment: individuals' relatives and ancestors, prenatal exposure, and early-life events. These epigenetic mechanisms may alter neurophysiological processes throughout an individual's life span via programming effects upon gene expression, sometimes in anticipation of particular life events. This signaling remains metastable, subject to perturbation by stochastic events, errors, environmental toxic agents, or epigenetic alteration that occurs as paternal age advances or during fetal adversity, endowing a causal susceptibility for schizophrenia (Dalman et al., 2006; Kaymaz et al., 2006; Perrin et al., 2010). The subsequent status of the regional structure and biomarkers affected is determined both by the predictable actions of the epigenetic processes and by the presence of random elements. The epigenetic state of an organism (or epigenome) describes a landscape of complex and plastic molecular events that may contribute links integrating genotype with phenotype including the characterization of epigenetic variation, in the form of *epialleles* over developmental processes, environments, under conditions of health, and disorder. Finer et al. (2010) have outlined the variable relationship of epialleles with phenotype, their detection, and influence on phenotype included (Rakyan and Beck, 2006). Neuropsychiatric disorders, such as schizophrenia and major depressive disorder, share genetic risk factors that have incomplete penetrance: an individual may possess a pathogenic allele without developing the disorder. Lahiri et al. (2009) have proposed the Latent Early-life Associated Regulation (LEARn) model, whereby latent changes during development in specific gene expressions occur. According to this notion, environmental agents cause long-lasting disturbances in gene regulation, initially at the early stages (infancy) but later expressing pathological outcomes. The LEARn model operates through the regulatory region (promoter) of the gene; it describes the induction of pathophysiology specifically through alterations in methylation and oxidation status within the promoter of specific genes.

The prodromal stages of schizophrenia during childhood, adolescence, and young adulthood may presage the dynamic pathophysiology of disturbances in brain maturation processes (Jaaro-Peled et al., 2009). As described previously, the indices of disease progression and symptom proliferation both complicate the array of gene involvement and provide a substrate of hazardous interactivity (Palomo et al., 2007). The neurodevelopmental-disorder notion posits that brain-development disturbances underlie the pathophysiology of psychiatric disorders (Tenyi et al., 2009). Rapoport and Gogtay (2010) have described the neuroanatomical development of patients with very early-onset schizophrenia displaying progressive loss of gray matter, delayed/disrupted white-matter growth, and a progressive decline in cerebellar volume; some of these structural deficits were shared by their healthy siblings. Global and regional cortical thinning in first-episode patients supporting a primary neurodevelopment disorder affecting the normal cerebral-cortex development in schizophrenia was observed (Crespo-Facorro et al., 2010). There exists much evidence that variations in disease-susceptibility genes influence higher brain functions, e.g., cognition and information processing (Keri, 2009; Tomppo et al., 2009). Genetic susceptibility genes, for affective disorders and schizophrenia-spectrum disorders, DISC1 and neureglin, have been identified (Krug et al., 2008; Schumacher et al., 2009). A role for DISC1 in regulating the formation and/or maintenance of primary cilia in the establishment of subtype-specific targeting of dopamine receptors to the ciliary surface has been identified (Marley and von Zastrow, 2010). Singh et al. (2010) have reported that DISC1 is regulated during embryonic neural progenitor proliferation and neuronal migration through an interaction with Disheveled-Axin (DIX) domain containing-1 (Dixdc1) gene, the third mammalian gene discovered to contain a DIX domain. Cohort studies have proved useful generally for observations on susceptibility genes: for example, multiple large cohort data indicate that gene loci on chromosome 6p22.1 are the most probable genes susceptible to schizophrenia (Shi et al., 2009; Stefansson et al., 2009). Walters et al. (2010) have shown in Irish samples that the ZNF804A genotype was associated with differences in episodic and working memory in patients but not in controls, thereby replicating results in the same direction in the German nationality samples. Additionally, in both samples, when patients with a lower IQ were excluded, the association between ZNF804A and schizophrenia strengthened. In a Chinese study, the association of DTNBP1 gene with some symptom factors of schizophrenia indicated allelic association of rs909706 with positive symptoms of schizophrenia, whereas the genotype C/G of rs2619539 has a negative symptom, lack of spontaneity, and flow of conversation (flatness of affect). In relation to negative symptom profiles, Tolosa et al. (2010), genotyping 27 single nucleotide proteins (SNPs) of *FOXP2* from cohorts of 293 schizophrenia patients and 340 healthy volunteers, reported a significant association between the

SNP rs2253478 and poverty of speech of the Manchester scale, with a higher degree of methylation in the left parahippocampal gyrus, compared with the right, in the patient group. Sun et al. (2010) concluded that susceptibility gene DTNBP1 variations were associated, putatively, with some symptoms of schizophrenia. Niwa et al. (2010) report the generation of an in utero animal model demonstrating that nonlethal deficits in early development may affect postnatal brain maturation and complex brain function in adulthood culminating in schizopsychotic disorders. They demonstrate that the transient knockdown of DISC1 in the pre- and perinatal stages of brain development, primarily in a lineage of pyramidal neurons of the prefrontal-cortex region, induces selective abnormalities in postnatal mesocortical dopaminergic maturation and functional abnormalities linked to disturbed postpubertal cortical neurocircuitry.

Candidate gene regulation for example, glutamic acid decarboxylase 67 (GAD67) and reelin, illustrates the evidence for epigenetic abnormalities in schizophrenia (Costa et al., 2009; Dong et al., 2005). Reduced GAD67, catalyzing decarboxylation of glutamate to GABA, is implicated in cognitive impairments in schizophrenia (Hashimoto et al., 2008; Lewis and Hashimoto, 2007). Reelin, involved in synaptic plasticity underlying cognition, is implicated in the reduced numbers of dendritic spines in postmortem schizophrenic brains (Grayson et al., 2005, 2006). During cerebellar development and in the modulation of adult synaptic function, reelin provides an extracellular-matrix protein synthesized in cerebellar granule cells that determines Purkinje-cell positioning. Maloku et al. (2010) have found a marked decrease of reelin expression in the cerebellum of schizophrenia and bipolar disorder patients that was unrelated to postmortem interval, pH, drugs of abuse, or to the presence, dose, or duration of antipsychotic medications. The promoters associated with reelin and GAD67 are downregulated as a consequence of DNA methyltransferase (DNMT)-mediated hypermethylation. Kundakovic et al. (2009) reported that other epigenetic drugs, histone-deacetylase (HDAC) inhibitors, activate these two genes with dose and time dependence comparable with that of DNMT inhibitors. Grayson et al. (2009) have presented the premise that coordinate regulation of multiple promoters expressed in neurons may be modulated through methylation. Although the identification of genes and promoters regulated epigenetically has increased steadily, few studies have examined methylation changes as a consequence of increased levels of dietary amino acids, e.g. methionine (MET). They have proposed that the MET mouse model may provide information regarding the identification of genes that are regulated by epigenetic perturbations. In addition to studies incorporating the reelin and GAD67 promoters, they have shown evidence that additional promoters expressed in select neurons of the brain are similarly affected by MET administration, implying the utility of epigenetically responsive promoters that uncover global methylation changes occurring in selected brain regions. MET is implicated in a drastic deterioration of schizophrenia symptoms (Wyatt et al., 1971; Erdelyi et al., 1978), possibly because of its conversion to *S*-adenosylmethionine (SAM) (Costa et al., 2002; Tremolizzo et al., 2002, 2005). Postmortem studies indicated increased levels of SAM (Guidotti et al., 2007) and elevated DNA methylation at the reelin promoter (Abdolmaleky et al., 2005). Tueting et al. (2010) indicate that the downregulation of spine density in L-methionine-treated mice may be due to the decreased expression of reelin and that sodium valproate, an antiepileptic drug that upregulates reelin and GAD67 mRNA/protein expression by reducing the methylation of their promoters (Dong et al., 2010), may prevent spine downregulation by inhibiting the methylation-induced decrease in reelin. Pertinent to these considerations, Gavin and Sharma (2010) have developed a model applying cultured peripheral blood mononuclear cells capable of probing pharmacologically epigenetic processes in humans. They have suggested that the capability of modifying appropriately chromatin structure may mediate the treatment response.

There is ample preclinical and clinical evidence that exposure to infection during the period of gestation influences the etiopathogenesis of schizophrenia (Archer, 2010a, 2010b; Brown, 2008, 2010; Brown et al., 2004a, 2004b, 2005; Gilmore and Jarskog, 1997; Mortensen et al., 2007), with epigenetic mechanisms involved as a critical determinant in disease predisposition (Bale et al., 2010). Even maternal immune-system activation may incur pathophysiology associated with neuropsychiatric disorders (Zuckerman and Weiner, 2005; Zuckerman et al., 2003). For example, in a study of the neurodevelopmental etiology in schizophrenia, Brown et al. (2010) have observed that prenatal infections, previously associated with schizophrenia in 26 patients with schizophrenia from a large and well-characterized birth cohort, were related to impaired executive-functioning performance, as measured on the Wisconsin Card Sorting Test and Trails B. The pattern of results suggested that cognitive set-shifting ability may be particularly vulnerable to this gestational exposure. A plethora of studies have served to indicate candidate infections that are linked with elevated schizophrenia risk, and the notion that individual susceptibility genes, such as DISC1 and neureglin, but also others (Cacabelos and Martínez-Bouza, 2010; Kamiya et al., 2005; Shen et al., 2008; Yolken et al., 2009), may act in concert with prenatal infections has been entertained (Clarke et al., 2009). Cannon et al. (1999) used a prospective, longitudinal design, involving 9,236 members of the Philadelphia cohort of the National Collaborative Perinatal Project screened for mental health service utilization in adulthood with chart reviews performed to establish diagnoses according to DSM-IV criteria, to study whether or not adverse obstetric experiences predict schizophrenia. From an epigenetic perspective, a deviant functional-developmental trajectory during the first 7 years of life among individuals who manifested schizophrenia as adults, was investigated. It was observed that risk for schizophrenia increased linearly with the number of hypoxia-associated obstetric complications but was unrelated to maternal infection during pregnancy or fetal growth retardation. Preschizophrenic (prodromal essentially) cases (and their unaffected siblings who were also cohort members) manifested cognitive impairment, abnormal involuntary movements and coordination deficits, and poor social adjustment during childhood. There was no evidence of intraindividual decline in any domain, but the preschizophrenic cases did show deviance on an increasing number of functional indicators with age. Finally, Brown and Derkits (2010) have presented several indices of interaction between in utero exposure to infection and genetic variants to illustrate the cellular and molecular mechanisms by which in utero exposure to infection contributes to schizophrenia risk. They recommend the adoption of translational approaches utilizing the observations derived from epidemiologic research for the integration of etiological clues with pathophysiological implications, for example, the laboratory evidence that prenatal infection exposure induces phenotypes expressing brain and behavior anomalies of schizophrenia. Translational strategies target the elucidation of those causal mechanisms through which normal neurodevelopmental processes are altered by prenatal insults in the context of epigenetic regulation (Archer et al., 2010a).

The International Schizophrenia Consortium (2009) has presented two analytic approaches assessing the extent to which common genetic variation underlies the risk of schizophrenia: a major histocompatibility complex and molecular genetic evidence for a substantial polygenic component for the risk of schizophrenia involving thousands of common alleles of very small effect. In a related manner, Chen et al. (2010) found that in the cardiomyopathy-associated 5 (CMYA5) gene, there were two nonsynonymous markers, rs3828611 and rs10043986, showing nominal significance in both the Clinical Antipsychotic Trials of Intervention Effectiveness and the molecular genetics of schizophrenia genome-wide association study supported by the genetic association information network samples. The schizophrenia phenotype remains variable between patients and throughout the course of disease, comprising positive symptoms (delusions and hallucinations); negative symptoms (lack of motivation and energy, social withdrawal, flatness of affect), often expressed as anhedonia and depression; and cognitive-core symptoms, with consequential social burdens and poor quality of life. Krebs et al. (2009) have outlined the determining epigenetic interplay between combinations of heritable and environmental factors in the disorder etiopathogenesis. Kim et al. (2010) have demonstrated that the histone deacetylase HDAC3 and HDAC4 genes have influenced the pathophysiology of schizophrenia in a Korean population. Thus, neurobiological evidence of altered DNA methylation, abnormal glutamatergic transmission, altered mitochondrial function, folate deficiency, and high maternal homocysteine levels all involve the one-carbon metabolism that influences epigenetic regulation.

Dendritic spines, small actin-rich protrusions on the surface of dendrites whose morphological and molecular plasticity play key roles in cognition, are dependent on both the form and function on the actin cytoskeleton. Dynamic interactions between microtubules and actin filaments within dendritic spines play important roles in dendritic spine plasticity (Dent et al., 2010), thereby contributing to their role under neuropsychiatric conditions (Deutsch et al., 2010). Dendritic spines, possessing a dynamic, developmentally regulated morphology (De Roo et al., 2008; Ethell and Pasquale, 2005), are responsive to different patterns of synaptic activity and are composed of large memory spines of synaptic strength and stability and thin plasticity spines responsive to potentiating stimulation (Bourne and Harris, 2007). Although dendritic spines form and mature as synaptic connections develop, they remain plastic even in the adult brain, where they can rapidly grow, change, or collapse in response to normal physiological changes in synaptic activity that underlie learning and memory processes. The physiopathological influences underlying neuropsychiatric disorders may affect dendritic spine morphology, shape, and number, adversely. For example, Pontrello and Ethell (2009) have described how actin-regulating proteins maintain the balance between F-actin assembly and disassembly that is needed to stabilize mature dendritic spines, and how changes in their activities may lead to rapid remodeling of dendritic spines. Herrick et al. (2010) have identified an interaction between spine-associated RapGAP, a postsynaptic protein that reorganizes actin cytoskeleton and drives dendritic-spine head growth, and PDLIM5/Enigma homolog, a postsynaptic density-95/Discs large/zona occludens 1-lin11/Isl-1/Mec3 family member, that is implicated in both major depression and schizophrenia (Horiuchi et al., 2006; Kato et al., 2005; Li et al., 2008; Liu et al., 2008). Herrick et al. (2010) demonstrated that PDLIM5 presence in the postsynaptic density promotes decreased dendritic-spine head size and longer filopodia-like morphology. Spine head size was linked to new PDLIM5 functions pertaining to shrinkage or expansion of dendritic spines as a neuropsychiatric biomarker for gene-environment interaction. The influence of actin cytoskeleton as an expression of epigenetic regulation may be illustrated further, albeit in the setting of mood disorders: examining the *disturbed cytoskeletal* theory of mood disorders, Nakatani et al. (2007) identified the gene *Cap1* (the gene that encodes the enzyme adenylyl cyclase-associated protein 1) as an important mediator of actin turnover and as a cogent quantitative trait gene for a depressiveness trait in mice, by combining the results of our prior genetic and current genome-wide expression analyses. Analysis of core actin-related gene expression in the frontal cortex of C57BL/6 (B6) (prone to depression) and C3H/He (C3) (resistant to depression) mice indicated that *Cap1* was downregulated at both transcript and protein levels in B6, whereas other differentially regulated genes included *cofilin1* and *profilin1* (upregulated in B6) and a Rho-family GTPase member (*Pak1*) (downregulated in B6). They also investigated the core actin-pathway components in human postmortem prefrontal cortices and observed a trend for *Cap1* reduction in the bipolar brains, suggesting an imbalance of actin dynamics toward actin depolymerization in mood disorders.

25.4 EPIGENETICS IN INTERVENTION

An understanding of the role of epigenetic defects or epi-mutations in dictating the susceptibility and predisposition for neuropsychiatric disorders may contribute towards the possible development of treatment drugs that aim at modifications of the epigenome (Wu et al., 2008; Yasuda et al., 2009). For example, the standard mood stabilizer, valproate (valproic acid [VPA]), has been shown to modulate the epigenome by inhibiting histone deacetylase (Machado-Vieira et al., 2010). Note, however, that maternal VPA exposure has permanent adverse effects upon neurological and behavioral development. These effects include delayed physical development, impaired olfactory discrimination, dysfunctional preweaning social behavior, and evidence from *in situ* hybridization assays of lower cortical expression of BDNF mRNA (Roullet et al., 2010). In humans, prenatal exposure to VPA can induce fetal valproate syndrome, which has been associated with autism. Nevertheless, the case of epigenetics in the therapeutic advances in cancer disorders may be illustrative: DNA methylation patterns have been found to be susceptible to changes in the response to environmental stimuli such as diet, toxins, or infections: under these conditions, the epigenome appears to be most vulnerable during early *in utero* development. Epigenetic programming of gene expression profiles is sensitive to the early-life environment in that both the chemical and social environment early in life could affect the manner by which the genome is programmed by the epigenome (Szyf, 2009). Aberrant DNA methylation changes have been detected in several diseases, particularly cancer, where genome-wide hypomethylation coincides with gene specific hypermethylation. Just as the concept of "staging" attributes its origins to disease progression in oncology, the effect of epigenetic interactions on DNA damage response and DNA repair and the involvement of epigenetic cross-talks in cancer formation, progression, and treatment (Murr, 2010) may open horizons as yet unexplored in neuropsychiatry (Archer et al., 2010b). Therapeutic strategies aimed at resetting the epigenetic state of dysregulated genes have been tested. Nevertheless, due to the complexity of epigenetic gene regulation, the first-generation drugs that function globally by inhibiting the cogs of epigenetic machinery may introduce severe side effects. A detailed understanding of how repressive chromatin states are established and maintained at specific loci appears to be fundamental for the development of more selective epigenetic treatment strategies in the future (Hahn et al., 2010). As described by Tost (2010), DNA methylation patterns can be used to detect cancer at very early stages, to classify tumors, and to predict and monitor the response to antineoplastic treatment (Altucci and Minucci, 2009). Similarly, the staging of prodromal schizophrenia, e.g., analyzing subsets of schizophrenia susceptibility genes also affecting schizotypy in nonpsychotic relatives (Fanous et al., 2007) or assessment of molecular-behavioral associations in Fragile X full mutation (Abrams et al., 1994) may facilitate prediction and disease-monitoring.

Müller and Dursun (2010), in discussing genetics, epigenetics, and infection as environmental factors, posit a major role for dysfunctional immune systems and draw attention to the as yet rudimentary therapeutic advantages of applying antiinflammatory agents (Akhondzadeh et al., 2007; Chaves et al., 2009a, 2009b, 2010). The role of inflammatory and neurodegenerative (I&ND) processes in depressive disorders emerges as a critical factor in accelerated neurodegeneration in the disorder (Maes et al., 2009). Thus, animal models of depression confirm the effects of multiple inflammatory-cytokines, oxygen radical damage, tryptophan catabolites, and neurodegenerative biomarkers in patients with depression. In addition, several vulnerability factors result in a predisposition to disease status by enhancing inflammatory reactions, for example, lower peptidase activities (dipeptidyl-peptidase IV), lower omega-3 polyunsaturated levels, and an increased gut permeability (leaky gut). Transgenic mouse models and endophenotype-based animal models seek to:

1. Discover new I&ND biomarkers, both at the level of gene expression and the phenotype, and elucidate the underlying molecular I&ND pathways causing depression.
2. Identify new therapeutic targets in the I&ND pathways.
3. Develop new anti-I&ND drugs for these targets (Maes et al., 2009).

In the search for treatment interventions for psychiatric disorders, the role of aberrant methylation alterations and chromatin remodeling contributes to an eventual derivation of therapeutic strategies. In this regard, histone methylation and microRNA expression emerged as potential therapeutic targets (Kelly et al., 2010). Finally, Narayan and Dragunow (2010) highlight the differences that may exist between humans and model systems (even those laboratory-derived) in seeking robust, reliable, and valid preclinical designs and procedures.

25.5 CONCLUSIONS

The significance of epigenetics for neuropsychiatric disorders unfolds with an increasingly labyrinthine sophistication. Several sites of epigenetic dysregulation, such as possible mechanisms of oligodendroglial abnormalities involving functional variations in oligodendroglia-related genes, epigenetic regulation of chromatin, dopamine system hyperactivity, may contribute to affective disorders, substance abuse, and schizophrenia (Sokolov, 2007). Klar (2010) and Armakolas et al. (2010) have proposed the Somatic Sister chromatid Imprinting and Selective chromatid Segregation (SSIS) model as a mechanism for brain hemispheric laterality specification by chromosome 11 sister chromatids differentiation, via epigenetics and their selective segregation, through centromeres, to affect an asymmetric cell division. This notion postulates expression of a hypothetical brain-laterality-determining *DOH1* gene in the chromatid inheriting the parental template. It is suggested that asymmetric cell development induces human brain hemispheric laterality specification, and schizopsychotic disorders may appear when brain hemispheric laterality is compromised (Klar, 2010). It appears essential that the special techniques developed to analyze the epigenome are currently applied to the investigation of the molecular basis of psychiatric etiopathogenesis in order to describe the contribution of gene-environment interactions to brain function (Isles and Wilkinson, 2008). In this regard, the methodological and logistic issues pertaining to postmortem brain-tissue epigenomic studies remain to be elucidated (Pidsley and Mill, 2010). The seriousness of early-life adversity cannot be overestimated: Vucetic et al. (2010) showed that dietary protein restriction to mouse dams throughout gestation induced a six-to-eight-fold enhanced expression in dopamine-related genes, illustrating eventual impact on the potential development of schizophrenia, substance use disorder, and attention-deficit hyperactivity disorder. In this context, Meaney and Ferguson-Smith (2010) have formulated a conceptual framework that integrates current notions on nucleotide sequence, chromatin remodeling, RNA signaling, and their respective interactions for the eventual derivation of and implications for a working model of experience-dependent phenotypic plasticity (Bagot and Meaney, 2010).

The present account examines epigenetic influences to inherited characteristics subjected to conditions of prenatal or early-life adversity that produce the eventual expressions of depressive disorders, schizophrenia-spectrum disorders, and such developmental disorders as Prader-Willi syndrome. The essential role of nutrition is central: epigenetic regulation encompasses alterations of genetic material that do not affect the DNA nucleotide sequence; these include DNA methylation patterns, chromatin structure, histone codes, and noncoding small RNAs (Kussman et al., 2010). DNA methylation is modified particularly around the time of birth, and early-life nutrition may affect health outcomes later in life. DNA methylation provides an epigenetic mechanism of gene regulation in neural development, function, and disorder conditions (Feng and Fan, 2009; Pregelj, 2009); the critical integration of these epigenetic processes accommodating those involving stress-regulating functions of the HPA axis with related behavioral profiles through which early-life trauma is linked to specific psychotic experiences is awaited (Read et al., 2009). These influences, epigenetic interactions on DNA damage response and DNA repair, may yet provide insights facilitating diagnosis and understanding of progression and intervention. Epigenetic changes often precede disease pathology, rendering these processes valuable diagnostic indicators for disease risk or prognostic indicators for disease progression. Integrating epigenetic regulatory processes into essential processes in developmental psychobiology defines the marking steps through which environmental agents, adverse or benign during the early life of the child, adjust the structural aspects of DNA (Meaney, 2010). The accomplishment of this physical expression provides the influence whereby perinatal environmental signals determine the phenotypes governing individual life span. Finally, the longitudinal-developmental study by Wong et al. (2010), measuring DNA methylation across the promoter regions of the dopamine receptor 4 (DRD4) gene, the serotonin transporter (SLC6A4/SERT) gene, and the X-linked monoamine oxidase A (MAOA) gene utilizing DNA sampled at ages 5 and 10 years in 46 MZ twin-pairs and 45 DZ twin-pairs, has shown that DNA-methylation differences are apparent already in early childhood, even between genetically identical individuals and that individual differences in methylation are not stable over time.

ACKNOWLEDGMENTS

The valuable suggestions of Rick Beninger are appreciated greatly.

REFERENCES

Abdolmaleky, H. M., Cheng, K. H., Russo, A., et al. (2005) Hypermethylation of the reelin (RELN) promoter in the brain of schizophrenic patients: A preliminary report. *Am. J. Med. Genet. B Neuropsychiatr. Genet.* 134B: 60–66.

Åberg, K., Saetre, P., Lindholm, E., et al. (2006). Human QKI, a new candidate gene for schizophrenia involved in myelination. *Am. J. Med. Genet. B Neuropsychiatr. Genet.* 141: 84–90.

Abrams, M. T., Reiss, A. L., Freund, L. S., et al. (1994). Molecular-neurobehavioral associations in females with the fragile X full mutation. *Am. J. Med. Genet.* 51: 317–327.

Akhondzadeh, S., Tabatabaee, M., Amini, H., et al. (2007). Celecoxib as adjunctive therapy in schizophrenia: A double-blind, randomized and placebo-controlled trial. *Schizophr. Res.* 90: 179–185.

Alt, S. R., Turner, J. D., Klok, M. D., et al. (2010). Differential expression of glucocorticoid receptor transcripts in major depressive disorder is not epigenetically programmed. *Psychoneuroendocrinology* 35: 544–556.

Altucci, L., Minucci, S. (2009). Epigenetic therapies in haematological malignancies: Searching for true targets. *Eur. J. Cancer* 45: 1137–1145.

Archer, T. (2010a). Effects of exogenous agents on brain development: Stress, abuse and therapeutic compounds. *CNS Neurosci. Ther.* PMID: 20553311.

Archer, T. (2010b). Neurodegeneration in schizophrenia. *Expert Rev. Neurother.* 10: 1131–1141.

Archer, T., Beninger, R. J., Palomo, T., et al. (2010a). Epigenetics and biomarkers in the staging of neuropsychiatric disorders. *Neurotox. Res.* 18: 347–366.

Archer, T., Kostrzewa, R. M., Palomo T., et al. (2010b). Clinical staging in the pathophysiology of psychotic and affective disorders: Facilitation of prognosis and treatment. *Neurotox. Res.* 18: 211–228.

Armakolas A., Koutsilieris, M., Klar, A. J. (2010). Discovery of the mitotic selective chromatid segregation phenomenon and its implications for vertebrate development. *Curr. Opin. Cell. Biol.* 22: 81–87.

Bagot, R. C., Meaney, M. J. (2010). Epigenetics and the biological basis of gene x environment interactions. *J. Am. Acad. Child Adolesc. Psychiatry* 49: 752–771.

Bale, T. L., Baram, T. Z., Brown, A. S., et al. (2010). Early life programming and neurodevelopmental disorders. *Biol. Psychiatry* 68: 314–319.

Beach, S. R., Brody, G. H., Todorov, A. A., et al. (2011). Methylation at 5HTT mediates the impact of child sex abuse on women's antisocial behavior: An examination of the Iowa adoptee sample. *Psychosom. Med.* 73: 83–87.

Behan, A. T., van den Hove, D. L., Mueller, L., et al. (2011). Evidence of female-specific glial deficits in the hippocampus in a mouse model of prenatal stress. *Eur. Neuropsychopharmacol.* 21: 71–79.

Berton, O., Nestler, E. J. (2006). New approaches to antidepressant drug discovery: Beyond monoamines. *Nat. Rev. Neurosci.* 7: 137–151.

Blakely, R. D., De Felice, L. J., Hartzell, H. C. (1994). Molecular physiology of norepinephrine and serotonin transporters. *J. Exp. Biol.* 196: 263–281.

Bocchio-Chiavetto, L., Bagnardi, V., Zanardini, R., et al. (2010). Serum and plasma BDNF levels in major depression: A replication study and meta-analyses. *World J. Biol. Psychiatry* 11: 763–773.

Boer, H., Holland, A., Whittington, J., et al. (2002). Psychotic illness in people with Prader-Willi syndrome due to chromosome 15 maternal uniparental disomy. *Lancet* 359: 135–136.

Bourne, J., Harris, K. M. (2007). Do thin spines learn to be mushroom spines that remember? *Curr. Opin. Neurobiol.* 17: 381–386.

Boyle, M. P., Brewer, J. A., Funatsu, M., et al. (2005). Acquired deficit of forebrain glucocorticoid receptor produces depression-like changes in adrenal axis regulation and behavior. *Proc. Natl. Acad. Sci. U.S.A.* 102: 473–478.

Brami-Cherrier, K., Lavaur, J., Pagès, C., et al. (2007). Glutamate induces histone H3 phosphorylation but not acetylation in striatal neurons: role of mitogen- and stress-activated kinase-1. *J. Neurochem.* 101: 697–708.

Brami-Cherrier, K., Roze, E., Girault, J. A., et al. (2009). Role of the ERK/MSK1 signalling pathway in chromatin remodelling and brain responses to drugs of abuse. *J. Neurochem.* 108: 1323–1335.

Brami-Cherrier, K., Valjent, E., Hervé, D., et al. (2005). Parsing molecular and behavioral effects of cocaine in mitogen- and stress-activated protein kinase-1-deficient mice. *J. Neurosci.* 25: 11444–11454.

Brideau, C. M., Eilertson, K. E., Hagarman, J. A., et al. (2010). Successful computational prediction of novel imprinted genes from epigenetic features. *Mol. Cell. Biol.* 30: 3357–3370.
Brockmann, H., Zobel, A., Schuhmacher, A., et al. (2011). Influence of 5-HTTLPR polymorphism on resting state perfusion in patients with major depression. *J. Psychiatry Res.* 45: 442–451.
Brown, A. S. (2008). The risk for schizophrenia from childhood and adult infections. *Am. J. Psychiatry* 165: 7–10.
Brown, A. S. (2011). The environment and susceptibility to schitzophrenia. *Prog. Neurobiol.* 93: 23–58.
Brown, A. S., Begg, M. D., Graven stein, S., et al. (2004a). Serologic evidence for prenatal influenza in the etiology of schizophrenia. *Arch. Gen. Psychiatry* 61: 774–780.
Brown, A. S., Derkits, E. J. (2010). Prenatal infection and schizophrenia: A review of epidemiologic and translational studies. *Am. J. Psychiatry* 167: 261–280.
Brown, A. S., Hooton, J., Schaefer, C. A., et al. (2004b). Elevated maternal interleukin-8 levels and risk for schizophrenia in adult offspring. *Arch. Gen. Psychiatry* 161: 889–895.
Brown, A. S., Schaefer, C. A., Quesenberry, C. P., et al. (2005). Maternal exposure to toxoplasmosis and risk for schizophrenia in adult offspring. *Am. J. Psychiatry* 162: 767–773.
Brown, A. S., Vinogradov, S., Kremen, W. S., et al. (2010). Prenatal exposure to maternal infection and executive dysfunction in adult schizophrenia. *Am. J. Psychiatry* 166: 683–690.
Butler, A. W., Breen, G., Tozzi, F., et al. (2010). A genomewide linkage study on suicidality in major depressive disorder confirms evidence for linkage to 2p12. *Am. J. Med. Genet. B Neuropsychiatr. Genet.* 1538: 1465–1473.
Cacabelos, R., Martínez-Bouza, R. (2010). Genomics and pharmacogenomics of schizophrenia. *CNS Neurosci. Ther.* PMID: 20718829.
Calabrese, F., Molteni, R., Cattaneo, A., et al. (2010). Long-term duloxetine treatment normalizes altered brain-derived neurotrophic factor expression in serotonin transporter knockout rats through the modulation of specific neurotrophin isoforms. *Mol. Pharmacol.* 77: 846–853.
Canli, T., Lesch, K. P. (2007). Long story short: The serotonin transporter in emotion regulation and social cognition. *Nat. Neurosci.* 10: 1103–1109.
Cannon, T. D., Rosso, I. M., Bearden, C. E., et al. (1999). A prospective cohort study of neurodevelopmental processes in the genesis and epigenesis of schizophrenia. *Dev. Psychopathol.* 11: 467–485.
Chaves, C., Marque, C. R., Chaudhry, I. B., et al. (2009a). Short-term improvement by minocycline added to olanzepine antipsychotic treatment in paranoid schizophrenia. *Schizophr. Bull. Suppl.* 1: 354.
Chaves, C., Marque, C. R., Trzesniak, C., et al. (2009b). Glutamate-*N*-methyl-D-aspartate receptor modulation and minocycline for the treatment of patients with schizophrenia: An update. *Braz. J. Med. Biol. Res.* 42: 1002–1014.
Chaves, C., de Marque, C. R., Wichert-Ana, L., et al. (2010). Functional neuroimaging of minocycline's effect in a patient with schizophrenia. *Prog. Neuropsychopharmacol. Biol. Psychiatry* 34: 550–552.
Chen, X., Lee, G., Maher, B. S., et al. (2010). GWA study data mining and independent replication identify cardiomyopathy-associated 5 (CMYA5) as a risk gene for schizophrenia. *Mol. Psychiatry* PMID: 20838396.
Chourbaji, S., Vogt, M. A., Gass, P. (2008). Mice that under- or overexpress glucocorticoid receptors as models for depression or posttraumatic stress disorder. *Prog. Brain Res.* 167: 65–77.
Clarke, M. C., Tanskanen, A., Huttunen, M., et al. (2009). Evidence for an interaction between familial liability and prenatal exposure to infection in the causation of schizophrenia. *Am. J. Psychiatry* 166: 1025–1030.
Conrad, M., Hubold, C., Fischer, B., et al. (2009). Modeling the hypothalamus-pituitary-adrenal system: Homeostasis by interacting positive and negative feedback. *J. Biol. Phys.* 35: 149–162.
Costa, E., Chen, Y., Dong, E., et al. (2009). GABAergic promoter hypermethylation as a model to study the neurochemistry of schizophrenia vulnerability. *Expert Rev. Neurother.* 9: 87–98.
Costa, E., Davis, J., Pesold, C., et al. (2002). The heterozygote reeler mouse as a model for the development of a new generation of antipsychotics. *Curr. Opin. Pharmacol.* 2: 56–62.
Cotter, D., Mackay, D., Chana, G., et al. (2002). Reduced neuronal size and glial cell density in area 9 of the dorsolateral prefrontal cortex in subjects with major depressive disorder. *Cereb. Cortex* 12: 386–394.
Cotter, D., Mackay, D., Landau, S., et al. (2001). Reduced glial cell density and neuronal size in the anterior cingulated cortex in major depressive disorder. *Arch. Gen. Psychiatry* 58: 545–553.
Craig, J. M. (2005). Heterochromatin: Many flavours, common themes. *Bioessays* 27: 17–28.
Crespo-Facorro, B., Roiz-Santiáñez, R., Pérez-Iglesias, R., et al. (2010). Global and regional cortical thinning in first-episode psychosis patients: relationships with clinical and cognitive features. *Psychol. Med.* 41: 1449–1460.
Dalman, C., Allebeck, P., Gunnell, D., et al. (2006). Infections in the CNS during childhood and the risk of subsequent psychotic illness: A cohort study of more than one million Swedish subjects. *Am. J. Psychiatry* 165: 59–65.

Davies, W., Isles, A. R., Humby, T., et al. (2007). What are imprinted genes doing in the brain? *Epigenetics* 2: 201–206.
Davies, W., Isles, A. R., Humby, T., et al. (2008). What are imprinted genes doing in the brain? *Adv. Exp. Med. Biol.* 626: 62–70.
De Roo, M., Klauser, P., Garcia, P. M., et al. (2008). Spine dynamics and synapse remodeling during LTP and memory processes. *Prog. Brain Res.* 169: 199–207.
Deltheil, T., Guiard, B. P., Guilloux, J. P., et al. (2008). Consequences of changes in BDNF levels on serotonin neurotransmission, 5-HT transporter expression and function: Studies in adult mice hippocampus. *Pharmacol. Biochem. Behav.* 90: 174–183.
Dent, E. W., Merriam, E. B., Hu, X. (2011). The dynamic cytoskeleton: Backbone of dendritic spine plasticity. *Curr. Opin. Neurobiol.* 21: 175–181.
Deutsch, S. I., Rosse, R. B., Schwartz, B. L., et al. (2010). Regulation of intermittent oscillatory activity of pyramidal cell neurons by GABA inhibitory interneurons is impaired in schizophrenia: Rationale for pharmacotherapeutic GABAergic interventions. *Isr. J. Psychiatry Relat. Sci.* 47: 17–26.
Doe, C. M., Relkovic, D., Garfield, A. S., et al. (2009). Loss of the imprinted snoRNA mbii-52 leads to increased 5htr2c pre-RNA editing and altered 5HT2CR-mediated behaviour. *Hum. Mol. Genet.* 18: 2140–2148.
Dong, E., Agis-Balboa, R. C., Simonini, M. V., et al. (2005). Reelin and glutamic acid decarboxylase67 promoter remodeling in an epigenetic methionine-induced mouse model of schizophrenia. *Proc. Natl. Acad. Sci. U.S.A.* 102: 12578–12583.
Dong, E., Chen, Y., Gavin, D. P., et al. (2010). Valproate induces DNA demethylation in nuclear extracts from adult mouse brain. *Epigenetics* 5: 8.
Erdelyi, E., Elliott, G. R., Wyatt, R. J., et al. (1978). *S*-Adenosylmethionine-dependent *N*-methyltransferase activity in autopsied brain parts of chronic schizophrenics and controls. *Am. J. Psychiatry* 135: 725–728.
Ethell, I. M., Pasquale, E. B. (2005). Molecular mechanisms of dendritic spine development and remodeling. *Prog. Neurobiol.* 75: 161–205.
Falkenberg, V. R., Gurbaxani, B. M., Unger, E. R., et al. (2011). Functional genomics of serotonin receptor 2A (HTR2A): Interaction of polymorphism, methylation, expression and disease association. *Neuromol. Med.* 13: 66–76.
Fanous, A. H., Neale, M. C., Gardner, C. O., et al. (2007). Significant correlation in linkage signals from genome-wide scans of schizophrenia and schizotypy. *Mol. Psychiatry* 12: 958–965.
Feng, J., Fan, G. (2009). The role of DNA methylation in the central nervous system and neuropsychiatric disorders. *Int. Rev. Neurobiol.* 89: 67–84.
Finer, S., Holland, M. L., Nanty, L., et al. (2011). The hunt for the epiallele. *Environ. Mol. Mutagen.* 52: 1–11.
Fiori, L. M., Turecki, G. (2010). Gene expression profiling of suicide completers. *Eur. Psychiatry* 25: 287–290.
Flattem, N. L., Blakely, R. D. (2000). Modified structure of the human serotonin transporter promoter. *Mol. Psychiatry* 5: 110–115.
Foley, D. L., Craig, J. M., Morley, R., et al. (2009). Prospects for epigenetic epidemiology. *Am. J. Epidemiol.* 169: 389–400.
Franklin, T. B., Mansuy, I. M. (2010a). The prevalence of epigenetic mechanisms in the regulation of cognitive functions and behaviour. *Curr. Opin. Neurobiol.* 20: 441–449.
Franklin, T. B., Mansuy, I. M. (2010b). Epigenetic inheritance in mammals: Evidence for the impact of adverse environmental effects. *Neurobiol. Dis.* 39: 61–65.
Franklin, T. B., Russig, H., Weiss, I. C., et al. (2010). Epigenetic transmission of the impact of early stress across generations. *Biol. Psychiatry* 68: 408–415.
Gavin, D. P., Sharma, R. P. (2010). Histone modifications, DNA methylation, and schizophrenia. *Neurosci. Biobehav. Rev.* 34: 882–888.
Gilmore, J. H., Jarskog, L. F. (1997). Exposure to infection and brain development: Cytokines in the pathogenesis of schizophrenia. *Schizophrenia Res.* 24: 365–367.
Grafodatskaya, D., Chung, B., Szatmari, P., et al. (2010). Autism spectrum disorders and epigenetics. *J. Am. Acad. Child Adolesc. Psychiatry* 49: 794–809.
Grayson, D. R., Chen, Y., Costa, E., et al. (2006). The human reelin gene: Transcription factors (+), repressors (–) and the methylation switch (+/–) in schizophrenia. *Pharmacol. Ther.* 111: 272–286.
Grayson, D. R., Chen, Y., Dong, E., et al. (2009). From trans-methylation to cytosine methylation: Evolution of the methylation hypothesis of schizophrenia. *Epigenetics* 4: 144–149.
Grayson, D. R., Jia, X., Chen, Y., et al. (2005). Reelin promoter hypermethylation in schizophrenia. *Proc. Natl. Acad. Sci. U.S.A.* 102: 9341–9346.
Groves, J. O. (2007). Is it time to reassess the BDNF hypothesis of depression? *Mol. Psychiatry* 12: 1079–1088.

Guiard, B. P., David, D. J., Deltheil, T., et al. (2008). Brain-derived neurotrophic factor-deficient mice exhibit a hippocampal hyperserotonergic phenotype. *Int. J. Neuropsychopharmacol.* 11: 79–92.

Guidotti, A., Ruzicka, W., Grayson, D. R., et al. (2007). *S*-Adenosyl methionine and DNA methyltransferase-1 mRNA overexpression in psychosis. *Neuroreport* 18: 57–60.

Guipponi, M., Deutsch, S., Kohler, K., et al. (2009). Genetic and epigenetic analysis of SSAT gene dysregulation in suicidal behavior. *Am. J. Med. Genet. B Neuropsychiatr. Genet.* 150B, 799–807.

Gurrieri, F., Accadia, M. (2009). Genetic imprinting: The paradigm of Prader-Willi and Angelman syndromes. *Endocr. Dev.* 14: 20–28.

Hahn, M., Dambacher, S., Schotta, G. (2010). Heterochromatin dysregulation in human diseases. *J. Appl. Physiol.* 109: 232–242.

Hamidi, M., Drevets, W. C., Price, J. L. (2004). Glial reduction in amygdale in major depressive disorder is due to oligodendrites. *Biol. Psychiatry* 55: 563–569.

Haroutunian, V., Katsel, P., Dracheva, S., et al. (2006). The human homolog of the QKI gene affected in severe dysmyelination "Quaking" mouse phenotype: Downregulated in multiple brain regions in schizophrenia. *Am. J. Psychiatry* 163: 1834–1837.

Hashimoto, T., Arion, D., Unger, T., et al. (2008). Alterations in GABA-related transcriptome in the dorsolateral prefrontal cortex of subjects with schizophrenia. *Mol. Psychiatry* 13: 147–161.

Heim, C., Mletzko, T., Purselle, D., et al. (2008). The dexamethasone/corticotrophin-releasing factor test in men with major depression: Role of childhood trauma. *Biol. Psychiatry* 63: 398–405.

Herrick, S., Evers, D. M., Lee, J. Y., et al. (2010). Postsynaptic PDLIM5/Enigma Homolog binds SPAR and causes dendritic shrinkage. *Mol. Cell Neurosci.* 43: 188–200.

Ho, D. H., Burggren, W. W. (2010). Epigenetics and transgenerational transfer: A physiological perspective. *J. Exp. Biol.* 213: 3–16.

Hogart, A., Leung, K. N., Wang, N. J., et al. (2009). Chromosome 15q11-13 duplication syndrome brain reveals epigenetic alterations in gene expression not predicted from copy number. *J. Med. Genet.* 46: 86–93.

Holm, V. A., Cassidy, S. B., Butler, M. G., et al. (1993). Prader-Willi syndrome: Consensus diagnosis criteria. *Pediatrics* 91: 398–402.

Holmes, A., Lit, Q., Murphy, D. L., Gold, E., et al. (2003a). Abnormal anxiety-related behavior in serotonin transporter null mutant mice: The influence of genetic background. *Genes Brain Behav.* 2: 365–380.

Holmes, A., Murphy, D. L., Crawley, J. N. (2003b). Abnormal behavioral phenotypes of serotonin transporter knockout mice: Parallels with human anxiety and depression. *Biol. Psychiatry* 54: 953–959.

Homberg, J. R., Olivier, J. D., Smits, B. M., et al. (2007). Characterization of the serotonin transporter knockout rat: A selective change in the functioning of the serotonergic system. *Neuroscience* 146: 1662–1676.

Horiuchi, Y., Arai, M., Niizato, K., et al. (2006). A polymorphism in the PDLIM5 gene associated with gene expression and schizophrenia. *Biol. Psychiatry* 59: 434–439.

Hsieh, J., Eisch, A. J. (2010). Epigenetics, hippocampal neurogenesis, and neuropsychiatric disorders: Unraveling the genome to understand the mind. *Neurobiol. Dis.* 39: 73–84.

International Schizophrenia Consortium (2009). Common polygenic variation contributes to risk of schizophrenia and bipolar disorder. *Nature* 460: 748–752.

Iritani, S., Tohgi, M., Arai, T., et al. (2006). Immunohistochemical study of the serotonergic neuronal system in an animal model of the mood disorder. *Exp. Neurol.* 201: 60–65.

Isles, A. R., Wilkinson, L. S. (2008). Epigenetics: What is it and why is it important to mental disease? *Br. Med. Bull.* 85: 35–45.

Isles, A. R., Davies, W., Wilkinson, L. S. (2006). Genomic imprinting and the social brain. *Philos. Trans. R. Soc. Lond. B Biol. Sci.* 361: 2229–2237.

Jaaro-Peled, H., Hayashi-Takagi, A., Seshadri, S., et al. (2009). Neurodevelopmental mechanisms of schizophrenia: Understanding disturbed postnatal brain maturation through neureglin-1-ErbB4 and DISC1. *Trends Neurosci.* 32: 485–495.

Jablonka, E., Raz, G. (2009). Transgenerational epigenetic inheritance: Prevalence, mechanisms, and implications for the study of heredity and evolution. *Q. Rev. Biol.* 84: 131–176.

Jewkes, R. K., Dunkle, K., Nduna, M., et al. (2010). Associations between childhood adversity and depression, substance abuse and HIV and HSV2 incident infections in rural South African youth. *Child Abuse Negl.* 34: 833–841.

Juruena, M. F., Cleare, A. J., Papadopoulos, A. S., et al. (2006). Different responses to dexamethasone and prednisolone in the same depressed patients. *Psychopharmacology* 189: 225–235.

Juruena, M. F., Cleare, A. J., Papadopoulos, A. S., et al. (2010). The prednisolone suppression test in depression: Dose-response and changes with antidepressant treatment. *Psychoneuroendocrinology* 35: 1486–1491.

Juruena, M. F., Gama, C. S., Berk, M., et al. (2009a). Improved stress response in bipolar affective disorder with adjunctive spironolactone (mineralocorticoid receptor antagonist): Case series. *J. Psychopharmacol.* 23: 985–987.

Juruena, M. F., Pariante, C. M., Papadopoulos, A. S., et al. (2009b). Prednisolone suppression test in depression: Prospective study of the role of HPA axis dysfunction in treatment resistance. *Br. J. Psychiatry* 194: 342–349.

Kalueff, A. V., Olivier, J. D., Nonkes, L. J., et al. (2010). Conserved role for the serotonin transporter gene in rat and mouse neurobehavioral endophenotypes. *Neurosci. Biobehav. Rev.* 34: 373–386.

Kamiya, A., Kubo, K., Tomoda, T., et al. (2005). A schizophrenia-associated mutation of DISC1 perturbs cerebral cortex development. *Nat. Cell Biol.* 7: 1167–1178.

Kappeler, L., Meaney, M. J. (2010). Epigenetics and parental effects. *Bioassays* 32: 818–827.

Kato, T., Iwayama, Y., Kakiuchi, C., et al. (2005). Gene expression and association analyses of LIM (PDLIM5) in bipolar disorder and schizophrenia. *Mol. Psychiatry* 10: 1045–1055.

Kaymaz, N., Krabbendam, L., de Graaf, R., et al. (2006). Evidence that the urban environment specifically impacts on the psychotic but not the affective dimension of bipolar disorder. *Soc. Psychiatry Psychiatry Epidemiol.* 41: 679–685.

Kelly, T. K., De Carvalho, D. D., Jones, P. A. (2010). Epigenetic modifications as therapeutic targets. *Nat. Biotechnol.* 28: 1069–1078.

Kelly, U. (2010). Intimate partner violence, physical health, posttraumatic stress disorder, depression, and quality of life in latinas. *West. J. Emerg. Med.* 11: 247–251.

Keri, S. (2009). Genes for psychosis and creativity: a promoter polymorphism of the neureglin 1 gene is related to creativity in people with high intellectual achievement. *Psychol. Sci.* 21: 1070–1073.

Kim, T., Park, J. K., Kim, H. J., et al. (2010). Association of histone deacetylase genes with schizophrenia in Korean population. *Psychiatry Res.* 178: 266–269.

Klar, A. J. S. (2010). A proposal for re-defining the way the aetiology of schizophrenia and bipolar human psychiatric diseases is investigated. *J. Biosci.* 35: 11–15.

Klempan, T. A., Ernst, C., Devela, V., et al. (2009a). Characterization of QKI gene expression, genetics, and epigenetics in suicide victims with major depressive disorder. *Biol. Psychiatry* 66: 824–831.

Klempan, T. A., Rujescu, D., Mérette, C., et al. (2009b). Profiling brain expression of the spermidine/spermine N1-acetyltransferase 1 (SAT1) gene in suicide. *Am. J. Med. Genet. B Neuropsychiatr. Genet.* 150B, 934–943.

Klempan, T. A., Sequeira, A., Canetti, L., et al. (2009c). Altered expression of genes involved in ATP biosynthesis and GABAergic neurotransmission in the ventral prefrontal cortex of suicides with and without major depression. *Mol. Psychiatry* 14: 175–189.

Kluge, M., Schüssler, P., Dresler, M., et al. (2011). Effects of ghrelin on psychopathology, sleep and secretion of cortisol and growth hormone in patients with major depression. *J. Psychiatry Res.* 45: 421–426.

Kouzarides, T. (2007). Chromatin modifications and their function. *Cell* 128: 707–719.

Krauss, M., Weiss, T., Langnaese, K., et al. (2007). Cellular and subcellular rat brain spermidine synthase expression patterns suggest region-specific roles for polyamines, including cerebellar pre-synaptic function. *J. Neurochem.* 103: 679–693.

Krebs, M. O., Bellon, A., Mainguy, G., et al. (2009). One-carbon metabolism and schizophrenia: Current challenges and future directions. *Trends Mol. Med.* 15: 562–570.

Krug, A., Markov, V., Leube, D., et al. (2008). Genetic variation in the schizophrenia-risk gene neureglin1 correlates with personality traits in healthy individuals. *Eur. Psychiatry* 23: 344–349.

Kubera, M., Obuchowicz, E., Goehler, L., et al. (2011). In animal models, psychosocial stress-induced (neuro) inflammation, apoptosis and reduced neurogenesis are associated to the onset of depression. *Prog. Neuropsychopharmacol. Biol. Psychiatry.* 35: 744–759.

Kundakovic, M., Chen, Y., Guidotti, A., et al. (2009). The reelin and GAD67 promoters are activated by epigenetic drugs that facilitate the disruption of local repressor complexes. *Mol. Pharmacol.* 75: 342–354.

Kunugi, H., Hori, H., Adachi, N., et al. (2010). Interface between hypothalamic-pituitary-adrenal axis and brain-derived neurotrophic factor in depression. *Psychiatry Clin. Neurosci.* 64: 447–459.

Kussmann, M., Krause, L., Siffert, W. (2010). Nutrigenomics: Where are we with genetic and epigenetic markers for disposition and susceptibility? *Nutr. Rev.* 68 (Suppl. 1): S38–S47.

Lahiri, D. K., Maloney, B., Zawia, N. H. (2009). The LEARn model: An epigenetic explanation for idiopathic neurobiological diseases. *Mol. Psychiatry* 14: 992–1003.

Lewis, C. M., Ng, M. Y., Butler, A. W., et al. (2010). Genome-wide association study of major recurrent depression in the U.K. population. *Am. J. Psychiatry* 167: 949–957.

Lewis, D. A., Hashimoto, T. (2007). Deciphering the disease process of schizophrenia: The contribution of cortical GABA neurons. *Int. Rev. Neurobiol.* 78: 109–131.

Li, B., Carey, M., Workman, J. L. (2007). The role of chromatin during transcription. *Cell* 128: 707–719.

Li, C., Tao, R., Qin, W., et al. (2008). Positive association between PDLIM5 and schizophrenia in the Chinese Han population. *Int. J. Neuropsychopharmacol.* 11: 27–34.

Lira, A., Zhou, M., Castanon, N., et al. (2003). Altered depression-related behaviors and functional changes in the dorsal raphe nucleus of serotonin transporter-deficient mice. *Biol. Psychiatry* 54: 960–971.

Liu, D., Diorio, J., Tannenbaum, B., et al. (1997). Maternal care, hippocampal glucocorticoid receptors, and hypothalamic-pituitary-adrenal responses to stress. *Science* 277: 1659–1662.

Liu, Z., Liu, W., Xiao, Z., et al. (2008). A major single nucleotide polymorphism of the PDLIM5 gene associated with recurrent major depressive disorder. *J. Psychiatry Neurosci.* 33: 43–46.

Luger, K., Richmond, T. J. (1998). The histone tails of the nucleosome. *Curr. Opin. Genet. Dev.* 8: 140–146.

Machado-Vieira, R., Ibrahim, L., Zarate, C. A., Jr. (2010). Histone deacetylases and mood disorders: Epigenetic programming in gene-environment interactions. *CNS Neurosci. Ther.* doi: 10.1111/j.1755-5949.2010.00203.x.

Maes, M., Yirmyia, R., Noraberg, J., et al. (2009). The inflammatory & neurodegenerative (I&ND) hypothesis of depression: Leads for future research and new drug developments in depression. *Metab. Brain Dis.* 24: 27–53.

Malik, S., Roeder, R. G. (2010). The metazoan mediator co-activator complex as an integrative hub for transcriptional regulation. *Nat. Rev. Genet.* 11: 761–772.

Maloku, E., Covelo, I. R., Hanbauer, I., et al. (2010). Lower number of cerebellar Purkinje neurons in psychosis is associated with reduced reelin expression. *Proc. Natl. Acad. Sci. U.S.A.* 107: 4407–4411.

Markopoulou, K., Papadopoulos, A., Juruena, M. F., et al. (2009). The ratio of cortisol/DHEA in treatment resistant depression. *Psychoneuroendocrinology* 34: 19–26.

Marley, A., von Zastrow, M. (2010). DISC1 regulates primary cilia that display specific dopamine receptors. *PloS One* 5: e10902.

McGowan, P. O., Sasaki, A., D'Alessio, A. C., et al. (2009). Epigenetic regulation of the glucocorticoid receptor in human brain associates with childhood abuse. *Nat. Neurosci.* 12: 342–348.

McGowan, P. O., Sasaki, A., Huang, T. C., et al. (2008). Promoter-wide hypermethylation of the ribosomal RNA gene promoter in the suicide brain. *PLoS One* 3: e2085.

McGowan, P. O., Szyf, M. (2010). The epigenetics of social adversity in early life: Implications for mental health outcomes. *Neurobiol. Dis.* 39: 66–72.

McGuffin, P., Perroud, N., Uher, R., et al. (2010). The genetics of affective disorder and suicide. *Eur. Psychiatry* 25: 275–277.

Meaney, M. J. (2010). Epigenetics and the biological definition of gene x environment interactions. *Child Dev.* 81: 41–79.

Meaney, M. J., Ferguson-Smith, A. C. (2010). Epigenetic regulation of the neural transcriptone: The meaning of the marks. *Nat. Neurosci.* 13: 1313–1318.

Molteni, R., Calabrese, F., Cattaneo, A., et al. (2009). Acute stress responsiveness of the neurotrophin BDNF in the rat hippocampus is modulated by chronic treatment with the antidepressant duloxetine. *Neuropsychopharmacology* 34: 1523–1532.

Molteni, R., Cattaneo, A., Calabrese, F., et al. (2010). Reduced function of the serotonin transporter is associated with decreased expression of BDNF in rodents as well as in humans. *Neurobiol. Dis.* 37: 747–755.

Mortensen, P. B., Nørgaard-Pedersen, B., Waltoft, B. L., et al. (2007). Early infections of *Toxoplasma gondii* and the later development of schizophrenia. *Schizophr. Bull.* 33: 741–744.

Müller, N., Dursun, S. M. (2011). Schizophrenia genes, epigenetics and psychoneuroimmunology: All make sense now. *J. Psychopharmacol.* 25: 713–714.

Murgatroyd, C., Patchev, A. V., Wu, Y., et al. (2009). Dynamic DNA methylation programs persistent adverse effects of early-life stress. *Nat. Neurosci.* 12: 1559–1566.

Murgatroyd, C., Wu, Y., Bockmühl, Y., et al. (2010a). Genes learn from stress: How infantile trauma programs us for depression. *Epigenetics* 5: 3.

Murgatroyd, C., Wu, Y., Bockmühl, Y., et al. (2010b). The Janus face of DNA methylation in aging. *Aging* 2: 107–110.

Murphy, D. L., Lesch, K. P. (2008). Targeting the murine serotonin transporter: Insights into human neurobiology. *Nat. Rev. Neurosci.* 9: 85–96.

Murr, R. (2010). Interplay between different epigenetic modifications and mechanisms. *Adv. Genet.* 70: 101–141.

Nadeau, J. H. (2009). Transgenerational genetic effects on phenotypic variation and disease risk. *Hum. Mol. Genet.* 18: R202–R210.

Nakatani, N., Ohnishi, T., Iwamoto, K., et al. (2007). Expression analysis of actin-related genes as an underlying mechanism for mood disorders. *Biochem. Biophys. Res. Commun.* 352: 780–786.

Narayan, P., Dragunow, M. (2010). Pharmacology of epigenetics in brain disorders. *Br. J. Pharmacol.* 159: 285–303.

Nicholls, R. D., Knepper, J. L. (2001). Genome organization, function, and imprinting in the Prader-Willi and Angelman syndromes. *Annu. Rev. Genomics Hum. Genet.* 2: 153–175.

Niwa, M., Kamiya, A., Murai, R., et al. (2010). Knockdown of DISC1 by in utero gene transfer disturbs postnatal dopaminergic maturation in the frontal cortex and leads to adult behavioral deficits. *Neuron* 65: 480–489.

Oberlander, T. F., Grunau, R., Mayes, L., et al. (2008). Hypothalamic-pituitary-adrenal (HPA) axis function in 3-month-old infants with prenatal selective serotonin reuptake inhibitor (SSRI) antidepressant exposure. *Early Hum. Dev.* 84: 689–697.

Olivier, J. D., Van Der Hart, M. G., Van Swelm, R. P., et al. (2008). A study in male and female 5-HT transporter knockout rats: An animal model for anxiety and depression disorders. *Neuroscience* 152: 573–584.

Olsson, C. A., Foley, D. L., Parkinson-Bates, M., et al. (2010). Prospects for epigenetic research within cohort studies of psychological disorder: A pilot investigation of a peripheral cell marker of epigenetic risk for depression. *Biol. Psychol.* 83: 159–165.

Palomo, T., Kostrzewa, R. M., Beninger, R. J., et al. (2007). Genetic variation and shared biological susceptibility underlying comorbidity in neuropsychiatry. *Neurotox. Res.* 12: 29–42.

Pariante, C. M., Lightman, S. L. (2008). The HPA axis in major depression: Classical theories and new developments. *Trends Neurosci.* 31: 464–468.

Perini, G., Tupler, R. (2010). Altered gene silencing and human diseases. *Clin. Genet.* 69: 1–7.

Perrin, M., Kleinhaus, K., Messinger, J., et al. (2010). Critical periods and the developmental origins of disease: An epigenetic perspective of schizophrenia. *Ann. N.Y. Acad. Sci.* 1204: E8–E13.

Perroud, N., Neidhart, E., Petit, B., et al. (2010). Simultaneous analysis of serotonin transporter, tryptophan hydroxylase 1 and 2 gene expression in the ventral prefrontal cortex of suicide victims. *Am. J. Med. Genet. B Neuropsychiatr. Genet.* 153B, 909–918.

Peters, A., Conrad, M., Hubold, C., et al. (2007). The principle of homeostasis in the hypothalamus-pituitary-adrenal system: New insight from positive feedback. *Am. J. Physiol. Regul. Integr. Comp. Physiol.* 293: R83–R98.

Philibert, R. A., Sandhu, H., Hollenbeck, N., et al. (2008). The relationship of 5HTT (SLC6A4) methylation and genotype on mRNA expression and liability to major depression and alcohol dependence in subjects from the Iowa adoption studies. *Am. J. Med. Genet.* 147B, 543–549.

Pidsley, R., Mill, J. (2011). Epigenetic studies of psychosis: Current findings, methodological approaches, and implications for postmortem research. *Biol. Psychiatry* 69: 146–156.

Pontrello, C. G., Ethell, I. M. (2009). Accelerators, brakes, and gears of actin dynamics in dendritic spines. *Open Neurosci. J.* 3: 67–86.

Post, R. M. (2010). Mechanisms of illness progression in the recurrent affective disorders. *Neurotoxicity. Res.* 18: 256–271.

Poulter, M. O., Du, L., Weaver, I. C., et al. (2008). GABAA receptor promoter hypermethylation in suicide brain: Implications for the involvement of epigenetic processes. *Biol. Psychiatry* 64: 645–652.

Pregelj, P. (2009). Neurobiological aspects of psychosis and gender. *Psychiatry Danub.* 21 (Suppl. 1): 128–131.

Rakyan, V. K., Beck, S. (2006). Epigenetic variation and inheritance in mammals. *Curr. Opin. Genet. Dev.* 16: 573–577.

Rapoport, J. L., Gogtay, N. (2011). Childhood onset schizophrenia: Support for a progressive neurodevelopmental disorder. *Int. J. Dev. Neurosci.* 29: 251–258.

Read, J., Benthall, R. P., Fosse, R. (2009). Time to abandon the bio-bio-bio model of psychosis: Exploring the epigenetic and psychological mechanisms by which adverse life events lead to psychotic symptoms. *Epidemiol. Psichiatr. Soc.* 18: 299–310.

Relkovic, D., Doe, C. M., Humby, T., et al. (2010). Behavioural and cognitive abnormalities in an imprinting centre deletion mouse model for Prader-Willi syndrome. *Eur. J. Neurosci.* 31: 156–164.

Renthal, W., Nestler, E. J. (2008). Epigenetic mechanisms in drug addiction. *Trends Mol. Med.* 14: 341–350.

Renthal, W., Nestler, E. J. (2009). Chromatin regulation in drug addiction and depression. *Dialogues Clin. Neurosci.* 11: 257–268.

Risch, N., Herrell, R., Lehner, T., et al. (2009). Transporter gene (5HTTLPR), stressful life events, and risk of depression: a mete-analysis. *JAMA* 301: 2462–2471.

Rossi-George, A., Virgolini, M. B., Weston, D., et al. (2009). Alterations in glucocorticoid negative feedback following maternal Pb, prenatal stress and the combination: A potential biological unifying mechanism for their corresponding disease profiles. *Toxicol. Appl. Pharmacol.* 234: 117–127.

Roth, T. L., Lubin, F. D., Funk, A. J., et al. (2010). Lasting epigenetic influence of early-life adversity on the BDNF gene. *Biol. Psychiatry* 65: 760–769.

Roullet, F. I., Wollaston, L., Decatanzaro, D., et al. (2010). Behavioral and molecular changes in the mouse in response to prenatal exposure to the anti-epileptic drug valproic acid. *Neuroscience* 170: 514–522.

Schroeder, M., Krebs, M. O., Bleich, S., et al. (2010). Epigenetics and depression: Current challenges and new therapeutic options. *Curr. Opin. Psychiatry* 23: 588–592.

Schumacher, J., Laje, G., Abou Jamra, R., et al. (2009). The DISC locus and schizophrenia: Evidence from an association study in a central European sample and from a meta-analysis across different European populations. *Hum. Mol. Genet.* 18: 2719–2727.

Sequeira, A., Turecki, G. (2006). Genome wide gene expression studies in mood disorders. *OMICS* 10: 444–454.

Sequeira, A., Gwadry, F. G., Ffrench-Mullen, J. M., et al. (2006). Implication of SSAT by gene expression and genetic variation in suicide and major depression. *Arch. Gen. Psychiatry* 63: 35–48.

Sequeira, A., Klempan, T., Canetti, L., et al. (2007). Patterns of gene expression in the limbic system of suicides with and without major depression. *Mol. Psychiatry* 12: 640–655.

Shen, S., Lang, B., Nakamoto, C., et al. (2008). Schizophrenia-related neural and behavioral phenotypes in transgenic mice expressing truncated DISC1. *J. Neurosci.* 28: 10893–10904.

Shi, J., Levinson, D. F., Duan, J., et al. (2009). Common variants on chromosome 6p22.1 are associated with schizophrenia. *Nature* 460: 753–757.

Singh, K. K., Ge, X., Mao, Y., et al. (2010). Dixdc1 is a critical regulator of DISC1 and embryonic cortical development. *Neuron* 67: 33–48.

Sokolov, B. P. (2007). Oligodendroglial abnormalities in schizophrenia, mood disorders and substance abuse: Comorbidity, shared traits, or molecular phenocopies? *Int. J. Neuropsychopharmacol.* 10: 547–555.

Stefansson, H., Ophoff, R. A., Steinberg, S., et al. (2009). Common variants conferring risk of schizophrenia. *Nature* 460: 744–747.

Sun, Y. H., Shen, Y., Xu, Q. (2010). DTNBP1 gene is associated with some symptom factors of schizophrenia in Chinese Han nationality. *Chin. Med. Sci. J.* 25: 85–89.

Szyf, M. (2009). The early life environment and the epigenome. *Biochim. Biophys. Acta* 1790: 878–885.

Tenyi, T., Trixler, M., Csabi, G. (2009). Minor physical abnormalities in affective disorders. *J. Affect. Disorder* 112: 11–18.

Tichomirowa, M. A., Keck, M. E., Schneider, H. J., et al. (2005). Endocrine disturbances in depression. *J. Endocrinol. Invest.* 28: 89–99.

Tolosa, A., Sanjuan, J., Dagnall, A. M., et al. (2010). FOXP2 gene and language impairment in schizophrenia: association and epigenetic studies. *BMC Med. Genet.* 11: 114–122.

Tomppo, L., Hennah, W., Miettunen, J., et al. (2009). Association of variants in DISC1 with psychosis-related traits in a large population cohort. *Arch. Gen. Psychiatry* 66: 134–141.

Tost, J. (2010). DNA methylation: An introduction to the biology and the disease-associated changes of a promising biomarker. *Mol. Biotechnol.* 44: 71–81.

Tremolizzo, L., Carboni, G., Ruzicka, W. B., et al. (2002). An epigenetic mouse model for molecular and behavioral neuropathologies related to schizophrenia vulnerability. *Proc. Natl. Acad. Sci. U.S.A.* 99: 17095–17100.

Tremolizzo, L., Doueiri, M. S., Dong, E., et al. (2005). Valproate corrects the schizophrenia-like epigenetic behavioral modifications induced by methionine in mice. *Biol. Psychiatry* 57: 500–509.

Tsai, S. J., Hong, C. J., Liou, Y. J. (2011). Recent molecular genetic studies and methodological issues in suicide research. *Prog. Neuropsychopharmacol. Biol. Psychiatry* 35: 809–817.

Tsankova, N. M., Berton, O., Renthal, W., et al. (2006). Sustained hippocampal chromatin regulation in a mouse model of depression and antidepressant action. *Nat. Neurosci.* 9: 519–525.

Tsankova, N., Renthal, W., Kumar, A., et al. (2007). Epigenetic regulation of psychiatric disorders. *J. Med. Invest.* 50: 25–31.

Tueting, P., Davis, J. M., Veldic, M., et al. (2010). L-Methionine decreases dendritic spine density in mouse frontal cortex. *Neuroreport* 21: 543–548.

Turner, J. D., Alt, S. R., Cao, L., et al. (2010). Transcriptional control of the glucocorticoid receptor: CpG islands, epigenetics and more. *Biochem. Pharmacol.* 80: 1860–1868.

Uher, R., McGuffin, P. (2008). The moderation by the serotonin transporter gene of environmental adversity in the aetiology of mental illness. *Mol. Psychiatry* 13: 131–146.

Uranova, N. A., Vostrikov, V. M., Orlovskaya, D. D., et al. (2004). Oligodendroglial density in the prefrontal cortex in schizophrenia and mood disorders: A study from the Stanley Neuropathology Consortium. *Schizophr. Res.* 67: 269–275.

Van Winkel, R., Esquivel, G., Kenis, G., et al. (2010). Genome-wide findings in schizophrenia and the role of gene-environment interplay. *CNS Neurosci. Ther.* 16: 2185–2192.

Vucetic, Z., Totoki, K., Schoch, H., et al. (2010). Early life protein restriction alters dopamine circuitry. *Neuroscience* 168: 359–370.

Walters, J. T., Corvin, A., Owen, M. J., et al. (2010). Psychosis susceptibility gene ZNF804A and cognitive performance in schizophrenia. *Arch. Gen. Psychiatry* 67: 692–700.

Wankerl, M., Wüst, S., Otte, C. (2010). Current developments and controversies: Does the serotonin transporter gene-linked polymorphic region (5-HTTLPR) modulate the association between stress and depression? *Curr. Opin. Psychiatry* 23: 582–587.

Warmuz-Stangierska, I., Baszko-Błaszyk, D., Sowiński, J. (2010). Emotions and features of temperament in patients with Addison's disease. *Endokrynol. Pol.* 61: 90–92.

Wawrzik, M., Unmehopa, U. A., Swaab, D. F., et al. (2010). The C15orf2 gene in the Prader-Willi syndrome region is subject to genomic imprinting and positive selection. *Neurogenetics* 11: 153–161.

Weaver, I. C. (2009). Epigenetic effects of glucocorticoids. *Semin. Fetal Neonatal Med.* 14: 143–150.

Weaver, I. C. G., Cervoni, N., Champagne, F. A., et al. (2004). Epigenetic programming by maternal behavior. *Nat. Neurosci.* 7: 847–854.

Whittington, J., Holland, A., Webb, T. (2009). Relationship between the IQ of people with Prader-Willi syndrome (PWS) and that of their siblings: Evidence for imprinted gene effects. *J. Intellect. Disabil. Res.* 53: 411–418.

Whittington, J., Holland, A., Webb, T., et al. (2004). Cognitive abilities and genotype in a population-based sample of people with Prader-Willi syndrome. *J. Intellect. Disabil. Res.* 48: 172–187.

Wichers, M. C., Myin-Germeys, I., Jacobs, N., et al. (2008). Susceptibility to depression expressed as alterations in cortisol day curve: A cross-twin, cross-trait study. *Psychosom. Med.* 70: 314–318.

Wichers, M., Schrijvers, D., Geschwind, N., et al. (2009). Mechanisms of gene-environment interactions in depression: Evidence that genes potentiate multiple sources of adversity. *Psychol. Med.* 39: 1077–1086.

Wilkinson, L. S., Davies, W., Isles, A. R. (2007). Genomic imprinting effects on brain development and function. *Nat. Rev. Neurosci.* 8: 832–843.

Woodcock, K. A., Oliver, C., Humphreys, G. W. (2009). A specific pathway can be identified between genetic characteristics and behaviour profiles in Prader-Willi syndrome via cognitive, environmental and physiological mechanisms. *J. Intellect. Disabil. Res.* 53: 493–500.

Wong, C. C., Caspi, A., Williams, B., et al. (2010). A longitudinal study of epigenetic variation in twins. *Epigenetics* 5: 6.

Wu, X., Chen, P. S., Dallas, S., et al. (2008). Histone deacetylase inhibitors up-regulate astrocyte GDNF and BDNF gene transcription and protect dopaminergic neurons. *Int. J. Neuropsychopharmacol.* 11: 1123–1134.

Wyatt, R., Termini, B. (1971). Part II. Sleep studies. *Schizophr. Bull.* 1: 45–66.

Yasuda, S., Liang, M. H., Marinova, Z., et al. (2009). The mood stabilizers lithium and valproate selectively activate the promoter IV of brain-derived neurotrophic factor in neurons. *Mol. Psychiatry* 14: 51–59.

Yolken, R. H., Dickerson, F. B., Fuller Torrey, E. (2009). Toxoplasma and schizophrenia. *Parasite Immunol.* 31: 706–715.

Zhang, T. Y., Meaney, M. J. (2010). Epigenetics and the environmental regulation of the genome and its function: Epigenetics and the environmental regulation of the genome and its function. *Annu. Rev. Psychol.* 61: 439–466: C1–C3.

Zhang, Y., Gu, F., Chen, J., et al. (2010). Chronic antidepressants administration alleviate frontal and hippocampal BDNF deficits in CUMS rat. *Brain Res.* 1366: 141–148.

Zuckerman, L., Weiner, I. (2005). Maternal immune activation leads to behavioral and pharmacological changes in the adult offspring. *J. Psychiatry Res.* 39: 311–323.

Zuckerman, L., Rehavi, M., Nachman, R., et al. (2003). Immune activation during pregnancy in rats leads to a postpubertal emergence of disrupted latent inhibition, dopaminergic hyperfunction, and altered limbic morphology in the offspring: A novel neurodevelopmental model of schizophrenia. *Neuropsychopharmacology* 28: 1778–1789.

26 Neurogenetics and Nutrigenomics of Reward Deficiency Syndrome

Kenneth Blum
University of Florida College of Medicine and McKnight Brain Institute
Gainesville, Florida

CONTENTS

26.1 WHAT IS THE BRAIN REWARD CASCADE?

Over half a century of dedicated and rigorous scientific research on the mesolimbic system provided insight into the addictive brain and neurogenetic mechanisms involved in man's quest for happiness. In brief, the site of the brain where one experiences feelings of well-being is the mesolimbic system. This part of the brain has been termed the *reward center.* Chemical messages, including serotonin, enkephalins, γ-aminobutyric acid (GABA), and dopamine (DA), work in concert to provide a net release of DA at the nucleus accumbens (NAc), a region in the mesolimbic system. It is well known that genes control the synthesis, vesicular storage, metabolism (Baker et al., 1994), receptor formation, and neurotransmitter catabolism (Hodge et al., 1996; Hodge and Cox, 1998). The polymorphic versions of these genes have certain variations that could lead to an impairment of the neurochemical events involved in the neuronal release of DA. The cascade of these neuronal events has been termed *brain reward cascade* (Blum et al., 1990) (Figure 26.1).

A breakdown of this cascade will ultimately lead to the dysregulation and dysfunction of DA. Because DA has been established as the *pleasure molecule* and the *antistress molecule*, any reduction in function could lead to reward deficiency and resultant aberrant substance-seeking behavior and a lack of wellness (Blum et al., 2000).

26.2 WHAT IS REWARD DEFICIENCY SYNDROME?

Homo sapiens are biologically predisposed to drink, eat, reproduce, and desire pleasurable experiences. Impairment in the mechanisms involved in these natural processes lead to multiple impulsive, compulsive, and addictive behaviors governed by genetic polymorphic antecedents. Whereas there are a plethora of genetic variations at the level of mesolimbic activity, polymorphisms of the serotonergic 2A receptor (5-HTT2a); serotonergic transporter (5-HTTLPR); DA D2 receptor (DRD2); DA D4 receptor (DRD4); DA transporter (DAT1); and the catechol-*O*-methyltransferase (COMT), monoamine-oxidase (*MOA*) genes, and other candidate genes predispose individuals to excessive cravings and resultant aberrant behaviors (Blum et al., 1996b).

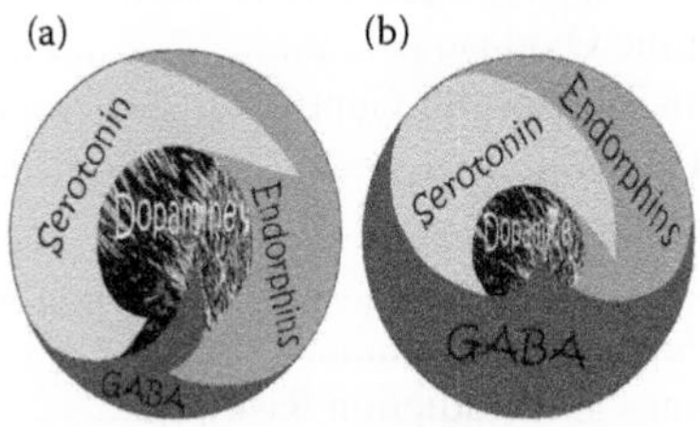

FIGURE 26.1 (See color insert.) (a) Schematic represents the normal physiologic state of the neurotransmitter interaction at the mesolimbic region of the brain. Briefly, in terms of the brain reward cascade first coined by Blum and Kozlowski (1990), serotonin in the hypothalamus stimulates neuronal projections of methionine enkephalin in the hypothalamus, which in turn inhibits the release of GABA in the substania nigra, thereby allowing for the normal amount of dopamine to be released at the nucleus accumbens (reward site of the brain). (b) Schematic represents the hypodopaminergic function of the mesolimbic region of the brain. It is possible that the hypodopaminergic state is due to gene polymorphisms as well as environmental elements, including both stress and neurotoxicity from aberrant abuse of psychoactive drugs (i.e., alcohol, heroin, cocaine, etc.). Genetic variables could include serotonergic genes (serotonergic receptors [5-HT2a]; serotonin transporter [5-HTLPR]); endorphinergic genes (mu *OPRM1* gene; proenkephalin [*PENK*] [*PENK* polymorphic 3′ UTR dinucleotide (CA) repeats]); GABAergic gene (*GABRB3*); and dopaminergic genes (*ANKK1* Taq A; *DRD2 C957T, DRD4 7R, COMT* Val/met substation, *MAOA-uVNTR*, and *SLC6A3 9* or *10R*). Any of these genetic and or environmental impairments could result in a reduced release of dopamine and or a reduced number of dopaminergic receptors.

In 1996, our laboratory first described reward deficiency syndrome (RDS) to define a common genetic variant, involving DA D2 receptor gene (*DRD2*) polymorphisms (Grandy et al., 1989; Smith et al., 1989) as a putative predictor of impulsive, compulsive, and addictive behaviors (Blum et al., 1996c). Thus, an umbrella term to describe the common genetic antecedents of multiple impulsive, compulsive, and addictive behaviors is RDS. Individuals possessing a paucity of serotonergic and/or dopaminergic receptors and an increased rate of synaptic DA catabolism because of high catabolic genotype of the *COMT* gene or high MOA activity are predisposed to self-medicating with any substance or behavior that will activate DA release, including alcohol, opiates, psychostimulants, nicotine, glucose, gambling, sex, and even excessive Internet gaming, among others (Comings and Blum, 2000). Use of most drugs of abuse, including alcohol, is associated with release of DA in the mesocorticolimbic system or reward pathway of the brain (Di Chiara and Impereto, 1988). Activation of this dopaminergic system induces feelings of reward and pleasure (Volkow et al., 2002; Eisenberg et al., 2007). However, reduced activity of the DA system (hypodopaminergic functioning) can trigger drug-seeking behavior (Volkow et al., 2001; Dackis et al., 1985). Variant alleles can induce hypodopaminergic functioning through reduced DA receptor density, blunted response to DA, or enhanced DA catabolism in the reward pathway (Hietala et al., 1994). Possibly, cessation of chronic drug use induces a hypodopaminergic state that prompts drug-seeking behavior in an attempt to address the withdrawal-induced state (Melis et al., 2005).

Acute utilization of these substances can induce a feeling of well-being, but unfortunately sustained and prolonged abuse leads to a toxic pseudofeeling of well-being, resulting in tolerance and disease or discomfort. Thus, low DA receptors caused by carrying the *DRD2* A1 allelic genotype result in excessive cravings and consequential behavior, whereas normal or high DA receptors result in low craving-induced behavior. In terms of preventing substance abuse or excessive glucose craving, one goal would be to induce a proliferation of DA D2 receptors in genetically prone individuals (Rothman et al., 2007). Experiments *in vitro* have shown that constant stimulation of the DA receptor system via a known D2 agonist in low doses results in significant proliferation of D2 receptors in spite of genetic antecedents (Boundy et al., 1995). In essence, D2 receptor stimulation signals negative feedback mechanisms in the mesolimbic system to induce mRNA expression, causing proliferation of D2 receptors. This molecular finding serves as the basis to naturally induce DA release to also cause the same induction of D2-directed mRNA and thus proliferation of D2 receptors in humans. This proliferation of D2 receptors in turn will induce the attenuation of craving behavior. In fact, this has been proven with work showing DNA-directed overexpression (a form of gene therapy) of the DRD2 receptors and significant reduction in both alcohol and cocaine craving-induced behavior in animals (Thanos et al., 2001, 2008).

These observations are the basis for the development of a functional hypothesis of drug seeking and drug use. The hypothesis is that the presence of a hypodopaminergic state, regardless of the source, is a primary cause of drug-seeking behavior. Thus, genetic polymorphisms that induce hypodopaminergic functioning may be the causal mechanism of a genetic predisposition to chronic drug use and relapse (Merlo and Gold, 2008). Finally, utilizing the long-term dopaminergic activation approach will ultimately lead to a common, safe, and effective modality in treating RDS behaviors, including substance use disorders (SUDs), attention-deficit hyperactivity disorder (ADHD), and obesity, among other reward-deficient aberrant behaviors.

26.2.1 Why Is DA the Pleasure Molecule?

As stated earlier, DA has been associated with pleasure, and it has been called the antistress molecule and the pleasure molecule (Blum et al., 2000; Comings et al., 1996b, Bau et al., 2000). When DA is released into the synapse, it stimulates a number of receptors (D1–D5), which results in increased feelings of well-being and stress reduction.

It is of particular interest that DA is known as the main neurotransmitter modulating the activation of the reward system of the brain. The D2 receptor has been associated with pleasure, and the

DRD2 A1 allele has been referred to as a reward gene (Blum et al., 1990). The *DRD2* gene has been one of the most widely studied in neuropsychiatric disorders in general and in alcoholism and other addictions in particular (Blum and Braverman, 2000). Grasping the mechanism of motivated behavior requires an understanding of the neural circuitry of rewards (Robbins and Everitt, 1996), otherwise called positive reinforcers. A positive reinforcer is operationally defined as an event that increases the probability of a subsequent response, and drugs of abuse are considered to be stronger positive reinforcers than natural reinforcers (e.g., food and sex) (Wightman and Robinson, 2002; Epping-Jordon et al., 1998; Cooper et al., 1995). The distinction between *natural rewards* and *unnatural rewards* is an important one. Natural rewards include satisfaction of physiological drives (e.g., hunger and reproduction), and unnatural rewards are learned and involve satisfaction of acquired pleasures such as hedonic sensations (Suhara et al., 2001) derived from alcohol and other drugs, as well as from gambling and other risk-taking behaviors (Wightman and Robinson, 2002; Hodge et al., 1996; Hodge and Cox, 1998).

The NAc, a site within the ventral striatum, is best known for its prominent role in mediating the reinforcing effects of drugs of abuse such as cocaine, alcohol, nicotine, food, and music. Indeed, it is generally believed that this structure mandates motivated behaviors, such as feeding, drinking, sexual behavior, and exploratory locomotion, which are elicited by natural rewards or incentive stimuli. A basic rule of positive reinforcement is that motor responses will increase in magnitude and vigor if followed by a rewarding event. Here I am hypothesizing that there is a common mechanism of action for the powerful effects that drugs, music, food, and sex have on human motivation. The human drive for the three necessary motivated behaviors, hunger, thirst, and sex, may all have common molecular-genetic antecedents that, if impaired, lead to aberrant behaviors. I hypothesize based on a plethora of scientific support that sexual activity is indeed like drugs, food, and music, which activate brain mesolimbic reward circuitry. Moreover, dopaminergic genes and possibly other candidate neurotransmitter-related genes and their polymorphisms affect both hedonic and anhedonic behavioral outcomes. I anticipate that future clinical studies involving genotyping of sex addicts will provide evidence for polymorphic associations with specific clustering of sexual typologies based on clinical instrument assessments. My associates and I are encouraging both academic and clinical scientists to embark on research coupling neuroimaging tools (i.e., functional magnetic resonance imaging [fMRI], quantitative electroencephalography [qEEG], and positron emission tomography [PET]), and natural dopaminergic agonistic agents (i.e., Synaptamine™) to systematically target specific gene polymorphisms and normalize hyper- or hyposexual response (Koob, 2008).

Hedonic *liking* for sensory pleasures is an important aspect of reward, and excessive liking of particular rewards might contribute to excessive consumption and to disorders such as RDS. Drug-microinjection studies have shown that opioids in both areas amplify the liking of sweet-taste rewards. Modern neuroscience tools such as Fos plume mapping have further identified hedonic hot spots within the accumbens and pallidum, where opioids are especially tuned to magnify the liking of food rewards. Hedonic hot spots in different brain structures may interact with each other within the larger functional circuitry that interconnects them (Peciña et al., 2006).

26.2.2 Where Are the Reward Structures of the Brain?

The neural circuitry for positive reinforcement involves multiple brain regions. Core regions constituting the brain reward pathway are located in the limbic system (Koob and Le Moal, 2001). Functions of the limbic system include monitoring of internal homoeostasis, mediating memory and learning, and contributing to emotions (Koob, 2000). The limbic system also drives important aspects of sexual behavior, motivation, and feeding behaviors. Primary areas of the limbic system include the hypothalamus, amygdale, sublenticular amygdale, hippocampus, septal nuclei, and anterior cingulate gyrus. Also important in the function of the limbic system are the NAc, ventral caudate nucleus, and the putamen (Kotter and Stephan, 1997). Other structures important in the brain reward system include the prefrontal cortex, substania nigra, periaqueductal gray matter, and the ventral tegmental area (Bowirrat and Oscar-Berman, 2005). It is of particular interest that DA is known as the main neurotransmitter modulating the activation of the reward system of the brain. See Konkle and Bielajew's 2004 review of functional neuroanatomical tools that have played an important role in proposing which structures underlie brain stimulation reward circuitry.

26.2.3 Why Is Dopamine the Key?

To reiterate, the D2 receptor has been associated with pleasure, and the *DRD2* A1 allele has been referred to as a reward gene (Blum et al., 1990). The *DRD2* gene has been one of the most widely studied on neuropsychiatric disorders in general and in alcoholism and other addictions, even in working memory and cognition (Blum and Braverman, 2000). With the advent of microarray analysis of the genome, much has been discovered about the role of genes and behavior. For example, array work has been accomplished by Li et al. (2008), who integrated 2,343 items of evidence from peer-reviewed publications between 1976 and 2006 linking genes and chromosome regions to addiction. Using single-gene strategies, microarray, proteomics, or genetic studies, they identified 1,500 human addiction-related genes. They developed KARG (http://karg.cbi.pku.edu.cn), the first molecular database for addiction-related genes with extensive annotations and a friendly Web interface. Then they performed a meta-analysis of 396 genes that were supported by two or more independent items of evidence to identify 18 molecular pathways that were statistically significantly enriched, covering both upstream signaling events and downstream effects. Five molecular pathways significantly enriched for all four different types of addictive drugs were identified as common pathways, which may underlie shared rewarding and addictive actions. These pathways included two new ones: GnRH signaling pathway and gap junction. They connected the common pathways into a hypothetical common molecular network for addiction. Interestingly two final pathways emerged: the glutamate pathway and the dopaminergic pathway.

26.2.4 RDS Is the Problem

It is well known that alcohol and other drugs of abuse (Cools et al., 1998), as well as most positive reinforcers (i.e., sex, food, gambling [Comings et al., 2001], aggression), cause activation and neuronal release of brain DA (Reuter et al., 2005), which can decrease negative feelings and satisfy abnormal cravings (Eshleman et al., 1994; Carboni et al., 2000; Gessa et al., 1985; Di Chiara and Impereto, 1988; Di Chiara, 1999, 2002; Di Chiara et al., 1999; Blum and Payne, 1991; Noble et al., 1994a; Adler et al., 2000; Hallbus et al., 1997; Koepp et al., 1998). A deficiency or absence of the D2 receptors then predisposes individuals to a high risk for multiple addictive, impulsive, and compulsive behaviors (Comings et al., 1995; Comings and Blum, 2000; Koob, 2003; Koob and Le Moal, 2001; Serý et al., 2006b). Although other neurotransmitters, e.g., glutamate, GABA (Dick et al., 2004), serotonin (Goldman et al., 1992), and enkephalins (Comings et al., 1999), may be important in determining the rewarding and stimulating effects of ethanol, DA may be critical for initiating drug use and for reinstating drug use during protracted abstinence (Gardner, 1997; Comings et al., 1999; Connor et al., 2002; Gordon et al., 2001; Rommelspacher et al., 1992).

The *DRD2* TaqIA polymorphism is associated with DA D2 receptor density that plays an important role in the context of reward. As cited above, persons carrying an A1 allele have a lower D2 receptor density and a higher risk to show substance abuse. One study was designed to investigate the influence of the *DRD2* TaqIA polymorphism and the selective D2 receptor agonist bromociptine on the activation of the reward system by means of fMRI. In a double-blind crossover study with 24 participants, Kirsch et al. (2006) found an increase of reward system activation from placebo to bromocriptine only in subjects carrying the A1 allele. Furthermore, only A1 carrier showed an increase of performance under bromocriptine. The results are interpreted as reflecting a specific sensitivity for DA agonists in persons carrying an A1 allele and may complement actual data and theories of the development of addiction disorders postulating a higher genetic risk for substance abuse in carriers of the A1 allele (Lawford et al., 1995).

When the mesocorticolimbic system DA reward system dysfunctions (potentially caused by certain genetic variants), the end result is RDS (Blum et al., 1996a) and subsequent drug-seeking behavior. In discussing RDS, we refer specifically to an insensitivity and inefficiency in the reward system (Blum et al., 1996a; Blum et al., 2000; Comings and Blum, 2000). RDS also encompasses the acquired need to escape or avoid negative effects created by repeated cycles of alcohol abuse (Koehnke et al., 2002) and dependence (Grant, B. F., 1994; Grant, K. A., 1994). The mesocorticolimbic dopaminergic pathway system plays an especially important role in mediating reinforcement by drugs of abuse, and it may be a common denominator for multiple addictions and a number of psychiatric disorders (Lowinson et al., 1997; Comings et al., 1991).

26.2.5 RDS Is Not Only the Problem; It Is the Disorder Phenotype

There is an ongoing lack of understanding that in order to carry strong informative genetic studies, the disease phenotype is RDS and not any one single addictive behavior. In an unpublished study, my group evaluated the potential association of DA D2 receptor gene (*DRD2*), DA D1 receptor gene (*DRD1*), DA transporter gene (*DAT1*), and DA β-hydroxylase gene (*DBH*) polymorphisms in reward deficiency syndrome subjects derived from two families. By demonstrating this association, we have illustrated the relevance of a generalized RDS behavior set as the phenotype. My group genotyped an experimental group of 55 subjects derived from up to five generations of two independent multiply affected families. Data related to RDS behaviors was collected on these subjects plus 13 deceased family members. Among the genotyped family members, 78% carried the *DRD2* Taql allele, 58% carried the *DAT1* 10/10 allele, 66% carried the *DBHB1* allele, and 35% carried either the *DRD1* A1/A1 or A2/A2 genotypes. The experimental positive rate for the *DRD2* Taql allele was significantly greater ($\chi^2 = 43.6$, $P < 0.001$), with an odds ratio of 103.9 (12.8, 843.2). All probands ($n = 32$) from Family A that were genotyped for the *DRD2* gene carried the TaqA1 allele. The experimental positive rate for the *DAT1* 10/10 allele was significantly greater ($\chi^2 = 6.0$, $P < 0.015$) with an odds ratio of 2.3 (1.2, 4.6). No significant differences were observed between the experimental and control positive rates of the *DBH*, *DRD1* A1/A1, or *DRD1* A2/A2 genotypes. These results confirm the putative role of dopaminergic polymorphisms in reward deficiency syndrome behaviors. This study demonstrates the importance of a nonspecific reward deficiency syndrome phenotype. Evaluating single subset behaviors (e.g., substance use disorder and Tourette's) of reward deficiency syndrome may lead to spurious results. Utilization of a nonspecific generalized reward phenotype may be a paradigm shift in future association and linkage studies involving dopaminergic polymorphisms and other neurotransmitter gene candidates.

26.2.6 Neurocommonality of Reward Dependence Behaviors

Support for the impulsive nature of individuals possessing dopaminergic gene variants is derived from a number of important studies illustrating the genetic risk for drug-seeking behaviors based on association and linkage studies implicating the subsequent alleles as risk antecedents having impact in the mesocorticolimbic system. Following the initial findings of a positive association of the Taql A1 of the *DRD2* gene and severe alcoholism (Blum et al., 1990), there have been a plethora of studies on this gene (3,100 general PubMed papers as of July 24, 2011) both positive (Blum et al., 1991, 1993; Comings et al., 1991, 1994; Smith et al., 1992; O'Hara et al., 1993; Parsian et al., 1991, 2000; Noble et al., 1991, 1994a, 1994b; Cloninger, 1991; Lawford et al., 1995, 1997; Neiswanger et al., 1995; Hietala et al., 1994; Ratsma et al., 2001; Hill et al., 1999; Bau et al., 2000; Connor et al., 2002; Laine et al., 1999, 2001; Ponce et al., 2003; Arinami et al., 1993; Kono et al., 1997; Goldman et al., 1993, 1997; Xu et al., 2004) and negative (Bolos et al., 1990; Gelernter et al., 1991; 1993; Cook et al., 1992; Turner et al., 1992; Parsian et al., 1991; Heinz et al., 1996;

Chen et al., 1996; Cruz et al., 1995; Gejman et al., 1994; Suarez et al., 1994; Goldman et al., 1997; Edenberg et al., 1998; Lu et al., 1996; Gebhardt et al., 2000). There are also reviews (Merikangas, 1990; Gorwood et al., 2000; Noble et al., 1998; Noble, 2000, 2003; Blum and Braverman, 2000; Bowirrat and Oscar-Berman, 2005; Barr and Kidd, 1993; Blum and Payne, 1991; Bouchard, 1994; Cadoret et al., 1990; Carey, 1994; Comings and Blum, 2000; Crowe, 1993; Grandy et al., 1991; Neiswanger et al., 1995; Oakley et al., 1991; Pato et al., 1993; Pickens et al., 1991; Plomin et al., 1994; Cook et al., 1992).

A number of studies have observed that the Taql A1 allele is associated with low DA D2 densities in alcoholics (Noble et al., 1991; Hietala et al., 1994). Moreover, other studies have confirmed that the striatal postsynaptic D2-receptor densities are low among alcoholics (Volkow et al., 1996). Studies of pre- and postsynaptic D2 receptors, as well as DA transporter densities among late-onset (Type 1) and violent (Type 11) alcoholics, have suggested an underlying dopaminergic defect (Tiihonen et al.1995; Gilman et al., 1998; Little et al., 1998; Kuikka et al., 1998, 2000). High DAT densities among violent Type 11 alcoholics were reported when compared with healthy controls (Repo et al., 1999), whereas late-onset Type 1 alcoholics has lower densities than healthy controls (Tiihonen et al., 1995). Another study using the highly selective radioligand PE2I technique (Kuikka et al., 1998) reported lower DAT densities among alcoholics compared with controls, but subtypes were not considered (Laine et al., 1999).

Regarding the polymorphic association, a major difficulty with an association of the *DRD2* TaqA1 allele with alcoholism is that the Taql A polymorphism is located more than 10 kb downstream from the coding region of the *DRD2* gene (Johnson, 1996), and a mutation at this site would not be expected to lead to any structural change in the DA receptor. The most likely explanation for an association is that the Taql A polymorphism is in linkage disequilibrium with an upstream regulatory element, or a 3′ flanking element, or another gene that confers susceptibility to RDS behaviors. Several linkage disequilibrium studies have found strong linkage disequilibrium between the Taql A1 allele and the Taq1B allele and the SSCP 1 allele (Blum et al., 1991; Hauge et al., 2001; Goldman et al., 1993; O'Hara et al., 1993; Johnson, 1996). As we have pointed out, the DA D2 receptor has been implicated extensively in relation to alcoholism, substance use disorder, nicotine dependence, anxiety, memory, glucose control, pathological aggression, pathological gambling, and certain sexual behaviors, all of which are RDS behaviors. The most frequently examined polymorphism linked to this gene is the Taql A restriction fragment length polymorphism, which has been associated with a reduction in D2 receptor density. In a recent study, within the 10-kb downstream region of the Taql A1 RFLP, Neville et al. (2004) identified a novel kinase gene named ankyrin repeat and kinase domain containing 1 (*ANKK1*) that contains a single serine/threonine kinase domain and is expressed at low levels in placenta and whole spinal-cord RNA. This gene is a member of an extensive family of proteins involved in signal transduction pathways. The *DRD2* Taq1A allele is a single-nucleotide polymorphism (SNP) that causes an amino acid substitution within the 11th ankyrin repeat of *ANKK1* (p. Glu713lYs), which, although unlikely to affect structural integrity, may affect substrate-binding specificity. If this is the case, then changes in ANKK1 activity may provide an alternative explanation for previously described associations between the *DRD2* gene and RDS behaviors (Neville et al., 2004). There are a number of papers that have shown the positive association of this gene and nicotine dependence. Huang et al. (2008) reported on the association of the rs2734849 polymorphism in the *ANKK1* gene as a functional variant in the causation of nicotine dependence indirectly affecting the D2 receptor density. Other studies have shown the relationship of both the *DRD2* gene and the *ANKK1* gene and executive network functioning (Fossella et al., 2008). However, recent studies from the laboratory of Hirvonen et al. (2009) show the C957T polymorphism of the human DA D2 receptor gene (*DRD2*) regulates DRD2 availability in striatum *in vivo*. Specifically, the T allele predicts high DRD2 availability in healthy volunteers (T/T > T/C > C/C). However, this finding was unexpected, because *in vitro* the T allele is associated with a decrease in DRD2 mRNA stability and synthesis of the receptor through a putative alteration in the receptor mRNA folding. To elucidate further how changes in

DRD2 density (B_{max}) and affinity (K_D) contribute to the differences in DRD2 availability between the C957T genotypes, they studied these parameters separately in a sample of 45 healthy volunteers. The subjects had two PET scans with [^{11}C]raclopride (high and low specific radioactivity scans) for the estimation of B_{max} and K_D and were genotyped for the C957T. Moreover, the role of the related and previously studied functional TaqIA polymorphism of the ankyrin repeat and kinase domain containing 1 (*ANKK1*) gene was reassessed for comparative purposes. The results indicate that the C957T increased binding potential by decreasing DRD2 K_D (C/C > C/T > T/T), whereas B_{max} was not significantly altered. These preliminary findings indicate that the C957T genotype-dependent changes in DRD2 availability are driven by alterations in receptor affinity and putatively in striatal DA levels. This mechanism seems to differ from that observed previously for the *ANKK1* gene TaqI A polymorphism, where the minor allele (A1) affects DRD2 availability predominantly by changing B_{max}. The hypothesis that the two SNPs may have independent effects on DA neurotransmission needs to be further tested.

Even in our first paper (Blum et al., 1990), the concept of the DA D2 receptor gene as a specific target for alcohol was appropriately dismissed by authors who suggested that they had found a nonspecific reward gene (Panagis et al., 1997). Moreover, the DRD2 TaqA1 allele has been also associated with sensitivity to stress and anxiety (Bau et al., 2000; Jonsson et al., 2003; Kreek and Koob 1998), and both symptoms have been related to sensitivity of presynaptic D2 receptors (Noble, 2000). The sensitivity is elevated in high-anxiety subjects compared with low-anxiety subjects. Furthermore, other RDS and related neurological and psychiatric disorders are also found to be associated with polymorphisms of the *DRD2* gene.

RDS refers to the breakdown of the reward cascade (Blum and Kozlowski, 1990) and resultant aberrant conduct caused by specific genetic and environmental influences (Rowe, 1986; Blum et al., 1996b, 1996c), especially in children and adults with ADHD and Tourette's syndrome (Ratsma et al., 2001; Romstad et al., 2003; Singer et al., 2002; Smith et al., 2003; Comings et al., 1995, 1996a, 1996b; Comings and Blum, 2000).

Studies at the University of Wisconsin (Clark and Grunstein, 2000), using identical twins raised in different families, who had parallel lives, showed that about half of human behavior (including aggression, sexuality, mental function, eating disorder, alcoholism, and drug abuse or generalized RDS) can be accounted for by genes. Very few behaviors depend upon a single gene. Complexes of genes (polygenic) drive most of our heredity-based actions, suggesting that genetic panels or algorithms organized into genetic indexes, such as Syn-R-Gene™ may be valuable clinically. Certainly abnormal functions of these brain systems can be due to specific genetic factors interacting with environmental factors such as abuse of various psychoactive substances, particularly alcohol and stimulants (Uhl et al., 1993). In this regard, it has been shown that these individuals may have a reduced number of DA D2 receptors (Noble et al., 1991; Hietala et al., 1994) and a high number of DA transporter sites (Tiihonen et al., 1995; Tupala et al., 2001a, 2001b, 2003). Certainly the finding of hypodominergic function as discovered in pathological gambling, obesity, and ADHD (Reuter et al., 2005; Comings et al., 1996a, 1996b, 2001), examples of RDS behaviors, helps us understand the potential driving force of some to induce activation of the DA system (Thut et al., 1997). Understanding the interaction of these components is likely to lead to better treatment.

26.3 HOW CAN THIS RDS PARADIGM BE USED IN TREATMENT? CAN YOU COUPLE GENE TESTING WITH REWARD CIRCUITRY MANIPULATION?

26.3.1 Diagnosis: Genetic Addiction Risk Score

As stated earlier, support for the impulsive nature of individuals possessing dopaminergic gene variants is derived from a number of important studies illustrating the genetic risk for drug-seeking behaviors based on association and linkage studies implicating the following alleles as risk antecedents having impact in the mesocorticolimbic system (Blum et al., 2010b) (Table 26.1).

TABLE 26.1
Proposed Addiction Gene Panel

Gene	Significant	Comment
ALDH2++	$P = 5 \times 10^{-37}$	With alcoholism and alcohol-induced medical diseases
ADH1B++	$P = 2 \times 10^{-21}$	With alcoholism and alcohol-induced medical diseases
ADH1C++	$P = 4 \times 10^{-33}$	With alcoholism and alcohol-induce medical diseases
DRD2+	$P = 1 \times 10^{-8}$	With alcohol and drug abuse
DRD4+	$P = 1 \times 10^{-2}$	With alcohol and drug abuse
SLC6A4++	$P = 2 \times 10^{-3}$	With alcohol, heroin, cocaine, methamphetamine dependence
HTRIB+	$P = 5 \times 10^{-1}$	With alcohol and drug abuse
HTRI2A+	$P = 5 \times 10^{-1}$	With alcohol and drug abuse
TPH+	$P = 2 \times 10^{-3}$	With alcohol and drug abuse
MAOA+	$P = 9 \times 10^{-5}$	With alcohol and drug abuse
OPRD1++	$P = 5 \times 10^{-1}$	With alcohol and drug abuse
GABRG2++	$P = 5 \times 10^{-4}$	With alcohol and drug abuse
GABRA2+	$P = 7 \times 10^{-4}$	With alcohol and drug abuse
GABRA6++	$P = 6 \times 10^{-4}$	With alcohol and drug abuse
COMT+	$P = 5 \times 10^{-1}$	With alcohol and drug abuse in Asians
DAT1+	$P = 5 \times 10^{-1}$	With alcohol and drug abuse in Asians
CNR1+	$P = 5 \times 10^{-1}$	With alcohol and drug abuse
CYP2E1++	$P = 7 \times 10^{-2}$	With alcohol liver disease
ANKKI++	$P = 5 \times 10^{-6}$	With alcohol and drug abuse

Source: Chen, T. J. H., Blum, K., Chen, L. C. H., et al., *Journal of Psychoactive Drugs,* 42, 108–127, 2011.

26.3.1.1 D2 DA Receptor Gene (*DRD2*)

The DA D2 receptor gene (*DRD2*), first associated by Blum et al. (1990) with severe alcoholism, is the most widely studied candidate gene in psychiatric genetics. The Taql A is a SNP (rs: 1800497) originally thought to be located at the 3′ untranslated region of the *DRD2* but now has been shown to be located within exon 8 of an adjacent gene, the ankyrin repeat, and kinase domain containing 1 (*ANKKI*). Importantly, although there may be distinct differences in function, Neville et al. (2004) suggest that the mislocation of the Taql A may be attributable to the *ANKKI* and the *DRD2* being on the same haplotype or the *ANKKI* being involved in reward processing through a signal transduction pathway. The *ANKKI* and the *DRD2* gene polymorphisms may have distinctly different actions with regard to brain function, as has been noted in recent experiments and fear-related conditioning in alcoholics (Dick et al., 2007; Huertas et al., 2010). Grandy et al. (1989) reported on the presence of the two alleles of the Taql A: A1 and A2. The presence of the A1+ genotype (A1/A1, A1/A2) compared with the A– genotype (A2/A2) is associated with reduced density (Noble et al., 1991; Pohjalainen et al., 1998). This reduction causes hypodopaminergic functioning in the DA reward pathway. Noble (2003), in reviewing the literature, concluded that the research supports a predictive relationship between the A1+ genotype and drug seeking behavior. This has also been discussed by my group (Blum et al., 1996b, 1996c), reporting that the presence of the A+ genotype using Bayesian analysis has a predictive value of 74% for a number of RDS behaviors. Other DRD2 polymorphisms such as the C (57T), a SNP (rs: 6277) at exon 7, also associates with a number of RDS behaviors including drug use (Duan et al., 2003; Hirvonen et al., 2004; Hill et al., 2008). Compared with the T– genotype (C/C), the T+ genotype (T/T, T/C) is associated with reduced translation of DRD2 mRNA and diminished DRD2 mRNA (Duan et al., 2003), leading to reduced DRD2 density (Hirvonen et al., 2004). Hill et al. (2008) have shown that the predictive relationship between the T+ allele and alcohol dependence results in hypoopaminergic function. Thus, the T+ is also a predictive risk allele.

The association of the *DRD2* A1 allele in alcoholism is well established. In a 10-year follow-up involving alcohol-dependent individuals, carriers of the *DRD2* A1 allele have a higher rate of mortality compared with carriers of the A2 allele (Berggren et al., 2010). There are 390 PubMed reports (as of June 5, 2010) providing significant support. The DA D2 receptor (DRD2) plays an important role in the reinforcing and motivating effects of ethanol. Several polymorphisms have been reported to affect receptor expression. The amount of DRD2 expressed in a given individual is the result of the expression of both alleles, each representing a distinct haplotype.

Most recently, Kraschewski et al. (2009) found that the haplotypes I-C-G-A2 and I-C-A-A1 occurred with a higher frequency in alcoholics ($P = 0.026$, odds ratio [OR]: 1.340; $P = 0.010$, OR: 1.521, respectively). The rare haplotype I-C-A-A2 occurred less often in alcoholics ($P = 0.010$, OR: 0.507) and was also less often transmitted from parents to their affected offspring (1 versus 7). Among the subgroups, I-C-G-A2 and I-C-A-A1 had a higher frequency in Cloninger 1 alcoholics ($P = 0.083$ and 0.001, OR: 1.917, respectively) and in alcoholics with a positive family history ($P = 0.031$, OR: 1.478; $P = 0.073$, respectively). Cloninger 2 alcoholics had a higher frequency of the rare haplotype D-T-A-A2 ($P < 0.001$, OR: 4.614) always compared with controls. In patients with positive family history haplotype I-C-A-A2 ($P = 0.004$, OR: 0.209), and in Cloninger 1 alcoholics, haplotype I-T-A-A1 ($P = 0.045$ OR: 0.460) was less often present. Kraschewski et al. confirmed the hypothesis that haplotypes, which are supposed to induce a low DRD2 expression, are associated with alcohol dependence. Furthermore, supposedly high-expressing haplotypes weakened or neutralized the action of low-expressing haplotypes.

26.3.1.2 D4 DA Receptor Gene (*DRD4*)

There is evidence that the length of the D4 DA receptor (DRD4) exon 3 variable number of tandem repeats (VNTR) affects DRD4 functioning by modulating the expression and efficiency of maturation of the receptor (Schoots and Van Tol, 2003). The 7-repeat (7R) VNTR requires significantly higher amounts of DA to produce a response of the same magnitude as other size VNTRs (Oak et al., 2000), this reduced sensitivity or DA resistance leads to hypodopaminergic functioning. Thus, 7R VNTR has been associated with substance-seeking behavior (Oak et al., 2000; McGeary et al., 2007). However, not all reports support this association (Lusher et al., 2001). Most recently Biederman et al. (2009), evaluating a number of putative risk alleles using survival analysis, revealed that by 25 years of age, 76% of subjects with a *DRD4* 7-repeat allele were estimated to have significantly more persistent ADHD compared with 66% of subjects without the risk allele. In contrast, there were no significant associations between the course of ADHD and the *DAT1* 10-repeat allele ($P = 0.94$) and 5-HTTLPR long allele. Their findings suggest that the *DRD4* 7-repeat allele is associated with a more persistent course of ADHD. Moreover, a study by Grzywacz et al. (2009) that evaluated the role of DA D4 receptor (DRD4) exon 3 polymorphisms (48-bp VNTR) in the pathogenesis of alcoholism found significant differences in the short alleles (two to five VNTR) frequencies between controls and patients with a history of delirium tremens and/or alcohol seizures ($P = 0.043$). A trend was also observed in the higher frequency of short alleles amongst individuals with an early age of onset of alcoholism ($P = 0.063$). The results of this study suggest that inherited short variants of DRD4 alleles (3R) may play a role in pathogenesis of alcohol dependence, and carriers of the 4R may have a protective effect for alcoholism risk behaviors. It is of further interest that work from Kotler et al. (1997) in heroin addicts illustrated that central dopaminergic pathways figure prominently in drug-mediated reinforcement including novelty seeking, suggesting that DA receptors are likely candidates for association with substance abuse in man. These researchers show that the 7-repeat allele is significantly overrepresented in the opioid-dependent cohort and confers a relative risk of 2.46.

26.3.1.3 DA Transporter Gene (*DAT1*)

The DA transporter protein regulates DA-mediated neurotransmission by rapidly accumulating DA that has been released into the synapse (Vandenbergh et al., 1998). The DA transporter gene (*SLC6A3* or *DAT1*) is localized to chromosome 5p15.3. Moreover, within three noncoding region of

DAT1 lies a VNTR polymorphism (Vandenbergh et al., 1998). There are two important alleles that may independently increase risk for RDS behaviors. The 9-repeat (9R) VNTR has been shown to influence gene expression and to augment transcription of the DA transporter protein (Michelhaugh et al., 2001). This results in an enhanced clearance of synaptic DA, yielding reduced levels of DA to activate postsynaptic neurons. The presence of the 9R VNTR has been linked to SUD (Guindalini et al., 2006), but not consistently (Vandenbergh et al., 2002). Moreover, in terms of RDS behaviors, Cook et al. (1995) were the first to associate tandem repeats of the DA transporter gene (*DAT*) in the literature. Although there have been some inconsistencies associated with the earlier results, the evidence is mounting in favor of the view that the 10R allele of *DAT* is associated with a high risk for ADHD in children and in adults alike. Specifically, Lee et al. (2007) found consistent support in several studies, the nonadditive association for the 10-repeat allele was significant for hyperactivity-impulsivity (HI) symptoms. However, consistent with other studies, exploratory analyses of the nonadditive association of the 9-repeat allele of *DAT1* with HI and oppositional defiant disorder symptoms were also significant.

26.3.1.4 Catechol-*O*-Methyltransferase

COMT is an enzyme involved in the metabolism of DA, adrenaline, and noradrenaline. The Val158Met polymorphism of the *COMT* gene has been previously associated with a variability of the COMT activity, and alcoholism. Serý et al. (2006a) found a relationship between the Val158Met polymorphism of the *COMT* gene and alcoholism in male subjects. They also found the significant difference between male alcoholics and male controls in allele and genotype frequencies ($P < 0.007$ and $P < 0.04$, respectively. No differences in genotype and allele frequencies of the 108 Val/Met polymorphism of *COMT* gene were observed between heroin-dependent subjects and normal controls (genotype-wise: $\chi^2 = 1.67$, $P = 0.43$; allele-wise: $\chi^2 = 1.23$, $P = 0.27$). No differences in genotype and allele frequencies of 900 Ins C/Del C polymorphism of *COMT* gene were observed between heroin-dependent subjects and normal controls (genotype-wise: $\chi^2 = 3.73$, $P = 0.16$; allele-wise: $\chi^2 = 0.76$, $P = 0.38$). Although there is still some controversy regarding the COMT association with heroin addiction, it was also interesting that the A allele of the Val/Met polymorphisms (–287 A/G) found by Cao et al. (2003) was found to be much higher in heroin addicts than controls. Faster metabolism results in reduced DA availability at the synapse, which reduces postsynaptic activation, inducing hypodopaminergic functioning. Generally Vandenbergh et al. (1997) and Wang et al. (2001) support an association with the Val allele and SUD, but others do not (Samochowiec et al., 2006).

26.3.1.5 Monoamine Oxidase A

Monoamine oxidase A (MAOA) is a mitochondrial enzyme that degrades the neurotransmitters serotonin, norepinephrine, and DA. This system is involved with both psychological and physical functioning. The gene that encodes MAOA is found on the X chromosome and contains a polymorphism (*MAOA-uVNTR*) located 1.2 kb upstream of the MAOA coding sequences (Shih, 1991). In this polymorphism, consisting of a 30-base pair repeated sequence, six allele variants containing either 2-, 3-, 3.5-, 4-, 5-, or 6-repeat copies have been identified (Zhu and Shih, 1997). Functional studies indicate that certain alleles may confer lower transcriptional efficiency than others. The three-repeat variant conveys lower efficiency, whereas 3.5- and 4-repeat alleles result in higher efficiency (Brummett et al., 2007). The 3- and 4-repeat alleles are the most common, and to date there is no consensus regarding the transcriptional efficiency of the other less commonly occurring alleles (e.g., 2-, 5-, and 6-repeat). The primary role of MAOA in regulating monoamine turnover, and hence ultimately influencing levels of norepinephrine, DA, and serotonin, indicates that its gene is a highly plausible candidate for affecting individual differences in the manifestation of psychological traits and psychiatric disorders (Shih et al., 1999). For example, recent evidence indicates that the *MAOA* gene may be associated with depression (Lee et al., 2010) and stress (Brummett et al., 2008). However, evidence regarding whether higher or lower *MAOA* gene transcriptional efficiency is positively associated with psychological pathology as been mixed. The low-activity 3-repeat allele of the *MAOA-uVNTR* polymorphism has been positively related to symptoms of antisocial personality (Lee et al., 2009) and cluster B personality disorders. Other studies, however, suggest that alleles associated with higher transcriptional efficiency are related to unhealthy psychological characteristics such as trait aggressiveness and impulsivity. Low MAO activity and the neurotransmitter DA are two important factors in the development of alcohol dependence. MAO is an important enzyme associated with the metabolism of biogenic amines. Therefore, Huang et al. (2007) investigated whether the association between the DA D2 receptor (*DRD2*) gene and alcoholism is affected by different polymorphisms of the MAO type A (*MAOA*) gene. The genetic variant of the *DRD2* gene was only associated with the anxiety, depression (ANX/DEP) ALC phenotype, and the genetic variant of the *MAOA* gene was associated with ALC. Subjects carrying the *MAOA* 3-repeat allele and genotype A1/A1 of the *DRD2* gene were 3.48 times (95% confidence interval = 1.47-8.25) more likely to be ANX/DEP ALC than the subjects carrying the *MAOA* 3-repeat allele and the *DRD2* A2/A2 genotype. The *MAOA* gene may modify the association between the *DRD2* gene and ANX/DEP ALC phenotype. Overall, Vanyukov et al. (2004) suggested that, although not definitive, variants in MAOA account for a small portion of the variance of SUD risk, possibly mediated by liability to early onset behavioral problems.

26.3.1.6 Serotonin Transporter Gene

The human serotonin (5-hydroxytryptamine) transporter, encoded by the *SLC6A4* gene on chromosome 17q11.1-q12, is the cellular reuptake site for serotonin and a site of action for several drugs with central nervous system effects, including both therapeutic agents (e.g., antidepressants) and drugs of abuse (e.g., cocaine). It is known that the serotonin transporter plays an important role in the metabolic cycle of a broad range of antidepressants, antipsychotics, anxiolytics, antiemetics, and antimigraine drugs. Saiz et al. (2009) found an excess of A-1438G and 5-HTTLPR L carriers in alcoholic patients in comparison with the heroin-dependent group (OR [95% confidence interval] = 1.98 [1.13–3.45] and 1.92 [1.07–3.44], respectively). The A-1438G and 5-HTTLPR polymorphisms also interacted in distinguishing alcohol from heroin-dependent patients (10.21 [4], $P = 0.037$). The association of –1438A/G with alcohol dependence was especially pronounced in the presence of 5-HTTLPR S/S, less evident with 5-HTTLPR L/S, and not present with 5-HTTLPR L/L. SCL6A4 polymorphism haplotypes were similarly distributed in all three groups. Moreover, Seneviratne et al. (2009) found that G allele carriers for rs1042173 were associated with significantly lower drinking intensity ($P = 0.0034$) compared to T-allele homozygotes. In HeLa cell cultures, the cells transfected with G allele showed a significantly higher mRNA and protein levels than the T allele-transfected cells. These findings suggest that the allelic variations of rs1042173 affect drinking intensity in alcoholics possibly by altering serotonin transporter expression levels. This provides additional support to the hypothesis that *SLC6A4* polymorphisms play an important role in regulating the propensity for severe drinking.

26.3.1.7 GABA Genes

GABA receptor genes have also received some attention as candidates for drug-use disorders as well as RDS. One reason for this is that DA and GABA systems are functionally interrelated (White, 1996). Research indicates that DA neurons projecting from the anterior ventral tegmental area to the NAc are tonically inhibited by GABA through its actions at the $GABA_A$ receptor (McBride et al., 1999). Most importantly, it has been shown that alcohol (Theile et al., 2008) or opioid (Johnson and North, 1992) enhancement of GABAergic (through $GABA_A$ receptor) transmission inhibits the release of DA in the mesocorticolimbic system. Thus, a hyperactive GABA system, by inhibiting DA release, could also lead to hypodopaminergic functioning. A dinucleotide receptor polymorphism of the GABA receptor β3 subunit gene (*GABRB3*) results in either the presence of the 181-bp G1 or 11 other repeats designated as non-G1 (NG1). Research indicates that NG1 is more prevalent in children of alcoholics (Namkoong et al., 2008). Moreover, Noble et al. (1998) associated NG1 with alcohol dependence, and Edenberg and Foroud (2006) associated a number of GABA receptor genes with alcoholism (Xuei et al., 2010).

26.3.1.8 Mu Receptor Gene

Opioid drugs play important roles in the clinical management of pain, as well as in the development and treatment of drug abuse. The mu opioid receptor is the primary site of action for the most commonly used opioids, including morphine, heroin, fentanyl, and methadone. The most prevalent SNP is a nucleotide substitution at position 118 (A118G), predicting an amino acid change at a putative *N*-glycosylation site. This SNP displays an allelic frequency of approximately 10% in a heroin population. The A118G variant receptor binds β-endorphin, an endogenous opioid that activates the mu opioid receptor, approximately three times more tightly than the most common allelic form of the receptor. Furthermore, β-endorphin is approximately three times more potent at the A118G variant receptor than at the most common allelic form in agonist-induced activation of G protein-coupled potassium channels. These results show that SNPs in the mu opioid receptor gene can alter binding and signal transduction in the resulting receptor and may have implications for normal physiology, therapeutics, and vulnerability to develop or protection from diverse diseases including the addictive diseases (Bond et al., 1998).

26.3.1.9 *PENK* Gene

Striatal enkephalin and dynorphin opioid systems mediate reward and negative influence, respectively, and are therefore relevant to addiction disorders. Nikoshkov et al. (2008) examined polymorphisms of proenkephalin (*PENK*) and prodynorphin (*PDYN*) genes in relation to heroin abuse and gene expression in the human striatum and the relevance of genetic dopaminergic tone, critical for drug reward and striatal function. Heroin abuse was significantly associated with *PENK* polymorphic 3' UTR dinucleotide (CA) repeats; 79% of subjects homozygous for the 79-bp allele were heroin abusers. Such individuals tended to express higher PENK mRNA than the 81-bp homozygotes, but PENK levels within the NAc shell were most strongly correlated to the COMT genotype. Control Met/Met individuals expressed lower PENK mRNA than Val carriers, a pattern reversed in heroin users. Upregulation of NAc PENK in Met/Met heroin abusers was accompanied by impaired tyrosine hydroxylase (TH) mRNA expression in mesolimbic DA neurons. Altogether, the data suggest that dysfunction of the opioid reward system is significantly linked to opiate-abuse vulnerability and that heroin use alters the apparent influence of heritable DA tone on mesolimbic PENK and TH function.

26.3.1.10 *DARP-32* Gene

Dopaminergic neurons exert a major modulatory effect on the forebrain. DA and adenosine 3',5'-monophosphate-regulated phosphoprotein (32 kDa) (DARPP-32), which is enriched in all neurons that receive a dopaminergic input, is converted in response to DA into a potent protein phosphatase inhibitor. Mice generated to contain a targeted disruption of the *DARPP-32* gene showed profound deficits in their molecular, electrophysiological, and behavioral responses to DA, drugs of abuse, and antipsychotic medication. The results show that DARPP-32 plays a central role in regulating the efficacy of dopaminergic neurotransmission (Fienberg et al., 1998). Other work showed that *in vivo*, the state of phosphorylation of DARPP-32 and, by implication, the activity of protein phosphatase-1 are regulated by tonic activation of D1, D2, and A2A receptors. The results also underscore the fact that the adenosine system plays a role in the generation of responses to DA D2 antagonists *in vivo* that has relevance to the treatment of RDS. Specifically, the D1 receptor agonist SKF82526 increased *DARPP-32* phosphorylation. In contrast, the D2 receptor agonist quinpirole decreased basal as well as D1 agonist-, forskolin-, and 8-bromo-cAMP-stimulated phosphorylation of DARPP-32. The ability of quinpirole to decrease D1-stimulated DARPP-32 phosphorylation was calcium-dependent and was blocked by the calcineurin inhibitor cyclosporin A, suggesting that the D2 effect involved an increase in intracellular calcium and activation of calcineurin. In support of this interpretation, Ca^{2+}-free/EGTA medium induced a greater than 60-fold increase in DARPP-32 phosphorylation and abolished the ability of quinpirole to dephosphorylate DARPP-32. The antipsychotic drug raclopride, a selective D2 receptor antagonist, increased phosphorylation of DARPP-32 under basal conditions and in D2 agonist-treated slices. The results of this study demonstrate that DA exerts a bidirectional control on the state of phosphorylation of DARPP-32 (Nishi et al., 1997). Thus, this gene has been considered by others as an important candidate for drug abuse (Nairn et al., 2004).

26.3.2 Combination of Genes and Addiction Risk

In general, inconsistencies in the literature involving association studies using single gene analysis prompted Conner et al. (2010) and others to evaluate a number of dopaminergic gene polymorphisms, as predictors of drug use in adolescents. As suggested by Bossert et al. (2005) we cannot ignore the importance of understanding the interaction of multiple genes and environmental elements in the neurochemical mechanisms involved in drug-induced relapse behavior. These investigators (using a drug-relapse model previously shown to induce relapse by reexposing rats to heroin-associated contexts) have found that after extinction of drug-reinforced responding in different contexts, reexposure reinstates heroin seeking. This effect is attenuated by inhibition of glutamate transmission in the ventral tegmental area and medial accumbens shell, two components of the mesolimbic DA system. This process enhances DA net release in the NAc. This fits well with Li's KARG addiction network map (Li et al., 2008).

Since the initial finding of Blum et al. (1990) showing the positive association of single-gene *DRD2* polymorphisms and severe alcoholism to date, the replication, although favorable, has been fraught with inconsistent results. This has been true for other complex behaviors as well (NCI-NHGRI Working Group on Replication in Association Studies, 2007). Moreover, when gene-gene and environment interactions are tested, the findings support the concept that complex gene relationships may account for inconsistent findings across many different single gene studies (Yang et al., 2008).

There are many different reasons for inconsistencies in trying to predict drug use including single-gene analysis, stratification of population, poorly screened controls, gender-based differences, personality traits, comorbidity of psychiatric disorders, positive and negative life events, and even neurocognitive functioning (Miller et al., 2010, Blum et al. 2010). Thus, instead of continuing to evaluate single-gene associations to predict future drug abuse, it occurred to us that we should embark on a study to evaluate multiple gene candidates especially linked to the brain reward cascade and hypodopaminergic functioning to gain a more complex but stronger predictive set of genetic antecedents. Our goal, albeit exploratory in nature, is to develop an informative panel to provide a means of stratifying or classifying patients entering a treatment facility as having low, moderate, or high genetic predictive risk based on a number of known risk alleles. We are coining the term *genetic addiction risk score* (GARS) for purposes of study identification and coupling this diagnostic tool with a putative natural dopaminergic agonist (Blum et al., 2010a, 2010b).

Most recently my group reported on an exploratory development of GARS. There is a need to classify patients at genetic risk for drug-seeking behavior prior to or upon entry to residential and or nonresidential chemical dependency programs. There are at least three practical reasons for such a diagnostic test:

1. Identifying those at risk prior to the onset of SUD providing early intervention and prevention of the negative outcomes from such use.
2. Removal of denial and guilt.
3. Genotype results could suggest different at-risk individuals, and programs could be tailored to a patient's risk profile.

My group has determined, based on a literature review, that there are seven risk alleles associated with six candidate genes that were studied in this patient population of recovering polydrug abusers. To determine the risk severity of these 26 patients, we calculated the percentage of prevalence of the risk alleles and provided an arbitrary severity score based on percentages of these alleles. Subjects carry the following risk alleles: *DRD2* = A1; *SLC6A3 (DAT)* = 10R; *DRD4* = 3R or 7R; *5-HTTlRP* = L or LA; *MAO* = 3R; and *COMT* = G. As depicted in Table 26.2, low severity (LS) = 1–36%, moderate severity (MS) = 37–50%, and high severity (HS) = 51–100%. We studied two distinct ethnic populations: group 1 consisted of 16 male Caucasian psychostimulant addicts, and group 2 consisted of 10 Chinese heroin-addicted males. Based on this model, the 16 subjects tested have at least one risk allele or 100%. Out of the 16 subjects, we found 50% (8) HS; 31% (5) MS; and 19% LS (3 subjects). These scores are then converted to a fraction and then represented as a GARS, whereby we found the average GARS to be: 0.28 low severity, 0.44 moderate severity, and 0.58 high severity, respectively. Therefore, using this GARS, we found that 81% of the patients were at moderate to high risk for addictive behavior. Of particular interest, we found that 56% of the subjects carried the DRD2 A1 allele (9 of 16). Out of the 9 Chinese heroin addicts (one patient was not genotyped) in group 2, we found 11% (1) HS, 56% (5) MS, and 33% LS (3). These scores are then converted to a fraction and then represented as GARS, whereby we found the average GARS to be: 0.28 low severity, 0.43 moderate severity, and 0.54 high severity, respectively. Therefore, using GARS, we found that 67% of the patients were at moderate to high risk for addictive behavior. Of particular interest, we found that 56% of the subjects carried the *DRD2* A1 allele (5/9) similar to group 1. Statistical analysis revealed that the groups did not differ in terms of overall severity (67 versus 81%) in these two distinct populations. Combining these two independent study populations reveals that subjects entering a residential treatment facility for polydrug abuse carry at least one risk allele (100%). We found 74% of the combined 25 subjects (Caucasian and Chinese) had a moderate to high GARS. Confirmation of these exploratory results and development of mathematical predictive values of these risk alleles are necessary before any meaningful interpretation of these results is to be considered (Blum et al., 2010a).

26.4 TREATMENT: NUTRIGENOMICS OF TREATING RDS

Based on neurochemical and genetic evidence, we (Blum et al., 2008) suggest that both prevention and treatment of multiple addictions, such as dependence on alcohol, nicotine, and glucose, should involve a biphasic approach. Thus, acute treatment should consist of preferential blocking of postsynaptic NAc DA receptors (D1–D5), whereas long-term activation of the mesolimbic dopaminergic system should involve activation and/or release of DA at the NAc site. Failure to do so will result in abnormal mood, abnormal behavior, and potential suicide ideation. Individuals possessing a paucity of serotonergic and/or dopaminergic receptors and an increased rate of synaptic DA catabolism because of high catabolic genotype of the *COMT* gene are predisposed to self-medicating with any substance or behavior that will activate DA release, including alcohol, opiates,

TABLE 26.2
Neuroadaptagen Amino-Acid Therapy

GARS Nutrient	Pathway
D-Phenylalanine	Opioid peptides
L-Phenylalanine	Dopamine
L-Tryptophan	Serotonin
L-Tyrosine	Dopamine
L-Glutamine	GABA
Chromium	Serotonin
R. rosea	COMT
Pyridoxine	Enzyme catalyst
Zehntose Metalosaceride™	Immune

psychostimulants, nicotine, gambling, sex, and even excessive Internet gaming. Acute utilization of these substances and/or stimulatory behaviors induces a feeling of well-being. Unfortunately, sustained and prolonged abuse leads to a toxic pseudofeeling of well-being, resulting in tolerance and disease or discomfort. Thus, a reduced number of DA receptors, caused by carrying the *DRD2* A1 allelic genotype, results in excessive craving behavior, whereas a normal or sufficient amount of DA receptors results in low craving behavior. In terms of preventing substance abuse, one goal would be to induce a proliferation of DA D2 receptors in genetically prone individuals. Although *in vivo* experiments using a typical D2 receptor agonist induce downregulation, experiments *in vitro* have shown that constant stimulation of the DA receptor system via a known D2 agonist results in significant proliferation of D2 receptors in spite of genetic antecedents. In essence, D2 receptor stimulation signals negative feedback mechanisms in the mesolimbic system to induce mRNA expression causing proliferation of D2 receptors. Our group (Blum et al., 2008) has proposed that D2 receptor stimulation can be accomplished via the use of Synaptamine™/Synaptose™/Endorphamine™, a natural but therapeutic nutraceutical formulation that potentially induces DA release, causing the same induction of D2-directed mRNA and thus proliferation of D2 receptors in the human. This proliferation of D2 receptors in turn will induce the attenuation of craving behavior. In fact, as mentioned earlier, this model has been proven in research showing DNA-directed compensatory overexpression (a form of gene therapy) of the DRD2 receptors, resulting in a significant reduction in alcohol-craving behavior in alcohol-preferring rodents. Utilizing natural dopaminergic repletion therapy to promote long-term dopaminergic activation will ultimately lead to a common, safe, and effective modality to treat RDS behaviors including SUD, ADHD, obesity, and other reward-deficient aberrant behaviors. This concept is further supported by the more comprehensive understanding of the role of DA in the NAc as a *wanting* messenger in the mesolimbic DA system (Davis et al., 2009). There are common genetic mechanisms responsible for both drug effects and subsequent seeking behavior. Past and current treatment of substance-seeking behavior, a subtype of RDS, is considered by most to be inadequate. Recently, we evaluated a complex named Synaptamine™ (KB220™) (Blum et al., 2007). The main difference with an older studied variant and the latest variant is the inclusion of a proprietary form of *Rhodiola rosea*, a known COMT inhibitor, to potentially enhance the activity of presynaptic released DA. In this regard, based on the current literature, we hypothesize that manipulation of COMT activity to influence the attenuation of substance seeking behavior is dependent upon gene polymorphisms. In this regard we hypothesize that carrying the LL genotype with low COMT activity should as theorized, increase the reward induced by substance-induced DA release, and may indeed increase the propensity to type 1 alcoholism and possibly other drugs that activate the dopaminergic system. Thus, when alcohol is present in low COMT LL genotype, increasing COMT activity, not inhibiting it should assist in the reduction of social consumption or abuse. Alternatively, under physiological conditions (no psychoactive substances present (e.g., alcohol) carrying the *DRD2* A1 allele with associated low D2 receptors should, as theorized, increase craving behavior because of a low or hypodopaminergic state, causing the individual to seek out substances that increase the release of DA for subsequent activation of unbound D2 sites in the NAc. Thus, in the absence of alcohol or other psychoactive drugs (DA releasers), especially during recovery or rehabilitation, decreasing, not increasing, COMT activity should result in enhanced synaptic DA as physiologically released, thereby proliferating D2 receptors while reducing stress, increasing well-being, reducing craving behavior, and preventing relapse. Based on this hypothesis, we believe that adding the COMT inhibitor *R. rosea* (as Rhodimin) to our amino-acid and chromium combination increases the potential for more targeted neurochemical rebalancing and enhanced relapse prevention.

Finally, we hypothesize that these data coupled together provide evidence that the combination of enkephalinase inhibition, neurotransmitter precursor loading, brain tryptophan enhancing, and COMT inhibition, as well as DNA analysis of the individual's genome, may be useful as an adjunct to therapy when used in outpatient recovery, specifically to assist in reducing craving behavior and preventing relapse (Table 26.2 contains a synaptamine list of ingredients with pathways affected).

26.4.1 Neuroadaptagen Amino-Acid Therapy

The proposed neuroadaptagen consists of amino-acid neurotransmitter precursors and enkephalinase-inhibition therapy called Synaptamine™, a patented formulation (U.S. patent 6,132,724 issued October 2000) in brain reward function. The basic patented (U.S. patent 6,132,734) formula included amino-acid precursors such as L-phenylalanine, L-tyrosine, L-tryptophan, 5-hydroxytryptophan, L-glutamine, a serotonin-concentrating substance, chromium, the enkephalinase inhibitor D-phenylalanine, a neurotransmitter synthesis promoter (vitamin B_6), and both methionine and leucine. The amounts of these ingredients will vary according to individualized assessment. Through a series of both neurogenetic and clinical experiments, it is becoming increasingly clear that this novel formulation is the first neuroadaptagen known to activate the brain reward circuitry. Ongoing research repeatedly confirms that the numerous clinical effects ultimately result in significant benefits for victims having genetic antecedents for all addictive, compulsive, and impulsive behaviors. These behaviors are all correctly classified under the rubric of RDS. My group is hereby proposing a novel addiction candidate gene map showing metaanalysis power (Table 26.2 and Figure 26.2). Other genes can also be added as more data become available, such as GABA Mu receptor PENK, Enkephalinase Genes (Blum et al., 1996b). The genotyping for the alcohol metabolizing genes may have utility in select populations (for example those of Asian and or American Indian descent).

The preliminary findings in the United States using qEGG and in China using fMRI regarding the effects of oral Synaptose Complex™ on activation of brain reward circuitry in SUD provide a novel framework that may ultimately lead to a biogenetic standard of care for RDS probands. It seems from this preliminary data, utilizing qEEG placebo controlled triple-blind crossover design here in the United Sates, that an increase in alpha and low beta brain waves (equal to 10 weeks of biofeedback training) may be important for treatment outcomes. These results are further confirmed by ongoing studies in China, where preliminary results of an fMRI at resting state and comparison of Synaptamine with placebo 2X2 design show activation of the caudate brain region and potentially a smoothing out of heroin-induced putamen abnormal connectivity. This study is in progress and awaits final analysis and replication.

Although there is support from several investigators for a higher likelihood of treatment response and compliance using dopaminergic agonist therapy (utilizing nutrigenomic principles) in carriers of the *DRD2* A1 allele compared with *DRD2* A2 allele genotype, the actual mechanism for positive clinical outcomes remains a mystery (Blum et al., 2009). For the first time in the history of this work involving dopaminergic genetics, Laakso et al. (2005) have provided a clue. Accordingly, the A1 allele of the TaqI restriction fragment-length polymorphism (RFLP) of the human DA D2 receptor gene (*DRD2*) is associated with a low density of D2 DA receptors in the striatum. Because of the important role of D2 autoreceptors in regulating DA synthesis, they aimed to examine whether subjects with the A1 allele have altered presynaptic DA function in the brain. They also studied the effects of two other *DRD2* polymorphisms, C957 T and –141C Ins/Del, which have been suggested to affect D2 receptor levels in brain. The relation between the Taq IA RFLP, C957 T, and –141C Ins/Del polymorphisms and striatal DA synthesis in 33 healthy Finnish volunteers was studied using PET scans and [18F]fluorodopa ([18F]FDOPA), a radiolabeled analog of the DA precursor L-DOPA. Heterozygous carriers of the A1 allele (A1/A2; 10 subjects) had significantly higher (18%) [18F]FDOPA uptake in the putamen than subjects without the A1 allele (A2/A2; 23 subjects). C957 T and –141C Ins/Del polymorphisms did not significantly affect [18F]FDOPA K_i values. These results demonstrated that the A1 allele of the DRD2 gene is associated with increased striatal activity of aromatic L-amino acid decarboxylase, the final enzyme in the biosynthesis of DA and the rate-limiting enzyme for trace amine (e.g., β-phenylethylamine) synthesis. The finding can be explained by lower D2 receptor expression leading to decreased autoreceptor function and suggests that DA and/or the trace amine synthesis rate is increased in the brains of A1 allele carriers to compensate. We are proposing that with an increased striatal activity of aromatic L-amino acid decarboxylase, the final enzyme in the biosynthesis of DA and the rate-limiting enzyme for trace

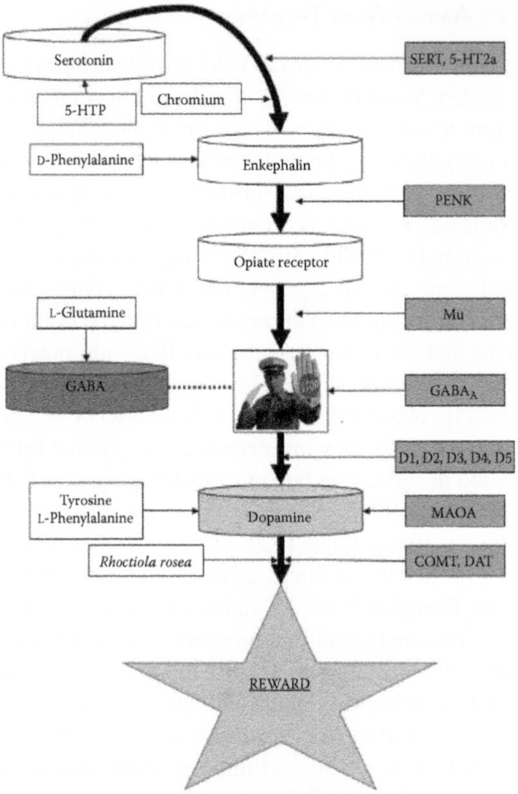

FIGURE 26.2 Understanding the relationship of neurotransmitters, candidate genes, and therapeutic nutrigenomic targets involvement in the brain reward cascade. Serotonin stimulates the neurons containing enkephalins in the hypothalamus of the mesocorticolimbic system of the reward pathway of the brain. The enkephalins that are released interact at postsynaptic neuronal terminals, thereby activating mu receptors at the substantia nigra region of the brain. This in turn inhibits the transmission of GAB$_A$, which in turn fine-tunes the release of dopamine and the nucleus accumbens via inhibitory regulatory control dopamine projection initiating at the ventral tegmental area of the brain. The appropriate amount of dopamine interacts at dopamine receptors, and under normal physiological conditions induces a feeling of well-being and reward. The synthesis, vesicular storage, receptor integrity, release, and synaptic catabolism are controlled by various well known candidate genes. The *SERT* gene is involved in serotonergic cellular transport by a reuptake mechanism; the 5-HT2a receptor is a major site for serotonin activation; the *PENK* gene controls the synthesis of enkephalins; the mu receptor is a major site for opioid or enkephalin activation; *GABAA* receptor gene is a major site for GABA activation; D2, D3, D4, and D5 genes are sites for dopaminergic activation (D2 being most prominent); MOAA is involved in the cellular catabolism of catecholamines including dopamine in the mitochondria; the DAT or dopamine transporter gene is responsible for clearing out excess synaptic dopamine by a reuptake mechanism; and the *COMT* gene is a catabolizing enzyme responsible for destroying excess dopamine in the synapse. There is mounting evidence especially from neuroimaging studies that the neuroadaptogen KB220-Z can interact at various neurotransmitter targets, causing a nutrigenomic solution affecting the net release and proliferation of dopamine D2 receptors. The nutrient 5-hydroxytryptophan (5-HTP) is involved in the synthesis of serotonin. Chromium salts (picolinate or polynicotinate) increase the sensitivity of 5-HT2a receptors, thereby reducing the need for high amounts of serotonin. D-Phenylalanine is a well-established inhibitor of carboxypeptidase A (enkephalinase), thereby increasing the quanta of enkephalins released into the synapse. L-Glutamate is involved in the synthesis of GABA, and it is carefully utilized to promote appropriate amounts of GABA. The rate-limiting step in the synthesis of dopamine is L-tyrosine. L-Phenylalanine is responsible for about 20% of dopamine synthesis in the doses utilized in Synaptose™ (KB220-Z). The Russian herb *R. rosea* has been shown to inhibit the action of the enzyme COMT. The therapeutic action of this nutrigenomic neuradaptogen results in a preferential release of dopamine with potential proliferation of D2 receptors in spite of genetic antecedents such as carrying the *DRD2* A1 allele.

amine (e.g., β-phenylethylamine), DA synthesis should occur with a more natural and less powerful agonistic compound relative to L-Dopa. This would support the use of Synaptamine complex (KB220), a precursor amino acid and enkephalinase therapy, as a safe DA agonist. It is postulated that a lower DA quanta release at presynaptic neurons in the NAc should result in an upregulation of postsynaptic D2 receptors in A1 carriers, which will ultimately result in a reduction of craving behavior.

This is the only known agent in the nutraceutical industrial space that on an acute basis normalizes persistent qEEG in serious psychostimulant abusers during protracted abstinence. Analysis from fMRI demonstrates that acute administration of this neuroadaptogen may even activate reward pathways deficient in D2 density in the mesolimbic system and warrants intensive investigation.

26.5 RELAPSE PREVENTION

Recent research on working memory and cognition is interesting and may be important in terms of recall and relapse prevention. The dopaminergic neurotransmitter system of the brain is involved in working memory and other cognitive functions. Studies suggest an important role for DA synthesis and uptake in modulation of human cognitive processes. Bolton et al. (2010) studied the association between polymorphisms in the *COMT* and *DRD2* genes and general cognitive ability in a secondary analysis of 2,091 men and women, aged 55–80 years living in Scotland. General cognitive ability g was derived from five cognitive tests of different domains. *COMT* was not associated with cognitive ability in this population. The *DRD2* C:C genotype of rs6277 was associated with decreased general cognitive ability g ($P = 0.003$), and *DRD2* rs1800497 heterozygotes had the lowest mean general cognitive ability g ($P = 0.007$). There was an indication of a potential interaction between the *DRD2* SNPs. This result may also offer a potential mechanism linking cognitive decline and substance-seeking behavior in ADHD probands. This finding takes on even more importance when you consider the role of DA in brain function and memory recall. DA is critical for reward-based decision making, yet dopaminergic drugs can have opposite effects in different individuals. This apparent discrepancy can be accounted for by hypothesizing an *inverted-U* relationship, whereby the effect of DA agents depends on baseline DA system functioning. Cohen et al. (2007) used functional magnetic resonance imaging to test the hypothesis that genetic variation in the expression of DA D2 receptors in the human brain predicts opposing dopaminergic drug effects during reversal learning. They scanned 22 subjects while they engaged in a feedback-based reversal learning task. Ten subjects had an allele on the Taq1A *DRD2* gene, which is associated with reduced DA receptor concentration and decreased neural responses to rewards (A1+ subjects). Subjects were scanned twice, once on placebo and once on cabergoline, a D2 receptor agonist. Consistent with an inverted-U relationship between the *DRD2* polymorphism and drug effects, cabergoline increased neural reward responses in the medial orbitofrontal cortex, cingulate cortex, and striatum for A1+ subjects but decreased reward responses in these regions for A1– subjects. In contrast, cabergoline decreased task performance and frontostriatal connectivity in A1+ subjects but had the opposite effect in A1– subjects. Further, the drug effect on functional connectivity predicted the drug effect on feedback-guided learning. Thus, there is individual variability in how dopaminergic drugs affect the brain reflects genetic disposition. These findings may help to explain the link between genetic disposition and risk for addictive disorders especially as it relates to appropriate decision making involving readdiction and relapse.

26.5.1 Deprivation Amplification Relapse Therapy

It is well known that after prolonged abstinence, individuals who use their drug of choice experience a powerful euphoria that often precipitates relapse. Although a biological explanation for this conundrum has remained elusive, we hypothesize that this clinically observed supersensitivity might be tied to genetic dopaminergic polymorphisms. Another therapeutic conundrum relates to the paradoxical finding that the dopaminergic agonist bromocriptine induces stronger activation of brain reward circuitry in individuals who carry the *DRD2* A1 allele compared with *DRD2* A2 allele carriers. Because carriers of the A1 allele relative to the A2 allele of the *DRD2* gene have significantly lower D2 receptor density, a reduced sensitivity to DA agonist activity would be expected in the former. Thus, it is perplexing that with low D2 density, there is an increase in reward sensitivity with the DA D2 agonist bromocriptine. Moreover, under chronic or long-term therapy with D2 agonists, such as bromocriptine, it has been shown *in vitro* that there is a proliferation of D2 receptors. One explanation for this relates to the demonstration that the A1 allele of the *DRD2* gene is associated with increased striatal activity of L-amino acid decarboxylase, the final step in the biosynthesis of DA. This appears to be a protective mechanism against low receptor density and would favor the utilization of an amino acid neurotransmitter precursor like L-tyrosine for preferential synthesis of DA. This seems to lead to receptor proliferation at normal levels and results in significantly better treatment compliance only in A1 carriers. We propose that low D2 receptor density and polymorphisms of the D2 gene are associated with risk for relapse of substance abuse, including alcohol dependence, heroin craving, cocaine dependence, methamphetamine abuse, nicotine sensitization, and glucose craving. With this in mind, my group has suggested a putative physiological mechanism that may help to explain the enhanced sensitivity following intense acute dopaminergic D2 receptor activation: denervation supersensitivity. Rats with unilateral depletions of neostriatal DA display increased sensitivity to DA agonists estimated to be 30–100-fold in the 6-hydroxyDA rotational model. Mild striatal DA D2 receptor proliferation may occur with the administration of a natural DA D2 agonist by increasing compromised D2 receptor density, especially in carriers of the *DRD2* A1 allele, and enhance DA release targeting D2 sensitization and, as such, attenuate relapse. This hypothesized mechanism is supported by clinical trials utilizing amino-acid neurotransmitter precursors, enkephalinase, and COMT enzyme inhibition, which have resulted in attenuated relapse rates in RDS probands (Blum et al., 2009). If future translational research reveals that DA agonist therapy reduces relapse in RDS, it would support the proposed concept, which we term deprivation-amplification relapse therapy. This term couples the mechanism for relapse, which is deprivation amplification, especially in *DRD2* A1 allele carriers with natural D2 agonist therapy utilizing amino-acid precursors and COMT-MOA enkephalinase inhibition therapy.

26.6 CONCLUSION

It is no surprise that it has taken over four decades to confirm and extend information about the critical role of DA and related genes and gene deficits in the etiology and risk for drug dependence. Hundreds of studies have been reported, and many are enabled by neuroscience, neuroimaging, and genetic advances. However, although DA theories have been reported, confirmed, replicated, and replicated again, changes have been slow to move from the bench to the bedside (Dackis and Gold, 1985).

Thus, acute pharmacological adjunctive treatment should consist of preferential blocking of postsynaptic NAc DA receptors (D1–D5), whereas long-term activation of the mesolimbic dopaminergic system should involve activation and/or release of DA at the NAc site. My associates and I have proposed that D2 receptor stimulation might be accomplished via the use of neuroadaptagen amino-acid therapy. Basic and clinical testing of such a compound is underway. Such a natural but therapeutic nutraceutical formulation that induces DA release could cause the induction of D2-directed mRNA and proliferation of D2 receptors in the human. This proliferation of D2 receptors in turn will induce the attenuation of craving behavior.

In fact, this model has been supported by research showing DNA-directed compensatory overexpression (a form of gene therapy) of the DRD2 receptors, resulting in a significant reduction in alcohol-craving behavior in alcohol-preferring rodents (Thanos et al., 2005). Still, 30 years from the advances in neuroscience of use, abuse, and dependence, most RDS patients are treated quite similarly to the way they would have been treated in the past. Efforts to integrate known neural mechanisms with other psychotherapeutic treatment options to combat relapse should be encouraged. It is well known that the addict in recovery, after prolonged abstinence, is particularly vulnerable to relapse. Individuals who use their drug of choice after abstinence experience a powerful euphoria that quickly can precipitate a full-blown relapse. A biological explanation has been hypothesized to explain this conundrum, whereby supersensitivity might be a result of a premorbid genetic hypodopaminergic trait. Although treatment with bromocriptine or bupropion has not appeared to be helpful in psychostimulant abuser outcomes because of potential downregulation of DA receptors, other approaches might be less provocative.

Prevention, diagnosis, treatment, and relapse-prevention tactics must be augmented by promoting rigorous 5-year outcome research in both out-patient and residential in-patient programs. After over four decades of pharmacological, neuroscience, and psychiatric research involving genetics and genomics, the field is poised to embark on large population studies incorporating these newer theories, especially as they relate to dopaminergic targeting of mesolimbic pathways. I encourage academic and clinical scientists to develop studies that will bridge science to recovery (Giordano and Blum, 2010), yielding novel standard of care guidelines to treat RDS (addictive, impulsive, and compulsive behaviors). Rather than simply choosing between pharmacological treatments that either block the high or maintain the drug state with less abusable agonists, we suggest a changing the focus to treating the underlying, premorbid vulnerabilities. This may be an approach that can help break the vicious cycle of use, abuse, abstinence, and relapse still so common today.

Utilizing exercise, pharmacological treatments, and/or natural dopaminergic repletion therapy to promote long-term dopaminergic activation could lead to a common, safe, and effective modality to treat SUD and other RDS disorders. This concept has been further explained by understanding the role of DA in the NAc as a *wanting* rather than *liking* messenger in the mesolimbic DA system. My group has further suggested that DSM diagnosis should include gene polymorphic testing using candidate gene analysis to assist in individualizing diagnosis, risk, and therapy. By classifying risk severity and striving to remediate these deficits with medication, diet, and exercise, we could improve future clinical trials. Conner et al. (2010), in a preliminary study, suggested that it is possible to identify children at risk for problematic drug use prior to onset of drug dependence by testing for hypodopaminergic genes and identifying personality and environmental elements as predictors.

Finally pharmacological therapies have had limited success because these powerful agents have focused on maintenance or interference with drug euphoria rather than correcting or compensating for premorbid DA system deficits. It is well known that powerful D2 agonists induce side effects following chronic use (Mizrahi et al., 2010).

ACKNOWLEDGMENTS

I would like to thank Margaret A. Madigan for editing the manuscript. The support of the staff of the Department of Psychiatry, University of Florida College of Medicine, Gainesville, Florida, LifeGen, Inc., and Path Foundation NY is much appreciated.

COMPETING INTERESTS

Kenneth Blum, PhD, is an officer and stockholder of LifeGen, Inc., which is the worldwide exclusive distributor of products related to patents concerning reward deficiency syndrome.

REFERENCES

Adler, C. M., Elman, I., Weisenfield, N., et al. (2000). Effects of acute metabolic stress on striatal dopamine release in healthy volunteers. *Neuropsychopharmacol.* 22: 545–550.

Arinami, T., Itokawa, M., Komiyama, T., et al. (1993). Association between severity of alcoholism and the A1 allele of the dopamine D2 receptor gene TaqI A RFLP in Japanese. *Biol. Psychiatry* 33: 108–114.

Baker, K. B., Halliday, G. M., Harper, C. G. (1994). Effects of chronic alcohol consumption on human locus coreruleus. *Alcohol. Clin. Exp. Res.* 18: 1491–1496.

Barr, C. L., Kidd, K. K. (1993). Population frequencies of the A1 allele at the dopamine D2 receptor locus. *Biol. Psychiatry* 34: 204–209.

Bau, C. H. D., Almeida, S., Hutz, M. H. (2000). The TaqI A1 allele of the dopamine D2 receptor gene and alcoholism in Brazil: Association and interaction with stress and harm avoidance on severity prediction. *Am. J. Med. Genet.* 96: 302–306.

Berggren, U., Fahlke, C., Berglund, K. J., et al. (2010). Dopamine D2 receptor genotype is associated with increased mortality at a 10-year follow-up of alcohol-dependent individuals. *Alcohol Alcohol.* 45: 1–5.

Biederman, J., Petty, C. R., Ten Haagen, K. S., et al. (2009). Effect of candidate gene polymorphisms on the course of attention deficit hyperactivity disorder. *Psychiatry Res.* 170: 199–203.

Blum, K., Braverman, E. R. (2000). Reward deficiency syndrome: A biogenetic model for the diagnosis and treatment of impulsive, addictive and compulsive behaviors. *J. Psychoactive Drugs* 32 (Suppl.): 1–112.

Blum, K., Kozlowski, G. P. (1990). Ethanol and neuromodulator interactions: A cascade model of reward. In Ollat, H., Parvez, S., Parvez, H. (Eds.), *Alcohol and behavior* (pp. 131–149). Utrecht, the Netherlands: VSP Press.

Blum, K., Payne, J. (1991). *Alcohol and the addictive brain.* New York: Simon and Schuster.

Blum, K., Braverman, E. R., Holder, J .M., et al. (2000). Reward deficiency syndrome: A biologic model for the diagnosis and treatment of impulsive, addictive, and compulsive behaviors. *J. Psychoactive Drugs* 32 (Suppl.): 1–112.

Blum, K., Braverman, E. R., Wood, R. C., et al. (1996a). Increased prevalence of the Taq I A1 allele of the dopamine receptor gene (DRD2) in obesity with comorbid substance use disorder: A preliminary report. *Pharmacogenetics* 6: 297–305.

Blum, K., Chen, A. L., Chen, T. J., et al. (2008). Activation instead of blocking mesolimbic dopaminergic reward circuitry is a preferred modality in the long term treatment of reward deficiency syndrome (RDS): A commentary. *Theor. Biol. Med. Model.* 12: 24.

Blum, K., Chen, T. J., Downs, et al. (2009). Neurogenetics of dopaminergic receptor supersensitivity in activation of brain reward circuitry and relapse: Proposing "deprivation-amplification relapse therapy" (DART). *Postgrad. Med.* 121: 176–196.

Blum, K., Chen, T. J., Meshkin, B., et al. (2007). Manipulation of catechol-*O*-methyl-transferase (COMT) activity to influence the attenuation of substance seeking behavior, a subtype of reward deficiency syndrome (RDS), is dependent upon gene polymorphisms: A hypothesis. *Med. Hypotheses* 69: 1054–1060.

Blum, K., Chen, T. J. H., Morse, S., et al. (2010a). Overcoming qEEG abnormalities and reward gene deficits during protracted Abstinence. In *Male psychostimulant and polydrug abusers utilizing putative dopamine D2 agonist therapy*: *Postgrad. Med.* 122: 214–226.

Blum, K., Cull, J. G., Braverman, E. R., et al. (1996b). Reward deficiency syndrome. *Am. Sci.* 84: 132–145.

Blum, K., Giordano, J., Morse, S., et al. (2010b). Genetic Addiction Risk Score (GARS) analysis: Exploratory development of polymorphic risk alleles in polydrug addicted males. *Int. J. Omics Biotechnol.* 1: 1–14.

Blum, K., Noble, E. P., Sheridan, P. J., et al. (1990). Allelic association of human dopamine D_2 receptor gene in alcoholism. *JAMA* 263: 2055–2060.

Blum, K., Noble, E. P., Sheridan, P. J., et al. (1991). Association of the A1 allele of the D2 dopamine receptor gene with severe alcoholism. *Alcohol.* 8: 409–416.

Blum, K., Noble, E. P., Sheridan, P. J., et al. (1993). Genetic predisposition in alcoholism: Association of the D2 dopamine receptor TaqI B1 RFLP with severe alcoholics. *Alcohol* 10: 59–67.

Blum, K., Sheridan, P. J., Wood, R. C., et al. (1996c). The D2 dopamine receptor gene as a determinant of reward deficiency syndrome. *J. R. Soc. Med.* 89: 396–400.

Bolos, A.M., Dean, M., Lucas-Derse, S., et al. (1990). Population and pedigree studies reveal a lock of association between the dopamine D2 receptor gene and alcoholism. *JAMA* 264: 3156–3160.

Bolton, J. L., Marioni, R. E., Deary, I. J., et al. (2010). Association between polymorphisms of the dopamine receptor D2 and catechol-*O*-methyltransferase genes and cognitive function. *Behav. Genet.* 40: 630–638.

Bond, C., LaForge, K.S., Tian, M., et al. (1998). Single-nucleotide polymorphism in the human mu opioid receptor gene alters beta-endorphin binding and activity: Possible implications for opiate addiction. *Proc. Natl. Acad. Sci. U.S.A.* 95: 9608–9613.

Bossert, J. M., Ghitza, U. E., Lu, L., et al. (2005). Neurobiology of relapse to heroin and cocaine seeking: An update and clinical implications. *Eur. J. Pharmacol.* 526: 36–50.

Bouchard, T. J. (1994). Genes, environment and personality. *Science* 9: 415–421.

Boundy, V. A., Pacheco, M. A., Guan, W., et al. (1995). Agonists and antagonists differentially regulate the high affinity state of the D2L receptor in human embryonic kidney 293 cells. *Mol. Pharmacol.* 48: 956–964.

Bowirrat, A., Oscar-Berman, M. (2005). Relationship between dopaminergic neurotransmission, alcoholism and reward deficiency syndrome. *Am. J. Med. Genet. B Neuropsychiatr. Genet.* 132B: 29–37.

Brummett, B. H., Boyle, S. H., Siegler, I. C., et al. (2008). HPA axis function in male caregivers: Effect of the monoamine oxidase-A gene promoter (MAOA-uVNTR). *Biol. Psychol.* 79: 250–255.

Brummett, B. H., Krystal, A. D., Siegler, I. C., et al. (2007). Associations of a regulatory polymorphism of monoamine oxidase-A gene promoter (MAOA-uVNTR) with symptoms of depression and sleep quality. *Psychosom. Med.* 69: 396–401.

Cao, L., Li, T., Liu, X. (2003). Association study of heroin dependence and catechol-*O*-methyltransferase gene. *Zhonghua Yi Xue Yi Chuan Xue Za Zhi* 20: 127–130.

Cadoret, R. J., Troughton, E., Bagford, J., et al. (1990). Genetic and environmental factors in adoptee antisocial personality. *Eur. Arch. Psychiatr. Neurol. Sci.* 239: 231–240.

Carboni, E., Silvagni, A., Rolando, M. T. P., et al. (2000). Stimulation of in vivo dopamine transmission in the bed nucleus of stria terminalis by reinforcing drugs. *J. Neurosci.* 20: 1–5.

Carey, G. (1994). Genetic association study in psychiatry: Analytical evaluation and a recommendation. *Am. J. Med. Genet. Neuropsychiatr. Genet.* 54: 311–317.

Chen, C. H., Chien, S. H., Hwu, H. G. (1996). Lack of association between TaqI A1 allele of dopamine D2 receptor gene and alcohol-use disorders in Atayal natives of Taiwan. *Am. J. Med. Genet.* 67: 488–490.

Chen, T. J. H., Blum, K., Chen, L. C. H., et al. (2011). Neurogenetics and clinical evidence for the putative activation of the brain reward circuitry by amino-acid precursor-catabolic enzyme inhibition therapeutic agent (a Neuroadaptagen): Proposing an addiction candidate gene panel map. *J. Psychoactive Drugs* 42: 108–127.

Clark, W. R., Grunstein, M. (2000). *Are we hardwired? The role of genes in human behavior*. New York: Oxford University Press.

Cloninger, C. R. (1991). D2 dopamine receptor gene is associated but not linked with alcoholism. *JAMA* 266: 1833–1834.

Cohen, M. X., Krohn-Grimberghe, A., Elger, C. E., et al. (2007). Dopamine gene predicts the brain's response to dopaminergic drug. *Eur. J. Neurosci.* 26: 3652–3660.

Comings, D., Johnson, P., Dietz, G., et al. (1995). Dopamine D2 receptor gene (DRD2) haplotypes and the defense style questionnaire in substance abuse, Tourette syndrome and controls. *Biol. Psychiatry* 37: 798–805.

Comings, D. E., Blum, K. (2000). Reward deficiency syndrome: Genetic aspects of behavioral disorders. *Prog. Brain Res.* 126: 325–341.

Comings, D. E., Comings, B. G., Muhleman, D., et al. (1991). The dopamine D2 receptor locus as a modifying gene in neuropsychiatric disorders. *JAMA* 266: 1793–1800.

Comings, D. E., Dietz, G., Johnson, J. P. M. (1999). Association of the enkephalinase gene with low amplitude P300 waves. *NeuroReport* 10: 2283–2285.

Comings, D. E., Gade, R., MacMurray, J. P., et al. (1996a). Genetic variants of the human obesity (OB) gene: Association with body mass index in young women psychiatric symptoms, and interaction with the dopamine D_2 receptor gene. *Mol. Psychiatry* 1: 325–335.

Comings, D. E., Gade-Andavolu, R., Gonzalez, N., et al. (2001). The additive effect of neurotransmitter genes in pathological gambling. *Clin. Genet.* 60: 107–116.

Comings, D. E., Muhlman, D., Ahn, C., et al. (1994). The dopamine D2 receptor gene: A genetic risk factor in substance abuse. *Drug Alcohol Depend.* 34: 175–180.

Comings, D. E., Wu, S., Chiu, C., et al. (1996b). Polygenic inheritance of Tourette Syndrome, stuttering, attention-deficit-hyperactivity, conduct and oppositional defiant disorder: The addictive and subtractive effect of the three dopaminergic genes-DRD2, DBH and DAT1. *Am. J. Med. Gen. Neuropsych. Genet.* 67: 264–288.

Conner, B. T., Hellemann, G. S., Ritchie, T. L., et al. (2010). Genetic, personality, and environmental predictors of drug use in adolescents. *J. Subst. Abuse Treat.* 38: 178–190.

Connor, J. P., Young, R. M., Lawford, B. R., et al. (2002). D2 dopamine receptor (DRD2) polymorphism is associated with severity of alcohol dependence. *Eur. Psychiatry* 17: 17–23.

Cook, B. L., Wang, Z. W., Crowe, R. R., et al. (1992). Alcoholism and the D2 receptor gene. *Alcohol. Clin. Exp. Res.* 16: 806–809.

Cook, E. H., Jr., Stein, M. A., Krasowski, M. D., et al. (1995). Association of attention-deficit disorder and the dopamine transporter gene. *Am. J. Hum. Genet.* 56: 993–998.

Cools, A. R., Gingras, M. A., Nijmegen J. (1998). High and low responders to novelty: A new tool in the search after the neurobiology of drug abuse liability. *Pharmacol. Biochem. Behav.* 60: 151–159.

Cooper, M., Frone, M., Russell, M., et al. (1995). Drinking to regulate positive and negative emotions: A motivational model of alcohol use. *J. Personality Social Psychol.* 69: 990–1005.

Crowe, R. R. (1993). Candidate genes in psychiatry: An epidemiological perspective. *Am. J. Med. Genet. Neuropsychiatr. Genet.* 48: 74–77.

Cruz, C., Camarena, B., Mejia, J. M., et al. (1995). The dopamine D2 receptor gene TaqI A1 polymorphism and alcoholism in a Mexican population. *Arch. Med. Res.* 26: 421–426.

Dackis, C. A., Gold, M. S. (1985). New concepts in cocaine addiction: The dopamine depletion hypothesis. *Neurosci. Biol. Rev.* 9: 469–477.

Dackis, C. A., Gold, M. S., Davies, R. K., et al. (1985). Bromocriptine treatment for cocaine abuse: The dopamine depletion hypothesis. *Int. J. Psychiatry Med.* 15: 125–135.

Davis, C. A., Levitan, R. D., Reid, C., et al. (2009). Dopamine for "wanting" and opioids for "liking": A comparison of obese adults with and without binge eating. *Obesity* 17: 1220–1225.

Di Chiara, G. (1999). Drug addiction as dopamine-dependent associative learning disorder. *Eur. J. Pharmacol.* 375: 13–30.

Di Chiara, G. (2002). Nucleus accumbens shell and core dopamine: Differential role in behavior and addiction. *Behav. Res.* 137: 75–114.

Di Chiara, G., Impereto, A. (1988). Drugs abused by humans preferentially increase synaptic dopamine concentrations in the mesolimbic systems of freely moving rats. *Proc. Natl. Acad. Sci. U.S.A.* 84: 1413–1416.

Di Chiara, G., Tanda, G., Bassare, V., et al. (1999). Drug addiction as a disorder of associative learning: Role of nucleus accumbens shell/extended amygdala dopamine. *Ann. N.Y. Acad. Sci.* 877: 461–485.

Dick, D. M., Wang, J. C., Plunkett, J., et al. (2007). Family-based association analyses of alcohol dependence phenotypes across DRD2 and neighboring gene ANKK1. *Alcohol. Clin. Exp. Res.* 31: 1645–1653.

Dick, D. M., Edenberg, H. J., Xuei, X., et al. (2004). Association of GABAG3 with alcohol dependence. *Alcohol. Clin. Exp. Res.* 28: 4–9.

Duan, J., Wainwright, M. S., Comeron, J. M., et al. (2003). Synonymous mutations in the human dopamine receptor D2 (DRD2) affect mRNA stability and synthesis of the receptor. *Hum. Mol. Genet.* 12: 205–216.

Edenberg, H. J., Foroud, T. (2006). The genetics of alcoholism identifying specific genes through family studies. *Addict. Biol.* 11: 386–396.

Edenberg, H. J., Foroud, T., Koller, D. L., et al. (1998). A family-based analysis of the association of the dopamine D2 receptor (DRD2) with alcoholism. *Alcohol. Clin. Exp. Res.* 22: 505–512.

Eisenberg, D. T., Campbell, B., Mackillop, J., et al. (2007). Season of birth and dopamine receptor gene associations with impulsivity, sensation seeking and reproductive behaviors. *PLoS One* 2: e1216.

Epping-Jordan, M. P., Markou, A., Koob, G. F. (1998). The dopamine D-1 receptor antagonist SCH 23390 injected into the dorsolateral bed nucleus of the stria terminalis decreased cocaine reinforcement in the rat. *Brain Res.* 784: 105–115.

Eshleman, A. J., Henningsen, R. A., Neve, K. A., et al. (1994). Release of dopamine via the human transporter. *Mol. Pharmacol.* 45: 312–316.

Fienberg, A. A., Hiroi, N., Mermelstein, P. G., et al. (1998). DARPP-32: Regulator of the efficacy of dopaminergic neurotransmission. *Science.* 281: 838–842.

Fossella, J., Fan, J., Liu, X., et al. (2008). Provisional hypotheses for the molecular genetics of cognitive development: Imaging genetic pathways in the anterior cingulate cortex. *Biol. Psychol.* 79: 23–29.

Gardner, E. L. (1997). Brain reward mechanisms. In Lowenson, J. H., Ruiz, P., Millman, R. B., Langrod, J. G. (Eds.), *Substance abuse: A comprehensive textbook* (pp. 51–58). Baltimore: Lippincott Williams & Wilkins.

Gebhardt, C., Leisch, F., Schussler, P., et al. (2000). Non-association of dopamine D4 and D2 receptor genes with personality in healthy individuals. *Psychiatr. Genet.* 10: 131–137.

Gejman, P.V., Gelernter, R.J., Friedman, E., et al. (1994). No structural mutation in the dopamine D2 receptor gene in alcoholism or schizophrenia: Analysis using denaturing gradient gel electrophoresis. *JAMA* 271: 204–209.

Gelernter, J., Goldman, D., Risch, N. (1993). The A1 allele at the D2 dopamine receptor gene and alcoholism: A reappraisal. *JAMA* 269: 1673–1677.

Gelernter, J., O'Malley, S., Risch, N., et al. (1991). No association between an allele at the D2 dopamine receptor gene (DRD2) and alcoholism. *JAMA* 266: 1801–1907.

Gessa, G. L., Mutoni, F., Coller, M., et al. (1985). Low doses of ethanol activate dopaminergic neurons in the ventral tegmental area. *Brain Res.* 48: 201–203.

Gilman, S., Koeppe, R. A., Adams, K. M., et al. (1998). Decreased striatal monoaminergic terminals in severe chronic alcoholism demonstrated with (+) [11C]dihydrotetrabenazine and positron emission tomography. *Ann. Neurol.* 44: 326–333.

Giordano, J., Blum, K. (2010). Probing the mysteries of recovery through nutrigenomic and holistic medicine: "Science meets recovery" through seminal translational research. *Prof. Counselor Mag.* 11.

Goldman, D., Brown, G.L., Albaugh, B., et al. (1993). DRD2 dopamine receptor genotype, linkage disequilibrium and alcoholism in American Indians and other populations. *Alcohol. Clin. Exp. Res.* 17: 199–204.

Goldman, D., Dean, M., Brown, G. L., et al. (1992). D2 dopamine receptor genotype and cerebrospinal fluid homovanillic acid, 5-hydroxyindoleacetic acid and 3-methoxy-4-hydroxyphenylglycol in alcoholisms in Finland and the United States. *Acta Psychiatr. Scand.* 86: 351–357.

Goldman, D., Urbanek, M., Guenther, D., et al. (1997). Linkage and association of a functional DRD2 variant (Ser311Cys) and DRD2 markers to alcoholism, substance abuse and schizophrenia in South-western American Indians. *Am. J. Med. Genet.* 74: 386–394.

Gordon, A. S., Yao, L., Jiang, Z., et al. (2001). Ethanol acts synergistically with a D2 dopamine agonist to cause translocation of protein kinase C. *Mol. Pharmacol.* 59: 153–160.

Gorwood, P. (2000). Contribution of genetics to the concept of risk status for alcohol dependence. *J. Soc. Biol.* 194: 43–49.

Grandy, D. K., Eubanks, J. H., Evans, G. A., et al. (1991). Detection and characterization of additional DNA polymorphisms in the dopamine D2 receptor gene. *Genomics* 10: 527–530.

Grandy, D. K., Litt, M., Allen, L., et al. (1989). The human dopamine D2 receptor gene is located chromosome 11 at q22-q23 identified a TaqI RFLP. *Am. J. Hum. Genet.* 45: 778–785.

Grant, B. F. (1994). Alcohol consumption, alcohol abuse and alcohol dependence: The United States as an example. *Addiction* 89: 1357–1365.

Grant, K. A. (1994). Emerging neurochemical concepts in the actions of ethanol at ligand-gated ion channels. *Behav. Pharmacol.* 5: 383–404.

Grzywacz, A,. Kucharska-Mazur, J., Samochowiec, J. (2009). Association studies of dopamine D4 receptor gene exon 3 in patients with alcohol dependence. *Psychiatr. Pol.* 42: 453–461.

Guindalini, C., Howard, M., Haddley, K., et al. (2006). A dopamine transporter gene functional variant associated with cocaine abuse in a Brazilian sample. *Proc. Natl. Acad. Sci. U.S.A.* 103: 4552–4557.

Hallbus, M., Magnusson, T., Magnusson, O. (1997). Influence of 5-HT1B/1D receptors on dopamine release in the guinea pig nucleus accumbens: A microdialysis study. *Neurosci. Lett.* 225: 57–60.

Hauge, X. Y., Grandy, D. K., Eubanks, J. H., et al. (2001). Detection and characterization of additional DNA polymorphisms in the dopamine D2 receptor gene. *Genomics* 10: 527–530.

Heinz, A., Sander, T., Harms, H., et al. (1996). Lack of allelic association of dopamine D1 and D2 (TaqIA) receptor gene polymorphisms with reduced dopaminergic sensitivity to alcoholism. *Alcohol. Clin. Exp. Res.* 20: 1109–1113.

Hietala, J., West, C., Syvalahti, E., et al. (1994). Striatal D2 dopamine receptor binding characteristics in vivo in patients with alcohol dependence. *Psychopharmacology* 116: 285–290.

Hill, S. Y., Hoffman, E. K., Zezza, N., et al. (2008). Dopaminergic mutations: Within-family association and linkage in multiplex alcohol dependence families. *Am. J. Med. Genet. B Neuropsychiatr. Genet.* 147B: 517–526.

Hirvonen, M., Laakso, A., Någren, K., et al. (2004). C957T polymorphism of the dopamine D2 receptor (DRD2) gene affects striatal DRD2 availability in vivo. *Mol. Psychiatry* 9: 1060–1061.

Hirvonen, M. M., Laakso, A., Någren, K., et al. (2009). C957T polymorphism of dopamine D2 receptor gene affects striatal DRD2 in vivo availability by changing the receptor affinity. *Synapse* 10: 907–912.

Hodge, C. W., Cox, A. A. (1998). The discriminative stimulus effects of ethanol are mediated by NMDA and GABA (A) receptors in specific limbic brain regions. *Psychopharmacology* 139: 95–107.

Hodge, C. W., Chappelle, A. M., Samson, H. H. (1996). Dopamine receptors in the medial prefrontal cortex influence ethanol and sucrose-reinforced responding. *Alcohol. Clin. Exp. Res.* 20: 1631–1638.

Huang, S. Y., Lin, W. W., Wan, F. J., et al. (2007). Monoamine oxidase-A polymorphisms might modify the association between the dopamine D2 receptor gene and alcohol dependence. *J. Psychiatry Neurosci.* 32: 185–192.

Huang, W., Payne, T. J., Ma, J. Z., et al. (2008). Significant association of ANKK1 and detection of a functional polymorphism with nicotine dependence in an African-American sample. *Neuropsychopharmacol.* 34: 319–330.

Huertas, E., Ponce, G., Koeneke, M. A., et al. (2010). The D2 dopamine receptor gene variant C957T affects human fear conditioning and aversive priming. *Genes Brain Behav.* 1: 103–109.

Johnson, K. (1996). *The dopamine D2 receptor as a candidate gene for alcoholism.* BS honors thesis, School of Community Medicine, University of New South Wales, Sydney, Australia.

Johnson, S. W., North, R. A. (1992). Opioids excite dopamine neurons by hyperpolarization of local interneurons. *J. Neurosci.* 12: 483–488.

Jonsson, E. G., Cichon, S., Gustavsson, P., et al. (2003). Association between a promoter dopamine D2 receptor gene variant and the personality trait detachment. *Biol. Psychiatry* 53: 577–584.

Kirsch, P., Reuter, M., Mier, D., et al. (2006). Imaging gene-substance interactions: The effect of the DRD2 TaqIA polymorphism and the dopamine agonist bromocriptine on the brain activation during the anticipation of reward. *Neurosci. Lett.* 405: 196–201.

Koehnke, M. D., Schich, S., Lutz, U., et al. (2002). Severity of alcohol withdrawal symptoms and the T1128C polymorphism of the neuropeptide Y gene. *J. Neural Transm.* 109: 1423–1429.

Koepp, M. J., Gunn, R. N., Lawrence, A. D., et al. (1998). Evidence for striatal dopamine release during a video game. *Nature* 393: 266–268.

Konkle, A. T., Bielajew, C. (2004). Tracing the neuroanatomical profiles of reward pathways with markers of neuronal activation. *Rev. Neurosci.* 15: 383–414.

Kono, Y., Yoneda, H., Sakai, T., et al. (1997). Association between early-onset alcoholism and the dopamine D2 receptor gene. *Am. J. Med. Genet.* 74: 179–182.

Koob, G. F. (2000). Neurobiology of addiction: Toward the development of new therapies. *Ann. N.Y. Acad. Sci.* 909: 170–185.

Koob, G. F. (2003). Alcoholism: Allostasis and beyond. *Alcohol. Clin. Exp. Res.* 27: 232–243.

Koob, G. F. (2008). Hedonic homeostatic dysregulation as a driver of drug-seeking behavior. *Drug Discov. Today Dis. Models* 5: 207–215.

Koob, G. F., Le Moal, M. (2001). Drug addiction, dysregulation of reward, and allostasis. *Neuropsychopharmacol.* 24: 97–129.

Kotler, M., Cohen, H., Segman, R., et al. (1997). Excess dopamine D4 receptor (D4DR) exon III seven repeat allele in opioid-dependent subjects. *Mol. Psychiatry* 2: 251–254.

Kotter, R., Stephan, K. E. (1997). Useless or helpful? The "limbic system" concept. *Rev. Neurosci.* 8: 139–145.

Kraschewski, A., Reese, J., Anghelescu, I., et al. (2009). Association of the dopamine D2 receptor gene with alcohol dependence: Haplotypes and subgroups of alcoholics as key factors for understanding receptor function. *Pharmacogenet. Genomics* 19: 513–527.

Kreek, M. J., Koob, G. F. (1998). Drug dependence: Stress and dysregulation of brain reward pathways. *Drug Alcohol. Depend.* 51: 23–47.

Kuikka, J. T., Baulieu, J. L., Hiltunen, J., et al. (1998). Pharmacokinetics and dosimetry of iodine-123 labeled PE2I in humans: A radioligand for dopamine transporter imaging. *Eur. J. Nucl. Med.* 25: 531–534.

Kuikka, J. T., Repo. E., Bergstrom, K. A., et al. (2000). Specific binding and laterality of human extrastriatal dopamine D2/D3 receptors in the late onset type 1 alcoholic patients. *Neurosci. Lett.* 292: 57–59.

Laakso, A., Pohjalainen, T., Bergman, J., et al. (2005). The A1 allele of the human D2 dopamine receptor gene is associated with increased activity of striatal l-amino acid decarboxylase in healthy subjects. *Pharmacogenet. Genomics* 15: 387–391.

Laine, T. P., Ahonen, A., Rasanen, P., et al. (1999). Dopamine transporter availability and depressive symptoms during alcohol withdrawal. *Psychiatry Res.* 90: 153–157.

Laine, T. P., Ahonen, A., Räsänen, P., et al. (2001). Dopamine transporter density and novelty seeking among alcoholics. *J. Addict. Dis.* 20: 91–96.

Laine, T. P. J., Ahonen, A., Torniainen, P., et al. (1999). Dopamine transporter increase in human brain after alcohol withdrawal. *Mol. Psychiatry* 4: 189–191.

Lawford, B. R., Young, R. M., Rowell, J. A., et al. (1995). Bromocriptine in the treatment of alcoholics with the D2 dopamine receptor A1 allele. *Nat. Med.* 1: 337–341.

Lawford, B. R., Young, R. M., Rowell, J. A., et al. (1997). Association of the D2 dopamine receptor A1 allele with alcoholism: Medical severity of alcoholism and type of controls. *Biol. Psychiatry* 41: 386–393.

Lee, S. S., Lahey, B. B., Waldman, I., et al. (2007). Association of dopamine transporter genotype with disruptive behavior disorders in an eight-year longitudinal study of children and adolescents. *Am. J. Med. Genet. B Neuropsychiatr. Genet.* 144B: 310–317.

Lee, S. Y., Hahn, C. Y., Lee, J. F., et al. (2009). MAOA-uVNTR polymorphism may modify the protective effect of ALDH2 gene against alcohol dependence in antisocial personality disorder. *Alcohol. Clin. Exp. Res.* 33: 985–990.

Lee, S. Y., Hahn, C. Y., Lee, J. F., et al. (2010). MAOA interacts with the ALDH2 gene in anxiety-depression alcohol dependence. *Alcohol. Clin. Exp. Res.* 34: 1212–1218.

Li, C. Y., Mao, X., Wei, L. (2008). Genes and (common) pathways underlying drug addiction. *PLoS Comput. Biol.* 4: e2.

Little, K. Y., McLaughlin, D. P., Zang, L., et al. (1998). Brain dopamine transporter messenger RNA and binding sites in cocaine users. *Arch. Gen. Psychiatry* 55: 793–799.

Lowinson, J., Ruiz, P., Millman, R., et al. (1997). *Substance abuse: A comprehensive textbook.* Baltimore, MD: William & Wilkins.

Lu, R. B., Ko, H. C., Chang, F. M., et al. (1996). No association between alcoholism and multiple polymorphism at the dopamine D2 receptor gene (DRD2) in three distinct Taiwanese populations. *Biol. Psychiatry* 39: 419–429.

Lusher, J. M., Chandler, C., Ball, D. (2001). Dopamine D4 receptor gene (DRD4) is associated with novelty seeking (NS) and substance abuse: The saga continues. *Mol. Psychiatry* 6: 497–499.

McBride, W. J., Murphy, J. M., Ikemoto, S. (1999). Localization of brain reinforcement mechanisms: Intracranial self-administration and intracranial place-conditioning studies. *Behav. Brain Res.* 101: 129–152.

McGeary, J. E., Esposito-Smythers, C., Spirito, A., et al. (2007). Associations of the dopamine D4 receptor gene VNTR polymorphism with drug use in adolescent psychiatric inpatients. *Pharmacol. Biochem. Behav.* 86: 401–406.

Melis, M., Spigra, S., Diana, M. (2005). The dopamine hypothesis of drug addiction: Hypodopaminergic state. *Int. Rev. Neurobiol.* 63: 101–154.

Merikangas, K. R. (1990). The genetic epidemiology of alcoholism. *Psychol. Med.* 20: 11–22.

Merlo, L. J., Gold, M. S. (2008). Special report: Frontiers in psychiatric research: Addiction research: The state of the art in 2008. *Psychiatric Times* 25: 52–57.

Michelhaugh, S. K., Fiskerstrand, C., Lovejoy, E., et al. (2001). The dopamine transporter gene (SLC6A3) variable number of tandem repeats domain enhances transcription in dopamine neurons. *J. Neurochem.* 79: 1033–1038.

Miller, D., Manka, M., Miller, M., et al. (2010). Acute intravenous synaptamine complex [KB220]™ variant "normalizes" abnormal neurological activity in protracted abstinence of alcohol and opiate patients using quantitative electroencephalographic (QEEG) and neurotransmitter genetic analysis: Part Two, pilot study with two case reports. *Postgrad. Med.* 122: 188–213.

Mizrahi, R., Houle, S., Vitcu, I., et al. (2010). Side effects profile in humans of (11)C-(+)-PHNO, a dopamine D(2/3) agonist ligand for PET. *J. Nucl. Med.* 51: 496–497.

Nairn, A. C., Svenningsson, P., Nishi, A., et al. (2004). The role of DARPP-32 in the actions of drugs of abuse. *Neuropharmacology* 47 (Suppl. 1): 14–23.

Namkoong, K., Cheon, K. A., Kim, J. W., et al. (2008). Association study of dopamine D2, D4 receptor gene, $GABA_A$ receptor beta subunit gene, serotonin transporter gene polymorphism with children of alcoholics in Korea: A preliminary study. *Alcohol.* 42: 77–81.

NCI-NHGRI Working Group on Replication in Association Studies (2007). Replicating genotype-phenotype associations. *Nature* 447: 655–660.

Neiswanger, K., Kaplan, B. B., Hill, S. Y. (1995). What can the DRD2/alcoholism story teach us about association studies in psychiatric genetics? *Am. J. Med. Genet.* 60: 272–275.

Neville, M. J., Johnstone, E. C., Walton, R. T. (2004). Identification and characterization of ANKK1: A novel kinase gene closely linked to DRD2 on chromosome band 11q23.1. *Hum. Mutat.* 23: 540–545.

Nikoshkov, A., Drakenberg, K., Wang, X., et al. (2008). Opioid neuropeptide genotypes in relation to heroin abuse: Dopamine tone contributes to reversed mesolimbic proenkephalin expression. *Proc. Natl. Acad. Sci. U.S.A.* 105: 786–791.

Nishi, A., Snyder, G. L., Greengard, P. (1997). Bidirectional regulation of DARPP-32 phosphorylation by dopamine. *J. Neurosci.* 17: 8147–8155.

Noble, E. P. (2000). The DRD2 gene in psychiatric and neurological disorder and its phenotypes. *Pharmacogenomics* 1: 309–333.

Noble, E. P. (2003). D2 dopamine receptor gene in psychiatric and neurologic disorders and its phenotypes. *Am. J. Med. Genet.* 116B: 103–125.

Noble, E. P., Blum, K., Ritchie, T., et al. (1991). Allelic association of the D2 dopamine receptor gene with receptor-binding characteristics. *Arch. Gen. Psychiatry* 48: 648–654.

Noble, E. P., Noble, R. E., Ritchie, T., et al. (1994a). D2 receptor gene and obesity. *Int. J. Eating Disorders* 15: 205–217.

Noble, E. P., Syndilko, K., Fitch, et al. (1994b). D2 dopamine receptor *Taq*1 A alleles in medically ill alcoholic and nonalcoholic patients. *Alcohol Alcohol.* 129: 729–744.

Noble, E. P., Zhang, X., Ritchie, T., et al. (1998). D2 dopamine receptor and GABA(A) receptor beta3 subunit genes and alcoholism. *Psychiatry Res.* 81: 133–147.

Noble, E. P., Blum, K., Ritchie, T., et al. (1991). Allelic association of the D2 dopamine receptor gene with receptor-binding characteristics. *Arch. Gen. Psychiatry* 48: 648–654.

Oak, J. N., Oldenhof, J., Van Tol, H. H. (2000). The dopamine D(4) receptor: One decade of research. *Eur. J. Pharmacol.* 405: 303–327.

Oakley, N. R., Hayes, A. G., Sheehan, M. J. (1991). Effect of typical and atypical neuroleptics on the behavioural consequences of activation by muscimol of mesolimbic and nigrostriatal dopaminergic pathways in the rat. *Neuropharmacology* 105: 204–208.

O'Hara, B. F., Smith, S. S., Bird, G., et al. (1993). Dopamine D2 receptor RFLPs, haplotypes and their association with substance use in black and Caucasian research volunteers. *Hum. Hered.* 43: 209–218.
Panagis, G., Nomikos, G. G., Miliaressis, E., et al. (1997). Ventral pallidum self-stimulation induces stimulus dependent increase in c-fos expression in reward-related brain regions. *Neuroscience* 77: 175–186.
Parsian, A., Cloninger, C. R., Zhang, Z. H. (2000). Functional variant in the DRD2 receptor promoter region and subtypes of alcoholism. *Am. J. Med. Genet.* 96: 407–411.
Parsian, A., Todd, R. D., Devor, E. J., et al. (1991). Alcoholism and alleles of the human D2 dopamine receptor locus: Studies of association and linkage. *Arch. Gen. Psychiatry* 48: 655–663.
Pato, C. N., Macciardi, F., Pato, M. T., et al. (1993). Review of the putative association of dopamine D2 receptor and alcoholism. *Am. J. Med. Genet. Neuropsychiatr. Genet.* 48: 78–82.
Peciña, S., Smith, K. S., Berridge, K. C. (2006). Hedonic hot spots in the brain. *Neuroscientist* 12: 500–511.
Pickens, R. W., Svikis, D. S., McGue, M., et al. (1991). Heterogeneity in the inheritance of alcoholism. *Arch. Gen. Psychiatry* 43: 19–28.
Plomin, R., Owen, M. J., McGuffin, P. (1994). The genetic basis of complex human behaviors. *Science* 264: 1733–1739.
Pohjalainen, T., Rinne, J. O., Någren, K., et al. (1998). The A1 allele of the human D2 dopamine receptor gene predicts low D2 receptor availability in healthy volunteers. *Mol. Psychiatry.* 3: 256–260.
Ponce, G., Jimenez-Arriero, M. A., Rubio, G., et al. (2003). The A1 allele of the DRD2 gene (Taq1 A polymorphisms) is associated with antisocial personality in a sample of alcohol-dependent patients. *Eur. Psychiatry* 18: 356–360.
Ratsma, J. E., van der Stelt, O., Schoffelmeer, A. N. M., et al. (2001). P3 event-related potential, dopamine D2 A1 allele, and sensation-seeking in adult children of alcoholics. *Alcohol. Clin. Exp. Res.* 25: 960–967.
Repo, E., Kuikka, J. T., Bergstrom, K. A., et al. (1999). Dopamine transporter and D2-receptor density in late-onset alcoholism. *Psychopharmacology* 147: 314–318.
Reuter, J., Raedler, R., Rose, M., et al. (2005) Pathological gambling is linked to reduced activation of the mesolimbic system. *Nat. Neurosci.* 8: 147–148.
Robbins, T. W., Everitt, B. J. (1996). Neurobiobehavioural mechanisms of reward and motivation. *Curr. Opin. Neurobiol.* 6: 228–236.
Rommelspacher, H., Raeder, C., Kaulen, P., et al. (1992). Adaptive changes of dopamine-D2 receptors in rat brain following ethanol withdrawal: A quantitative autoradiographic investigation. *Alcohol* 9: 355–362.
Romstad, A., Dupont, E., Krag-Olsen, B., et al. (2003). Dopa-responsive dystonia and Tourette syndrome in a large Danish family. *Arch. Neurol.* 60: 618–622.
Rothman, R. B., Blough, B. E., Baumann, M. H. (2007). Duel dopamine/serotonin releasers as potential medications for stimulant and alcohol addictions. *AAPS J.* 9: E1–E10.
Rowe, D. C. (1986). Genetic and environmental components of antisocial behavior: A study of 265 twin pairs. *Criminology* 24: 513–532.
Saiz, P. A., Garcia-Portilla, M. P., Florez, G., et al. (2009). Differential role of serotonergic polymorphisms in alcohol and heroin dependence. *Prog. Neuropsychopharmacol. Biol. Psychiatry* 33: 695–700.
Samochowiec, J., Kucharska-Mazur, J., Grzywacz, A., et al. (2006). Family-based and case-control study of DRD2, DAT, 5HTT, COMT genes polymorphisms in alcohol dependence. *Neurosci. Lett.* 410: 1–5.
Schoots, O., Van Tol, H. H. (2003). The human dopamine D4 receptor repeat sequences modulate expression. *Pharmacogenomics J.* 3: 343–348.
Seneviratne, C., Huang, W., Ait-Daoud, N., et al. (2009). Characterization of a functional polymorphism in the 3' UTR of SLC6A4 and its association with drinking intensity. *Alcohol. Clin. Exp. Res.* 33: 332–339.
Serý, O., Didden, W., Mikes, V., et al. (2006a). The association between high-activity COMT allele and alcoholism. *Neuro. Endocrinol. Lett.* 27: 231–235.
Serý, O., Drtílková, I., Theiner, P., et al. (2006b). Polymorphism of DRD2 gene and ADHD. *Neuro. Endocrinol. Lett.* 27: 236–240.
Shih, J. C. (1991). Molecular basis of human MAO A and B. *Neuropsychopharmacol.* 4: 1–7.
Shih, J. C., Chen, K., Ridd, M. J. (1999). Monoamine oxidase: From genes to behavior. *Annu. Rev. Neurosci.* 22: 197–217.
Singer, H. S. (2010). Treatment of tics and Tourette syndrome. *Curr. Treat. Options Neurol.* 12: 539–561.
Smith, K. M., Daly, M., Fischer, M., et al. (2003). Association of the dopamine beta hydroxylase gene with attention deficit hyperactivity disorder: Genetic analysis of the Milwaukee longitudinal study. *Am. J. Med. Genet.* 119: 77–85.
Smith, M., Wasmuth, J., McPherson, J. D. (1989). Cosegregation of an 11q22.3-9p22 translocation with affective disorder: Proximity of the dopamine D2 receptor gene relative to the translocation breakpoint. *Am. J. Hum. Genet.* 45: A220.

Smith, S. S., O'Hara, B. F., Persico, A. M., et al. (1992). Genetic vulnerability to drug abuse: The D2 dopamine receptor Taq I B1 restriction fragment length polymorphism appears more frequently in polysubstance abusers. *Arch. Gen. Psychiatry* 49: 723–727.
Suarez, B. K., Parsian, A., Hampe, C. L., et al. (1994). Linkage disequilibria at the D2 dopamine receptor locus (DRD2) in alcoholics and controls. *Genomics* 19: 12–20.
Suhara, T., Yasuno, F., Sudo, Y., et al. (2001). Dopamine D2 receptors in the insular cortex and the personality trait of novelty seeking. *Neuroimage* 13: 891–895.
Thanos, P. K., Michaelides, M., Umegaki, H., et al. (2008). D2R DNA transfer into the nucleus accumbens attenuates cocaine self-administration in rats. *Synapse* 62: 481–486.
Thanos, P. K., Rivera, S. N., Weaver, K., et al. (2005). Dopamine D2R DNA transfer in dopamine D2 receptor-deficient mice: Effects on ethanol drinking. *Life Sci.* 77: 130–139.
Thanos, P. K., Volkow, N. D., Freimuth, P., et al. (2001). Overexpression of dopamine D2 receptors reduces alcohol self-administration. *J. Neurochem.* 78: 1094–1103.
Theile, J. W., Morikawa, H., Gonzales, R. A., et al. (2008). Ethanol enhances GABAergic transmission onto dopamine neurons in the ventral tegmental area of the rat. *Alcohol. Clin. Exp. Res.* 32: 1040–1048.
Thut, G., Schultz, W., Roelcke, U., et al. (1997). Activation of the human brain by monetary reward. *Neuroreport* 8: 1225–1228.
Tiihonen, J., Kuikka, J., Bergstrom, K., et al. (1995). Altered striatal dopamine re-uptake site densities in habitually violent and non-violent alcoholics. *Nat. Med.* 1: 654–657.
Tupala, E., Hall, H., Bergstrom, K., et al. (2003). Dopamine D2 receptors and transporters in type 1 and 2 alcoholics measured with human whole hemisphere autoradiography. *Hum. Brain Mapp.* 20: 91–102.
Tupala, E., Hall, H., Bergstrom, K., et al. (2001a). Dopamine D(2)/D(3)-receptor and transporter densities in nucleus accumbens and amygdala of type 1 and type 2 alcoholics. *Mol. Psychiatry* 6: 261–267.
Tupala, E., Kuikka, J.T., Hall, H., et al. (2001b). Measurement of the striatal dopamine transporter density and heterogeneity in type 1 alcoholics using human whole hemisphere autoradiography. *Neuroimage* 1: 87–94.
Turner, E., Ewing, J., Shilling, P., et al. (1992). Lack of association between an RFLP near the D2 dopamine receptor gene and severe alcoholism: Clinical Center for Research on Alcoholism, San Diego Veterans Affairs Medical Center, CA. *Biol. Psychiatry* 31: 285–290.
Uhl, G., Blum, K., Noble, E., et al. (1993). Substance abuse vulnerability and D2 receptor genes. *Trends Neurosci.* 16: 83–88.
Vandenbergh, D. J. (1998). Molecular cloning of neurotransmitter transporter genes: Beyond coding region of cDNA. *Methods Enzymol.* 296: 498–514.
Vandenbergh, D. J., Bennett, C. J., Grant, M. D., et al. (2002). Smoking status and the human dopamine transporter variable number of tandem repeats (VNTR) polymorphism: Failure to replicate and finding that never-smokers may be different. *Nicotine Tob. Res.* 4: 333–340.
Vandenbergh, D. J., Rodriguez, L. A., Miller, I. T., et al. (1997). High-activity catechol-*O*-methyltransferase allele is more prevalent in polysubstance abusers. *Am. J. Med. Genet.* 74: 439–442.
Vanyukov, M. M., Maher, B. S., Devlin, B., et al. (2004). Haplotypes of the monoamine oxidase genes and the risk for substance use disorders. *Am. J. Med. Genet. B Neuropsychiatr. Genet.* 125B: 120–125.
Volkow, N. D., Chang, L., Wang, G. J., et al. (2001). Low level of brain dopamine D2 receptors in methamphetamine abusers: Association with metabolism in the orbitofrontal cortex. *Am. J. Psychiatry* 158: 377–382.
Volkow, N. D., Fowler, J. S., Wang, G. J. (2002). Role of dopamine in drug reinforcement and addiction in humans: Results from imaging studies. *Behav. Pharmacol.* 13: 355–366.
Volkow, N. D., Wang, G. J., Fowler, J. S., et al. (1996). Decreases in dopamine receptors but not in dopamine transporters in alcoholics. *Alcohol. Clin. Exp. Res.* 20: 1594–1598.
Wang, T., Franke, P., Neidt, H., et al. (2001). Association study of the low-activity allele of catechol-*O*-methyltransferase and alcoholism using a family-based approach. *Mol. Psychiatry* 6: 109–111.
White, F. J. (1996). Synaptic regulation of mesocorticolimbic dopamine neurons. *Annu. Rev. Neurosci.* 19: 405–436.
Wightman, R. M., Robinson, D. L. (2002). Transient changes in mesolimbic dopamine and their association with "reward." *J. Neurochem.* 82: 721–735.
Xu, K., Lichterman, D., Kipsky, R. H., et al. (2004). Association of specific haplotypes of D2 dopamine receptor gene with vulnerability to heroin dependence in distinct populations. *Arch. Gen. Psychiatry* 61: 567–606.
Xuei, X., Flury-Wetherill, L., Dick, D., et al. (2010). GABRR1 and GABRR2, encoding the GABA-A receptor subunits rho1 and rho2, are associated with alcohol dependence. *Am. J. Med. Genet. B Neuropsychiatr. Genet.* 153B: 418–427.

Yang, B. Z., Kranzler, H. R., Zhao, H., et al. (2008). Haplotypic variants in DRD2, ANKK1, TTC12, and NCAM1 are associated with comorbid alcohol and drug dependence. *Alcohol. Clin. Exp. Res.* 32: 2117–2127.

Zhu, Q. S., Shih, J. C. (1997). An extensive repeat structure down-regulates human monoamine oxidase A promoter activity independent of an initiator-like sequence. *J. Neurochem.* 69: 1368–1373.

Section V

Future Perspective

27 Future Perspective
Paving a Path to Optimization of Health

Debmalya Barh
Institute of Integrative Omics and Applied Biotechnology
Nonakuri, India

Margaret A. Madigan
LifeGen Inc.
San Diego, California

Kenneth Blum
University of Florida College of Medicine and McKnight Brain Institute
Gainesville, Florida

CONTENTS

27.1 HISTORY

Although the future looks bright, following the successes of the Human Genome Project and a decade of arduous research from world class scientists, there is still a long way to go. To many, the Human Genome Project has failed so far to provide the medical miracles that it promised in 2000. In that year, leaders of the Human Genome Project announced completion of the first rough draft of the human genome. One of the predictions was that follow-up research could pave the way to personalized medicine within 10 years. Accordingly, few medical applications have emerged, although important insights have already revolutionized medical research, including psychiatric genetics. Although some leading geneticists argue that a key strategy known as the *common variant hypothesis* for seeking medical insights into complex diseases such as addiction is fundamentally flawed, others say that the strategy is valid.

The obvious way to make progress was to sequence the full genome of diseased and healthy individuals with powerful computers and develop a catalog of genetic variation by identifying DNA variations that turned up in patients with the given disease but not in control subjects. Using sophisticated techniques, many scientists embarked on large-scale studies, known as genome-wide association studies (GWAS) that relied on landmarks in DNA known as single-nucleotide polymorphisms or SNPs, to uncover gene variants important in disease. These concepts led to the development of HapMap, the 1000 Genomes Project, the Human Microbe Project, and the Cancer Genome Atlas, which have been made available as online databases for researchers in the field. Thus, in the past 5 years genome-wide association studies have looked at hundreds of thousands of common SNPs in the genomes of tens of thousands of individual subjects and controls in search of SNPs linked to common disease.

This strategy has revealed important clues and uncovered pathways for a number of common diseases such as addictive behaviors, schizophrenia, type 2 diabetes, Alzheimer's disease, and hypertension. Although the human genome has been cracked, allowing scientists to look at the entire compliment of common genetic variants, it is astounding that this discovery has not led to the hoped-for major breakthroughs. Moreover, David Botstein of Princeton University (Hall, 2010), discussing the HapMap, stated, "It had to have (been) done. If it had not been tried, no one would have known that it didn't work." He called the $138 million HapMap a "magnificent failure." Walter Bodmer, who first proposed the genome project, also believes that the common variant hypothesis is a dead end and suggests that the vast majority of common variants have shed no light on the biology of disease (Hall, 2010).

Gresham et al. (2008) have suggested that DNA microarrays present an alternative way to study the differences between closely related genomes by looking at the main forms of genomic variation (amplifications, deletions, insertions, rearrangements, and base-pair changes). However, it is plausible that new approaches should consider re-exploring the older candidate approach instead of relying exclusively on GWAS for an answer.

The current argument over the common variant hypothesis suggests at least one way forward for solving what many are calling the missing heritability problem, that is, to search for rare genetic variants (Manolio et al., 2009). McClellan and King et al. (2010) suggest that complex diseases may be produced by many different mutations on many different genes and that high-impact genes are rare, are relatively recent in the gene pool, and could lead researchers to disease-related molecular pathways and, as a consequence, new therapies. An example that illustrates this paradigm shift is the research of Hobbs and Cohen, who performed intensive medical workups for cardiac disease on 3,500 residents of Dallas County and compiled a detailed database of individual physical traits to identify people with particularly dramatic phenotypes (Horton et al., 2007; Cohen et al., 2006). They then focused on those with either extremely high or low numbers for high-density lipoproteins (HDL) or for low-density lipoproteins (LDL). Aware that at least three genes are implicated in rare cholesterol metabolism disorders, they compared DNA sequences from both groups and identified rare variants linked to the extremely depressed HDL levels. Their prize discovery was that their analysis of the *PCSK9* sequence found two mutations that silenced the gene and correlated with the low LDL levels. Drug companies are already testing molecules that perturb the molecular pathway or shut off the *PCSK9* gene (Horton et al., 2007; Cohen et al., 2006).

Another example of discovery in genomic research is that the vast areas of DNA that do not code for proteins once thought of as junk are now known to conceal important regulatory regions that produce small bits of RNA that can interfere with gene expression. This epigenetic RNA does not change the DNA sequence but can influence gene expression and can be modified by environmental factors and be passed on to offspring (Nadeau, 2009). As researchers catch up within different aspects of this exciting field, we find that the more we uncover, the greater is our appreciation of the complexity of biological systems and the field we refer to herein as *integrative omics*.

27.2 SCIENTIFIC AND ETHICAL STANDARDS

As scientists we must transcend the morality-influenced laws against stem-cell research, for example, by continually building public trust through scholarly reporting on the importance of this potentially life-saving research and eliminating unwanted fear. Indeed, stem-cell therapy has been used in the successful treatment of myelodysplastic syndrome and leukemia. Meanwhile, within the challenge to the viability of these cell lines as treatment for degenerative pathology remains the need to identify transcription factors that signal these cells to perform their new roles, for example, as endocrine cells.

Another ethical issue involves encouraging the avoidance of animal models (replace, reduce, and refine) by increased use of *in vitro* models for toxicogenomic assays. A surprising result from the survey carried out in 2010 by Pettit et al. is that *in vitro* studies using cell lines, primary cultures, and organ cultures are presently favored in genomic analysis. The reasons include reduced costs, increased screening speed using smaller amounts of material, and simplicity of data extraction and interpretation, because data interpretation can be straightforward when the research questions focus on receptor or pathway activation or inhibition. However, the authors warn that *in vitro* models may lack the complexity of *in vivo* models required for research questions such as those involving bioactivation or complex metabolic cascades.

Following the successful application of pharmacogenomic analysis to the identification of a number of molecular mechanisms affecting an individual's drug response, barriers to the use of genetic testing and analysis in personalized medicine include: the cost, the lack of patient education about the importance of genetic testing in response to drug treatment, and the acceptability of gene testing to the individual. Ways to assess and communicate risk information obtained from association and drug-response studies have been neglected to date. This communicative lack has been recently addressed in the United States by the Genetic Information Nondiscrimination Act of 2008 (enforced November 2009). Although this new law provides some guidelines to protect an individuals' right not to be penalized because of their genome, most laypeople do not understand all the related caveats and might still shy away from important genetic testing. Hence, communication of risk information in diagnostic medicine also demands attention to the processes involved in the production of knowledge and the human values embedded in scientific practice, for example, why, how, by whom, and to what ends are association and linkage studies conducted and standards developed (or not).

27.3 TRANSLATIONAL RESEARCH

Genomic biomarkers are an objective measure or indicator of normal biologic processes, pathogenic processes, or pharmacologic responses to a therapeutic intervention, and integrative omics technologies are relatively new biomarker discovery tools. Biomarkers, both at the level of phenotype and gene expression can be used to elucidate the underlying molecular pathways and determine disease conditions such as type 2 diabetes cardiac channelopathies, AIDS/HIV, and psychiatric illnesses.

27.3.1 Genotype-Phenotype Association Studies

Genotype-phenotype association studies are a centerpiece of translational research in omics science. Association studies decisively influence which genetic loci become genetic tests in the clinic or products in the genetic test marketplace. This is being developed for highly complex mental disorders such as substance use disorder, attention-deficit hyperactivity disorder, binge eating, and schizophrenia and for even general happiness. The challenge here involves the interpretation of large number of identified gene polymorphisms that are typically observed in the distribution of quantitative traits in genetic association studies. They may have a polygenic/multifactorial phenotype that may overlap, leading to spurious results.

27.3.2 Gene Expression Biomarkers

The utility of gene expression biomarkers has been acknowledged for many years (Van Leeuwen et al., 1986). Until recently the use of single technologies to determine biomarkers has been the focus of the majority of efforts. However, changes can be detected concerning gene or protein expression or on metabolic profiles, so that a more promising and predictive approach has been to integrate methodologies to discover and evaluate reproducible biomarkers for disease, risk assessment, or drug toxicity.

The authors of this book have elucidated the potential of various omics technologies in regard to genomic-biomarker research. Gene-expression profiling can be used to determine which genes are differentially expressed in disease conditions; proteomic analysis holds the promise of creating noninvasive tests that monitor biomarkers in body fluids; toxicogenomics has the potential to speed up clinical trials, clinical monitoring, and toxicity assessment in drug development, detect the presence of environmental contaminants, and improve compound selection (Pettit et al., 2010); whereas metabolic biomarkers, in the low-molecular-weight range, may contain disease-specific information (Hu et al., 2009) and are offered as highly specific metrics for testing therapeutic efficacy while not being comprised of any of the drug metabolites. The products of these emerging omics technologies that are reaching the market include companion diagnostics and targeted drug discovery for devastating diseases and complex mental conditions from cancer to cardiovascular, autoimmune diseases, and diabetes and to reward deficiency syndrome (obsessive, compulsive, and impulsive disorders) and other psychiatric diseases, such as schizophrenia and manic-depressive illness.

27.4 THE FUTURE

The chapters of this book include applicable discussions of the challenges faced by multiomics scientists involved in integrative omics research. Some of these relevant and perplexing questions provide unique opportunities for future research, as explored below.

27.4.1 Evolutionary Systems Biology

During the past decade, microevolution of intermediary metabolism has become an important new research focus at the interface between metabolic biochemistry and evolutionary genetics. Increasing recognition of the importance of integrative omic studies in evolutionary analysis, the rising interest in evolutionary systems biology, and the development of various omics technologies have all contributed significantly to this developing interface. Some of these studies have provided a new perspective on important evolutionary topics that have not been investigated extensively from a biochemical perspective (hybrid breakdown and parallel evolution).

Recent studies have provided new data that augment previous biochemical information, resulting in a deeper understanding of evolutionary mechanisms (allozymes and biochemical adaptation to climate, life-history evolution, marine biology, and the genetics of adaptation). Finally, other studies have provided new insights into how the function or position of an enzyme in a pathway influences its evolutionary dynamics, in addition to providing powerful experimental models for investigations of network evolution. One potential result of this new integrative approach involves the complex comprehension of, for example, microevolution. Such microevolutionary studies of metabolic pathways will undoubtedly become increasingly important in the future because of the central importance of intermediary metabolism in organismal fitness, the wealth of biochemical data being provided by various omics technologies, and the increasing influence of integrative and systems perspectives in biology (Zera, 2011).

27.4.2 Bioinformatics

Bioinformatics (computational tools and omic databases available on the Internet) have fueled the growth of the biological sciences and integrative omics in particular. The time and costs of biological research have been reduced, but there are challenges in the future, including adequate and accessible data storage, overcoming limitations to computing powers, and a lack of algorithms sophisticated enough to reflect the complexity of biological systems. In addition, sophisticated algorithms take longer to run, so that presently there is a trade-off between accuracy and pace/cost based on a lack of computational power. Bioinformatic analysis also relies on the quality of the experimental data. For example, incorrect sequence alignment will affect outcome.

Another challenge is that in a model, if the sample size is small and has too many parameters, the outcome may fit the original data but cannot be generalized to predict the outcome of other independent data. Validation of the entire data-analysis process and validation of independent datasets can be used to overcome this problem, although a lack of standardization in the development of public databases remains a barrier to effective sharing of microarray data. Finally, bioinformatic analysis relies on the quality of data interpretation. Development of the next generation of trained human resources in bioinformatics is required in order to affect the quality of data analysis now and in the future.

27.4.3 Functional Genomics

Cellular functions are very complex and involve a multitude of biochemical processes—not just genetic processes (DNA) but epigenetic processes such as transcriptional, translational, and allosteric regulation as well. The application of integrated multiomics, which include genetic, proteomic, and metabolic approaches, has already become the key to successes in deciphering some regulatory mechanisms and complex metabolic networks in a variety of cellular organisms (Hegde, White et al., 2003; Mootha, Bunkenborg, et al., 2003; Ray, Mootha, et al., 2003; Alter and Golub, 2004).

As an example, scientists have recently embraced this new integrative approach of exploiting various aspects of functional genomics in complex mental disorders such as bipolar disorder. Progress in understanding the genetic and neurobiological basis of bipolar disorder(s) has come from both human studies and animal-model studies. Until recently, the lack of concerted integration between the two approaches has been hindering the pace of discovery or, more exactly, constituted a missed opportunity to accelerate our understanding of this complex and heterogeneous group of disorders. Le-Niculescu et al. (2007) has helped overcome this *lost-in-translation* barrier by developing an approach called convergent functional genomics. The approach integrates animal-model gene-expression data with human genetic linkage/association data, as well as human tissue (postmortem brain and blood) data. This Bayesian strategy for cross-validating findings extracts meaning from large datasets and prioritizes candidate genes, pathways, and mechanisms for subsequent targeted, hypothesis-driven research.

27.4.4 Toxicogenomics

Toxicogenomics will be applied to in a variety of areas of safety and risk assessment including the future evaluation of chemical compounds for compound selection and for product development as well as xenobiotic exposures and, at early preclinical stages, the identification of drug-target and off-target effects. The identification of species differences in toxicity and biomonitoring of populations should be followed by assessment of risk in sensitive/vulnerable populations.

The necessity of exploiting bioinformatics, understanding toxicity of biological substrates, and using a systems biological approach has been clearly documented. The National Academy of Sciences publication *Toxicity testing in the 21st century: A vision and a strategy* proposes a paradigm shift in toxicology from current animal-based testing towards the application of emerging technologies, i.e., assays based on human cells or nonmammalian models, high-throughput testing, omics approaches, systems biology, and computational modeling. These technologies can be used to identify how chemicals interact with cellular response networks and alter them to toxicity pathways. According to the authors, such a new paradigm will provide a better scientific understanding and more adequate data to predict the adverse effects of chemicals on human health (van Vliet, 2010). In light of this, the results from the recent survey carried out by Pettit et al. (2010) showing that *in vitro* studies using cell lines, primary cultures, and organ cultures are presently favored in genomic analysis are not surprising. In addition to the cost of using genetic profiles obtained with microarrays for toxicogenomics, the challenges include the integration, validation, and storage of data (National Research Council, 2007). According to the National Institute of Environmental Health Science (NIEHS) Health and Environmental Sciences Institute (HESI), the most critical aspects of technology development that will drive toxicogenomics in the next decade will be as follows:

1. New sequencing technologies that offer the prospect of cost-effective individual whole-genome sequencing and comprehensive genotype analysis.
2. Array-based whole-genome scanning for variations in individual genes, known as SNPs, will dramatically increase throughput for genotyping in population studies.
3. Advances in nuclear magnetic resonance and mass spectrometry instrumentation will enable high-sensitivity analyses of complex collections of metabolites and proteins and quantitative metabolomics and proteomics.
4. New bioinformatic tools, database resources, and statistical methods will integrate data across technology platforms and link phenotypes and toxicogenomic data (Pettit et al., 2010).
5. The role of clinical research will expand in environmental health sciences.
6. Environmental toxicants will be used to understand basic mechanisms in human biology.
7. Integrated environmental health research programs will be developed to address the cross-cutting problems in human biology and human disease.
8. Community-linked research will be improved and expanded.
9. Sensitive markers of environmental exposure, early (preclinical) biological response, and genetic susceptibility will be developed.
10. The next generation of environmental-health scientists will be recruited and trained.
11. The development of partnerships between the NIEHS and other divisions of the National Institutes of Health, national and international research agencies, academia, industry, and community organizations will be fostered to improve human health.

27.4.5 Nanotechnology

It is noteworthy that the delivery of safe, effective future pharmaceutical/nutraceutical agents may include nano-sized particles. Although these delivery systems promise high absorption, there is emerging concern about safety. In fact, the National Toxicology Program Nanotechnology Safety Initiative, a broad-based research program to look at how nanoscale materials interact with biological systems, evaluates their toxicological properties and addresses potential health hazards associated with their manufacture and use. Presently, there is intense interest in future research and development of nanotechnology. One of us (.K.B) along with others is exploring the use of novel liposomal-colloidal nontoxic nano-sized particles in the range of 100–900 nm (average, 300 nm) as a novel delivery system for products having central nervous system function (Chen et al., 2011).

27.4.6 Tumorigenesis

In studying the molecular genetics of human cancers, a combination of whole-genome sequencing techniques with *N*-ethyl-*N*-nitrosourea mutagenesis in genetically homogenous mice populations has become the tool of choice for the identification of new gene functions. The barrier to the development of new drugs and novel diagnostic and therapeutic procedures against cancer is in understanding the reasons underlying the mechanism of the role reversal involved in tumor-suppressor gene function, where inactivation causes cancer, and overexpression can be tumorigenic.

27.4.7 Omics and Imaging Modalities

In their chapter "Safety in Diagnostic Imaging Techniques Used in Omics," Visaria et al. point out that more work to enhance understanding of the physiological responses activated during the imaging procedures must be done to better protect those using imaging technologies.

27.5 CONCLUSION

27.5.1 The Role of Academic, Industrial, and Governmental Organizations in the Future of Omics Research

We as scientists must begin to think out of the box and, while taking pride in individual accomplishments, continue to develop scientific interactions through publications, conferences, consortium group discussions, and media transparency. Our unified goals are to embrace the sustainability of the omics fields and potentially suggest legislative initiatives to create a multidisciplinary oversight body, at arm's length from conflict of interests; to carry out independent, impartial, and transparent innovation analyses and prospective technology assessment; and to highlight the important health benefit breakthroughs in the field that will garnish widespread public support and ultimately lead to new found funding sources. In this regard the work of Pettit et al. (2010), "Current and Future Applications of Toxicogenomics: Results Summary of a Survey from the HESI Genomics State of Science Subcommittee," is a great example of the value to the field of these political and organizational processes.

The survey was carried out in response to a lack of application of toxicogenomic data within organizations. The multisector survey was designed to address issues and perspectives that are important for the current and future challenges to the translation of toxicogenomic data to facilitate discussions for decision making by policy makers.

Respondents (112) from industry, academia, and government included researchers and managers. They answered questions on the online survey about areas including technology used, organizational capacity and resource allocation, experimental approaches, data storage and exchange, perceptions of benefits, hurdles, and future expectations. These results are available as a public resource and a basis for efforts to ensure that the utmost benefit can be obtained from both the research results and the resources available for toxicogenomic studies.

Another good example of is the NIEHS. One of their goals is the development of the next generation of multiomic scientists required, in order to affect the quality of data analysis now and in the future. Ideally academia and industry working through scientific organizations can provide proactive programs that offer incentives and provide experience to attract the brightest science students from science fairs and scientific meetings. An example is the summer research program held by the Federation of American Societies for Experimental Biology for 2,000 students from schools and colleges funded through the American Recovery and Reinvestment Act. The report describes the very positive experiences for both the students and the science educators who participated (Deschamps, 2011).

27.5.2 Omics and the Future

Information presented in this compendium indicates areas for future investment in omics research and the technologies used for omics analysis. Since the completion of the Human Genome Project, the application of omics technologies has promised to unravel disease processes and pave the way for real personalized medicine including accurate diagnosis and genetically based therapeutics. Omics has been expected to revolutionize medical practice (Taylor, 2007; Weber and Eng, 2008). The reality has been that the field of omics, in becoming integrative and driven by bioinformatics, has grown exponentially and accomplished much; however, this work has not been without great difficulty, and many barriers and challenges remain. However, given time, hard work, and a bit of luck, these disciplines will lead to specific disease and environmental-biomarker discoveries, individualized care, prevention and treatment of human diseases, and identification of disease biomarkers and will finally make a recommendation for personalized medicine. The hard work of scientific enquiry is exemplified in the chapters of this book where omics technologies and methodologies have been characterized and integrated to study diseases in human and animal models.

> Hypothesis-driven research and discovery-driven research (through omic methodologies) are complementary and synergistic. Modern screening technologies speed up the discovery process and give broader insight into both biochemical events that follow the exposure to harmful agents e.g. chemical substances, ionizing radiation or electromagnetic fields and diagnosis of various pathological conditions. Moreover, it highlights the response to therapy at a molecular level. Omics technologies have the advantage of containing methods for investigations of DNA-, RNA- and protein level as well as changes in the metabolism.
>
> **Nalinini Reghavarhari, PhD**

For us, the editors, this has been an exciting but intense experience. The chapters in this compendium will serve as a definitive genesis of the global work accomplished by many omic scientists. Understanding the problems beseeched by the complexity of the subject matter, although somewhat difficult to unravel, will ultimately lead to new techniques, diagnostics, and genomic solutions–both pharmaceutical and nutraceutical in composition. We as scientists will look forward to these advancements and the elimination of legal impediments in the fruitful years ahead. We encourage the pioneers of this field to continue their arduous and sometimes serendipitous work, always looking toward the future paving the path to the optimization of health. The real question, albeit a cliché, is, "Where do we go from here?"

REFERENCES

Alter, O., Golub, G. H. (2004). Integrative analysis of genome-scale data by using pseudoinverse projection predicts novel correlation between DNA replication and RNA transcription. *Proc. Natl. Acad. Sci. U.S.A.* 101: 16577–16582.

Chen, T. J. H., Blum, K., Bowirrat, A., et al. (2011). Neurogenetic and clinical evidence for the putative activation of the brain reward circuitry by an amino-acid precursor-enkephalinase inhibition therapeutic agent (a Neuroadaptagen™): Proposing an addiction candidate gene panel map. *J. Psychoactive Drugs* 42: 108–127.

Cohen, J., Boerwinkle, E., Mosley, H, Jr., Hobbs, H. (2006). Sequence variation in PCSK9, low LDL, and protection against coronary heart disease. *N. Engl. J. Med.* 354: 1264–1272.

Deschamps, A. M. (2011). Energizing and investing in the future of science. *FASEB*. Available at http://www.faseb.org.

Gresham, D., Dunham, M. J., Botstein, D., et al. (2008). Comparing whole genomes using DNA microarrays. *Nat. Rev. Genet.* 9: 291–302.

Hall, S. S. (2010). Revolution postponed. *Scientific American* 303: 60–67.

Hegde, P. S., White, I. R., et al. (2003). Interplay of transcriptomics and proteomics. *Curr. Opin. Biotechnol.* 14: 647–651.

Horton, J. D., Cohen, J. C., Hobbs, H. H. (2007). Molecular biology of PCSK9: Its role in LDL metabolism. *Trends Biochem. Sci.* 32: 71–77.

Hu, L., Ye, M., Zou, H. (2009). Recent advances in mass spectrometry-based peptidome analysis. *Expert Rev. Proteomics* 6: 433–447.

Jepsen, K. J., Courtland, G. H., Nadeau, J. H. (2010). Genetically determined phenotype covariation networks control bone strength. *J. Bone Miner. Res.* 25: 1581–1593.

Le-Niculescu, H., McFarland, M. J., Mamidipalli, S., et al. (2007). Convergent functional genomics of bipolar disorder: From animal model pharmacogenomics to human genetics and biomarkers. *Neurosci. Biobehav. Rev.* 31: 897–903.

Manolio, T. A., Collins, F. S., Cox, N. J., et al. (2009). Finding the missing heritability of complex diseases. *Nature* 461: 747–753.

McClellan, J., King, M. (2010). Genetic heterogeneity in human disease. *Cell* 141: 210–217.

Mootha, V. K., Bunkenborg, J., Olsen, M., et al. (2003). Integrated analysis of protein composition, tissue diversity, and gene regulation in mouse mitochondria. *Cell* 115: 629–640.

Nadeau, J. H. (2009). Transgenerational genetic effects on phenotypic variation and disease risk. *Hum. Mol. Genet.* 18: 202–210.

National Research Council Committee on Toxicity Testing and Assessment of Environmental Agents (2007). *Toxicity testing in the 21st century: A vision and a strategy*. SBN: 978-0-309-10988-8. Available at http://www.nap.edu.

Pettit, S., des Etages, S. A., Mylecraine, L., et al. (2010). Current and future applications of toxicogenomics: Results summary of a survey from the HESI Genomics State of Science Subcommittee. *Environ. Health Perspect.* 118: doi:10.1289/ehp.0901501.

Ray, H. N., Mootha, V. K., Boxwala, A. A. (2003). Building an application framework for integrative genomics. *AMIA Annu. Symp. Proc.* 981.

Taylor, D. L. (2007). Past, present, and future of high content screening and the field of cellomics. *Methods Mol. Biol.* 356: 3–18.

Van Leeuwen, B. H., Evans, B. A., Tregear, G. W., et al. (1986). Mouse glandular kallikrein genes. Identification, structure, and expression of the renal kallikrein gene. *J. Biol. Chem.* 261: 5529–5535.

van Vliet E. (2010). Current standing and future prospects for the technologies proposed to transform toxicity testing in the 21st century. *ALTEX* 28: 17–44.

Weber, F., C. Eng (2008). Update on the molecular diagnosis of endocrine tumors: Toward -omics-based personalized healthcare? *J. Clin. Endocrinol. Metab.* 93: 1097–1104.

Zera, A. J. (2011). Microevolution of intermediary metabolism: Evolutionary genetics meets metabolic biochemistry. *J. Exp. Biol.* 214: 179–190.

Index

C

F

G

H

M

O

P

Q

R

S